T0255652

CAMBRIDGE LIBRARY COLLECTION

Books of enduring scholarly value

Botany and Horticulture

Until the nineteenth century, the investigation of natural phenomena, plants and animals was considered either the preserve of elite scholars or a pastime for the leisured upper classes. As increasing academic rigour and systematisation was brought to the study of 'natural history', its subdisciplines were adopted into university curricula, and learned societies (such as the Royal Horticultural Society, founded in 1804) were established to support research in these areas. A related development was strong enthusiasm for exotic garden plants, which resulted in plant collecting expeditions to every corner of the globe, sometimes with tragic consequences. This series includes accounts of some of those expeditions, detailed reference works on the flora of different regions, and practical advice for amateur and professional gardeners.

A Dictionary of the Economic Products of India

A Scottish doctor and botanist, George Watt (1851–1930) had studied the flora of India for more than a decade before he took on the task of compiling this monumental work. Assisted by numerous contributors, he set about organising vast amounts of information on India's commercial plants and produce, including scientific and vernacular names, properties, domestic and medical uses, trade statistics, and published sources. Watt hoped that the dictionary, 'though not a strictly scientific publication', would be found 'sufficiently accurate in its scientific details for all practical and commercial purposes'. First published in six volumes between 1889 and 1893, with an index volume completed in 1896, the whole work is now reissued in nine separate parts. Volume 1 (1889) opens with the prefatory matter, along with lists of works consulted, contributors and abbreviations. It contains entries from *Abaca* (a name in the Philippines for Manila hemp) to *Buxus* (a genus of evergreen shrubs).

Cambridge University Press has long been a pioneer in the reissuing of out-of-print titles from its own backlist, producing digital reprints of books that are still sought after by scholars and students but could not be reprinted economically using traditional technology. The Cambridge Library Collection extends this activity to a wider range of books which are still of importance to researchers and professionals, either for the source material they contain, or as landmarks in the history of their academic discipline.

Drawing from the world-renowned collections in the Cambridge University Library and other partner libraries, and guided by the advice of experts in each subject area, Cambridge University Press is using state-of-the-art scanning machines in its own Printing House to capture the content of each book selected for inclusion. The files are processed to give a consistently clear, crisp image, and the books finished to the high quality standard for which the Press is recognised around the world. The latest print-on-demand technology ensures that the books will remain available indefinitely, and that orders for single or multiple copies can quickly be supplied.

The Cambridge Library Collection brings back to life books of enduring scholarly value (including out-of-copyright works originally issued by other publishers) across a wide range of disciplines in the humanities and social sciences and in science and technology.

A Dictionary of the Economic Products of India

Volume 1: Abaca to Buxus

George Watt

CAMBRIDGE
UNIVERSITY PRESS

University Printing House, Cambridge, CB2 8BS, United Kingdom

Published in the United States of America by Cambridge University Press, New York

Cambridge University Press is part of the University of Cambridge.
It furthers the University's mission by disseminating knowledge in the pursuit of education, learning and research at the highest international levels of excellence.

www.cambridge.org
Information on this title: www.cambridge.org/9781108068734

This edition first published 1889
This digitally printed version 2014

ISBN 978-1-108-06873-4 Paperback

Selected botanical reference works available in the CAMBRIDGE LIBRARY COLLECTION

al-Shirazi, Noureddeen Mohammed Abdullah (compiler), translated by Francis Gladwin: *Ulfáz Udwiyeh, or the Materia Medica* (1793) [ISBN 9781108056090]

Arber, Agnes: *Herbals: Their Origin and Evolution* (1938) [ISBN 9781108016711]

Arber, Agnes: *Monocotyledons* (1925) [ISBN 9781108013208]

Arber, Agnes: *The Gramineae* (1934) [ISBN 9781108017312]

Arber, Agnes: *Water Plants* (1920) [ISBN 9781108017329]

Bower, F.O.: *The Ferns (Filicales)* (3 vols., 1923–8) [ISBN 9781108013192]

Candolle, Augustin Pyramus de, and Sprengel, Kurt: *Elements of the Philosophy of Plants* (1821) [ISBN 9781108037464]

Cheeseman, Thomas Frederick: *Manual of the New Zealand Flora* (2 vols., 1906) [ISBN 9781108037525]

Cockayne, Leonard: *The Vegetation of New Zealand* (1928) [ISBN 9781108032384]

Cunningham, Robert O.: *Notes on the Natural History of the Strait of Magellan and West Coast of Patagonia* (1871) [ISBN 9781108041850]

Gwynne-Vaughan, Helen: *Fungi* (1922) [ISBN 9781108013215]

Henslow, John Stevens: *A Catalogue of British Plants Arranged According to the Natural System* (1829) [ISBN 9781108061728]

Henslow, John Stevens: *A Dictionary of Botanical Terms* (1856) [ISBN 9781108001311]

Henslow, John Stevens: *Flora of Suffolk* (1860) [ISBN 9781108055673]

Henslow, John Stevens: *The Principles of Descriptive and Physiological Botany* (1835) [ISBN 9781108001861]

Hogg, Robert: *The British Pomology* (1851) [ISBN 9781108039444]

Hooker, Joseph Dalton, and Thomson, Thomas: *Flora Indica* (1855) [ISBN 9781108037495]

Hooker, Joseph Dalton: *Handbook of the New Zealand Flora* (2 vols., 1864–7) [ISBN 9781108030410]

Hooker, William Jackson: *Icones Plantarum* (10 vols., 1837–54) [ISBN 9781108039314]

Hooker, William Jackson: *Kew Gardens* (1858) [ISBN 9781108065450]

Jussieu, Adrien de, edited by J.H. Wilson: *The Elements of Botany* (1849) [ISBN 9781108037310]

Lindley, John: *Flora Medica* (1838) [ISBN 9781108038454]

Müller, Ferdinand von, edited by William Woolls: *Plants of New South Wales* (1885) [ISBN 9781108021050]

Oliver, Daniel: *First Book of Indian Botany* (1869) [ISBN 9781108055628]

Pearson, H.H.W., edited by A.C. Seward: *Gnetales* (1929) [ISBN 9781108013987]

Perring, Franklyn Hugh et al.: *A Flora of Cambridgeshire* (1964) [ISBN 9781108002400]

Sachs, Julius, edited and translated by Alfred Bennett, assisted by W.T. Thiselton Dyer: *A Text-Book of Botany* (1875) [ISBN 9781108038324]

Seward, A.C.: *Fossil Plants* (4 vols., 1898–1919) [ISBN 9781108015998]

Tansley, A.G.: *Types of British Vegetation* (1911) [ISBN 9781108045063]

Traill, Catherine Parr Strickland, illustrated by Agnes FitzGibbon Chamberlin: *Studies of Plant Life in Canada* (1885) [ISBN 9781108033756]

Tristram, Henry Baker: *The Fauna and Flora of Palestine* (1884) [ISBN 9781108042048]

Vogel, Theodore, edited by William Jackson Hooker: *Niger Flora* (1849) [ISBN 9781108030380]

West, G.S.: *Algae* (1916) [ISBN 9781108013222]

Woods, Joseph: *The Tourist's Flora* (1850) [ISBN 9781108062466]

For a complete list of titles in the Cambridge Library Collection please visit:
www.cambridge.org/features/CambridgeLibraryCollection/books.htm

A

DICTIONARY

OF

THE ECONOMIC PRODUCTS OF INDIA.

BY

GEORGE WATT, M.B., C.M., F.L.S.,

PROFESSOR, BENGAL EDUCATIONAL DEPARTMENT,
ON SPECIAL DUTY WITH THE GOVERNMENT OF INDIA, DEPARTMENT OF REVENUE
AND AGRICULTURE.

ASSISTED BY NUMEROUS CONTRIBUTORS.

IN SIX VOLUMES.

VOLUME I.

Abaca to Buxus.

Published under the Authority of the Government of India,
Department of Revenue and Agriculture.

CALCUTTA:
PRINTED BY THE SUPERINTENDENT OF GOVERNMENT PRINTING, INDIA.
1889.

PREFACE.

THE following are the circumstances which led to the publication of the Dictionary of Economic Products. In 1877 the Agricultural Department of the North-Western Provinces was required to provide a collection of products for the Paris Exhibition, and again, in 1880, for the Melbourne Exhibition. Early in 1883 the Imperial Department of Agriculture was called upon for a third collection for the Exhibition at Amsterdam. Requisitions were at the same time received from the Governments of Italy and Belgium for sample collections of commercial products. The collections above referred to were all made, under the direction of the undersigned, by Babu Trailokhya Nath Mukharji, now the officer in charge of the Exhibition Branch in the Agricultural Department of the Government of India, and their formation led to the gradual compilation of a list of the more important Economic Products of India, which was illustrated by a series of samples or specimens arranged in glass-fronted tin cases designated the "Index collection." But the list was far from being complete, and was necessarily wanting in scientific detail and arrangement. Matters stood thus when a request was made to the Imperial Government by the Government of Bengal for assistance and co-operation in providing a collection of Economic Products for the Calcutta Exhibition of 1883–84. The opportunity was taken to obtain from the Government of Bengal, for the purpose of securing the scientific arrangement of the collection, the services of Dr. George Watt, of the Bengal Educational Department, who had been originally sent out to India as Professor of Botany, and who had already done useful work in the field of

A Y 18673. 300.—7/89. I. 1192.

botanical research. The Bengal Government had, at the same time, formed the intention of bringing into scientific order a valuable collection of Bengal products at the Provincial Economic Museum in Calcutta.

Dr. Watt, with the assistance of Babu Trailokhya Nath Mukharji, devoted himself, during 1883, to the combined duty, which he was called upon to perform, of enlarging and arranging both the Imperial and Provincial lists and collections. The results were exhibited in the Economic Court at the Calcutta Exhibition, and formed a great stride in advance of all that had been previously done. The time, however (less than a year), allowed for the preparation of the Calcutta collection was too short for a full compilation of the facts and statistics which were necessary for the proper investigation and description of each product. Nor was the index collection itself sufficiently full. These circumstances, as well as the likelihood of having to provide a collection for the London Exhibition of 1886, led to the retention of Dr. Watt's services by the Imperial Department for the purpose of preparing as complete a "Dictionary of Economic Products" as might be possible, with the information already existing in the Agricultural Office; Babu Trailokhya Nath Mukharji being at the same time entrusted with the duty of enlarging the "Index collection" of specimens.

It is needless to explain that the work upon which Dr. Watt has been thus engaged is one which it would have been in any case necessary for the Agricultural Department to carry out independently of any call which was made upon it in connection with Exhibitions. At the same time the utility of Exhibitions in forwarding the performance of the duty should not escape notice. Now that the work has reached the stage of a compilation of existing facts and statistics up to date, it will become the further duty of the Department to make, with the assistance of the Agricultural Departments of the Provinces, such investigations as may be necessary for obtaining the fuller information, which is still wanting, about many of the Economic Products of the country, as well as to maintain and publish periodical statistics of the production of or the traffic in the more important articles. It will become a question for consideration whether a second and more complete edition of the present work should not be published by the Department when sufficient material has been collected, and the present opportunity is taken to invite all who may be interested in the matter to contribute any information which may serve to correct or supplement the contents of the first edition.

E. C. BUCK,

Secretary to the Government of India,
Department of Revenue and Agriculture.

January 1*st*, 1889.

A Y 18673.

PREFACE.

THE circumstances which led to the publication of the present work having been indicated by Sir Edward Buck, it may not be deemed out of place for me to offer a few words of explanation regarding the manner in which I have endeavoured to accomplish the task committed to me. The present edition professes to be an approximately complete *résumé* of the opinions of Indian authors, and of extensive official and private enquiries, regarding the Economic Products of India. Care has been taken to show the sources from which the more important facts have been obtained, and to give, in most cases, the entire list of works consulted in the preparation of the account of each product. I have chosen to adopt this course, even in dealing with facts so well known that they might legitimately have been published without acknowledgment. It is hoped that on this account the "Dictionary of the Economic Products of India" will be found a useful work of reference, and that it may form the nucleus of an extended and systematic enquiry into the productive resources of the Indian Empire. The limited time at my disposal has almost precluded original and personal investigation of critical questions, and therefore, except in so far as ten years of botanical research in India have supplied me with the means of rapidly correcting misunderstandings, the opinions of authors, even when apparently conflicting, have been placed side by side. One of the objects I have kept in view has been to remove the confusion and ambiguity due to wrong or antiquated botanical names being associated with economic products. If I have succeeded, an important step will have been taken in the right direction; and the Dictionary, though not a strictly scientific publication, will, I trust, be found sufficiently accurate in its scientific details for all practical and commercial purposes.

I have had to keep in view a twofold purpose; *viz.*, on the one hand, to supply scientific information which may be useful to the administrative officer; and, on the other, to meet the requirements of the reader in search of definite information regarding Indian economics. It may perhaps convey a not inadequate idea of the scope of the Dictionary to say that with this double purpose in view it is hoped that something may have been done to advance the material interests of India, and to bring the trade and capital of the West into more direct contact with the resources of the Empire.

With regard to botany, Sir J. D. Hooker's *Flora of British India* has been taken as the standard of scientific names and synonyms, and the reference to that invaluable work will be found in the first line following the name of each species. The botanic diagnosis of the plants from Ranunculaceæ to Acanthaceæ has been entirely derived from the *Flora of British India.* For the plants which fall into the natural orders after Acanthaceæ, a larger number of authors have had to be consulted; the scientific names used in this portion have been derived chiefly from Bentham and Hooker's *Genera Plantarum*; De Candolle's *Prodromus*; Roxburgh's *Flora of British India*; The Linnæan Society's publications; Brandis's *Forest Flora*; Kurz's *Forest Flora of British Burma*; Thwaites' *Enum. Ceylon Plants*; Dalzell and Gibson's *Bombay Flora*; Stewart's *Panjáb Plants*; and Mr. Gamble's recent and most useful *Manual of Indian Timbers.* Until the authors of the *Flora of British India,* who are associated with Sir J. D. Hooker, have completed their account of the plants of India, it will be next to impossible for Indian authors to obviate a repetition of some of the errors of the older botanical writers. This is chiefly due to the fact that the libraries and collections available in India are too poor to admit of much literary and critical botanical research; moreover, work of this nature was not contemplated in connection with the present edition of the Dictionary.

A list of the principal authors from whom economic facts have been compiled will be found on page xiii. The economic products which belong to the Animal and Mineral Kingdoms have been but very imperfectly touched upon. It is hoped that these products may, however, receive more attention in a future edition, and the reader is, for the present, referred to the publications of the Geological Survey for detailed information about the Ores and Minerals of India. The majority of the brief notices regarding minerals which are here published have, at the request of the Revenue and Agriculture Department, been kindly supplied by the Superintendent of the Geological Survey.

It may be explained that, with the permission of the Government of India, some 300 copies of each of my "Catalogues of the Economic Products of India" (Calcutta International Exhibition) were issued to officers of all Departments throughout India for additions and corrections. Of these a considerable number have been duly returned, and the Dictionary, incorporating as it does the new economic facts which have come to light through this combined enquiry, may be viewed as an improved and enlarged edition of the Catalogues. I have endeavoured to give as much prominence as possible to the information thus obtained, and the notes supplied will accordingly be found under each article following the symbol § and bearing an abbreviation of the name and address of each contributor.

A full list of all contributors will be found on page xxiii; but I must here express the very great obligation I am under to them all, and severally, for the liberal and invaluable aid they have given me.

With regard to the spelling of vernacular names, it may be stated that it has been thought unwise to depart from the method adopted by the authors from whom this work has been chiefly compiled. To correct the names given to the same object in the numerous languages and dialects of India would of necessity require the co-operation of many persons acquainted with these languages. At the same time the economic plants known to local authorities under certain vernacular names would have to be botanically determined. As this could not be accomplished in the limited time at my disposal, an effort has been made to indicate the long vowels by a diacritic mark, and it is hoped this will enable the reader to pronounce the majority of the vernacular names correctly. In a future edition greater care will doubtless be observed and the vernacular names will be revised and confirmed.

It may be explained that the vernacular names are given with reference to their provincial distribution rather than with regard to the language to which they actually belong; thus BENG. may mean simply that the word is in use in Bengal but need not be Bengali.

The alphabetical arrangement of the Dictionary is based upon the scientific names of the animals, plants, and minerals. This has been accepted as at once the most convenient and satisfactory standard, since it secures uniformity. With regard to large commercial products obtained from more than one species, such as silk, the subject has been treated collectively, instead of being broken up into a number of sections under the scientific names of the insects which yield the various kinds of silk. This should cause no difficulty, however, as the numerous cross references will serve to direct attention to the heading under which the detailed accounts are to be found.

On the margin a number for each product or object has been given. It is hoped that these numbers may not only prove useful for museum purposes, but that they may also afford a convenient clue for correspondence regarding the products. To avoid using an inconveniently large number of figures, the numbers for each letter of the alphabet will be found to commence anew. The index will contain all the European and Vernacular names. It will give the corresponding scientific name and will enumerate the known and described products, referring the reader to the marginal number for each; thus—Gum, A. 756, or Fibre, B. 399. It is hoped that in this way the index may even prove useful as an independent work upon the names given to the Economic Products; it will contain over 30,000 vernacular words.

I desire to acknowledge in this place the invaluable services of the following gentlemen: Sir J. D. Hooker, late Director, Royal Gardens, Kew, for kindly permitting me to consult him upon difficult botanical points; Professor W. T. Thiselton Dyer, Director, Kew, for many valuable additions and corrections; Dr. George King, Superintendent, Royal Botanic Gardens, Calcutta, for the liberty to utilise freely the resources of the Herbarium and Library attached to the Gardens; Mr. C. B. Clarke, Inspector of Schools, Calcutta, for identifying doubtful plants; and Mr. J. F. Duthie, Superintendent, Botanic Gardens, Saharanpur, for notes and suggestions; Dr. H. Trimen, Director of the Botanic Gardens, Ceylon, for many important additions, more particularly with reference to Ceylon Botany. I am also specially indebted to Mr. E. T. Atkinson, Accountant General, Bengal, for many valuable additions to the proof-sheets of this work, and for having placed at my disposal manuscripts containing many interesting notes and original observations. I am likewise greatly indebted to the officers in charge of the Provincial Agricultural Departments, both for much personal assistance and for the prompt manner in which they have uniformly responded to my solicitations for the aid of local specialists.

To Dr. Charles Rice of New York, Dr. W. Dymock of Bombay, and Dr. Moodeen Sheriff of Madras, my best thanks are due for kindly consenting to revise the proofs of the present edition with the view of correcting the vernacular names and of adding to the information. I may be pardoned for quoting a passage from a private letter from Dr. Moodeen Sheriff as showing the liberal way in which he has co-operated with me: "In revising the vernacular names I am not solely depending upon my '*Supplement to the Pharmacopœia of India*,' but am consulting many other works and making fresh enquiries. This, together with my experience, which is much greater than before, will, I hope, enable me to accomplish satisfactorily the work entrusted to me." I deeply regret the death of Dr. U. C. Dutt, of Serampore, the able author of "*The Materia Medica of the Hindus*" which has deprived me, in the greater portion of the work, of the assistance which he was so eminently qualified to render. Dr. Dutt undertook to supply a series of notes regarding the plants enumerated in the Glossary to his Materia Medica, and to revise the Sanskrit names in the bulk of the work. The latter duty has been generously undertaken by Dr. Rice, and as the result I have received from New York many valuable additions and corrections which cannot fail to prove valuable to students of Oriental literature. These distinguished scholars are not, however, in any way responsible for the accuracy of the vernacular names throughout the work, since many additions have been made subsequent to their kind supervision.

My acknowledgments are specially due to Dr. Cuningham, late Surgeon-General with the Government of India, for having encouraged the numerous medical officers throughout India to supply the series of notes which constitute a most important feature of the work.

The trade statistics have been furnished by Mr. J. E. O'Conor, Assistant Secretary, Department of Finance and Commerce, Government of India. In supplying these figures Mr. O'Conor has rendered me a most important service; at the same time he has obligingly offered many other suggestions and corrections.

Further, I would acknowledge the great personal interest taken in the work by Mr. E. J. Dean, Superintendent of Government Printing, India. Working with the staff of printers who have but an imperfect knowledge of the English language, Mr. Dean has shown, as I believe will be readily admitted, that he can produce an elaborate work of this nature in a manner worthy of a high-class European press.

GEORGE WATT.

CALCUTTA,
January 1st, 1889.

LIST OF WORKS CONSULTED.

The following is a list of the more important works, journals, reports, or other publications which have been either directly or indirectly consulted in the preparation of the "Dictionary of the Economic Products of India." As far as possible, the rule has been followed of quoting original authors, and an effort has been made to show the source from which every statement has been derived. It is hoped that this fact may be accepted as a decided advantage, but it has somewhat encumbered the work with references.

AINSLIE, W.; Materia Medica of Hindustan, 1813.
„ Materia Indica, 1826.
ANDERSON, GEORGE; The Palms of Sikkim. Journal of the Linnæan Society, 1868.
„ „ An Enumeration of the Indian Acanthaceæ: Journal of the Linnæan Society, 1866.
ANDERSON, JOHN; Report on Expedition to Western Yunan, 1871.
AITCHISON, J. E. T.; Catalogue of the Punjab and Sind Plants, 1869.
ARCHER; Economic Botany.
ATKINSON, E. F. T.; Economic Products of the North-West Provinces, 1881; Gums and Gum-Resins, 1876; Himalayan Districts, being Vol. X of the Gazetteer, N.-W. P., 1882; and N.-W. P. Gazetteer.
BAKER AND HOOKER (W. J.); Genera Filicum, 1842.
BALFOUR, E.; Cyclopædia of India and of Eastern and Southern Asia, 1857-73.
„ Timber Trees, &c., 1870.
BALL, V.; Economic Geology, being Part III., Geology of India, 1881.
BALL; Cultivation and Manufacture of Tea in China, 1848.
BATTEN, J. H.; Tea Cultivation in Kumaon, 1878.
BEAVAN, R.; Hand-book of the Fresh-water Fishes of India, 1877.
BEDDOME, R. H.; Flora Sylvatica of the Madras Presidency, in 4 vols., 1869-73.
„ „ The Ferns of Southern India, 1863-64.
„ „ The Ferns of British India, 1866-70.
„ „ Species Filicum, 1846-64.
„ „ Synopsis Filicum, 1868.
„ „ Hand-book of the Ferns of British India, 1883.
BEE-KEEPING IN INDIA; Papers published by the Revenue and Agricultural Department, 1883.
BELLEW, H. W.; Journal of a Political Mission to Afghanistan, 1863.
BENTHAM, GEORGE; Flora Hongkongensis, 1861.
„ „ Flora Australiensis, 4 vols., 1863-73.
„ „ Genera and Species of Labiatæ, 1832-36.
BENTHAM AND HOOKER; Genera Plantarum, 3 vols., 1867-83.
BENTLEY AND TRIMEN; Medicinal Plants, 4 vols., 1880.
BERKELEY, M. J.; Intrcduction to Cryptogamic Botany, 1857.
BIDIE, GEORGE; Report on the Nilgiri Loranthaceous Plants, 1874.
„ „ Raw Products of Southern India (Paris Exhibition Catalogue), 1878.
BIRDWOOD, GEORGE; Catalogue of the Economic Products of the Presidency of Bombay, 2nd Edition, 1865.
„ „ Monograph Burseraceæ in Transactions, Linnæan Society. Vol. XXVII.
„ „ Industrial Arts of India, 1880.
„ „ Portfolio of Industrial Art, 1881.
BLYTH, EDWARD; Catalogue, Bengal Asiatic Society of the Mammalia, Birds, Feline Animals, Reptiles, &c., 1863.
BOISSIER; Flora Orientalis, 1867.

BOOT; Illustrations of the genus Carex (Cyperaceæ), 1858-67.
BRANDIS, D.; Forest Flora of the North-West and Central India, 1874.
BRITISH PHARMACOPŒIA, 1868.
BRITISH MANUFACTURING INDUSTRIES; a series of 15 vols., edited by G. P. Bevan, 1876 to 1878.
BROWN, J.; The Forester, 1871.
BUCHANAN, F.; Journey through Mysore and Kanara, 1857.
BUCK, E. C.; Dyes and Tans of the N.-W. Provinces, 1878.
BURMANN; Flora Indica, 1768.
BURRELL; India Tea Culture, 1877.
BUTTER, D.; Medical Topography of the Southern Districts of Oudh, 1839.
CALVERT'S Dyeing and Calico-printing; Ed. by Stenhouse and Groves.
CAMPBELL, GEORGE; Ethnology of India, in the Bengal Asiatic Society's Journal, 1866.
" " The Non-Hindú Tribes of the Borders of Hindústan, in the Journal of the Ethnological Society, 1867; and the Races of India, 1869.
CAMPBELL, A.; The Lepchas, in Journal, Ethnological Society, 1869.
CASSELS, W. R.; Cotton Culture in the Bombay Presidency, 1862.
CATHCART AND HOOKER; Illustrations of Himálayan Plants, 1855.
CHRISTY, T.; New Commercial Plants, Parts I. to VIII., 1878—85.
CHURCH, A. H.; Food, 1882.
CHURCHILL'S Technological Hand-books, Vol. on Dyeing and Calico-printing, 1884.
CLARKE, C. B.; Compositæ Indicæ and Commelynaceæ and Cyrtandraceæ, 1876.
" " Ferns of Northern India, 1880.
CLEGHORN, HUGH; Forests and Gardens of Southern India, 1861.
CLEGHORN, J.; Notes upon the Pines of the N.-W. Himálaya, 1866.
CLIFFORD; Memorandum on the Timber of Bengal, 1862.
COIN, R. L. DE; Cotton and Tobacco, 1864.
COLEMAN; Mythology of the Hindús, 1832.
COLLINS; Report on Caoutchouc of Commerce, 1872.
COOKE, M. C.; Gums and Gum-Resins; Oil and Oil-seeds, 1874.
" Hand-book of British Fungi, in 2 vols., 1871.
" Report on Diseased Leaves of Coffee and other Plants, 1876.
COOPER, E.; Forest Culture and Eucalyptus Trees, 1876.
CROOKE, W.; A Practical Hand-book of Dyeing and Calico-printing, 1874.
CUNNINGHAM; Ladakh, Physical, Statistical, and Historical, 1854.
DALTON, E. T.; Ethnology of Bengal; and Wild Tribes of Central India, in Journal of the Ethnological Society, 1867.
DALZELL AND GIBSON; Bombay Flora, 1861.
DAVIES, R. H.; Report on the Trade and the Resources of the Countries on the North-West Frontier of British India, 1862.
DAVIES, R. H.; Coal and Iron in the Punjáb, 1859.
DAY, F.; Fishes of India, 2 vols., 1876-78.
DECANDOLLE; Prodromus, 17 vols., 1854-73.
" Monographiæ, Phanerogamarum, Vols. I. to III., 1878-83.
DECANDOLLE, A.; Origin of Cultivated Plants (Intern. Sc. Series), 1884.
DELOUREIRO; Flora Cochin Chinensis, 1793.
DEY, K. L.; The Indigenous Drugs of India, 1867.
DIGBY, W.; Famine Campaign in South India, 1878.
DOLLARD, W. M.; Medical Topography of Kali Kumaon and Shore Valley, 1840.
DON, D.; Prodromus Floræ Nepalensis, 1825.
DONOVAN, E.; Natural History of the Insects of India, Ed. by J. O. Westwood, 1842.
DOUGLAS, J. C.; Hand-book of Bee-keeping for India, 1884.
DRURY, H.; Useful Plants of India, 1873.
" A Hand-book of the Indian Flora, 1864-69.

DUTHIE AND FULLER; Field and Garden Crops of the North-Western Provinces, 2 vols., 1883.
DUTHIE, J. F.; Grasses of the North-Western Provinces, 1883.
DUTT, U. C.; Materia Medica of the Hindús, 1877.
DYMOCK, W.; Materia Medica of Western India, 1883.
EDGEWORTH; Catalogue of the Plants found in the Banda District, 1847.
„ Florula Mallica, Journal of the Linneæan Society, Vol. VI., 1862.
ELLIOT, W.; Flora Andhrica, 1859.
„ Mammalia of the Deccan, in the Journal of the Madras Literary Society, 1839.
ELLIOT, H. M.; Glossary of Indian Terms, 1860.
ELLISON; Hand-book of the Cotton Trade, 1858.
ENDLICHER; Genera Plantarum, 1836-47.
„ Synopsis Coniferarum, 1847.
Famine Commission Report, 1880-85.
FAULKNER, A.; Dictionary of Commercial Terms, with their Synonyms in various languages, Bombay, 1856.
FAYRER, J.; The Thanatophidia of India, being a description of the Venomous Snakes of the Indian Peninsula, 1872.
FIRMINGER, T. A. C.; Manual of Gardening for Bengal, 1869.
FLEMING, JOHN; Catalogue of Indian Medicinal Plants, &c., in Transactions of the Asiatic Society, 1810.
FLÜCKIGER AND HANBURY; The Pharmacographia, 2nd Ed., 1879.
FORSYTH; The Highlands of Central India, 1871.
FORTUNE, R.; Report on the Tea Plantations in the North-West Provinces, 1851.
„ A Visit to the Tea Districts in China and India, 1852.
FOSBERRY; The Mountain Tribes of the North-West Frontiers of India, 1869.
FRASER; Journey on the Himálayan Mountains, 1820.
FULLER, J. B.; Agricultural Primer for India, 1881.
GAMBLE, J. S.; List of the Trees and Shrubs of the Darjeeling District, 1878.
„ Manual of Indian Timbers, 1881.
GAZETTEERS CONSULTED—

1. Imperial Gazetteer by W. W. Hunter, in 9 vols., 1881 (a new edition in Press).
2. Statistical Account of Bengal, by W. W. Hunter, in 20 Vols., 1875.
3. Bombay Gazetteer, Vols. II.-VI., X.-XVI., XXI.-XXIII., by J. M. Campbell, 1877-1884.
   Bombay Gazetteer, Vol. VII., by F. A. H. Elliot, 1883.
   „ „ Vol. VIII., by J. W. Watson, 1884.
4. North-Western Provinces, I.-VI., X., by E. F. T. Atkinson, 1874-82.
5. Oudh, in 3 vols., 1877.
6. Central Provinces, by C. Grant, 1870.
7. British Burma, 2 vols., by H. R. Spearman, 1880.
8. Rajputana (unauthenticated), 1879.
9. Sindh, by A. W. Hughes, 1876.
10. North Arcot, by A. F. Cox, 1881.
11. Cuddapah, by J. D. B. Gribble, 1875.
12. Godavery, by H. Morris, 1878.
13. Kistna, by G. Mackenzie, 1883.
14. Madura, by J. H. Nelson, 1868.
15. Nellore, by J. A. C. Boswell, 1873.
16. Salem, in 2 vols., by H. LeFauer, 1883.
17. Tanjore, by T. Ven Kassami Raw, 1883.
18. Trichinopoly, by L. Moore, 1878.
19. Orissa, by W. W. Hunter, 1872.
20. Mysore, in 3 vols., by L. Rice, 1877.
21. Coorg, by Rev. G. Richter, 1870.

22. Ajmere, by J. D. LaTouche, 1875.
23. Ulwar, by Major P. W. Powlett, 1878.
24. Marwar, Mallani, and Jessulmere, by Major C. K. M. Walter, 1877.
25. Mathura, by T. S. Growse, 1874.
26. Manipur, by R. Brown, 1873.

GEOGHEGAN, J.; Silk in India, 1880.
GORKOM, K. W.; Cultivation of Chinchonas in Java, 1870.
Geological Survey of India: Records, Memoirs, and Reports.
GORDON, G.; Pinetum, 1875.
GRAHAM; Catalogue of the Bombay Plants, in Thompson's Records of General Science, 1839.
GRANT-DUFF, M. E.; Notes on an Indian Journey, 1876.
GRAY, ASA; Structural Botany, Part I., 1st Edition, 1842; 5th Edition, 1857; and 6th Edition, 1881.
GRIFFITH, W. Itinerary Notes on Plants collected in India, 1837-48.
„ Icones Plantarum Asiaticarum, 1847-51.
„ The Palms of British East India, 1850.
„ Posthumous Papers, 1845.
„ Report on the Tea Plant of Upper Assam, 1836.
GÜNTHER, A.; Manual of Fishes, 1880.
„ Reptiles of British India, 1864.
HAMILTON, F. (Buchanan); The Fishes of the Ganges, 1852.
„ „ Statistical Account of Behar.
HANBURY, D.; Science Papers, 1876.
HANNAY, S. E.; Observations on the Timber Trees of Upper Assam, Agri.-Horticultural Society of India, 1845.
HANNAY, S. E.; Sketch of the Singphos published by Government, 1847.
HARKNESS; Account of a Singular Aboriginal Tribe on the Nilgiris, 1832.
HAWKES, H. P.; Report on the Oils of Southern India shown at the Madras Exhibition, 1855.
HELFER; Provinces of Ye, Tavoy, and Mergui on the Tenasserim Coast, 1839.
HENDERSON AND HOME; Lahore to Yarkand, 1873.
HIERN; Monograph of Ebenaceæ, Cambridge Philological Society's Transactions, 1873.
HOLTZAPPFEL; Descriptive Catalogue of the Woods commonly employed in Mechanical and Ornamental Arts, 1852.
HOME; Report on the Vegetation of the Andaman Islands, 1874.
HONIGBERGER, J.; Thirty-five years in the East, 1852.
HOOKER AND THOMSON; Flora Indica, Vol. I., 1855.
HOOKER, J. D.; Flora of British India, Vols. I., II., III., and Parts I. and II. of Vol. IV., 1872-85.
„ Rhododendrons of Sikkim-Himálaya, 1849.
„ Illustrations of Himálayan Plants, 1856.
„ Flora of the British Islands, 3rd Edition, 1884.
„ Himalayan Journals, 1854.
HOOKER, W. AND BAKER; Synopsis Filicum, 1874.
HOOKEK, W.; Musci Exotici, 1818-20.
HORSFIELD AND MOORE; Catalogue of Mammalia in the East India Museum, London, 1851.
„ „ Do. Birds Do. 1854-58.
„ „ Lepidopterous Insects, 1857.
HUC; Journey through Tartary, Thibet, and China, 1852.
„ Chinese Empire, 1855.
HUGON, T.; The Silkworms of Assam, Journal, of the Asiatic Society of Bengal, 1857.
HUNTER, ALEXANDER; Report on the Fibrous Materials shown at the Madras Exhibition, 1855.
HUNTER, W. W.; Statistical Account of Bengal, 20 vols.
„ Imperial Gazetteer, 9 vols.
HUTCHINS, D. E.; Report on Measurements of Growth of Australian Trees grown in the Nilgiris, 1882.

HUTTON, T.; Notes on the Bombycidæ, Supplement, Gazette of India, September 2nd, 1871.
HUXLEY; Ethnology and Archæology of India, Journal of the Ethnological Society, 1869.
INDIAN PHARMACOPŒIA; *See* WARING.
INDIAN FORESTER.
IRVINE, R. H.; General Medical Topography of Ajmere, 1841.
„ „ A short account of the Materia Medica of Patna, 1848.
JACQUEMONT; Voyage dans l'Inde, Vols. I.-V., 1841-45.
„ Letters from India, 1839.
JAMES, A. G. F. E.; Indian Industries, 1880.
JAMESON, W.; Report on Tea Cultivation in the North-West Provinces, 1857.
„ Cultivation and Manufacture of Flax in the North-West Provinces.
„ „ „ „ Tea in Kumaun and Garhwál, 1843-45.
„ Suggestions for the importation of Tea-makers, Implements, and Seeds from China to the N.-W. Provinces, 1852.
JAMESON, W.; Formation of Plantations in Canal Banks, 1876.
JENNINGS, S.; Orchids and how to grow them in India, 1875.
JERDON, T. C.; The Birds of India, 1862-64.
„ „ „ Ed. by Godwin-Austen, 1877.
„ The Mammals of India, 1874.
JERVIS; Indian Weights and Measures, 1836.
JOHNSTON, T. W.; Chemistry of Common Life, edited by Church, 1880.
JOHNSON'S; Gardeners' Dictionary, with a revised Supplement by N. E. Brown, 1882.
JOURNALS and other Scientific Periodicals of which the following may be specially mentioned:—
Agri.-Horticultural Society of India.
„ „ of Madras.
Asiatic Society of Bengal.
Linnæan Society, London.
Pharmaceutical Society, London.
&c., &c.
KEAN, A. H.; Ed. by Sir R. Temple: Asia, with Ethnological Appendix, 1882.
KEARN, J. F.; Tribes of South India, 1860.
KEIGHLY, E. F.; How to rear Silkworms in the Punjáb, 1884.
KERR, H. C.; Report on the Cultivation and Trade in Jute in Bengal, 1874.
„ Report on the Cultivation and Trade in Gánja in Bengal, 1877.
KEW Reports from 1877 to 1883; also Museum Catalogue and Guide to the Gardens, 1883.
KING, G.; Manual of Cinchona Cultivation in India, 1880.
KURZ; Report on the Vegetation of the Andaman Islands, 1870.
„ Preliminary Forest Report of Pegu, 1875.
„ Forest Flora of British Burma, in 2 vols., 1877.
„ Additions to our knowledge of Burmese Flora and other Papers in Journal of Asiatic Society of Bengal.
„ Bamboo and its Uses, in *The Indian Forester.*
LAMBERT, K. K.; Description of the Genus Pinus, 1828-37.
LASLETT; Timber and Timber Trees, 1875.
LATHAM, R. G.; Descriptive Ethnology, 1859.
LEMAOUT and DECAISNE; General System of Botany; translated by Mrs. Hooker and edited by Sir J. Hooker. Second Edition, 1876.
LEWIN; the Hill Tracts of Chittagong, 1869.
„ Wild Races of South-Eastern India.
„ A Fly on the Wheel, 1885.
LIOTARD, L.; Materials in India suitable for the Manufacture of Paper, 1880.
„ Introduction of Carolina Rice into India, 1880.
„ Memorandum on Dyes of Indian Growth and Production, 1881.

LIOTARD, L.; Note on the Tea Industry in the N.-W. Provinces and Punjáb, 1882.
„ Note on Tea Trade with Thibet, 1883.
„ Hop Culture in India, 1883.
„ Note on Nankin Cotton in India, 1883.
„ Memorandum on Silk in India, Part I, 1883.
LINDLEY AND MOORE; Treasury of Botany, 1877.
LINDLEY; Vegetable Kingdom, 1872.
LINDLEY, J.; Genera and Species of Orchidaceous Plants, 1830-38.
LISBOA, J. C.; Useful Plants of the Bombay Presidency (proof copy) 1884.
LOARER; Proceedings of a Committee on the Process for Converting Oil into Vegetable Wax, 1858.
LOUDON; An Encylopædia of Trees and Shrubs, 1842.
LOUIS, J. A. H.; Sericulture in Bengal, 1882.
LOW; Sarawak, its Inhabitants and Productions, 1848.
LOWE, E. J.; Ferns, British and Exotic, 1872.
MACDONNELL, A. P.; Report on the Food-grain Supply of Bengal, 1876.
MACLEOID, D. A.; Medical Topography of Assam, 1837.
MACKENZIE, A; Hill Tribes of the North-Eastern Frontier of Bengal, 1884.
MACPHERSON; Reports upon the Khonds of Ganjam and Cuttack, 1842.
MADDEN; Observations on some of the Pines and other Coniferous Trees of the Northern Himálaya and on Himálayan Coniferæ; Journ. Agri.-Hort. Soc. of India, Vols. IV. and VII.; also many scattered papers on the Flora of Kumaon and other parts of the Himálaya, especially in Journal, Asiatic Society of Bengal, Vols. XV., 79; XVI., 226, 596; XVII., 349 and XVIII., 603: In Agri.-Horticultural Society of India, Vols. IV., VII., & VIII., 1845-50.
MARKHAM, C. R.; Travels in Perua and India, 1862.
MARSHMAN; Cotton in Dharwar.
MARSHMAN, J. C.; Notes on the Cultivation of Tea in Assam.
MASON; Burmah, its People and Natural Products, 1860.
MATHIEU; Flore Forestière, 1877.
MCCANN, H.; Dyes and Tans of Bengal, 1883.
MCCLELLAND; Indian Cyprinidæ, Asiatic Researches, XIX., 1839.
„ Isinglass Fishes; Journal, Asiatic Society, 1839; Calcutta Journal of Natural History, Vols. II. and III., 1842-43.
MCCLELLAND; The Indian Botanical Works of W. Griffith arranged by —, 1848-54.
MCCOSH; Topography of Assam, 1837.
MEDLICOTT, J. G.; The Cotton Hand-book of Bengal, 1862.
MILLBURN, W.; Oriental Commerce, 1813.
MILLS, J.; Diseases of Cattle in India (Plain Hints on), 1883.
MIQUEL; Annales Musei Bot. Lugduno, Batavi, Amsterdam, Vols. I. to IV., 1863-69.
MITTEN; Mosses of Northern India.
MOERMAN, T.; The Ramie (Rhea), 1874.
MOHUN LAL; Travels in the Panjáb, Afghánistan, Turkhistan. Bokhara, England, &c., 1846.
MONEY; The Cultivation and Manufacture of Tea, 1883.
MOODEEN SHERIFF; Supplement to the Pharmacopœia of India, 1869.
MOON; Catalogue of the Indigenous and Exotic Plants of Ceylon, 1824.
MOORCROFT AND TREBECK, ; Travels in Peshâwar, Cabul, and Bokhara, 1841.
MOORE, F.; Synopsis of the Silk-producing Insects of India, 1859.
„ Bombyces, 1859.
„ Adolia Lepidoptera, 1859.
MORTON'S Cyclopædia of Agriculture, 1875.
MOSA, P.; Coal Fields of Asia, 1877.
MUELLER, F. VON; Select Extra-Tropical Plants for Industrial Culture, 1881.
MUKHARJI, T. N.; Amtersdam Exhibition Catalogue of Indian Exhibits, 1883.
MUNRO; A Monograph of the Bambusaceæ; Transactions of the Linnæan Society, 1868.

MURCHISON, C.; Chaulmoogra Seeds of India, in Edinburgh New Philosophical Journal.
MURRAY, J. A.; Plants and Drugs of Sind, 1881.
„ „ Vertebrate Zoology of Sind, 1884.
NEWBERY; Descriptive Catalogue of the Economic Woods of Victoria, 1877.
NEWSPAPERS, Indian, Letters and Editorial Articles upon Economic Products.
NICHOLSON, EDWARD; Indian Snakes, 1874.
NIETNER; Observations on the Enemies of the Coffee Trees of Ceylon, 1861.
NOTES ON ECONOMIC PRODUCTS,—*viz.*, on Flax, Eucalyptus, Carob, Calophyllum, Malachra, and Cardamom, published by Department of Revenue, Agriculture and Commerce, 1881.
O'CONOR, J. E.; Report on the Production of Tobacco in India, 1873.
„ Note on Colophyllum inophyllum and other products, 1881.
„ Annual Review of the Accounts of the Sea-borne Foreign Trade of British India, 1878-1885.
„ Annual Statement of the Trade and Navigation of British India with Foreign Countries, Parts I., II, and III., 1878-1885.
„ Vanilla, its Culture in India, 1881.
OFFICIAL Correspondence of Revenue and Agricultural Department and important files on Economic Products furnished by the Home Department, such as the Correspondence regarding a proposed new edition of the Pharmacopœia of India; also Forest Department Correspondence.
OLDHAM, THOMAS; Geology of the Khasi Hills, 1854.
„ Papers on the Coal of the Nerbudda Valley, Tenasserin Provinces, and Thayet-Myo (No. 10), Government of India Records, 1856.
OLIVER; Flora of Tropical Africa, Vols. I., II., 1868-71.
O'SHAUGHNESSY; The Bengal Dispensatory, 1842.
„ Bengal Pharmacopœia, 1844.
OSWALD; Catalogue, Madras Exhibition, 1857.
OUCHTERLONY; Mineralogical Report of Nellore, Cuddapah, 1841.
PACKARD, A. S.; Guide to the study of insects, 1870.
PAXTON; Botanical Dictionary. New Edition, 1885.
PEMBERTON; Report on the Eastern Frontier of British India, 1835.
PELLY; Report on the Tribes around the shores of the Persian Gulf, 1874.
PHARMACY; Year-book of, 1874-1884.
PIDDINGTON, H.; English Index of the Plants of India, 1832.
PORTER, Tropical Agriculturist, 1833.
POSTAN; Observations on Sindh, 1843.
POTTINGER; Travels in Beluchistan and Sindh, 1816.
POWELL, BADEN H.; Punjab Products, Vols., I., II., 1868-72.
PRINSEP, H.T.; Thibet, Tartary, and Mongolia, 1851.
REH; Report of Committee for investigating into the cause of deterioration of Land by Reh, 1878.
REPORTS; (1) Administration, General and Revenue—for each Province, 1874-84.
„ (2) Internal and External Trade, 1874-85.
„ (3) Forest Administration for the Provinces.
„ (4) Botanic Gardens of Calcutta and Saharanpur.
„ (5) Agriculture Department.
„ (6) Curator of Ancient Monuments in India.
„ (7) Sanitary Commissioner.
„ (8) Famine Commission.
„ SPECIAL; (9) Official and Descriptive Catalogue of the Madras Exhibition of 1857, 1857.
„ „ (10) Papers regarding the Tobacco Monopoly in the Presidency of Fort St. George, and the proposed levy of a tax on tobacco in the other presidencies, 1870.
„ „ (11) Annual Reports on Sericulture in Dehra Dún in the Government Proceedings.

REPORTS, SPECIAL, (12) Parliamentary Papers on Cotton Cultivation in Bengal, 1857.
" " (13) Calcutta International Exhibition, 1883-84.
" " (10) Area and Out turn of the Cotton Crop, N.-W. Provinces., Sel. Rec., N.-W. Provinces, IV., VI., 1869-75.
" " (11) Reports of the Cotton Commissioner with the Government of India.
" " (12) Reports of the Manchester Cotton Supply Association from 1858.
RHEEDE; Hortus Indicus Malabaricus, 1678-86.
RICHARDSON, G. G.; Corn and Cattle-Producing Districts of France.
RIBBENTROP; Hints on Arboriculture in the Panjáb, 1874.
RILEY, C. V.; The Locust Plague, 1877.
ROBERTSON, W. R.; Agriculture for Schools of South India, 1880.
" Report on Agricultural condition of Nilgiri and Coimbatore Districts, 1881.
ROBINSON. P.; Synopsis of Tobacco Cultivation and Manufacture, 1872.
ROBINSON, W.; Account of Assam, 1841.
RONDOT, NATALIS, M.; Green Dye of China, 1860.
RÖEPSTORFF, FR. A. DE; Vocabulary of Dialects spoken in Nicobar and Andaman Islands, 1875.
ROTTLER; Indian Cyperacæ, 1773.
ROXBURGH; Flora Indica, edited by Carey and Wallich, 1820-24.
" " " " " " 1832, and reprinted by C. B. Clarke, 1874.
" Plants of the Coromandel Coast, 1795-1819.
" An account of the Tusseh and Arrindy Silkworms, in Transactions, Linnæan Society, VII., 1802.
ROYLE, J. FORBES; Illustrations of the Botany of the Himálayan Mountains, 1839.
" " The Fibrous Plants of India, 1855.
" " Essay on the Productive Resources of India, 1840.
" " Production of Isinglass along the Coasts of India, with a notice of its Fisheries.
" " Arts and Manufactures of India.
" " Culture of Cotton in India, 1851.
" " Measures for the improved cultivation of Cotton in India, 1859.
" " Report on the progress of the Culture of the China Tea Plant in the Himálayas from 1835 to 1847, 1849.
RUMPH; Herbarium Amboinense, 1741-45.
RUSSELL, P.; The Fishes of the Coromandel Coast, 1803.
SAKHÁRÁM ARJUN; Catalogue of the Bombay Drugs, 1879.
SCHLICH; Report on the Pyin Kado Forests of Arakan and Rangoon, 1870.
SCHONBERG, BARON E. V.; Travels in India and Kashmir, 1853.
SCHWARZ, R. C. VON.; Financial Prospectus of the Chanda Iron-works, 1882.
SEEMAN, BERTHOLD; Popular History of the Palms and their Allies, 1856.
SEN, RAMSHUNKER; Report on Agricultural Statistics in Jessore, 1874.
SHORTT; Ethnology of Jeypore in Journ. Ethn. Soc., 1867.
" Hill Ranges of Southern India, 1870-71.
" Rude Tribes " 1870.
" Leaf Festival, Madras, 1865.
SIMMONDS, P. L.; Tropical Agriculture, 1877.
" Waste Products, 1876.
SIRR, H. C.; Ceylon and the Singhalese; their History, Government, Religion, Antiquities, &c., 1850.
SKINNER; Indian and Burman Timbers, 1862.
SLEEMAN; Journey through Oudh, 1858.
SMITH, J.; Dictionary of Economic Plants, 1882.
SMITH, R. M.; Persian Art.
SPEEDE, G. F.; The Indian Hand-book of Gardening, 1842.

SPONS' ENCYCLOPÆDIA; 1880.
STERNDALE, R. A.; Natural History of the Mammalia of India, 1884.
STEWART, J. L.; Punjab Plants, 1869.
STIVEN, W. S.; Paper-making in India, Sel. Rec. N.-W. Provs., III.
STOCKS, J. E.; Botany of Scinde, 1841-59.
STRETTELL, G. W.; Note on Caoutchouc obtained from Chavannesia esculenta, 1874.
 „ Ficus elastica in Burma, 1876.
TALÍFE-SHARÍFÍ (translated by G. Playfair), 1823.
TAYLOR, J.; Medical Topography of Dacca in Bengal, 1840.
TEMPLE, RICHARD; Aboriginal Tribes of Central Provinces, 1866.
TENNENT, J. E.; Natural History of Ceylon, with Narrative and Anecdotes illustrative of the habits and instincts of the Mammalia, Birds, Reptiles, Fishes, Insects, &c., 1861.
THACKER and HALLEN; More Deadly Forms of Cattle-diseases in India, 1880.
THEOBALD, W.; Catalogue of Reptiles in the Museum of the Asiatic Society of Bengal, 1868.
 „ The Reptiles of Burma, Journ., Linnæan Soc., Zoology, Vol. X, 1868.
 „ Recent Shells (Cat. Beng. As. Soc.), 1860.
 „ Land and Fresh-water Shells of India, 1860.
THOMAS, H. S.; The Rod in India, 1881.
 „ Report on Pisciculture in South Canara, 1870.
THOMPSON; Report on Insects destructive to Woods and Forests, 1868.
THOMSON; Western Himálaya and Thibet, 1852.
THWAITES; Enumeratio Plantarum Zeylaniæ, 1864.
Tobacco Planting and Manufacturing in South India, 1880.
TOD; Travels in Western India, 1839.
 „ „ in Rájputana.
ULFAZ UDWIYEH; by Noureddeen Mohammed Abdulah Shirazy (translated by Francis Gladwin), 1793.
United States Dispensatory, 15th Ed., 1883.
 „ Reports of the Commissioner of Agriculture up to 1884.
URE; Dictionary of Arts, Manufactures, and Mines, 1872-78.
VIGNE; A visit to Cabul, Ghuzni, and Afghánistan, 1840.
 „ A personal Narrative.
 „ Kashmir, 1842.
VOIGT; Hortus Suburbanus Calcuttensis, 1845.
WACE, E. G.; Punjab wheat, 1884.
WAGNER; Chemical Technology, edited by Crookes.
WALKER, C.; Report on Government Chinchona Plantations, 1878.
WALLICH; List of the Woods of Nepál and Goalpara, in Jury Reports of the Exhibition of 1851-1852.
WALLICH; Plantæ Asiaticæ Rariores, Vols. I. to III., 1830-32.
WALTON, W.; Cotton in Belgaum and Kaládgi Districts, Bombay, 1880.
WARDLE, T.; The Wild Silks of India, 1881.
WARING; Pharmacopœia of India, 1868.
 „ Bazar Medicines, 1874.
WATSON, J. FORBES; Cotton in India, 1859.
 „ A classified and Descriptive Catalogue of the Indian Department of the International Exhibition of 1862, 1862.
 „ Index to names of Eastern Plants and Products, 1868.
 „ Textile Manufactures and Costumes of India, 1866.
 „ List of Indian Products, 1872.
 „ Report on the Preparation and Uses of Rheea Fibre, 1875 and 1884.
WATT, A.; Soap-making, 1884.
WATT, G.; Catalogue of the Economic Products of India shown at the Calcutta International Exhibition—Part I., Gums and Resins; Part II., Dyes and Tans; Part III., Fibres; Part IV., Oils; Part V., Medicines; Part VI., Foods; and Part VII., Timber; 1883.

WEDDELL, H. A.; Quinquinas, Notes on, 1871.
WHEAT PRODUCTION and TRADE; Published by Revenue and Agricultural Department, 1883.
WHEAT of PANJÁB; *See* WACE.
WIGHT; Icones Plantarum Indiæ Orientalis, 1840-53.
„ Illustrations of Indian Botany, 1838-53.
„ The Timber Trees of the Madras Presidency.
„ Contributions to the Botany of India, 1834.
„ Illustrations of Indian Botany, principally of the Southern parts of the Peninsula, 1831.
WIGHT AND ARNOTT; Prodromus Floræ Peninsulæ Indiæ Orientalis, 1834.
WILCOX; Asiatic Researches in the Bor-Kampti Country, Vol. XVII., 436.
WILSON, H. H.; Glossary of Indian Terms, 1855.
WINTER; Six months in British Burmah, 1858.
WISE; Hindu Medicine.
WOLFF; Mission to Bokhara, 1845.
WOOD-MASON, J.; Report on the Tea-mite and Tea-bug of Assam, 1884.
YARKAND; Lahore to Yarkand, by Henderson and A. O. Hume, 1873.
„ Second Mission to—, 1875.
„ Scientific Results of the Second Mission to—, XI. Parts, 1878-1885

LIST OF OFFICERS

who have rendered assistance, having contributed original communications regarding Indian Economic Products.

AITCHISON, M.D., C.I.E.; Brigade Surgeon J. E. T., Simla (at present with the Afghán Boundary Commission).
ANDERSON, M.B.; Surgeon J., Bijnor, North-Western Provinces.
ANDERSON; Honorary Surgeon Peter, Zillah Guntur, Madras.
ASHWORTH; Apothecary J. G., Kumbakonam, Madras.
ATKINSON, C.S.; E. F. T., Accountant-General, Bengal.
BANERJI; Assistant Surgeon Nil Ratton, Etawah.
BARKER, M.D.; Surgeon R. A., Doomka, Santal Parganas.
BARREN; Surgeon W. A., Belgaum, Bombay.
BASU; Assistant Surgeon Shib Chunder, Bankipore.
BASU; Surgeon D., Faridpore, Bengal.
BATESON; Brigade Surgeon R. S., Umballa City.
BEECH; Surgeon-Major Lionel, Cocanada.
BENSLEY; Surgeon-Major E. C., Rajshahye.
BHATTACHARJI; Assistant Surgeon Nogendra Nath, Somastipore, Tirhoot State Railway.
BHATTACHARJI; Assistant Surgeon Shib Chandar, Chanda, Central Provinces.
BHIMBHÁI KIRPÁRÁM; Personal Assistant to the Director of Agriculture, Bombay.
BIDIE, M.B., C.I.E.; Deputy Surgeon-General George, Madras.
BONAVIA, M.D.; Surgeon-Major E., Etawah.
BORILL; Surgeon Edward, Motihari, Champarun.
BRANDER, M.B., F.R.C.S.E.; Surgeon Edward S., Rungpore.
BRIDGES, C.S.; J. E., Director of Agriculture, British Burma.
BROWNE, M.D.; Surgeon S. H., Hoshangabad, Central Provinces.
BUCHANAN; A. M., Forest Department, Rangoon.
CALTHROP, M.D.; Surgeon-Major C. W., Morar.
CAMERON; J., Superintendent, Botanic Gardens, Bangalore, Mysore.
CAMERON, M.D.; Surgeon L., Nuddea.
CAMPBELL; Rev. A., Santal Missionary, Free Church of Scotland, Pokhuria Govindpur, Chutia Nagpur.
CARTER, M.D. (LOND.), I.M.D.; Brigade-Surgeon H. V., Bombay.
CHUNA LALL; 1st Class Hospital Assistant, Jubbulpore.
CLARKE, M.A.; C. B., Inspector of Schools, Assam.
COLDSTREAM, C.S.; W. C., Commissioner, Hissar.
COOK; Surgeon-Major Henry David, Calicut, Malabar.
CORNISH, F.R.C.S., C.I.E.; Surgeon-General William Robert, Madras.
CROMBIE; Surgeon A., Dacca.
CROOKSHANK; Col. A. C. W., Deputy Secretary, Military Department, Government of India.
DARRAH, C.S.; H. Z., Director, Agricultural Department, Assam.
DASS; Assistant Surgeon Bhugwan, Rawal Pindi, Panjáb.
DETATHAM, M.D. (BRUS.), M.R.C.P. (LOND.), M.R.C.S.; Surgeon-Major H., Ahmednagar Bombay.
DEY, C.I.E.; Surgeon K. L., Calcutta.
DEY; Assistant Surgeon J. N., Jeypore.
DIMMOCK, M.R.C.S., L.R.C.P. (LOND.); Surgeon H. P., Shikárpur, Bombay.
DOBSON, M.B.; Surgeon-Major Andrew Francis, Bangalore.
DRIVER; H. P., Ranchi, Chutia Nagpur.

DRUMMOND, C.S.; J. R., Assistant Commissioner, Simla.
DUTHIE; J. F., Superintendent, Government Botanic Gardens, Saharanpore.
DUTT; Assistant Surgeon Nobin Chunder, Durbhunga.
DUTT, M.D.; Surgeon R. L., Pubna.
DUTT; Civil Medical Officer U. C., Serampore.
DYER, C.M.G.; Prof. W. T. Thiselton, Assistant Director, Royal Botanic Gardens, Kew, London.
DYMOCK, B.A.; Surgeon-Major W., Medical Store-keeper, Bombay.
EMERSON; Surgeon G. A., Calcutta (at present with the Suakim Field Force).
EVERS, M.D.; Surgeon B., Wardha.
FARQUHAR, M.D.; Surgeon-Major W., Ootacamund.
FINUCANE, C.S.; M., Director of Agriculture, Bengal.
FITZPATRICK, M.D.; Surgeon-Major J. F., Coimbatore.
FORSYTH, F.R.C.S. (EDIN.); Civil Medical Officer W., Dinajpore, North Bengal.
FRENCH, M.D., F.R.C.S.; Surgeon-Major J. G., Bankipore.
FULLER, C.S.; J. B., Director, Agricultural Department, Central Provinces.
GHOLAM NABI; Assistant Surgeon.
GHOSE, M.D., M.R.C.S.; Surgeon K. D., Khoolna.
GHOSE; Surgeon K.D., Bankura, Bengal.
GHOSE; Assistant Surgeon Nundo Lal, Bankipore
GHOSE; Assistant Surgeon T. N., Meerut.
GILLIGAN; Surgeon W. A., Durbhunga.
GRAY; Surgeon R., Lahore.
GUPTA; Assistant Surgeon Ram Chunder, Bankipore.
GUPTA, M.B.; Surgeon-Major Bankabehari, Pooree.
HAZLETT; Surgeon-Major H. J., Salem, Madras.
HILL; Surgeon H. W., Mánbhúm.
HOUSTON; Surgeon-Major J. McD., Travancore; and JOHN GOMES, Esq., Medical Store keeper, Trivandrum.
HUNTER, M.R.C.S. (LOND.), L.S.A.; Surgeon-Major G. Y., Bombay.
HUTCHINSON, M.D.; Deputy Surgeon-General R. F., Morar.
JASWANT RAI; Assistant Surgeon, Mooltan.
JAYAKAR; Surgeon-Major A. S. G., Muskat.
JOHNSON; Surgeon-Major E. R., Simla.
JOUBERT, M.B.; Surgeon C. H., Darjeeling, Bengal.
JOYNT, M.D.; Brigade-Surgeon C., Poona.
KAMELL; Apothecary H.
KENNEDY, C.S.; T. J., Assistant Commissioner, Hazára.
KING, M.B., LL.D.; Surgeon-Major George, Superintendent, Botanical Gardens, Seebpore, Calcutta.
KING, M.B.; Surgeon W. G., Kurnool, Madras.
KINSLEY; Honorary Surgeon P., Chicacole, Ganjam, Madras.
LANCASTER, M.B.; Surgeon John, Chittore.
LANGLEY, M.R.C.S. (LOND.), B.A.; Brigade-Surgeon E. H. R., Bombay.
LEE; Surgeon W. A., Mangalore.
LEVINGE; Surgeon-Major E., Rajahmundry, Godavery District, Madras.
LIOTARD, L.; Department of Finance and Commerce, Government of India, Calcutta.
MACLEOD, M.D.; Surgeon Roderick, Gya.
MAHOMED; Hospital Assistant Lal, Hoshangabad, Central Provinces.
MAITLAND, M.B.; Surgeon J., Madras.
MANSON; F.B., Deputy Conservator of Forests, Ranchi, Lohardaga.
MASANI; Surgeon H. D., Karáchi.
MATHEW; Surgeon R. G., Mozufferpore.
MCCALMAN, M.B., M.D., C.M. (ABER.); Surgeon H., Ratnágiri, Bombay.

McCloghry; Surgeon James, Jakobabad, Sind.
McConaghey, M.D.; Surgeon John, Shahjahanpore.
McKenna; Surgeon-Major C. J., Cawnpore.
Meadows; Surgeon C. J. W., Burrisal.
Meer Comer Ali; Native Doctor, Bhognipore Division, Etawah.
Misr; Narain, Hoshangabad, Central Provinces.
Modeen Sheriff, Khan Bahadur; Honorary Surgeon, Triplicane.
Moir; Surgeon-Major W., Meerut.
Mokund Lal; Surgeon, Agra.
Morgan; Brigade Surgeon W. H., Cochin.
Morris; Honorary Surgeon Easton Alfred, Tranquebar.
Mukerji, M.B., C.M.; Civil Medical Officer U.C., Dinajpore.
Mukerji, Surgeon Anund Chunder, Noakhally.
Mukerji; Surgeon-Major P. N., Cuttack, Orissa.
Mukerji; T. N., Revenue and Agricultural Department, Calcutta.
Mullane; Surgeon J., Gauhati, Assam.
Mullen; Surgeon J. Ffrench, Saidpur, N. B. State Railway, Bengal.
Murray, M.B.; Surgeon R. D., Burdwan.
Nanney; Surgeon-Major Lewis Charles, Trichinopoly.
Nariman, L. M. (Bo.), L. R. C. S. (Ed.), L. S. A. (Lond.); Surgeon K. S., Broach, Bombay.
Nehal Sing; Assistant Surgeon, Saharunpore.
Nolan, M.A., M.D. (Dub.); Surgeon-Major W., Bombay.
Norman; Apothecary J., Chatrapore, Ganjam.
North; Surgeon-Major John, Bangalore.
O'Conor; J. E., Assistant Secretary, Department of Finance and Commerce, Government of India.
Oliver; J. W., Deputy Conservator of Forests, Tharrawaddy, British Burma.
Ozanne, C.S., M.R.A.C.; E. C., Director of Agriculture, Bombay.
Parakh, L.R.C.P., M.R.C.S. (Lond.); Surgeon D. N., Bombay.
Parker; Surgeon-Major C. R. G., Bangalore.
Parker, M.D.; Surgeon Joseph, Poona.
Peacocke, L.R.C.S., L.M.K., Q.C.P. (Ir.); Surgeon J. C. H., Shikárpur, Bombay.
Penny, M.D.; Surgeon J. C., Amritsur.
Perry; Surgeon F., Jullunder City, Panjáb.
Peters; Surgeon-Major C. T., South Afghánistan.
Picachy; Surgeon D., Purneah.
Pilcher; Surgeon-Major J. G., Howrah.
Pitcher; Major D. G., Assistant Director, Agricultural Department, North-Western Provinces.
Poynder; Surgeon G. F., Roorkee.
Price; Surgeon G., Shahabad.
Ratton, M.D.; Surgeon-Major J. J. L., Medical College, Madras.
Reid; Surgeon-Major J., Fort William, Calcutta.
Rennie; Surgeon S. J., Cawnpore.
Rice, Ph.D.; Charles, New York, America.
Rice, M.D.; Brigade Surgeon W. R., Jubbulpore.
Richards; Surgeon Vincent, Goalundo, Bengal.
Robb, M.D., C.M. (Aber.); Surgeon-Major J., Ahmedabad, Bombay.
Robb; Surgeon R., Hazaribagh.
Robinson; Surgeon-Major Mark, Coorg.
Ross; Surgeon George Cumberland, Delhi.
Roy; Assistant Surgeon Devendro Nath, Campbell Medical School, Sealdah, Calcutta.
Roy; Surgeon-Major G. C., Birbhúm.
Russell, M.D.; Surgeon C. M., Sarun.
Russell; Surgeon-Major E. G., Superintendent, Asylums, Calcutta.

RUTHNUM MOODELIAR, Native Surgeon T., Chingleput, Madras.
SAISE, D.S.C., F.G S.; W., Manager, East Indian Railway Company's Collieries, Giridhi.
SAKHÁRÁM ARJUN RÁVAT, L.M.; Assistant Surgeon, Bombay.
SANDERS; Surgeon-Major E., Chittagong, Bengal.
SANGKY, M.B.; Surgeon-Major E. H. R., Bombay.
SEN; Assistant Surgeon Bolly Chand, Campbell Medical School, Sealdah, Calcutta.
SHIRCORE; Brigade Surgeon S. M., Moorshedabad.
SHOME; Assistant Surgeon Doyal Chunder, Campbell Medical School, Sealdah, Calcutta.
SMEATON, C.S.; D. M., Director, Agricultural Department, North-Western Provinces and Oudh.
SMITH; Deputy Surgeon-General C., Hyderabad Subsidiary Force, Secunderabad.
STEVENSON; W. H., Missionary, Free Church of Scotland, Santal Mission, Pachumba, Giridhi, Chutia Nagpur.
STEWART; Surgeon-Major W. D., Cuttack.
SUKHIA, L M.S.; Assistant Surgeon Nádarshah Hormasji, Bombay.
TALBOT; W. A., Conservator of Forests, Yellapur, Kanara.
TEMPLE; Captain R. C., Cantonment Magistrate, Umballa.
THOMAS; Surgeon-Major J. Byers, Masulipatam, Kistna.
THOMAS; Surgeon W. F., Mangalore.
THOMPSON, M.D., C.I.E.; Surgeon-Major D. R., Madras.
THORNTON, B.A., M.B.; Brigade Surgeon J. H., Monghyr (at present with the Suakim Field Force).
TILLERY; Mr., Jeypore.
TOMES; Surgeon A., Midnapore.
TRIMEN; Dr. H., Director, Botanic Gardens, Ceylon.
UMMEGUDIEN; Native Doctor V., Mettapolliam, Madras.
WACE; Lieutenant-Colonel E. G., Commissioner of Settlements and Agriculture, Panjáb.
WARD; Surgeon G. J., Mhow.
WARD; Apothecary Thomas, Madanapalle, Cuddapah.
WARDEN; Surgeon C. J. H., Professor of Chemistry, Medical College, Calcutta.
WATSON; Brigade-Surgeon George Alder, Allahabad.
WOODROW; G. M., Lecturer, College of Science, Poona.
WESTCOTT; Surgeon S., Dinapore.
WILSON; Surgeon William, Bogra.
WILSON, C.S.; W., Director, Revenue Settlement and Agriculture, Madras.
ZORAB; Surgeon-Major John M., Balasore, Orissa.

LIST OF ABBREVIATIONS.

Authors' names are abbreviated and printed in italics after the name given by them to an animal or plant. This is necessary, since the same object may have come to receive more than one name, and may even have been referred to two distinct genera—a result either due to ignorance of what had already been published, or in consequence of a difference of opinion regarding the nature of the species. The ambiguity thus caused through the existence of more than one name for the same object, or through the application of the same name to two distinct objects, is mitigated by affixing the author's name in full, if this be composed of only one syllable, or (when of more than one syllable) some convenient abbreviation to represent the author's name. The abbreviation once adopted by one author is never assumed by a second author who may chance to bear the same, so that *Wall,* is accepted by all botanists in the world to stand for Dr. Wallich, one of the most distinguished of Indian botanists.

The multiplicity of synonyms for the same object is one of the most perplexing evils—and an unavoidable evil—which besets systematic science. The changing of names is not, however, the result of fancy, but is a necessary consequence of improved knowledge and extended acquaintance with the forms of life. The difficulty of synonymy experienced by Indian students is scarcely felt in Europe, where the names given to the animals and plants are thoroughly established. This explanation is offered because of the very general complaint against the changes which have, within the past few years, been made in the names given to Indian animals and plants. It should be recollected, that the natural history of an empire, like that of India, cannot be worked out in a century, and that any attempt to systematise the scattered publications which have appeared from time to time must result in the suppression of names which have come incorrectly into existence. The following list gives the principal abbreviations adopted for authors' names :—

Abbreviations of Names of Botanists and Botanical Authors.

Ach. = Acharius.
Adans. Adanson.
Afz. Afzelius.
Ag. Agardh.
C. Ag. C. A. Agardh.
J. Ag. / *Ag. f.* } J. G. Agardh, son.
Ait. Aiton.
Aitch. Aitchison.
All. Allioni.
Amm. Amman.
Anders. Andersson of Stockholm.
Anders. Anderson.
Andr. Andrews.
Andr. B. Repos. The Botanists' Repository, by Henry Andrews.
Andrz. Andrzejowski.
Ann. mus. Annales du Museum d'Histoire Naturelle.
Aresch. Areschoug.
Arn. = Arnott.
Arrh. Arrhenius.
Asa Gray. Asa Gray.
Asch. Ascherson.
Asiat. Reser. Asiatic Researches.
Aubl. Aublet.

B. Cab. Botanical Cabinet.
B. Mag. Botanical Magazine.
B. Misc. Botanical Miscellany.
B. Reg. Botanical Register.
Bab. Babington.
Bail. Baillon.
Baker. Baker.
Balb. Balbis.
Balbis. Balbis (John Baptist), a French Professor of Botany
Baldw. Baldwin.
Balf. Balfour.
Banks. Herb. Banksian Herbarium

Barn = Barnéoud.
Barr. Barrelier.
Bart. Benj. Smith Barton.
W. Bart. W. P. C. Barton, nephew.
Bartl. Bartling (Th. Fr.) Ord. Nat. Plantarum.
Bartr. John Bartram.
Bartr. f. Wm. Bartram.
Bauh. Bauhin.
J. Bauh. Bauhin (Johannes).
Beauv. Palisot de Beauvois.
Bedd. Beddome.
Benj. Benjamin.
Benn. J. J. Bennett.
A. Benn. A. W. Bennett.
Benth. or Bth. Bentham.
Berg. Bergius.
Berk. M. J. Berkeley.
Berkh. Berkhey.
Berken. Berkenhout.
Berland. Berlandier.
Bernh. Bernhardi, a German botanist.
Bert. Bertero.
Bertol. Bertoloni.
Bess. Besser.
Bieb. Marschall von Bieberstein.
Bigel. Jacob Bigelow.
Bisch. Bischoff.
Bivon. Bivona (a Sicilian botanist).
Bl. or *Blume.* Blume.
Bl. Bijdr. Blume (C. L.), Bijdragen tot de flora von Nederlandsche Indië.
Bl. Cat. Blume's Catalogue.
Bœhm. Bœhmer.
Boerh. Bœrhaave.
Boiss. Boissier.
Boland. Bolander.
Bong. Bongard.
Booth. Dr. Booth.
Bonpl. Bonpland.
Bork. Borkhausen.
Bosc. Bosc, a French botanist, &c.
Borsz. Borszcow.
Brack. Wm. D. Brackenridge.
Brebis. Brebisson.
Bref. Brefeld.
Brew. & Wats. W. H. Brewer and Sereno Watson.
Brid. Bridel.
Brong. Brongniart.
Brot. Brotero.
Brouss. = Broussonet.
Br. or *R. Br.* Robert Brown.
Br. or *N.E. Br.* N. E. Brown.
P. Br. Patrick Browne.
Brunf. Brunfels.
Ham. Buchanan Hamilton.
Buch. Dr. F. Hamilton (formerly Buchanan).
Buch. Herb. Buchanan's Herbarium.
Buch. MSS. Buchanan's Manuscripts.
Buckl. Buckley.
Bull. Bulliard.
Burch. Burchell (Wm.), a South African traveller and botanist.
Burm. Burman.
Burm. Afric. Burmann's (Johannes) African Plants.
Burm. Zeyl. Thesaurus Zeylanicus, by J. Burmann.
Burm. Ind. Flora Indica, by N.L. Burmann.
Buxb. Buxbaum.
Cam. Camerarius.
Camb. } Cambessedes.
Cambes. }
Campd. Campdera.
Cand. De Candolle, usually DC.
Casp. Caspary.
Cass. Cassini.
Catesb. Catesby.
Cav. Cavanilles.
Cerv. Cervantes.
Cham. Chamisso.
Chapm. A. W. Chapman.
Chav. Chavannes.
Chois. Choisy.
Clarke or *C. B. C.* C. B. Clarke.
Clayt. Clayton.
Clus. Clusius.
Colebr. Colebrooke (H. T.), a well known English writer on Indian plants.
Collad. Colladon.
Colm. Colmeiro.
Comm. Commelin.
Corn. Cornuti.
Coss. Cosson.
Cunn. Cunningham, A. or J.
Curt. Wm. Curtis.
M. A. Curt. M. A. Curtis.
Cyril. Cyrilli, an Italian botanist.
Dalech. Dalechamps.
Dalib. Dalibard.

Dalz. =	Dalzell.
Dalz. & Gibs.	Dalzell and Gibson.
Darl.	Darlington.
DC. / *DeC.*	A. P. DeCandolle.
A. DC.	Alphonse DeCandolle, son.
Cas. DC.	Casimir DeCandolle, grandson.
Decne.	Decaisne.
Deless.	Delessert.
Dennst.	Dennstedt.
Desc.	Descourtilz.
Desf.	Desfontaines.
Desj.	Desjardins.
Desmar.	Desmazières.
Desmoul.	Desmoulins.
Desv.	Desvaux.
Dicks.	Dickson, an English cryptogamic botanist.
Diesb.	Diesbach.
Dieter.	Dieterich.
Dietr.	Dietrich.
Dill.	Dillenius.
Dillw.	Dillwyn.
Dod.	Dodonæus (Dodœns).
Don.	D. Don.
D. Don.	Don (D.) Prodromus floræ Nepalensis.
G. Don.	Don (G.) in Miller's Dictionary.
Donn. Hort. Cantab.	Donn. Hortus Cantabrigiensis, 1796.
D'Orb.	D'Orbigny.
Dorst.	Dorstenius.
Dougl.	Douglas.
Drej.	Drejer.
Dryand.	Dryander.
Dub.	Dubois, a French botanist.
Dufr.	Dufresne.
Duham.	Duhamel du Monceau.
Dumort.	Dumortier.
Dun.	Dunal.
Eat.	Amos Eaton.
D. C. Eat.	D. C. Eaton, grandson.
Edgew.	Edgeworth.
Edw.	Edwards.
Ehren.	Ehrenberg.
Ehret.	Ehret (Geo. Dion).
Ehrh.	Ehrhart.
Eichl.	Eichler.
Eiseng.	Eisengrein.
Ell.	Elliott.
Endl.	Endlicher.
Engelm. =	Engelmann.

Engl.	Engler.
Eschs.	Eschscholtz.
Eschw.	Eschweiler.
Ettingsh.	Ettingshausen.
Falk.	Falkner.
Fendl.	Fendler.
Fenill.	Feniller, a Chilian botanist.
Feuil.	Feuillée.
Fingerh.	Fingerhuth.
Fisch.	Fischer.
Fl. Dan.	Flora Danica.
Forsk.	Forskāl.
Forst.	Forster.
Fourn.	Fournier.
Fresen.	Fresenius.
Freyc.	Freycinet.
Frœl.	Frœlich.
Gærtn.	J. Gærtner.
Gærtn. f.	C. T. Gærtner.
Gamble.	J. S. Gamble.
Gardn.	Gardner.
Garid.	Garidel.
Gasp.	Gasparrini.
Gaud.	Gaudin.
Gaudich.	Gaudichaud.
Germ.	Germain.
Gesn.	Gesner.
Gilib.	Gilibert.
Ging.	Gingins de Lassaraz.
Gis.	Giseke.
Gled.	Gleditsch.
Gleich.	Gleichen.
Glox.	Gloxin.
Gmel.	J. G. Gmelin.
C. Gmel.	C. C. Gmelin of Baden.
S. Gmel.	S. G. Gmelin.
Godr.	Godron.
Gæpp.	Gœppert.
Good.	Goodenough.
Grah.	Graham (Dr. J.), Professor of Botany at Edinburgh.
Grah. Cat. B. J. pl.	Graham's (J.) Catalogue of Bombay plants.
Gren.	Grenier.
Grev.	Greville.
Griff.	Griffith (Dr. W.)
Griseb.	Grisebach.
Grœn.	Grœnland.
Gron. / *Gronov.*	Gronovius.
Guett.	Guettard.
Guib. =	Guibord.

Guillem.	Guillemin.
Guimp.	Guimpel.
Gunn.	Gunnerus.
Guss.	Gussone.
Habl.	Hablizl, a Russian botanist.
Hagenb.	Hagenbach.
Hall.	Haller.
Ham.	Hamilton.
Ham.	Hamilton (Dr. Francis), formerly Buchanan.
Hanb.	Hanbury.
Hanst.	Hanstein.
Hartm.	Hartmann.
Hartw.	Hartweg.
Harv.	Harvey.
Hass.	Hassall.
Hassk.	Hasskarl.
Hausm.	Hausmann.
Haw.	Haworth.
Hb. Madr.	Herbarium (Madras), by Drs. Klein, Heyne, and Rottler.
Hebens.	Hebenstreit.
Hedw.	Hedwig.
Hegelm.	Hegelmaier.
Hegetsch.	Hegetschweiler.
Heist.	Heister.
Heldr.	Heldreich.
Helw.	Helwing.
Hemsl.	Hemsley.
Henck.	Henckel.
Henfr.	Henfrey.
Hensl.	Henslow.
Herb.	Herbert.
Herm.	Hermann.
Hild.	Hildebrand.
Hochst.	Hochstetter.
Hoffm.	G. F. Hoffmann.
H. Hoffm.	Hermann Hoffmann.
Hoffmanns.	Hoffmannsegg.
Hofm.	Hofmeister.
Hohen.	Hohenacker.
Holmsk.	Holmskiold.
Homb.	Hombron.
Hook.	Wm. J. Hooker.
Hook. f.	J. D. Hooker, son.
H. f. and T. T.	Hooker, f. and T. Thomson.
Hopk.	Hopkirk.
Hornem.	Hornemann.
Hornsch.	Hornschuch.
Horsf.	Horsfield.
Houst.	Houston.
Houtt.	Houttuyn.
Huds.	Hudson.
Hueb.	Huebener.
Humb.	Humboldt.
H. B. K.	Humboldt, Bonpland, and Kunth.
Jack.	Jack (Dr. William), a most distinguished botanist.
Jacq.	N. J. Jacquin.
Jacq. f.	J. F. Jacquin, son.
J. St. Hil.	Jaume St. Hilaire.
Jord.	Jordan.
Jungh.	Junghuhn.
Juss.	A. L. Jussieu.
Adr. Juss.	Adrien Jussieu, son.
Kæmp.	Kæmpfer.
Karst.	Karsten.
Kaulf.	Kaulfuss.
Kindb.	Kindberg.
King.	G. King.
Kirschl.	Kirschleger.
Kit.	Kitaibel.
Koch.	Koch, Professor at Erlang.
Kœlr.	Kœlreuter.
Koen.	Koenig (J. Gerard), a Danish botanist and pupil of Linnæus.
Korth.	Korthals.
Kostel.	Kosteletzky.
Kremp.	Krempelhuber.
Kromb.	Krombholz.
Kth.	Kunth, an eminent Prussian botanist.
Kuetz.	Kuetzing.
Kunze.	Kunze, a German Cryptogamic botanist.
Kurz.	S. Kurz.
L. or *Linn.*	Linnæus.
Labill.	La Billardière.
Læst.	Læstadius.
Lag.	Lagasca.
Lall.	Lallement.
Lam	Lamarck (Monnet de La Marck).
Lam. fl. Fr.	Lamarck (J. Be Monet de) Flore Francaise.
Lam. fl.	Lamarck (J. B. Monet de) Illustration des genres.
Lamb.	Lambert.
Lamour.	Lamouroux.
Langsd.	Langsdorf.
La Peyr.	La Peyrouse.
La Pyl.	La Pylaie.

Laws. = Lawson.
Laxm. Laxmann, a writer on Siberian plants.
Ledeb. Ledebour.
Lehm. Lehmann.
Lem. Lemaire.
Lesch. Leschenault de la Tour.
Lesq. Lesquereux.
Less. Lessing.
Lestib. Lestiboudois.
Lév. Léveillé.
L'Her. L'Heritier.
L'Hérm. L'Herminier.
Liebm. Liebmann.
Lightf. Lightfoot.
Lilij. Lilijeblad.
Lindb. Lindberg.
Lindbl. Lindblom.
Lindenb. Lindenberg.
Lindh. Lindheimer.
Lindl. Lindley.
Linn. Linnæus. Also *L.*
Linn. f. C. Linnæus, son.
Lockh. Lockhart (W.)
Lodd. Loddiges.
Lœfl. Lœfling.
Lœs. Lœselius.
Lois. Loiseleur-Delongschamps.
Loud. Loudon.
Lour. Loureiro.
Ludw. Ludwig.
Lumn. Lumnitzer.
Lyngb. Lyngbye.

Macf. Macfadyen.
Macgil. MacGillivray.
Magn. Magnol.
M. Bieb. Marschall von Bieberstein.
Marsh. Marshall (Humphrey).
Mars. Marsili.
Mart. Martius.
Mass. Massalongo.
Mast. Masters.
Maxim. Maximowicz.
Med. Medikus or Medicus.
Meisn., *Meissn.* Meisner or Meissner.
Meneg. Meneghini.
Menz. Menzies.
Mert. Mertens.
Metten. Mettenius.
Mich. Micheli.
Michx., *Mx.* André Michaux.
Michx. f. F. A. Michaux, son.

Midden. = Middendorff.
Mill. Philip Miller.
Mill. J. John S. Mueller or Miller.
Miq. Miquel.
Mirb. Mirbel.
Mitch. John Mitchell.
Mitt. Mitten.
Moc. Moçino.
Moen. Moench, a German botanist.
Molk. Molkenboer.
Mont. C. Montagne.
Moq. Moquin-Tandon.
Moric. Moricand.
Moris. Morison.
Morr. Morren.
Moug. Mougeot.
Muell. Arg. J. Mueller of Argau.
F. Muell. Ferdinand Mueller.
O. Muell. Otto Mueller of Denmark.
Muhl. Muhlenberg.
Munt. Munting.
Murr. J. A. Murray.
A. Murr. Andrew Murray.

Nacc. Naccari.
Næg. Nægeli.
Naud. Naudin.
Neck. Necker.
Nees or *N. ab E.* C. F. Nees von Esenbeck.
T. Nees. T. F. L. Nees von Esenbeck.
Nestl. Nestler.
Newb. Newberry.
Newm. Newman.
Nœgg. Nœggerath.
Nois. Noisette.
Nord. Nordstedt.
Not. Notaris.
Nutt. Nuttall.
Nyl. Nylander.
Nym. Nyman.

Œd. Œder.
Œrst. Œrsted.
Oliv. Olivier.
D. Oliv. D. Oliver.
Orb. A. or C. d'Orbigny.
Orph. Orphanides.
Ort. Ortega.
Oudem. Oudemans.

P. de. Beauv. Palisot de Beauvois.
Pall. Pallas.
Panz. Panzer.
Park. Parkinson.

Parl. = Parlatore.
Pasq. Pasquale.
Pav. Pavon.
Perl. Perleb.
Pers. Persoon.
Philib. Philibert.
Planch. J. E. Planchon.
G. Planch. Gustave Planchon.
Pluk. Plukenet.
Plum. Plumier, Lat. Plumerius.
Pœpp. Pœppig.
Poir. Poiret.
Poit. Poiteau.
Poll. Pollich.
Post. Postels.
Pourr. Pourret.
Pringsh. Pringsheim.
Pritz. Pritzel.
Putter. Putterlich.

Rabenh. Rabenhorst.
Radlk. Radlkofer.
Raf. Rafinesque-Schmaltz.
Rasp. Raspail.
Red. Redouté.
Reich. Reichard.
Reichenb. H. G. L. Reichenbach.
Reichenb. f. H. G. Reichenbach, son.
Reinw. Reinwardt.
Reiss. Reisseck.
Retz. Retzius.
Reut. Reuter.
Rheed. auct. Rheede, author of Hortus Malabaricus.
Rich. L. C. Richard.
Rich. f.
A. Rich. } Achille Richard.
Richards. John Richardson.
Richt. Richter.
Ridd. Riddell.
Riv. Rivinus.
Rœhl. Rœhling.
Rœm. J. J. Rœmer.
M. J. Rœm. M. J. Rœmer.
Rœm and Sch. Rœmer and Schultes.
Rœp. Rœper.
Rohrb. Rohrbach.
Roscoe. Roscoe.
Rostk. Rostkovius.
Roth. Roth (A. W.)
Rothr. Rothrock.
Rottb. Rottbœll.
Rottl. Rottler.
Roum. Roumegère.

Roxb. = Roxburgh.
Roxb. H.Bengl. Roxburgh's Hortus Bengalensis.
Royle. Royle.
Royle. Ill. Royle's Illustrations.
Roy. Royen.
Rudb. Rudbeck.
Rupr. Ruprecht.
Sacc. Saccardo.
Sadl. Sadler.
St. Hil. A. Saint Hilaire.
Salisb. Salisbury.
Salm-Dyck. Prince Jos. Salm-Rifferschied-Dyck.
Sauss. Saussure.
Schimp. Schimper.
Schk. Schkuhr.
Schlecht. Schlechtendal.
Schleich. Schleicher.
Schomb. Schomburgh.
Schrad. Schrader.
Schreb. Schreber.
Schueb. Schuebeler.
Schult. Schultes.
Schultz.
Bip. } C. H. Schultz, Bipontinus (Zweibrucken).
Schum. Schumacher.
Schnitzl. Schnitzlein.
Schwægr. Schwægrichen.
Schwein. Schweinitz.
Schweinf. Schweinfurth.
Schwend. Schwendener.
Scop. Scopoli.
Seem. Seemann.
Sendt. Sendtner.
Seneb. Senebier.
Ser. Seringe.
Seub. Seubert.
Sibth. Sibthorp.
Sieb. Sieber.
Sieb. Siebold.
Sim. Simmonds.
Sm. Smith (J. E.)
Soland. Solander.
Sow. Sowerby.
Spenn. Spenner.
Spreng. Sprengel.
Sternb. Sternberg.
Steud. Steudel.
Stev. Steven.
Stocks Stocks.
Sull. Sullivant.
Sw. Swartz.
Swt. Sweet (R.)

Tabern.	=	Tabernæmontanus (J. T.)
Targ.		Targioni-Tozetti.
Telf.		Telfair.
Ten.		Tenore.
Thib.		Thibaud deChauvalon.
Thoms.		Thomas Thomson.
Thonin		Thonin.
Thore		Thore.
Thuill.		Thuillier.
Thunb.		Thunberg.
Thurb.		Thurber.
Thurm.		Thurman.
Tod.		Todaro.
Torr.		Torrey.
Torr. & Gr.		Torrey & A. Gray.
Tourn.		Tour'nefort.
Trans. Linn. Soc.		Transactions of the Linnæan Society.
Tratt.		Trattinick.
Traut.		Trautvetter.
Trev.		Treviranus.
Trim.		H. Trimen.
Trin.		Trinius.
Tuck.		Tuckerman.
Turcz.		Turczaninow.
Turn.		Turner.
Turp.		Turpin.
Tussc.		Tussac.
Vahl.		Vahl (Prof. Martin) of Copenhagen.
Vaill.		Vaillant.
Veill.		Veillard or Vieillard.
Vauch.		Vaucher.
Vent.		Ventenat.
Vill.		Villars, or Villar.
Vis.		Visiani.
Vittad.		Vittadini.
Viv.		Viviani.
Vog.		T. Vogel.
Wahl.	=	Wahlenberg.
Wahlst.		Wahlstedt.
Walds.		Waldstein.
Wall.		Wallich.
Wall. Cat.		Wallich's Catalogue.
Wallm.		Wallman.
Wallr.		Wallroth.
Walp.		Walpers.
Walt.		Walter.
Wang.		Wangenheim.
Warm.		Warming.
Watt.		G. Watt.
Wats.		P. W. Watson.
H. C. Wats.		H. C. Watson.
S. Wats.		Sereno Watson.
Web.		Weber.
Wedd.		Weddell.
Weinm.		Weinmann.
Welw.		Welwitsch.
Wender.		Wenderoth.
Wendl.		Wendland.
Wight. icon.		Wight's Icones.
Wight. Ill.		Wight's Illustrations of Indian Botany.
W. & A.		Wight and Arnott.
Wiks.		Wikstrom.
Wildb.		Wildbrand.
Willd.		Willdenow.
Willk.		Willkomm.
Wils.		Wilson.
Wimm.		Wimmer.
Wisliz.		Wislizenus.
With.		Withering.
Woodv.		Woodville.
Wulf.		Wulfen.
Zanard.		Zanardini.
Zetterst.		Zetterstedt.
Zucc.		Zuccarini.
Zuccag.		Zuccagini.

DICTIONARY

OF

THE ECONOMIC PRODUCTS OF INDIA.

Abaca, a name in the Philippine Islands for Manilla Hemp—**Musa textilis**, which see.

Abele, the Dutch name for the White Poplar, see **Populus alba.**

ABELIA, *R. Br.; Gen. Pl.,* Vol. II., 4.*

1

A genus of shrubs (belonging to the Natural Order CAPRIFOLIACEÆ), containing in all only five species: distributed from Kashmir to China and Japan. They are characterised by having opposite ex-stipulate *leaves*. *Calyx* adnate to the ovary, *lobes* elongate. *Corolla* regular or nearly so, funnel-shaped, with five short, rounded teeth. *Style* long. *Stigma* capitate. *Ovary* 3-celled, two cells being many-ovuled but early aborted, the other developing. *Fruit* elongated, coriaceous, 1-seeded.

This genus was named by **R. Brown** after the Chinese explorer, **Mr. Clarke Abell.** The species belonging to it are more objects of ornament than of utility. **A. floribunda** from Mexico has purple-red flowers, and **A. rupestris** from China has pale rose-coloured flowers.

There is only one species met with in India.

Abelia triflora, *R. Brown; Fl. Br. Ind.,† III., 9.*

2

Vernacular Names used in different parts of India.—*Adei, paktawar,* PUSHTO (TRANS-INDUS); *Cheta búta,* JHELUM VALLEY; *Ban bakhara, salanker,* CHENAB VALLEY; *Dalúng, kút sái,* RAVI VALLEY; *Zbang, matsbang, peni,* SUTLEJ VALLEY; *Munri, gogatti, kumki,* KUMAON.

Habitat.—A large shrub, met with in Safedkoh and the Suliman Range, North-West Himálaya, between 4,000 and 10,000 feet; also from Kashmír to Kumaon.

Properties and Uses—

Structure of the Wood.—Hard, close, and even-grained. Weight 65 lbs. per cubic foot.

TIMBER 3

Not used, except for walking-sticks.

ABELMOSCHUS, *Medik.; Gen. Pl., I., 208.*

4

A generic name formerly given to a group of species now referred to the genus **Hibiscus.** They are characterised by having an elongated, spathaceous calyx. The word Abelmoschus is derived from the Arabic signifying musk-seeded, in allusion to the odour possessed by the seeds of **Hibiscus Abelmoschus,** *Linn.*

* By Gen. Pl. is meant Bentham and Hooker's Genera Plantarum.
† Fl. Br. Ind. means Sir J. D. Hooker's Flora of British India.

Abelmoschus esculentus, *W.&A.; O'Shaughnessy, Beng. Disp, .217.*

Syn. used by old authors for **Hibiscus esculentus,** *Linn.*, which see.

A. ficulneus, *W. & A.* (as in Drury).

Syn. for **Hibiscus ficulneus,** *Linn.*, which see.

A. moschatus, *Mœnch.*

Syn. used by old authors for **Hibiscus Abelmoschus,** *Linn.*, which see.

A. strictus (as in Voigt).

Syn. for **Hibiscus ficulneus,** *L.*, which see.

5

ABIES, *Juss.; Gen. Pl., III., 441.*

A genus of lofty evergreen monœcious trees, belonging to the Natural Order CONIFERÆ, containing some 18 species, widely distributed throughout the colder temperate regions. *Leaves* single, spirally arranged, or in two rows, needle-shaped or narrow, linear. *Male—catkins* single, in the axils of the leaves; *anther-cells* two. *Ovules* inverted, in pairs at the base of the carpillary scales. *Cones* ripening the same year, terminal or lateral, erect or pendulous. *Seeds* oily, winged.

The members of the genus **Abies** are popularly known as the FIRS in contradistinction to the PINES (*Pinus*). **Professor W. T. Thiselton Dyer** has kindly drawn my attention to the fact that the genus **Abies**, as viewed by Indian authors, has, in the *Genera Plantarum*, been broken up into the following: **Picea,** *Link.*, **Tsuga,** *Carr.*, **Pseudotsuga,** *Carr*, and **Abies,** *Juss.* It was found impossible, however, to effect the rearrangement which this necessitates, owing to the *Genera Plantarum* having only reached my hands after a great part of vol. I. was in type. The correct genera of the Indian species will be found indicated under each.

In India there are three important species:—

6

Abies dumosa, *Loudon;* (in *Gen. Pl.* referred to **Tsuga**).

THE INDIAN HEMLOCK SPRUCE.

Vern.—*Changathasi dhúp, thingia, thingáni súla* (*Gamble*) or *tingúri-salla* (*Atkinson*), NEPAL; *Tangshing* or *tungsing*, BHUTIA; *Dárma*, KUMAON; *Tanghing, Surain*, N.-E. KUMAON (*Duthie*); *Semadung* (*Gamble*) or *semadúng (Atkinson), chemdang*, LEPCHA.

References.—*Brandis, For. Fl., 52; Gamble, Man. Timb., 408.*

Habitat.—A large tree met with in North-East Kumaon, Nepal, and Sikkim, between 8,000 and 10,500 feet. In Kumaon 3,650 acres are under this tree.

Botanic Diagnosis.—*Cones* 1 inch long, occurring on the ends of the branches. *Leaves* white underneath.

Properties and Uses—

RESIN. 7

Resin.—Little or nothing is known regarding its resinous properties.

TIMBER. 8

Structure of the Wood.—Soft, with a slight pinkish tinge. Weight 27 to 29 lbs. per cubic foot.

Used in Sikkim for shingles. It is suitable for planking and rough furniture.

DOMESTIC. Bark. 9

Domestic Uses.—The bark is used for roofing.

10

A. excelsa, *DC.;* (in *Gen. Pl.* referred to **Picea excelsa,** *Link.*).

THE SPRUCE FIR OF EUROPE, *Eng.;* POIX DE BOURGOGNE OU DES VOSGES, *Fr.;* FICHTENHARZ, TANNENHARZ, *Ger.*

Habitat.—A noble tree found in the mountains of Central Europe, in Norway, Sweden, and Russia; introduced into India.

Botanic Diagnosis.—This species much resembles the Himálayan **A. Smithiana.**

Properties and Uses—

RESIN. 11

Description of the Resin.—A yellowish-brown, opaque substance, which naturally exudes from the bark, hard and brittle when cold, strongly adhesive, has an agreeable, aromatic odour, especially when heated.

Chemical Composition.—§ "The resins obtained from the Coniferæ, according to **Maly**, are all similar. The oleo-resinous exudations known as TURPENTINE consist of an amorphous resin $C^{44}H^{62}O^{4}$ mixed with an essential oil or hydrocarbon of the composition $C^{10}H^{16}$. Burgundy Pitch is obtained from **Picea excelsa**; it has been deprived more or less of its essential oil by evaporation, but it is not a turpentine. The so-called Turpentines possess their essential oil; they are obtained from **Pinus Pinaster** (in France) and **Pinus australis** (in America). If exposed to water or moisture the amorphous resin contained in the turpentines combines chemically to a certain extent with the water and becomes opaque and crystalline, being transformed into a substance having the character of an acid. When crude, soft turpentine is distilled, nearly the whole of the OIL OF TURPENTINE passes over while the resin (*rosin* or *colophony*) remains. If this substance still contains a little water it is known as *yellow rosin*; if completely deprived of water, it is *transparent rosin*; and by a continued application of heat it becomes *black rosin*. The crude turpentine which concretes upon the stem is termed in France *galipot* or *barras*. The American concrete crude turpentine is known in trade as common *Frankincense* or *gum Thus*. OIL OF TURPENTINE is distilled from the liquid crude turpentine, collected in *boxes* or artificial cavities cut on the trunk of the tree. This liquid substance is technically called *Dip*. The first year's flow from a new tree is called *Virgin Dip*: this yields the best quality of turpentine oil and of rosin. From the wood of Coniferæ a crystalline glucoside *Coniferin* has been isolated by **Rube**."

Turpentine. Galipot. 12

Gum thus. 13

Oil of Turpentine. 14

Rosin or Colophony. 15

"The essential oils of the various coniferous resins vary considerably. Of the turpentines the two most important varieties commercially are the French and the American; that chiefly used in England being the American. Canada Balsam and Strassburg Turpentine are also well-known coniferous products. Hydro-carbons similar to those obtained from the Coniferæ are also derived from Rutaceæ, Myristiceæ, Lauraceæ, Umbelliferæ, and even from some Labiatæ." (*Surgeon C. J. H. Warden, Prof. Chemistry, Medical College Hospital, Calcutta.*)

MEDICINE. Burgundy Pitch. 16

Medicine.—The resinous exudation from the stem of **Picea excelsa** is officinal in the *Pharmacopœia of India*, and is used as a stimulant and rubefacient; always applied in the form of plaster. It is known to English writers as BURGUNDY PITCH, although that term, as used by French authors and popularly in many English works, has a wider significance, being applied to the turpentines of other Coniferæ.

Abies Smithiana, *Forbes*; (in *Genl. Pl.* referred to Picea? Morinda, *Link.*).

17

THE HIMÁLAYAN SPRUCE.

Vern.—*Wesha, 'bajúr*, AFG.; *Kachal, kachan*, HAZARA, KASHMIR; *Rewari, rái, ban lúdar, sangal, salla, salie, sarei, káulí, roi, rág, ráo, bang rê, krok*, PB. HIMALAYA generally; *Tos*, RAVI; *Rau, raiang, re, ro*, SUTLEJ; *Rái*, JAUNSAR; *Kandre, re, rhái, ráo, rái, khutrau (Brandis*

§ Information specially contributed to this work by the authors whose names follow each paragraph.

and Gamble) or *kudrau (Atkinson), riálla* or *rái ála, rágha, morinda, kail, káluchilu, kilu,* GARHWAL, KUMAON; *Raiyang,* KANAWAR; *Seh-shing,* BHUTIA. *Landar, anandar,* names used at timber depôts.

References.—*Brandis, For. Fl., 527; Gamble, Man. Timb., 407; Stewart's Pb. Pl., p. 219; Atkinson, Gaz., N.-W. P., X., p. 834; Madden, Jour. Agri.-Hort. Soc. Ind., VII, 87; Baden Powell, I., 564.*

Habitat.—A lofty tree met with in the North-West Himálaya between 7,000 and 11,000 feet; in Sikkim and Bhután in the inner valleys between 7,000 and 10,000 feet; and in the mountains of Afghánistan, Kafiristan, and Gilgit.

Botanic Diagnosis.—*Cones* 4-6 inches long, occurring at the ends of the branches, drooping, pale green when young; *scales* persistent. *Leaves* stiff, sharp, 4-sided, green, spirally arranged; when young, crowded into pendulous, tail-like twigs.

An elegant tree, growing rapidly in moist localities where not under too much shade.

RESIN. 18

Resin.—It yields a resin, of no importance.

TIMBER. 19

Structure of the Wood.—White, with a reddish or brown tinge, a little harder than that of **A. Webbiana.** The inner belt of annual rings soft and spongy. Weight, on an average, 30 lbs. per cubic foot.

The wood is extensively used locally, *e.g.*, in Simla, for packing-cases, rough furniture, and planking, and sometimes for shingles. It crackles and sends out sparks in burning, and is consumed very quickly, but it is in much demand for charcoal.

DOMESTIC. Bark 20 Leaves. 21

Domestic Uses.—The *bark* is used for roofing shepherds' huts, and the *leaves* are collected by the hill-people as a manure, and they are also stored for winter use as a litter for cattle.

22

Abies Webbiana, *Lindl.*; (in *Gen. Pl.* referred to **Picea Webbiana,** *Loud.*).

THE HIMÁLAYAN SILVER FIR.

Syn.—ABIES WEBBIANA, *Lindley*; A. SPECTABILIS, *Spach.*; ABIES DENSA, *Griffith*; PINUS TINCTORIA, *Wall*; P. WEBBIANA, *Wall.*

Var. PINDROW, *Loudon*: the flat horizontal branches of this Western form give it a very distinctive effect from the variety Webbiana proper met with in Sikkim and Bhután.

Vern.—*Palúdar, rewari,* JHELAM; *Bádar, búdar, túng,* KASHMIR; *Dhúnu, rág, rail, pe, re, salle, sara,* CHAMBA; *Tos,* KULU; *Spun, pun, krok, kalrei,* KANAWAR; *Bharda, thanera,* SHALE; *Burla, pindrau, pindrai,* HATTU; *Kúdrom,* MATIYANA; *Burúl, búrra, búldu,* BHAJJI; *Kalrai, satrai, chúr,* KOTKAI; *Raho, row, chilrow, kilaunta,* CHOR; *Morinda,* JAUNSAR; *Bang, dodhma rágha, teliya* or *chíli rágha,* SOUTH-EASTERN GARHWAL; *Chilrao,* CENTRAL GARHWAL; *Morunda,* NORTH-WEST GARHWAL; *Raunsla* or *rái salla,* KOSI RIVER; *Rágha, ráo rágha, ransla, raisalla,* KUMAON; *Wúman, wumbusing (Mr. Duthie),* BYANS; *Gobria sulah,* NEPAL; *Dumshing,* BHUTIA.

References.—*Brandis, For. Fl., 528; Gamble, Man. Timb., 408; Atkinson, Gaz., N.-W. P., X., 837; Madden, Jour. Agri.-Hort. Soc. Ind., VII., 96.*

Habitat.—A lofty, evergreen tree, met with in the Himálaya, from the Indus to Bhután; in the North-West Himálaya, between 7,000 and 13,000 feet; in the inner ranges of Sikkim and Bhután, between 9,000 and 13,000 feet; in the outer ranges not below 10,000 feet.

Botanic Diagnosis.—*Cones* lateral, erect, 4-6 inches long, solitary or clustered, dark blue when young; *scales* deciduous. *Leaves* flat, narrow, linear, spirally arranged, but spreading in one plane so as to appear distichous.

RESIN. 23

Resin.—It yields a white resin, which is sometimes used medicinally in India.

§ "Hakims affirm that the resin of **Abies Webbiana,** mixed with oil of roses, when taken internally, produces intoxication. This mixture is used externally for headache, neuralgia, &c." (*Surgeon G. A. Emerson, Calcutta.*)

DYE. Young Cones. 24

Dye.—Mr. Duthie, Superintendent of the Botanic Garden, Saharanpur, has drawn my attention to the fact that **Veitch,** in his *Manual of Coniferæ,* states that "a beautiful violet dye is extracted from the young cones" of this plant. It is remarkable that neither **Stewart, Brandis,** nor **Gamble** alludes to this dye, while in *Gordon's Pinetum* occurs the following: "It is called *Rai-sulla* (fragrant fir) and *Gobrea-sulla* (fragrant or indigo fir) by the Gorkhalis on account of an indigo or purple pigment being extracted from the young cones." It would be exceedingly interesting to have this dye properly confirmed by fresh information, and specimens of the dye-stuff and cloths dyed by this process, also information as to the extent this curious dye is actually used by the hill tribes of India.

MEDICINE. Leaves. 25

Medicine.—The dried leaves of this plant (*Tálíspatra,* HIND. and BENG., *Tálísapatra,* SANS.) are regarded as carminative, expectorant, stomachic, tonic, and astringent; useful in phthisis, asthma, bronchitis, and catarrh of the bladder. The powdered leaves are often given along with the juice of **Adhatoda Vasica** and honey, and a confection called *tálísádya churna* is prepared from the *tálíspatra* along with pepper, ginger, bamboo manna, cardamoms, cinnamon, and sugar. The *tálíspatra* also enters into the preparation of numerous complex prescriptions. (*U. C. Dutt's Hin dú Mat. Med.*) In **Ainslie** and the earlier writers on Indian Economic Botany, *tálíspatrie, tálísapatra,* DUK. and HIND., and *tálísha, vidara,* SANS., are the vernacular names for the dried leaves and twigs of **Flacourtia Cataphracta,** the *paniyala* of Bengal. (*Ainslie, II., 407.*)

In his *Manual of Indian Timbers,* p. 17, **Mr. Gamble** gives *tálíspatri* as the Hindi name for **Flacourtia Cataphracta,** *Roxb.,* and this is also the name given by **Babu T. N. Mukharji** in his *Amsterdam Catalogue.* **Surgeon U. C. Dutt** informs me that he is of opinion **Abies Webbiana** is the *tálíspatra* of the ancient Sanskrit writers, and that specimens of the drug which he submitted to Dr. King were found to be the leaves and twigs of this plant. It seems difficult to account, however, for a man of **Dr. Ainslie's** ability mistaking the ovate leaf of a **Flacourtia** for the needle-shaped leaves of a Fir, and, having few or no authors to compile from, he must have personally identified the plants of which he wrote, and ascertained locally that **Flacourtia** was the *tálísha* of the shops of South India.

The Hindú Doctors of Behar, according to **Dr. F. Hamilton,** use an infusion of *tálíspatra* in the treatment of hoarseness. It is probable that, in different parts of India, the dried leaves of various plants receive the name of *tálíspatra,* provided they are found useful in the treatment of coughs. It seems likely, however, that the leaves of **Abies Webbiana** are the original or true *tálíspatra.*

Dr. Dymock states that the *tálíspatra* of the Bombay shops (also called *Barmi*) consists of the leaves and young shoots of **Taxus baccata,** *Linn.* While admitting that the *tálíspatr* of the ancients has not been identified for certain, he quotes under **Taxus** the properties and mode of prescribing the *tálíspatr* as given by **Surgeon U. C. Dutt,** an author who refers it to **Abies.** The importance of this observation lies in the fact that the therapeutic properties of **Abies** and **Taxus** are widely distinct, and therefore these distinct plants, one would imagine, could not possibly be used for the same purposes. (See **Taxus.**) Since writing the above, however, I have seen in Calcutta a specimen of a drug called *tálíspatra,* which I believe to be the leaves of **Taxus,** and **Mr. Woodrow** says that it is the leaves of that plant which are sold in Poona as the *tálíspatr.* The whole subject is thus exceedingly obscure.

MEDICINE.

The description of the *tálíspatra* in some of the older books on Indian medicinal plants would agree very well with the leaves of a **Cinnamomum,**—much better, in fact, than with those of an **Abies. Surgeon Moodeen Sheriff** gives *tálíshapatri* as the Tamil and Telegu names for **C. Tamala,** *Nees* and also the Arabic and Persian for the leaves of that plant. He may be quite right in this opinion, modern usage having appropriated the name to **Abies** and **Taxus.** But one only of the plants referred to in this critical note can be the true *tálíspatra* since they have such distinct properties. It seems clear, however, that in different parts of the country different plants are prescribed as *tálíspatr,* and it is quite probable that none of them are the true *tálíspatra* of the Sanskrit writers.

26

Special Opinions.—§ "*Tálíspatra* (leaves of **Abies Webbiana)** are sold in all *Baniahs'* shops in Bengal, Behar, and the N.-W. Provinces. I do not think any other leaf is used in these provinces under the name of *Tálíshapatri.*" (*Surgeon U. C. Dutt, Serampur.*) "The *Tálíspatra* of the Bombay shops, also called *Barmi,* consists of the leaves and young shoots of **Taxus baccata.** *Many of the shoots bear the male flowers of that plant.* If this is the source of the *Tálíspatra* of Sanskrit writers, we ought to have a Sanskrit name *Talish* for the tree, and the Hindi name would probably be similar." (*Surgeon-Major W. Dymock, Bombay.*) "The same argument is equally against **Taxus baccata,** since in no Indian language is the tree called *tálísha,* nor by any name that could be called a derivative from that word." (*G. Watt.*) "The Bhuj Hakim uses the dried leaves when a carminative action is needed, dose half to one drachm internally." (*Surgeon W. Barren, Bhuj, Cutch, Bombay.*)

27

"The juice of the fresh leaves is used as a family medicine in fevers, acting as an antiperiodic, for infants dose 5-10 drops in water or mother's milk. It is also prescribed in affections of the chest and during dentition. In Bengal it is given as a tonic after parturition." (*Surgeon J. McConaghey, Shahjahanpur.*)

FODDER. Leaves. 28

Fodder.—In tracts near the Jhelum the twigs and leaves are cut and stored for winter use as fodder and litter for cattle. (*Brandis.*) That cattle could eat the dried leaves of this plant seems incredible and at strange variance with their reputed medicinal properties.

TIMBER. 29

Structure of the Wood.—White, soft. The inner zone of each annual ring is soft and spongy. Weight about 29 lbs. per cubic foot.

The wood is not durable when exposed to the weather, but seems to last well as shingles in Sikkim, whence it is sometimes exported to Thibet for roofing. At Murree, shingles are said to last eight to ten years, and in Kulu three to six. In Kanawár and Lahoul it is much used for construction. Very little information exists regarding the rate of growth.

DOMESTIC. Bark. 30

Domestic Uses.—The bark is used for roofing shepherds' huts, and it is also made into troughs for the salt given to the sheep grazing on the higher Himálaya.

ABIR.

31

Abir (sometimes called *Phák,* Beng., or *Phaku,* Ass.), or the white perfumed powder which is mixed with the red *Gulál* powder and used at the *Holi* festival.

Dr. McCann, in his *Dyes and Tans,* publishes, from the records of the Bengal Economic Museum, as the practice adopted in Mymensing, the following description:—

Bengal Powder with *bakram.* 32

1st.—"The *shati* is washed and pounded in a *dhenki.* The powder is then put into an earthen vessel full of water and allowed to rot. The water is afterwards poured off, and the powder dried. It is then mixed with the juice extracted from *bakram* wood. This turns it red, and it is then called *Abír* or *Holi* powder. *Shati* is gathered for this purpose in the

ABIR.

month of *Poos* (December-January)." There is no mention of alum in the above, but without that substance the colour could not be produced.

The practice which seems to prevail in most other parts of India is to prepare the two powders quite distinct from each other, and to mix them as required.

Bengal Powder with aniline. 33

2nd.—The Bengal *Holi* powder is prepared from **Curcuma Zedoaria,** *Roscoe* (or common flour or arrow-root), Sappan-wood, and alum. "In Bengal aniline magenta is now largely used to colour the flour obtained from **Zedoaria,** and being cheaper has almost superseded the older preparations." (*Mr. T. N. Mukharji, Calcutta.*) **Dr. Buchanan Hamilton** says that in some parts of Bengal the yellow dye obtained from Bixa Orellana is used as a *gulál.*

Dr. Dymock has favoured me with the following list (Nos. 3 to 7) of *Abír* powders:—

Bombay Powder. 34

3rd.—"The Bombay *Holi* powder or *Gulál* is made of flour coloured with Sappan-wood and alum." (*Surgeon-Major W. Dymock, Bombay.*)

White Abir. 35

4th.—A whitish *Abír* made from the following:—

Andropogon muricatus. **Hedychium spicatum.**
Santalum album. **Sorghum vulgare** (flour).

Ghisi Powder. 36

5th.—The buff-coloured Hindi *Gulál,* known as *Ghisi,* contains, in addition to the above, the following:—

Cerasus (Prunus) Mahaleb. **Artemisia Sieversiana** (imported).
Cedrus Deodara. **Curcuma Zedoaria.**
Cloves. Cardamoms.

Bukka or Deccan Abir. 37

6th.—Deccan *Abír* or *Bukká* is of a black colour, and in addition to all the above contains the following:—

Aquilaria Agallocha.

Costus Root (**Saussurea Lappa,** *C.B.C.,* formerly *Aucklandia Costus, Falc.*); Jatamansi; Liquid Storax.

Jain Abir. 38

7th.—The *Abír* of the Jains is of a pale yellow colour. It is called *Vasakshep;* it is made of—

Santalum album. Saffron; Camphor (Borneo); Musk.

Pulas Gulal. 39

8th.—**Voigt** in his *Hortus Sub. Calc.,* states that **Trapa bispinosa** (the Singara nut) is used as an *Abír.* "During the *Holi* festival a red dye is made of it mixed with the yellow dye procured from the flowers of **Butea frondosa.**" **Drury** apparently quotes this paragraph. I presume the flour of the Singara nut is simply used in place of rice or wheat flour and coloured with the *gulál.* My friend **Major Pitcher** informs me, however, that in Oudh the flour from the Singara nut is used during the *Holi.* **Professor W. T. Thiselton Dyer** also writes me that a specimen of *Abír,* composed of Singara-nut flour, coloured with **Butea frondosa,** was received by the Kew authorities from the late Indian Museum. It would thus appear that the use of Singara-nut flour is more general than was thought. What peculiar merit it is supposed to possess I have been unable to discover. The use of **Butea frondosa** as *Gulál* is, however, referred to by several authors. See **Butea.**

Absorbent—(L. *ab,* from; *sorbeo,* I suck up). Applied adj. and subs. to anything which absorbs acidity in the intestinal canal, or blood and other fluids, from any part of the body. For list of Indian Absorbents, see "Drugs."

40

ABROMA, *Jacqr.; Gen. Pl., I., 224.*

A genus of trees or shrubs (belonging to the Natural Order Sterculiaceæ), containing in all two or at most three species, natives of tropical Asia. *Leaves* cordate, angled. *Petals* purple-coloured, concave at the base, with

An Important Fibre.

a stipulate, ovate blade. *Andrœcium* tubular, *stamens* 1-seriate, *anthers* marginal, 2-4 between each staminode. *Capsule* membranous, 5-angled and 5-winged. *Seeds* numerous.

The word Abroma is derived from ἀ and βρῶμα—unfit for food.

41

Abroma augusta, *Linn.; Fl. Br. Ind., I., 375.*

ABROME, *Fr.;* ABROME, *Ger.*

Vern.—*Ulatkambal,* BENG., HIND., and CUTCH.

References.—*Roxb., Fl. Ind., Ed. C.B.C., 510; Gamble, Man. Timb., 45; Atkinson, Gaz., N.-W. P., X., 792; Drury, U. P., 2; U. S. Dispens., Ed. 15th, 1559.*

Habitat.—A small bush, widely distributed (native or cultivated) throughout the hotter parts of India.

Botanic Diagnosis.—A large spreading shrub, with leaves and branches softly hairy. *Sepals* lanceolate, almost free to the base, and nearly as long as the petals; *fruit* becoming more than twice the length of the persistent calyx. It flowers most profusely during the rains, and the seeds ripen in the cold season.

Properties and Uses—

FIBRE. Bark. 42

Fibre.—The bark of the twigs yields a much-valued fibre, which deserves to be more generally known. It might be used with advantage as a substitute for silk. The plant yields three crops a year. **Roxburgh** says of this plant: "The bark abounds with strong, white fibres, which make a very good substitute for hemp, and as the plant grows so quickly as to yield two, three, or even four crops of cuttings annually fit for peeling" it may with advantage be cultivated for its fibre. It is a large and more easily cultivated plant than jute or *sunn,* and is a perennial, while the produce is as great if not greater, and the fibre as easily separated. With all these facts to recommend it, it seems remarkable that we should have had to urge the claims of this fibre for over 50 years without its having attained the high commercial rank which its great beauty, softness, cheapness, and durability deserve. There are many purposes to which jute is put nowadays, which Abroma could meet with greater acceptance, and which it most certainly would, by this time, have supplied but for the unparalleled early success of jute. The future seems likely to raise Abroma, however, to a position of great importance.

"To separate the root-bark from the shoots, maceration in stagnant water, from four to eight days, answers well during the warmer parts of the year, while three times as much is scarce sufficient during the cold season: indeed, the process is scarcely practicable then; besides, the fibres are generally weakened by prolonged maceration. The fibres being naturally very white and clean, they do not require to be cleansed. **Dr. Roxburgh** states that, in its native state, without being dressed in any way, it is about one tenth part stronger than *sunn,* and in that state much more durable in water. A cord of its fibre bore 74 lbs., when *sunn* broke with 68 lbs." (*Royle's Fibrous Plants of India, p. 267.*)

There can be no doubt that sooner or later the trade in jute, having lost the enormous profits obtained during its early history, will subside into an old and established industry. The demand for some new fibre to give life to the progressive textile industries will then turn to **Abroma, Agave, Yucca, Abutilon,** and a few others; but to the owners of jute mills and jute machinery, **Abroma** seems the most likely to prove the new outlet for enterprise.

MEDICINE. Root-bark. 43

Medicine.—The bark of the root is an emmenagogue, which does not appear to have received the attention it deserves. In the *Indian Medical Gazette* for 1872, **Mr. Bhoobun Mohun Sircar** gives an account

MEDICINE.

of the uses of this drug, specially recommending it in the treatment of dysmenorrhœa. It seems to be the fresh viscid sap from the root-bark which possesses the properties attributed to this plant. Half a drachm is given at a dose. (*Am. Jour. Med. Science, July 1873, p. 276.*)

Dr. Kirton, in a correspondence with the Home Department, Government of India, says: "Fresh root, beat into paste with water, dose 1 drachm in dysmenorrhœa."

44

Special Opinions.—§ "Is a valuable medicine in dysmenorrhœa. The way in which it is usually given is, the fresh root of the plant made into a paste with black pepper. The medicine is given about a week before the menstruation begins and is continued till it commences. I have seen it prove very efficacious in some cases; especially the congestive form of the disease." (*Surgeon R. Macleod, M.D., Gya.*) "Two ounces of dried bark boiled in one pint of water will make a good decoction. An ounce thrice daily acts well in cases of dysmenorrhœa." (*Surgeon R. L. Dutt, M.D., Pubna.*) "Have never used it as an emmenagogue, but the infusion of fresh leaves and stems in cold water is demulcent and very efficacious in gonorrhœa." (*Surgeon C. J. W. Meadows, Burrisal.*) "A well-known emmenagogue in Bengal: said by some native doctors to possess antispasmodic properties." (*Surgeon J. McConaghey, Shahjahanpur.*) "Useful in dysmenorrhœa." (*Surgeon J. Anderson, Bijnor.*)

45

"The slender roots of this plant are useful in the congestive and neuralgic varieties of dysmenorrhœa. It regulates the menstrual flow and acts as a uterine tonic. It should be given during menstruation, 1½ drs. of the fresh root for a dose, with black pepper, the latter acting as a stomachic and carminative." (*Brigade Surgeon J. H. Thornton, Monghyr.*) "I have tried the green tender roots and also the bark of roots in two-drachm doses with a few grains of pepper just to cover the bland taste of it in congestive dysmenorrhœa with excellent results. It is to be given as soon as the vague sense of discomfort and weight in the loins begins and should be continued till the flow appears." (*Assistant Surgeon Devendro Nath Roy, Sealdah, Calcutta.*)

46

"I used the root-bark in three cases of dysmenorrhœa, in which it was vaunted as a specific, but without any effect." (*Surgeon Shib Chunder Bhattacharji, Chanda, Central Provinces.*) "Babu B. M. Sarkar of Calcutta is selling pills made of this as a successful remedy for dysmenorrhœa, and I have heard of good effects from it. Requires a trial." (*Surgeon K. D. Ghose, Khulna.*) "Action tonic." (*Surgeon W. Barren, Bhuj, Cutch, Bombay.*) "Has never failed in my hands in speedily relieving painful dysmenorrhœa." (*Surgeon B. Evers, M.D., Wardha.*) "In my personal experience, I know the root is very efficacious in cases of amenorrhœa as an emmenagogue." (*Assistant Surgeon Shib Chunder Basu, Bankipur.*) "Reported to be an emmenagogue; said to remove sterility in cases depending on dysmenorrhœa, but it has failed in two cases in my hand." (*Assistant Surgeon Bolly Chand Sen, Sealdah, Calcutta.*)

47

ABRUS, *Linn.; Gen. Pl. I., 527.*

A genus of climbing shrubs (belonging to the Sub-Order PAPILIONACEÆ, of the Natural Order LEGUMINOSÆ) comprising 5 species, cosmopolitan in the tropics. *Leaves* equally pinnate; *leaflets* numerous and deciduous. *Calyx* campanulate, teeth short. *Corolla* much exserted, standard adhering below to the staminal tube. *Stamens* 9, united into one bundle, the uppermost and tenth one being abortive. *Pod* dehiscent (exposing the red shining seeds), not jointed.

The word Abrus is derived from the Gr. ἁβρός, graceful, either in allusion to the graceful, delicate leaves, or elegant, shining seeds.

ABRUS precatorius.

Indian Wild Liquorice.

In India there are three species, closely allied and economically viewed as but forms of one species. These may be briefly distinguished thus:—

48 **A. precatorius,** *Linn.* Pod oblong, turgid, 3-5 seeded.

49 **A. pulchellus,** *Wall.* Pod linear, flat, incurved, 9-12 seeded.

50 **A. fruticulosus,** *Wall.* Pod linear, oblong, flattish, 4-6 seeded.

51

Abrus precatorius, *Linn.; Fl. Br. Ind., II., 175.*

INDIAN OR WILD LIQUORICE ROOT, *Eng.;* LIANE À RÉGLISSE, RÉGLISSE D'AMÉRIQUE, *Fr.*

Vern.—*Gaungchi, rati, chirmiti,* HIND.; *Ratak, labri,* PB.; *Gunjá, ghungchi,* BOMB.; *Gumchí,* DUK.; *Chanoti,* GUJ.; *Ratí,* CUTCH; *Maspati,* NEPAL; *Kunch, gunch, chun-hati,* BENG.; *Kawet;* SANTAL; *Látuwani,* ASS.; *Gunja, gunjá, krishnala, kakachinchi,* SANS.; *Aainu-ddík,* ARAB.; *Chashmkhurós,* PERS.; *Gundumani,* TAM.; *Ghurie-ghénzá* TEL.; *Ywe-guwe* or *gyin-ywe,* BURM.

References.—*Roxb., Fl. Ind., Ed. C.B.C., 544; Drury's U. P., p. 3; Brandis, For. Fl., 139; Gamble, Man. Timb., 121; C. I. H. Warden, in Year Book Phar. 1882, p. 211; Flück, and Hanb., Pharmacog., 2nd ed., 188; Bentl. and Trim., Med. Pl., 25; Dymock's Drugs of W. Ind., p. 183; Stewart's Pb. Pl., p. 50; Atkinson, Gaz., N.-W.P., X., pp. 7, 24, 767.*

Habitat.—A beautiful climber, met with all along the Himálaya ascending to altitude 3,000 feet, and spreading through the plains of India to Ceylon and Siam.

Botanic Diagnosis.—There are three principal varieties under this species described by **Roxburgh**:—

1st—With rose-coloured flowers and red seed, with black eye.

2nd—With dark-coloured flowers and black seed, with white eye.

3rd—With white flowers, white seed.

§ "The white variety as seen here has no black eye." (*Dr. Dymock, Bombay.*)

Properties and Uses—

MEDICINE. Root. 52

Medicine.—THE ROOT, **Ainslie** and **O'Shaughnessy** say, "is a perfect substitute for liquorice in every way." Modern authors differ from them in this opinion. According to Sanskrit writers it is emetic and useful in poisoning. **Dr. Bidie** says: "The Abrus root has little or no saccharine taste, and would form a very indifferent substitute for liquorice." Liquorice root is largely imported into India and extensively used in native medicine, and is probably often sold under the same vernacular name as that given to the Abrus root. **Dr. Dymock,** speaking on this subject, says: "I consider the root to bear very little resemblance to liquorice, either as regards appearance or qualities; as pointed out by **Dr. Moodeen Sheriff,** the leaves are by far the sweetest part of the plant, and from them a tolerable extract may be made." An alkaline ash is prepared from the plant.

53

§ **Special Opinions.**—"The root of the third variety, as described by **Roxburgh,** is used for gonorrhœa. A quantity equal to a drachm in weight is pounded, and the expressed juice mixed with sugar-candy." (*Surgeon H. McCalman, M.D., Bombay.*) "At Poona *Safed Gunja* is considered the best variety, and accordingly under this name **Glycerrhiza glabra** is sold in the bazars. The root of **Abrus precatorius** appears to have fallen much out of use of late." (*G. M. Woodrow, Prof., College of Science, Poona.*) "An infusion of the roots is used for procuring abortion in Hoshiarpur District, Panjáb." (*W. C. Coldstream, Esq., Commissioner, Hissar.*)

54

"Further experience confirms the view that the root of this plant is not a substitute for liquorice, and that the article sold in the bazars as 'Indian liquorice' is not the root of **Abrus precatorius.** I would therefore strike out everything connecting liquorice with **Abrus.**" (*Deputy Sur-*

MEDICINE.

geon-General G. Bidie, C.I.E., Madras.) "The powdered root, with ginger and other carminatives, is much used for coughs, including hooping-cough: Dose—5 to 7 grains. A cold infusion of the root is also used in leucorrhœa." (*Narain Misser, Hoshangabad, Central Provinces.*)

55

"Not equal to liquorice, but a fair substitute; used chiefly in the form of infusion, ½ oz. of the root-bark to 10 oz. water: dose one to two ounces." (*Apothecary Thomas Ward, Madanapalle, Cuddapah.*) "I often use a decoction of bazar liquorice in cough mixtures, and find it very useful, especially in the bronchitis of children." (*Surgeon-Major J. J. L. Ratton, M.D., Salem.*)

Leaves. 56

THE LEAVES.—§ "If the leaves are steeped in warm mustard oil, and applied over the seat of pain in rheumatism, much benefit will be derived." (*Surgeon W. Wilson, Bogra.*)

"The leaves, warmed over the fire, are applied to painful swellings, over which a little warm castor-oil is first smeared. In this way they are said to be antiphlogistic in their action." (*Hon. Surgeon P. Kinsley, Chicacole, Ganjam, Madras.*) "Juice of fresh leaves, mixed with some bland oil, and applied externally, seems to relieve local pain." (*Surgeon-Major Bunkabehari Gupta, Púrí.*)

Seeds. 57

THE SEEDS.—In his *Sanskrit Materia Medica* **Surgeon U. C. Dutt** says the seeds are "used internally in affections of the nervous system and externally in skin diseases, ulcers, affections of the hair." They are pounded and made up with mercury, sulphur, *ním* seeds, hemp leaves, and cotton seeds, &c.

58

§ "The seeds are used as a purgative, but in large doses are an acrid poison, giving rise to symptoms resembling those of cholera. The poisonous property is generally believed to be in the red covering of the seed." (*Surgeon-Major A. S. G. Jayakar, Arabia.*)

"When boiled with milk it is said to have a very powerful tonic action on the nervous system. Dose of the powder boiled with milk, one to three grains. The powder, when administered internally, *uncooked*, acts as a strong purgative and emetic." (*Surgeon W. Barren, Bhuj, Cutch, Bombay.*) "The white seeds are made into confection with several other drugs and used as tonic. A *Vaid* tells me that the roots are similarly used." (*Surgeon-Major J. Robb, Ahmedabad.*) "Used in granular chronic conjunctivitis causes slight inflammation and absorption of the granulation." (*Surgeon H. D. Masani, Karachi.*)

59

"If the colour of the seeds is sufficient to constitute a variety, there are more than 10 varieties of **A. precatorius**; some seeds are entirely black, white, yellow or rosy, and such seeds are more used in medicine than those described by **Dr. Roxburgh** as having an eye of a different colour. Taken internally by women, the seeds of **A. precatorius** disturb the uterine functions and prevent conception. For the latter purpose, 4 to 6 seeds are swallowed every day in 2 doses for several days after each menstruation. The white and black seeds are preferable to those of other colours." I am aware of one successful case under this treatment. The white seeds, again, are considered deobstruent or repellant. Rubbed up with water and applied to swellings, they succeed in some slight cases." (*Hon. Surgeon Moodeen Sheriff, Khan Bahadur, Madras.*)

60

"I presume that this plant is the one referred to by **Major Ramsay**, of the Bengal Police, in his book, Detective Foot-prints, Bengal, 1874-81, London, Army and Navy Co-operative Society, 1882. He calls the plant *Korgani*; his description at page 44 agrees. The seed is used for killing cattle. Consult the reference." (*Surgeon L. Cameron, Nuddea.*) "The powdered seeds are taken as snuff in cases of violent headache arising from cold." (*Mr. T. N. Mukharji, Calcutta.*) "The boiled seeds are

Opinions regarding the Gunja.

CHEMISTRY.

said to possess powerful aphrodisiac properties." (*Surgeon-Major J. M. Honston, Travancore, and Civil Apothecary John Gomes, Travandrum.*)

"The seeds rubbed down with a small quantity of water (paste) is used for contusions, &c., to reduce the inflammation." (*Surgeon W. A. Lee, Mangalore.*)

61

"The root is a good substitute for imported liquorice. The seeds are poisonous and used by the *chamars* for poisoning cattle; they are poisonous when mixed with the blood, but not so when taken internally. I treated a case at the Bankipore Hospital who died with symptoms of nervous excitement. On *post mortem* examination no trace of the drug could be detected, only there was a suppurative spot with inflammation all around it, and the brain was highly congested." (*Asst. Surgeon Bolly Chand Sen, Calcutta.*)

62

"Prescribed as a general tonic, and, mixed with milk and cardamoms, natives use it as an aphrodisiac. In large doses it is emetic. A mixture consisting of vinegar 2 parts, sugar 3 parts, and lime-juice 1 part, acts as an antidote to the poison—dose 67 grains." (*Surgeon J. McConaghey, Shahjahanpur.*) "The seeds, mixed with the roots and cocoanut milk, are given in hæmorrhoids according to **Rheede.**" (*Surgeon H. W. Hill, Mánbhúm.*)

63

"The seeds, when decorticated and finely ground, are used for treatment of pannus cornea and for granular lids with good effect. They cause a true purulent ophthalmia, and in cases where too vigorously used in diphtheritic exudation on the conjunctiva. In mild cases of pannus and granular lids the treatment need not be carried beyond the purulent stage. But in severe cases diphtheritic exudation must be caused before good results are procured. A 3 per cent. solution, prepared by steeping the decorticated seeds in cold water for 24 hours, brushed over the reversed lids three times a day, will cause purulent ophthalmia. In bad cases a 5 per cent. infusion is required. Moderate ulceration of the cornea is not a contra-indication for the use of this remedy; on the other hand the ulceration speedily improves." (*By a Surgeon.*)

Off. Preparation.—An extract of **Abrus** root is regarded as officinal in the Indian Pharmacopœia, dose *ad libitum.*

64

Description.—The root is woody, tortuous, and much-branched, about ¼ inch in thickness. Section not broken up by medullary rays into wedge-shaped blocks as in the true liquorice. Bark thin, reddish brown; wood yellowish white. Odour, when broken, peculiar and disagreeable. Taste at first bitter, then sweet.

65

Chemical Composition.—In 1882, **Dr. de Wecker** directed attention to the fact that the **Abrus precatorius** seeds, used in the form of a lotion, were capable of producing purulent inflammation of the conjunctiva. He advanced the theory that this was due to the presence of a bacillus, and stated that if an infusion of the seeds be sterilized, it is no longer capable of exciting inflammation: if the bacillus be cultivated separately, it will set up the factitious ophthalmia. That when inoculations were carried very far, a transmission to the lymphatic glands took place, causing suppuration and erysipelatous symptoms, as well as a distinctly febrile condition. (*Jour. Ph. Soc.*)

66

It has been observed by **Dr. Nicholson** that Abrus seeds are sometimes found on prisoners in jail. Sudden attacks of ophthalmia in jails may perhaps often be factitious. The action of the Abrus seeds in establishing purulent inflammation is at all events well known. According to **Dr. de Wecker,** the ophthalmia set up by Abrus seeds disappears in ten days or a fortnight without any therapeutic intervention or danger to the cornea, and he therefore recommends a lotion, prepared from 155 grains of the decorticated and powdered seeds, macerated in 17 ounces of cold water for 24 hours, and then filtered, in the treatment of certain eye diseases in which it may be necessary to produce purulent conjunctive inflammation. The theory that this infusion of the seeds of **Abrus precatorius** owes its power of producing ophthalmia to the action of a bacillus, has by this time no doubt been abandoned even by its first propounders, **Messrs. de Wecker** and **Sattler. Warden** and **Waddell** have conclusively proved that the Abrus poison or *abrin* is a proteid closely allied to ordinary albumen, that it is insoluble, and is present not only in the seeds, but also in the stem and root of the **Abrus precatorius.** They further show that *abrin* produces its poisonous effect when injected subcutaneously, but not when ingested. (*Practitioner, XXXII., 135, 437.*)

CHEMISTRY.

My attention has been drawn to some interesting information regarding the chemistry of this plant, but too late unfortunately, to be taken advantage of. See *The Lancet*, July and August 1884.

67

§ "It has been proved that the action is not, as supposed, due to a bacillus, but to some peculiar chemical constituent not yet isolated; see papers in *American Druggist*, 1884, pp. 103 and 105. An alcoholic dialysate of **Abrus** was found very active; hence bacilli can certainly not be the cause." (*Dr. Charles Rice, New York.*)

FOOD. Seeds. 68

Food.—The seeds, known vulgarly as Crab's eyes, are said to be used in Egypt as an article of food, and are wholesome when boiled; if eaten in any quantity, however, they produce violent headache.

When injected under the skin they are poisonous. Heat, with moisture, however, destroys completely this poisonous principle, hence when cooked they become inert, or if simply eaten the heat and moisture of digestion destroys the poisonous property. A temperature of 100° C. will destroy completely the activity of the poison.

POISON. Sui. 69

Poison.—**Dr. Warden**, experimenting with the poisonous property of the seeds, found that half a seed rubbed down with a small quantity of water and injected into the thigh of a full-grown cat produced fatal effects in 24 hours. He succeeded in isolating an acid which he called Abric Acid; it proved to be non-poisonous. While the seeds used in this manner are highly poisonous, it is remarkable, as stated above, that when boiled they may be eaten.

70

Special Opinions.—§ "The seeds are used by sweepers and *chamars* to poison cattle. From the throat of a bullock suffering from symptoms of poisoning resembling narcotism, I once extracted a bamboo containing bruised *ratak* seeds, loosely tied with hair. The animal died the same day. Iron needles, soaked in a paste made from the ground seeds, are also used. I once extracted a black ball made of wax and *ratak* seeds from the vagina of a cow." (*Asst. Surgeon Bhagwan Dass, Rawal Pindi.*)

"The seeds are powdered and made into a little cylinder called *sui*, almost exactly resembling a silver nitrate point. A puncture is made in the skin and the *sui* pushed in. The animal soon becomes uneasy, and if discovered and the *sui* cut out, the animal may live." (*Surgeon-Major C. W. Calthrop, M.D., Morar.*) "The powdered seed is made up into a paste and formed into short sharp pieces (called *sui* or needles) used by cattle-poisoners. The *sui* is inserted under the skin of the animal, causing inflammation and death." (*W. C. Coldstream, Esq., Commissioner, Hissar, Panjáb.*) "The pulp of the seed, made into small needles called '*sutarí*,' is used for poisoning cattle in Behar. I gave evidence in a case of this nature before the Sessions Judge of Patna." (*Asst. Surgeon Shib Chunder Basu, Bankipur.*)

DOMESTIC. Seeds as a weight 71

Domestic Uses.—The small, shining, red seeds, *rati* or *ghungchi*, are largely used by the Indian goldsmiths as weights, each weighing about 1·75 grains; in the Panjáb they are regarded as of correct weight when

DOMESTIC.

equal to about eight grains of *bansmatti* rice (*Mr. Coldstream*). It is stated that the famous *Koh-i-núr* diamond was first weighed by the *rati*, a word which, by some authors, is supposed to have given origin to the jewellers' carat (*kírát,* ARAB.). The carat is the twenty-fourth part of an ounce or $3\frac{1}{6}$ grains; this approximately would be equal to two *rati* seeds. The climber, with its open, withered fruits, exposing the scarlet *rati* (or crab's eye) seeds, is twisted round leafy boughs to simulate the holly at Christmas time in English stations.

A Necklace. 72
As Ornaments. 73
A Rosary. 74
75

The *rati* are also extensively used for necklaces, ornaments for the ears, and to decorate small boxes, baskets, &c. The fact of their being used as rosaries doubtless suggested the specific name *precatorius.*

76 **Abrus precatorius,** *Linn.*, var. **melanospermus,** *L.* (as in Voigt's Hort. Sub. Calc., p. 228.)

Syn. for **A. pulchellus,** *Wall,* which see under generic note.

Absinthe—An intoxicating liquor, largely consumed by the French, prepared in Europe from one or two species of Wormwood, but chiefly from **Artemisia Absinthium,** *Linn.*, which see.

Abstergent—A term applied in Pharmacy to a substance such as Fuller's earth, which has a cleansing property. The Soap-nut (**Sapindus Mukorossi,** *Gærtn.*) is an excellent abstergent or detergent, and so also are the pods of **Acacia concinna.** See under "Soap Substitutes" for other natural abstergents.

ABUTA, *Aubl.*

77 **Abuta rufescens,** *Aubl.; DC. Prod., Vol. I., 103.*

Said by **O'Shaughnessy** to be considered Ceylon as an excellent stomachic, and used for the same purpose as Cissampelos. The plant is a native of Cayenne and Guiana, and if used in Ceylon must be imported.

78 ABUTILON, *Gærtn.; Gen. Pl., I., 204.*

A genus of herbs or under-shrubs (belonging to the Natural Order MALVACEÆ), containing about 70 species, all inhabitants of the tropical or subtropical regions. *Leaves* softly tomentose, cordate, angled, or palmately lobed. *Bracteoles* none. *Staminal-tube* divided at the apex into numerous filaments. *Carpels* 5 or more, entire (*i.e.*, not divided by a false partition), when ripe separating from the axis. *Styles* as many as the carpels.

The word Abutilon is said by some authors to be an ancient Greek name for the Mulberry tree, and to be given to this genus in allusion to the resemblance of the foliage. **Dr. Rice** writes me, however, that the word does not occur in old Greek authors, but appears to have been first used by **Avicenna.** The members of this genus are annual or perennial low bushes, growing gregariously and forming clumps in the jungles. There are 10 or 12 Indian species, most of which are very abundant in the plains, and yield beautiful white fibres. The flowers of **A. esculentum,** a Brazilian species, are edible. The leaves of all contain a large quantity of mucilage, and are used in the same manner as the Marsh-mallow of Europe.

The allied genera are readily separated from **Abutilon** by the number and position of the ovules. In **Althæa** they are solitary, ascending; in **Sida** solitary, pendulous; while in **Abutilon** there are two or more ovules in each carpel, one ascending and another pendulous from the top and bottom of the carpel.

The following are the more important Indian species of Abutilon:—

79 **Abutilon asiaticum,** *G. Don,* and **A. indicum,** *G. Don,* MALVACEÆ, are two species so nearly allied botanically, that from an economic point

of view they may be regarded as one and the same. The former is characterised by having the upper surface of the leaf rugose, the under velvety, with the carpels scarcely longer than the calyx. The latter, by the leaves being covered on both surfaces with closely-felted white down, the carpels being also longer than the calyx.

Abutilon asiaticum, *G. Don; Fl. Br. Ind., I., 326.*

80

COUNTRY MALLOW.

Syn.—SIDA ASIATICA, *Linn.*, belongs to the former species.

Vern.—*Kangahi*, or *kanghi*, *jhampi*, HIND.; *Petári*, BENG.; *Kangori*, *chakra-bhenda*, *Petári*, MAR.; *Tutti* (or *tuthi nar*), *perun-tutti*, TAM.; *Tuttura-benda*, *nugu-benda chettu*, TEL.

A. asiaticum, *G. Don*, is chiefly met with in Western India and Ceylon, while **A. indicum,** *G. Don*, is widely distributed throughout tropical India, to Prome, Pegu, and Ava (wanting in Malacca). They are annual or perennial bushes, frequenting roadsides, banks of rivers, &c., especially in the vicinity of villages. Their curious fruit, consisting of a whorl of pretty carpels, has apparently suggested many of the designs in jewellery made in Eastern India. They blossom and seed all the year, and when not insect-eaten, their graceful velvety leaves contrast elegantly with their yellow flowers.

FIBRE Stems. 81

Fibre.—The stems contain a good fibre, suitable for cordage. (See remarks under **A. Avicennæ.**) These exceedingly abundant wild plants deserve attention as paper-yielding fibres.

For further particulars see **A. indicum.**

Abutilon Avicennæ, *Gærtn.; Fl. Br. Ind., I., 327.*

82

INDIAN MALLOW; AMERICAN JUTE.

Syn.—SIDA ABUTILON, *Willd. in Roxb. Fl.*

References.—*Roxb., Fl. Ind., Ed. C.B.C., 518; Voigt's Hort. Sub. Calc., 114; Christy, New Comm. Pl., 33-34.*

Habitat.—A native of North-West India, Sind, Kashmír, and distributed to North Asia and westward to South Europe and North America. It is said to be also met with in Bengal, but **Roxburgh** first reared it from seeds received from China under the name of *King-ma.* In Bengal it would therefore seem to be introduced or met with in cultivation only.

Botanic Diagnosis.—*Leaves* orbicular, cordate, with a long point. *Peduncles* solitary, axillary shorter than the *petiole. Sepals* free or nearly so. *Petals* yellow, hardly exceeding the sepals. *Carpels* 15-20, becoming ultimately divergent, and much exceeding the sepals.

Properties and Uses—

FIBRE. Stems. 83

Fibre.—Considerable attention has of late years been directed to the fibre produced from this species; in the United States vast quantities are being prepared over the region from Ohio to Missouri. "It is pronounced superior to Indian jute and finer than Manilla hemp. It takes readily any colour, and its natural lustre displays more in the aniline dye than in any other—a great advantage over Indian jute, which is antagonistic to cheap bleaching and dyeing." "It is stated that an acre of ground will produce 5 tons of **Abutilon** stalks, and about 20 per cent. of pure fibre is obtained after preparation. Considered superior to jute fibre as imported; the long fibre is fully equal in value to Calcutta prime jute, and Philadelphia rope-manufacturers have already offered to buy any quantity at the highest market price for jute." (*Christy.*) This is exceedingly important, and points to the advisability of a thorough examination of this and other Indian species with special reference to their fibres. It is recommended to be sown broadcast, the yield from good soil being 4 tons an acre of dry stalks.

84

Abutilon graveolens, *W. & A.; Fl. Br. Ind., I., 327.*

Syn.—SIDA GRAVEOLENS, *Roxb.;* SIDA HIRTA, *Lam., DC. Prod.*
Vern.—*Barkanghi,* HIND. and BENG.
References.—*Roxb., Fl. Ind., Ed. C.B.C., 518.*

Habitat.—North-West Provinces, Sindh, Nilgiri Hills, and Ceylon; distributed to Beluchistan, Java, tropical Africa, and Australia.

Botanic Diagnosis.—Whole plant covered with clammy pubescence and spreading hairs. *Leaves* orbicular, cordate, acuminate. *Peduncles* as long as the petioles. *Carpels* 20 or more, rounded and hairy, not awned. The whole plant has an unpleasant smell.

Properties and Uses—

FIBRE. Stems. 85 MEDICINE. Leaves. 86 Roots. 87 Seeds. 88

This is chiefly a native of the North-West Provinces and Sind, extending along the western side of India to the Nilgiri Hills and Ceylon. **Kurz** says it is also met with in Pegu, frequenting uncultivated places. This species was first obtained by **Dr. Roxburgh** from seeds sent him from Cawnpore. From its prevalence in the North-West Provinces, and the fact that it contains a much larger proportion of mucilage than any of the preceding species, this is most probably the plant which yields the *Kangai* medicine used in these Provinces. It may be employed for the same purposes as the preceding, the stems yielding fibre, the leaves, roots, and seeds medicine.

89

A. indicum, *G. Don; Fl. Br. Ind., I., 326.; Wight, Ic., t. 12.*

Syn.—A. ASIATICUM, *W. & A.,* non SIDA ASIATICA, *Linn.;* SIDA INDICA, *Linn.;* SIDA POPULIFOLIA, *Lam.;* SIDA BELVERE, *L'Her.*

Vern.—*Kanghí, kanghani, jhampi,* HIND.; *Potári,* BENG.; *Miru baha,* SANTAL, *Masht-ul-ghoul, deishar,* ARAB.; *Darakhte-shanah,* PERS.; *etári, Pmadni, Kangori, kangoi, chakra-bhenda,* BOMB. and DUK.; *Dábali,* GUJ.; *Balbij,* CUTCH; *Khápato,* SIND; *Petári, túp-kadi,* GOA; *Tutta, uram,* MALAY; *Shrimudrigida,* KAN.; *Tutti, perun-tutti,* TAM.; *Tuttura-benda, nugu-benda, tutiri-chettu,* TEL.; *Anoda-gaha,* SINGH.; *Bon-kho-e, bon-khoye, tha-ma-chók,* BURM. The seeds are known as *balbij,* BOMB.

References.—*Drury's U. P., 4; Roxb., Fl. Ind., Ed. C.B.C., 518; Atkinson's N.-W. P. Gaz., X., 724-791; Stewart's Pb. Pl., p. 21; Dalz. and Gib.'s Bomb. Fl., 18.*

Habitat.—A small shrub, common throughout the hotter parts of India.

Botanic Diagnosis.—*Leaves* cordate, entire or toothed, closely felted with white down on both surfaces. *Sepals* ovate, acute. *Carpels* 15-20, when ripe longer than the calyx, glabrescent, truncate, or shortly awned, awns spreading. *Flowers* yellow, opening in the evening. *Peduncles* longer than the petioles, joined near the top.

Properties and Uses—

FIBRE. Stems. 90

Fibre.—The STEMS contain a good fibre, suitable for cordage. (See remarks under **A. Avicennæ.**)

MEDICINE. Leaves. 91

Medicine.—The LEAVES yield a mucilaginous extract, used as a demulcent.

Root. 92

From the ROOTS an infusion is prepared and given in fevers as a cooling remedy; said also to be useful in the treatment of leprosy.

Seeds. 93

The SEEDS are considered laxative and demulcent, and are given in the treatment of coughs. They are generally known by the name of *Balbíj,* especially in Western India; they cost about R6 per maund. The bark is astringent and valuable as a diuretic. The bark, roots,

MEDICINE.

leaves, and seeds are used in native medicine, and said to be useful in chest affections. An infusion of the leaves or of the roots is prescribed in fevers as a cooling medicine (*Ainslie*). The seeds are used as a laxative in piles. In Bombay, they are supposed to be laxative and demulcent, and, like the leaves, are very mucilaginous. **A. indicum** and **A. asiaticum** are used indiscriminately, if not also one or two other species. A decoction of the bark, leaves, and seeds together has been long used by the Hindús on account of its mucilaginous and diuretic properties, much in the same way as the Marsh-mallow of Europe. (*Dymock, Mat. Med., Western India.*)

94

"§ A decoction is used as a mouth-wash in cases of toothache and tender gums. Boiled milk, whisked with the fibrous twigs, coagulates; the fluid obtained on decantation is by hakims regarded as efficacious in hæmorrhoids when given internally." (*Surgeon G. A. Emerson, Calcutta.*)

95

"The seeds are reckoned aphrodisiac. The leaves are cooked and eaten in cases of bleeding piles. The infusion of root is useful in strangury and hæmaturia." (*Surgeon-Major D. R. Thompson, M.D., C.I.E., Madras.*) "Dose, half to 2 drachms, laxative and demulcent, invariably used in combination with other purgative medicines." (*Surgeon W. Barren, Bhuj, Cutch, Bombay.*) "A decoction of the leaves is used in gonorrhœa and chronic bronchitis." (*J. Norman, Chatrapore, Ganjam.*) "A mucilaginous decoction used in gonorrhœa and inflammation of the bladder." (*Surgeon J. Anderson, Bijnor.*) "The seeds are burned on charcoal and the recta of children affected with thread-worm are exposed to the smoke. It is said to be a very rapid and certain cure, but I have never tried it myself." (*Surgeon-Major C. W. Calthrop, Morar.*)

96

Abutilon muticum, *G. Don; Fl. Br. Ind., I., 327.*

Syn.—SIDA TOMENTOSA, *Roxb., Fl. Ind., Ed. C.B.C., 518; Dalz. & Gib.'s Bomb. Fl., 18.*

Habitat.—An erect annual, native of rubbish heaps, road-sides, hedges, &c., where the soil is good; met with in the North-West Provinces and Western Peninsula.

FIBRE. Bark. 97

Fibre.—Yields a fibre.

98

A. polyandrum, *Schlecht; Fl. Br. Ind., I., 325; Atkinson, Gaz, N.-W. P., X., 791; Dalz. and Gibs., Bomb. Fl., 17.*

Syn.—SIDA POLYANDRA, *Roxb.; Fl. Ind., Ed. C.B.C., 516.*
Vern.—*Velai-thuthi,* TAM.?

Habitat.—A native of the North-West Provinces, tropical Himálayas, up to altitude 3,000 to 4,000 feet, Western Peninsula, Nilgiris, and Ceylon.

FIBRE. Bark. 99

Fibre.—It yields a long, silky fibre, resembling hemp.

100

ACACIA, *Willd.; Gen. Pl., I., 594.*

A genus of spiny or prickly shrubs or trees belonging to the Natural Order LEGUMINOSÆ and constituting the most characteristic group of the Sub-Order MIMOSEÆ. It comprises in all some 430 species, of which the foliiferous are cosmopolitan to the tropics, and the phyllodineous (comprising two thirds of the genus) almost restricted to Australia. The genus is characterised by having small *flowers* aggregated into rounded or elongated heads. Each flower or rather *floret* has its *calyx* and *corolla* regular and valvate. The *stamens* are indefinite and free.

In India there are 18 species, chiefly distributed throughout the plains, two species ascending the hills to altitude 5,000 feet. The following are those of economic interest.

Indian Gum Arabic.

101

Acacia arabica, *Willd.; Fl. Br. Ind., II., 293,* LEGUMINOSÆ.

INDIAN GUM-ARABIC TREE.

Syn.—MIMOSA ARABICA, *Lamk.;* ACACIA VERA, *Willd.;* A. NILOTICA, *Desf.*

Vern.—*Babúl, babur, bábla, kikar,* HIND., BENG., PB.; *Vabbúla, barbara,* SANS. (*Dr. Rice says these are doubtfully Sanskrit words or only Sanskritized*); *Ummughílán,* ARAB.; *Kháré-mughílán,* PERS.; *Karúvelum, karú-veylam, kar-vaila* (by Cleghorn), TAM.; *Túma, nella túma, nallatumma kara* (by Cleghorn), TEL.; *Gobli, karrijáli,* MYSORE; *Jáli,* KAN.; *Bábhula, káli-kikar, ráma-káti,* BOMB.; *Bával,* GUJ.; *Babbar, babhúla, kálikikar,* SIND; *Bábúl,* C. P.; *Gabur,* SANTAL; *Babola,* MAL. (S. P.); *Hnan-lón-kyaing?* BURM. (*contributed by A. M. Buchanan*); *Sant,* AFRICA. IN KANARESE, *mara* (tree), *chukke* (bark), *chipo* (leaves), are often given as port-fire; *Jáli-mara*=Jáli-tree.

References.—*Voigt, Hort. Sub. Calc., 262; Stewart, Pb. Pl., 50; Brandis, For. Fl., 180; Drury, U. Pl., 4; Gamble, Man. Timb., 151; Flück. & Hanb., Pharmacog., 1879, 233; Bent. & Trim., Med. Plants, II, 94; U. S. Dispens., 15th Ed., 11; Year Book, Pharm., 1873, 52; 1874, 280; 1881, 191; and 1882, 152; Atkinson, Him. Distts. (X., N.-W. P. Gaz.), 781; Gums and Resins, p. 7; Baden Powell, Pb. Prod., I., 345, 364, and 471.*

Habitat.—Panjáb to Behar, Western Peninsula, and Ceylon. Cultivated throughout the greater part of India save in the moist humid regions on the coast and in the extreme north-west beyond the Jhelum. One of the commonest plants of the Deccan, but, except where planted, it is rare in the Panch Mahál; it covers most parts of Surat and Gujarát. Is common in the Upper Godaveri, and is the most abundant plant of the Shewan district, Sind. An experiment to introduce the tree into Kashmír failed. It is also absent from Assam and the greater part of the warm moist districts of Eastern Bengal, British Burma, and Mánipur. It becomes smaller in stature on approaching the coast, attaining its greatest prevalence and most prolific condition in lower and middle Sind. It is not indigenous to the Panjáb, although plentiful, attaining a girth of 5 or 6 feet in the submontane districts; nor is it indigenous to Madras, although it grows plentifully near villages and on waste lands, especially on black cotton soil. Very common in East and South Mysore. It prefers a dry to a moist soil, and seems to avoid the influence of the sea. The tree is never leafless, but the fresh foliage appears in February to April. (*Brandis; Stewart.*)

Botanic Diagnosis.—An erect shrub or tree, with straight spines. *Leaves* composed of from 6-12 pinnæ and 20 to 40 leaflets. *Flowers* in rounded heads, axillary; supported upon short *peduncles* with the *bracts* above the middle. *Pod* stalked, straight, sub-indehiscent, persistently grey-downy, with sutures deeply indented between the seeds; seeds 8-12.

This species belongs to the series **Gummiferæ globiferæ** or arborescent Acacias with globular, axillary flower-heads, and straight long spines. This series contains the following species:—

A. arabica, A. eburnea, A. Farnesiana, A. Jacquemontii, A. leucophlœa, and **A. planifrons,** which compare.

For series **Gummiferæ spicatæ** see **A. Catechu,** and for series **Vulgares** see **A. concinna.**

Properties and Uses—

SYNOPSIS OF PRODUCTS AND ECONOMIC PARTS OF THE PLANT.

1. It yields a GUM, in India used by the calico-printer and for other industrial purposes; as a medicine by the natives, and to a certain extent as an indifferent substitute for true gum arabic. In times of scarcity it also constitutes an important article of food.

Babúl Gum.

2. THE BARK is largely used by the Indian tanners in the preparation of leather, and also as a dye. It is a valuable astringent medicine, extensively used by the natives, and in Indian European medical practice as a substitute for oak bark. It is also used to flavour native spirits.

3. THE POD has recently attracted much attention as a tanning material, containing a high percentage of tannin, and imparting a good and uniform colour to the leather. From the immature pods, by expression and inspissation, an extract was formerly prepared. This was known to the ancients as *Acaciæ veræ succus*, and was highly prized by the Greek medicinal writers. From them the virtues of this Acacia extract passed doubtless to the Arabs, and, at the present day, a drug known as *akaka* or *akakia* (or *acacia*) is regularly imported into Bombay from Turkey and Persia, and kept by all Mahomedan druggists in India.

4. THE LEAVES are used as a tan and dye; they are often eaten; and constitute an important fodder in times of scarcity.

5. THE TIMBER is highly valued because of its hardness and durability. One of the principal trees in the Panjáb for the rearing of lac insects (see "Lac").

THE GUM.

GUM. Babul. 102

Vern.—*Babúl-ki-gónd, kíkar* (or *kikar*) *ki-gónd*, HIND.; *Bábla-átá, bablá-gónd, babúl-gónd*, BENG.; *Káli-kikar-ki-gónd*, DUK.; in TAM. *pishin*, and in TEL. and MAL. *pasha* are added to the respective vernacular terms for the tree to denote the gum. According to **Moodeen Sheriff**, *kála barbúra-niryásam* is the Sanskrit for this substance. *Samghul*, HINDI, ARAB. **Dr. Rice** writes that *kála barbúra niryásam* is only a modern attempt to render in Sanskrit. It means the juice of the black *barbúra*, whatever the latter may be.

Dr. Dymock says: "There appears to be no mention of Gum Arabic in Sanskrit works." The Arabic and Persian name *Samgh-i-Arabi* given to this gum is more correctly the name for the True Gum Arabic than for the Indian Gum Arabic, which is sometimes called *babúl-gónd*. Commercially the gum from **A. arabica** is known as Morocco or Brown Barbary Gum or Mogador Gum—*Gum Gattie* according to **Atkinson**. The name Indian Gum Arabic must be carefully distinguished from the commercial term East Indian Gum Arabic. Many authors use these incorrectly as synonymous.

This gum is a tolerable substitute for the True Gum Arabic, but the mucilage is weak, and the red colour often objectionable. It exudes chiefly in March and April, each tree yielding about 2 lbs. In the bazars it occurs in the form of irregular and broken tears, agglutinated in masses, each tear about half an inch in size, and of a brown or red to light straw colour.

For the chemical composition of Gum Arabic, see **A. Senegal.**

TRADE AND COMMERCE.

External.—The subject of the identification of the true Gum Arabic of Europe is one still involved in considerable obscurity. The term Gum Arabic is of course incorrect, little or no gum ever having come from Arabia. Several species of Acacia yield gums, belonging to the Arabic series, and these, mixed together or distinct from each other, but more or less adulterated with foreign matter, reach Europe, and are by the dealers classed into geographical varieties, to which, according to their purity, are attributed certain commercial values. The principal English supply is from Egypt; France obtaining its gum from Senegal. The so-called East Indian gum, which is exported from Bombay to England, to the continent of Europe, and to America, is first imported into Bombay from the Red Sea ports. Except as adulterating this so-called East Indian gum, it is doubtful if

GUM.

any Arabic gum of Indian origin ever finds its way to Europe. The impurities of the East Indian gum are said to be chiefly a substance resembling Bassora (readily distinguished by its insolubility in water), and a resinous substance belonging to the turpentine series. It would seem that the subject of our Arabic gums has not hitherto obtained the attention which it deserves. The various species of **Acacia**, wild or under cultivation, should be critically examined and their respective gums carefully collected; the area under each, and the present and prospective supply of gum, accurately recorded. It would, of course, be of little importance to attempt to calculate the probable number of trees scattered here and there over the country near villages—the forest tracts or areas where gum-yielding trees form the prevailing feature are those of importance. Were this to be done, and a report published and forwarded to Europe along with a complete set of samples, a future trade would, for certain, develop,—a trade which would soon check the anomalous importation of African gum, and become a new source of revenue to India and her people. When it is recollected that all the Egyptian and African species which yield the true Gum Arabic of European commerce are either wild in India or are extensively cultivated or naturalised, it is remarkable that no effort should have been made on the part of India to compete with Egypt and Africa. The commercial gums known as MOROCCO, MOGADOR, or BROWN BARBARY gums are derived from **A. arabica,** *Willd.* They are described as consisting of tears or broken pieces of a light dusky-brown tint. When dry they are permeated by cracks and are very brittle. They are perfectly soluble in water. (For further information consult **A. Catechu** and True Gum Arabic under **A. Senegal.**)

Internal.—While our Indian gums are little if at all exported to Europe, *babúl-gónd* constitutes an important article of internal trade. But even in this trade there is a great deal of uncertainty, which a little careful investigation might easily enough remove. It seems probable that the finer qualities of *babúl-gónd* sold in our bazars may prove to be Senegal gum or white sennar (the true commercial Gum Arabic), the confusion in calling *babúl gum* Gum Arabic having arisen in all probability from the fact of the scientific name **Acacia arabica** having been given to this species. The *babúl-gónd* or *kikar-gónd* of the bazars is rarely pure, being mixed with mechanical impurities, and adulterated with other gums. The following are the gums generally mixed with each other and sold as *babúl-gónd:* the true Gum Arabic obtained in Sind from **A. Senegal** (the *khor,* SIND, or *kumta,* RAJPUTANA), **A. Catechu** gum (the *khair*), **A. Farnesiana** (the Cassie or *Vilayati-babúl*), **A. lenticularis** (*khair*), **A. modesta,** (*phuláhi*), and several gums obtained from various species of **Albizzia.**

MANUFACTURES AND PREPARATIONS.

Babúl-gónd is extensively used by the calico-printers. For certain colours a mixture of this gum, with that obtained from **Anogeissus latifolia,** the *dhawa*, is regarded as most serviceable, and this mixture is likewise used to stiffen dyed fabrics and to give them a polish. It is also said to be used extensively in precipitating the indigo fecula. With turmeric (*huldí*) dye the *Dhawa-gónd* is used alone, while with madder the *babúl-gónd* is regarded as most efficacious.

(For the " medicinal " uses and properties of this gum and for its value as an article of "food," see the remarks under those sections.)

THE TAN.

TAN. Bark. 103

THE BARK (*babúl-ki-chhál,* HIND., or *kas* or *sák* in N.-W. P.) is a powerful astringent, and is one of the tanning substances most extensively used in India. There seems no reason why this might not compete with the Australian wattle-bark if once made properly known; its cheapness, as compared with the wattle, would more than compensate for a slight inferiority in quality.

TAN.

It is obtained by felling the trees 8-10 years old, and cutting them up into pieces 2½ by 3 feet in length, the bark being removed when green. To do this the logs are beaten until the bark is removable, the wood being sold as fuel. The bark in appearance consists of reddish-brown slabs, hard and rough, with longitudinal fissures, the inner surface being smooth and fibrous. Recent experiments made by **Mr. W. Evans**, one of the best practical authorities upon tanning, in his laboratory at Taunton, have shown this bark to contain 18·95 per cent. of a beautiful cream-coloured tannin, when precipitated with gelatin. There is at present no exportation of *babúl* bark to speak of, but there seems every chance of a future trade developing. **Mr. Christy**, in his *New Commercial Plants*, suggests that the bark, pods, and twigs should be used in the preparation of a tanning extract which apparently he recommends should be prepared in India, and in that condition, instead of as bark, the *babúl* tan should be exported. In a correspondence with the Provincial Agricultural Departments the Government of India obtained much new information regarding the prospects of a future trade in this bark. In that correspondence **Mr. J. S. Gamble**, Conservator of Forests, Northern Division, Madras, suggested that if grown for its bark, **Acacia arabica** "would probably be found most profitable," if treated "in coppice and cut over every 8 to 10 years." "It would therefore require to be planted close, say, 10′ × 10′, or, still better, propagated by broadcast sowing": "the wood could, at the time of cutting, be also easily sold at good rates in any locality away from the large forests where *babúl* is grown."

Pods.
104

THE PODS (*babúl-séngrí*) are also used as a tan, imparting a buff colour to the leather. If it is desired to cultivate the tree on account of its pods, **Mr. Gamble** recommends that it should be planted at first 15 feet apart and thinned out to 30 feet. A proper amount of air and sunlight is essentially necessary for the full development of the fruit. This would give about 48 trees an acre.

An enquiry has recently been received by the Revenue and Agricultural Department of the Government of India from the Berlin Leather Trades Association, through Messrs. Fischer & Co., of Bombay, regarding what appears to be the *babúl* pods. The Association writes that a small pod known under the nam eof *blablab* (? *báblá*, the Bengali for **Acacia arabica**) formerly imported from the East Indies and partly also from the African coast, has of late years entirely dropped out of the market. On being boiled it gave a dull decoction of a light grey-yellowish colour. This colour became evenly diffused, and was specially serviceable for sheep-skin. "It is earnestly desired to recover again this tan and dye-stuff." I have taken the liberty to publish the substance of this correspondence, since by doing so this may lead to the resuscitation desired. There seems little doubt that the so-called *blablab* referred to by the Berlin Leather Trades Association is the pods of the common *babúl*. It is chiefly as a leather dye-stuff that the Berlin Association wishes to recover the pods. Speaking of these pods **Mr. Christy**, in a letter to the *Tropical Agriculturist, Vol. 2, p. 989*, states that they had been submitted to a chemical examination and were found to contain 60 per cent. of tannin. The seeds are worthless and should be rejected. He also adds that by means of this tanning agent a beautiful light-coloured leather was produced. Valonia yields only 25·2 per cent. of tannin and sells for £18 per ton. When once the *babúl* pods come into the market as a recognised tan, **Mr. Christy** anticipates they will fetch £40 a ton. If this statement be correct it is remarkable that a large trade

TAN.

should not already exist in *babúl* pods, and still more so that such a trade apparently did exist but has now died out. **Mr. Buck,** in his *Dyes and Tans of the North-West Provinces,* says: "The babúl pods are used for tanning in the villages of the Cawnpore district, and when nothing better is obtainable the leaves are employed as a make shift." But of the bark he says "it is the commonest and most effective tanning agent" used in these Provinces.

Leaves. 105

THE LEAVES alone are, in some parts of the country, used as a tan, but chiefly as a substitute for the bark. By the native tanners in India the bark is regarded as a much more powerful tan than the pods—a fact somewhat at variance with **Mr. Christy's** anticipations.

THE DYE.

DYE. Pods. 106

As a dye-stuff the boiled pods are often used, especially in the North-West Provinces, constituting the black colour known as *siyah-bhura;* this, with a subsequent application of a solution of sulphate of iron, changes the black into shades of grey ranging to dark brown, giving origin to the colour known as *Agrai-khaki.* Both colours are fast: but the iron is most objectionable. In combination with the barks of **Acacia Catechu** and **Butea frondosa** rich brown colours are also produced. The pods and bark, with alum as the mordant, yield dark brown shades approaching to black. In Upper India the pods are largely used by the dyers both as a direct source of colour and as an accelerator to other dyes. In addition to the local supply, which, in the warm, dry regions of the central and northern tracts of India, is very considerable, large quantities are brought from the forests, Kumaon alone sending some 30 cwts., valued at R400. The price in Calcutta of the dried pods is from 2 to 3 annas a ℔.

§ "The bark furnishes a dye largely used by the natives." (*Surgeon S. H. Browne, M.D., Hoshangabad, Central Provinces.*)

THE FIBRE.

FIBRE. Bark. 107

THE BARK of the slender twigs yields a fibre, which is used in the Panjâb for the manufacture of paper. It is also made into coarse ropes.

MEDICINAL PRODUCTS.

MEDICINE. Gum. 108

The true Gum Acacia is used in the preparation of the mucilage; the *Pharm. Ind.* says that the gum of **Feronia Elephantum,** *Corr.*, is a better substitute for this than the gum from **A. arabica.** Gum is used as a demulcent, and along with water as a vehicle for bismuth, oxide of zinc, or other insoluble substance and also in lozenges.

109

§ "I do not agree with this remark; the gum of **Feronia Elephantum,** is much inferior in various respects to that of **A. arabica.**" (*Deputy Surgeon-General G. Bidie, M.B, C.I.E., Madras.*) "Gum Acacia is also administered to recently-delivered women as a tonic." (*Surgeon H. McCalman, M.D., Bombay.*) "The gum is largely used in the form of a mucilage in diarrhœa and dysentery." (*Bhagwan Das, Rawal Pindi.*) "Some native hakims say the gum is very useful in diabetes mellitus, as the gumis not converted into sugar. The bark and seeds burnt and powdered areused as a tooth-powder." (*Surgeon Emerson, Calcutta.*) "The powdered gum is useful, combined with quinine, in fever cases complicated with diarrhœa and dysentery." (*Brigade Surgeon Shircore, Moorshedabad.*) "The mucilage in its simple form used as an injection has been found to allay the irritation of gonorrhœa, and to lessen the discharge. It has also been found to allay rectal irritation in the diarrhœa and dysentery of children, when given as an injection." (*Peter Anderson, Guntur, Madras Presidency.*)

"**Acacia arabica** mucilage I have used in cases of cystitis with good effect." (*Surgeon-Major J. J. L. Ratton, M.D., Salem.*) "An infusion

MEDICINE.

110

or decoction of the bark is used as gargle for sore-throat and stomatitis. The juice of the tender leaves is dropped into the eye for epithora and conjunctivitis. The gum is fried in *ghí* and made into sweetmeat for the use of women in childbed." (*Surgeon-Major J. Robb, Ahmedabad.*)

"Decoction of bark useful in chronic dysentery and diarrhœa as an astringent enemata. Useful gargle in spongy, bleeding gums and mercurial salivation. Excellent mucilage from gum used in gastro-intestinal irritation." (*Surgeon Shib Chunder Bhattacharji, Chanda, Centl. Prov.*)

Bark. 111

THE BARK is a powerful astringent; recommended to be made officinal. It occurs in coarse, fibrous pieces of a deep reddish colour. It may be used in external applications as a substitute for oak-galls. It has been found a valuable remedy in prolapsus ani, as an external applicant in leucorrhœa, and has been recommended as a poultice for ulcers attended with sanious discharge. (*Pharm. Ind.*) As a substitute for oak-bark, *babúl* is now issued to the Government hospitals and dispensaries in India. I am informed that the powdered dry bark dusted over sores or ulcers on the lips of horses is one of the best cures for these troublesome affections. A similar powder is used in the Panjáb in the treatment of snake-bite.

112

§ "I have frequently used the decoction of Babla bark as a substitute for oak-bark for vaginal injections. It might take the place of imported oak-bark." (*Surgeon C.H. Joubert, Darjíling.*)

"In addition to the ordinary uses of the gum and the bark, I have frequently used the decoction of the bark as an astringent injection in different forms of leucorrhœa and found it to be more efficacious and less irritating than the alum and zinc injection generally used." (*Doyal Chunder Shome.*) "Useful as a gargle in relaxed sore-throat." (*Surgeon J. Anderson, Bijnor.*) "The infusion is useful as an injection in whites." (*Surgeon-Major G. Y. Hunter, Karachi.*)

"Babúl Bark } Mango " } . . . . . . $1\frac{1}{2}$ oz. each.
Water . . . . 20 oz.

Boil for half an hour, filter, and make a gargle: said to be used in mercurial salivation." (*Surgeon J. Parker, M.D., Poona.*) "A decoction of the bark is used by women in menorrhagia and leucorrhœa, and is said to be efficient. The decoction is also used in caries of the teeth as a mouth-wash. The young leaves in doses of 2 drachms are used in gonorrhœa with good effect." (*Narain Misser, Hoshangabad, Central Provinces.*)

113

"The bark of the **Acacia arabica,** combined with the bark of the Banyan tree **(Ficus bengalensis),** after being infused, has been frequently used by me as a gargle in relaxed sore-throat with excellent effect. Strong infusion of the above barks has also been used locally in cases of excessive bleeding from hœmorrhoids with very good effect. The dissolved gum with bismuth is very effectual in checking diarrhœa arising from intestinal irritation." (*Honorary Surgeon Easton Alfred Morris, Negapatam.*)

Pods 114

THE DRIED PODS reduced to a powder are sometimes given internally as an astringent.

§ "The powder of the tender legumes of **A. arabica** is astringent and demulcent, and has a beneficial influence over diarrhœa and dysentery. Its usefulness is much enhanced by the combination of some preparation of opium. It is generally more useful for children than adults." (*Honorary Surgeon Moodeen Sheriff, Khan Bahadur, Madras.*)

The extract Akakia 115

AN EXTRACT prepared from the young pods of this and allied species constituted the famous *Akakia* of the ancient Greeks. This extract was at one time much extolled, and its virtues were doubtless ultimately made

MEDICINE.

known to India through the Arabs. At the present day it is largely imported into India from Turkey and Persia, and under the name *akaka* or *akakia* or *akhákhíyá* (acacia) is sold by all Mahomedan druggists. According to **Dymock** "it should be heavy, hard, and have an agreeable odour; small fragments held between the eye and the light should be of a bottle-green colour; when seen in bulk it appears black. It is considered to be cold and dry, astringent, styptic, and tonic, and is used internally and externally in relaxed conditions of the mucous membranes; also as a collyrium in purulent conjunctivitis and chronic congestion of the vessels of the conjunctiva. Applied as a lotion, it is said to improve the complexion. With the white of an egg it is applied to burns and scalds. Powdered it arrests hæmorrhage; in short, it is used in all cases in which an astringent is applicable."

Decoction from Bark. 116

A DECOCTION of the bark mixed with milk, according to **Mr. Baden Powell** (*Panjáb Products, Vol. I, p. 345*), is in Delhi and Lahore evaporated into the *akakia* juice. The term *akakia* is apparently incorrectly applied to this decoction, but the preparation is nevertheless an interesting adaptation of the ancient *akakia* or *akhákhíyá*. It is said to be made into dark flat cakes with a sweet astringent taste. It is regarded as acting as a demulcent and astringent, and is said to be prescribed in coughs. There is probably some mistake regarding the source of these cakes. It is difficult to see how they could be sweet if prepared from the bark, unless sweetened through some action upon the milk.

117

§"I have often used the decoction of the bark as an injection in chronic dysentery with relaxed state of the rectum, when, with tolerably healthy motions, a little mucus is passed. The injection reduces the quantity of the mucus, soothes the irritability, and gives tone to the mucous membrane. I have also used the decoction in acute congestion of throat with success." (*Surgeon D. Basu, Furridpur.*) "The bark contains a large quantity of tannin; a decoction of it, as a local application, is most useful in cases of prolapsus uteri and of prolapsus ani, and in other uterine and vaginal affections of an asthenic nature." (*Brigade Surgeon S. M. Shircore, Moorshedabad.*) "The decoction of the bark is used as an astringent in diarrhœa and dysentery, also as an injection in gleet and leucorrhœa, and as a wash for hæmorrhagic ulcers, also as a gargle in affections of the mouth and throat. The tender leaves are sometimes used in dysentery as well as in diarrhœa." (*Brigade Surgeon J. H. Thornton, B.A., M.B., Monghyr.*) "A decoction of the bark is used as an astringent gargle in ulcers of the mouth. The tender growing tops rubbed into a paste with a little sugar and water is given morning and evening as a demulcent to allay irritation in acute gonorrhœa." (*Surgeon-Major Bankabehari Gupta, Púrí.*)

Leaves. 118

THE TENDER LEAVES beaten into a pulp are given in diarrhœa as an astringent (*U. C. Dutt*). Mixed with the leaves of the pomegranate this pulp is also given in gonorrhœa.

119

§"The young leaves are beaten up with black pepper and sugar, and given in hæmatemsis." (*Surgeon-Major C. W. Calthrop, M.D., Morar.*) "The powdered gum is lightly fried in *ghí* and used as an aphrodisiac. The tender leaves, fried in *ghí*, are used as a poultice over the eyelid to remove chronic congestion of the conjunctiva. The bark is said to possess antisyphilitic properties." (*Surgeon-Major D. R. Thompson, M.D., C.I.E., Madras.*) "The tender leaves, bruised and mixed with human milk, is used in conjunctivitis as a poultice, or the juice mixed with milk is dropped into the eye. The burnt bark mixed with salt and burnt almond shell is used as a tooth-powder in Southern India." (*Surgeon-Major John Lancaster, M.B., Chittore.*)

The Babúl—Food, Fodder, Timber.

FOOD.

FOOD. Gum. 120

The gum is highly nutritious, and to a limited extent forms an article of food, largely so in times of scarcity; in fact, there are few trees more valuable to the cultivator than the babúl. It yields his most valuable timber while luxuriating on the poorest waste lands, and even in seasons of drought it is evergreen. Its bark forms a useful domestic medicine, and along with the leaves and pods it is also used in dyeing and tanning. The leaves are a never-failing source of fodder, and the gum an article of food; each tree yielding about 2 lbs. In times of scarcity even the ground bark mixed with the seeds of the **Sesamum orientale** may be used for food.

Bark. 121

FODDER.

FODDER. Pods. 122

The green pods with tender shoots and leaves are given as fodder to cattle, sheep, goats, and camels; and are specially valuable for this purpose during a season of drought when other fodder fails. In the drought of 1877-78 the road-side trees in the North-West Provinces were denuded of leaves for this purpose, and this resource saved numbers of cattle. In ordinary seasons goats are largely fed upon the pods; hence, in all probability, the rapidity with which the plant becomes diffused over the country, springing up self-sown on the banks of tanks, rubbish heaps, and walls. It is remarkable that sheep and goats not only eat but eat greedily a substance which is stated to contain so much of tannin. **Balfour** in his *Cyclopædia*, however, says it is only the seed that is given to sheep, but his statement is at variance with that of all other authors, the whole pod being given.

Seed. 123

THE TIMBER.

TIMBER. 124

Structure of the Wood.—Sapwood large, whitish; heartwood pinkish-white, turning reddish-brown on exposure, hard, mottled with dark streaks. It consists of darker and lighter coloured bands of an equal width. Weight about 54 lbs.

Implements. 125

It is very durable if well seasoned. Used extensively for wheels, well-curbs, sugar and oil presses, rice-pounders, ploughshares, agricultural implements, and tool handles; in fact, for all purposes for which a bent hard wood is required. In Sind it is largely used for boat-building, rafters, and for fuel; also, occasionally, for railway sleepers. One of the most valuable timbers for tent-pegs.

The lighter-coloured sapwood of this tree is subject to be attacked by white-ants, the heartwood much less so if properly seasoned. A good tree will sell in the Panjáb for as much as R30 or more: one of the best trees for broad-cast sowings in the reclamation of waste lands, because independent of rain. It is said that seasoned wood will only float for a few days. The Sind and the Madras Railway Companies refuse to use babúl as a fuel from an idea that the pyroligneous acid injures the boilers. It is, however, largely used for this purpose by railways in other parts of India.

Fuel. 126

DOMESTIC USES.

DOMESTIC. Spirits. 127

The bark of the roots is used to flavour native spirits and to assist the fermentation of the sugar. The bark is also stated to be used as a substitute for soap. The green pods are made into ink. The young thorny twigs are universally used for temporary dry fences to protect certain crops, and large bundles of the boughs are used by the fishermen as decoys. In the rivers of Assam these decoys end in cone-shaped baskets placed on the margins of the streams sloping down water. Every now and then they are raised from the water, and the fish, unable to escape through the decoy in time, are caught in the basket or amongst the

Bark. 128

Pods. 129

Dry Hedges. 130

ACACIA arabica.

Domestic uses of, and re-afforestation with, Babúl.

Fish Decoys. 131
Fishing-hooks. 132
Tooth-brushes. 133

thorny twigs. In the tanks and lakes of Bengal similar decoys are submerged, and alongside of these, or imbedded between passages in the decoy, traps of various forms are placed. The sharp spines are made into the common fishing-hook used in Bengal. A portion of the twig is removed along with the spine, the string being tied to the hook so that the spine points upwards. Young fresh twigs are in Bengal used as tooth-brushes.

§ "The bark is extensively used in the preparation of country liquor." (*Bhagwan Das, Rawal Pindi.*)

134

"The bark is largely used for the distillation of rum. Before distillation takes place it is steeped in water with molasses for a few days. For loose teeth and tender gums, the young stems are used largely by the natives for rinsing the teeth daily." (*Assistant Surgeon J. N. Dey, Jeypore.*)

CULTIVATION AND RE-AFFORESTATION.

Improvement of waste lands.

The extended cultivation of this species and of the allied gum-yielding Acacias cannot but be attended with the most beneficial results to the country. As an agent to improve sterile tracts of country, or to arrest the destructive development of the efflorescence known as *reh*, the most sanguine expectations may be entertained. It should not be cultivated longer than has been found necessary to neutralise the salt, because *babúl* is an exhaustive crop. In fact, the moment *babúl* has taken a hold of the soil the improvement may be regarded as established, and should the soil be desired for other purposes the *babúl* should be early removed. There are extensive tracts of country highly suited for the cultivation of the gum Acacias, and tracts, too, where this crop would prove far more productive than the futile efforts which are now being spent upon them. The extended cultivation of the *babúl* as a hedge to protect the fields and roads would greatly help to avert the dangers of a season of temporary scarcity both to men and cattle. **Cleghorn** says "it is of rapid growth, and requires no water, flourishing in dry arid plains, and especially in black cotton soil, where other trees are rarely met with."

§ "One of the best subjects for planting in poor soils or exposed situations. During long periods of intense drought it continues to flourish." (*J. Cameron, Esq., Bangalore.*) "It is said that the seed collected from the litter of goats germinates more freely, and such seed is sometimes collected for sowings. If sown thick in a road, well, or ditch in the rains, it will, under favourable circumstances, come up as a wood-hedge. It is one of the most useful trees for broadcast sowings; the seed germinates so easily, and it is so hardy and independent of rain. In the drier districts it should be planted as a thick belt along all roads to form a roadside avenue. There would then be a vast supply of fodder available when rain failed. This is being attended to in this division." (*W. C. Coldstream, Esq., C.S., Hissar, Panjáb.*)

The seed is gathered in April, and by the native cultivator it is coated with cow-dung and kept in that condition until July, when it is sown at once into the spot where it is intended the tree should be allowed to grow. It requires no further care except to protect the young seedling from being browsed upon by animals. **Stewart** states that in the upper parts of the Gangetic Doáb it is raised by cuttings. **Major Pitcher,** Agricultural Secretary to the Chief Commissioner of Oudh, writes me that the best mode of propagating *babúl* is to fold sheep previously fed on *babúl* seed on the field where it is desired to plant the trees; or to carry the dung to the field and plough it in, when the plants will spring up like corn. He also informs me that **Dr. Bonavia** tried an experiment with a number of trees to discover which could withstand repeated swamping. All died except the *babúl*, which continues to flourish on the banks of the Gúmti, where

RE-AFFORESTATION.

they are annually flooded. The *babúl* would, therefore, prove most useful upon embankments to break the force of water in storms or during floods.

Gamble writes that it is an extremely useful tree in the re-afforestation of waste lands, being associated with **Albizzia Lebbek, Balanites ægyptiaca, Parkinsonia aculeata,** the tamarind, margosa, and wood-apple. These plants constitute the most prevalent association of trees met with in the warmer parts of India. The *bábul* is ready for barking in eight to ten years. It is then cut down and renewed by fresh seedlings, or the ground, after this season of repose and leaf-manuring, is brought under other forms of cultivation. A full-sized tree, eight to ten years old, will yield half a ton of bark. The gum-yielding property increases with age, the gum exuding naturally from the bark, or accelerated by artificial scars. **Stewart** mentions a tree at a Mussulman shrine close to Lahore said to be over 100 years old, which is popularly reputed to have shed blood when sacrilegiously cut by the Sikhs.

135

Acacia Catechu, *Willd.; Fl. Br. Ind., II., 295; Roxb., Cor. Pl., t. 175.*

This is the plan t which yields—

CATECHU, CUTCH, CATECHU NIGRUM, PEGU CATECHU, *Eng.;* CACHORE, CACHOU, *Fr.;* KATECHU, *Germ.;* CATECU, CATEIN, CALTO, *It.;* CATECU, *Sp.;* CATCH, *Port.*

Sometimes called Terra Japonica, a name which more correctly means Gambier.

§ "Perhaps it would be better not to call this *the true catechu,* as the **C. pallidum** is now the officinal variety in British Pharmacy." (*Deputy-Surgeon General George Bidie, M.B., C.I.E., Madras.*) "The name **Catechu pallidum** has always been given to Gambier, and the word Catechu was originally applied to Cutch, which is officinal in America." (*G. W.*)

Syn.—MIMOSA CATECHUOIDES, *Roxb.;* ACACIA CATECHUOIDES, *Wall.;* A. POLYACANTHA, *Willd.;* A. SUNDRA, *Bedd.*

Vern.—*Khair,* or *khair-babúl, katha,* HIND., DUK.; *Khayer, kuth,* BENG.; *Khaiyar,* SANTAL; *Khoira, koir, kat,* ASS.; *Vódalam, vodalai, karangalli, bágá, wodalior, wodahalle* (by Cleghorn), (? *kasku katti*), *wothalay,* TAM. *Podala-manu, kaviri sandra, nalla sandra,* TEL.; *Khadira,* SANS.; *Khoiru,* URIYA; *Kagli, kagali, tare* (in MYSORE), KAN.; *Khaderi, khaira, khera,* BOMB., MAR.; *Kher,* SURAT, BARODA, GUJ.; *Khair,* C. P.; *Ratkihiri,* SINGH.; *Sha,* BURM.

"The khadira tree is mentioned in the Vedas, where it is used as a simile for 'strength,'" &c. (*Dr. Ch. Rice, New York.*)

References.—*Brandis, For. Fl., 186; Stewart, Pb. Pl., 52; Kurz, For. Fl. Burm., I., 422; Gamble, Man. Timb., 153; Roxb., Fl. Ind., Ed. C.B.C., 423; Madden, Jour. Asiat. Soc. Beng., vol. XVII., part I., p. 563; Royle's Ill., 182; Baden Powell, Pb. Prod., 345; Atkinson, Him. Dists., vol. X., N.-W. P. Gaz., p. 775; Balfour's Cycl., I.; Spons' Encycl.; Bombay Gazetteer, vols. IV., 24; VI., 13; X., 130; XI, 71, 72, 415; XII, 25; XIII, 160; British Burma Gaz., I, 128, 134, 415; C. P. Gaz., 118, 503; Oudh Gaz., vol. I., p. v.*

Habitat.—Common in most parts of India and Burma, extending in the Sub-Himálayan tract westward to the Indus and eastward to Sikkim, ascending to altitude 5,000 feet. **Mr. J. W. Oliver** reports that trees 70 to 80 feet high, with a girth of 8 to 9 feet, are not uncommon in the North Tharrawaddy Reserves. In Burma it grows in the dry forests all over the plains of Pegu; rare in the savannah forests; common in the northern part of the Irrawaddy (*Burm. Gaz., I., 128*). In the Central Provinces it is plentiful in the forests of Bilaspur, Chánda, and Raipur. It is remarkable that in Raipur the natives seem to be ignorant of its value, no attempt, as far as is known, having ever been made to extract the Catechu

Catechu Gum.

(*C. P. Gaz., 419*). In Gonda, Oudh, it is abundant, also in the forests of the Upper Godaveri, spreading through the forests of Chutia Nagpur to the North-West Provinces. In the Bombay Presidency it is most abundant in the forests of Ahmedabad, Broach, the Panch Mahâls, Surat, and Baroda. In Madras the tree is by no means uncommon, but the natives appear to know nothing of its properties. It is also met with in some parts of Mysore, where it is even common.

Botanic Diagnosis.—A moderate-sized, deciduous tree, with dark brown, much-cracked bark, and short-hooked spines in pairs. *Leaves* composed of from 40 to 80 pinnæ and 60 to 100 leaflets. *Flowers* white or pale yellow, peduncled, spikes in the axils of leaves; *rachis* downy. *Corolla* 2 to 3 times the tomentose *calyx*. *Pod* straight, strap-shaped, narrow, thin, dark brown.

This belongs to the series **Gummiferæ spicatæ** or arborescent Acacias, with spiked flower-heads and short recurved spines (except **Latronum**):—

A. Catechu, A. ferruginea, A. Latronum, A. lenticularis, A. modesta, A. Senegal, A. Suma, and **A. Sundra,** which see and compare with above diagnostic characters.

For series **Vulgares** see **A. concinna.**

THE GUM.

GUM. 136

This plant yields a pale-yellow gum, often occurring in tears one inch in diameter, generally less than half an inch in size. It is sweet to the taste and soluble in water; it forms a strong mucilage, and is a better substitute for true gum arabic than babúl gum, with which it is generally mixed and sold as *babúl gónd*. The tears are mostly bright coloured, varying to dark amber. They occur chiefly in broken pieces, very much resembling large brown sugar, the fragments being cracked and granular.

A pure sample of *Khair gónd* was, in 1873 sent from Chánda in the Central Provinces to London for report and valuation. It was pronounced as ordinary gum arabic and valued at 20*s*. to 25*s*. a cwt. The Deputy Commissioner, in sending the sample, remarked that by a little care on the part of the merchant, a large supply of gum, equal in purity to the sample supplied, might easily enough be procured, and that the ordinary mixed gum arabic of Chánda sold for about R5-4 a cwt. landed in Bombay. The cost of collecting the gum in the forests was put down at R2-8 a maund (80 ℔), carriage from forest to Bombay R1-4 a maund, and this gum might easily realise in London 20*s*., *i.e.*, twice what it cost landed at Bombay. In the Ahmedabad district and Gujarát this gum is largely collected, as also *bavál* gum, by the Bhils and sold either for grain or money or eaten by the poorer classes (*Bomb. Gaz., IV., 24*).

History of Catechu Gum.—No fresh effort appears to have been made to bring this gum to the notice of Europe, nor so to organise its collection that comparatively pure consignments might continuously find their way into the market. It is reported to be much superior to babúl gum, and it would seem a large trade might easily enough be established; confidence in the supply, however, must be created before any extended trade is possible. Indian gums hold so low a position, at present, in the commercial world, that they require to be freed from an undeserved stigma. To accomplish this the most likely course would be to collect authentic samples as types upon which, in Europe and in India, all experiments, reports, and administrative assistance or encouragement might be based. It must be borne in mind, however, that vernacular names of bazar products are most untrustworthy, far more so than the knowledge which every uneducated native of India possesses in the plants of his district. Adulteration and substitution is practised very largely in India; and in different districts, still more so in different provinces, widely dissimilar products are sold under the same vernacular names. The type samples require, therefore, to be collected from the individual plants and carefully examined both as to their physical appearance and chemical properties. So little is actually known regarding the subject of gums (even in Europe) that at present it is next to impossible to determine the botanical origin of a sample not accompanied with its geographical history, and even then a mere commercial approximation is all that can be arrived at. To remove this ignorance the greatest care is necessary, and it would be advisable to collect the type samples of each species of gum from one individual tree, and, if possible, say, during each of the great climatic seasons of the district—cold season, hot season, and during the rains. Such collections would enable questions of specific variation to be determined. It would also remove all doubt as to identity were a flowering twig of the tree to be dried and preserved so as to admit of botanical determination, together with a branch or log of the tree with gum exuding from it. The latter is necessary to afford some data by which the actual source or formation of each gum might be determined, for it is possible that the same tree may yield more than one gum.

Within the past few years the plants which yield the true commercial gum arabic have been pretty well determined, and it is important to note that all these are either wild or cultivated in India. To enable India to come into the field of competition with Egypt and Africa, the merits of the Indian gums must first be made known by some accurate and reliable mode such as has been indicated. When the gums obtained from Indian trees have been chemically and botanically examined and commercially tested, those most deserving of encouragement would be discovered, and we should be able to indicate to the merchant the provinces and districts where these were most easily and plentifully procurable. We should be able also to supply the merchant with the tests by which he could protect himself against adulteration and admixture. The possession of such a knowledge would, in time, check fraudulent action, and it would also act as the most powerful impulse to future progression, for, when freed of impurities, there does not seem any reason why a large trade should not be established in Indian gums. Nothing is so fatal to the development of such a trade, however, as adulteration and admixture, and it would seem that it is this that has prevented India from taking an important position as a gum-supplying country. **Dr. Cooke,** in his report on gums produced in India, says: "It is scarcely necessary to add that the gum must be unmixed. Two gums—it even may be that both are soluble—might not dissolve equally in rate or freely coalesce; this would be a disadvantage; but when two or more gums are mixed, of which one is insoluble in whole or in part, the whole sample is greatly deteriorated and reduced to almost the rate of the insoluble admixture. Hence then, commercially, it is of importance that gum should be (1) entirely derived from the same species of tree, (2) as light and uniform in colour as possible, and (3) free from all foreign admixture." In our Indian forests it frequently happens that two or more gum-yielding trees are found growing together, and the gums from these are indiscriminately collected together, the mixture being often reduced to a powder and sold in that condition with the object of preventing detection of the admixture and adulteration with sand or other non-gummy substances.

THE EXTRACT CATECHU OR CUTCH.

137

Vern.—*Kat* or *Kath, katthá,* HIND.; *Kát,* MAR.; *Kátho,* GUZ.; *Káshu-katta, káshu,* TAM.; *Kánchu,* TEL.; *Katti, káshi-katti,* MAL.; *Kachu,* KAN.; *Kaipu,* CINGH.

History of Catechu.—At the present day, by far the most important

ACACIA Catechu.

The Extract Catechu or Cutch.

CUTCH.

product of **Acacia Catechu** is the resinous extract (Catechu) obtained by boiling down a decoction obtained from chips of the heartwood. The practice of preparing this extract has been handed down from remote periods. The Sanskrit authors mention the drug, and **Barbosa,** in his description of the East Indies, published in 1514, mentions what is, in all probability, this drug under the name *Cacho.* He states that it was at that time exported from Cambay to Malaccā. *Cacho* is apparently the Kanarese word *Káchu* now applied to it. It is in fact probable that the word Catechu is a modern Latin derivative from the South Indian name, and that from South India the product was first exported. Some authors, however, say that it is derived from the Cochin Chinese word *Caycau.* One of the Tamil names for the plant is *Kati, Kuti,* or *Cate,* and the second half of the word may have been derived from *Chuana,* to drop or distil. Whatever may be the origin of the word Catechu, it would save much ambiguity if it could be restricted to the extract from **Acacia Catechu** instead of being made popularly to include one or two other substances such as Gambier, a word of Malayan origin signifying bitter, and applied to a purely Malayan product (see **Uncaria Gambier**). It is quite true that both these astringents contain the same chemical properties, but they are obtained from widely different plants and manufactured in countries separated from each other. In our *Trade and Navigation Returns* the exportation appears as "Cutch and Gambier," from which one would naturally infer that both Cutch (or Catechu) and Gambier were exported from India, the relative proportions of which had not been determined. I am informed, however, by my friend Mr. J. E. O'Conor that this practice is a remnant of the time when the Straits Settlements returns were published with those of India. At present, therefore, by "Cutch and Gambier" is meant in all probability chiefly Catechu; a small amount only of Gambier is re-exported.

From the time **Barbosa** wrote in 1514 we have no further mention of this substance until 1574, when **Garcia de Orta** gave a complete account of the plant and the process of preparation of the extract, describing it under its Tamil name *Cate* (*Kati* or *Kuti*). It was not, however, until the 17th century that Catechu attracted the attention of Europe. It was then supposed to be a natural earth, and as it reached Europe by way of Japan (being simply re-exported from that island) it received the name of Terra Japonica. At this period, or shortly after, Gambier also found its way to Europe, and was indiscriminately with Catechu called Terra Japonica. **Cleyer** exploded the mineral notion regarding Catechu, and in 1685 republished **Garcia de Orta's** account of the preparation of the extract, and declared it to be of Indian origin, the best quality coming from Pegu and other sorts from Surat, Malabar, Bengal, and Ceylon.

Catechu (from **Acacia Catechu**) was received as an officinal drug into the *London Pharmacopœia* of 1721. It was officinal in the *British Pharmacopœia* of 1864, but has since been discarded and Gambier retained, both in Great Britain and India, as the officinal form of the drug Catechu. In the *United States Pharmacopœia* Catechu (from **A. Catechu**) is retained, however, as officinal and Gambier discarded.

138

Interesting Ethnological Facts connected with the Catechu Industry.—Before passing to discuss the chemical properties and modes of preparation of Cutch (Kutch) and *Káth,* the digression may be regarded as not altogether inappropriate to say something of the race of people who, from time immemorial, have made it their sole occupation to prepare this extract. A brief history of this nature it is thought may help to throw some light upon questions connected with the early history of the drug, and may prove interesting to those who have not the opportunity of consulting the voluminous gazetteers and official records from which it is extracted. In his *Himálayan Districts* (which constitutes Vol. X. of *N.-W. P. Gazet-*

teer) **Mr. Atkinson** says: "The men employed are of the *Dom* caste, and are called *Khairís*, from the vernacular name of the tree."

In the *Bombay Gazetteer*, Vol. XIII., an interesting account of the *Káthkarís* is given, of which the following synopsis contains the most interesting features. It is stated that the *Káthkarís*, or makers of *káth*, are believed to have entered the district of Thána from the north and to have been originally settled in the *Gujarát Athávisi*, the present district of Surat. According to their story, they are descended from the monkeys which the god *Rám* took with him in his expedition against the demon-king of Ceylon. They are darker and slimmer than other forest tribes. They have no peculiar language of their own, but in conversation they have a tendency to reduce words and shorten speech, and uniformly endeavour to get rid of the personal, not the tense, inflections of the verbs. The women are strong and healthy, and pass through child-birth with little trouble or pain. They are said, sometimes when at work in the fields during the rains, to retire behind a rice-bank and give birth to a child, and, after washing it in the cold water, to put it under a teak-leaf rain-shade and go back to their work. They are divided into two sections—*Sons* or *Maráthás* and *Dhors*. The former do not eat cow's flesh, and are accordingly allowed to draw water from the village well. They are also more or less a settled tribe. Some of them still make *káth* or catechu, but from the increase of forest conservancy the manufacture is nearly confined to private *inám* villages and to forests in Native States. When they go to the forests to make Catechu, they hold their encampment sacred, and let no one come near without giving warning. Before they begin their wood-cutting, they choose a tree, smear it with red-lead, offer it a cocoanut, and, bowing before it, ask it to bless their work. The Catechu is made by boiling the heart-juice of the *khair* tree, straining the water and letting the juice harden into cakes. The *Káthkarís* will never go in for regular cultivation; they eat rats and monkeys, and live chiefly upon jungle produce, or by theft, stealing from fields and barns.

In the tenth volume of the *Bombay Gazetteer, p. 48*, it is further stated that the "*Káthkarís*, a wild forest tribe in the Ratnágiri District, who subsist almost entirely by hunting, now that their more legitimate occupation of preparing Catechu, *káth*, has been interfered with, habitually kill and eat monkeys, shooting them with bows and arrows. In order to approach within range, they are obliged to have recourse to stratagem, as the monkeys at once recognise them in their ordinary costume. The ruse usually adopted is for one of the best shots to put on a woman's robe, *sári*, under the ample folds of which he conceals his murderous weapons. Approaching the tree on which the monkeys are seated, the disguised *shikári* affects the utmost unconcern, and busies himself with the innocent occupation of picking up twigs and leaves. Thus disarming suspicion, he is enabled to get a sufficiently close shot to render success a certainty."

"In the villages of Navágám, about 7 miles north-east, Gángadia, 11 miles south, and Nelsa, about 9 miles south-west of Dohad, every year on the day after *Holi* (April) a ceremony called the *chul* or hearth takes place. In a trench seven feet by three and about three deep, *kher* **(Mimosa Catechu)** logs are carefully and closely packed till they stand in a heap about two feet above ground. The pile is then set on fire and allowed to burn to the level of the ground. The village *Bhangia* or sweeper breaks a cocoanut, kills a couple of fowls, and sprinkles a little liquor near the pile. Then after washing their feet, the sweeper and the village headman walk barefoot hurriedly across the fire. After this strangers come to fulfil vows, and giving one anna and a half cocoanut to the sweeper and the other half cocoanut to the headman, wash their feet, and, turning to the left, walk over the pile; the fire seems to cause none of

CUTCH.

them any pain." (*Bombay Gazetteer, III., 310.*) At the village of Chosala about 7 miles north of Dohad, a stream runs into a cave, and on this spot an image of *Mahádev*, under the name of *Kedáreshvar*, has been set up. The place is sacred to the Bhils.

Many other similar ceremonies and sacred practices might be mentioned, showing that the preparation of Catechu dates back to the remotest antiquity. The tree is sacred to *Mangala* or *Kárttikeya*, one of the Hosts of Heaven. It receives special worship and is often mentioned in the Vedas.

Forms of the Catechu Extract.

139

There are three substances, all very similar in chemical composition, derived from this plant: (1) Dark Catechu or *Cutch* used for industrial purposes; (2) Indian Pale Catechu or *Kath*, a crystalline substance eaten in *pán* or used medicinally by the Hindús, prepared from the decoction; (3) Keersal, a crystalline substance found imbedded in the wood.

Cutch. 140

1.—Preparation of Catechu or Cutch. The trees are regarded as mature when about a foot in diameter. They are then felled and cut up into blocks two to three feet in length. In some parts of the country the natives "test whether the tree will pay to cut by making a small notch in its heartwood. Trees between 25 and 30 years old are regarded as best suited for the manufacture, and are said to yield more or less *káth* according to the number of the white lines in the heartwood." (*Bombay Gazetteer, VII., p. 35.*) The bark and the outer white wood is removed and rejected. The red heartwood is then cut up into small chips (generally about a square inch in size). In Báriya, Gujarát, the trees are not felled, but the branches lopped and the extract prepared from them, and it is stated that in some parts of the country the unripe fruits and leaves are also used. The furnace in most frequent use is of a curious construction: it is built over, leaving a number of small openings into which the earthen pots are placed, in which the chips are boiled down into a decoction. The process is somewhat varied in different parts of the country, the departures, however, being in minor details. After being boiled, the red liquid obtained is poured over fresh chips and boiled again, and when a decoction of sufficient strength has been made, this is poured, either into larger earthen pots, or, as in Pegu, into iron bowls and boiled down into the consistency of a black paste.

141

In Baroda "the men, after removing all the sapwood and a little of the heartwood, cut it into thin chips about a square inch in size. These chips are boiled in small earthen pots with water. When sufficiently charged with *káth* the water is poured into two pots and allowed to go on boiling. The infusion in the two pots is poured into a wooden trough one yard long and eighteen inches broad, and a woman strains it through a piece of blanket about a foot square. Sitting on the ground she dips the blanket into the infusion, stirs it about, and holding it as high as she can, wrings it into the trough. This process goes on for about two hours, after which the trough is covered with a lid of split bamboos and the sediment is allowed to subside. The water is then poured off and the *káth* cut into small cakes and left to dry. On account of the destruction it causes to trees, *káth* manufacture has been stopped in the Navsári forests." (*Bombay Gaz., VII., 35.*) "In Báriya, Gujarát, during February and the three following months, *káth*-making gives employment to a large number of Kolis and Náikdás. Branches stripped of their bark are cut into small, three or four-inch, pieces and boiled in earthen pots till only a thick sticky decoction remains. A narrow pit five or six feet deep is dug and a basketful of the extract placed over the pit's mouth, the water soaks into the earth, and the refuse remains in the basket, leaving the *káth* in the pit. The

CUTCH.
142

extract is then taken out and dried on leaves in the sun." *Bombay Gaz., VI., 13.*)

In the *British Burma Gazetteer* the process is described as follows: "Three men generally work together, one cutting down the trees (*sha* or **Acacia Catechu**) and driving the buffaloes that drag them to the site of the furnace, one clearing off the sapwood and cutting the heartwood into chips, and the third attending to the fires. The chips are put into four-gallon cauldrons, which are filled up with water, and the whole is boiled for twelve hours. When the water is reduced to one half, the chips are taken out and the liquid placed in large iron pans and again boiled and stirred till it attains the consistency of syrup; the pans are then taken off the fire and the stirring continued till the mass is cool, when it is taken out and spread on the leaves, arranged in a wooden frame, and left for the night; in the morning it is dry and ready to be cut up into pieces for the market. The chips are boiled down twice, but there is not much extracted by the second boiling. There was formerly no restriction on the felling of the trees and the supply was getting exhausted; now no tree can be felled without permission, and a fee of R5 is charged for each cauldron used.

The following interesting correspondence, which has been obligingly placed at my disposal by Dr. Schlich, Inspector General of Forests, will be found to give important facts regarding the manufacture of Cutch as practised in Pegu. The Conservator of Forests in Oudh asks the Conservator of British Burma the following questions:—

Enquiry.—"In my circle the season for manufacturing Cutch extends to only three months in the year, and the rates levied here from catechu-makers are 12 annas a pot, capable of holding about 3 gallons of water or liquid substance. The rate has been raised this year from 9 to 12 annas per earthen pot, which even seems to be a very low price compared with what you get in Pegu. I should feel much obliged if you would kindly give me the following information:—

"(1) The length of the season during which cutch is manufactured in your circle;
"(2) The outturn per season per cauldron of 20 gallons;
"(3) The process of manufacturing as conducted in Pegu;
"(4) The price per maund of catechu on the spot;
"(5) The distance of the market from the forests where the cutch is sold to retail dealers and others;
"(6) What percentage of cutch is obtained from a maund of heartwood."

Answer.—"(1) Cutch is manufactured in this circle from 1st June to 31st March; but the months from December to March (inclusive) are those in which the manufacture is most energetically carried on. In April and May, and in the drier parts in March even, scarcity of water stops the work; while in the rainy season carts cannot ply, and boilers have difficulty in provisioning themselves and disposing of their cutch.

"(2) The outturn of a cauldron per season depends on such a variety of circumstances—the duration of the season, the quality of the trees, their proximity to the boiling-place, and, above all, the working days of the party—that an average cannot be struck. It may be 2,000 lbs. only, or it may reach, and even exceed, 6,000 lbs., for a cauldron of 20 gallons (those in use have a capacity of 12 gallons).

"(3) Mr. Carter, Deputy Conservator, Tharrawaddy, well describes the process of manufacture as follows:—

"For the working of one cauldron three men are necessary; but if a larger number of cauldrons are employed, there is some saving of labour. Of the three men one man is employed in felling the trees and dragging

CUTCH.

them, by means of cattle, to the cutch boiling-place. The second clears the logs of sapwood and cuts the heartwood into chips. The third attends to the fires and the boiling process. The chips are put into earthen pots, which are filled with as many chips as they will contain, then water is poured in until the pots are nearly full. The pots (which have a capacity of about 3 gallons) are then placed on the fire and boiled for about 12 hours, in which time the water is reduced to about one half the original quantity. For one cauldron, 20 to 25 of these earthen pots are employed. The cauldron is nearly filled from these pots; and when the extract in the cauldron is reduced to about one half, the cauldron is again filled from the pots, and this is repeated until the pots are emptied. The boiling process is generally accelerated by the employment of a large earthen pot, which is set up near the cauldron, and is filled at the same time as the cauldron and kept boiling, the extract from the small pots being constantly added as that in the larger pot is reduced. The cauldron is then filled from the large pot, instead of from the small ones. The Burmans call this large pot the *Ye-ni-o,* or red-water pot. The extract from the pots having all found its way into the cauldron, the boiling is continued and the liquid is stirred until it attains the consistency of syrup, and fills only about one fifth of the cauldron. The cauldron is then removed from the fire and stirred with a piece of wood, shaped like a paddle, for 4 hours or more, by which time the mass has obtained a greater consistency and is cool enough to be handled. It is then placed in a mould, like a brick mould, and is left to cool. This generally happens at night, and by next morning the result is a brick-like mass of cutch weighing 36 to 44 lbs.

"The stirring business, which takes place after the cauldron is removed from the fire, is more of a beating up, and I have never been able to ascertain what the object or effect of the process is. Cooks differ, too, in the amount of beating up that is desirable, some being satisfied with half an hour's application." The outturn of one cauldron of 12 gallons in 24 hours, when properly worked, is fairly constant at the figure given by Mr. Carter.

"(4) Cutch was worth, last year, R438 to R558 per maund (equivalent to R15, R20, and R25 per 100 viss), on the edge of the forest, according to the distance from the Irrawaddy river or the railway.

"(5) The above rates correspond to some 40, 25, and 15 miles from the markets on the railway and river, where the price was R30 per 100 viss (365 lbs.), or R658 per maund.

"(6) Regarding the amount of cutch yielded by heartwood, no reliable data are available. The yield has been stated at from 3 to 10 per cent. in weight.

"For practical purposes, I believe a ton of timber in the round may be taken to yield 250 to 300 lbs. of cutch."

By whatever process the Catechu is prepared, the final drying and hardening takes place by exposure to the sun and atmosphere.

Commercial Forms of Catechu.—In the Dún (North-West Provinces) it is then thrown (*a*) into moulds of clay, forming small squarish pieces, or (*b*) into moulds formed of leaves. In other parts of the country it is thrown upon a cloth covered with ashes of cow-dung, and (*c*) either allowed to harden into irregular slabs, or, when soft, it is cut (*d*) into blocks by means of a string. In Pegu it is manufactured into (*e*) great masses a cwt. in weight. These blocks are composed of layer upon layer of catechu, of succeeding preparations separated by leaves. As the block form enters commerce it is generally broken into (*f*) pieces which may be readily distinguished from the other forms through the presence of the dried leaves

143 (*a*)
144 (*b*)
145 (*c*)
146 (*d*)
147 (*e*)
148 (*f*)

CUTCH.

dividing the layers. In Bombay it is described by **Dr. Hamilton** as formed into (*g*) rounded balls of the size of an orange. These are probably after-preparations, for in Thána it is made into (*h*) cakes.

149 (*g*)
150 (*h*)

Description of Commercial Catechu met with in Europe.—It occurs in great masses surrounded by leaves, or broken into small blocks, in balls, cubes, or irregular-shaped pieces. In colour it is externally of a rusty brown, internally a dirty orange to dark liver colour, in some cases almost black, in others port-wine coloured. It is inodorous, with an astringent and bitter taste, followed by a sense of sweetness. It is brittle and breaks with a fracture more or less resinous and shining. The pale form *kath* is grey-coloured, porous, and under the microscope is seen to be composed of agglutinated masses of needle-shaped crystals.

Kath.
151

2.—THE CRYSTALLINE SUBSTANCE KNOWN AS *Kath* (in some parts of the country (Bombay) pronounced *Káth*) or the PALE CATECHU OF INDIA.—*Kath* or Pale Catechu is the restricted name given in Northern India to a grey crystalline substance prepared from a concentrated decoction of **A. Catechu** wood by placing in it a few twigs and allowing the decoction to cool. The twigs are removed and the crystalline substance collected. Whether the liquid is rejected or afterwards boiled down to produce a poor quality of dark catechu or cutch has not been ascertained. As sold in the bazars this crystalline substance occurs either in irregular pieces or in square blocks similar to the dark orange-brown homogeneous cubes of catechu. This is the substance eaten by the natives in their *pán*, and which imparts with lime the red colour to the lips. It is apparently never exported to Europe; the name *Kath*, while chiefly applied to it, is in some parts of India erroneously applied to Cutch also. *Kath* and Cutch have by Europeans been mistaken for the same substance, but the former is much purer chemically than the latter, and it may be owing to the fact of Cutch being the form exported to Europe, that Catechu has lost the former position it held as an astringent medicine. It seems probable that the preparation of *Kath* may be a secondary process from the Cutch, since its direct preparation from the original decoction has only been observed at Kumaon, although the substance is universally used in *pán* all over India. This subject deserves to be thoroughly investigated, and the merits of *Kath* and its process of preparation made known. The dark and the pale forms of *Khadira* were both well known to the Sanskrit writers, but in later times they seem to have been confused with each other.

The process of preparation of *Kathá* or *Káth* is described by **Madden**: "One portion of the *Khairis* is constantly employed in cutting down the best trees, and for these they have to search far in the jungles; only those with an abundance of red heartwood will answer. This is chopped into slices a few inches square. Under two large sheds are the furnaces,—shallow, and with a slight convex clay roof pierced for twenty ordinary-sized earthen pots. This operation takes place in about an hour and a half. The liquor resembles thin light port, and the *kathá* crystallizes on leaves and twigs thrown into it for the purpose. Each pot yields about a seer of an ashy white colour. The work is carried on for twenty out of the twenty-four hours, by relays of women and children; the men merely preparing the wood, which, after being exhausted, is made use of as fuel."

Keersal.
152

3.—KEERSAL OR KHERSAL.—From the wood of **Acacia Catechu** is occasionally obtained a pale crystalline substance known as *Khersál*. The woodmen, when cutting up the timber for fuel, sometimes come across this substance and carefully collect it, since it is much valued as a medicine by the Hindús, and fetches a high price. **Dr. Dymock** (*Mat. Med.*,

The Chemistry of Catechu.

CUTCH.

Western Ind., *232*) says of it: "Keersal or catechuic acid is obtained from cavities in the wood, and occurs in small irregular fragments like little bits of very pale catechu mixed with chips of reddish wood." "In the forests near Báriya, Gujarát, this substance is collected and is regarded as a valuable cure for coughs." (*Bom. Gaz.*, *VI.*, *13.*)

153

§ "Cutch is in large, irregular cakes; the characteristic squares are Gambier. There are five kinds of Catechu in the Indian markets:—

(*a*) Kath in irregular fragments, eaten with pán.
(*b*) Cutch of commerce, in large, irregular cakes, soft internally.
(*c*) Cutch from Singapore, in lozenges, almost colourless.
(*d*) Gambier in characteristic squares.
(*e*) Areca-nut cutch, rarely met with." (*Surgeon-Major W. Dymock, Bombay.*)

CHEMICAL COMPOSITION.

154

References.—*Flück. & Hanb., Pharmacog., 1879, 243; U. S. Dispens., 15th Ed. 379; Dymock's Mat. Med. of W. Ind., 232; Bentl. & Trim., Med. Pl., II., 95; Crooke's Dyeing and Calico Printing, 490-98; Etti, Monatsh. Chem., II., 547, and Liebig's Ann., 186, p. 327; Liebermann and Tauchert, Ber. der deutsch. Chem. Ges., XIII., 694; Nees von Esenbeck, Ann. der Chem. und Pharm., I., 343; Zwenger, Ann. der Chem. und Pharm., XXXVII., 320; Hagen Ann. der Chem. und Pharm., XXXVII., 336; Van Delben and Kraut, Ann. der Chem. und Pharm., CXXVIII., 283; Schützenberger and Rack, Bullet. Soc. Chimiq. de Paris, 1865, p. 5; Meyer, Jour. de Pharm., 1870, p. 479; Watt's Dict. Chem., I., 816; Gmelin's Chem., XV. (1862), 515; Hist. des Plantes (Monogr. des Rosacées, 1869), I., 415; Year-Book of Pharm., 1881, p. 61; and 1882, p. 84.*

§ "Catechu contains a variety of tannic acid called *Mimotannic acid*, which is soluble in water, and *Catechu* or *Catechuic* acid, which is insoluble. Mimotannic acid differs from tannic acid in yielding a greenish-grey precipitate with ferric chloride, and by not producing pyrogallic acid when heated. The destructive distillation of Cutch yields *Pyrocatechin*. Quercetine is stated to be contained in Cutch. This principle is the yellow crystallizable substance to which the bark of **Quercus tinctoria**, *Oliver*, owes its colour." (*Dr. C. J. Hislop Warden.*)

The chemistry of the Catechus has occupied the attention of chemists for some time back, but as yet the views and conclusions arrived at are somewhat conflicting, and the subject may be regarded as still involved in considerable obscurity. The brief chemical note (above) which my friend **Dr. Warden** has supplied, may be regarded as an abstract of all that is known. In his "*Science Papers*" **D. Hanbury** suggests that the process by which the various kinds of Cutch, Catechu, and Gambier are obtained should be carefully studied by persons who have the opportunity of doing so on the spot, that the trees yielding each of the forms of these substances should be accurately recorded; for, he adds, "we wish to identify the trees with the respective extracts." It would seem that our ignorance upon these important points may have much to do with the conflicting chemical results which at present exist regarding the composition of Cutch. There are at least two if not three distinct products obtained from each of the Cutch-yielding trees, and it is just probable these may have been experimented upon indiscriminately by the chemists of Europe. It would be but in keeping with other instances of two or more species (still more so of members of different natural orders), yielding approximately the same product, to find that the trees which afford the Cutch of commerce produce substances chemically dissimilar. Some such explanation may be found in the future to account for a certain number of the conflicting opinions which at present exist regarding the chemical composition of Cutch and its derivatives. A similar example may be mentioned in the fact that **Aconitum**

CUTCH.

Napellus yields a different alkaloid from **A. ferox,** although both species have hitherto been used in the preparation of Aconitia.

Catechuic Acid. 155

Pegu Catechu, "when immersed in cold water, turns whitish, softens, and disintegrates, a small proportion of it dissolving and forming a deep-brown solution. The insoluble part is Catechin in minute acicular crystals." (*Flück. and Hanb., Pharmacog., 243.*) When the crude Cutch of commerce is subjected to a dry heat of 110°, or 100°, in an atmosphere of hydrogen, it fuses and becomes transparent, losing 4 to 5 per cent. of its weight. It melts at 140° without further loss of water. On ignition there is left 3 to 4 per cent. of ash. If pure it should be completely soluble in boiling hot water, the solution precipitating the insoluble crystals of catechuic acid on cooling. Ether extracts from Cutch its catechin or catechuic acid, so that by precipitation from a hot solution, or by means of ether, this substance may be separated for chemical or industrial purposes.

Mimotannic Acid. 156

In addition to catechin, Cutch contains, however, other two substances, *viz.*, Mimotannic acid and a gummy extractive principle. Mimotannic acid is soluble in cold water, and by simple maceration may therefore be removed from Cutch. The solution will be observed to be of a thick chocolate colour. If heated to the boiling point it is rendered quite transparent, becoming turbid on cooling. With this solution ferric chloride gives a dark-green precipitate, which will immediately change into purple on the addition of cold water, or of an alkali.

Gummy Principle. 157

Catechuic and Mimotannic acids are present in Cutch in about equal proportions. The effect of heat upon Cutch and its compounds is most important, and, as pointed out by **Etti,** the chemical changes effected by heat afford the most likely explanation of the discordance of authors as to the formula for Catechin. According to **Liebermann,** confirmed by **Etti's** re-examination of the substance, the formula for Catechuic acid or Catechin is $C_{18} H_{18} O_8$. If a piece of Cutch be first heated in a crucible and then macerated it will be found to be completely soluble in cold water. This is explained by **Etti** as due to the formation of soluble anhydrides from Catechin, thus:

$$2 (C_{18} H_{18}O_8) = 2H_2 O + C_{36} H_{32}O_{14}.$$

Catechu-tannic Acid. 158

The compound thus produced is known as *Catechu-tannic acid,* and is completely soluble in cold water. By a further loss of water at 190°-200° this becomes $C_{36}H_{30}O_{13}$. Under the influence of heat the anhydride that is first formed is $C_{36}H_{34}O_{15}$, an insoluble, brownish red, amorphous powder, a substance soluble in alcohol and precipitated in crystals by lime-water. These compounds, if formed, in varying proportions, in a piece of Catechu, would greatly tend to produce conflicting chemical formulæ in the results of different experiments, and a piece of Catechu, which is found to be completely soluble in cold water, should be regarded as inferior in quality (injured through heat) and most probably adulterated by the trader.

For some time **Gautier** regarded the Catechin of Gambier as quite distinct from that obtained from Catechu, but in his more recent publications he admits them as identical. He now corrects his formula, $C_{19} H_{18}O_8$ which he published as expressing Catechin (adopted in *Flück. and Hanb. Pharmacog.*), into $C_{19}H_{20}O_8$ and suggests for this compound the name of *Methylcatechin.*

The soluble Catechu-tannic compounds constitute the active astringent principle of the drug and the tanning and dyeing property for which it justly holds so high a position for industrial purposes.

159

Preparation of Pure Catechin.—Etti directs that Catechu should be dissolved in about eight times its own weight of boiling water, and the liquid, after being strained through a cloth, should be set aside for some days until the insoluble Catechin subsides. This should then be collected and

CUTCH.

placed under a screw-press, being thereafter dissolved in a sufficient amount of dilute alcohol and the filtered solution shaken up in ether. The ether is next removed by distillation, and the crystals obtained washed repeatedly in pure distilled cold water. It is then found to exist in the form of almost colourless crystals.

160

Adulteration and Detection of Catechu.—**Meyer** regards ether as the best reagent for this purpose. Whether it has been partially heated or not, the whole of the Catechu-tannic compounds may be abstracted from a given weight of pulverised Cutch by repeated treatment with ether, about 53 per cent. of the original weight being thus removed. The dried residue should thus weigh about 47 per cent., the excess over this being adulterants. The chief substances used for adulteration are sand, clay, sugar, starch, and dried blood. On ignition pure Cutch should leave a residue of 3 to 4 per cent. It should be completely soluble in boiling hot water; if soluble in cold water, it may be suspected of impurities or of having been injured by heat.

TRADE FORMS OF CUTCH OR CATECHU.

From Burma. 161

(*a*) *Burma.*—It is largely prepared at Pegu, in the districts of Prome and Thayetmo; it is in fact, next to teak wood, the most valuable product of the forests of Burma. **Dr. Brandis**, in his "Forest Administration Report for 1875-76," says that more than half of the Catechu exported comes from territories beyond the frontier of Assam and British Burma. The total number of trees felled during the year 1869-70 is stated to have been 284,198 in Pegu. The total earnings of a cutch-maker, in a good season, is about R70. The Pegu season is from November to March; very little is made from July to October, and hardly any from April to May. (*Official Correspondence with the Government of India.*) Pegu Catechu, as this form is commercially known, is in the London market regarded as the most valuable, and according to *Spons' Encycl.*, p. 1983, it fetches 21*s.* to 42*s.* a cwt.; but, according to quotations published by the *Tropical Agriculturist* for 1882-83, the market value for it is 25*s.* to 37*s.* a cwt. The manufacture of Pegu Cutch in the year 1869-70 afforded, in the Prome and Thayetmo districts alone, employment for 4,000, and with their families, a total population of about 16,000 persons, yielding an article of commerce worth on the spot about three lakhs of rupees. (*Indian Agriculturist, November 1882.*)

Bengal. 162

(*b*) Next in importance is placed the so-called Bengal Catechu. This, with the exception of the supply from Chutia Nagpur, seems to come chiefly from Nepal, the Terai forests of the North-West Provinces, Oudh, and of Behar. The Kumaon form seems to be entirely *Kath*, while in the Dún forests and from the Keri Pass, **Mr. Buck** reports that Cutch is made in these forests into cubes and cakes. A most important trade exists in Cutch between Gonda and Calcutta. **Dr. McCann**, in his *Dyes and Tans of Bengal*, compiled from the correspondence and records of the Bengal Economic Museum, reports that about 1,000 maunds of Catechu is annually consigned from Hazaribagh in Chutia Nagpur to Calcutta, where it sells at from R8 to R12 a maund. He also states that the local price is from R5 to R7 a maund, while **Dr. Schlich** says: "It is sold in the bazars of Chutia Nagpur at R2-8 per maund." It may be here remarked that **Dr. McCann** seems to have attached too great importance to the consumption of Gambier, for he informs us that "the imports are nearly altogether Gambier, which is imported for making *pán*. None of this is re-exported, except a small quantity to the Mauritius, &c., for the *pán* of the Hindú coolies there." (*Dyes and Tans of Bengal, p. 129.*) The question of the

Chutia Nagpur. 163

CUTCH.

imports and exports under the heading of "Cutch and Gambier" to and from Bengal is certainly obscure, but it is probable the exports, which are very considerable, relative to the entire industry, are chiefly Cutch. The imports include the Cutch coasting trade, principally from Burma (equal to more than half the entire industry), and a comparatively small quantity of Gambier imported from the Straits Settlements and consumed in India as the officinal form of the drug Catechu; "only a very small amount of this imported article is eaten in *pán*." Nearly the whole of the Cutch imported into Calcutta from Rangoon is re-exported to the United States, where it is largely used as a brown dye and as a drug. It seems probable, also, that a considerable quantity of the imported Catechu is the product of **Areca Catechu,** re-exported as the true article. In the Calcutta bazars four kinds are said to be sold: (*a*) *Belguti*, 4*d*. a pound; (*b*) Pegu, 6*d*. a pound; (*c*) Gánti, 5*d*. a pound; and (*d*) *Janakpur*, 5½*d*. a pound (*Amst. Exhib. Cat., T. N. Mukharji*). Bengal or Calcutta Cutch fetches on an average from 20*s*. to 30*s*. in the London market. On application to Calcutta for samples of the above I received from *Babu T. N. Mukharji*, the following: *1st, Papri*, a sample of Gambier in cubes; *2nd, Janakpuri*, apparently a sample of the so-called red Cutch of Upper Burma. This sample is pale pink-brown in irregular masses with fragments of wood. It has the colour and texture of Gambier, but does not exist in cubes. *3rd, Pegu,* irregular pieces marked with leaves, dark glassy brown. This is unmistakeably Cutch. *4th, Tele*, irregular masses, yellow on the outside, glassy brown within—Cutch. *5th, Belguti*, rounded pieces marked with leaves uniformly brown-black—Cutch. There would thus appear to be very considerable variations in the supply met with in the Calcutta market; but there seems no doubt that a certain amount of Gambier is regularly sold as Cutch; the sample called *Papri* is most distinctly Gambier, and appears to be the Singapore article; it occurs in the characteristic regular cubes.

From Calcutta. 164 (*a*) 165 (*b*) 166 (*c*) 167 (*d*)

(*c*) Bombay Cutch, in the *U. S. Dispensatory, 15th Ed.*, is said to yield a higher percentage of tannin than Bengal Cutch, but it is commercially almost unknown outside the Presidency. **Birdwood** describes four forms of it:—

Bombay. 168

(1) Kauchú of Dharwar—flat, round cakes, two inches in diameter and one thick; dark brown in colour, and preserved in *bajrí* husks.

Dharwar. 169

(2) South Konkan, covered with paddy husks.

Konkan. 170

(3) Khandesh, in angular grains, pale earthy brown internally, darker externally.

Khandesh. 171

(4) Surat, in irregular lumps from the size of a hazel-nut to a walnut.

Surat. 172

Dr. Dymock gives the local value of the Surat Cutch as R20 per maund of 37½ lbs. This is the fine *kath* used with *pán supari*; common *kath* is from 4½ to 5 rupees a maund.

(*d*) The Cutch of Madras is in all probability purely the product of **Areca Catechu. Dr. Moodeen Sheriff** says that the natives of Madras do not know that **Acacia Catechu** yields Cutch, although the tree is common.

Madras. 173

Up to 5th August 1875 imported Cutch and Gambier was subject to a duty of 7½ per cent.; after that date the duty was reduced to 5 per cent., and on the 9th March 1882 removed altogether.

TRADE RETURNS OF CATECHU.

174

* Cutch is exported in mats, bags, or boxes; the following table gives

* This presumes that the exports which appear in the *Trade and Navigation Returns* under the heading of Cutch and Gambier mean chiefly Cutch.

ACACIA Catechu.

Trade Returns of Catechu.

CUTCH.

the total exports of this important substance from India for the past five years:—

Years.	Weight in Cwt.	Value in Rupees.
1879-80	222,123	28,13,994
1880-81	316,077	42,22,527
1881-82	198,897	25,80,840
1882-83	246,506	30,52,434
1883-84	302,302	35,32,000

Of these amounts the following are the exports from Bengal and British Burma respectively:—

Years.	Bengal.	British Burma.
1879-80	67,757	154,290
1880-81	99,155	216,678
1881-82	57,747	141,013
1882-83	12,131	200,780
1883-84	68,885	230,005

The following analysis of the exports of Cutch shows the Provinces from which shipped and the countries to which consigned for the year ending 31st March 1884:—

Presidency from which exported.	Weight in Cwt.	Value in Rupees.	Country to which exported.	Weight in Cwt.	Value in Rupees.
Bengal	68,885	9,15,504	U. Kingdom	121,898	13,43,789
Bombay	3,263	33,582	Egypt	51,284	5,82,814
Madras	149	2,814	St. Helena	33,020	3,89,920
British Burma	230,005	25,80,100	U. States	67,566	9,01,513
			Straits	21,887	2,33,503
			O. Countries	6,647	80,461
Total	302,302	35,32,000	Total	302,302	35,32,000

THE DYE.

DYE. 175

A solution of Catechu is, by the action of lime or of alum, changed into a dull red colour, which constitutes a fairly good dye, and is used for that purpose in some parts of India; the extract may be used or the heartwood broken up and boiled with the lime. With salts of copper and salammoniac, Catechu gives a permanent bronze brown, much

DYE. 176

used by the calico printers of India. This colour is deepened by the use of perchloride of tin, with the addition of copper nitrate. In Dinajpur the red expectoration from chewing the *pán* is preserved and used as an auxilliary in dyeing *eri* silk. **Dr. McCann**, in his *Dyes and Tans of Bengal,* mentions a dye-combination, and **Mr. Buck**, in his *Dyes of the North-West Provinces,* adds several others in most of which lime constitutes the metallic agent. The rationale of these dyes lies in the fact that, under the influence of oxidising agents (chiefly metallic salts) the soluble Catechu compounds are converted into insoluble and thus permanent dyes. By the calico printers of Upper India 2 lbs. of Catechu are boiled in 3 gallons of water. To this solution is added 1 lb. of shell lime, and the mixture set aside for 12 hours. The surface coloured liquid is skimmed off and preserved as the printing "standard." In this case the oxidisation has taken place, or nearly so, before the colour is printed on the fabric. In Europe this is never done, the dye-solution containing soluble Catechin and Gum is printed on the fabric and the oxidisation accomplished within the tissue. This is a much more effectual and permanent process. The dyed fabric would in time become oxidised by exposure to the air, but the process is completed more rapidly by exposing the fabric to steam, or much more expeditiously by the still more modern process of passing it through a solution of bichromate of potash. The oxy-salts of copper along with salammoniac are also sometimes used for this purpose, and at one time enjoyed a high reputation. Milk of lime is selected as the oxy-salt when the colours employed in the prints, such as blue, are naturally fixed by that agent.

There are several "standards" adopted by the European calico-printer containing Catechu, of which the following are the more important:

Brown Standard. 177

BROWN STANDARD.—Water 50 gallons, Catechu 200 lbs. Boil for six hours, add acetic acid 4½ gallons, and make up to 50 gallons by adding water. Allow to stand for two days, thereafter decant the clear solution, heat to 54° C., and add salammoniac 96 lbs.; dissolve and allow to settle for 48 hours. Decant the clear portion and thicken with 4 lbs. of gum Senegal per gallon.

Madder Brown. 178

MADDER BROWN STANDARD to resist heavy purple colours.—Catechu ½ lb.; salammoniac, ¼ lb.; lime-juice at 8° Tw., 1 quart; nitrate of copper at 8° Tw., 2½ oz.; acetate of copper, 1½ oz.; gum Senegal, 1 lb. (*Spons' Encycl. 840*).

THE TAN.

TAN. 179

As a tan Catechu extract does not hold a very high position owing to the colour it imparts to the skin. It is said to contain from 45 to 55 per cent. of tannin, or about 10 per cent. less than divi-divi pods and 20 per cent. less than gall-nuts.

FIBRE.

FIBRE. 180

The Kew Museum catalogue describes a sample of fibre said to be prepared from the bark of this plant. I can find no record of fibre being prepared in India from the *Khair* tree.

THE MEDICINE CATECHU.

181

Vern.—*Kath, katthá,* HIND., BENG.; *Kath, kach,* PB.; *Káth,* BOMB., MAR., and SIND; *Kattakambu, káshu, káshu-katti,* TAM.; *Kánchu,* TEL.; *Katti, kashi-katti,* MAL.; *Káchu,* KAN.; *Kaipu,* CINGH.; *Kátho,* GUJ.; *Katu,* SWAHILI; *Khadira khadirasára,* SANS.

References.—*Pharm. Ind., 62; U.S.Dispens., 15th Ed.,379; Flück. & Hanb., Pharmacog., 1879, 244; Bentl. and Trimen, Med. Plants, 95; Royle's Mat. Med., Ed. Harley, 640; Dymock's Mat. Med., W. Ind., 230; U. C. Dutt's Mat. Med., Hindús, 158; O'Shaugh. Disp., 302; Bidie's Cat. Prod.*

Medicinal Properties of Catechu.

MEDICINE.

Madras at Paris Exhib., 1878, p. 6; Murray's Drugs of Sind, p. 138; Birdwood's Bomb. Prod., 26; Stewart's Pb. Pl., 52; Moodeen Sheriff's Sup. to Pharm. Ind., 20; Ainslie's Mat. Ind., I., 63 and 590.

PROPERTIES AND THERAPEUTIC USES.

MEDICINE Extract. 182

THE RESINOUS EXTRACT is a powerful astringent, and may be used where most other astringents are indicated.

Internally it is useful in diarrhœa with pyrosis, depending upon a relaxed state of the mucous membrane. (*Pharm. Ind.*) Recommended to be given to adults in the form of a simple powder along with honey, 15-20 grains, or for dysentery in larger doses up to one drachm. It holds the reputation of being useful in intermittent fevers and scurvy. "A small piece held in the mouth and allowed slowly to dissolve is an excellent remedy in relaxation of the uvula and the irritation of the fauces and troublesome cough which depend upon it." (*U. S. Disp.*) The Hindú physicians recommend a piece of Catechu, rubbed with oil, to be kept in the mouth in hoarseness. (*U. C. Dutt.*) Catechu, boiled down in five times its weight of water, to one eighth, then flavoured with nutmeg, camphor, and betel-nut and made into balls of a convenient size, is directed to be kept in the mouth for affections of the gums, palate, tongue, and teeth.

Injection. 183

Ainslie cautions the free use of Catechu in ordinary diarrhœa until the full extent of the complications with the liver have been ascertained.

"*Locally* it holds a high reputation in ptyalism, ulceration, and sponginess of the gums, relaxation of the uvula, hypertrophy of the tonsils, &c." (*Pharm. Ind.*) An injection of the aqueous solution is often used in leucorrhœa and atonic menorrhagia. "In obstinate gonorrhœa, gleet, and leucorrhœa we have found it highly beneficial." (*U. S. Disp.*) A useful injection for severe hæmorrhage after confinement. By Hindú physicians it is much used both internally and externally in skin diseases: "a decoction of Catechu is used as a wash for inflamed parts and ulcers." (*U. C. Dutt.*) **Stewart** says it is used in the Panjáb externally "in ointment for itch, syphilis, and burns." "Chronic ulcerations, attended by much fœtid discharges, are frequently speedily benefited by the use of an ointment composed of the fine powder and lard; and in obstinate cases with the addition of sulphate of copper. In prolapsus ani and protruding piles, Catechu, with lard and opium, has been found of great service; bathing or fomenting with an infusion of Catechu is also beneficial." (*Murray, p. 138.*) Recommended as a dentifrice in combination with powdered charcoal, peruvian bark, myrrh, areca-nut burned to charcoal, powdered almond shell, and many other combinations in which the Catechu exercises the chief influence.

Ointment. 184 Dentifrice. 185 Powder. 186

Opinions of Medical Officers.—§ "As a styptic in hæmorrhage the powdered extract has been found useful if sprinkled over the wound." (*Civil Surgeon, Aligarh.*) "Reported to be beneficial as a local application in primary syphilitic sores." (*Dr. Parker, Poona.*) "A useful application to sore nipples and as a preventive against the ill effects of nursing, may be used in infusion as a wash, for some weeks before confinement." (*Surgeon-Major Hunter, Karachí.*) "Mixed with aromatics it is used by natives in melancholia; powdered and mixed with water it is used in conjunctivitis. Hakims state that it will produce abortion, but that, at the same time, it is useful for women who are barren but are desirous of having offspring (?)" (*Dr. Emerson, Calcutta.*) "Powdered and rubbed up with sulphate of copper and yolk of eggs it is a common application to cancers in East Africa." (*Surgeon-Major Robb, Bombay.*) "A mixture of Catechu and Myrrh, called *Káthbol,* is very generally given to women after confinement as a tonic and to promote the secretion of milk." (*Surgeon-Major W. Dymock, Bombay.*)

187

Kathbol. 188

Medicinal Properties of Catechu.

MEDICINE. 189

"It is used internally in congestion of the fauces, sore-throat, hæmoptysis, diarrhœa, and dysentery. Externally, it is used as a wash in hæmorrhagic ulcers, and with burnt areca-nut in soft chancres. It is believed to be anaphrodisiac and to cause impotence when used in excess." (*Brigade Surgeon J. H. Thornton, B.A., M.B., Monghyr.*) "It is a valuable astringent. Internally, the extract, with opium or other medicines, is preferable in a pillular form; dose gr. 5 to 15 grains. The tincture is less efficacious. Externally as a gargle and as a wash it is very serviceable where an astringent is wanted. The tincture is a good application to sore or spongy gums." (*Surgeon R. L. Dutt, M.D., Pubna.*) "A most efficient and useful astringent, largely used in Charitable Dispensaries and Hospitals." (*Brigade Surgeon S. M. Shircore, Moorshedabad.*) "Much used in cleaning the tongue and the gums of infants, and is a preventive to the formation of ulcers." (*Surgeon-Major J. M. Zorab, Balasore.*)

Tincture. 190

"Astringent and tonic in diarrhœa, in combination with aromatics, such as cinnamon and nutmeg." (*Surgeon C. M. Russell, Sarun.*) "Pale catechu is also used in soft chancre, after it has been softened in water and made into a paste." (*Surgeon Anund Chunder Mookerjee, Noakhally.*) "The only tact worth recording about this well-known drug, is its supposed anaphrodisiac properties. It is taken in doses of from ten to twenty grains (the powder being simply mixed with water) by Hindú widows with a view of suppressing sexual desire." (*Surgeon R. G. Mathew, Mozufferpur.*) "Catechu, with areca-nut slightly toasted and pounded into an impulpable powder, is in common use by the natives for sponginess of the gums, but the prolonged use of it darkens the teeth." (*Hon. Surgeon Easton Alfred Morris, Negapatam.*) "Combined with the seeds of **Bonducella** and with sulphate of iron it is used for strengthening the gums." (*Surgeon-Major John Lancaster, M.B., Chittore.*)

191

"Dose five to twenty grains as an astringent, one to four grains as an expectorant. It is used in bronchial affections with sugar-candy and turmeric. Frequently prescribed in diarrhœa as an astringent, also as an astringent lotion in conjunctivitis and to ulcers. It is supposed to have an analogous action to that possessed by ergot on the womb, when prescribed with myrrh. It increases the secretion of milk after delivery." (*Surgeon W. Barren, Bhuj, Cutch, Bombay.*) "The powder is useful in otorrhœa; it is also made up into an ointment with *ghí* and applied to cancers." (*Surgeon James McCloghey, Poona.*)

192

OFF. PREPARATIONS.

"**Infusion.**—160 grains in 10 fl. ounces. Dose one to two fluid ounces.

193

"**Tincture.**—2½ ounces in 1 pint. Dose one and a half to two fluid drachms. A valuable adjunct to Mistura Cretæ and other astringent mixtures.

194

"**Compound Powder.**—Dose fifteen to thirty grains. A valuable aromatic astringent." (*Pharm. Ind.*)

195

KEERSAL OR KHERSAL.—The crystalline form of Catechuic acid, found naturally in crevices of the wood, fetches a high price in India as a drug, and is regarded as a valuable cure for coughs. (*See page 35.*)

196

FOOD.

FOOD Pan. 197

The chief product of this tree is *Kath* and Cutch, obtained by boiling down a decoction from the chopped wood, say for 20 hours continuously. In the preparation of *kath* twigs are placed in the boiling liquid and upon these crystals of the substance generally known as *kath* are deposited. Both *Kath* and Cutch are commercially designated Catechu, but the

FOOD.

former is regarded as purer than the latter, and is largely used as an ingredient in the *pán* or the betel-leaf preparation which the natives of India are so fond of chewing. The *Kath* is reduced to a fine powder, a little of which is smeared on the *pán* leaf, together with some white lime and crushed betel-nuts. It is the *Kath*, in combination with the lime, which gives the teeth and lips the red colour so characteristic of Hindús. Continued use blackens the teeth. The people of Assam very seldom eat *Kath* with *pán*, as they consider it too rich for them. (*Mr. Darrah, Assam.*)

THE TIMBER.

TIMBER. 198

Structure of the Wood.—Sapwood yellowish white; heartwood either dark or light red, extremely hard. The wood seasons well, takes a fine polish, and is extremely durable. **Cleghorn** says the wood of this plant is less hard and durable than that of other Acacias. It is not attacked by white-ants or by teredo.

199

It is used for rice-pestles, oil and sugarcane crushers, agricultural implements, bows, spear and sword handles, and wheelwrights' work. In Burma it is employed for house-posts, and very largely as firewood for the steamers of the Irrawaddy Flotilla. The felling of Cutch trees for the purpose of fuel should, however, be altogether prohibited, although the wood is greatly admired for its high heating powers. It is much valued in Broach for posts which have to be driven into the ground. The fuel of the dead *khair* is much valued by goldsmiths. In North India it is made into charcoal, and is one of the best woods for that purpose. It has been found good for railway sleepers, and it is probably only the smallness of the tree and the consequent waste in cutting up, that has prevented its more general use. A cubic foot of the wood weighs about 70 lbs.

Several other plants yield Catechu, such as **A. Suma, Areca Catechu,** and **Uncaria Gambier,**—see **Catechu.**

Acacia Campbellii, *Arn.*

Syn. for **A. planifrons,** *W. & A.*

A. chrysocoma, *Miq.*

Syn. for **A. tomentosa,** *Willd.*

200

A. concinna, *DC.; Fl. Br. Ind., II., 296.*

Syn.—MIMOSA CONCINNA, *Willd.; Roxb., Fl. Ind., Ed. C.B.C., 424.*

Vern.—*Rithá, kochi,* HIND.; *Ban-rithá,* BENG.; *Toldúng,* LEPCHA; *Saptalá,* SANS.; *Aila, rassaul,* OUDH; *Sika, shika,* BOMB., *Sikekái,* MAR., DUK.; *Shiká kái,* MAR.; *Chikakai,* GUJ.; *Shika,* TAM.; *Chikaya, shikáya, gogu,* TEL.; *Sigé, sige* (the unripe fruit being known as *kayi*), KAN.; *Ken bwon, kinbun, subóknwé* (or *su-kwot-nwé* or *soop-wotnway*), BURM.

Rita or *ritha* is the Hind. for the Soap-nut, **Sapindus Mukorosi,** *Gærtn.* As this name is also given to the detergent legumes of **Acacia concinna,** *DC.*, the two bazar products require to be carefully distinguished.

References.—*Brandis, For. Fl., 423; Gamble, Man. Timb., 150.*

Habitat.—A common, prickly, scandent bush, common in the tropical jungles throughout India; in Bengal flowering during the rains. Very common in East and Central Mysore planted as a hedge.

Botanic Diagnosis.—*Prickles* abundant, minute, hooked. *Leaves* with 12-16 pinnæ and 30-50 leaflets; *stipules* and *bracts* cordate, ovate. *Flowers* in copiously panicled, globose, yellow heads; *panicles* with densely downy branches, the lower springing from the axils of the leaves, the upper subtended by copious, membranous, subpersistent *bracts*. *Corolla* a little longer than the calyx. *Pod* thick, succulent, strap-shaped, straight, 3-4 by ¾ inches, depressed between the seeds, the broad sutures narrowed to a short stalk; when dry, shrivelled and rugose with slightly waved sutures, when young hairy.

This species belongs to the series **Vulgares.** Climbers with copious scattered prickles and flowers in globose heads forming panicles:—

A. concinna, A. Intsia, and **A. pennata.**

Properties and Uses—

DYE. 201

Dye.—Balfour says the bark is used for dyeing and tanning fishing-nets in South India.

§ "The bark is imported into Bombay from Kanara for this purpose." (*Surgeon-Major W. Dymock, Bombay.*)

"Turmeric and the leaves of this acacia afford a beautiful green dye (*Mason*)." (*I. N. Pickard, Esq., Burma For. Dept.*)

DETERGENT. Pods. 202

Soap.—"A considerable trade is carried on in some parts of India with the saponaceous legumes of this species." (*Roxburgh.*) The thick fleshy pods are used for washing the hair (*Gamble*). The nut of the *Shikákái* is used in the Kolába district, Bombay, instead of soap (*Bombay Gaz., XI., 26*). In Kanara they sell at R12 to 20 for 560 lbs; every other year comes a bumper crop (*Bombay Gaz., XV., Part I., 60*).

The legumes of several other species of **Acacia** are also used for this purpose, being regarded as efficacious in destroying vermin. To **Ainslie** must be attributed all our information regarding these pods, very little having been obtained regarding them since he wrote.

MEDICINE. Pods. 203

Medicine.—The pods are largely used by the natives of India externally as a detergent, and internally they are deobstruent, and, according to **Ainslie,** are also expectorant. **Ainslie** recommends the drug to be prescribed in the form of an electuary, in a dose about the size of a small walnut, every morning for three successive days. **Dr. Dymock** gives the value in Bombay as R1½ to 1¾ a maund of 37½ lbs.

204

§ "In South India the pods and leaves are used as an aperient in bilious affections. (*Ainslie.*)" (*Surgeon-Major W. Dymock, Bombay.*) "The pods of **A. concinna** are a mild cathartic, nauseant, and emetic. As a cathartic it is superior to senna, but it is rather nauseous and disagreeable in taste and smell. Like senna, it is not an efficient purgative when used alone, but a very good adjunct to other purgatives, as sulphate of magnesia. The pods are also a pretty good emetic in jaundice not depending upon obstruction." (*Honorary Surgeon Moodeen Sheriff, Khan Bahadur, Madras.*) "A popular household remedy for promoting the growth of hair and removing dandriff from the scalp, a decoction of pods (½ an ounce to the pint of water) being used as a hair-wash. In small doses the pods act as a tonic, but in large and repeated doses they have purgative and emetic properties assigned to them." (*Surgeon-Major J. M. Honston, Travancore; and John Gomes, Medical Store-keeper, Travandrum.*)

Hair-wash. 205

Leaves. 206

"Very young leaves, ground up with a little salt, tamarind, and a few chillies, are used by the natives as a chutney with their food when they suffer from biliousness. I have seen it act as a laxative producing one or two copious motions, deeply tinged with bile; it is also a detergent." (*Honorary Surgeon Easton Alfred Morris, Negapatam.*) "Powdered leaves, in form of infusion, act as a mild laxative; can be used as a substitute for Senna indica, but less powerful in action. Tender leaves are used by natives in the form of chutney in bilious affections, with successful results. (*Surgeon E. W. Savinge, Rajamundry, Godavery District.*) "The tender leaves made into a decoction are used as an aperient." (*Surgeon-Major John Lancaster, M.B., Chittore.*) "The tender leaves are subacid and make a good chutney. The pods are in daily use for washing purposes." (*Native Surgeon Ruthnam T. Moodelliar, Chingleput, Madras.*) "An infusion made from the pods is given to check malarious fevers. The tender leaves made into infusion, or ground down into a paste, are used to prevent flatulence, and to act as a mild laxative." (*Dr. Lee, Mangalore.*) "The pods of this species are largely used for washing the hair in Madras." (*Deputy Surgeon-General G. Bidie, M.B., C.I.E., Madras.*) "The pods are used in the form of an ointment in skin diseases." (*Surgeon J. Parker, M.D., Poona.*)

FOOD. The Leaves. 207

Food.—The leaves are pleasantly acid, and they are sometimes used by the Hindús as a substitute for tamarind, and are made into chutney.

DOMESTIC. Pods. Soap. 208

Domestic Uses.—The Hindús, according to **Drury**, use the (legumes) pods to mark the forehead. This statement requires confirmation. They are largely used as a substitute for soap, especially to wash the hair. "The *sige kayi* or soap-nut is planted for village hedges in the East, but it grows wild in Manjarabad and Belur." (*Mysore Gaz., 291.*)

209

Acacia dealbata, *Link.; Fl. Br. Ind., II., 292.*

THE SILVER WATTLE.

References.—*Brandis, For. Fl., 180; Gamble, Man. Timb., 155; Benth., Fl. Austr., II. 415; Mueller's Extra-Tropical Plants, 4.*

Habitat.—A tree, spreading rapidly by numerous root-suckers, indigenous in New South Wales, Victoria, and Tasmania, introduced on the Nilgiris, and now naturalised since 1840. Experimentally cultivated in the Panjáb.

TIMBER. 210

Structure of the Wood.—The wood is moderately hard, light brown, but warps considerably. It is extensively used in Australia for timber.

According to **Mueller** this is placed as a variety under **A. decurrens,** *Willd.*, the Black Wattle. "It prefers for its habitation humid river-banks, and attains there a height of sometimes 150 feet, supplying a clear and tough timber, used by coopers and other artisans, but principally serving as select fuel of great heating power."

TAN. Bark. 211

Tan.—"The bark of this variety is much thinner and greatly inferior to the Black Wattle in quality, yielding only about half the quantity of tanning principle. It is chiefly employed for lighter leather. This tree is distinguished from Black Wattle by the silvery or rather ashy hue of its young foliage; it flowers early in spring, ripening its seed in about five months, while the Black Wattle blossoms late in spring or at the beginning of summer, and its seeds do not mature before about 14 months." (*Baron von Mueller, Select Extra-Tropical Plants, 4.*)

Introduced in the Nilgiris, where "A very curious fact has been observed about the wattle tree. In 1845 and up to about 1850, the trees flowered in October, which corresponded with the Australian flowering time, but about 1860 they were observed to flower in September; in 1870 they flowered in August; in 1878 in July; and here this year, 1882, they have begun to flower in June, this being the spring month here, corresponding with October in Australia." It is very curious that the tree takes nearly 40 years "to regain its habit of flowering in the spring," *i.e.*, to become perfectly acclimatised. (*Ind. For., VIII., 26.*)

212

A. decurrens, *Willd.*

THE BLACK WATTLE.

Habitat.—The eastern part of South Australia, through Victoria and New South Wales to the southern part of Queensland.

The Black Wattle.

A small or middle-sized tree. The bark constitutes the tanner's wattle-bark. It is rich in tannin, and this fact, together with the many uses of the gum derived from the tree, make this one of Australia's most valuable plants.

TAN. Bark. 213

TAN.—In England the price of wattle-bark runs from about £8 to £11; in Melbourne about £5 a ton. "It varies, so far as experiments made in my laboratory have shown, in its contents of tannin from 30 to 54 per cent. in bark artificially dried. In the mercantile bark the percentage is somewhat less, according to the state of its dryness, it retaining about 10 per cent. moisture; 1⅓ lb. of black wattle-barkgives 1 lb. of leather, whereas 5 lbs. of English oak-bark are requisite for the same results, but the tannic principle of both is not absolutely identical. Melbourne tanners consider a ton of black wattle-bark sufficient to tan 25 to 30 hides; it is best adapted for sole leather and other so-called heavy goods. The leather is fully as durable as that tanned with oak-bark, and nearly as good in colour. Bark carefully stored for a season improves in tanning power 10 to 15 per cent. From experiments made under the author's direction it appears that no appreciable difference exists in the percentage of tannin in wattle-bark, whether obtained in the dry or in the wet season. The tannin of this Acacia yields a grey precipitate with the oxide salts of iron, and a violet colour with sub-oxides; it is completely thrown down from a strong aqueous solution by means of concentrated sulphuric acid. The bark improves by age and desiccation, and yields about 40 per cent. of Catechu, rather more than half of which is tannic acid. Bichromate of potash, added in a minute quantity to the boiling solution of Mimosa tannin, produces a ruby-red liquid, fit for dye purposes; and this solution gives, with the salts of sub-oxide of iron, black pigments, and with the salts of the full oxide of iron, red-browndyes. As far back as 1823, a fluid extract of wattle-bark was shipped to London, fetching then the extraordinary price of £50 per ton, one ton of bark yielding 4 cwts. of extract of tar consistence (Simmons), thus saving much freight and cartage. For Cutch or Terra Japonica the infusion is carefully evaporated by gentle heat. The estimation of tannic acid in Acacia barks is effected most expeditiously by filtering the aqueous decoction of the bark after cooling, by evaporating and then re-dissolving the residue in alcohol and determining the weight of the tannic principle obtained by evaporating the filtered alcoholic solution to perfect dryness."

"The cultivation of the black wattle is extremely easy, being effected by sowing either broadcast or in rows. Seeds can be obtained in Melbourne at about 5*s.* per lb., which contains from 30,000 to 50,000 grains. They are known to retain their vitality for several years. Seeds should be soaked in warm water before sowing. Any bare, barren, unutilised place might most remuneratively be sown with this Wattle Acacia; the return would be in from five to ten years. Full-grown trees, which supply also the best quality, yield as much as 1 cwt of bark." I have taken the liberty to extract almost the entire article published by **Baron von Mueller, K.C.M.G.**, in his exceedingly valuable work "*Select Extra-Tropical Plants,*" thinking it was certain to prove most useful to persons experimenting in India with the cultivation of the wattle or with its most valuable tanning bark. The variety **mollis (A. mollisima,** *Willd.*) is the most plentiful form in Victoria, and this is also admitted to be the most powerful tanning agent. It grows rapidly, and in addition to the bark and gum which it affords, the timber is much valued, chiefly as fuel.

TIMBER 214

A fuller account of this plant and of the other trees yielding the commercial product will be found under the name WATTLE-BARK.

Acacia dumosa, *W. & A.*

Syn. for **A. Latronum,** *Willd.*

215

Acacia eburnea, *Willd.; Fl. Br. Ind., II., 293.*

Syn.—MIMOSA EBURNEA, *Roxb., Fl. Ind., Ed. C.B.C., 421.*

Vern.—*Marmati*, MAR.

References.—*Brandis, For. Fl., 183; Gamble, Man. Timb., 151.*

Habitat.—A short or small deciduous tree, met with in Sind, Suliman Range, Berar, Deccan, and South India.

Botanic Diagnosis.—General habit of **A. arabica.** *Leaves* with 4-10 pinnæ and 12-16 leaflets. *Flowers* in rounded yellow heads in the axils of undeveloped leaves; *peduncles* densely grey downy, with involucre about the middle. *Pod* narrow linear, straight, rigidly coriaceous, dehiscent, glabrous, with slightly repand sutures.

Compare with **A. arabica** and the other members of this series.

Properties and Uses—

TIMBER. 216

Structure of the Wood.—Hard, yellowish white; splits in drying. Weight 52 lbs. per cubic foot.

A. elata, *Wall.* Syn. for **Albizzia procera,** *Benth., var.* **elata.**

A. Farnesiana, *Wall.* Syn. for **A. planifrons,** *W. & A.*

217

A. Farnesiana, *Willd.; Fl. Br. Ind., II., 292; Wight, Ic., t. 300.*

THE CASSIE FLOWER, *Eng.*

Syn.—MIMOSA FARNESIANA, *Linn.*

Vern.—*Vilayati kikar, vilayati bábúl, pissi babui, gú-kikar, gand-bábúl, gúh-babúl*, HIND., DUK.; *Gúya bábúla*, BENG.; *Gabur*, SANTAL; *Gubábhul*, MAR.; *Talbaval*, GUJ.; *Knebáwal*, SIND; *Vedda vala, puj-velam*, TAM.; *Kusturi, piktúmi, oda sale, murki tumma, naga-túmma, kampu-túmma*, TEL.; *Jáli*, KAN.; *Hnanlóng yaing* (or *Huanlóngyaing*), *nanlon-kyaing* or *nan-loon-gyaing*, BURM.

Moodeen Sheriff seems to think that because the name *Kastúri* or *kastúri-tumma*, TEL., is inappropriate for this plant, it is incorrectly applied to it.

References.—*Roxb., Fl. Ind. Ed. C.B.C., 421; Brandis, For. Fl., 185; Gamble, Man. Timb., 150; Baillon, II., 41; Mueller, Select Extra-Tropical Plants, 4; Smith's Dict., 3; Piesse on Perfumery, 106; Hanbury's Science Papers, 151-152; Atkinson's Gums and Gum-Resins, 9.*

Habitat.—*The Flora of British India* regards this small tree as indigenous to India, "cosmopolitan in the tropics, but often cultivated." It is common enough everywhere in India and Burma, growing freely by self-sowing. Its strong-scented, yellow flower-heads perfume the atmosphere very pleasantly. Cultivated in Europe and most successfully at Cannes. It is abundant in the valley of the Dead Sea, where it is covered with the scarlet flowers of the parasite **Loranthus acaciæ**, giving the effect as if on fire.

Botanic Diagnosis.—An erect shrub or low tree, with straight spines, flowering in the cold season. *Flowers* in rounded heads, axillary, fragrant, bright yellow; supported upon *peduncles* which are crowded in the nodes of the leaves, and having a whorl of bracts like an involucre at the apex. *Pod* thick, swollen or fleshy, cylindrical, more or less curved or hooked, glabrous, and having straight sutures. *Seeds* biserial.

Allied to **A. planifrons,** *W. & A.*, a tree of the Western Peninsula, with umbrella-like spreading branches and flower-heads in clusters in the axils of mature bracts. Compare with **A. arabica.**

Properties and Uses—

GUM. 218

Gum.—The gum is collected in Sind; *Bomb. Gaz., XV., Part I., p. 60*, says the gum exudes from the trunk in considerable quantities. **Waring** states that it is considered superior to gum arabic in the arts and as a medicine. **Murray** remarks that it is used to adulterate gum arabic. It is very desirable that its peculiar properties should be investigated: adulteration has, in all probability, prevented it from becoming better known. **Mr. Baden Powell** (*Pb. Pr. I.*, 345) describes it "as dark, conchoidal masses, translucent and transparent at the edges. Some pieces are much whiter."

GUM. 219

§ "The distilled flowers yield a delicious perfume, and the gum is generally considered useful." (*J. C. Hardinge, Esq., Rangoon.*)

"A fair substitute for gum arabic. A decoction may be used internally in diarrhœa and externally where an astringent is required." (*Surgeon R. L. Dutt, M.D., Pubna.*)

Dye and Tan.—Christy, in his *New Commercial Plants*, includes the bark of this tree among the Indian tans. It is not in much demand for this purpose in India, but is reported to be sometimes used in Dacca mixed with salts of iron. It gives an inky dye. The pods are also used in some parts of Bengal as a dye-stuff. (*Dr. H. McCann.*)

DYE. Bark 220 PODS. 221 TAN. 222

Perfumery.—The round yellow heads constitute the Cassie flowers so much used in European perfumery. With the development of the art of perfumery in India, this plant should prove a source of wealth. It grows freely without any care whatever, and should it ever be cultivated there cannot be a doubt but that it would prove a great success. **Piesse** says that the European practice is to sow the seeds in beds, the best plants being left and the doubtful ones removed. In the third year they are two to three feet in height, and are then planted out into the fields, each tree receiving about 12 square feet. Before planting into their final places the ground is recommended to be well ploughed and manured and dug to the depth of 4 to 6 feet. The locality chosen for cultivation should be exposed to the sun. After the third year the trees produce flowers. A full-grown tree is calculated to yield 2 lbs. weight of flowers, valued at from three to four pence a lb., the acre under Cassie cultivation in Europe thus giving £30 to £40. **Hanbury,** in his "Science Papers," gives the value of Cassie flowers in Cannes as "five to six francs the kilogramme."

PERFUMERY. Flower-heads. 223

The plant is wild in most parts of Bengal, and its cultivation might, if used as a hedge plant, or if scattered through the fields devoted to garden produce, be most profitable. The flowers are a certain source of wealth. The gum seems likely to hold in the future a much higher position than in the past. The pods yield dye, and the leaves, in times of scarcity, would prove an important addition to the fodder sources of the country. **Piesse** says: "I cannot leave Cassie without recommending it more specially to the notice of perfumers and druggists, as an article well adapted for the manufacture of essences for the handkerchief and pomades for the hair. When diluted with other odours, it imparts to the whole such a true flowery fragrance, that it is the admiration of all who smell it, and has not a little contributed to the great sale which certain proprietary articles have attained. The Cassie perfume retains its fragrance for a long time and is hence most useful for sachets. For this purpose a good combination is Cassie heads 1 lb., orris-root 1 lb." The perfume Cassie should not be confounded with Cassia or Cinnamon.

224

Medicine.—The bark is astringent, and is often used as a substitute for **Acacia arabica** bark. **Mr. Baden Powell** says: "the pods contain a balsamic liquid."

MEDICINE. The Bark. 225

§ "It is used as an adjunct to aphrodisiacs in the treatment of spermatorrhœa." (*Surgeon-Major C. W. Calthrop, M.D., Morar.*) "The bark is used as an astringent in the form of a decoction, strength 1 to 10 of water. The tender leaves, bruised in a little water and swallowed, are said to be useful in gonorrhœa, dose ½ ounce." (*Surgeon James McCloghey, Poona.*)

Leaves. 226

Structure of the Wood.—White, close-grained, hard and tough. Weight 49 lbs. per cubic foot.

TIMBER. 227

In the Panjáb usually grown as a fence, for which purpose it answers well.

DOMESTIC. 228

Domestic Uses.—§ "The plant is supposed to be obnoxious to rats and snakes, and is accordingly planted as a protection against the injury caused by these animals burrowing in the embankments." (*Rev. A. Campbell, Santal Mission, Pachumba.*)

229

Acacia ferruginea, *DC.; Fl. Br. Ind., II., 295; Bedd., Fl. Sylv., t. 51.*

Syn.—MIMOSA FERRUGINEA, *Roxb.; Fl. Ind., II., 561.*

Vern.—*Khour*, NEPAL; *Kaiger*, PANCH MAHALS; *Son khair*, BERAR; *Kar, khair*, GOND; *Pándhrá khair*, MAR.; *Teóri|khair*, BHIL; *Banni*, KAN.; *Shimai-velvelam, shemi-velvel*, TAM.; *Ana-sandra, wúni*, TEL.

References.—*Brandis, For. Fl., 185; Kurz, Burm. For. Fl., I., 423; Gamble, Man. Timb., 153.*

Habitat.—Found in Northern Bengal, Central and South India, and Gujarát.

Botanic Diagnosis.—A large deciduous tree with brown bark; *spines* short hooked, in pairs; *leaves* composed of 6-12 pinnæ and 20 to 40 leaflets. *Flowers* in peduncled spikes in the axils of the leaves; *rachis* glabrous; *corolla* 2-3 times the campanulate and glabrous *calyx*. *Pod* straight, strap-shaped, 3-4 inches long, veined, the upper suture winged, distinctly stalked, and 4-6-*seeded.*

Compare with **A. Catechu.**

Properties and Uses—

GUM. 230

Gum.—It yields a good gum, similar to gum arabic.

MEDICINE. 231

Medicine.—The bark possesses astringent properties.

TIMBER. 232

Structure of the Wood.—Sapwood large; heartwood olive-brown, extremely hard, harder than **A. Catechu.** Weight 70 lbs. per cubic foot. A fine timber, but little used. **Beddome** says it is used for building carts, and for agricultural implements.

A. Hookeriana, *Zippel.*

Syn. for **A. concinna,** *DC.*

A. indica, *Desv.*

Syn. for **A. Farnesiana,** *Willd.*

233

A. Intsia, *Willd.; Fl. Br. Ind., II., 297.*

Syn.—MIMOSACÆSIA, *Roxb.;* M. INTSIA, *Linn.*

Vern.—*Arhai-ka-bél*, SUTLEJ; *Katrar*, KUMAON; *Kondro-janum*, SANTAL; *Kundaru*, KOL; *Harrari*, NEPAL; *Payir rik, ngraem rik*, LEPCHA; *Korinta, kórendam*, TEL.; *Chilári*, MAR.

References.—*Kurz, Burm. For. Fl., I., 423;* **A. cæsia,** *W. & A.; Brandis, For. Fl., 189;* **M. cæsia,** *Roxb., Fl. Ind., II., 565.*

Habitat.—A large climber found in the Sub-Himálayan tract from the Chenab eastward (ascending to 4,000 feet), throughout India and Burma.

Botanic Diagnosis.—*Prickles* minute hooked. *Leaves* with 12-16 pinnæ and 16-24 ligulate-oblong leaflets. *Flowers* in globose yellow heads, panicled; *bracts* minute, lanceolate. *Pod* dry, thin, straight, strap-shaped, glabrous, smooth, 4-6 inches long by $\frac{3}{4}$ to $1\frac{1}{4}$ broad, cuneately narrow to a short stalk.

Var. cæsia, *W. & A.* Leaflets 40-60, not more than $\frac{1}{12}$ to $\frac{1}{8}$ inch broad, obtuse, with a minute point. Western Himálaya, 3,000 feet; Sikkim, 5,000 feet.

Var. oxyphylla, *Grah., sp.* Leaflets 40-50, more membranous than in the preceding, and acutely pointed.

Compare with **A. concinna.**

Properties and Uses—

Dye.—The bark or the fresh leaves of this plant are said to be used as an auxiliary or astringent in dyeing with morinda or lac, giving brightness. (*McCann.*) The bark is also used as a substitute for soap to wash the hair. (*Gamble.*)

DYE. Bark and Leaves. 234 SOAP SUBSTITUTE. Bark. 235

Medicine.—§ "The flowers are used by Santal women in deranged courses." (*Rev. A. Campbell, Santal Mission, Pachumba.*)

MEDICINE. Flowers. 236

Structure of the Wood.—White, soft, porous.

TIMBER. 237

Acacia Jacquemontii, *Benth.; Brandis, For. Fl., 183.*

238

Vern.—*Hausa,* AFG.; *Kinkar, babúl, bamúl, babbil,* PB.; *Rátobával,* GUJ.

References.—*Fl. Br. Ind., II., 293; Gamble, Man. Timb., 150.*

Habitat.—A small bushy, thorny shrub, met with on the east flank of the Suliman Range, ascending to 2,500 and at times to 3,200 feet; on the outer Himálaya near the Jhelum to about the same elevation, on the Panjáb plains, in Sind, and on the banks of the Nerbudda. Common in ravines and dry water-courses in Rájputana and North Gujarát. (*Brandis.*)

Botanic Diagnosis.—An elegant shrub with polished stems and straight, polished, and slender spines. *Leaves* composed of 6-8 pinnæ and 12-16 leaflets. *Flowers* in rounded axillary heads, yellow, sweetly-scented. *Pod* thin, flat, broad, ligulate, dehiscent, glabrous, grey, with straight, sutures; 2-3 inches long. *Seeds* 5-6.

Allied to **A. eburnea,** *Willd.*, which has narrow pods, and to **A. tomentosa,** *Willd.*, which has purple heads. Compare with **A. arabica.**

Properties and Uses—

Gum.—§ "The var. *baonli* (*Rájputana*), a bush with straight slender branches, common in dry, sandy water-courses, yields a small quantity of gum resembling that from **Acacia arabica.**" (*E. A. Fraser, Rájputana.*)

GUM. 239

Tan.—§ "The bark is used in tanning, and gives a brown or black colour." (*E. A. Fraser, Rájputana.*)

TAN. The Bark. 240

Spirits.—The bark of the root is used in the distillation of spirits; the branches are lopped, and the leaves, thrashed out with sticks, are used as fodder. (*Brandis.*)

Spirits. 241 FODDER. The Leaves. 242

Domestic Uses.—The polished stems and thorns and the sweetly-scented yellow flowers make this bush an object of much beauty and interest (*Ráj. Gaz.*, 29). Might with advantage be extensively cultivated as a hedge plant and its flowers collected for perfumery purposes.

DOMESTIC. Hedge. 243

A. Latronum, *Willd.; Fl. Br. Ind., II., 296; Wight., Ic., t. 1157.*

244

Vern.—*Bhes,* HIND.; *Devbábhul,* MAR.; *Paki-tuma,* TEL.; *Donn mulli-na-jali,* KAN.; *Hote-jali,* MYSORE.

References.—*Brandis, For. Fl., 180; Gamble, Man. Timb., 149.*

Habitat.—A thorny shrub found in South India, forming gregarious thickets.

Botanic Diagnosis.—A small tree with umbrella-like head and brown glabrous branches; *spines* in pairs, long, straight. *Leaves* composed of 6-10 pinnæ and 20-30 leaflets. *Flowers* in pedunculate spikes, abundantly produced from the nodes of leafless branches; *corolla* $\frac{1}{12}$ inch long and 3 or 4 times the minute campanulate and glabrous *calyx.* *Pod* oblong, thin, flat, somewhat recurved.

In the nature of its spines this species departs from the character of the series, having more the form of the spines of **A. arabica** (series **Globiferæ**). See **A. Catechu.**

Properties and Uses—

FIBRE. Bark. 245

Fibre.—It is said to yield a good fibre. Bark dark brown, dotted with white. (*Brandis.*)

TIMBER. 246

Structure of the Wood.—Useful for tent-pegs. (*Bomb. Gaz., XV., Part I., p.60.*)

247

Acacia lenticularis, *Ham.; Fl. Br. Ind., II., 296.*

Vern.—*Khin,* KUMAON.

References.—*Brandis, For. Fl., 186; Gamble, Man. Timb., 150.*

Habitat.—A small tree of the Siwaliks and of Kumaon, extending to the Rájmahal Hills in Bengal, to Central and South India and Burma.

Botanic Diagnosis.—*Spines* in pairs, short, hooked or recurved. *Leaves* with 4-8 pinnæ and 12-16 leaflets. *Spikes* very dense, shortly pedunculate, 3-4 inches long. *Corolla* twice the length of the campanulate *calyx. Pod* straight, 6-8 inches long by ¾ broad, thin, flat, opaque, venulose, with both sutures thickened and winged.

Compare with **A. Catechu.**

GUM. 248

Gum.—Yields an Acacia gum.

249

A. leucophlœa, *Willd.; Fl. Br. Ind., II., 294; Bedd., Fl. Sylv., t. 48.*

Syn.—MIMOSA LEUCOPHLŒA, *Roxb.*

Vern.—*Safed kikar, verú, raunj, karír, nimbar, ringa, rinj, rohani, jhind,* HIND.; *Safed-bábúl,* BENG.; *Sharáb-ki-kikar, hivar,* DUK.; *Goira,* URIYA; *Shvéta-barbúra-vrikshaha,* (modern) SANS.; *Safed kikar,* PB.; *Arinj,* RAJ.; *Raundra, runjra,* BANSWARA; *Renuja,* BIJERAGOGARH; *Tumma, reúnja, rinja,* GOND.; *Hewar,* C. P., MAR.; *Haribával,* GUJ.; *Hívar, pánharyá bábhuliche jháda,* MAR.; *Vel-velam, vel-vel, vevay-lam,* TAM.; *Tella-tuma,* TEL.; *Bili-jáli, togral naibela,* or *nayi béla* (*Gaz. Mysore*), *vel-vaila, bilijali topál,* KAN.; *Katu, vela maram, andara,* SINGH.; *Tanaung,* BURM.

References.—*Brandis, For. Fl., 184; Kurz, Burm. For. Fl., I., 421; Gamble, Man. Timb., 152; Roxb., Fl. Ind., Ed. C.B.C., 421; Atkinson, Gums and Gum-resins, p. 9.*

Habitat.—Found in the plains of the Panjáb from Lahore to Delhi, and in all forest tracts of Central and South India, Rájputana, and Burma (in the dry forests of Prome). Seems indifferent to climatic conditions.

Botanic Diagnosis.—A large, deciduous tree, with short, straight, and white spines. *Leaves* composed of 12-24 pinnæ and 30 to 60 leaflets. *Flowers* in small, rounded, yellow heads, aggregated into terminal *panicles,* which, when fully expanded, are a foot long and broad, and densely tomentose. *Pod* sessile, narrow ligulate-falcate, thin, flat tomentose, with straight sutures. The inflorescence of this species is its most characteristic feature.

Compare with **A. arabica.**

Properties and Uses—

GUM. 250

Gum.—The gum yielded by this plant is used in native medicine; it somewhat resembles gum-bassora, and received that name from **Ure.**

DYE. Leaves and Bark. 251

Dye.—The leaves and bark are used in dyeing, and give a black colour. The bark is also used for dyeing in Burma and gives a red colour, but mixed with other barks gives black. (*Prof. Romanis, Rangoon.*)

Dr. Schlich (*C. P. Forest Adm. Report, 1883, p. 45*) says he found the bark of *Hewar* being prepared for export, but he seems to have omitted to note the economic use for which it was to be exported.

FIBRE. Bark. 252

Fibre.—A coarse, tough fibre is prepared from the bark, much valued for fishing-nets and ropes (*Bom. Gaz., XII., 25; also XV., Part I., p. 60*).

Dr. Brandis says by steeping the bark in water for four or five days and beating it, a tough fibre may be obtained, which is used for making nets and coarse cordage.

Medicine.—The bark partakes more or less of the astringent properties of **A. arabica.** In an official correspondence with the Government of India, it is recommended that this drug should be excluded from further trial.

MEDICINE. Bark. 253

Food.—Dr. Brandis says that the young pods and seeds are eaten, and even the bark in times of scarcity is ground and mixed with flour. The latter is used to assist in preparing spirits from sugar and palm-juice, to precipitate, by the tannin which it contains, the albuminous substances in the juice and to facilitate the fermentation. It flavours at the same time the spirits, and is supposed to increase the amount of alcohol. "This fact was first ascertained and pointed out by **Mr. Broughton,** the Quinologist on the Nilgiris." (*Deputy Surgeon-General G. Bidie, M.B., C.I.E., Madras.*) The fruit is largely collected for fodder in the Panjáb.

FOOD. Bark. 254

Spirits 255

FODDER. Fruit. 256

§ "In the South Mahratta Country the bark is used in the distillation of a spirit, in consequence of which the trees are farmed on account of Government." (*Surgeon-Major W. Dymock.*)

"The legumes are called *Padiá,* which are pickled in Kathiawar." (*Asst. Surgeon Sakharam Arjun Ravat, L.M., Girgaum, Bombay.*)

Structure of the Wood.—Sapwood large; heartwood reddish brown, with lighter and darker streaks; extremely hard. It seasons well and takes a good polish; is strong and tough, but often eaten by insects. It is brittle, makes good posts, but bad planks. (*Bomb. Gaz., XII., 25.*) It makes an excellent fuel. When seasoned a cubic foot weighs about 55 lbs. (*Bomb. Gaz., XV., Part I., p. 60.*)

TIMBER. 257

Acacia megaladena, *Desv.;* Syn. for **A. pennata,** *Willd.*

A. melanoxylon, *R. Br.*

258

THE AUSTRALIAN BLACK-WOOD.

References.—*Benth., Fl. Aust., II., 415; Mueller's Select Extra-Tropical Plants, 6; Brandis, For. Fl., 180; Gamble's Man. Timb., 155; Kew Museum Cat., p. 56.*

Habitat.—A large tree met with in New South Wales, Victoria, Tasmania, and South Australia; introduced on the Nilgiris since 1840 and now completely naturalised. Also being grown in the hills of the Panjáb, Kumaon, and Sikkim.

Botanic Diagnosis.—*Leaves* seen only in young trees and then bipinnate, generally abortive, and represented by phyllodia. *Flowers* in globose, compact heads, on short axillary racemes.

Properties and Uses—

Structure of the Wood.—Hard and durable; heartwood dark-brown, beautifully mottled, soft, shining, and even-grained. Weight 41 to 48 lbs. per cubic foot in Australia; 36 lbs. on the Nilgiris.

TIMBER. 259

It is used in Australia for cabinet-work, coach-building, railway carriages, and agricultural implements; on the Nilgiris chiefly for firewood. Regarded as one of the best Australian woods, being easily cut into veneers. "It takes a fine polish, and is considered as almost equal to walnut." (*Mueller.*)

260

A. modesta, *Wall.; Fl. Br. Ind., II., 296.*

261

Syn.—MIMOSA DUMOSA, *Roxb., Fl. Ind., II., 559,* and probably M. OBOVATA, *l.c., 561.*

Vern.—*Kántosariyo,* GUJ.; *Palosa,* AFG.; *Phulahi,* PB.

References.—*Brandis, For. Fl., 185; Gamble, Man. Timb., 152; Roxb., Fl. Ind., Ed. C.B.C., 646; Atkinson, Gums and Gum-resins, p. 8.*

Habitat.—Found in the Suliman and Salt ranges, Sub-Himálayan tract between the Indus and the Sutlej, and in the northern part of the Panjáb plains. It is in fact one of the characteristic trees of the Panjáb.

Botanic Diagnosis.—A moderate-sized tree; *spines* in pairs, short hooked. *Leaves* with 4-6 pinnæ and 6-8 leaflets. *Flowers* in peduncled spikes 2-3 inches long, not very dense. *Corolla* greenish coloured, $\frac{1}{12}$ inch long, twice the length of the glabrous campanulate *calyx*. *Pod* 2-3 inches long by $\frac{3}{8}$, glabrous, glossy, venulose, straight strap-shaped, narrowed into a short peduncle.

Compare with **A. Catechu.**

Properties and Uses—

GUM. 262

Gum.—It yields a gum which occurs in the form of small, round, smooth, subtranslucent and very characteristic tears. I found what appeared to be this gum being used by the Lucknow calico-printers under the name of *bábúl*. It is quite tasteless.

MEDICINE. The Gum. 263

Medicine.—The gum, which is used in native medicine, is stated by **Bellew** to be regarded by the people of the Pesháwar valley as restorative.

FODDER. The Leaves. 264

Fodder.—The leaves and fallen blossoms are collected for cattle fodder.

TIMBER. 265

Structure of the Wood.—Sapwood large, white, perishable; heartwood dark brown, with black streaks, extremely hard—harder than that of **A. Catechu.** A most beautiful wood, strong and durable. Weight about 70 lbs. (In *Bomb., Gaz. VII., p, 31*, it is stated to weigh only 53-56 lbs.)

Valuable for cart-wheels, sugarcane-crushers, Persian water-wheels, and agricultural implements.

DOMESTIC. Tooth-brushes. 266

Domestic Uses.—§ "The delicate green twigs are used by the natives of the Panjáb in the form of tooth-brushes. The bark, though slightly unpleasant at first, imparts subsequently a pleasant and sweet taste to the mouth." (*Asst. Surgeon Bhagwan Das, Rawal Pindi.*)

Acacia paludosa, *Miq.*; Syn. for A. pennata, *Willd.*

267

A. pennata, *Willd.*; *Fl. Br. Ind., II., 297; Bot. Mag., t. 3408.*

Syn.—MIMOSA PENNATA, *Roxb.*; M. TORA, *Roxb.*; ACACIA MEGALADENA, *Desv.*; A. PENNATA, *Dalz. & Gib.*

Vern.—*Agla, awal,* KUMAON; *Thembi,* MAR.; *Biswúl,* HIND.; *Kundaru,* KÓL.; *Arar,* KHARWAR; *Undaru,* SANTAL; *Gurwa,* MAL. (S. P.); *Arfu,* NEPAL; *Tol rik,* LEPCHA; *Súyit,* BURM.

References.—*Brandis, For. Fl., 189; Kurz, Burm. For. Fl., I., 424; Gamble, Man. Timb., 155.*

Habitat.—A large, climbing shrub, found in Oudh, Kumaon, Nepal, Eastern Bengal, Burma, and South India.

Botanic Diagnosis.—*Prickles* fewer and less hooked than in **A. concinna** and **A. Intsia.** *Leaves* with 16-30 pinnæ and 80-100 leaflets, rigidly coriaceous, very narrow, densely crowded. *Flowers* in heads forming panicles; *bracts* minute, lanceolate. *Corolla* $\frac{1}{12}$ inch, slightly exceeding the glabrous *calyx*. *Pod* dry, thin, glabrous, dehiscent; 6-8 inches by $\frac{3}{4}$ to $1\frac{1}{4}$, strap-shaped, distinctly stalked. There are three or four varieties of this species.

Compare with **A. concinna.**

Properties and Uses—

TIMBER. 268

Structure of the Wood. Reddish, porous, moderately hard, with large vessels and numerous medullary rays. (*Gamble, List Darj.*) Weight 50 lbs. per cubic foot.

269

A. planifrons, *W.&A.*; *Fl. Br. Ind., II., 293; Roxb., Cor. Pl., t. 199.*

THE UMBRELLA THORN.

Vern.—*Salé, sal,* TEL.

References.—*Brandis, For. Fl., 575; Gamble, Man. Timb., 150.*
Habitat.—Western Peninsula.

Botanic Diagnosis.—A small, gregarious tree, with flat, umbrella-like spreading branches; *branches* glabrous, but with grey lenticular dots. *Leaves* with 10-16 pinnæ and 16-24 leaflets. *Flowers* in globular heads in clusters from the axils of branchlets. *Pod* glabrous, narrow, ligulate, turgid, with straight sutures, indehiscent; distinguished from **A. eburnea** by being shorter and crooked. Compare with **A. arabica.**

Properties and Uses—

TIMBER. 270

Structure of the Wood.—Hard and strong; heartwood red, sapwood white.

Used for agricultural implements and as fuel.

Acacia pycnantha, *Bth.*

271

THE GOLDEN OR GREEN WATTLE.

Habitat.—Victoria and South Australia.

TAN. Bark. 272

"This tree, which attains a maximum height of about 30 feet, is second perhaps only to **A. decurrens** in importance for its yield of tanner's bark; the quality of the latter is even sometimes superior to that of the *Black wattle*, but its yield is less, as the tree is smaller and the bark thinner. It is of rapid growth, content almost with any soil, but is generally found in poor, sandy ground, near the sea-coast, and thus is also important for binding rolling sand. Experiments instituted by me have proved the artificially dried bark to contain from 30 to 45 per cent. tanning principle, full-grown sound trees supplying the best quality. The aqueous infusion of the bark can be reduced by boiling to a dry extract which, in medicinal and other respects, is equal to the best Indian Catechu, as derived from **Acacia Catechu** and **A. Suma.** It yields about 30 per cent., about half of which or more is Mimosa-tannic acid. This Catechu is also of great use for preserving against decay articles subject to exposure in water, such as ropes, nets, fishing-lines, &c." "**A. pycnantha** is also important for its copious yield of gum." "The wood, though not of large dimensions, is well adapted for staves, handles of various instruments, and articles of turnery, especially bobbins." (*Baron von Mueller, Extra-Tropical Plants.*) It is remarkable that the Catechu-like products of Australia are all apparently made from a decoction of the bark instead of the heartwood, while in India the bark is rejected. It is probable that a combination of both practices would be more remunerative.

A. rupestris, *Stocks,* is by *Fl. Br. Ind.* reduced to **A. Senegal,** *Willd.*, which see.

A. Senegal, *Willd.; Fl. Br. Ind., II., 295.*

273

Syn.—A. RUPESTRIS, *Stocks,* as in *Brandis, For. Fl., 185;* A. VEREK, *Guill. et Perrot;* MIMOSA SENEGAL, *Linn.;* M. SENEGALENSIS, *Lam.*
Vern.—*Khor,* SIND; *Kúmta,* RAJ.; *Verek,* in WEST, and *Hashab* in EAST AFRICA.
References.—*Brandis, For. Fl., 185; Rájputana Gazetteer, 29; Flück. and Hanb. Pharmacog., 1879, 233.*

Habitat.—Chiefly found in Sind and Ajmere; abundant in West Africa, north of the River Senegal. It is also found in South Nubia, Kordofan, and in the region of Atbura in East Africa.

Botanic Diagnosis.—A low tree with grey bark, flexuose and glucose branches; *spines* strong, short, sharp-hooked, often 3-nate, two lateral and one below the petiole. *Leaves* with 6-10 pinnæ and 16-28 small ligulate leaflets and finely downy raches about one inch long, with a gland at the base and between the upper pair of pinnæ. *Flowers* in peduncled spikes; *spikes* 2-3 inches long and not crowded. *Corolla* yellow, twice the length of the campanulate, glabrous, deeply-toothed *calyx*. *Pod* 3 by ¾ inch, thin, grey, indehiscent, straight, strap-shaped, with a strong, fibrous, marginal midrib, and constricted between the seeds.

Compare with **A. Catechu.**

THE COMMERCIAL GUM ARABIC.

GUM. 274

In Sind and Rájputana the gum from this tree is collected, but unfortunately it is sold in a mixed condition with the gums from other species of **Acacia.** This fact, in all probability, accounts for its superiority not having been recognised by the natives of India. It would be exceedingly interesting to have the gum carefully collected from Indian trees and chemically examined along with authentic samples from the African plant. The gum from the same species apparently varies considerably under diversified climatic conditions, for from Africa widely different gums are exported to Europe, and these meet the demands of distinct markets, but are apparently obtained from one and the same plant. **Flückiger** and **Hanbury** describe five or six African gums as regularly imported into Europe, of which the gum from **A. Senegal** is the most frequent, abundant, and valuable constituent.

Senegal Gum. 275

1st.—**Gum Senegal, or the Gum of A. Senegal,** the *Verek* of the Negroes.—To the French colony of Senegal (on the west coast of Africa) this is a most important product. The trade was first established by the Dutch and Portuguese, but the French afterwards monopolised it, and planted the colony of Senegal, having St. Louis and Portendic as the chief ports for the exportation of the gum. The tree from which this gum is obtained was first accurately described by **Adanson** in 1788. It is collected by the Moors after the close of the rains in November (when the wind sets in from the desert) up till July. The gum is found to exude in greatest abundance during the dry desert winds, and most frequently at the bifurcations of the branches. **Mr. M. C. Martius** has also observed that the production of the gum is stimulated by the growth of the parasite **Loranthus senegalensis.** The principal supply is to the north of the River Senegal, or about 16° N. latitude. The gum is shipped chiefly to Bordeaux, the quantity annually imported into France being from 1½ to 5 millions of kilogrammes. It is usually of a yellowish to reddish colour, occurring in larger lumps than Turkey gum, roundish or oval, or even elongated, pulverisable, and less brittle than Turkey gum. It is not very much used in England, and may be distinguished from Turkey or Kordofan gum by the absence of the numerous fissures so characteristic of the latter, the masses being in consequence firmer and less easily broken.

Turkey Gum. 276

2nd.—**Kordofan or Turkey Gum.**—This is also ascertained to be a pure form of the gum from **A. Senegal,** the *Hashab* of East-Central Africa. It comes from the mountainous tracts of Kordofan on the Upper Nile (between Nubia and Sennaár) and almost in the same latitude as Senegal, although across the vast continent of Africa to the north-eastern division. It is calculated that about 30,000 cwts. of this gum are annually collected in Kordofan. The most valuable kind of Kordofan gum comes from the province of Dejara and is known as *Hashabi*. This is generally conveyed down the Nile to Egypt and thence exported to Europe. It occurs in rounded lumps, often as large as a walnut, or in irregular broken pieces, pure white, very much fissured, specially upon the surface. This is the gum most frequently used for medicinal purposes, and may, in fact, be regarded as the true officinal Gum Arabic of England, India, and America. It is in fact the only gum which should be used for medicinal purposes. It

The True Gum Arabic.

is chiefly imported into Europe from Alexandria and the neighbouring ports, hence the commercial name Turkey gum.

Suakim Gum. 277

3rd.—**Suakim Gum.**—There are, in addition to the above, several inferior forms of gum obtained from North-East-Central, and North-East Africa, exported from Alexandria, and occasionally met with in the London marts. Amongst these may be mentioned the Sennaár gum known as *Hashab-el-Jesire;* the gum from the eastern territories of the Blue Nile and from the mountain tracts between Khartoum and Berber. From being exported from Suakim to Alexandria these are collectively known as Suakim gum, Talca or Talka gum. Suakim Gum is supposed to be obtained from **A. Seyal,** *var.* **fistula,** and **A. stenocarpa.** It generally exists in the form of a powder or in a semi-pulverulent state, owing to its being very brittle. It is a very inferior variety of gum.

Barbary Gum. 278

4th.—**Barbary or Morocco or Mogador Gum.**—This brown-coloured gum is obtained from Morocco and the northern provinces, or brought to Mogador from Fezzan, or by caravans from Timbuctu. It is now pretty well determined to be the produce of **A. arabica,** *Willd.* (**A. Nilotica,** *Desf.*). The tree is said to bear the vernacular name of *attaleh,* and the gum is reported to be collected when the weather is hot and dry (July and August). It has a faint smell, and, when fresh, constantly produces a crackling noise. It is usually of a brownish colour and found in small angular or broken pieces.

East Indian Gum. 279

5th.—**East Indian Gum.**—This is perhaps the most abundant gum in England and America, and is applied to industrial purposes. It is generally a mixture of gums, and chiefly from the following species: **A. Senegal,** *Willd.*, **A. stenocarpa,** *Hochst.*, **A. fistula,** *Schweinf.*, and **A. arabica,** *Willd.* (=**A. Nilotica,** *Desf.*). It is imported into Bombay from the Red Sea ports, from Aden, and the east coast of Africa. It is re-exported to Europe from Bombay under the name of Indian or East Indian gum. **Dr. Dymock**, speaking of Bombay, says there are "two kinds met with in this market, *viz.*, '*maklai,*' in large, round tears or vermicular pieces, white, yellow or reddish, much like gum Senegal, but more fissured (it derives its name from Makalla); and '*maswai,*' in angular fragments and vermicular pieces, fissured, white, yellow, or reddish, which derives its name from the port of Massowa. Both of these are good, soluble gums, and if carefully sorted not much inferior to Kordofan gum." In the year 1872-73, Bombay imported 14,352 cwt. of East Indian gum valued at R2,29,627, and re-exported 4,625 cwt. valued at R78,898. Ten years later the imports into Bombay were only 8,691 cwt. valued at R1,13,028.

Cape Gum. 280

6th.—**Cape Gum.**—This is produced at the Cape Colony from **A. horrida,** *Willd.*, the *Dúrnbúm* or *Witledúrn* of the Colonists—the commonest tree of South Africa. The gum is pale yellow or amber-brown, and is regarded as inferior.

Wattle Gum. 281

7th—**Australian Gum or Wattle Gum.**—This is chiefly derived from **A. pycnantha,** *Benth.*, and **A. decurrens,** *Willd.* It occurs in hard, elongated, or globular pieces, varying in colour from dark amber to pale yellow. It is very adhesive and is said not to be liable to crack. The bark is very astringent, and seems to impart this property to the gum.

From the preceding sketch of the commercial substances known as gum arabic, it will be seen that they vary very much, but that the Kordofan form of the gum from **A. Senegal** is the purest and most valuable, and is the Gum Arabic of Pharmacy.

282

History of Gum Arabic.

Gum arabic would appear not to have been known to the Sanskrit authors. The Persian and Arabic writers describe it under the name of *Samgh-i-arabi* (*Dymock*). From the very remotest antiquity gum was

The True Gum Arabic.

known to the Egyptians. It is frequently mentioned by the ancient writers, and there are numerous representations both of the plant and of the gum itself. The Egyptian fleet brought gum from Aden in the early part of the 17th century B.C. The word *Kami* is the original of the Greek Κόμμι, whence, through the Latin *gummi*, the English word gum was derived. (*Flück. and Hanb., Pharmacog.*) Gum was used by the Arabian physicians, but in the 12th century it was apparently unknown in Europe; it first reached Europe in 1340 A.D., through Italian merchants trading with Egypt and Turkey, and by the Portuguese in 1449 it became a regular article of trade from the west coast of Africa.

283

Chemical Composition.

Important Characters.—§ "Gum Acacia or *Arabic* is a type of gum found in the juice of various plants, and especially in the genus Acacia. Chemically gum Acacia consists of *Arabic Acid* (C_{12} H_{22} O_{11}) in combination with lime, potash, and magnesia. By the action of dilute sulphuric acid *Arabin* is converted into *Arabinose*, a crystalline principle, which is sweet, and which has the same composition as grape-sugar. By the action of dilute nitric acid, mucic acid and saccharic acids are obtained. Gum arabic is soluble in water but insoluble in alcohol. Another variety of gum is known of which gum tragacanth may be mentioned as a type, which does not dissolve in water, but merely swells up to a soft gelatinous mass. Gums of this latter class contain a principle named *Bassorin* or *Tragacanthine*. Gums sometimes resemble resins physically, but are distinguished by dissolving or softening in water, and by being insoluble in alcohol, while resins are unaffected by cold water but are more or less completely soluble in alcohol." (*Surgeon C. J. H. Warden, Prof. of Chemistry, Calcutta.*)

Gum dissolves slowly in an equal weight of water, without affecting the thermometer, and forms a thick glutinous liquid which possesses a distinctly acid reaction. This property is but slightly accelerated at higher temperatures. Gum is insoluble in alcohol and most other liquids. An aqueous solution of gum, if poured into glycerine, becomes intimately mixed, and this mixture may be evaporated to a thick jelly without any separation taking place. Dry lumps of gum are, however, insoluble in glycerine. Gum undergoes no change by age if kept dry, but if prepared with warm water its disposition to sour is increased. The solution does not, however, ferment upon the addition of yeast, but chalk and cheese start in it a fermentation which gives origin to lactic acid and alcohol, but not to mannite or glycerine.

Arabic Acid. 284

To separate *Arabic Acid* or *Arabin*, acidulate slightly a solution of gum with hydrochloric acid, and add alcohol, when the Arabin will be precipitated. Calcium chloride, upon a dialyser, will also separate this substance from the acidulated solution. An Arabin solution differs from a solution of gum in not being precipitated by alcohol, but if the Arabin precipitate obtained from the acidulated solution be removed by means of a filter and dried, it will be found that it has lost its solubility and cannot be dissolved even in boiling water. It has by the action of heat been changed into metagummic acid, a substance identical with *Cerasin* found in beet and in cherry gum, or in the series of the Tragacanth or Bassora gums.

Upon a chemical examination of the different kinds of commercial gum, **Masing** found that their botanical sources could not by that means be ascertained. He also observed that the value of a gum is better judged of from its solubility than from its colour. He noted that the percentage of ash varied but little, while the degree of alkalinity varied considerably, being chiefly due to lime. **Fremy** first called the attention of chemists to the

Alkalis of Gum. 285

peculiar relation between the organic substance of a gum and its mineral ash. The importance of his investigations lay in their physiological bearings upon the relation of the gum to the plant itself. If Arabin, which has by heat been rendered insoluble (converted into metagummic acid), be next subjected to the influence of an alkali, it is at once converted into a soluble gummy substance, which differs in no essential from natural gum. It is thus concluded that gum is a salt of lime, with Arabic acid or a mixture of such alkaline salts—magnesia and potash being frequently found in gum ash in addition to lime. Gum is thus viewed as a salt containing an overwhelmingly larger proportion of the organic acid than of the alkaline base, since 3 per cent. is about the largest proportion of lime detected. Although small, the percentage of alkali is, however, exceedingly constant—a fact which would seem to justify **Fremy's** conclusion that the formation of gum by the plant depends upon an important function, and is not accidental, the product eliminated being most probably the organic acid, which, on escaping from the structure of the plant, obtains its alkalis from the cell wall, and is thus reduced to the saline condition.

Ash of Gum. 286

Dr. Graeger found that gum dried in the air contains 85·25 per cent. of organic matter, 3·15 of ash, and 11·60 of water. The ash was found in three experiments to contain the average of 48 per cent. of lime, 18 of magnesia, and 34 of potash. It still remains to be explained why the Bassora group of gums merely swell, when placed in water, instead of being dissolved.

Detection of Gum.—"Neutral acetate of lead does not precipitate gum arabic mucilage; but the basic or sub-acetate forms, even in a very dilute solution, a precipitate of definite constitution." (*Flück. and Hanb. Pharmacog.*) A gum solution is rendered turbid by silicates, borates, and ferric salts, but it is unaffected by silver salts, mercuric chloride, and iodine. Acted upon by nitric acid, mucic acid is produced from gum, and also a little oxalic acid.

Kiliani has recently shown that *Arabinose,* the sweet substance obtained from gum arabic by the prolonged action of dilute sulphuric acid, is identical with lactose obtained from milk sugar. Gum may be distinguished from dextrine by the following tests:—

1st.—It contains no dextro-glucose, a substance present in dextrine and recognised by the copper test (Fehling's solution).

2nd.—Gum contains a lime compound detected by the milky action of oxalic acid.

3rd.—Gum gives a yellow precipitate with ferric salts.

287

Substitutes and Adulterations.—The *Indian Pharmacopœia* recommends the gum of **Feronia Elephantum,** *Corr.,* as a good substitute. It forms small rounded tears, transparent, frequently stalactitic, colourless, yellow, or reddish. It is soluble in two parts of water, forming a tasteless mucilage and much stronger than a gum arabic solution of the same proportions. It is chemically, however, considerably different from true gum arabic. It is precipitated by neutral acetate of lead or caustic baryta, but not by potash; in this reaction it resembles Tragacanth, which, unlike gum arabic, yields an abundant precipitate with neutral acetate of lead.

288

The gum from **Prosopis glandulosa,** *Torrey,* a tree successfully introduced into the North-West Provinces by the Department of Agriculture and Commerce (nearly allied to the Panjáb species, **P. spicigera,** *Linn.*), "yields a gum sometimes used in America as a substitute for gum arabic, and known there as Mezquit Gum." (*Dr. Charles Rice, New York.*)

The adulterations are chiefly mixtures of other less valuable gums, and, indeed, so frequently is this the case that for pharmaceutical purposes it is desirable that the gum should be picked and assorted and each fragment cleansed from mechanical impurities. Flour or starch is often mixed with powdered gums, but this may readily be detected by the blue reaction with iodine. *Dextrin* is also often used as an adulterant, but this may be at once detected by the tests already given. Sand and other non-gummy substances are sometimes mixed to increase the weight, taking advantage of the presumption that these will be viewed as mechanical and accidental.

THE DYE.

DYE. Gum. 289

Large quantities of gum arabic are used for giving lustre to crape and silk, and for thickening colours and mordants in calico-printing; for suspending tannate of iron in the manufacture of ink and blacking.

MEDICINAL USES.

MEDICINE. The Gum. 290

The gum is used in medicine as a demulcent and emollient. Taking advantage of its viscidity it is used *externally* to cover inflamed surfaces, such as burns, sore nipples, &c., and it blunts the acrimony of irritating matters by being blended with them. The powdered gum has also been found useful in checking hæmorrhage from leech-bites, and when blown up the nostrils it arrests severe epistaxis. *Internally* it has been found useful in inflammations of the gastric and intestinal mucous membrane. If held in the mouth in the form of a special preparation, the gum is found serviceable in allaying cough, thus affording relief. Its influence as a demulcent is supposed to extend even to the urinary organs. Gum has also been recommended as a substitute for amylaceous food in diabetes, since it is not converted into sugar, but it does not appear to have been attended with any appreciable benefit.

Acacia sirissa, *Buch.;* Syn. for **Albizzia Lebbek,** *Benth.*

A. speciosa, *Willd.;* Syn. for **Albizzia Lebbek,** *Benth.*

291

A. Suma, *Kurz, Mss. in Brandis' For. Fl., 187; Fl. Br. Ind., II., 294; Bedd., Fl. Sylv., t. 49.*

Syn.—MIMOSA SUMA, *Roxb.*

Vern.—*Sai-kanta,* BENG.; *Kumtia,* PERTABGARH; *Dhaula khejra* (White Acacia), BANSWARA; *Gorado,* MANDEVI; *Sonkairi,* DANGS; *Tella sandra,* TEL.; *Mugalisoppu* (IN MYSORE), *banni mara, mugli,* KAN.

Habitat.—Common in Bengal, Behar, the Western Peninsula, Ava, and Ceylon.

Botanic Diagnosis.—A medium-sized tree, with *white bark* and downy branchlets; *spines* in pairs, short-hooked. *Leaves* with 20-40 pinnæ and 60 to 100 leaflets; *rachis* ½ feet long, densely downy, with a large basal gland and several glands between the upper pinnæ. *Corolla* nearly white, scarcely exceeding the canescent *calyx. Pod* 3-4 inches by ½ to ¾ inch, thicker than in **A. Catechu;** veined, distinctly beaked, strap-shaped, narrowed suddenly into a stalk ½ to ¾ inch long.

Compare with **A. Catechu.**

CATECHU. 292

Gum.—The extract Catechu is said to be made from the heartwood of this tree (*Brandis, 188*).

TAN. The Bark. 293

Tan.—The bark is peeled off and used as a tan (*Brandis*).

TIMBER. 294

Structure of the Wood.—The wood resembles that of **Acacia Catechu,** but has smaller and more numerous pores, and finer and more numerous medullary lays.

295

A. Sundra, *DC.; Fl. Br. Ind., II., 295; Bedd., Fl. Sylv., t. 50.*

Vern.—*Lál khair,* MAR.; *Kempu khairada, shemi,* KAN.; *Nalla sandra, sandra, darisanchai,* TEL.; *Karangalli, bága,* TAM.; *Banni* (in MYSORE), KAN.

Habitat.—Found in the Western Peninsula, Ceylon, and Upper Burma.

Botanic Diagnosis.—The *Flora of British India* remarks: "This is scarcely more than a variety of **A. Catechu,** from which it differs by its fewer leaflets and pinnæ, and by the total absence of pubescence," and in "the dark-brown colour of its branchlets."

Compare with **A. Catechu.**

Properties and Uses.—

Gum.—It yields Catechu of good quality. CATECHU. 296

Structure of the Wood.—Dark red, rather close grained, durable, very heavy, not attacked by insects. Much like **A. Catechu,** and when seasoned weighs about 80 lbs. per cubic foot. (*Bomb. Gaz., XV., Part 1., p. 60.*) TIMBER. 297

Cleghorn says it is "used for posts and rice-pestles. The supply is rather large and abundant, but the wood is not generally to be obtained in the market in planks of any size. At Guntur, **Mr. Rohde** states that posts 5 feet long are procurable at R12 per 100. These are well suited for fencing, though the non-elastic nature of the wood is unfavourable to the holding of nails driven into it. The natives regard it as the most durable wood for posts in house-building." (*Cleghorn's For. and Gard. S. Ind., 223.*) 298

Acacia tomentosa, *Willd.; Fl. Br. Ind., II., 294.* 299

Habitat.—A small tree of the Western Peninsula and Ceylon; very common in the Panch Mahál̈s and Gujarát, where it is known as *anjar.*

A. umbraculata, *Wall.;* Syn. for **A. Latronum,** *Willd.*

A. vera, *Willd.,* see **A. arabica,** *Willd.;* and for the true Gum Arabic, see **A. Senegal,** *Willd.* 300

A. vera may be described as the hypothetical species to which the true gum arabic was attributed before the plants which yielded that product were definitely determined.

ACALYPHA, *Linn.; Gen. Pl., III., 311.* 301

A genus of shrubs (belonging to the Natural Order EUPHORBIACEÆ) having alternate, ovate, 3-5 or pinnivened *leaves,* with long petioles; often in cultivation variously coloured or marked, chiefly in shades of yellow to dark red. *Flowers* in axillary simple racemes or spikes, apetalous and monœcious or diœcious. *Male calyx* usually 4-partite, valvate. *Stamens* indefinite (rarely 8), attached to an elevated receptacle; *filaments* free, compressed, attenuate at the apex; *anthers* inserted below the apex, cells often free. *Female flowers* in spikes, hid within the axils of bracts, solitary or 2-3, cymose, sessile. *Bracts* much varied in form, usually dentate, variously evolute, and in most species accrescent, more or less covering the fruit. *Female calyx* 3-4-partite, subchavate. *Ovary* 3-locular, *cells* (two anterior) 1-ovuled. *Style* 3, distinct, or shortly connate at the base. *Capsules* 3, often echinate or rugose. (Compare with *Baillon, V., 212.*)

A large genus, comprising some 220 species, mostly American, but more or less distributed over all tropical and sub-tropical countries. In India there are some six or eight unimportant species. The name of the genus was originally *Acalépha,* from ἐκαλος, unpleasant; and ἡφη, touch, or ἀκαληφη, the nettle.

Acalypha brachystachya, *Horn, Hort.; DC. Prod., XV., II., 870.* 302

Syn.—A. CONFERTA, *Roxb.; Fl. Ind., Ed. C.B.C., 686;* A. CALYCIFORMIS, *Wall. Cat.*

Habitat.—A small bush, little over a foot in height, met with on the Himálaya and the Nilgiri Hills.

Botanic Diagnosis.—*Leaves* round, cordate, long-petioled, three-nerved. *Spikes* sessile, aggregated together; *male flowers* minute, purplish, forming a head within the exterior involucre; *female flowers* two or three within each involucre. *Bracts* crowded, sessile, proliferous.

303

Acalypha ciliata, *Müll.-Arg.; DC. Prod., XV., II., 873; Roxb., Fl. Ind., Ed. C.B.C., 686.*

Habitat.—A common annual plant, throughout the plains of India, most plentiful in the Western Peninsula, where it almost takes the place of **A. indica,** *Linn.*, to which it is nearly allied.

304

A. fruticosa, *Forsk.; DC. Prod., XV., II., 822; Kurz, For. Burm., II., 397;* **A. amentacea,** *Roxb., Fl. Ind., Ed. C.B.C., 686;* **A. betulina,** *Retz.; Ainslie. Mat. Ind., II., 388; Drury, Us. Pl., 10.*

THE BIRCH-LEAVED ACALYPHA.

Vern.—*Sinni-marum,* TAM.; *Chinni-ká-jhar,* DUK.; *Chinni-áka, chinni,* TEL.

Habitat.—A bush, 4-8 feet high, leaf-shedding; met with in South India—Madras, Pondicherry, Mysore, and the Carnatic; Ceylon; Burma—frequent in the tropical forests of Pegu Yomah and Martaban up to 2,000 feet in altitude; Moluccas.

Botanic Diagnosis.—*Leaves* ovate-oblong, deeply serrate, acute to long acuminate, 3-or almost 5-nerved, tomentose beneath, on a long petiole. *Flowers* minute, green, clustered, sessile, forming slender puberulous spikes, occurring singly or 2-3 above the scars of fallen leaves; the *female flowers* at the very base of the spike or on separate small axillary spikes. *Styles* simple, many-cleft, about 2-3 times longer than the floral bracts. Flowering time the beginning of the hot season.

MEDICINE. Leaves. 305

Medicinal Properties and Uses.—The LEAVES "are much esteemed by the native practitioners, who prescribe them as a grateful stomachic in dyspeptic affections and in cholera; they are, besides, considered as attenuant and alterative, and are accordingly administered when it is necessary to correct the habit." "The dose of the infusion of the leaves" as ordered by the *Vytians,* "is half a tea-cupful twice in the day." (*Ainslie, Mat. Ind., II., 388.*)

306

A. indica, *Linn.*

Syn.—A. SPICATA, *Forsk.;* A. CILIATA, *Wall. Cat., No. 7779;* and A. CANESCENS, *Wall. Cat., No. 7785.*

Vern.—*Kuppi, khokali,* or *khokli,* HIND., BOMB.; *Khokli, khájoti,* MAR.; *Vanchhi kánto,* GUJ.; *Muktajuri, shwet busunta, murkanta,* BENG.; *Indra-maris,* URIYA; *Arittamunjayrie?* SANS.; *Kuppaimeni,* TAM.; *Kuppai-chettu, murkanda-chettu* or *múrúkonda* (GODAVARI), *pappanti, marupindi, harita-manjiri,* TEL.; *Chalmari, kúppi,* KANARA; The *Cupameni* of Rheed, Mal., X., 161, t. 81, 83; *Kúpamenya,* CINGH.

References.—*Roxb., Fl. Ind., Ed. C.B.C., 685; DC. Prod., XV., II., 868; Pharm. Ind., 205; Ainslie, Mat. Ind., II., 161; O'Shaugh. Disp., 562; Balf. Cycl., I., 19; Bidie, Cat. Paris Exh., 1878, 42; Dymock, Mat. Med. W. Ind., 587.*

Habitat.—A small annual shrub (1-2 feet in height) occurring as a troublesome weed in gardens and road-sides throughout the plains of India, flowering all the year.

Botanic Diagnosis.—*Leaves* scattered, ovate-cordate, 3-nerved, serrate, smooth, about 2 inches long and $1\frac{1}{3}$ broad; *petiole* as long as the blade. *Spikes* axillary, generally single, peduncled, erect, as long as the leaves, many-flowered, crowned by a cross-shaped body, the base of which is surrounded with a three-leaved *calyx.* From the base of this cross-shaped body issues a *style* having a *stigmatic* fringe. *Male flowers* numerous, crowded around the apex of the spike.

MEDICINE. Roots. 307 Leaves. 308 Shoots. 309

Medicinal Properties and Uses.—"The ROOTS, LEAVES, and TENDER SHOOTS are all used in medicine by the Hindús. The powder of the dry leaves is given to children in worm cases, also a decoction prepared from the leaves with the addition of a little garlic. The juice of the same part of the plant, together with that of the tender shoots, is occasionally mixed with a small portion of margosa oil, and rubbed on the tongues of infants for the purpose of sickening them and clearing their stomachs of viscid phlegm. The *hakims* prescribe the *koopamaynee* in consumption." (*Ainslie, Mat. Ind., II., 161.*) "The leaves with garlic are regarded as anthelmintic; mixed with common salt the leaves are applied externally in scabies, and the juice rubbed up with oil is used externally in rheumatism." (*Balf. Cycl.*) According, to **Rheede** the root is used as a purgative on the Malabar Coast. (*Hort. Mal., X., 161.*) This property "is confirmed by **Dr. H. E. Busteed**, who has used it as a laxative for children." A contributor in Dacca informs me he uses it as a laxative, and in an official correspondence with the Government of India, **Rai Kanai Lal De, Bahadur**, includes the *muktajhuri* amongst emetics. In Bombay "the plant had a reputation as an expectorant, hence the native name *khokli* (cough)". (*Dymock, Mat. Med. W. Ind., 588.*) "**Dr. George Bidie** furnishes the following remarks: 'The expressed juice of the leaves is in great repute, wherever the plant grows, as an emetic for children, and is safe, certain, and speedy in its action. Like Ipecacuanha, it seems to have little tendency to act on the bowels or to depress the vital powers, and it decidedly increases the secretion of the pulmonary organs. Probably an infusion of the dried leaves or an extract prepared from the green plant, would retain all its active properties. The dose of the expressed juice, for an infant, is a teaspoonful.'" (*Pharm Ind.*) A decoction of the leaves is given in earache; a cataplasm of the leaves is applied as a local application to syphilitic ulcers, and as a means of relieving the pain of snake-bite. (*Drury.*) According to **Nimmo** the roots "attract cats quite as much as those of valerian." (*Voigt, 160; Treasury of Botany.*)

310

§ "Much used by Mahomedan practitioners in treating cases of acute mania in early stage. The fresh juice (℥i) with (6 gr.) chloride of sodium dissolved in it and dropped in both nostrils every morning, followed by cold shower-baths for three mornings regularly, proves highly successful. Thus it is supposed by them to act as a 'brain purge,' so called probably owing to a quantity of mucus and other matter escaping from the nostrils immediately after the application of the above recipe. I have given it internally; it acts as an anthelmintic and laxative." (*Surgeon E. W. Savinge, Rajamundry, Godavery District.*) "Juice of the fresh plantemetic, laxative; dose one to four drachms, according to age. Fresh leaves ground into a paste, made into a ball, to the size of a large marble and introduced into the rectum, very useful in relieving obstinate constipation of children." (*Apothecary Thomas Ward, Madanapalle, Cuddapah.*) "The juice or the bruised leaf is applied to the skin to allay the irritation caused by the bite of the centipede." (*Surgeon Ruthnam T. Moodelliar, Chingleput, Madras Presidency.*)

311

"The juice of the fresh leaves mixed with lime is applied topically in painful rheumatic affections." (*Surgeon-Major John Lancaster, M.B., Chittore.*) "Used in scabies and ringworm, also internally as a carminative." (*Surgeon-Major J. J. L. Ratton, M.D., Salem.*) "The root possesses purgative properties; the leaf-juice is a safe, useful emetic, especially adapted for children." (*Surgeon-Major J. M. Honston, Travancore, and John Gomes, Travandrum.*) "The juice of the fresh plant is given to children as an emetic in ℥½ to ℥i doses." (*Apothecary J. Norman, Chattrapur, Ganjam.*)

312

"This plant is called in Kanara *chalmarí* as well as *kúppí* (the latter word means a 'heap,' the plant being found in waste places and rubbish heaps). The natives use it in congestive headaches: a piece of cotton is saturated with the expressed juice and inserted into each nostril, relieving the head symptoms by causing hæmorrhage from the nose. The powder of the dry leaves is used in bed sores and wounds attacked by worms. In asthma and bronchitis I have employed it with benefit both in children and adults.

313

"**Mode of preparation.**—Macerate 3 oz. of the fresh leaves, stalks, and flowers, with a pint of spirits of wine, in a closed jar for 7 days, occasionally agitating the same. Strain, press, filter, and add sufficient spirits of ether to make one pint.

314

"**Physiological effects.**—In small doses it is expectorant and nauseant; in large doses emetic.

315

"**Dose.**—Minims 20 to 60, frequently repeated during the day in honey." (*Surgeon-Major E. H. R. Langley, Bombay.*) "One drachm of the expressed juice of the fresh leaves is an easy and rapid emetic in children. The bruised leaves are useful as an application to maggot-eaten sores." (*Surgeon W. D. Stewart, Cuttack.*)

316

"The root, bruised in hot water, is employed as a cathartic, and the leaves as a laxative in decoction mixed with common salt. The leaves are used in scabies, and mixed with chunam in other cutaneous diseases (*Drury*)." (*Surgeon H. W. Hill, Mánbhúm.*)

ACAMPE, *Lindl.; Gen. Pl., III., 579.*

317

Acampe papillosa, *Lindl.*; ORCHIDEÆ.

Syn.—SACCOLABIUM PAPILLOSUM, *Lindl.*

Vern.—*Kánbher*, MAR.; *Rásná, gandhanákuli*, SANS. The Drug, *Rásná*, HIND., BENG., BOMB.

MEDICINE.
318

Medicine.—This plant is said by **U. C. Dutt** to be used indiscriminately with the **Vanda Roxburghii**, the roots of both constituting *rásná*, BENG. and SANS., also *gandhanakulí*. **Acampe** is a native of the coast of Burma and South India, and is not met with in Bengal.

Acampe differs from **Vanda** in having small brittle flowers, with a lip adnate to the edges of the column, sepals and petals thick, concave; racemes short, rigid, crowded upon a short simple peduncle. **Dr. Dymock** includes both the above plants as yielding the *rásná*, and says the bazar drug comes from Kathiawar. The comparatively limited distribution of **A. papillosa** as compared with **Vanda Roxburghii** should assist in determining which of these plants is the true *rásná*.

It seems probable that the roots of two or three distinct orchids are indiscriminately used as *rásná*,—*i.e.*, **Acampe papillosa, Vanda Roxburghii,** and **Vanda Wightiana. U. C. Dutt** gives as follows a popular prescription for rheumatism, in which *rásná* is one of the ingredients:—

319

"*Rásná-panchaka.*—Take of *rásná, gulancha, devdáru*, ginger, and the root of the castor-oil plant in equal parts, and prepare a decoction in the usual way. This is apparently a popular prescription for rheumatism, being mentioned by most writers."

320

§ **Special Opinions.**—"This orchid is very common in the Konkan; its roots are considered to have cooling properties, and are used medicinally as *Rásná.*" (*Surgeon-Major W. Dymock, Bombay.*) "I found the **Acampe papillosa** in the mango-groves of Malda. It is a common parasite on the mango tree, and flowers in the rainy season, when it can readily be distinguished from allied species." (*U. C. Dutt, Civil Medical Officer, Serampore.*)

321

"It is said to be a specific for acute rheumatism. It is invariably given internally as a substitute for Sarsaparilla. There are three preparations in use. The first is called ***Rásná panchak***, and is prepared by boiling together equal parts of ***rásná, dewadára***, ginger, ***garula*** and castor root, and water 8 oz. This decoction is prescribed extensively for cases of acute rheumatism. The second preparation, or ***Rásná suptaka***, has seven ingredients, and is given to cases of lumbago, sciatica, and neuralgia. Prepared by boiling together equal quantities of ***rásná, gokharu***, castor root, ***dewadára, poonarnawá, goilwel***, pulp of ***babawa***, and ginger. The third preparation is named the great ***Rásná*** (***Bará rásná***) and is considered a specific remedy for rheumatic and nervous affections, paralysis, secondary syphilis, and uterine diseases. It has the peculiar power of preventing abortions and miscarriages. The ingredients of the decoction are ***rásná*** two parts, ***dhumasa, chiknamula*** one part, castor root one, ***dewadára*** one, ***kuchora vekhanda*** one, ***adulasa*** one, ginger one, small ***hirda*** one, ***chuwak*** one, ***nagar-motha*** one, ***punarnawá gulawel*** one, ***wardhárá*** one, ***budishep*** one, ***gokhroo asandha*** one, ***ativish*** one, pulp of ***bahawa, sutawari*** one, ***pimplee*** one, ***kolista*** one, coriander one, ***ringnee*** one, and ***moti ringnee*** one." (*Surgeon W. Barren, Bhuj, Cutch, Bombay.*)

ACANTHACEÆ; *CXXII., Gen. Pl.*

322

The name of a large and important Natural Order of herbs and shrubs (rarely trees), comprising about 1,500 species, almost exclusively inhabitants of the tropical and warm-temperate regions of the Old World. Indeed, the gregarious, sub-bushy or herbaceous under-vegetation of damp tropical forests may be regarded as at least one-third composed of Acanthaceous plants. Over 400 species are met with in India, 154 of which, according to *Flora of British India*, are referred to the genus STROBILANTHES. Only a few species occur in the temperate zones and none in the alpine or arctic regions.

This Natural Order belongs to the Cohort PERSONALES or plants, having a monopetalous, hypogynous, and chiefly 2-labiate *corolla*; *stamens* generally 4, two long and two short (*didynamous*). *Ovary* 1-2, rarely 4-celled. *Fruit* capsular. In this cohort are placed the following Natural Orders: SCROPHULARINEÆ, VERBENACEÆ, LENTIBULARIEÆ, OROBANCHACEÆ, GESNERACEÆ, BIGNONIACEÆ, ACANTHACEÆ, and PEDALINEÆ.

Diagnostic characters of the Acanthaceæ.—*Leaves* opposite, exstipulate. *Flowers* aggregated into compact, bracteated *spikes* (sometimes solitary as in some THUNBERGIEÆ). *Corolla* imbricate or twisted in æstivation, bilabiate, rarely sub-regular. *Stamens* often reduced to two. *Ovary* 2-celled; *ovules* one or more in each cell, 1 to 2, seriate, anatropous. *Capsule* dehiscent, loculicidal; *valves* often elastically recurving and carrying the seeds attached to the septa. *Seeds* 2-seriate along the *septum*, and each (except in THUNBERGIA) seated upon a sharp, up-curved, hook-like process from the placenta (called the *retinacula*), ovoid or compressed; *testa* smooth or warted, rarely hispid. *Albumen* none.

This Order has its nearest affinities in SCROPHULARINEÆ, BIGNONIACEÆ, and VERBENACEÆ. The bracteated spike, contorted æstivation, and the presence of retinacula will unerringly separate it, however, from these Orders. Indeed, the dark-greenish blue, and more or less glabrous condition of the foliage, with many approximately parallel veins, when taken along with the bracteated spikes, will be found to possess something so characteristic and impressive, that if this feature be once carefully observed it is not again readily mistaken.

For the analysis and diagnostic characters of the genera and species of this Order reference should be made to the *Flora of British India, Vol. IV., 388*, from which the following classification into Tribes has been extracted for the convenience of the general reader:—

Tribe 1, Thunbergieæ.—Scandent or twining. *Calyx* minute, annular or 10-15-toothed. *Corolla-lobes* twisted in bud. *Ovules* 2 in each cell, collateral, capsule beaked; retinacula 0.

Genus—Thunbergia.

Tribe 2, Nelsonieæ.—*Corolla-lobes* imbricate in bud. *Ovules* many, superimposed in two rows in each cell. *Seeds* small, seated on minute papillæ, not on hard retinacula, obscurely albuminous.

Genera.—Elytraria, Nelsonia, Ebermaiera, Ophiorrhiziphyllum.

Tribe 3, Ruellieæ.—*Corolla-lobes* twisted to the left in bud. Seeds on retinacula. *Sepals* 5 or 4, with one larger. *Anthers* usually 2-celled; *cells* parallel or one a little below the other. *Style* 2-fid, one lobe often suppressed.

Sub-tribe 1, POLYSPERMEÆ.—*Ovules* 3 to 12 in each cell. *Capsule* normally 6 or more seeded.

Genera.—Cardanthera, Hygrophila, Nomaphila, Ruellia, Echinacanthus, Æchmanthera, Hemigraphis, Stenosiphonium.

Sub-tribe 2, TETRASPERMEÆ.—*Ovule* 2 in each cell. *Capsule* 4 or fewer seeded.

Genera.—Strobilanthes, Calacanthus, Calophanes, Dædalacanthus, Phaylopsis, Petalidium.

Tribe 4, Acantheæ.—*Corolla-tube* short; upper lip obsolete, lower 3-lobed. *Ovules* 2 in each cell. *Retinacula* curved, hardened.

Genera.—Blepharis, Acanthus.

Tribe 5, Justicieæ.—*Corolla-lobes* imbricate in bud. *Retinacula* curved, hardened. *Anthers* 2-1-celled; cells often spurred at the base, one frequently placed much above the other. *Style* shortly equally 2-fid or sub-entire.

Sub-tribe 1, ANDROGRAPHIDEÆ.—*Ovules* 3-10 in each cell. *Capsule* normally 6 or more seeded.

Genera.—Andrographis, Haplanthus, Gymnostachyum, Phlogacanthus, Cystacanthus, Diotacanthus.

Sub-tribe 2, BARLERIEÆ.—*Ovules* 2-1 in each cell. *Corolla-lobes* 5, sub-equal. *Stamens* 4, of which 2 are small or obsolete, or 4-1-celled.

Genera.—Barleria, Neuracanthus, Crossandra.

Sub-tribe 3, ASYSTASIEÆ.—*Ovules* 2 in each cell. *Corolla-lobes* 5, sub-equal. *Sepals* 5, small, sub-equal. *Stamens* 4 or 2; anther-cells 2, sub-equal, parallel, muticous.

Genera.—Asystasia, Eranthemum, Codonacanthus.

Sub-tribe 4, EUJUSTICIEÆ.—*Ovules* 2 in each cell. *Corolla* distinctly 2-lipped.

Genera.—Lepidagathis, Phialacanthus, Monothecium, Clinacanthus, Hypœstes, Rungia, Dicliptera, Peristrophe, Justicia, Adhatoda, Rhinacanthus, Dianthera, Ptysiglottis, Sphinctacanthus, Ecbolium, Graptophyllum.

323

ACANTHUS, *Linn.; Gen. Pl., II., 1090.*

A genus of herbs belonging to the ACANTHACEÆ; characterised by having the upper lip of the *corolla* obsolete, lower 3-lobed. Anterior *filaments* without an excurrent process. *Ovules* 2 in each cell.

It contains 7 Indian species, none of which are of any great economic value; all are confined to the warm forests of the Eastern Peninsula. **A. spinosus** is found in Italy, Spain, and the south of France, and is supposed to have suggested the idea of the decoration of columns in the style now known as Corinthian architecture.

324

Acanthus ilicifolius, *Linn.; Fl. Br. Ind., IV., 481; Wight, Ic., t. 459.*

Vern.—*Harcuch kanta*, BENG.; *Harikusa*, SANS.; *Kaya*, BURM.

Habitat.—A common plant, growing everywhere near the coast, from the Sunderbuns to Malacca, and one of the most characteristic plants in that region. It makes its appearance in the swamps around Calcutta, Dum-Dum, &c.

It does not appear to be put to any economic purpose, although it covers many miles of country.

ACER.

325

A genus of trees with opposite, simple, or palmately lobed, exstipulate *leaves* (belonging to the Natural Order SAPINDACEÆ). *Flowers* regular, polygamous, formed into terminal and lateral racemes or corymbs. *Calyx* usually 5-partite, imbricate, deciduous. *Petals* of the same number as the sepals, or absent, shortly clawed, and without scales. *Stamens* 4-12, usually 8, inserted on the glabrous disk; generally shorter in the hermaphrodite than in the male flowers. *Ovary* laterally compressed, 2- (rarely 3-) lobed and celled; cells 2-ovuled. *Fruit* an indehiscent double samara. *Seeds* exalbuminous.

A genus containing about 40 to 50 species, found in Europe, Asia, and North America, chiefly in the temperate zones. The name Acer is derived from the Latin *Acer,* sharp or pointed (aigu, *Fr.*; acute, eager, *Eng.*). A name applied in all probability by the Latin people to the members of this genus in allusion to the form of the leaves. This is the genus of the maple, the sycamore, and the plane-tree of English authors. **A. Pseudo-platanus** is a native of Germany, Switzerland, Austria, and Italy, now largely cultivated in England; it is the greater maple, sycamore, or plane-tree. **A. campestre** of England is the common maple. **A. saccharinum,** *Wang,* is the common sugar-maple of the Northern United States and of Canada. In addition to yielding sugar, **A. rubrum,** the Swamp maple of Pennsylvania, gives from its bark a dark-blue dye made into ink.

MAPLE-SUGAR (a form of Cane-sugar).

326

References.—*Spons' Encycl., 1902; Flück. and Hanb., Pharmacog., 721; Smith's Dictionary of Economic Plants; Smith on Foods, 256; U. S. Dispens., 15th Ed., 1256; Simmonds' Comm. Prod. Veg. Kingdom, p. 205.*

It would appear that either none of our Indian species of Acer yield maple-sugar, or that this property is quite unknown to the natives of our hill tracts, where various species of maple are plentiful. This subject seems worthy of a little attention, for there does not seem the slightest reason why the sugar-yielding species could not be introduced into India. If they were found to take naturally to the soil and climate of Indian subalpine regions, they might supply the poor hill tribes with the little-known luxury of sugar; they are, however, of no commercial value. Speaking of maple-sugar, *Spons' Encycl.* says: "In sections of the United States where it has not been exterminated, the manufacture of sugar and syrup from it is a remunerative adjunct to other farming industries, occupying a period in which little other farm work can be pursued. The apparatus for collecting the sap and manufacturing the sugar involves a very small investment; the fuel consumed usually consists of the prunings of the maple grove, which is benefited thereby; and at least 90 per cent. of the gross return is net profit."

The census of Pennsylvania for 1870 gives the following figures for maple-sugar as manufactured in that State: 1850, 2,326,525 lbs.; 1860, 2,768,965 lbs.; and in 1870, 1,545,917 lbs.; and the United States as a whole are said to manufacture about 40,000,000 lbs. annually, the Indians manufacturing some 30,000,000 lbs. in addition for their own consumption. Of this amount Vermont yields about 10,000,000 lbs. and New York

The Species of Indian Maple.

a somewhat larger quantity. Canada manufactures about 10,000,000 lbs. annually.

In addition to the manufacture of sugar a large quantity of maple sap is consumed in the form of molasses.

327

In Nebraska an equally good saccharine product is obtained from **A. Negundo,** *L.* **(Negundo aceroides,** *Mœnch*). From some investigations made in Illinois, with reference to the value of this tree as a supply of sugar, it was found (1) that trees 5 years old commence to yield, and that an ordinary tree will yield more than one of equal size of the true maple-sugar tree; (2) that the sap is richer in sugar, the yield being 2·8 per cent. to the weight of sap; (3) that the sugar produced is whiter than that from the sugar-maple. These facts should recommend themselves to planters, and it is probable this tree would succeed better in India than any of the preceding species.

There are 13 species of ACER met with in India, grouped into the following sections by the *Flora of British India*:—

[*Note.*—Those marked * will be found described further on.]

SECTION I.—LEAVES UNDIVIDED.

† *Leaves with 3 basal nerves.*

* 1. **A. oblongum,** *Wall.*—Kashmír, Sikkim, Bhután.
2. **A. niveum,** *Blume.*—Leaves quite entire, white, glaucose beneath. *Cymes* lax, flowered, glabrous, *cell* not angular.—Assam, Moulmein, Sumatra.
* 3. **A. lævigatum,** *Wall.*—Simla, Sikkim, Khásia Hills, &c.

†† *Leaves with 5 basal nerves.*

* 4. **A. sikkimense,** *Miq.*—Sikkim, Bhután.
* 5. **A. Hookeri,** *Miq.*—Sikkim, Bhután.
6. **A. stachyophyllum,** *Hiern.*—*Leaves* serrate. *Carpels* 1¾ to 2 in. long. *Cells* angular.—Sikkim, Bhután.

SECTION II.—LEAVES 3-LOBED.

7. **A. isolobum,** *Kurz.*—*Leaves* deeply 3-lobed and nerved, glabrous, shining, acutely serrate, lobes lanceolate, acuminate,—Pegu.
8. **A. pentapomicum,** *J. L. Stewart.*—Leaves 3-lobed, with tufts of hairs in the axils of the 3-5 nerved lobes, ovate, obtusely serrate. Peduncles fascicled.—Kashmír to Kumaon.

SECTION III.—LEAVES 5-LOBED AND NERVED (EXCEPT 3-LOBED FORM, UNDER **A. villosum var. Thomsoni.**)

* 9. **A. cæsium,** *Wall.*—Kashmír to Nepal.
* 10. **A. villosum,** *Wall.*—Kashmír to Nepal.
    * **Var. Thomsoni,** *Miq., Sp.*—Bhután to Mánipur.
* 11. **A. caudatum,** *Wall.*—Chumba to Sikkim.

SECTION IV.—LEAVES 7- TO 5-LOBED AND NERVED.

* 12. **A. Campbellii,** *Hook.*—Sikkim to Mánipur.
* 13. **A. pictum,** *Thunb.*—Kashmír to Bhután.

328

Acer cæsium, *Wall.; Fl. Br. Ind., I., 695;* SAPINDACEÆ.

THE INDIAN MAPLE.

Vern.—*Trekhan, tarkhana, tilpattar, mandar, kauri, kalindra, salima, kanzal,* PB.; *Kanshin,* THIBET; *Jerimu, shumanjra,* SIMLA; *Kilu, kainshing,* KUMAON.

References.—*Brandis, For. Fl., 111; Gamble, Man. Timb., 100.*

Habitat.—A large, deciduous tree, found in the North-West Himálaya from the Indus to Nepal, between 7,000 and 11,000 feet.

Botanic Diagnosis.—*Leaves* palmately 5-lobed, pale, glaucose beneath and almost quite glabrous, except a few short hairs on the veins. *Cymes* corymbose, appearing after the leaves and becoming nearly as long. *Fruit* black and quite glabrous, carpels $1\frac{3}{4}$ to 2 in. long, wings venose, somewhat diverging; cells angular, black.

A large tree, often 80 feet in height; twigs red or bluish, laterally compressed, flowering in April, and the fruit ripening in October.

TIMBER 329

Structure of the Wood.—White or pale cream-coloured, with brown bands,porous, close-grained, less mottled than that of **A. caudatum,** soft to moderately hard; annual rings distinct. Weight about 40 lbs. per cubic foot.

DOMESTIC. 330

Scarcely used. Drinking-cups are sometimes made of it by the Thibetans.

Acer Campbellii, *Hook. f. & Th.; Fl. Br. Ind., I., 696.*

331

Vern.—*Kabashi*, NEPAL; *Daom, yatli*, LEPCHA; *Kilok*, BHUTIA.

Reference.—*Gamble, Man. Timb., 100.*

Habitat.—A large deciduous tree, found in the Sikkim Himálaya, above 7,000 feet in altitude. This is the chief Maple of the North-East Himálaya.

Botanic Diagnosis.—*Leaves* beautifully green, 5-7-lobed, sub-membranous, glabrescent, except in the axils of the 5-7 nerves, petioles red. *Cymes* pyramidal or elongated, sub-glabrous, appearing with the leaves.

TIMBER. 332

Structure of the Wood.—Greyish white, moderately hard, shining, close-grained. Annual rings marked by a thin line. Weight 38 lbs. per cubic foot.

333

It is extensively used for planking and for tea-boxes. It reproduces freely either by seed or by coppice, and plays an important part in the regeneration of the hill forests.

A. caudatum, *Wall.; Fl. Br. Ind., I., 695.*

334

Vern.—*Kanela, kandaru, kanjara*, SIMLA; *Khansing, kabashi*, NEPAL; *Yalishin*, BHUTIA.

References.—*Brandis, For. Fl., 112; Gamble, Man. Timb., 100.*

Habitat.—A moderate-sized, deciduous tree, with dark-brown bark, flowering in March and April; met with in the Himálaya from the Chenab to Bhután, between 7,000 to 11,000 feet.

Botanic Diagnosis.—*Leaves* 5-lobed, serrate, nearly glabrescent except in the axils of the nerves; *lobes* caudate, the two basal ones small. *Racemes* short. *Carpels* $\frac{3}{4}$ to $1\frac{1}{3}$ inch long. *Fruit* nearly glabrous; *wings* pink, or at length ferruginous, *front* sinuous, crenulate.

TIMBER. 335

Structure of the Wood.—White, with a faint pink tinge, shiny, compact, moderately hard, sometimes with small masses of heartwood near the centre. Annual rings distinct. Weight 43 lbs. per cubic foot.

A, cultratum, *Wall.*

A Synonym used by *Baden Powell, Pb., Prod., I., 566.* See **A. pictum,** *Thunb.*

A. Hookeri, *Miq.; Fl. Br. Ind., I., 694.*

336

Vern.—*Lal kabashi*, NEPAL; *Palé*, LEPCHA.

Reference.—*Gamble, Man. Timb., 99.*

Habitat.—A deciduous tree, found in Sikkim and Bhután, above 7,000 feet. Plants with copper-coloured foliage are not uncommon about Darjíling.

Botanic Diagnosis.—*Leaves* undivided, finely duplicate-serrate, ovate caudate-acuminate, base 5-nerved, cordate, both sides green and sub-glabrate. *Racemes* puberulent, simple, nearly equal and appearing with the leaves. *Fruit* glabrous; *carpels* $\frac{3}{4}$ to $\frac{7}{8}$ inch long; *wings* venose, widening above and divergent, *back* slighly curved.

TIMBER. 337

Structure of the Wood.—Grey. Weight 37 lbs. per cubic foot.

338

Acer lævigatum, *Wall.; Fl. Br. Ind., I., 693.*

Vern.—*Saslendi, cherauni, thali kabashi,* NEPAL; *Tungnyok,* LEPCHA.

References.—*Brandis, For. Fl., 110; Gamble, Man. Timb., 99 Kurz.; For. Fl. Burm., I., 289.*

Habitat.—A deciduous tree, found in the Himálaya from the Jumna eastward to Bhután, between 5,000 and 9,000 feet; in the Khásia Hills, and in Tenasserim.

Botanic Diagnosis.—*Leaves* undivided, quite entire or minutely serrate, when young ovate-oblong acuminate, glabrous, penninerved, reticulate, green on both surfaces, base rounded three-nerved. *Cymes* panicled, glabrous, appearing with the leaves. *Fruit* glabrous; *carpels* 1 to $1\frac{1}{4}$ inch long; *wings* venose, slightly diverging, widened above, *back* curved.

A handsome tree with broad, oval crown, flowering in April, and ripening its fruit in July and August. Bark smooth, yellowish, or dark ash-coloured.

TIMBER. 339

Structure of the Wood.—White, shining, hard, close-grained. Weight 43 lbs. per cubic foot.

DOMESTIC. 340

Used for planking and tea-boxes. Much used in Nepal for building.

341

A. oblongum, *Wall.; Fl. Br. Ind., I., 693.*

Vern.—*Mark,* PB.; *Mharengala, patangalia, kirmòli,* N.-W. P.; *Pugila, buzimpála,* NEPAL.

References.—*Brandis, For. Fl., 110; Gamble, Man. Timb., 99.*

Habitat.—A moderate-sized, deciduous tree, found in the Himálaya from the Jhelum eastward to Bhután, up to 6,000 feet in altitude.

Botanic Diagnosis.—*Leaves* undivided, quite entire, oblong-ovate acuminate, penninerved, silvery glaucous beneath, base obtuse, 3-nerved. *Cymes* panicled, appearing with the leaves. *Fruit* glabrous; *carpels* 1 to $1\frac{1}{4}$ inch long; *wings* venose, diverging, contracted below, *back* nearly straight; cells woody, angular, clothed inside with white hairs.

Never leafless, the mature foliage of a deep dark-green colour. A gregarious tree, flowering in February to April, the fruit ripening in June to November.

TIMBER. 342

Structure of the Wood.—Light reddish-brown, moderately hard, close-grained. Annual rings faintly marked. Weight 45 lbs. per cubic foot.

DOMESTIC. 343

Used for agricultural implements and drinking-cups.

344

A. pictum, *Thunb.; Fl. Br. Ind., I., 696.*

Syn.—A. LÆTUM, *C. A. Mey.*; A. TRUNCATUM, *Bunge*; A. CULTRATUM, *Wall., Pl. As. Rar.*

Vern.—*Tilpattar, trekhan, tarkhana, kakkru, kanzal, kánjar, jarimu, laur,* PB.; *Trikunda* or *trikanna,* MURREE; *Mandal, maner,* CHUMBA; *Trekhan,* HAZARA; *Tian,* KANÁWÁR; *Kanchli,* N.-W. P.; *Dhadonjra,* SIMLA.

References.—*Brandis, For. Fl., 112; Baden Powell, Pb. Prod., I., 566; Gamble, Man. Timb., 101.*

Habitat.—A beautiful, moderate-sized tree, met with in the temperate Himálaya from Kashmír, altitude 4,000 to 6,000 feet, to Bhután, altitude

9,000 feet. A widely diffused species, being distributed to Japan, China, the Caucasus, Armenia, and North Persia; the type in these widely different countries constantly preserving itself. It is the most abundant Maple in the Himálaya, flowering in April, and the fruit ripening in July and August.

Botanic Diagnosis.—*Leaves* 5-7-lobed, glabrescent, except hairy tufts on the axils of the basal nerves, cordate, lobes lanceolate caudate, entire, green on both sides. *Cymes* corymbose, appearing with the leaves. *Fruit* glabrous; *carpels* 1⅛ to 1¾ inch long, divaricating almost in one line; *wings* sinuous, venose, *back* arcuate; *cells* compressed.

Properties and Uses—

Medicine.—The KNOTS on the stem are made into the curious water-cups supposed by some of the hill tribesto have a medicinal influence over the water. The LEAVES are said to yield an acrid juice in Kanáwár, which blisters the hands; but in most other parts of the Himálaya they are lopped off as fodder.

MEDICINE. Cup. 345 Leaves. 346

Fodder.—The branches are lopped for fodder.

FODDER. 347

Structure of the Wood.—Pinkish white, soft to moderately hard, close-grained, fairly strong and elastic. Weight 41 lbs. per cubic foot.

TIMBER. 348

It is used for the construction of ploughs, bedsteads, and poles to carry loads. Thibetan drinking-cups are made of the knotty excrescences; in fact this is the species most frequently used for this purpose.

Acer sikkimense, *Miq.; Fl. Br. Ind., I., 694.*

349

Vern.—*Palegnyok*, LEPCHA.
References.—*Gamble, Man. Timb., 99.*

Habitat.—A small tree, found in the hills in Sikkim and in Bhután, from 7,000 to 9,000 feet, and in the Mishmi mountains.

Botanic Diagnosis.—*Leaves* undivided, minutely serrate, ovate, cuspidate, cordate, penninerved, with 5 basal nerves, sub-coriaceous, glabrous, green on both sides. *Racemes* spicate and glabrate, appearing with the leaves. *Fruit* glabrous; *carpels* ⅝ to ¾ inch long; *wings* venose, diverging, widened above; *back* straight or slightly curved; *cells* not angular.

Structure of the Wood.—Shining, grey; annual rings distinct. Weight 37 lbs.

TIMBER. 350

A. Thomsoni, *Miq.; Fl. Br. Ind., I., 695.*

351

Syn.—A. VILLOSUM, *Wall.*; var. THOMSONI, in *Hook. Fl. Ind., I., 695.*
Vern.—*Kabashi*, NEPAL.
References.—*Brandis, For. Fl., 109.*

Habitat.—A large tree, often 150 feet in height, found in the hills in Sikkim and in Bhután, altitude 4,000 feet.

Botanic Diagnosis.—By the *Flora of British India,* this is reduced to a mere variety under **A. villosum.** In Mánipur I compared this with **A. villosum** carefully, side by side, and I regard **Thomsoni** as a distinct and well-marked species, which should be placed in the section with undivided leaves having 5 basal nerves. *Leaves* ovate, cordate, acute, entire or 3-angled on the apex, but never 5-lobed or even 5-angled; a foot or more long, thick, coarse, glabrous.

Structure of the Wood.—Greyish white, soft, and very brittle. A tree, 150 feet in height, hewn down by me in the Koupra forest, Mánipur, having a stem 80 feet without branches, was shattered to pieces by the fall. Weight 44 lbs. per cubic foot.

TIMBER. 352

353

A. villosum, *Wall.*; *Fl. Br. Ind., I., 695.*

Vern.—*Karendera*, SIMLA; *Kikalnshing*, N.-E. KUMAON.

References.—*Brandis, For. Fl., 111; Gamble, Man. Timb., 100.*

Habitat.—A distorted, deciduous tree, 80 feet in height, found in the North-West Himálaya from the Jhelum to Nepal, between 7,000 and 9,000 feet, and in Mánipur (on the eastern frontier of Assam), altitude 6,000 feet.

Botanic Diagnosis.—*Leaves* 3-5-lobed, cordate, 5-nerved, lobes ovate or lanceolate, serrate or repand, thin, shining, membranous, glabrous or tomentose below (the latter condition being simply the younger or less exposed condition upon the same tree). *Racemes* branched or simple, pubescent, appearing a little before the leaves. *Fruit* puberulent, brownish; *carpels* 1½ to 2¼ inches long, diverging; *wings* venose, margins often crenulate; *back* rather curved; *cells* angular, nervose.

A handsome tree with the stem-bark grey; flowering in February and March, the leaves appearing in May and the fruit ripening in June.

FODDER. 354

Fodder.—Leaves lopped for fodder.

TIMBER. 355

Structure of the Wood.—White, moderately hard, close-grained, beautifully mottled and shining; annual rings distinct. Weight 38 lbs. per cubic foot. Not used.

ACETUM.

356

Acetum.

VINEGAR, *Eng.*; VINAIGRE, *Fr.*; ESSIG, *Germ.*; ACETO, *It.*; VINAGRE, *Sp.*

Vern.—*Khall*, ARAB.; *Sirkah*, PERS.; *Sirka*, HIND.; *Kádi*, TAM.; *Shukta*, SANS.

References.—*Spons' Encyl., 2038; U. S. Dispens., 15th Ed., 18; Smith on Foods, 230; Balfour's Cycl., V., under "Vinegar"; Baden Powell's Pb. Prod., I., 312; Pharm. Ind., 267; Year-Book of Pharm. 1875, 46; 1876, 35, 376; 1878, 159, 160, 174; 1882, 119; The Vinegar Plant, Treasury of Botany; Smith's Dictionary of Economic Plants, &c.*

Vinegar is an acid liquid, used largely as a food auxiliary, and to preserve certain articles of food, as a medicine, and for industrial purposes. It is produced (*a*) by what is known as acetous fermentation of a mixture of malt and unmalted grain (forming malt vinegar), and (*b*) by the oxidation of white or red wine (forming the white and red wine vinegar). Chemically, vinegar is a dilute solution of acetic acid ($C_2 H_4 O_2$) with certain organic substances derived from and peculiar to the source from which it is derived. "The finest artificial table vinegar is prepared by mixing 15 parts of glacial acetic acid (of at least 99 per cent.) with 235 parts of water and one part of alcohol (94 per cent.) or by adding to 250 parts of diluted acetic (6 per cent.) 1 part of alcohol. If allowed to stand a few weeks, the mixture will develop enough acetic ether to impart to it a fine flavour; it may be coloured with caramel, hut is much nicer and cleaner without it. A popular prejudice in favour of malt, cider, and other similar vinegars should be gradually removed. (*Dr. D. R. Squibb.*)" (*Dr. Charles Rice, New. York.*) Instead of being made from malt vinegar it is also largely derived from artificial glucose ($C_6 H_{12} O_6 H_2 O$) and cane-sugar or molasses ($C_{12} H_{22} O_{11}$); the substance thus produced is identical with malt vinegar.

PREPARATION OF VINEGAR.

THE GERMAN PROCESS.

MALT VINEGAR. 357

Malt Vinegar.—There are various processes for producing vinegar from malt, the practice chiefly prevailing in England being that known as the "German process." The grain (wheat, barley, oats, rice, maize, or other grain) is broken and the husk thereby got rid of. This is necessary, since the phosphoric compounds of the husk retard the process, and the presence of the husk in the vats would add an unnecessary amount of vegetable matter. The unmalted grain must be well dried over a kiln previous to crushing, in order that many of the glutinous and albuminoid compounds of the grain may be destroyed. If this be not done the vinegar will not keep well. The crushed grain is now conveyed to the "Mash tun" and there mixed intimately with water at 77° (170° Fahr.). After an hour the wort is conveyed to the boilers and well boiled for the purpose of coagulating the albumen. It is then mixed with a certain proportion of malt grain, which has the affect of converting the insoluble starch into glucose. Malt grain alone canhot be used, as it contains much more *diastase* than is necessary for the conversion of its starch into *glucose.* The superabundance of *diastase* would produce secondary and putrefactive fermentation. It is therefore necessary to mix it with unmalt grain which has previously been roasted on a kiln to destroy its glutinous and albuminoid compounds. These compounds compose the protoplasm from which the *diastase* is derived.

During the germination of a grain a portion of the protoplasm becomes converted into a chemical substance known as *diastase.* This substance has the curious property of converting the insoluble starch contained in the grain into a soluble saccharine compound, namely, first into *dextrine,* and then into artificial *glucose.* In the economy of plant life this curious action allows the embryo plant to obtain a soluble food from the seed, enabling it thereby to produce and develope a root which soon supplies the further wants of the plant. The proportions in which the malt and unmalt grain should be used will depend very much upon the nature of the grain, and can be determined by experience or experiment only.

358

The Chemistry of Vinegar.—Up to this stage the manufacture of vinegar is almost identical with beer-brewing. The mass, by the chemical action described, having been converted into glucose, is run into the fermenting vat, and yeast added, when fermentation at once commences. Unlike beer-brewing, it is now the object of the vinegar-maker to convert the whole of the glucose into alcohol. This is done by forced and repeated additions of yeast; the glucose being converted into alcohol and carbonic acid eliminated:—

Glucose. **Alcohol.**

$$(C_6H_{12}O_6) = (2C_2H_6O) + 2CO_2.$$

When the entire volume has been converted into an alcoholic liquid it is pumped into vats and allowed to clear itself of dead yeast and cloudiness by subsidence. It is then passed through a filter into the acetifier. It is the process by which this alcoholic liquid is at this stage converted into acetic acid (or vinegar) that constitutes the difference between the various processes of vinegar-brewing. In the German process, for a period of about six weeks, the liquid is being constantly pumped from the bottom of the acetifier or vinegar-generator and made to fall down again in a finely divided condition, by passing through small openings in a wooden lid placed over the top of the generator, thereafter to trickle through a layer of specially prepared or purified beech twigs or charcoal. While trickling through these contrivances the alcoholic liquid becomes oxidised and reduced thereby to vinegar. This is accomplished by two atoms of hydrogen separating from the alcohol and forming with one atom of oxygen from the air a molecule of water. By this separation of hydrogen the alcohol is reduced to the inflammable ethereal liquor known as aldehyde (or alcohol dehydrogen). By the absorption of an atom of oxygen from the air this compound is next formed into acetic acid :—

$$\underset{\text{Alcohol.}}{C_2H_6O} + \underset{\text{Oxygen.}}{O} = \underset{\text{Aldehyde.}}{C_2H_4O} + \underset{\text{Water.}}{H_2O}$$

$$\underset{\text{Aldehyde.}}{C_2H_4O} + \underset{\text{Oxygen.}}{O} = \underset{\text{Acetic acid.}}{C_2H_4O_2}$$

The whole process of making malt vinegar occupies about two months, after which time it is stored in vats for the purpose of cleansing, colouring, &c.

The French Process.

359

According to **Pasteur** the process of acetification is not spontaneous oxidation, caused through the simple exposure of an alcoholic liquid to the complete influence of the oxygen of the air, but is due to the oxidising influence of the growth of a mycoderm closely allied to dry rot (see Vinegar Plant) within the vinegar generator. The chemical actions indicated are quite correct, but the initial or starting agent in these changes is the growth of the vinegar plant upon the oxidising apparatus. In support of this theory, it may be stated that, in the absence of light, both fermentation and acetification is greatly retarded, light being necessary for the growth of the yeast and vinegar plants.

Vinegar may be clarified by throwing about a tumblerful of boiling milk into fifty gallons of the liquid and stirring the mixture. This operation has the effect at the same time of rendering red vinegar pale. At one time it was thought necessary to add free sulphuric acid to vinegar in order to stop further chemical changes and to destroy the mycoderm which might still be surviving within the liquid. The law formerly authorised 1 part of acid in 1,000 of vinegar. The necessity for acid is the confession of defective preparation, chiefly due to the whole of the glucose not having been changed into alcohol, or to too much diastase having been formed in the wort. According to the "Food and Drug Adulteration Act," the addition of sulphuric acid, in however small a proportion, constitutes an adulterant. Accordingly, **Pasteur** recommended that the mycoderm should be systematically sowed and the alcoholic liquid carefully added, until complete acetification be accomplished. The mycoderm exists in two forms (which may be distinct species), both of which have the power of acetification. In the one it consists of extremely small globules (micrococci), arranged and adhering together in contiguous rows (sometimes, in addition, enveloped in a glue-like mass). In the other the mycoderm is made up of rod-like forms (bacilli).

This has been called the French Process, and it was originally applied practically in 1869 by **Beton-Laugier**, and more recently perfected by **Mr. Emanuel Warm.**

360

The Quick German Process.

This is chiefly practised on the Continent, and consists in taking alcohol and water instead of malt. Alcohol being free of duty (on the Continent), this can be done at a price which admits of competition with the ordinary malt vinegar. The pleasant aromatic odour of the malt vinegar is obtained by mixing a small quantity of the brewed wort with the alcohol and water.

WINE-VINEGAR. White. 361

Wine-Vinegar or **Acetum Gallicum.**—This is made from either white or red wine giving origin to white and red wine vinegar. This is chiefly prepared from wines which have shown a tendency to become sour. It is the sugar present in the wine which acts as the ferment, when wine naturally undergoes acetous fermentation. In the same way vinegar is made from ciders molasses, &c.

Red. 362

To prepare wine-vinegar, casks capable of holding 100 gallons are arranged in rows in a shed kept at the temperature of 75° to 85°. The casks are open at the top to allow free access of air. A small quantity of boiling vinegar is added to each cask, and every eight or ten days a few gallons are added until the casks are two-thirds full. After 10 to 14 days the acetous fermentation is complete. A few gallons of vinegar are withdrawn from each and wine added, and so the process is carried on. Upon the inside of the older casks a crop of the mother-of-vinegar plant will be found, and when once established this greatly accelerates the fermentation.

363

Wine-vinegar is said to always contain a little aldehyde. It is nearly one-sixth stronger than pure malt vinegar.

Chemical Tests.—Vinegar should have the sp. gr. of 1·017 to 1·019. If aldehyde be present (sometimes the case in wine-vinegar), its presence may be at once detected by Trommer's test, for having, like glucose, the power of absorbing oxygen, the red sub-oxide of copper will be precipitated from Fehling's test solution proving the presence of aldehyde. The most dangerous impurities are copper, lead, and even tin, derived through carelessness from the apparatus used. These metals will at once be detected through the black precipitate thrown down in the vinegar on the addition of sulphuretted hydrogen gas, and by no precipitate being formed on being boiled with common salt. Vinegar should also be devoid of free sulphuric acid. If 10 minims of a 10% solution of chloride of barium be added to one fluid ounce of vinegar, and the resulting precipitate (if any) removed by filtration, the further addition of the barium solution would give no further precipitation if the amount of sulphuric acid formerly authorised (1 in 1,000) were present. It must not, however, be overlooked that this test would not only throw down sulphuric acid (if present), but also any sulphate which might exist in the vinegar. The presence or absence in vinegar of free sulphuric acid may be conveniently demonstrated by saturating a piece of white paper or of loaf-sugar with the vinegar. On evaporating the vinegar, the paper or sugar will become charred should the acid be present. (For charring tests, see *Year-Book of Pharm., 1878, 174.*) **Dr. A. Jorissen** (in *Journ. de Pharm., d'Anvers 1881, 233*) describes a new and interesting mode of detecting mineral acids in vinegar. To a mixture of one drop of gurjun oil and 25 drops of glacial acetic acid, one drop of vinegar is added, and after agitation, four to six drops of ordinary acetic acid is added. No reaction takes place if the vinegar be free from mineral acids, but if these be present, a violet colour is produced, which does not disappear on the addition of alcohol.

In commerce the strength of vinegar is determined by the number of grains of dry carbonate of soda required to neutralise 1 fluid ounce. This is he practice in England, but in the U. S. America carbonate of potash is used. The strength of vinegar may be accurately determined by means of a standard solution of bicarbonate of potash, one fluid ounce of vinegar becoming saturated with 35 grains of the potash salt. After filtration the remaining liquid should now be quite free from acidity; if not, this would prove the presence in the vinegar of other acid substances, the presence of which was disguised by the acetic acid.

Vohl (in *Ber. der deutsch. Chem., Ges., November 1877*) has designed a simple contrivance to determine the amount of acetic acid in vinegar, consisting of a flask provided with a $CaCl_2$ tube, closed by a caoutchouc stopper, through which passes a glass rod terminating in a platinum hook, and supporting a tube of sodium bicarbonate. The apparatus is weighed alone, the vinegar added, and after weighing, the bicarbonate is lowered into the liquid. The resultant CO_2, after being entirely removed by suction, is determined by the loss of weight, and the acetic acid calculated therefrom. A more rapid mode of arriving at the same conclusion has been published by **Dr. Jehn** (*Ber. der deutsch Chem., Ges., Dec. 10th, 1877*). This consists in introducing 10 c.c. of the vinegar into a flask containing an excess of sodium bicarbonate, from which the liberated CO_2 is conducted into a second jar filled with water. The fall of the level of water displaced by the carbonic acid will indicate the volume of gas liberated, and from this the percentage of acetic acid contained in the 10 c.c. of vinegar may be determined. The jar may be so graduated as to indicate at once the percentage composition, or the displaced water may be conducted into an accurately graduated cylinder, indicating by the volume of water the same result. (*Archiv. der Pharm., May 1877; Year-Book of Pharm., 1878, 160; U. S. Dispens., 15th Ed., 20.*)

Adulterations.—In addition to sulphuric acid, copper, lead, and tin are commonly met with.

364

Indian Vinegar.

§ "The acidity of vinegar is due to the presence of acetic acid. An impure variety, pyroligneous acid, is a product of the destructive distillation of wood. Glacial acetic acid, the variety which crystallises below 63° Fahr., is obtained by converting the crude acid into a salt, and distilling the purified and dried acetate with concentrated sulphuric acid. Acetic acid is also largely obtained by the oxidation of alcohol. An aqueous solution of pure alcohol will not yield acetic acid on exposure to air, but when mixed with certain easily changeable organic materials, it undergoes the so-called acetous fermentation. But this oxidation can also be brought about by inorganic oxidising agents. According to **Pasteur** the first described variety of fermentative action is due to the presence of a mycoderm, the *Mycoderma aceti*, which, when developed in large quantity, is commonly called "the mother of vinegar." The acidification of wines thus yields "white" and "red" wine vinegar, while beer or malt infusions produce the "brown" malt vinegar of commerce. When solutions of malt or saccharine substances are used for the manufacture of vinegar, the sugar is first converted into alcohol by yeast, and then on exposing the so-formed dilute alcohol to air, oxidation of the alcohol occurs with the formation of acetic acid. As a rule vinegar contains about 5 per cent. of acetic acid.

"In India, vinegar is made from toddy, the fresh juice of the **Borassus flabelliformis** and **Cocos nucifera,** and also from the inspissated juice, jaggery, by dissolving it in water and exposing the solution to air in earthen jars. The dried flowers of the **Bassia latifolia,** *mohwa,* infused in water, also yield a saccharine liquid which readily undergoes the acetous fermentation, and produces an excellent vinegar. Regarding the strength of country vinegars, according to **Dr. Lyon's** experiments, the amount of acetic acid ought to be at least from 4 to 5 per cent. in toddy vinegar. Many samples have, however, been found to contain less than 3 per cent., and in one case the acidity was as low as one and a half per cent." (*Surgeon C. J. H. Warden, Calcutta.*)

The manufacture of vinegar may almost be said to be common to every district in India, especially in Mahomedan centres; the Hindús use very little except as medicine. The alcoholic liquid is placed in enormous earthen jars half imbedded in the soil and left until the acetification has been accomplished. The article produced is very inferior in quality to European vinegar, and is entirely consumed within the country, being largely used in the preparation of pickles, &c. **Mr. Baden Powell** says:

"The vinegar obtained from sugar-cane juice is generally a poor stuff, and does not contain more than 2 per cent. acetic acid; but at some places it is made well, especially at Delhi. A large number of bottles of vinegar sold in the country, with the ticket and capsule of 'Crosse and Blackwell,' are in reality bottles which once contained the real article, but when emptied, are refilled with country vinegar, and sold a little cheaper under the above name. I have seen, however, really excellent vinegar from Peshâwar which was made from grapes; it was quite fit for table use."

There appears to be no export trade in vinegar, and the extent of the import trade cannot be determined, since it is included under the heading of Oilman's Stores.

MEDICINAL USES OF VINEGAR.

MEDICINE. 365

§ "In India, vinegar is made from rice, sugar, various fruits, &c., and is largely used by hakims as a medicinal agent. In the pure state it is escharotic. Applied externally, mixed with sweet oil or water, it is largely used for congestive headaches, and in sunstroke. In catarrh it is used like smelling-salts. As a vapour bath it is useful in reducing the high temperature of fevers. It is extensively made use of by hakims for the destruction of ectozoa and entozoa. The vapour of vinegar applied to the ear is beneficial in earache and deafness. In dyspepsia with foul breath the natives use it internally mixed with salt, and this same mixture, combined with alum, is employed as a dentifrice and astringent for bleeding gums. For sore-throat it is used as a gargle mixed with hot water. Diluted solutions are given as cooling draughts and to quench thirst. Highly prized by natives for reducing obesity. The vinegar made from grapes, mixed with salt, is a local application to the bites of mad dogs; it is also a much-prized remedy for ringworm. A weak solution is applied to burns and scalds. Combined with sulphur it is said to be beneficial in chronic rheumatism and gout. Mixed with sweet oil it is applied locally over rheumatic and stiff joints. If used for any length of time it is an anaphrodisiac." (*Surgeon G A. Emerson, Calcutta.*)

"Vinegar (a wine glassful) is recommended by **Dr. Grigg*** in cases of *post partum* hœmorrhage, but its action is so rapid that he refrains from using it or permitting its use before the placenta is expelled, for fear of causing a retention of that body and making its removal difficult. (See *British Medical Journal, January 12, 1884, p. 56.*)" (*Brigade Surgeon W. H. Morgan, Cochin*).

"I have found it effectual (as recently suggested in *Br. Med. Journal*) in *post partum* hœmorrhage, in dose of a W. F. wine-glassful." (*Surgeon-Major W. Farquhar, M.D., Ootacamund.*) "A solution of borax in vinegar is much used for ringworm in dispensary practice." (*Asst. Surgeon Jaswant Rai, Multan.*)

"Vinegar, prepared from toddy obtained from the Palmyra palm fermentation, is very effectual in checking troublesome hiccough; a diluted solution of vinegar has been frequently used by me to check the intolerable itching in some forms of herpes and with very good effect." (*Hon. Surgeon Easton Alfred Morris, Negapatam.*)

"Mahwa vinegar is particularly useful as a diaphoretic, especially when neutralised by carbonate of ammonia, a cheap and valuable kind of Liquor Ammoniæ Acetatis." (*A Contributor.*) "Also used in cases of cholera, diluted with water, in the form of a drink; and for this purpose, it is prepared from sugar-cane juice by keeping it exposed to the sun,

* Physician to Queen Charlotte's Hospital.

till fermentation ensues." (*Assistant Surgeon Anund Chunder Mukerji, Noakhally.*) "The efficacy of vinegar gargle in sore-throat is considerably increased by the addition of a few grains of powdered capsicum." (*Brigade Surgeon S. M. Shircore, Moorshedabad.*)

366

ACHILLEA, *Linn.; Gen. Pl., II., 419.*

A genus of pubescent herbs (belonging to the Tribe ANTHEMIDEÆ of the Natural Order COMPOSITÆ), comprising some 50 species, distributed throughout the north temperate regions of the globe, one species being met with on the Himálaya from Kashmir to Kumaon. *Leaves* alternate, narrow, serrulate or pinnatisect. *Flower-heads* small, corymbose, heterogamous and rayed, or homogamous and disciform (or rayless). *Ray-flowers* few, male, rarely neuter, ligulate, short, white, pinkish or yellow. *Disk-flowers* hermaphrodite, tube terete or compressed and 2-winged, base often produced over the top of the achene, limb 5-fid. *Involucre bracts* few, seriate, appressed; *receptacle* flat or elevated, paleaceous. *Anther-cells* obtuse at the base (not produced into tails). *Pappus* absent. *Achenes* oblong or obovoid, dorsally compressed, glabrous, with two cartilaginous wings.

The name **Achillea** is given because of its being supposed to be the plant with which Achilles cured the wounds of his soldiers. By the ancients the aromatic plant used for this purpose received the name of ἀχίλλειος.

367

Achillea millefolium, *Linn.; Fl. Br. Ind., III., 312.*

MILFOIL OR YARROW, *Eng.*; HERBE AUX CHARPENTIERS, MILLEFEUILLE, *Fr.*; SCHAFGARBE, SCHAFRIPPE, *Germ.*; MILLEFOGLIE, *It.*; YERBA DE SAN JUAN, *Sp.*

By the older English writers, this plant received the name of NOSE-BLEED, because the leaves, if inserted in the nostrils, were supposed to cause bleeding to take place.

Vern.—*Rojmari*, BOMB.; *Biranjasif*, CUTCH. **Stewart** says this is one of the plants sold in the bazars under the names *Momádrú chopándiga*, KASHMIR; *Búi máderán*, AFG.

References.—*Dymock's Mat. Med. W. Ind., 356; U. S. Dispens., 15th Ed., 1560; Bentl. & Trim., III., 153; Waring, Pharm. Jour. Ser. 2, V., 504; Proc. Amer. Pharm. Assoc., 1862, 113; Watt's Dict. Chem., I, 36.*

Habitat.—A native of North Asia, Europe, and America, and of the Western Himálaya, from Kashmír to Kumaon; altitude 6,000 to 9,000 feet. Common on the hills a little to the north of Simla.

Botanic Diagnosis.—A small, herbaceous plant, $\frac{1}{2}$ to $1\frac{1}{2}$ feet in height; glabrous or pubescent, with leaves 2-6 inches long. *Leaves* oblong-lanceolate, 3-pinnatisect, minutely divided into linear, dentate, mucronate segments. *Flower-heads* corymbose ovoid, shortly pedunculate. *Achenes* shining.

Properties and Uses—

MEDICINE. Leaves. 368 Flower-heads. 369

Medicine.—The LEAVES and FLOWER-HEADS are used medicinally as an aromatic stimulant (see **Artemisia vulgaris**). They are also used as a tonic, and in medicated vapour baths for fever.

In Scotland at the present day a warm decoction of the fresh leaves is regarded as a family specific against the colds and other ailments common to childhood. This plant once held a creditable position amongst British drugs, and its recent introduction into the American Pharmacopœia may have the effect of reviving its use in England. It might, with great advantage, be added to our list of Indian indigenous drugs. Formerly it was much used in England as "a vulnerary, and was given internally for the suppression of hæmorrhages and of profuse mucous discharges. It was employed also in intermittents and as an antispasmodic in flatulent colic and nervous affections. Its hot infusion is used as an emmenagogue in France, and also in suppression of the lochia; it is sometimes employed in low exanthematous fevers with difficult eruption. In these cases it probably acts as a stimulant sudorific, as do most aromatic herbs. In some parts of Sweden it is employed as a substitute for hops in the preparation of beer, which it is thought to render more intoxicating." (*U. S. Dispens., Ed. 15th, 1560.*) (For this curious property compare remarks under **Absinthe** and **Artemisia Absinthium,** *Linn.*)

§ "Carminative, dose 5 to 30 grains. (*Surgeon W. Barren, Bhuj, Cutch.*)

Chemical Composition.—**Von Planta-Reichenau** found in this species as also in **A. moschata** (the *Iva* of Europe), a bitter, aromatic, bluish-green, volatile oil, *Ivaol* ($2C_{12}H_{20}O$), a substance faintly resembling oil of peppermint and a peculiar nitrogenous principle *Moschatin* ($C_{21}H_{27}NO_7$) (*Ann. Chem. Pharm., CLV., 145*). The aromatic property is strongest in the flowers and the astringency in the leaves. **Zanon** found in addition to the volatile oil and *moschatin*, a third substance which has been called *Achillein,* a compound which has been determined to be composed of $C_{20}H_{38}N_2O_{15}$ (*Ann. Chem. Pharm., LVIII., 21, and CLV., 1870*), a bitter principle soluble in water, but with difficulty in absolute alcohol. **S. de Luca,** experimenting with **A. Ageratum,** has found similar results. This plant, when rubbed between the hands, gives out an aromatic camphorous odour, and if distilled in a current of steam, furnishes an essential oil, the composition of which has been found to be $C_{26}H_{44}O_8$ (*Year-Book, Pharm., 1876, 43; 1881, 156*).

370

Dr. Dymock says that the flowers of **A. Santolina,** *Stocks* (? *Linn.*), are used in the Bombay Presidency, and are known as *biranjasif, bui-mád-eran,* PERS. This plant is apparently imported into Bombay from Persia, where it is called *dermeneh,* or vulgarly *varek* or *yoshen.* It is not a native of Egypt.

Domestic Uses.—A large number of species and cultivated forms of **Achillea** are met with in gardens in Europe, many of them forming highly ornamental foliage clumps for border and bed cultivation. They are propagated by root divisions, cuttings, and seeds.

Ornamental foliage. 371

Achilleinum—A spirit, distilled from **Achillea millefolium,** is used by the Italians in intermittent fever.

Achilleinum spirit. 372

ACHIMENES, *P. Br. Hist. Jam., 271; Gen. Pl., II., 998.*

373

A genus of elegant, villose herbs (belonging to the Natural Order GESNERACEÆ). They are all favourites of the modern gardener, and his art has perhaps done more to multiply the cultivated forms of this than any other genus of similar size. They are occasionally met with in our Indian orchid houses, but require great care: they are natives of tropical America. The Natural Order GESNERACEÆ is sub-divided into three great sections or sub-orders. The Indian indigenous examples belong to the CYRTANDREÆ (characterised by having ex-albuminous seeds contained within a contorted capsule or berry). The ACHIMENES belong to the sub-order characteristic of America, *viz.*, GESNERACEÆ, recognised by having albuminous seeds, contained within a capsular fruit, semi-inferior or inferior.

In **Achimenes** the *flowers* are large, axillary, variously coloured. *Corolla* tubular, straight or curved, often obliquely dilated at the mouth. *Anthers* in fours, connivent or coherent, included within the corolla, the *filaments* being attached to the base of the tube. *Disk* annular, entire or 5-lobed. *Ovary* inferior, cohering to the base of the calyx; *style* elongated; *stigma* dilated and concave or sometimes distinctly 2-lipped. Chiefly

Ornamental plants. 374

tomentose herbs with spherical rhizomes, from which they are readily propagated.

By some authors the genus has been broken up into various genera, of which perhaps **Tydæ** is the most deserving of an independent position. The following may be mentioned as a few of the more important species generally met with in cultivation : **A. coccinea,** especially *var.* **major ; A. Escherii ; A. floribunda elegans ; A. formosa ; A. grandiflora ; A. Jayii ; A. longiflora ; A. Mountfordii ; A. patens ; A. pedunculata ;** &c. They are perennials, the leaves dying annually. When the old rhizomes commence to give off-shoots, these should be collected and planted six in a pot in a soil composed of equal parts of loam, leaf-mould, and sand.

375

ACHRAS, *Linn. ; Gen. Pl., II., 657.*

A genus comprising only one or two species (belonging to the Natural Order SAPOTACEÆ). By *DeCandolle's Prodromus* this genus was reduced to Sapota, but Bentham and Hooker in their *Genera Plantarum* have restored it. The genus may be briefly defined as having the *flowers* clustered in the axils of the leaves, 6-merous, with leafy *staminodes* inserted upon the lower part of the corolla, and nearly as large as the petals. *Stamens 6 ; filaments* complanate-subulate, opposite the petals. *Fruit* superior, 8-10 celled, with one large, erect, albuminous seed in each cell.

376

Achras Sapota, *Linn. ; Fl. Br. Ind., III., 534.*

The SAPODILLA PLUM of the WEST INDIES ; the BULLY TREE, or NEESBERRY.

Syn.—MIMUSOPS MANILKARA, *Don.*

Vern.—*Sapotá,* HIND. and BENG. ; *Chikalí,* BOMB. ; *Shímai-eluppai,* TAM. ; *Síma-ippa,* TEL. ; *Kumpole,* KANARA ; *Chakchakóti-kajhár,* DUK. ; *Twotta pat,* BURM.

The Tamil name should not be confused with the names applied to MIMUSOPS and to BASSIA.

Habitat.—Introduced from America, and now cultivated throughout India. Roxburgh speaks of the Chinese specimens not flowering in the Calcutta Botanic Gardens, while the West Indian plants growing alongside of them were doing so. It would thus appear that the plant apparently reached India by way of China as well as direct from the West Indies.

Fruit. 377

An evergreen tree, with dome of dark-green shining leaves. The fruit is about the size of a hen's egg, and much of the same shape, dark brown, with a sort of mealy surface. Pulp greenish brown, with generally 2 or 3 seeds developed. When ripe, and just before it becomes over-ripe, the fruit is often very pleasant, although little eaten by Europeans in India. Commonly sold in the streets of Calcutta, about the beginning of the hot season, under the name of *Mangosteen,* a fruit which, at first sight, it somewhat resembles. The absence of the sessile peltate stigma, so characteristic of the apex of the Mangosteen fruit, will at once remove the delusion, and expose the impostor, who makes a large profit by selling his fruit under a false name. "A fine evergreen tree, producing delicious fruit." (*Baron Von Mueller.*)

MEDICINE. Bark. 378

Medicine.—The bark was formerly regarded as a good and useful substitute for Cinchona.

TIMBER. 379

Structure of the Wood.—Kurz says : "Wood uniformly brown, close-grained, rather light, hard ; valued in South America for the shingles of corn-houses." Wood "reddish brown, hard, heavy, and very durable (Bullet or Bully-wood of Central America and the West Indies.)" (*Brandis.*) "Wood is dull red, short, but straight in the grain and very dense." (*Bomb. Gaz., XV., Part I., 61, Kanara.*)

ACHYRANTHES, *Linn.; Gen. Pl., III.*, 35.

380

A genus of herbs or small shrubs (belonging to the Natural Order AMARANTACEÆ), found in the tropical and sub-tropical regions of the Old World.

Leaves opposite, petiolate, ovate oblong or lanceolate, entire. *Flowers* hermaphrodite, deflexed and compressed to the rachis, arranged in elongated, loose, terminal, simple or paniculate spikes, white or coloured. *Bracts* 3, the central large, in contact with the flower, the lateral spreading, spiny. *Perianth* of 4 to 5, green, rigid coriaceous; *sepals* sub-equal, subulato-lanceolate, aristato-acuminate, glabrous or pilose. *Stamens* 5, rarely 2 or 4; *filaments* subulate above the base, membranous, united into a cup; *anthers* oblong, 2-locular. *Ovary* oblong, sub-compressed, glabrous; *style* filiform; *stigma* capitate; *ovule* one, suspended. *Seeds* inverted, oblong; *testa*, thin, coriaceous; *arillus* absent; *albumen* farinaceous, *radical*, erect.

The name **Achyranthes** is derived from ἄχυρον, chaff, and ἄνθος, a flower or blossom, in allusion to the appearance of the flowers. There are in all about 12 species in the world.

Achyranthes alternifolia, *Linn.*; AMARANTACEÆ.

Syn. for **Digera muricata**, *Mart.*, which see.

A. aquatica, *Roxb.*; by **Wallich** formed into the genus CENTROSTACHYS, but by *Genera Plantarum* has been again reduced to ACHYRANTHES.

381

A. aspera, *Linn.*

382

THE PRICKLY CHAFF-FLOWER.

Vern.—*Apáng*, BENG.; *Apáng*, ASS.; *Latjirá, chichra, chirchira, chirchitta*, HIND.; *Aghàda*, BOMB., MAR.; *Apámárga, ághàta, apángaka*, SANS.; *Utta-réni, antisha, apa márgamu*, TEL.; *Ná-yurivi*, TAM.; *Kutri*, PB.; *Kataláti*, MAL; *Utráni-gida, uttaráne*, KAN.; *Aghedo*, GUJ.; *Atkumah*, ARAB.; *Khâre-vázhún*, PERS.; *Kiva-lá-mon, kune-lá mon*, BURM.

References.—*DC. Prod. XIII., II. 314; U. C. Dutt, Sans. Mat. Med., 221; Dymock, Mat. Med. W. Ind., 538; Roxb., Fl. Ind., Ed. C.B.C., 226; Voigt's Hortus Suburb. Calc., 318; Drury's Us. Pl., II; Pharm. Ind., 184.*

Habitat.—A shrub, 3-4 feet high; found all over India, ascending to 3,000 feet in altitude. A troublesome weed in gardens.

Botanic Diagnosis.—Stem erect, striated. *Leaves* ovate-obtuse, acuminate, base cuniate, petiole short, pubescent from a coat of long simple hairs.

Properties and Uses—

The Ash as a Mordant.—The ashes of this plant are used as an alkali in dyeing.

DYE. 383

Medicine.—THE WHOLE PLANT has astringent and diuretic properties assigned to it. Of the former property little is known for certain, but it is said to be successfully used in native practice in the treatment of menorrhagia and diarrhœa. It is reported to be used by the women of Bengal to produce abortion, and holds a high reputation for this purpose, but it most probably acts mechanically, the prickly flowering spikes being inserted into the uterus. A correspondent informs me that if THE JUICE of the plant be injected into the os uterus labour pains will be set up rapidly. On the other hand A DECOCTION of the plant is highly spoken of as "a laxative and promoter of secretions; it is used in combination with other medicines of its class in ascites and anasarca." (*U. C. Dutt, Civil Medical Officer, Serampore.*) **Dr. Bidie** says: "Various English practitioners agree as to its marked diuretic properties in the form of a decoction." **Dr. Cornish** reports favourably, having found it efficacious in the treatment of dropsy. It possesses valuable medicinal properties as a pungent

MEDICINE. 384 Seeds. 385 Spikes. 386 Decoction. 387

and laxative, and is considered useful in dropsy, piles, boils, eruptions of the skin, &c. THE SEEDS and LEAVES are considered emetic, and are useful in hydrophobia and snake-bites. (*T. N. Mukharji's Amsterdam Catalogue.*)

As a POWDER the dried plant is given to children for colic, and also as an astringent in gonorrhœa (*Stewart's Panjáb Plants*).

Major Madden says that the FLOWERING SPIKES are regarded as a protective against scorpions, the insects being paralysed through the presence of a twig. **Dr. Shortt** reports on its use as an external applicant in the treatment of the bites of insects, and **Mr. Turner** calls attention to it as a remedy in snake-bite (*Pharm. Indica*).

Ash. 388

THE ASH yields a large quantity of potash, rendering it useful in the arts as well as in medicine. Mixed with orpiment it is used externally in the treatment of ulcers, and of warts on the body (*U. C. Dutt, Civil Medical Officer, Serampore*). Sesamum oil and the ash (*apamarga taila*) are used in the treatment of disease of the ear, being poured into the meatus. As an ash, however, there seems no reason to think it possesses any virtues other than those of the simple alkali of our shops.

The seeds are given in cases of hydrophobia and snake-bite, the juice of the flowering spikes for scorpion-bite, and the ashes of the plant have been successfuly used in dropsy. (*Bomb. Gaz., VI., 14.*) "In Western India the juice is applied to relieve toothache." (*Surgeon-Major W. Dymock, Bombay.*)

The drug is best administered in the form of a decoction prepared by boiling two ounces of the fresh plant in a pint and a half of water till reduced to one pint, then straining. Of this the dose is two fluid ounces or more, should the diuretic operation be desired.

389

Special Opinions.—§ "The seeds and also leaves relieve the pain of scorpion-bites and allay the irritation of boils and of pleurodynia. The juice of the leaves is useful in snake-bite. The dried leaves are smoked in asthma and they are also used internally in dropsy: dose 4 grains." (*Surgeon J. McConaghey, Shahjahanpore.*) "Is found highly useful in the treatment of general dysentery. (*Pharm. Ind., 184.*)" (*Surgeon H. McCalman, M.D., Bombay.*)

"No protection against scorpions. I have tried the flowery spikes, the root and branches, but the paralysing effect did not follow; the insects ran busily so soon as the pressure of the twig or root, &c., was removed." (*Surgeon B. Evers, M.D., Wardha.*)

"This is found useful as a diuretic in dropsical affections, and has been freely used in the hospital here in combination with other diuretics and tonics. "*Preparations:* Take of root with stems and leaves one ounce, water ten ounces, boil for fifteen minutes in a covered vessel and strain. *Dose*—From one and a half to two fluid ounces twice or thrice a day." (*Apothecary J. G. Ashworth, Kumbakonum.*) "The fresh leaves bruised into a paste and mixed with black pepper and garlic are used as an antiperiodic in the form of pills. It is given before the attack comes on." (*Surgeon-Major John Lacaster, M..B., Chittore.*) "Decoction of the entire plant is useful as a diuretic in dropsical affections; rubbed into a paste with a little water it forms a favourite application to stings of wasps, bees, and other insects." (*Assistant Surgeon Shib Chunder Bhuttacharji, Chanda, Central Provinces.*)

390

"In simple anasarca I have found it of marked benefit." (*Surgeon-General William Robert Cornish, F.R.C.S., Madras.*) "The tender leaves ground into a paste with a little sugarcandy, and with the addition of a little butter fried to a proper consistence it is useful in the early stage of dysentery." (*Surgeon-Major D. R. Thompson, M.D., C.I.E., Madras.*)

"The extract is used in cases of dropsy and gonorrhœa mixed with

MEDICINE.

the extract of white *rati*." (*Surgeon W. Barren, Bhuj, Cutch.*) "The seeds are supposed to impair appetite and are given in the form of *congee* for excessive hunger." (*Surgeon-Major J. Robb, Ahmedabad.*)

"It is useless in snake-bite." (*Surgeon V. Richards, Goalundo, Bengal.*) "I have tried this largely in the shape of a strong decoction in two-ounce doses in dropsy, especially of malarial origin, and found good results as a diuretic." (*Assistant Surgeon Debendro Nath Roy, Sealdah, Calcutta.*)

"If the bark of the root is mixed with an equal quantity (say five grains) of black pepper, it may be given in cases of intermittent fever with good results." (*Surgeon W. Wilson, Bogra.*) "The fresh juice of the leaves thickened into an extract by exposure to the sun and then mixed with a little opium, may be beneficially applied to primary syphilitic sores." (*Surgeon-Major Bankabehari Gupta, Púrí.*)

"Several Native practitioners use the decoction as a diuretic in gonorrhœa and dropsical affections." (*Surgeon-Major R. L. Dutt, M.D., Pubna.*)

Achyranthes ferruginea, *Roxb.;* Syn. for **Psilotrichum ferrugineum,** *Endl.*

A. incana, *Roxb.;* Syn. for **Ærua javanica,** *Juss.*, which see.

A. lanata, *Linn.;* Syn. for **Ærua lanata,** *Forsk.*, which see.

A. lappacea, *Linn.;* Syn. for **Desmochæta atropurpurea,** *DC.*, but by the *Genera Plantarum* it has been reduced to the genus **Papalia,** which see.

A. nodiflora, *Linn.;* Syn. for **Allmannia nodiflora,** *R. Br.*, which see.

A. prostrata, *Linn.;* Syn. for **Papalia prostrata,** *Mart.*, which see.

A. scandens, *Roxb.;* Syn. for **Ærua scandens,** *Mart.*, which see.

A. triandra, *Roxb.;* Syn. for **Alternanthera sessilis,** *R. Br.*, which see.

ACID.

391

Acids are chemical compounds of two or more elements, of which Hydrogen is generally one. A few compounds are known as acids, however, which do not contain Hydrogen, such as Silicic Acid (SiO_2) and CO_2 popularly known as Carbonic Acid. They may be referred to two great sections—acids which contain oxygen and hydrogen in combination with one or more elements, and acids which do not contain oxygen. They generally possess an acid taste, turn blue litmus into red, and neutralise the basic oxides, forming salts. When basic oxides form with water soluble hydroxides, these compounds are alkaline in their reactions and antagonistic to acids: they transform the acid red of litmus into the alkaline blue. An acid may be defined as a hydrogen compound, which has the power of combining with basic oxides to form salts by the partial or complete displacement of its hydrogen. In the commoner salts the basic principle is a metallic oxide. Conversely, salts are binary compounds or compounds composed of two principles, *viz.*, a basic oxide and an acid.

The reader is referred to works on chemistry for an account of the properties and uses of acids, a good many of which are now being prepared in India, and a large import trade exists in others.

ACIPENSER.

Acipenser huso, *Linn.;* PISCES.

392

THE SOURCE OF ISINGLASS.

ISINGLASS. 393

Vern.—*Machchhi-ká-sirish,* HIND., DUK.; *Ghirriyus-samak,* ARAB.; *Sireshame-mahi,* PERS.; *Min-vajjaram,* TAM.; *Chepa-vajra mú,* TEL.

The name of the fish is *Sek máhr;* isinglass *is Serishom—Sek máhr.* (*Dr. C. Rice, New York.*)

A sturgeon inhabiting the Caspian and Black Seas.

The swimming bladder or sound is cut up into shreds forming the Isinglass of commerce. This is insoluble in cold water, but when boiled it is completely soluble, and on cooling forms a beautiful jelly. Fifteen grains of Isinglass are sufficient to form a consistence to one ounce of water.

MEDICINE. 394

Medicine.—It is demulcent and nutritive. It is also used as a test to distinguish gallic from tannic acid, the latter becoming yellow. As a substitute for isinglass see **Gracilaria lichenoides.**

A very extensive trade is done from India in what is known, in our Trade and Navigation Returns, as "Fish Maws and Shark-fins." During the year 1883-84 the exports under this head amounted to 1,612,014 cwts., valued at R11,23,254. How much of this could be said to be isinglass is impossible to ascertain. The imports are chiefly from Arabia, and the bulk of the exports are to China. The finer qualities of fish maws doubtless find their way into the isinglass market.

For further information see under "Fish Maws."

395

§ "Indian isinglass, which is of very good quality, is got from a species of Polynemus. (*See Royle's pamphlet and Day's Fishes.*)" (*Deputy Surgeon-General G. Bidie, Madras.*) "In the last July sales of Mincing Lane 380 packages of East Indian isinglass were offered, comprising 192 cases Penang, 127 cases Bombay and Kurrachee, and 61 cases Saigon. Bombay *tongue,* good to fine, fetched from 3*s.* 5*d.* to 3*s.* 11*d.* per lb. The qualities in the market are: *1st Leaf, 2nd Tongue, 3rd Pipe, 4th Purse.* The *Purse* is worth 1*s.* to 1*s.* 6*d.* per lb. only. A reference to the Mincing Lane Reports will show that the isinglass of commerce is practically all East Indian (at any rate as far as the London market is concerned). The sales are sometimes over 500 packages at one sale. In the Indian Trade Returns isinglass is entered as "Fish Maws;" it is obtained from various kinds of large fish. The trade names depend upon the shape when dried. Russian and Brazilian isinglass would seem to have found another market." (*Surgeon-Major W. Dymock, Bombay.*)

396

ACONITUM, *Linn.; Gen. Pl., I., 9.*

A genus of perennial herbs (belonging to the Natural Order RANUNCULACEÆ and the Tribe HELLEBOREÆ), comprising about 180 species, inhabitants chiefly of the northern temperate zones, some seven species being met with on the Himálaya.

Leaves **alternate, palmi-partite, rarely entire.** ***Flowers*** **irregular, racemed, blue, purple, white, or yellow.** ***Sepals*** **5-petaloid, the** ***helmet*** **or posterior sepal convex or vaulted, the others flat, the two anterior ones being narrower than the lateral.** ***Petals*** **2-5, the two posterior ones stalked, (*clawed*)** ***limb*** **hooded and formed somewhat in the shape of a hammer, concealed within the helmet; the three lower (or anterior petals) small or obsolete.** ***Stamens*** **many.** ***Follicles*** **3-5, sessile.** ***Seeds*** **many, with a spongy, rugose, or wrinkled** ***testa.***

The word Aconitum is ἀκόνιτον, the classical Greek name being derived most probably from ἄκων, a dart, from its having been used to poison darts.

In connection with the subject of the poison used by the Aka hill tribes on the frontier of Assam, I had recently some correspondence with **Dr. Warden**, Professor of Chemistry, Calcutta. The roots which furnish the poison were identified by me as those of a species of Aconitum, most probably indigenous to the mountains bordering on Assam; in other words, they were not the Nepal Aconite which finds its way over the greater part of India. I had also the pleasure of examining a root said to be used by the Akas as an antidote against aconite poisoning. This proved to be the classical Costus root. In a private correspondence upon this subject with **Dr. Dymock** of Bombay, I received a most interesting letter drawing my attention to the numerous references to this same property having been attributed to the Costus root (**Saussurea Lappa**). I take the liberty to extract a passage from this correspondence :—

"Regarding Costus, it was regarded by the Greeks, Romans, and Muhammadans as an antidote to poisons generally. **Dioscorides** says of it, *πινόμενος δὲ ὀυγγίας βπλῆθος, ἐχιοδήκτοις βοηθεῖ*; and again, *μίγνυται δέ καὶ μαλάγμασι και ἀντιδότοις.*" *Celsus* 5·23, *de antidotis* has three receipts, in all of which Costus is an ingredient.

Mir Muhammad Hossein in the *Makhzan* says :—

يکمثقال آن با خمر و انسنتين ترياق سموم و جاذب آنها بسوى ظاهر جلد وجهت رفع سم افعي و عقرب و رتيلا و امثال اينها از سموم قتاله نافع ،

i.e., 'one miskal of it, with wine and wormwood, is an antidote to poisons and draws them to the surface of the skin. It is useful to counteract the poison of the viper, scorpion, and tarantula, and other deadly poisons.' Practically, the antidotes for aconite are diffusible stimulants: costus is a stimulant, and is given as such in cholera by the natives."

The subject of Aconite seemed deserving of a thorough investigation, both with the view of establishing a trustworthy supply of uniform quality for medicinal purposes, and, if possible, of checking the indiscriminate way in which the drug is placed within the reach of persons desiring to use it for criminal purposes. Accordingly, I recently addressed the Government of India, in the Department of Revenue and Agriculture, on this subject, submitting, along with some of the more interesting facts brought to light in connection with the Aka arrow poison, a suggestion to form a Commission of Enquiry. I take the liberty to republish a few passages from that communication :—

"The genus of plants which yields aconite belongs to the poisonous Natural Order RANUNCULACEÆ. The members of that genus are exclusively confined, as far as India is concerned, to the alpine and subalpine regions, chiefly occurring on the Himálaya from Nepal westward to Kashmír. There are in all seven Indian species known to botanists, with two or three varieties under two of these species. One contains no aconitia, *viz.*, **A. heterophyllum**, and is largely eaten as a vegetable or mild tonic. This species is perfectly well known. Of the other species, some are poisonous, others not, and even some of the varieties of one species are poisonous, while other varieties are not. The poisonous forms have never been accurately identified,and the result is, that of a given weight of the root sold in our druggists' shops, a certain percentage frequently contains no aconitia whatever; indeed, an entire consignment may be perfectly inert. This uncertainty renders the use of aconite objectionable, its action not being constant; while it is for many diseases, such as certain forms of malarial fever, and all skin diseases, the most valuable drug known. It is very much to be regretted that so valuable a medicine should thus suffer in consequence of ignorance, and I would, therefore, strongly recommend that a Commission of Enquiry be instituted with the object of determining the following points :—

"*1st.*—The scientific determination of the various species of **Aconitum.** Dried specimens of each species in flower should, for this purpose, be collected, with their roots attached, so that the characters of the roots of each individual species or variety may be determined and clearly described. Were this done, it might be possible to recognise the various forms of aconite sold in the bazars.

"*2nd.*—To chemically determine the average amount of aconitia in each species, and perform a series of experiments in the cultivation of the various species with the object of ascertaining if there is a fixed average percentage, or whether different alkaloids or different proportions of alkaloids are characteristic of certain species. In other words, ascertain whether the presence or absence of aconitia is a specific peculiarity or a formation which may or may not take place within the tissue of an individual plant.

"*3rd.*—To endeavour to ascertain whether the percentage of aconitia can be increased by cultivation; and, if possible, to establish some source from which a constant supply might be obtained. I am not aware whether the Medical Store Department imports its aconite or not; but there does not seem the slightest reason why India might not supply the world with this most valuable medicine, and supply it in such a manner as to do away with the uncertainty which centres around the use of the drug at the present day. Perhaps it might be that the extermination of inert forms from certain limited racts of the higher Himálaya would be all that might be necessary, allowing thereby the aconitia-yielding forms to naturally become more prevalent. All the species are exceedingly plentiful on the alpine Himálaya from Nepal westward to Kashmír, one or two species finding their way eastward to Assam through Sikkim and Bhután.

"*4th.*—To ascertain the areas and statistics of production of the different aconite roots at the present time and the chief centres of exportation.

"In support of these recommendations, I beg to quote one or two of the numerous appeals which have appeared in European medical publications. **Dr. Cook**, in the *Year-Book of Pharmacy for 1873, page 21*, writes as follows: 'In this instance of an important drug involved in mystery, we see the necessity for some official medium through which to prosecute inquiry. Individual effort is insufficient, and the only effectual mode should emanate from the Government of India, to ascertain the areas and statistics of production of the different aconite roots in the Himálaya,—their value commercially on the spot, the native names applied to the different kinds, and the plants producing them properly and satisfactorily identified. Then, as a consequence, the different varieties will be analysed, and the value of each determined according to the amount of alkaloid present. This is but one out of scores, perhaps hundreds, of instances, in which information is required of a special character on the products of our vast Indian Empire, and which no private effort is capable of obtaining.' **Dr. E. R. Squibb** also writes in the same volume: 'Although but few drugs are apparently more cheaply and easily obtained than aconite root, yet perhaps in no other is there so great an amount of uncertainty, many parcels having been found to be comparatively worthless in a medical point of view.'"

In the *Admiralty Manual of Scientific Inquiry* (and republished in *Hanbury's Science Papers*, 187) occurs the following significant interrogation, which, strange to say, remains unanswered: "*Aconite root* has been imported in considerable quantities from India. In what district is it collected, and from what species of *Aconitum?*" This admission of want of definite information regarding the source of Indian aconite was made in 1871 by **Professor Oliver**, and the late distinguished scholar and pharmacologist **Daniel Hanbury**, and it has still to be answered before we can be said to possess any trustworthy data upon which to base a definite and accurate knowledge of what may be justly called India's most valuable indigenous drug.

The following are the principal Indian species with the information which can be gathered regarding each from works on Indian Economci Science:—

Aconitum ferox, *Wall.; Fl. Br. Ind., I., 28.*

397

INDIAN ACONITE.

Vern.—*Bish, (bikh), bis,* (derived from the Sanskrit *Visha*), *bachnak, mítházahar, singyá-bis, singyá, téliyá-bis, bachhnág,* HIND; *Kat bísh,* or simply *bísh,* BENG.; *Bish,* ASS.; *Visha, vatsanábha,* SANS.; *Bish,* ARAB.; *Bishnág,* PERS.; *Bachnág,* MAR.; *Vachnág* or *vachha-nág,* GUJ.; *Buchnága,* CUTCH; *Vasha návi,* TAM.; *Vasanábhi, nábhi,* TEL.; *Valsanábhi,* MAL.; *Vasanábhi,* KAN.; *Vachanabhi,* CINGH.

It is probable that, under the above vernacular names, the majority of the poisonous forms of aconite are sold by our native druggists. Most authors, however, agree in regarding these as more properly belonging to this species. **Dr. Moodeen Sheriff** regards *Jadvár* as the only safe generic name for the species of medicinal aconite. The names *Singyi* or *singyá-bis* and *mítha-zahar* are applied to two forms of Aconite generally referred to this species. (See below.)

References.—*Pharm. Ind., 3 and 434; Flück. & Hanb., Pharmacog.; Bentl. & Trim., Med. Pl., 5; O'Shaughnessy's Bengal Disp., 165; Dymock's Mat. Med., W. Ind., 1; Drury's Us. Pl., 12; Royle's Mat. Med., Ed. by Harley, 776; Groves, in Year-Book of Pharm., 1870; 1873, p. 500; Moodeen Sheriff, Supp. Pharm. of Ind., 25, 32, 265; Cooke, in Pharm. Jour., Ser. 3, Vol. III., p. 563; Duquesnel, De l'Aconitine Crystal, Paris; and Pharm. Jl., Ser. 3, Vol. II., pp. 602, 623, and 662; Alder-Wright, Year-Book of Pharm., Report on the Aconites, 1st article in 1875, p. 514; 2nd, 1876, p. 531; 3rd, 1877, p. 444; 4th, 1878, p. 483; 5th, 1879, p. 417; 6th, 1880, p. 455; and 1881, p. 140.*

Habitat.—Temperate sub-alpine Himálaya, from Sikkim to Garhwál, altitude 10,000 to 14,000 feet.

Botanic Diagnosis.—*Stem* erect, 3-6 feet high, simple below. *Leaves* 3-6 inches, rounded, palmately 5-fid, cut into irregularly indented lobes. *Inflorescence* a terminal dense-flowered raceme, with only one or two branches below; *bracts* at the base of the peduncle leaf-like, smaller upwards, cut or lobed, two small ones usually about half way up the pedicel. *Flowers* on long erect stalks thickened above and glandularly pubescent, large, pale dirty-blue. *Helmet* about twice as long as high, vaulted, with a short sharp beak. *Follicles* 5, erect, usually densely villous and transversely wrinkled. *Seeds* having the testa pitted or plaited.

Differs from **A. Napellus** chiefly in the less divided leaves, denser flowered racemes, and shorter beak to the helmet.

MEDICINAL PROPERTIES.

MEDICINE. Singyi. 398

Medicine.—The mass of the root sold in Indian druggists' shops as Aconite is derived from this species, but several others are no doubt used as substitutes or adulterants. **Dr. Bidie** says the root of **Methonica superba** is used as an adulterant in Madras.

Description of the Root.—Fusiform, 2-5 inches long and ½ to 1½ broad at the top. As sold in the bazars of India, it is often broken through the middle as if from carelessness in digging up, shrivelled longitudinally, and marked here and there with the scars of small detached rootlets. The colour of the dry root as sold by druggists is blackish brown; when broken the fractured surface is compact, hard, horny, somewhat translucent, and of various shades of brown. In the rainy season it becomes moist, and when handled stains the fingers brown. This is the *Singyi* or *singyá-bis* (horny poison) of the Hindús, the *sringí* poison of the Sanskrit authors.

Mitha-Zahar. 399

Dr. Dymock (as also most other writers) has described another form of Indian aconite attributed to this species, *viz.*, with white spongy roots, and

Indian Aconite.

Chemistry.

known in Bombay as "Lahore *Bachnáb.*" This is the *mithá-zahar* (sweet poison) of the older writers. It is generally 1 to 1½ inches long, tapering, compressed, rough and wrinkled. Brown externally, much paler internally than the preceding. This form is said by some authors to contain more aconitia than the brown horny variety; by others it is regarded as decidedly inferior in strength. It is also devoid of the peculiar smell of the former, which resembles hyraceum or castor. It seems probable that this smell, as also the peculiarity of staining the fingers when moist, is due to some mode of drying and rendering the root proof against the attacks of insects, rather than a specific property of the root. For this purpose the roots are often dried over a fire, boiled in cow's urine or milk, preserved in oil, &c. A similar practice prevails in Japan, where the roots of an aconite are said to be preserved in child's urine.

The specific identity of these two forms appears open to grave doubt, however, and, indeed, the chemical nature, structural characteristics, and the association with Lahore would seem to suggest that the white spongy root was much more likely to be obtained from **A. Lycoctonum**—a species plentiful on the north-west Himálaya from Kumaon to Kashmír—than from **A. ferox.** The latter is the characteristic species of the eastern Himálaya, and nowhere occurs west of Garhwál. Being the root of a temperate plant, the "Lahore *Bachnáb*" most probably comes from Kashmír and the surrounding mountains (where **A. Lycoctonum** at altitude 7,000 to 10,000 feet, and **A. Napellus** at 10,000 to 15,000, are very plentiful); indeed, the latter species is one of the commonest plants on all the higher Himálaya from Kumaon westward, but is not met with in the region of **A. ferox,**—the eastern Himálaya.

In European commerce all the Indian forms of aconite are classed as forms of **A. ferox.** This seems an unfortunate mistake, the more so, since it is by no means the most plentiful, and certainly not the most accessible, species. It may be quite wrong to limit the dark-coloured form to **A. ferox,** but it seems only natural to expect that, should the suggestion to form a Commission of Enquiry into Indian aconite be acted upon, we shall be able to refer the numerous forms met with in commerce to distinct species. Indeed, we shall very likely discover that **A. Napellus** supplies a far larger proportion of our Indian Aconite than we had any idea of before.

CHEMICAL COMPOSITION.

400

The roots of this species contain, relatively to the roots of the other species of aconite, a much larger amount of *Pseudo-aconitine* ($C_{36}H_{49}NO_{12}$) (or *Nepaline*), a much smaller quantity of *Aconitine* ($C_{33}H_{43}NO_{12}$), and an amorphous alkaloid analogous to, but not identica with, that found in **A. Napellus.** This alkaloid is non-crystalline, and yields non-crystalline salts. **Wright** considers pseudo-aconitine as nearly related to the opium alkaloids narceine, narcotine, and oxy-narcotine, which, like pseudo-aconitine, all give rise to derivatives of dimethyl-proto-catechuic acid. Napelline (or Nepaline, or pseudo-aconitine) was discovered originally by **Hübschmann** in 1857. It is a white powder, in the form of transparent needle-crystals, with a bitter burning taste, having a strong alkaline reaction, forming with difficulty crystallizable salts. It is readily soluble in water, alcohol, or chloroform, but insoluble in ether. According to **Prof. Flückiger** the most characteristic features of pseudo-aconitine, separating it from Aconitine proper, is the absence of bitterness and its ready solubility in water. The physiological properties of this alkaloid have been carefully investigated by **Bœhm and Ewers,** who state that it exerts an influence similar to that of aconitine, except that it is more powerful. It has entered largely into the composition of English commercial acon-

MEDICINE.

itine (aconitia) owing to the Nepal aconite being so largely used in the preparation of that alkaloid. (Conf. with chemistry of the compounds derived from **A. Napellus.**)

Until the suggestion contained above regarding the black, horny, and white spongy roots has been proved correct by original investigation of the living plants, it would, perhaps, be unwise and prove confusing to depart too far from the universally accepted notion that they are both forms of **A. ferox.** **Dr. Royle,** in his *Materia Medica* (*Ed. J. Harley*), says: "**Dr. Headland** found in several experiments, the results of which were uniform, that while from 54 to 56 grains of aconitia could be obtained from one pound of the horny root, 88 to 96 grains were extracted from a pound of the friable root."

MEDICINAL USES.

The root is commonly regarded as much more powerful than that obtained from **A. Napellus.** On this account it is chiefly recommended in the manufacture of preparations to be used externally. Experimenting with the poisonous properties of the aconites, **Ewers** discovered that the root of **A. ferox** is much more virulent than **A. Napellus.** By the natives of India the *bish* (as sold in the bazars) is used extensively, both externally and internally, but owing to the want of definite knowledge regarding the species which afford this drug, much of what has been written by the older authors, or, indeed, can at present be written regarding it, may have to be rearranged and placed under other species.

It is a very effective medicine in various diseases, acting as a narcotic sedative, regarded as heating and stimulant, useful in fever, cephalalgia, affections of the throat, dyspepsia, and rheumatism. *Bish* appears to have been known to the Hindú doctors from the earliest ages. It is much used as an external application, the root being formed into a paste (*lép*) and spread upon the skin in neuralgia, boils, &c. Internally, it is chiefly used in the treatment of chronic intermittent fevers (*Dymock*). Europeans use it as a substitute for true aconite. In a recent correspondence between the leading members of the Indian Medical Department and the Government of India, several Provincial Committees and distinguished officers recommended that this root should be substituted for **A. Napellus** in the officinal preparations. The Bombay Committee recommended that this should be done chiefly with external remedies, adding that "it must not be used for internal administration in the same doses, the alkaloid *Nepaline* being much more powerful than the aconitine of **A. Napellus.**" Its therapeutic uses are also defined "externally to relieve neuralgia, rheumatism, gout, &c.; internally, to control the action of the heart when increased by disease, and to relieve pain in rheumatism. In native practice *Bachanága* is used in combination with cinnabar, sulphur, borax, and aromatics, in extremely small, almost homœopathic, doses, in intermittent fevers and common coughs, with considerable success." (*Brigade Surgeon H. V. Carter, President, and Surgeon-Major W. Dymock and Assistant Surgeon Sakhárám Arjun, Members of Committee, Bombay, 20th May 1879.*) Before, however, the bazar aconite is substituted for the aconite of European commerce, it seems highly desirable that the identity of the Indian forms be thoroughly established, and, if possible, some arrangement made by which a full and uniform supply of the best Indian article may be ensured.

Special Opinions.—§ "The root is very useful in the form of liniment in cases of neuralgia and muscular rheumatism." (*Surgeon S. H. Browne, M.D., Hoshangabad, Central Provinces.*) "Tonic and antiperiodic (dose $\frac{1}{60}$ to $\frac{1}{30}$ grain). Used by hakims in the form of pills called *anand* *chairawa*, which are made up of sulphide of mercury, *buchnága, tankana, khar, pimplee*, and mucilage (gum)." (*Surgeon W. Barren, Bhuj, Cutch.*)

"I have tried it and found it utterly useless in cobra poisoning. Indeed, without having any influence whatever upon the lethal effects of cobra poison, in fact it certainly very much increased the severity of one of the most marked symptoms of cobra poisoning, *viz.*, salivation. It would appear to be an aphrodisiac of some power. I believe it to be a very poor substitute for true aconite." (*Surgeon V. Richards, Goalundo, Bengal.*)

"An oil is extracted from the root and used for rheumatism as an external application." (*Surgeon-Major J. Robb, Ahmedabad.*)

"Used by natives in fevers attended with constipation; it enters in the composition of purgative pills, containing cinnabar and Indian calomel, otherwise called '*Rasacurpooram.*'" (*Surgeon-Major J. F. Fitzpatrick, M.D., Coimbatore.*)

"The fresh root is given in small quantities internally in gonorrhœa." (*Surgeon-Major D. R. Thompson, M.D., C.I.E., Madras.*)

"Found growing wild in Kalahandi on hilly places, used by the Khonds to poison their arrows." (*Assistant Surgeon Shib Chunder Bhattacharji, Chanda, Central Provinces.*)

"**Aconitum ferox.**—There are several varieties of aconite root met with in Southern India, the most common of which are those which, according to their colour or taste, are known as the *white* or *sufed-bachnág*, the *reddish brown* or *lál-bachnág*, the *black* or *kála-bachnág*, and the *sweet* or *míthá-zahar*. The white and reddish-brown varieties can be used internally. They are very useful as sedative, nervine, and alterative tonics in medicinal doses, but a virulent poison in large ones. A few years ago I took the white variety myself in small quantities, and found that its internal use is not attended with more danger than that of the European drug (**A. Napellus**). Since that period I have employed it very extensively in my practice, and do not hesitate in saying that it is one of the most useful medicines in India. Its beneficial influence over diabetes is very remarkable, the immoderate flow of urine beginning to diminish from the very day of its use, with a proportionate decrease in the saccharine matter. Its control over spermatorrhœa and incontinence of urine is equally great. It has lately been found useful in some cases of paralysis and leprosy. The advantages of this drug over all other varieties of the Indian Aconite root are that it is not only much milder but also more certain and uniform in its actions. The white and hard variety which I am speaking of is quite different from the white and spongy variety mentioned in some books. I have also used the reddish-brown variety pretty extensively, and with almost the same results. The above roots are best used in the form of powder with some inert or farinaceous substance, as follows: Take of the white or the red variety of the aconite root in powder, one ounce; arrowroot or wheat-flour, seven ounces. Mix them thoroughly, pass the powder through a fine sieve and rub it lightly in a mortar and keep it in a bottle. The roots can also be employed in the form of tincture, but the powder I have just described was so convenient and cheap, and proved so successful, that I did not think it necessary to resort to any other form. The dose of the powder is from two to six grains, gradually increased, three times in the twenty-four hours; the average and usual dose being four grains." (*Honorary Surgeon Moodeen Sheriff, Khan Bahadur, Madras.*)

"Said by hakims to be useful in large doses, along with stimulants, in cases of snake-bite and scorpion-stings; it is aphrodisiac. Very useful for reducing the temperature in fevers." (*Surgeon G. A. Emerson, Calcutta.*)

"Antiperiodic, alterative, and expectorant, used as a nervine tonic in cases of paralysis. Used as an external application to chronic sores.

Natives prepare it by boiling the root in milk for half an hour, repeating this process seven times, and afterwards pulverise. This process is said to reduce the poisonous effects of the drug. Dose $\frac{1}{20}$ to $\frac{1}{16}$ of a grain." (*Surgeon J. McConaghey, Shahjahanpur.*) "A very useful anodyne liniment is prepared by heating an ounce of coarsely-powdered aconite root in half a seer of linseed oil." (*Assistant Surgeon Mokund Lal, Agra.*) "Said to be used in bronchitis and asthma; no personal experience." (*Surgeon J. Parker, Poona.*)

"While serving at Buxa, Bhután, a woman attempted to poison her husband by means of the root given in a curry. The symptoms were well marked, but he recovered." (*Surgeon L. Cameron, Nuddea.*) "A tincture of this drug acts like the true aconite, hence it is admissible in inflammations and fevers like the European drug." (*Surgeon-Major R. L. Dutt, M.D., Pubna.*) "Useful not only in chronic but in acute intermittent fever, during the hot stage, also in continued fever, as well as in all neuralgic affections. (*Brigade Surgeon S. M. Shircore, Moorshedabad.*)

Aconitum heterophyllum, *Wall.; Fl. Br. Ind., I., 29.*

401

Syn.—A. CORDATUM, *Royle;* A. ATEES, *Royle.*

Vern.—*Atis, atviká,* HIND., BOMB.; *Ataicha, ativishá,* SANS.; *Vajjeturkí,* PERS.; *Ati-vadayam,* TAM.; *Ati-vasa,* TEL.; § "*Mohand-i-gúj saféd, hong-i-saféd,* KASHMIR; *A'is,* BHOTE" (*Dr. Aitchison*); *Súkhihari, chitijari, patrís,* or *patís, bonga,* PB.; *Atavishni-kalí, ativish* or *ativakh,* GUJ.; *Atavish,* MAR. (*Assistant Surgeon Sakharam Arjun Ravat, Bombay*); *Ativista,* CUTCH.

Moodeen Sheriff cautions the use of the term *ati-visha* (*ati,* great, and *visha,* a poison) as applied to this plant, and thinks that it should be restricted to the poisonous forms. The Telegu name *Ati-vasa* (*Ati,* great, and *vasa,* the sweet-flag) is given in allusion to its supposed resemblance to the rhizomes of **Acorus Calamus.** The Arabic word *Jadvár* is the only safe one in ordering the non-poisonous forms of aconite, much safer than the Hindustani *Nirbisí* (*Nir,* free from, *bisí,* poison), because of the latter having been applied by modern usage to many other things (*conf.* **Curcuma.**)

References.—*Royle's Ill., 56, t. 13; Bentl. & Trim. Med. Pl., 7; Flück. and Hanb., Pharmacog., 14; Pharm. of Ind., 4; O'Shaughnessy's Beng. Disp., 167; Dymock's Mat. Med., W. Ind., 4; Alder Wright,* in *Year-Book Pharm., 1879, 422; Agri.-Hort. Soc. of Ind. (1857), XVI., Sec. I., p. 311.*

This is apparently **Caltha Nirbisa,** *Ham.,* and **Nirbisia Hamiltonii,** *Don;* it is most probably the species of Aconite to which the vernacular *Nirbisia* belongs. **Hamilton** says it is in Nepal called *Nirbishí* or *Nirbechí.* (See the concluding para. upon adulteration.)

Habitat.—West temperate Himálaya, from Kumaon to Hasora, altitude 8,000 to 15,000 feet, very plentiful in the neighbourhood of Simla; very common on the Sach Pass, Chumba, along with **A. Napellus.** Altitude 7,000 to 15,000 feet.

Botanic Diagnosis.—*Stem* erect, leafy, 1-3 feet, simple or branched from the base, glabrous below, puberulous above. *Leaves* 2-4 inches, broad ovate or orbicular-cordate, more or less 5-lobed and toothed (inciso-crenate), acute or obtuse; upper shortly stalked or sessile, not lobed, amplexicaul, thick, bright green, pale beneath, lower long-petioled. *Raceme* often panicled, many-flowered. *Flowers* on long rufous-pubescent peduncles, more than an inch long, bright blue or yellow-greenish with purple veins; *helmet* shortly beaked, half as high as long. *Follicles* 5, downy. *Seeds* angled but with a smooth testa. In form of leaf it varies considerably, hence the specific name **heterophyllum.**

Description of the Root.—Ovoid-conical, tapering to a point from $\frac{1}{2}$ to $1\frac{1}{2}$ inch long, and $\frac{1}{8}$ to $\frac{1}{2}$ or more thick. Externally light ash-coloured, wrinkled and marked with the scars of the fallen rootlets, with a rosette of scaly rudimentary leaves on the top. A transverse section shows the root to consist of a pure white, friable, amylaceous substance, marked towards the centre by from 4 to 7 concentrically arranged yellow dots corresponding to the ends of the fibro-vascular bundles which traverse the root longitudinally. In taste the root is simply bitter with no acridity. It has no odour, and may be distinguished from other roots by its bitterness and absence of tingling sensation when a small portion is eaten.

402

CHEMICAL COMPOSITION.

The composition of this root has not by any means been thoroughly established, but it may be stated that **Broughton** first succeeded in separating what appears to be its active principle, *viz.*, the amorphous alkaloid to which he gave the name *Atisine* and assigned the formula $C_{46}H_{47}N_2O_5$. The exact composition of this alkaloid cannot, however, be said to have been determined as yet, and in the *Year-Book of Pharmacy for 1879*, **Wright** suggests a correction upon **Broughton's** formula as likely to be soon established. He percolated a powdered dry root with alcohol containing a little tartaric acid, and evaporating the percolate, he obtained ultimately **Broughton's** alkaloid atisine. This was uncrystallizable, but with hydrochloric acid and gold chloride, he obtained a crystalline hydrodichloride $C_{22}H_{31}NO_2HCl$, $AuCl_3$ from which he suggests that $C_{22}H_{31}NO_2$ may prove nearer the correct formula for atisine than that given by **Broughton.**

Aconitine has not been found in the *atís* root, or only in very minute quantities, and atisine is not poisonous; it tastes intensely bitter, without the slightest tendency to produce the tingling characteristic of the aconite alkaloids. **Wasowicz** states that it contains the following: (1) a fat, probably a mixture of oleic, palmitic, and stearic glycerides; (2) aconitic acid; (3) an acid related to tannic acid; (4) cane-sugar; (5) vegetable mucilage; (6) pectous substances; (7) atisine; and (8) starch. It contains $\frac{8}{100}$ of 1 per cent. atisine.

MEDICINE.
Grey Roots.
403
White Roots
404

MEDICINAL USES.

The ROOT is pleasantly bitter, and is regarded as a valuable, mild antiperiodic, aphrodisiac, and tonic, checking diarrhœa. It may be administered internally with safety owing to the absence of Aconitia or other poisonous properties. It is specially useful in convalescence after fever. As a tonic the dose is 5 to 10 grains, three times daily, and as an antiperiodic from 20 to 30 grains of the powdered root every three or four hours.

Special Opinions.—§ "The white or common variety of the root of **A. heterophyllum** is a very useful antiperiodic and antipyretic, but to ensure its best effects it is required to be administered in its full medicinal doses, which are, according to my own experience, from one to two drachms. It is quite safe up to two drachms and a half. In smaller doses (twenty to forty grains) it is a good tonic, but its action as an antiperiodic is very feeble." (*Honorary Surgeon Moodeen Sheriff, Khan Bahadur, Madras.*)

"This drug has never produced any substantial benefit in my hands. I am satisfied it possesses no curative value. I have abandoned the use of it. Any benefit derived while it was used might with equal accuracy be set down to *time* and the *vis mericatrix naturæ*, of which I had ample evidence in my practice." (*Brigade Surgeon W. R. Rice, M.D., Jubbulpore.*)

"I have found *atís* an uncertain antiperiodic, although acting well in some cases, especially in mild agues. It requires to be administered in large doses, frequently repeated, during the intervals of the fever." (*Surgeon S. H. Browne, M.D., Hoshangabad, Central Provinces.*)

MEDICINE.

"Formerly was largely used in Hyderabad (Deccan) as an antiperiodic in mild fevers." (*Deputy Surgeon-General G. Bidie, C.I.E., Madras.*)

"I used this in Government out-door dispensaries in large doses (20 to 40 grains) as an antiperiodic in simple intermittent fever, but cannot speak very favourably of it." (*Assistant Surgeon Debendra Nath Roy, Sealdah, Calcutta.*)

405

"Valuable antiperiodic, formerly largely used in out-door dispensary in intermittent fevers. *Dose* of powdered root half a drachm." (*Assistant Surgeon Shib Chunder Bhattacharji, Chanda, Central Provinces.*) "Antiperiodic and tonic, *dose* 1 to 5 grains of the powder." (*Surgeon W. Barren, Bhuj, Cutch.*) "I have largely used it in ten-grain doses as an antiperiodic in intermittent fevers and as a febrifuge in five-grain doses in slight cases of fevers with benefit, but I could not depend on it in cases of remittent fever." (*Assistant Surgeon Bollye Chand Sen, Teacher of Medicine.*) "White *atís*, if from ⅛ to 2 grains of ipecacuanha is added to each dose, is useful as a febrifuge." (*Surgeon-Major John North, Bangalore.*) "*Atís* powder used as an antiperiodic. It does not cause sickness." (*Surgeon J. Ffrench Mullen, M.D., Saidpur.*)

"Used in chronic diarrhœa and dysentery, with other astringents." (*Surgeon-Major J. J. L. Ratton, M.D., Salem.*) "Given internally it is said to be useful in boils." (*Surgeon G. A. Emerson, Calcutta.*) "Is a fairly good febrifuge, can be obtained in the bazars, and may be used when other bitter drugs are not available." (*Assistant Surgeon Nehal Singh, Saharanpur.*) "This is a good febrifuge and obtainable in the bazars of Umballa, Panjáb." (*Brigade Surgeon R. Bateson, Umballa.*) "Appears to be efficacious in mild intermittent fever, but less so than the cinchona alkaloids." (*Assistant Surgeon Jaswant Rai, Mooltan.*)

406

"Febrifuge and tonic, used as a substitute for quinine. *Dose* of the powder 10 grs. with 3 grs. of *Heera kus* (Ferri. sulph.)." (*Surgeon C. M. Russell, Sarun.*) "An indifferent antiperiodic, used in dispensary practice on account of its cheapness; very much inferior to cinchona preparations." (*Surgeon G. Price, Shahabad.*) "In mild fever 5 to 10 grain doses of the powdered root have been found antiperiodic. In convalescence it may be given with advantage, in combination with iron, ginger, &c. The infusion of root has been tried, but not found so efficacious as the powder." (*Surgeon E. S. Brander, Rungpur.*) "I tried this medicine extensively in the epidemic fever of Burdwan and elsewhere. It certainly possesses antiperiodic virtues in larger doses, 30 to 40 grains. It can only be used in mild intermittent fevers when cinchona alkaloids or arsenic are not procurable." (*Surgeon-Major R. L. Dutt, M.D., Pubna.*) "Very efficacious in the acute stage of dysentery, with febrile symptoms. Good for ordinary malarious fever in doses of 10 grains, given every three hours, during the remission." (*Dr. Forsyth, Civil Medical Officer, Dinajpore.*) "*Atís*, in doses of gr. xxx, three times a day, is a useful antiperiodic in intermittent and other periodic fevers. It is also a valuable tonic in cases of debility." (*Brigade Surgeon J. H. Thornton, B.A., M.B., Monghyr.*)

The numerous opinions received regarding this drug are fairly represented by those published above. The remainder are so uniformly to the same purpose, if not exactly in the same words, that it has not been considered necessary to publish more than the selection given.

Adulterants. 407

Adulteration and Substitution.—The root is said by **O'Shaughnessy** to be adulterated with that of **Asparagus sarmentosus** (*satamulí*). Two kinds of the root are met with in the market—(*a*) grey, shrivelled tubers, larger and longer than (*b*) white, the daughter off-shoots broken from the former. The latter fetch the best price. They are slightly scarred from the abrasion of rootlets, are generally 2 inches long, with a thin tap-like extremity, often bifurcated. They should break with a short, starchy fracture, presenting a white surface (*Dymock*). The *atís* is eaten fresh by the hillmen of Kanáwár as a mild tonic.

408

Dr. Buchanan, who first made known the various forms of Aconite, referred them to the genus **Caltha,** but **Don** early corrected this mistake by forming them into a new genus to which he gave the unhappy name Nirbisia (the antidote) in honour of the vernacular name *Nirbisí* applied to one of **Dr. Buchanan's** plants. **Wallich** subsequently referred these plants to their correct genus, **Aconitum.** Much confusion still exists as to the true *Nirbisí,* for it is by no means clear that it is a pure synonym for *Jadvár,* the generic name for the non-poisonous forms of the Aconite root. The following plants have been also mentioned as bearing the vernacular name *Nirbisí:* **Curcuma aromatica,** *Salisb.,* **C. Zedoaria,** *Roxb.,* which **Colebrooke** regarded as the Zedoary of the ancients from its synonyms being *Jádwár* and *Zadwár.* **Dr. Royle** states that the roots of **Delphinium denudatum,** *Wall.* (**D. pucifiorum,** *Royle*), bear the name *Nirbisí.* In **Dr. Dymock's** *Glossary of the Bombay Plants and Drugs, Nirbisee* is given as the Deccan name for **Cissampelos Pareira. Dr. Dymock** has, however, drawn my attention to the fact that **Prof. Rudolph Roth,** the distinguished Sanskrit scholar, has identified the roots of **Kyllingia monocephala,** *Linn.,* as the *Nirvisha* of the Sanskrit writers. This agrees with **Roxburgh's** remark under **Kyllingia,** where he gives the Bengali of this plant as *Swetagothubi,* remarking that *Nirbishee,* its fragrant aromatic root, is accounted an antidote to poison. **Dr. Moodeen Sheriff, Khan Bahadur,** distinguishes between the words *Nir-bisí* (a synonym for *Jadvár*) and the Sanskrit expression *Nir-visham* or *Nirvisha,* which expression he says means antidote. He concludes his remarks by urging that great care should be shown in prescribing the forms of Aconite under their vernacular names, and he regards *Jadvár* as the only name which can with safety be used for the non-poisonous forms.

409

Dr. Aitchison, in a note published under **A. palmatum,** says that the word *Nirbisí* is in the Panjáb applied to a poisonous form of aconite root. By most authors it is applied to **A. heterophyllum** only; by others (as it would appear with more correctness) to the non-poisonous forms of aconite (**A. heterophyllum** is known as *atís*), the non-poisonous forms being regarded as antidotes to poisons generally, hence the name *Nir-bisí.* It seems probable, however, that this name has become associated with many antidotes, and may, indeed, have originally been applied to a quite distinct plant, such as the roots of **Kyllingia.** (Conf. with non-poisonous forms of **A. Napellus**).

410

Aconitum luridum, *H. f. & T.; Fl. Br. Ind., I., 28.*

Habitat.—Sikkim, altitude 14,000 feet.

Botanic Diagnosis.—*Stem* erect, simple. *Leaves* palmately 5-fid below, the middle segments cuneate-ovate, 3-fid, coarsely crenate. *Racemes* ½ to 1 foot, simple, pedicels short. *Flowers* dull red; *helmet* with a long, straight beak, and broad, dome-like, dorsal prominence. *Seed* with a smooth testa. No information regarding the economic uses of this species.

411

A. Lycoctonum, *Linn.; Fl. Br. Ind., I., 28.*

Habitat.—The temperate Western Himálaya from Kashmír to Kumaon, altitude 7,000 to 10,000 feet. *Distrib.*—Europe and North Asia.

Botanic Diagnosis.—*Stem* erect, much-branched, 3-6 feet, glabrous or pubescent. *Leaves* 6 to 10 inches diameter, palmately and deeply 5-9-lobed; *lobes* cuneate-ovate, sharply cut, lower on long peduncles, upper sessile. *Racemes* long-branched, tomentose; *bracts* minute. *Flowers* pale yellow or dull purple, variable in size; *helmet* with a short beak and long cylindrical or conical dorsal prominence. *Follicles* 3, spreading. Seeds with plaited testa.

No definite information exists in India regarding the economic uses of this species; but it has been suggested under **A. ferox** that this may afford part of the Indian aconite.

MEDICINE. Japanese Aconite. 412

Medicine.—By most botanists in Europe this is regarded as the species to which the various forms of Japanese aconite **(A. japonicum,** *Thunb.*) are most nearly related. It was first chemically analysed by **Hübschmann,** who is said to have extracted two alkaloids. These he named *Acolyctine,* a white powder insoluble in ether but soluble in water and alcohol, *Lycoctonine,* a crystallizable substance, very soluble in alcohol. but only slightly so in water and ether. He subsequently came to regard acolyctine as identical with napelline. **Prof. Flückiger** regards this alkaloid as quite distinct from aconitine and pseudo-aconitine, and much less poisonout than either. **Dragendorff** has shown that **Hübschmann's** two alkaloids are but decomposition products of the true alkaloids present in the root of this species. He gave the name *Lycaconitine* to the alkaloid soluble in ether; it differs from every other known alkaloid, being richer in nitrogen. The alkaloid extracted by chloroform **Dragendorff** called *Myoctonine;* it is quite distinct from *acolyctine.* (*Pharm. Journ., 1884, 104.*)

At the same time **Wright** considers that he has discovered a new base from Japanese aconite, which he has called *Japaconitine* ($C_{66}H_{88}N_2O_{21}$), which is said to be even more poisonous than the aconite alkaloids.

It is probable that this species and also **D. paniculatum** (both frequent upon the Alps) yield much of the aconite of European commerce.

Aconitum Napellus, *Linn.; Fl. Br. Ind., I, 28.*

413

This is the true MONKS'-HOOD OR WOLVES'-BANE ACONITE.

Vern.—*Dudhiabish, kátbísh, míthá-zahar, tilia cachang, mohri,* KASHMIR and PANJAB HIMALAYAN NAMES. (§ *"The root of this plant is in Kashmír called Ban-bal-nág."* (*Surgeon-Major J. E. T. Aitchison, Simla.*) *Vasa nabhi* (*Surgeon-Major E. Levinge, Rajmundry*), TEL.; *Dudhio vachhanág* (*Assistant Surgeon Sakharam Arjun Ravat*), GUJ.

References.—*Pharm. Ind., 1; Flück. and Hanb., Pharm., 8; Bentl. & Trim., Med. Pl., 6; U. S. Dispens., Ed. 15th, 126; Bentley in Pharm. Jour., XV., 18 Ser., 449; Royle, Mat. Med., Ed. Harley, 773; Groves, Year-Book of Pharm., 1873, 500; 1874, 507; Wright's Reports on his experiments with Aconite, Year-Book of Pharm., 1875, 514; 1876, 531; 1877, 444; 1878, 483; 1879, 417; 1880, 455; 1881, 24; 1882, 223.*

Habitat.—Temperate alpine Himálaya, from 10,000 to 15,000 feet (Sach Pass, Chumba, *Watt*), ascending in stunted alpine forms to the highest limit of vegetation in the North-West Provinces. *Distrib.*—Temperate and Arctic Europe, Asia, and America.

Botanic Diagnosis.—A herbaceous perennial with short fleshy roots. *Stem* erect, simple, 2-4 feet high, smooth, green, slightly hairy above. *Leaves* variable in size, on long petioles, spreading, deeply, palmately, cut into 5 or 3 segments; *segments* linear, deeply and irregularly multifid, dark green above, pale beneath. *Racemes* simple, few or many flowered, sometimes with one or two smaller racemes at the base; *bracts* entire or 3-fid. *Flowers* large (¾ to 1 inch), stalked, erect; pedicels downy, thickened at the end with two small bracts on the apex close to each flower, bright blue or dull greenish-blue. *Helmet* shallow, 3 times as long as high, tapering to a slender beak. *Follicles* 3-5, generally hairy. *Seeds* with a smooth testa.

The *Flora of British India* refers the forms of this species to four varieties, as follows:—

414

Var. 1, Napellus proper: *Stem* 2-3 feet, leafy; *raceme* dense-flowered. *A poisonous form.*

Monks'-hood Aconite.

415

Var. 2, rigidum : *Stem* 2-3 feet, few-leaved; *leaves* firm, sub-coriaceous, with spreading, falcate, sharp teeth; *racemes* lax, few-flowered, tomentose. *A poisonous form.*

Syn.—A. DISSECTUM, *Don, Prod. Nepal, 197 ; Royle, Ill., 54;* A. FEROX, *Wall., Plant. As. Rar., t. 41.*

416

Var. 3, multifidum : *Stem* 6-12-inch, erect or decumbent, few-leaved, *leaves* 1 to 2 inches in diameter, many-lobed to the base; *lobes* cut into linear segments; *racemes* lax, few or many flowered. *Eaten by Bhotias.*

Syn.—A. MULTIFIDUM, *Royle, Ill., 56 ;* A. OLIGANTHEMUM, *Kern.*

417

Var. 4, rotundifolium : Like var. 3, but leaves not divided to the base. *Eaten by the Bhotias.*

Syn.—A. ROTUNDIFOLIUM, *Kar. ;* A. TIANSCHANICUM, *Osk. & Rupr.*

Bikhma. 418

Nirbisi. 419

The above sub-division into poisonous and non-poisonous varieties has been republished here in the hope that persons who may have the opportunity of examining the forms of this plant may be able to confirm the curious fact that some of the varieties of a species are poisonous while others are wholesome. I strongly suspect that the *Bikhma* of Nepal, referred, by most authors, to **A. palmatum**, may prove one of the edible forms of this species, and that it may be found to be the same as the *Nirbisí* or *Jadvár* of other authors. **Dr. Buchanan**, in his account of the Kingdom of Nepal, includes *Bikhma* as one of the four forms of aconite met with by him, **Dr. Dymock** says it reaches Bombay from the North, and that its value is R2-6 per lb.; that it is a valuable tonic of bitter quality, but resembling the *atís* root in its chemical properties and therapeutical actions. See also **A. palmatum** and **A. heterophyllum.**

Properties and Uses—

MEDICINE. Root. 420

Description of the European Medicinal Root.—Usually from 2 to 4 inches long and ½ to 1 inch thick at the top, tapering, shrivelled longitudinally, and beset with the prominent scars of fallen rootlets; externally blackish brown, internally composed of a whitish, farinaceous substance. It breaks with a short fracture, is sometimes hollow in the centre. The transverse section of a sound root shows a pure white central pith, many-sided, with a fibro-vascular bundle at each of its angles. The fibro-vascular tissue is devoid of true ligneous cells, its tissue being chiefly composed of uniform parenchyma, loaded with starch granules. A minute portion chewed causes prolonged tingling and numbness of the lips and tongue. It tastes at first sweet, but soon becomes alarmingly acid.

In the fresh state the root has the odour of the radish, a peculiarity which disappears on drying. It has been mistaken for horse-radish, and is said to have caused the death of many persons who have eaten it by mistake for that root. This accident could only occur in winter when the leaves have faded and the roots been dug up leafless. But even then there should be no mistake. The root-stock of the horse-radish is much larger than the Monks'-hood; it does not taper like the latter plant, is pale-yellow coloured, and the crown marked by transverse scars indicating the positions of the old leaves. The aconite has not the sharp pungency of the horse-radish, and the scrapings will be observed to turn rapidly red, while the tingling sensation of the lips on biting the root should prevent fatal accidents.

It has been found by experiment that the proper season to dig up the root for medicinal purposes is in autumn when the plant is leafless. The two new roots occur on either side of the old one. The tincture and principal medicinal preparations used in European practice are prepared from the root.

MEDICINE. Leaves. 421 Herb. 422

Leaves and Herb.—In European practice the fresh leaves, and, indeed, the whole herb, are also used as medicine. The inspissated juice forms the *Extract of Aconite* of our European druggists. This preparation, which is somewhat uncertain in its action, is sometimes prescribed to relieve the pains in rheumatism, inflammatory and febrile affections, neuralgia, and heart-disease. Aconite herb was introduced into the London Pharmacopœia in 1788.

Chemical Composition.—Under **A. ferox** and **A. heterophyllum** a good deal has been already said regarding the results of recent chemical analyses of the various aconites. It may be stated that we are on the eve of dispensing with the aconite drugs now in use, which, owing to the uncertainty of the root used in their preparation, could never be depended upon. The chemical constituents of the more important aconites have recently been finally determined and their active principles or alkaloids extracted in a definite and crystalline form. The officinal aconite preparations of the future may be expected to contain a chemically fixed amount of the alkaloid, and their reactions will thus be perfectly trustworthy. For this invaluable result we are mainly indebted, in the first instance, to **Groves** and **Duquesnel**, perfected by **Alder—Wright** and his collaborateurs.

Formerly, by means of rectified spirit, water, ammonia, ether, and sulphuric acid, an amorphous powder was extracted from the roots of **A. Napellus**, known as Aconitia. But as the result of the researches of the distinguished chemists whose names are associated with this subject, it has been shown that the substance extracted from **A. Napellus** consists of two distinct compounds, *viz.*, *Aconitine* (proper)—a crystallizable alkaloid—and two non-crystalline bases, of which *Picraconitine* may be mentioned.

423

Aconitine or Aconitia, $C_{33}H_{43}NO_{12}$—This compound crystallizes in an anhydrous condition; melts at 183°C. It is dehydrated by heating with acids, more particularly with tartaric acid, forming *Apo-aconitine* $C_{33}H_{41}NO_{11}$. On saponification with alkali it splits up into benzoic acid and the base *Aconine* $C_{26}H_{39}NO_{11}$.

One of the most important features of the chemistry of aconite is, that unless the amount of aconitine present in the root be relatively to the picraconitine very considerable, it is impossible to obtain the crystalline form of aconitine. Even when it is possible to produce the crystallization, a certain amount of aconitine is always held in solution through the agency of the amorphous base, much as alkaline salts prevent the complete crystallization of sugar. It is also important to add that, after repeated crystallization, aconitine always retains mechanically a certain amount of the amorphous base. This can be completely got rid of by transforming the aconitine into a salt and by regenerating the alkaloid from this salt after being thoroughly freed from the mother liquor. In this way chemically pure and crystalline aconitine may be obtained. The purity may be determined by the melting point being 183° to 184°C., or by allowing an acidulated and ethereal solution, to which carbonate of soda has been added, to slowly evaporate. If the entire mass crystallizes, it is pure, but if the last drop dries into a varnish, picraconitine must have been present.

Dr. Wright concludes his most valuable report on the aconites by saying: "The questions now remaining to be solved are essentially of a pharmaceutical and manufacturing nature, and as such somewhat out of the province of the scientific chemical investigator; these questions being simply the determination of the circumstances (as to soil, climate, age of plants, &c.) which influence the relative proportions between the crystallizable aconitine and the non-crystalline bases naturally accompanying it; so that the plants most suitable for the extraction of the alkaloids may be known; and the elaboration of the best method of separating the crystallizable from the amorphous substances on a large scale." (*Year-Book of Pharm., 1881, p. 27.*)

424

Medicinal Preparations.—It is not necessary to enter into the properties of the aconite of European commerce. It is too well known to require to be treated of here, and no definite information can be obtained regarding the **A. Napellus** of Indian origin. It is enough to have briefly indicated the modern advances which have been made, and alongside of these to show the part which India must play before its aconite can either attain a larger commercial position or take the place, in Indian medical practice, of the imported article. It is more than likely that a large proportion of the aconitine found in the so-called **A. ferox** of India may be due to the exports of that drug having consisted of the roots of at least four species—**A. ferox, A. Lycoctonum, A. Napellus, and A. palmatum**—indiscriminately mixed together.

425

§ "I have found **Aconitum Napellus** an invaluable drug in the reduction of the temperature of sun-stroke and pneumonia. As an external application, I have found it to be a most useful anodyne in facial neuralgia." (*Surgeon J. Parker, Poona.*)

426

Aconitum palmatum, *Don.; Fl. Br. Ind., I., 28.*

Habitat.—A perennial herb, chiefly of the eastern temperate Himálaya, extending from Garhwál to Sikkim, the Mishmí Hills, and along the north-eastern lofty ranges forming the frontier of Assam, to Mánipur, and to the higher peaks of the Nagá Hills, extending into Northern Burma.

Botanic Diagnosis.—*Stem* leafy, erect, simple, 2-5 feet high, glabrous. *Leaves* reniform, deeply 5-lobed, 4-6 inches in diameter, sinus shallow; *segments* cuneate-ovate, deeply and sharply cut; *petioles* long. *Panicles* few-flowered. *Flowers* large, greenish blue, on long pedicels; *helmet* much vaulted, shortly beaked, rather higher than broad. *Follicles* 5, 1-1½ inches long, glabrous. *Seeds* with a plaited testa.

MEDICINE. Root. 427

Medicine.—No definite information can at present be given regarding the roots of this plant. It was found plentiful on the loftier peaks north of Mánipur, where its roots are said to be poisonous. Samples of the roots sent from the Aka country (on the frontier of Assam), as those said to afford the arrow poison used by these wild hill tribes, seemed to agree with roots from Mánipur; and from the imperfect descriptions which could be obtained from other sources, it is probable that they are the roots of **A. palmatum.** The Aka roots on being chemically analysed by **Dr. Warden** were found to be poisonous.

428

§ "*Jadwár-khatái* is the name in Leh for the root of an aconite (probably **A. ferox**) that is imported from Nepal *viá* Lhassa. It is called in the Panjáb *Nirbisí,* by Bhoteas in Leh *Bongá,* and by the Yarkandis *Farfí:* it is poisonous. It is administered in cases of poisoning and in severe illness such as cholera, and is carried as a talisman about the person." (*Surgeon-Major J. E. T. Aitchison, Simla.*) It is probable that the above remark, written by **Dr. Aitchison** on the proof copy of this work under **A. palmatum,** has little reference to that species; but, like most other facts regarding Indian aconites, it is quite impossible to determine the species referred to. **Dr. Aitchison** may probably be correct in attributing it to **A. ferox,** but the Panjáb name *Nirbisí* (*Nir,* free from, and *bisí* poison) is somewhat at variance with its being poisonous.

"**Dr. Dymock** *(Mat. Med., West. Ind.)* suspects the *Bikhma, bishma,* HIND., *Wakhma,* BOMB., to be the root of **A. palmatum,** which would therefore be a non-poisonous species. The root is very bitter and contains a well-defined bitter alkaloid; it has no poisonous properties."

In Sikkim the natives consider the root of **A. Palmatum** as not poisonous. (*Surgeon-Major G. King, Calcutta.*)

ACORUS, *Linn.; Gen. Pl., III., 999.*

429

A genus of aquatic herbaceous perennials (belonging to the Natural Order AROIDEÆ), comprising a number of forms which may, with advantage, be reduced to two species. *Spike* not enclosed by a spathe—the spathe forming a leaf appearing to continue the growth of the axis like a long ordinary leaf, so that the spike seems to arise upon the side of a leaf and near the middle. *Flowers* all hermaphrodite, composed of six green perianth leaves; six stamens opposite the perianth segments; and a three-celled ovary with sessile stigma.

Acorus Calamus, *Linn.; Roxb., Fl. Ind., Ed. C.B.C., 296.*

430

THE SWEET-FLAG.

Vern.—*Bach, ghor* or *gor bach,* HIND.; *Bach,* BENG., ASS.; *Vekhand, gandhílovaj, godá vaj,* GUJ.; *Vekhand,* MAR.; *Vekhanda,* CUTCH; *Gand-ki-lakri, vach,* DUK.; *Vachá ugra gandhaha, shadgranthá,* SANS.; *Vaj,* ARAB.; *Agre-turkí,* PERS.; *Bariboj, warch,* PB.; *Vahi,* KASH.; *Vashambu,* TAM.; *Vasa, wasa, vadaja,* TEL.; *Vashanpa,* MAL.; *Bajé,* KAN.; *Linhe,* BURM.

In connection with the Telugu name *Vasa,* it may be noted (on the authority of **Moodeen Sheriff**) that the word *Ati-vasa,* which means greater Vasa, is applied to the root of **Aconitum heterophyllum,** and that *Ati-visha,* or greater poison, is the name for **A. ferox.** These two names must not therefore be confounded with *Vasa,* **Acorus Calamus.**

Habitat.—A semi-aquatic perennial, with indefinitely branched rhizomes; a native of Europe (?) and North America. Cultivated in damp, marshy places in India and Burma, altitude 3,000 to 6,000 feet; exceedingly common in Mánipur and the Nagá hills, often a weed of cultivation spreading apparently from the walls dividing the fields. Originally a native of Asia and probably introduced into Europe. Some difference of opinion prevails as to whether this is the **Calamus Aromaticus** of the Greeks which **Royle** regards as an Andropogon, but it seems probable that this was the plant.

Botanic Diagnosis.—*Rhizome* indefinitely branched, creeping in mud, with stout joints and large-leaf scars, cylindrical or somewhat compressed, about ¾ inch in diameter, smooth, pinkish or pale green, the leaf-scars brown, white and spongy within. Gives off below numerous straight rootlets. *Leaves* few, distichously alternate, forming erect tufts at the extremities of the rhizome, tapering into long acute points, entire, smooth; scapes arising from the outer leaves. All parts, but especially the rhizome, aromatic.

Properties and Uses—

OIL. 431 Perfumery. 432

Oil and Perfumery.—An essential oil is obtained from the leaves, which is used in England by perfumers in the manufacture of hair-powder. From the rhizome a pale or dark-yellow oil, with the strong penetrating odour of the root, and an aromatic, bitter, burning, camphoraceous flavour (due to the presence of a glucoside known as *Acorin*) is obtained by distillation.

The *volatile* oil and *acorin* may be said to be the two substances to which Sweet-flag owes its properties.

MEDICINE. Root-stock. 433

Medicine.—The aromatic rhizome or root-stock is considered emetic in large doses, and stomachic and carminative in smaller doses. (*U. C. Dutt, Civil Medical Officer, Serampore.*) It is a simple useful remedy for flatulence, colic, or dyspepsia, and a pleasant adjunct to tonic or purgative medicines. It is also used in remittent fevers and ague by the native doctors, and is held in high esteem as an insectifuge, especially for fleas. In **Voigt's** *Hortus Suburbanus Calcuttensis* occurs the

ACORUS Calamus.

Opinions regarding the Sweet-flag.

MEDICINE.

following (taken from *Thomson's Mat. Med.*): "The root has been employed in medicine since the time of Hippocrates. By the moderns it is successfully used in intermittent fevers, even after bark has failed and it is certainly a very useful addition to cinchona. It is also a useful adjunct to bitter and stomachic infusions." In European practice this medicine is not much if at all used. The rhizome is sold to a small extent by chemists in England, and in Scotland it is regarded useful to clear the throat before taking part in any public performance; for this purpose a small piece is chewed for a few minutes. It holds, however, an important position amongst the drugs regularly prescribed by Indian doctors. The Sweet-flag is the only plant which can be said to be taken by the Nagás as medicine, and it is also much valued by the Manipuris, especially in the treatment of coughs or sore-throat.

434

Opinions of Medical Officers.—§ "In Meerut the rhizome, with *bhang* and *ajowain* in equal parts, is powdered and used as a fumigation in painful piles." (*Surgeon-Major W. Moir* and *Assistant Surgeon T. N. Ghose, Meerut.*) "I found the root extremely useful in the dysentery of children, and also in bronchitic affections—*vide Ind. Med. Gazette* for February 1875, page 39, for further particulars." (*Surgeon B. Evers, M.D., Wardha.*)

"Aromatic, bitter, stimulant; useful as an expectorant in bronchitis.

"As a stomachic in flatulency in the form of infusion.

Bruised root	1 oz.
Boiling water	14 „

"Dose: 1 ounce and a half thrice daily." (*Surgeon C. M. Russell, Sarun.*)

"The root, rubbed up with water or spirit, is used as a counter-irritant to the chest in the catarrh of children. It is generally supposed that the smell is disliked by the cobra, on which it produces a narcotic effect. For this reason it is cultivated near dwellings and chewed by snake-catchers." (*Surgeon H. McCalman, M.D., Ratnagiri, Bombay.*) "*Bach* is commonly used to allay distressing cough. I use it much for this purpose, with excellent results. A small piece of the dried root-stock kept in the mouth acts better than many cough lozenges. It produces a warm sensation in the mouth and a beneficial flow of saliva." (*Surgeon-Major R. L. Dutt, M.D., Pubna.*) "Used as a tonic and stomachic. Combined with chiretta, is used by natives for intermittent fever, also in dysentery (especially of native children)." (*Surgeon H. W. Hill, Mánbhúm.*) "In finely-powdered 10-grain doses taken internally with warm milk, allays the tingling sensation of the throat in catarrhal sore-throat." (*Assistant Surgeon Devendro Nath Roy, Calcutta.*) "I have myself used it in coughs and sore-throats with some success. It seems to stimulate the mucous membrane and the salivary glands, the result being an increased secretion and relief of dryness of throat and harassing dry cough. I used to chew a small piece now and again." (*Surgeon D. Basu, Faridpur.*) "Useful in dysentery; a decoction is made from the bruised root. Dose one ounce and a half." (*Dr. W. Forsyth, Civil Medical Officer, Dinajpore.*) "Is a tonic and stomachic, useful in cases of dyspepsia, loss of appetite, and debility." (*Brigade Surgeon J. H. Thornton, Monghyr.*)

435

"The rhizome is emetic, nauseant, antispasmodic, carminative, stomachic, stimulant, and insecticide. As an emetic it is more nauseant and depressent than Ipecacuanha, and it is therefore useful in most of the diseases in which the latter is indicated, including dysentery. It is one of the two vegetable drugs in this country which act efficiently as emetics in so small a dose as 30 grains. It should not be used in more than 35 grains, and in 40 grains its action is very violent and obstinate. It

MEDICINE.

is a good remedy in asthma, to relieve which, it should be first used in pretty large or nauseant doses (15 to 20 grains) and then repeated every 2 or 3 hours in smaller or expectorant doses (10 grains) till relieved. Among other diseases which are most benefited by this drug are bronchial catarrh, hysteria, neuralgia, and some forms of dyspepsia. The rhizome can also be used in the form of a tincture or an infusion." *(Hony. Surgeon Moodeen Sheriff, Khan Bahadur, Madras.)*

436

"It is used in Madras as a flea-powder and is very effective." *(Deputy Surgeon-General G. Bidie, C.I.E., Madras.)* "A time-honoured domestic remedy for cough and fever, especially in children, even of the tenderest age, given, grated into a paste, with the milk of the mother. Is administered in the same form in colic of children. Has been used here in the form of infusion (1 in 20) in bronchial catarrh and febricula. In the former disease its efficacy is increased in combination with infusion of liquorice root (1 in 20). It is also supposed to destroy fleas, for which purpose an infusion of the root is used." *(Surgeon-Major Lewis Charles Nanney, Trichinopoly.)* "Useful as an external application on the abdomen of children suffering from flatulent colic. The root is burnt to cinder, mixed either with cocoanut or castor oil and smeared over the abdomen." *(Hony. Surgeon Peter Anderson, Guntúr, Madras Presidency.)* "A carminative, tonic, and insectifuge. Root is slightly burnt and powdered, grains 2 to 10 for a dose. An infusion sprinkled in infected places drives away vermin." *(Surgeon-Major A. F. Dobson, Bangalore.)* "Used largely by the natives in the flatulent colic of infants. The rhizome is roasted over a light, and a small portion, rubbed down with human milk, is given internally, and also used as a paste over the umbilicus." *(Hony. Surgeon Easton Alfred Morris, Negapatam.)* "Antispasmodic and sedative to the nervous system, used for colic of young children both externally and internally." *(Surgeon-Major Henry David Cook, Calicut, Malabar.)* "The root-stock is burnt to charcoal, then pulverised: 10 to 20 grains of this powder mixed with water is given to counteract the effect of croton. Is considered as an antidote in cases of croton-poisoning." *(Surgeon W. A. Lee, Mangalore.)* "Rhizome powder, dose 20 to 40 grains; infusion (1 oz. to 10 oz. boiling water), dose 1 to 2 oz. (*Pharm. Ind.*) Stomachic and carminative, insecticide." *(Apothecary Thomas Ward, Madanapalle, Cuddapah.)* "Is common in Southern India; it is antispasmodic and carminative, often used for children, also applied externally on the abdomen to expel flatus. It is used to keep moths from woollen goods and fleas from rooms, &c." *(Surgeon Mark Robinson, Coorg.)* "Used internally in the shape of decoction for children, as a carminative, dose grains v.; also externally in the form of paste applied to the abdomen in tympanitis." *(Surgeon-Major J. J. L. Ratton, M.D., Salem.)* "Is given internally by first burning the end of the root and rubbing it down in milk (as a vehicle) for flatulence, &c. It is also applied externally over the abdomen for flatulence in infants." *(Surgeon-Major Lionel Beech, Cocanada.)*

"The burnt root acts as an astringent in infantile diarrhœa." *(Assistant Surgeon Ruthnam T. Moodelliar, Chingleput, Madras Presidency.)* "In 3-grain doses it is very effectual in relieving the colic of small children." *(Surgeon-Major John North, Bangalore.)*

437

"The rhizome is largely used in North Bengal in coughs and sore throats; a few thin slices are given to chew, having been slightly warmed before the fire; it is more efficacious than cough lozenges in relieving the irritation of the throat. It is also used as a carminative in dyspepsia. In Western India it is used externally as an application on bruises and rheumatism rubbed up with the spirits made from the Cashew-nut fruit." *(Surgeon-Major C. T. Peters, South Afghánistan.)*

"Stimulant 1 to 5 grs., emetic one to two scruples. It is used in cases of colic and dyspepsia; when applied externally in the form of a paste over the head it relieves headache; when rubbed over the nose, it arrests the progress of influenza and bronchial catarrh." *(Surgeon W. Barren, Bhuj, Cutch, Bombay.)* "A small piece kept in the mouth and its juice swallowed relieves cough and tickling of the throat; it produces salivation and an agreeable warmth." *(Assistant Surgeon Shib Chunder Bhuttacharji, Chanda, Central Provinces.)*

Chemical Composition.—§ "The dried rhizome yields, according to the editors of the *Pharmacographia*, 1·3 per cent. of a neutral yellowish essential oil, of an agreeable odour, which **Curratow** has shown contains Tæpene. **Faust** has isolated a bitter semi-fluid, nitrogenous glucoside, *acorin*, while **Fluckiger** and **Hanbury** have obtained a very bitter crystalline principle." (*Surgeon C. J. Hislop Warden, Prof. of Chemistry, Calcutta.*)

Spirit. 438

Spirit.—It is stated that a considerable amount of the rhizomes of this plant are used in flavouring gin, beer, &c. For this purpose the market is said to be chiefly supplied from the river-banks in Norfolk. Formerly the leaves of the Sweet-flag were spread over the floors of churches and cathedrals (especially in Norfolk) upon great occasions, the pressure of the foot causing a pleasant odour. The rhizome is said to be used in India in the preparation of an aromatic vinegar.

Vinegar. 439

Trade.—**Dr. Dymock** remarks that the drug imported into Bombay comes chiefly from the Persian Gulf; it brings about R3 a maund of 37½ seers. There is a very considerable trade in this article done in Calcutta.

ACROCARPUS, *W. & A.; Genl Pl.*

A genus containing only a single species (belonging to the sub-order MIMOSÆ, of the Natural Order LEGUMINOSÆ).

440

Acrocarpus fraxinifolius, *Wight; Fl. Br. Ind., II., 292; Wight, Ic., t. 254.*

RED OR PINK CEDAR (of tea-planters).

Vern.—*Mandania*, NEPAL; *Mad ling*, LEPCHA; *Malai-kone*, TINNEVELLY; *Kalinji*, NILGHIRIS; *Kilingi*, BURGHERS; *Hantige, belanjihavulige*, KAN.

Habitat.—A lofty, deciduous tree, found in the Eastern Himálaya and lower hills down to Chittagong, ascending to 4,000 feet; also in South India and Burma.

TIMBER, 441

Structure of the Wood.—Sapwood white; heartwood light red, moderately hard. Weight 39 lbs. per cubic foot.

DOMESTIC USES. 442

Used by planters in Darjiling for tea-boxes and planking, in the Wynaad for building and furniture, and in Coorg for shingles.

In the *Tropical Agriculturist* for May 1883, some interesting information is given regarding wood for tea-boxes. **Mr. Bruce** writes: "I have used this timber more perhaps than any other for tea-boxes and tea-house furniture in general, and if it has been well seasoned it is as good a wood as could be procured for the purpose."

443

ACROSTICHUM, *L.; Syn. Fil., 399.*

Acrostichum (Stenochlœna) scandens, *Willd.*; FILICES.

A common fern in the warmer parts of Ceylon.

Fibre.—Dr. Trimen informs me that ropes are made in Ceylon from this plant.

ACTÆA, *Linn.; Gen. Pl., I., 9.*

444

A genus of herbaceous perennials (belonging to the Natural Order RANUNCULACEÆ), comprising only two species, inhabiting the cold temperate regions of Europe, North Asia, and North America.

Leaves alternate, ternately compound. *Flowers* small, regular, in short crowded racemes. *Sepals* 3-5, unequal, petaloid. *Petals* 4-10, small, spathulate or wanting. *Stamens* many, slender. *Carpel* 1, many-ovuled; *stigma* sessile, dilated. *Fruit* a many-seeded berry.

(Compare with the allied genus CIMICIFUGA, which will be found to differ chiefly in the longer racemes (3-8 inch), and dry, dehiscent capsule, instead of a succulent berry.)

Actæa spicata, *Linn.; Fl. Br. Ind., I., 29.*

445

THE BANEBERRY.

References.—*U. S. Dispens., 15th Ed., 1560.*

Habitat.—Temperate Himálaya, from Bhután to Hazára. *Dist.*—Europe, North Asia, North America.

Botanic Diagnosis.—*Leaflets* ½-2 inches, ovate-lanceolate, entire or 3-lobed, acutely serrate. *Flowers* ¼ inch diameter, white. *Berry* black in the European and Himálayan form; white and red in the American. The two American forms are in popular scientific works treated as distinct species—**A. alba** and **A. rubra.** The berries are very poisonous.

Properties and Uses—

MEDICINE. The Root. 446

Medicine.—The drug which in Europe and America is prescribed under the name of Tinctura Actæa racemosæ is prepared from **Cimicifuga racemosa** and not from a species of **Actæa. Stewart** remarks regarding **Actæa spicata**: "I have found no trace of its being used or dreaded" by the hill people on the Panjáb Himálaya. It would be interesting to know whether this be correct; for it is curious that so useful a plant should have escaped the notice of the natives of India. Canadian doctors administer the root in snake-bite; and it is said to be attended with much success in the treatment of nervous diseases, rheumatic fever, chorea, and lumbago. **Mr. Frederick Stearns** describes the root as violently purgative. The berries were formerly used internally for asthma and scrofula, and externally for skin complaints. Baneberry Root is largely exported into Europe and used to adulterate the root of **Helleborus niger,** but the former may readily be distinguished on section by the presence of radiating medullary bands, while **Hellebore** has an entire or undivided substance. An infusion of **Actæa** root is changed into black on adding a solution of persulphate of iron acting upon the tannic acid of the **Actæa.** No such change is effected upon an infusion of **Hellebore.**

§ "**Actæa racemosa.**—A tincture of the root is a powerful nerve sedative, and will often relieve severe neuralgia when all other drugs fail." (*Dr. S. Westcott.*) "It would be worth while to try this drug as a substitute for **Actæa racemosa,** which I find very serviceable in chronic rheumatism and uterine disorders." (*Surgeon-Major R. L. Dutt, M.D., Pubna.*) It seems probable that by **Actæa racemosa** is meant **Cimicifuga racemosa,** and in that case the above medical opinions should be transferred to the latter species. See **Cimicifuga and Helleborus.**

ACTINIOPTERIS, *Link.; Syn. Fil., 246.*

447

A genus of ferns (FILICES) belonging to the tribe ASPLENIEÆ. *Sori* linear, elongated, submarginal; *indusium* the same shape as the sorus and folded over it; placed one on each side of the narrow segments of the frond and opening towards the midrib.

448 **Actiniopteris dichotoma,** *Bedd.; Clarke's Ferns, N. Ind., in Trans. Lin. Soc., 1880.*

Syn.—A. RADIATA, *Link.*; ACROSTICHUM DICHOTOMUM, *Forsk.*
Vern.—*Mor-pankhi, mor-pach,* N.-W. P.; *Mápursika,* BOMB.

Habitat.—Common throughout India on the lower hills of the peninsula. Very characteristic of the Nilgiris up to altitude 2,000 feet, and of Kumaon and West Nepal; rare in the plains of India, occurring, as at Agra, Delhi, and Moradabad, in crevices of rocks and in old masonry.

Botanic Diagnosis.—An exceedingly pretty fern, like a miniature palm. *Fronds* fan-shaped, 1 to 1½ inches in breadth, composed of numerous dichotomous segments.

MEDICINE. 449

Medicine.—Used as an anthelmintic (*Atkinson*).

§ "Very common on old walls in the Deccan; used as a styptic." (*Surgeon-Major W. Dymock, Bombay.*)

450

ACTINODAPHNE, *Nees; Gen. Pl., III., 160.*

A genus of trees or bushes (belonging to the Natural Order LAURINEÆ) comprising 50 species, of which 9 or 10 are Indian, inhabiting the warm, moist forests of the lower hills.

Leaves sub-opposite or clustered at the ramifications and tips of the branches, thick coriaceous, pennivened. *Flowers* diœcious, sessile clustered in sessile fascicles, in bud enclosed by imbricate caducous scales. *Perianth* with a short tube broken into 6 sub-equal leaves. *Male flowers* with 9 perfect stamens arranged in three rows of three each, the innermost having a gland on either side of the filament; *anthers* all introrse, 4-loculate. *Ovary* immersed in the cup-shaped tube of the calyx; *style* tapering; *stigma* dilated. *Fruit* a berry placed in the disk or cup of the perianth.

Very little of importance can be said regarding the Indian species of this genus. The following are those best known:—

451 **Actinodaphne angustifolia,** *Nees; Wight's Ic., t. 1841.*

Syn.—LITSÆA ANGUSTIFOLIA, *Bl.*
Vern.—*Samkoh,* ASS.; *Boltanaro,* GÁRO; *Tabongdeing,* MAGH.; *Shwoaygjo, nalingjo,* BURM.

Habitat.—An evergreen tree, with the leaves rusty-tomentose beneath; met with in Eastern Bengal, South India, and Burma.

452 **A. Hookeri,** *Meissn.*

Syn.—A. LANCEOLATA, *Dalz.*
Vern.—*Pisa,* BOMB.

Habitat.—A small tree or shrub of Sikkim, and of the Eastern and Western Gháts of South India and in Kanára and Sattára, and particularly at Mahableshwar.

Medicine.—"A cold infusion of the leaves is mucilaginous, and is used in urinary disorders and in diabetes. The oil of the seeds, *Pisa-tila,* is used as an external application to sprains; it is of a reddish colour and has a fatty odour." (*Surgeon-Major Dymock, Mat. Med., 554.*)

A. obovata, *Hook. f.*

Vern.—*Muslindi,* NEP.; *Pohor,* LEPCHA; *Laiphauzeh,* MECHI; *Cherritinga,* ASS.

Habitat.—A tall tree (with large 3-nerved leaves) occurring in the outer Sikkim Himálaya, Assam, Khásia Hills, and Sylhet.

Adamia versicolor, *Fortune;* **A. cyanea,** *Wall.,* and **A. chinensis,** *Gard.*
Synonyms for **Dichroa febrifuga,** *Lour.,* which see.

Adam's Apple—A name sometimes applied to the Lime or Lemon.

Adam's Needle, see **Yucca gloriosa.**

ADANSONIA, *Linn.; Gen. Pl., I., 209.*

454

A genus (belonging to the Natural Order MALVACEÆ and the Tribe BOMBACEÆ), containing in all only two species, one met with in tropical Africa, the other in Australia; the former is cultivated in India. *Leaves* digitate. *Calyx* 5-cleft, leathery. *Petals* 5, exceeding the sepals, adnate below to the stamens. *Style* divided into 5-10 branches; *stigmas* radiating. *Fruit* oblong, woody, indehiscent.

A genus named after **Adanson**, a celebrated French traveller, who lived in Senegal from 1749 to 1754.

Adansonia digitata, *Linn.; Fl. Br. Ind., I., 348.*

455

THE BAOBAB TREE; the SOUR GOURD, or the MONKEY BREAD TREE OF AFRICA.

Vern.—*Gorakhchincha* or *gorakh chintz, choyari chinch* (The horse's tamarind, *gorakha-amli*), BOMB.; *Gor-amli chora, górak-amali,* or *gorakh-amli,* HIND.; *Gorakha-amli, bukha,* GUJ.: *Gorakh chinch,* CUTCH; *Kalpbriskh,* or *kalbriskh,* AJMERE, DELHI; *Háthí khatyán* (the plant), DUK.; *Vilayti-imlí,* MORAR; *Anai-puli, paparapuli* (the plant, *ánai-puliya-marram*), TAM.; *Hujed,* ARAB.

Dr. Dymock says the Bombay name *Gorakh* is derived from the name of a celebrated Hindú ascetic, who probably taught his disciples under this tree. Gorgkh and his disciple Machindar were well-known *Sadhus.*

Habitat.—This is one of the largest and longest-lived trees in the world. Trunk short, thick, often found 30 feet in diameter; branches spreading.

Cultivated, to a small extent, in some parts of India, but deserves to be extended; originally introduced by Arab traders, who call it *Habhabú.* It is chiefly met with in Bombay, being plentiful on the coast. "Four or five venerable specimens are in the Futtehpore district." The abandoned capital, Mandoo, near Indore, is overrun with Adansonias as other ruins are with the *pípul.*" (*R. T. H., Morar.*) "Pretty common about Madras; at one time it was proposed to cultivate it on account of the fibrous material in its bark." *(Deputy Surgeon-General G. Bidie, Madras.)* "Specimens are to be seen at Lucknow and at Allahabad." (*Brigade Surgeon G. A. Watson, Allahabad.*) It is also being experimentally cultivated in the Sunderbuns. There is a good specimen in the Barrackpore Park, and a small one on the Calcutta Maidan, a little beyond the Cathedral. In Africa it is said to extend through the continent from Senegal to Abyssinia. It has also been introduced into the West Indies. **Humboldt** speaks of this tree as "the tree of a thousand years," "the oldest organic monument of our planet." **Adanson** made a calculation to show that a tree 30 feet in diameter was over 5,000 years of age. He saw two trees 5 to 6 feet in diameter, on the bark of which were cut European names, one dated in the 14th and another in the 15th century. **Livingstone** says: "I would back a true *Mowana* (the name given to this tree in the neighbourhood of Lake Ngami) against a dozen floods, provided you do not boil it in salt water, but I cannot believe that any of those now alive had a chance of being subjected to the experiment of even the Noachian deluge."

Properties and Uses—

GUM. 456

Gum.—§ "The bark when wounded yields a large quantity of white semi-fluid gum, which is odourless and tasteless, and has an acid reaction.

The Baobab.

Under the microscope, in addition to amorphous matter, a considerable number of minute bodies, with sharp projecting rays, are visible. The ash contains a large quantity of lime. Gum Baobab is insoluble in water, and appears to be allied to gum tragacanth." (*Surgeon C. J. H. Warden, Prof. of Chemistry, Calcutta.*)

FIBRE. 457

Fibre.—The bark yields a strong, useful fibre.

In Senegal it is made into ropes and woven into cloth. The hard outer bark is first chopped away, and the inner bark stripped off in large sheets. These are beaten with sticks to remove the pithy matter. The fibre is then sun-dried and pressed into bales. Small trees yield finer and softer fibre than large ones. The Africans use the fibre for making rope, twine, and sacking; in India, elephant saddles are made from it. The fibre imported into England from Portuguese West Africa readily sold at £9 to £15 a ton. It produces an exceedingly strong paper, suitable for bank-notes, and has received much attention. The slow growth of the tree, and the careful cultivation and shading it requires while young, renders it, however, a precarious source of paper fibre. (*Spons' Encycl.*)

Chemical Composition.—§ "From the bark **Walz** extracted a non-nitrogenous principle, which crystallizes in needles and prisms, and which he named *Adansonia*. The root contains a red colouring matter, soluble in water and in absolute alcohol. From its aqueous solution it is deposited as a red powder." (*Surgeon C. J. H. Warden, Prof. of Chemistry, Calcutta.*)

MEDICINE. Pulp. 458 Leaves. 459 Bark. 460

Medicine.—The FRUIT has a mucilaginous PULP, having a pleasant, cool, subacid taste, like cream of tartar; "a good refrigerant in fever." (*Bomb. Gaz., VI., 14.*) Used in Africa in dysentery. LEAVES dried and powdered constitute the "Lalo" of Africans, used to check excessive perspiration. The BARK is antiperiodic. "A useful substitute for quinine in low fever." (*Bom. Gaz., VI., 14.*)

MEDICINE. Wood. 461 Seed. 462

The pulp is used in Bombay with butter-milk in diarrhœa and dysentery. "The WOOD is said to possess antiseptic properties." (*Bom. Gaz., XIII., Part I., 24.*) The SEEDS are said to possess febrifugal properties.

In a recent correspondence with the Government of India regarding the desirability of producing a revised edition of the *Indian Pharmacopœia*, it was proposed this plant should be excluded from the new edition. The *U. S. Dispensatory, Ed. 15th*, says of it, however, that "the leaves and the bark abound in mucilage." "**Dr. Duchassaing**, of Guadaloupe, West Indies, and **M. Pierre**, of France, commend the bark highly as an antiperiodic. It is said to be acceptable to the stomach, and to produce no other observable physiological effect than increase of appetite, increased perspiration, and perhaps diminished frequency of pulse. An ounce may be boiled in a pint and a half of water to a pint, and the whole taken in a day." (*U. S. Disp., Ed. 15th, 1561; Jour. de Pharm. 2e Ser. XIII., 412 and 421.*)

In the *Pharm. India* it is stated that, according to **Dr. R. F. Hutchinson**, its action is not due to any astringent property which it possesses, but to its virtues as a refrigerant and diuretic. **Dr. Gibson** hinks that the properties of this tree are well deserving of attention.

§ "The pulp is said to be a useful external application in skin diseases —no personal experience." (*Surgeon J. Parker, Poona.*) "Useful astringent in diarrahœa and dysentery, dose 1 to 20 grains." (*Surgeon W. Barren, Bhuj, Cutch, Bombay.*) "Decoction used in bilious headaches. (*Surgeon-Major J. J. L. Ratton, M.D., Salem.*)

FOOD. The Fruit. 463

Food.—The fruit, which varies in size and shape, is frequently 12 inches long, or only as large as a lemon, and resembles a gourd; contains many brown seeds, is somewhat acid, and makes a cooling and refreshing drink.

It is also eaten by the natives. **Major Pedley**, in his expedition in search of Mungo Park, lived almost exclusively on it for twelve days. In Gujarát the fishermen eat the leaves with their food, and consider them cooling. In Senegal the negroes use the bark and leaves powdered as a condiment.

Leaves. 464

Structure of the Wood.—Light, soft, and porous, made into rafts to support fishermen in tanks. It is readily attacked by fungi.

TIMBER. 465

Domestic Uses.—Owing to the softness of the wood the stems of the Baobab trees are often excavated into living houses. **Livingstone** describes one of these excavated trunks as sufficient to allow 30 men to lie down. The bodies of men denied the honour of a burial are often in Africa suspended within these houses and soon become perfectly dry and converted into mummies without the necessity of being embalmed. The ash of the fruit and bark boiled in oil is used as soap by the negroes. The dry fruits are used as floats by the Indian fishermen.

DOMESTIC USES. **Living houses.** 466 **Soap.** 467 **Floats.** 468 **The Fruit.** 469

ADELIA.

Adelia castanicarpa, *Roxb.* Syn. for **Chætocarpus castaneæcarpus,** EUPHORBIACEÆ, which see.

A. cordifolia, *Roxb.* Syn. for **Macaranga cordifolia,** *Müll.-Arg.*, which see.

A. neriifolia, *Roxb.* Syn. for **Homonoya riparia,** *Lour.*, which see.

ADENANTHERA, *Linn.; Gen. Pl., I., 590.*

470

A genus of trees or shrubs, without prickles (belonging to the Natural Order LEGUMINOSÆ, sub-order MIMOSEÆ), comprising in all some 4 species, spread throughout the tropics of the Old World, 2 being natives of India.

Leaves bipinnate. *Flowers* in spikes, minute, white, hermaphrodite. *Calyx* campanulate, equally toothed. *Petals* valvate, cohering only at the very base. *Stamens* 10, all free, equalling the corolla. *Seeds* scarlet.

The word Adenanthera is derived from the Gr. ’αδὴν, an acorn or gland, and ἀνθηρός, a flower.

Adenanthera aculeata, *Roxb.*, see **Prosopis spicigera,** *Linn.;* LEGUMINOSÆ.

A. pavonina, *Linn.; Fl. Br. Ind., II., 287; Wight, Ill., t. 84 (80).*

471

RED WOOD (sometimes called RED SANDAL WOOD).

Vern.—*Rakta kanchan, rakta-kambal, ranjana* (sometimes also called *Rakta-chandan*, a name more correctly applied to **Pterocarpus santalinus**), BENG.; *Chandan*, ASS.; *Bir-mungara*, SANTAL; *Anai-gundumani*, TAM.; *Bandi gurivenda, pedda-guriginga*, TEL.; *Manjati*, MAL.; *Vál, thorligunj*, MAR.; *Bari-gumchi, hatti-gumchi*, DUK., GUJ.; *Manjadi*, KAN.; *Madateya*, CINGH.; *Gung*, MAGH.; *Ywaygyee*, or *ywegyi*, BURM.; *Rechedá*, AND.

Habitat.—A large, deciduous tree, met with in Bengal, South India, and Burma.

Botanic Diagnosis.—*Leaves* compound, with 8-12 pinnæ and 12-18 obtuse leaflets. *Racemes* short, peduncled, 2-6 inches long; *seeds* bright scarlet.

In Ceylon a nearly allied species occurs known as **A. bicolor,** with 6-8 pinnæ, and leaflets acute. *Seeds* half black, half red.

Properties and Uses—

Gum.—*Spons' Encyclopœdia* mentions a gum obtained from this plant and known as *madatia*.

GUM. 472

Dye.—The wood is sometimes used as a dye, but chiefly as a substitute for the true red sandal wood.

DYE. 473

OIL. 474

Oil.—The seeds yield an oil.

MEDICINE. Seeds. 475

Medicine.—The powder made from the SEEDS is said to be a useful external application, hastening suppuration. § "A grateful application over boils, soothing the burning pain and hastening recovery. It is also used to cure prickly heat." (*R. N. Gupta.*) "The emulsion, made by rubbing the seeds on a stone with water, forms a cooling external application useful in headache and in the early stages of inflammation." (*Surgeon J. Anderson, Bijnor.*) "*Vála* powder mixed with honey is used in colic—no personal experience." (*Surgeon-Major R. L. Dutt, M.D., Pubna.*) "A decoction is used by *Kobirages* in rheumatic affections. The powdered seed rubbed with water is used to disperse boils." (*Surgeon J. Parker, Poona.*) "Hakims use the powder in gonorrhœa." (*Surgeon-Major W. Moir* and *Assistant Surgeon T. N. Ghose, Meerut.*)

Leaves. 476

A decoction is made from the LEAVES in South India, and given as a remedy for chronic rheumatism and gout. If used for any length of time it is said to be anaphrodisiac. It is regarded as useful in hœmorrhage from the bowels and hæmaturia.

Wood. 477

§ "The decoction is very useful as an astringent and tonic in atonic diarrhœa and dysentery." (*Assistant Surgeon Bhagwan Das, Rawal Pindi.*) "The wood, powdered and mixed with water, is said to be useful when applied to the forehead in cases of headache from over-exertion or exposure." (*Surgeon-Major C. W. Calthrop, Morar.*) "The red wood rubbed on stone with a little water and applied to the body is a certain temporary cure for prickly heat." (*Surgeon-Major Henry David Cooke, Calicut, Malabar.*) "Used as an emetic, in 60-grain doses, in warm water." (*Surgeon-Major J. J. L. Ratton, M.D., Salem*) "It is used as an external application in orchitis." (*Surgeon-Major J. F. Fitzpatrick, M.D., Coimbatore.*)

FOOD. 478

Food.—The seeds are sometimes eaten as an article of food.

TIMBER. 479

Structure of the Wood.—Heartwood red, hard, close-grained, durable, and strong.

The timber is used in South India for house-building and cabinet-making purposes.

§ "This is sometimes confounded with the **Pterocarpus santalinus**; but the latter yields the red sandal-wood of commerce, which is largely exported from South India." (*Deputy Surgeon-General G. Bidie, Madras.*)

DOMESTIC USES. The seeds as weights. 480 As Necklaces. 481 Forming Cement. 482 Tilak Paste. 483

Domestic Uses.—The bright scarlet seeds are used as weights, each being about 4 grains; they are also strung and made into necklaces. Powdered and beaten up with borax, they give a good cement. The red paste (*tilak*) made by rubbing the wood upon a moist stone is used by the Brahmins to colour the forehead after bathing.

ADHATODA, *Nees; Gen. Pl., II., 1112.*

A genus of sub-herbaceous bushes (belonging to the Natural Order ACANTHACEÆ, and the tribe JUSTICIEÆ), comprising in all some six species, distributed through tropical India, south tropical Africa, and Brazil.

Leaves opposite, entire, arising from swollen nodes. *Flowers* purple or white, crowded into a bracteated spike, sub-sessile, each flower having three bracts, the outer largest and persistent. *Calyx* campanulate, 5-fid, lobes lanceolate. *Corolla-tube* short; limb 2-labiate, the posterior lip erect, the anterior broad recurved, 3-fid. *Stamens* 2, each with 2 large diverging anther-cells, one much higher than the other. *Ovules* 2 in each cell, *placenta* not rising elastically from the base of the capsule.

The word Adhatoda is derived from the Tamil name for the Indian species.

Adhatoda Vasica, *Nees ; Fl. Br. Ind., IV.*

484

Syn.—JUSTICIA ADHATODA, *Linn.*

Vern.—*Arúsá, arushá, adalsá, adulasá, adulaso, adársa,* HIND., BOMB.; *Bákas, vásaka,* BENG.; *Adulsa,* MAR.; *Aduso,* GUJ.; *Arátórá,* DUK.; *Bhekkar,* JHELUM; *Basúti,* BEÁS; *Bekkar,* SALT-RANGE; *Tora bajja,* TRANS-INDUS; *Bashangarús,* KUMAON; *Árus, vásaka, vajidantakaha atarusha,* SANS.; *Rús,* OUDH; *Basung,* URIYA; *Adhatodai,* TAM.; *Adasara,* or *addasaram,* TEL.; *Atalótakam,* MAL.; *Teeshæ,* NAGÁ; *Kath, alesi,* NEPÁL.

Bánsa, vásá, bahikat, (?) HIND., in *Baden Powell's Pb. Prod., I., 365*; he also gives the vernacular names *Behikar, bhekar, p. 565.*

Habitat.—A small, sub-herbaceous bush, often gregarious, found everywhere in Bengal and in the Sub-Himálayan tracts, ascending to 4,000 feet in altitude.

Properties and Uses—

Dye.—A yellow dye, obtained from the leaves by boiling, is used for dyeing coarse cloth. It gives a greenish-blue when combined with indigo. This property is not apparently known to the Nagás, who cultivate the plant to shade the approaches to their villages. I repeatedly asked if they prepared a dye from it, and was told that they did not, but that they used the stems for divining.

DYE. 485

Medicine.—The LEAVES and the ROOT of this plant are considered a very efficacious remedy for all sorts of coughs, being administered along with ginger. "The medicine was considered so serviceable in phthisis that it was said no man suffering from this disease need despair as long as the *vásaka* plant exists." (*U. C. Dutt, Civil Medical Officer, Serampore.*) It is often administered along with honey, the fresh juice or a decoction with pepper being made into a cough mixture. Dr. Irvine gives the dose as ½ oz. to 1 oz. of the decoction, and states that the price of the drug in Patna was 1 anna per lb. in the year 1848. The *Pharm. India* states that strong testimony has been given in favour of the remedial properties of this plant drawn from personal experience, in the treatment of chronic bronchitis, asthma, &c., when not attended with febrile action. The FLOWERS and the FRUIT are bitter, aromatic, and antispasmodic. The FRESH FLOWERS are bound over the eyes in cases of ophthalmia. "The flowers, leaves, and root, but especially the flowers, are supposed to possess antispasmodic qualities." "They are bitterish and sub-aromatic, and are administered in infusion and electuary." (*Ainslie.*) "The leaves are used as a cattle medicine." (*Gamble.*)

MEDICINE. Leaves. 486 Root. 487

Fruit. 488

Flowers. 489

Special Opinions.—§ "Leaves recently dried are also *smoked* in cases of asthma; they produce very beneficial effects. Hospital Assistant Gopal Chunder Gangooly, of the Noakhally Dispensary, who is subject to asthma, has used the leaves in this form and testifies to their property." (*Assistant Surgeon Anund Chunder Mukerji, Noakhally.*) "The leaves made into cigarettes are used in asthma; they act as an antispasmodic." (*Brigade Surgeon J. H. Thornton, Monghyr.*) "Decoction of fresh leaves was found to be very useful in bronchial catarrh." (*Surgeon C. J. W. Meadows, Burrisal.*) "The dried bark, when smoked, relieves asthmatic fits; strong decoction of it in subacute bronchitis did not produce, in my hand, much benefit as an expectorant." (*Assistant Surgeon Devendro Nath Roy, Calcutta.*) "There are two varieties, one with red and the other with white flowers. The first is medicinally much more important. An infusion is used in bronchitis and consumption. A fluid extract of the leaves and flowers would be a desirable preparation for trial." (*Surgeon-Major R. L. Dutt, M.D., Pubna.*)

Leaves smoked. 490

Bark. 491

"Excellent expectorant, dose 1 to 20 grains in chronic bronchitis and asthma." (*Surgeon W. Barren, Bhuj, Cutch, Bombay.*) "A fomentation

MEDICINE.

with a strong decoction of the leaves is considered efficacious in rheumatic pains, and in neuralgia. Also useful in reducing swellings." *(Hony. Surgeon P. Kinsley, Chicacole, Ganjam Dist., Madras Presidency.)* "This shrub is very common in Mysore, and the powdered root is used by native doctors in cases of malarial fever." *(Surgeon-Major John North, Bangalore.)* "The juice of the leaves has been used as an excellent expectorant when combined with some native medicines; also used by native doctors as a diuretic in dropsical affections attended with anæmia." *(Surgeon-Major J. F. Fitzpatrick, M.D., Coimbatore.)* "Decoction is used to quench thirst in fever." *(Surgeon-Major J. J. L. Ratton, M.D., Salem.)*

"The juice of the leaves is used for diarrhœa and dysentery. It is considered especially useful in hæmoptysis and bleeding in dysentery." *(Surgeon-Major J. Robb, Ahmedabad.)* "Is a useful refrigerant in fever, given as decoction." *(Surgeon-Major John Lancaster, Chittore.)*

FODDER. 492

Fodder.—Not browsed by any animals, except occasionally by goats.

TIMBER. 493

Structure of the Wood.—White, moderately hard.

The timber of the thicker stems is used for gunpowder-charcoal and as a fuel for brick-burning.

§ "Though only a shrub, it is valuable, as yielding a good charcoal for gunpowder. Specimens of the wood may be got an inch in diameter. It is quite the characteristic plant of the lower hills." *(Baden Powell, Pb. Prod., I., 565.)*

DOMESTIC USES. 494

Domestic Uses.—The stems are used in the Nagá Hills for divining and to foretell omens. The twig is held in the left hand and rapidly cut into thin slices, an incantation being repeated all the while; the prognostications are based upon the number of times the heart-shaped, dark, central wood turns towards or away from the operator. The idea of medicine seems scarcely to have occurred to the Nagá, and he does not appear to attribute to this plant any virtues other than those described.

"The *rús* is extensively employed in the construction of the fascine-like supports of mud wells. The smaller branches are exceedingly pliant and are worked round and round in a sort of neat triple plait. The leaf is held to possess high qualities as a manure, and is scattered over the fields just before the rainy season commences. It is then worked into the soil with the plough, and left to decay with the moisture and thus form mould. As fuel it is almost exclusively used in the process of boiling down the cane-juice, and is collected into large heaps some days prior to the cutting down of the sugar-cane." *(Oudh Gaz., III., 72.)*

Manure. 495

Fuel. 496

497

ADIANTUM, *Linn.; Syn. Fil., 113.*

A genus of Ferns belonging to the Tribe PTERIDEÆ, recognizable from all the other ferns (except some LINDSAYÆ) by the texture and one-sidedness of their segments; veins bifurcating but not anastomosing (except in the small section HEWARDIA). *Sori* marginal, varying in shape from globose to linear, sometimes confluent. *Indusium* of the same shape as the sorus, being a modification of the margin of the leaf thrown over the sorus; it is free from the frond except at the edge. *Capsules* attached to the under-surface of the indusium.

A large genus, having its head-quarters in tropical America, comprising some 80 species, 9 of which are met with in India.

498

Adiantum Capillus-Veneris, *Linn.; Syn. Fil., 123.*

THE MAIDEN-HAIR FERN.

Vern.—*Dúmtúli*, KASHMIR; *Kirwatzei (bisfáij)*, TRANS-INDUS; *Parsha warsha* (a corruption of *para-siyávashán*, Moodeen Sheriff) SALT-RANGE (Stewart); *Mubáraka*, KUMAON (Atkinson); *Pursha, hansráj*, *mubáraka*, HIND.; *Shir* or *shair-ul-jin*, ARAB. (*Murray's Drugs of Sind*); *Sír siá-peshâne*, PERS. (**Dr. Rice**, NEW YORK); *Hanspadi*, GUJ.

Habitat.—A graceful, delicate fern of damp places, in rocks, walls, or wells, found chiefly in the Western Himálaya, ascending to altitude 8,000 feet, but found also far to the east in the valley of Mánipur, extending to the mountains of the Burma-Mánipur frontier and to Chittagong. It is common in the Panjáb, descending even to the plains, where it is found in wells and damp places. "This plant is quite common in South India—see *Beddome's Ferns*." *(Deputy Surgeon-General G. Bidie, C.I.E., Madras.)* **Mr. C. B. Clarke**, in his *Ferns of Northern India*, gives its distribution as "Malabaria, Bombay to Ceylon (rare); from Kábul to England and Morocco; in tropical and temperate Africa and America; Queensland."

Botanic Diagnosis.—*Frond* usually 2-pinnate; segments ½-1 inch broad, the base cuneate, the outer edge rounded, deeply lobed from the circumference towards the centre, the lobes often again bluntly crenate; *petiole* near the centre. *Sori* roundish or obreniform.

Properties and Uses—

Sirop de Capillaire. 499

Smith's Economic Dictionary (1882) states that this is the plant used in the preparation of the so-called *Sirop de Capillaire* of Europe. This syrup is largely used in Italy and Greece in the treatment of chest complaints. **A. pedatum,** *Linn.*, is also extensively used for this purpose, being exported from Canada. **Dr. Dymock** draws my attention to the fact that **A. pedatum** is the French officinal plant, and that **A. Capillus-Veneris** is allowed as a substitute only. The former is a common North-West Himálayan plant. *Sirop de Capillaire* is imported into India, but might be prepared in the country to an unlimited extent, since at least four species of Adiantum are exceedingly common plants, especially **A. caudatum.**

MEDICINE. Fronds. 500

Medicine.—It is more than likely that the bulk of Adiantum sold medicinally in India is the true maiden-hair fern, **A. Capillus-Veneris,** although most writers on Indian drugs attribute this to **A. venustum.**

Dr. Irvine says: "*Hans Raj, shair-ul-jin*, Venus's hair, grows at Patna, but brought from Nepal; used as heating and febrifuge. Dose 20 to 30 grains, price 5 annas per lb." In the Panjáb the leaves, along with pepper, are administered as a febrifuge, and in South India, when prepared with honey, they are used in catarrhal affections.

It is probable, however, that the officinal root sold in the Panjáb bazars under the name of *Báisfaij* is a species of **Polypodium,** which see.

Adiantum caudatum, *Linn.; Syn. Fil., 115.*

501

Vern.—*Adhsarita-ka-jari, kanghai ? gunkiri*, PB.; *Mayúrashikhá*, SANS.; *Mayurshika*, CUTCH; *Mylekondai* ? TAM.

Habitat—An exceedingly common plant in many parts of India, Bengal, N.-W. Provinces, the Panjáb, Madras, Bombay, &c., covering nearly every old wall in shady places, fronds rooting at the tip and thus forming new plants.

Botanic Diagnosis.—*Fronds* simply pinnate, tomentose, often elongated into a tail which generally roots at the tip. *Segments* (or pinnæ) ½-¾ inch long by ¼ inch broad, dimidiate, nearly sessile, the lower edge straight and horizontal, the upper rounded, more or less cut. *Sori* roundish.

Properties and Uses—

Fronds. 502

Ainslie says that in the Island of Bourbon the fronds of this species, as also of the preceding, are used in the preparation of *Sirop de Capillaire*. **Mr. Baden Powell** (*Pb. Prod., I., 384*) associates this with **A. venustum** and other species, and gives them the vernacular names *par-i-siyá-*

washán and *hansuráj*. He adds: "An astringent and aromatic; said to be emetic in large doses; also tonic and febrifuge. This is the fern which is used in making 'Capillaire' Syrup." It would be interesting to know if a syrup was actually prepared in the Panjáb from this plant, or if **Mr. Baden Powell** has simply associated the Panjáb plants with the European drug. (See **A. Capillus-Veneris** and **A. pedatum**, the plants used in Europe in the manufacture of the *Sirop de Capillaire.*)

MEDICINE. 503

Medicine.—§" Used externally as a remedy for skin diseases." *(Surgeon W. Barren, Bhuj, Cutch, Bombay.)* "Said to be useful in diabetes." *(Surgeon-Major D. R. Thompson, M.D., C.I.E., Madras.)*

504

Adiantum flabellulatum, *Linn.; Syn. Fil., 126.*

Habitat.—Is very common in some parts of India. It was found plentiful in the oak and mixed forests of Mánipur.

Botanic Diagnosis.—Nearly allied to **A. pedatum**, smaller in size, but much thicker or coriaceous. *Scales* on the rhizome long, linear, chestnut-coloured; *rachis* often hairy, repeatedly dichotomous; *segments* glabrous, more or less rounded and toothed, the lower edge nearly straight; *sori* $\frac{1}{10}$ inch broad.

MEDICINE. 505

Medicine.—At Chuttuck I was told by a Mánipuri sepoy that the root was used medicinally.

506

A. lunulatum, *Burm.; Syn. Fil., 114; Hk., Ic. Pl., t. 191.*

Vern.—*Kálí-jhánt*, BENG., HIND.; *Mubárak, rájahans* or *hansráj, kansaráj*, BOMB.; *Hansráj*, GUJ.; *Ghodkhúrí*, MAR. (Horse's hoof, on account of the shape of the leaflets and the arrangement of the spores on under side like the horse-shoe.—*Surgeon-Major W. Dymock, Bombay.*)

Habitat.—This is unquestionably the commonest and most widely-spread Adiantum in India. In Bengal, every hedgerow and old brick-wall is covered with it, also the rocks and banks of the lower hills throughout the greater part of India; ascends to 4,000 feet; in damp glades it often becomes 2 feet in length, rooting as in **A. caudatum**, *Linn.*

Botanic Diagnosis.—*Frond* simply pinnate (in this respect allied to **A. caudatum**). *Rachis* naked, polished, dark brown. *Segments* glabrous, $\frac{3}{4}$ to $1\frac{1}{2}$ inch long by $\frac{1}{2}$ to 1 inch broad, subdimidiate, the lower edge nearly in a line with or oblique to the petiole, the upper edge rounded and usually more or less lobed.

Properties and Uses—

MEDICINE. Fronds. 507

While this and the preceding species are plentiful everywhere throughout Bengal, they do not seem to be collected for medicinal purposes, and it is probable that the fern root (*hansráj*) to be had in Calcutta native druggists' shops is imported and not procured locally. **Dr. Dymock** (*Mat. Med., W. Ind., p. 760*) seems to regard this as one of the species used medicinally in Bombay. (See under **A. venustum.**)

DYE. 508

Dye.—This and the preceding ferns, and probably also several other species, form ingredients in certain dye recipes.

Medicine.—§ "In Gujarát this is known as *Kálo-Hansráj*. It is extensively used in the treatment of children for febrile affections. The leaves are rubbed with water and given with sugar. It is worked up with ochre and applied locally for erysipelous affections. It is called *Kálo Hansráj*, probably on account of the black colour of the stalks." *(Surgeon-Major J. Robb, Ahmedabad.)*

"Demulcent, dose of the decoction one to two ounces. Used externally as a cooling lotion in cases of erysipelas." *(Surgeon W. Barren, Bhuj, Cutch, Bombay.)* "Very common in Madrás, but not, so far as I know, used in native medicine." *(Deputy Surgeon-General G. Bidie, C.I.E., Madras.)*

Adiantum pedatum, *Linn ; Syn. Fil., 125.* 509

Habitat.—North-West Himálaya from Garhwál to Sikkim; nowhere very plentiful.

Botanic Diagnosis.—*Frond* herbaceous glaucose and glabrous, with shining naked rachis once dichotomous, main divisions flabellately branched and somewhat scorpioid on either side, central pinnæ 6-9 inches long and 1 to 1½ in. broad.

Properties and Uses—

This is the French officinal species used in the preparation of the *Sirop de Capillaire.* The *United States Dispensatory* says of it: "An indigenous fern, the leaves of which are bitterish and aromatic, and have been supposed to be useful in chronic catarrhs and other pectoral affections. A European species, known by the vulgar name, is the **A. Capillus-Veneris**, which has similar properties, though feebler."

A. venustum, *Don ; Syn. Fil., 125.* 510

Vern.—*Par-e-siyá-washán, hansráj,* PERS., HIND. (in the BAZARS); *Shirul-jinn, shirul-jibal,* ARAB. The *Makhzan* gives *Káli-jhánp* or *Jhant* as the Hindi name of this plant. In Bombay it is chiefly known as *mubárak. Mayirsikki,* TAM.

Habitat.—A fern found in the Himálaya up to 8,000 feet in altitude, and, chiefly in the North-West, extending to Afghánistan; exceedingly plentiful in the fir forest north of Simla, often forming for miles the most characteristic under-vegetation.

Sirop de Capillaire. 511

Botanic Diagnosis.—*Fronds* 3 to 4 times pinnate. *Rachis* slender, polished, naked; *segments* rigid, prominently veined and toothed, upper edge rounded, lower cuneate into the petiole; *sori* 1 to 3, large, roundish, placed in a distinct hollow on the upper edge.

Properties and Uses—

MEDICINE. Fronds. 512

Medicine.—It possesses astringent and aromatic properties, is emetic in large doses; it is also tonic, febrifuge, and expectorant. This remark is given by **Mr. Baden Powell** in his *Panjáb Products* under **A. caudatum, A. venustum,** and other species, and it is probable that if all the preceding are not actually used indiscriminately, or as substitutes for each other, in different districts, they might easily be so, since they seem all to possess the same properties. **Stewart** says that "in Chumba it is pounded and applied to bruises, &c., and the plant appears to supply in the Panjáb most of the officinal *hansráj* which is administered as an anodyne in bronchitis, and is considered diuretic and emmenagogue."

Dr. Dymock describes the drug obtained from Adiantum under the joint names of A. **venustum,** *Don,* and **A. lunulatum,** *Spr.* The former plant is confined, however, to the North-West Himálaya, never descending below 3,000 feet in altitude, and is in fact much more temperate in its likings than any of the other supposed medicinal Adiantums. It has never been collected in Bombay, while the latter is plentiful, and, indeed, is one of the most abundant ferns in India, but is almost confined to the plains, or warm moist valleys of the lower hills. There can be no mistaking these two species,—the simply pinnate fronds of **A. lunulatum,** with segments (or pinnæ) sometimes as much as one inch in breadth, is quite unlike the tripinnate frond of **A. venustum,** with its rounded, deltoid, and cuneate-toothed and strongly-veined segments. There should be no difficulty in separating these two plants, but it would seem more than probable that they are never sold mixed together, although they may, in different parts of India, be substituted for each other. **A. Capillus-Veneris** is much more likely to be mixed with **A. venustum,** since it is very plentiful in most parts of India (as, for example, in the Panjáb plains, in wells) and very much resembles **A. venustum.** Speaking of these two species collectively, **Dr. Dymock,** however, says: "The Native physicians consider the maiden-hair to be deobstruent and resolvent, useful for clearing the *primæ viæ* of bile, adust bile, and phlegm, also pectoral, expectorant, diuretic, and emmenagogue. Used as a plaster it is considered to be discutient, and is applied to chronic tumours of various kinds. The Persian name is *Parsiawashan.* In Arabic it has many names: the best known are *Shir-ul-jinn* and *Shir-ul-jibal* (fairy's hair or mountain hair)." (*Dymock's Mat. Med., W. Ind.*)

§ "This is imported into Bombay from Persia as *Parisiyahwashan,* but the people here often call it *Mubaraka* and *Hansráj.* Native writers on Materia Medica do not distinguish between the species of Adiantum. The *Makhzan* gives *Káló-jhánt* as the Hindi for *Parisiyahwashan.*" (*Surgeon-Major W. Dymock, Bombay.*)

"It is recommended by Hakims for hydrophobia. It is resolvent, and is also used for the prevention of hair from falling. For internal use it is given in the form of a syrup." (*Assistant Surgeon J. N. Dey, Jeypore.*)

"A vapour bath medicated by a decoction from this plant is regarded useful in fever." (*Surgeon G. A. Emerson, Calcutta.*) "Very useful as a mild tonic, especially during convalescence from fever." (*Surgeon J. Anderson, Bijnor.*)

513

ADINA, *Salisb.; Gen. Pl., II., 30.*

A genus of trees or shrubs (belonging to the Natural Order RUBIACEÆ), comprising in all some six species, distributed through tropical Asia and America, four being met with in India. Adina is referred to the Tribe NAUCLEÆ, having the flowers collected into dense globose heads. *Corolla* funnel-shaped; *stigma* simple. It is placed in the section of the Tribe said to have the ovaries free or nearly so.

The genus may be diagnosed thus: *Leaves* having large caducous stipules. *Flowers bracteate, densely crowded in solitary or panicled heads. Calyx-tube angled, 5-lobed. Corolla* funnel-shaped, with a long tube; throat glabrous, lobes 5-valvate. *Stamens* 5, on the mouth of the corolla; *filaments* short. *Ovary* 2-celled; *style* filiform; *stigma* capitate or clubbed; *ovules* numerous, imbricated upon a pendulous placenta in each cell. *Capsule* of 2 dehiscent cocci, many-seeded.

514

Adina cordifolia, *Hook. f. & Bth.; Fl. Br. Ind., III., 24; Cor., Pl. I., t. 53.*

Syn.—NAUCLEA CORDIFOLIA, *Roxb., Fl. Ind., Ed. C.B.C., 172.*

Vern.—*Haldu, hardu, kadámi, karam,* HIND.; *Bangka, keli-kadam, petpuria, da-kóm,* BENG.; *Hardua, hardú (haldi in Gazetteer),* C. P.; *Kúrumba, komba sanko,* KOL; *Karám,* SANTAL; *Bara kuram,* MAL., (S. P.); *Karam,* NEPAL; *Tikkoe,* BAHRAICH and GONDA; *Hardu, paspu, kurmi,* GOND; *Holonda,* URIYA; *Shangdong,* GARO; *Roghu, keli-kadam,* ASS.; *Manjakadambe,* TAM.; *Daduga, betta-ganapa, bandaru, dúdagú, paspu kandi, paspu kadimi,* TEL.; *Arsintega,* MYSORE; *Hedde, yettéga-pettega, arsanatéga, yettada, ahuau,* KAN.; *Hedú,* MAR.; *Haladhwán,* GUJ.; *Kolong,* CINGH.; *Thaing,* MAGH; *Dhárákadamba,* SANS.; *Hnaw* or *hnaubeng, nhingpen* or *nhan-ben,* BURM.

Habitat.—A large, deciduous tree, found in the Sub-Himálayan tract from the Jumna eastward, ascending to 3,000 feet in altitude and extending throughout the moister regions of India, Burma, and Ceylon. It is common in the Western Peninsula, especially in the forests of the Ratnagiri and Thána Districts of the Konkan, and in the forests of Surat and of Baroda, in Gujarát; from thence it extends south into the forests of Mysore, is plentiful in the forests of the Upper Godaveri and of Bhandara in the Central Provinces. Is common in the mixed leaf-shedding forests all over Burmá from Chittagong and Ava to Pegu and Martabán.

Botanic Diagnosis.—*Leaves* with petiole 2 to 3 inches long, orbicular-cordate, abruptly acuminate, pubescent beneath; *stipules* orbicular or oblong. *Peduncles* 1 to 3 axillary and one-headed. *Heads* of flowers ¾ to 1 inch diameter; *bracts* small towards the apex; *flowers* yellow.

MEDICINE. 515

Medicine.—§ "The small buds ground with round pepper are sniffed into the nose in severe headache." *(Rev. A. Campbell, Santal Mission, Pachumba.)* "Roots used as a medicine in Assam." *(H. Z. Darrah, Esq., Assam.)*

TIMBER. 516

Structure of the Wood.—Yellow, moderately hard, even-grained. No heartwood, no annual rings. It seasons well, takes a good polish, and is durable, but somewhat liable to warp and crack. Weight 40 to 50 lbs.

It is good for turning, and is extensively employed in construction, for furniture, agricultural implements, opium boxes, writing-tablets, gun-stocks, combs, and occasionally for dug-out canoes.

The *Bomb. Gaz., XIII., Part I., 24,* says: "this is a large, handsome tree; logs often more than 30 feet long; from durability in water they are much prized for fishing-stakes." In the *Mysore Gaz., I., p. 48,* occurs the following regarding this plant: "Wood like that of the box-tree; very close-grained, light and durable, but soon decays if exposed to wet."

Adina Griffithii, *Hook.f.; Fl. Br. Ind., III., 24.* 517

Habitat.—Khásia Mountains, altitude 3,000 feet.

Botanic Diagnosis.—*Leaves* shortly petioled, elliptic-oblong or obovate, shortly acuminate. *Heads* 1¼ inch diameter; *corolla* glabrous; *bracteoles* short, stiff, conical spines.

A. polycephala, *Benth.; Fl. Br. Ind., III., 25.* 518

Habitat.—Sylhet and the Khásia Hills, Chittagong, Tenasserim, and Moulmein.

Botanic Diagnosis.—*Leaves* shortly petioled, lanceolate, caudate-acuminate, glabrous. *Heads* in trichotomously-branched panicles.

A. sessilifolia, *Hook. f. & Bth.; Fl. Br. Ind., III., 24.* 519

Syn.—Nauclea sessilifolia, *Roxb.;* Nauclea sericea, *Wall.*

Vern.—*Kúm,* Beng.; *Kúmkoi,* Chakma; *Thaing,* Magh; *Teinkala, thitpayoung,* Burm.

Habitat.—A small tree of Chittagong and Burma. In Chittagong it is perhaps the only gregarious tree, being commonly found on flat places on the banks of rivers.

Botanic Diagnosis.—*Leaves* sessile, oblong, base cordate, tip rounded, glabrous. *Heads* 1½ inches diameter, silky. *Corolla* shaggy.

TIMBER. 520

Structure of the Wood.—Hard, yellow-brown. Weight 55 lbs.

Firewood. 521

Used in Chittagong for building purposes and firewood.

Adul Oil of Travancore. OIL 522

Was forwarded to the Great Exhibition of 1851. The oil is medicinal, but the botanical name of the plant from which it is obtained has not as yet been discovered.

ADONIS, *Linn.; Gen. Pl., I., 5.* 523

A genus of herbaceous annuals or perennials, found chiefly as weeds of cultivation in the temperate regions; they belong to the Natural Order Ranunculaceæ. There are in all only some three or four species, of which three are met with on the western alpine Himálaya from Kumaon to Kashmír. None are found east of that region. They are botanically interesting as belonging to the tribe Anemoneæ, although they possess a distinct calyx and corolla, the latter having 5 to 15 non-nectariferous petals. They are not known to be of any economic value, except that they are often met with in cultivation in Europe.

524

ÆCHMANTHERA, *Nees; Gen. Pl., II., 1088.*

A small genus of hairy shrubs (belonging to the Natural Order ACANTHACEÆ and the Tribe RUELLIEÆ), containing in all only two species, one met with on the Himálaya and the other in the Khásia Hills.

Leaves broad elliptic, acute, crenate, often viscid. Clusters of *flowers* sessile, scattered on the branches of the trichotomous cyme. *Corolla* tubula rventricose, nearly straight, widened suddenly near the middle, glabrous but with two hairy lines on the palate within; *segments* 5, sub-equal, rounded, twisted to the left in bud, pale violet or purplish. *Stamens* 4, didynamous, included longer filaments hairy; *anthers* two-celled, *cells* muticose; *connective* excurrent at the tip or not. *Ovary* densely hairy at the apex; *stigma* large, simple; *ovules* 4-6 in each cell. *Fruit* seeded to the bottom (a character which at once separates this genus from RUELLIA).

525

Æchmanthera leiosperma, *Clarke; Fl. Br. Ind., IV., 429.*

Habitat.—Jaintia and Khásia Hills, altitude 3,000 feet.

Mr. C. B. Clarke, in the *Flora of British India*, says of this curious plant that, except that the seeds are glabrous, and when wetted not discoid, the plant is hardly distinguishable from **Æ. tomentosa.**

No information regarding its economic uses.

526

Æ. tomentosa, *Nees, var.* **Wallichi**; *Fl. Br. Ind., IV., 428.*

Syn.—Æ. WALLICHI, *Nees*; Æ. GOSSYPINA, *Nees.*

Vern.—*Patrang, ban marú,* CHUMBA.

Habitat.—A small shrub, met with in the temperate Himálaya, from Kashmír to Bhután, altitude 3,000 to 5,000 feet.

Botanic Diagnosis.—*Leaves* hairy, elliptic, acute; *petiole* 1½ inches long. *Anthers* oblong, connective, not excurrent. *Seeds* densely hairy, the hairs starting out when wetted.

FIBRE. 527 Bees. 528

Fibre.—In **Dr. Stewart's** *Panjáb Plants* occurs the following note regarding this plant: "**Madden** states that bees are particularly fond of its flowers, and **Jameson** mentions that a kind of cloth is made from the tomentum of the leaf."

ÆGIALITIS, *R. Br.; Gen. Pl., II., 624.*

529

Ægialitis annulata, *R. Br.*; PLUMBAGINEÆ.

Syn.—Æ. ROTUNDIFOLIA, *Roxb., Fl. Ind., Ed. C.B.C., 278*; Æ. ANNULATA, *Kurz, in Journ., As. Soc.*; Æ. ROTUNDIFOLIA, *Prest. Bot. Bermeck.*

Habitat.—A small, evergreen treelet, with pale yellow, sessile flowers; found in the tidal forests of the Sunderbans, Chittagong, Arakan, Burma, and the Andaman Islands.

TIMBER. 530

Structure of the Wood.—Very curious, resembling that of the monocotyledons. It consists of a soft pithy substance, with scattered white pore-bearing wood, resembling fibro-vascular bundles, but quite distinct in character.

ÆGICERAS, *Gærtn.; Gen. Pl., II., 648.*

531

Ægiceras majus, *Gærtn.; Fl. Br. Ind., III., 533; Wight, Ill., t. 146;* MYRSINEÆ.

Vern.—*Halsi, khalshi,* BENG.; *Bútayet,* BURM.; *Kánjlá,* MAR.; *Chawir,* SIND.

Habitat.—A small, evergreen tree, met with in the coast forests and tidal creeks of the Western Coast, Bengal, Burma, and the Andaman Islands.

TIMBER. 532

Structure of the Wood.—Hard, close-grained. No annual rings. Weight 40 lbs.

Used for firewood and in Jessore in the construction of native huts.

533

ÆGLE, *Corr.; Gen. Pl., I., 306.*

A genus (belonging to the Natural Order RUTACEÆ) comprising two or three trees inhabiting tropical Asia and Africa.

Leaves alternate, 3-foliolate; *leaflets* membranous, subcrenulate. *Flowers* large, white, in axillary panicles. *Stamens* 30 to 60, inserted round an inconspicuous disk; *filaments* short, subulate; *anthers* elongated, erect. *Ovary* ovoid from a broad axis; *cells* 8-20, peripheral; *ovules* many in each cell, 2-seriate. *Fruit* large, globose, 8-15-celled, 6- to many-seeded, rind woody. *Seeds* within membranous cells, buried in the aromatic pulp, oblong compressed; *testa* woolly and mucous; *cotyledons* thick, fleshy; *radicle* pointing away from the *hilum.*

The name Ægle is in allusion to one of the Hesperides, whose orchard bore golden fruit. Marmelos is the Portuguese for quince. By the mediæval writers this was called "Marmelos de Benguala," or Bengal Quince.

534

Ægle Marmelos, *Corr.; Fl. Br. Ind., I., 516; Wight, Ic., t. 16.*

THE BAEL OR BEL FRUIT TREE; THE BENGAL QUINCE.

Syn.—CRATÆVA MARMELOS, *Linn.*; C. RELIGIOSA, *Ainslie.*

Vern.—*Bél, si-phal, siriphal,* HIND.; *Bela, bel, vilva,* BENG.; *Bel,* ASS.; *Bela, bila,* BOMB.; *Bel,* MAR.; *Bil,* GUJ.; *Bila, katori,* SIND.; *Sriphal, bilva, malura, bilvaphalam, balva,* SANS.; *Safarjale-hindi, shul,* ARAB., PERS.; *Lohagasi,* KOL.; *Auretpang,* MAGH.; *Vilva-pazham,* TAM.; *Marédu, maluramu, bilvapandu, patir,* TEL.; *Maika, mahaka,* GOND.; *Kúvalap-pazham,* MAL.; *Corvalum, bela,* KURKU; *Bilapatri,* or *Bel-patri,* KAN.; *Okshit, ushitben,* BURM.; *Bélli,* SINGH.

The *Bilva, mabura,* or *matura,* of the ancients. **Roxburgh** says a small variety is called *Shriphula* in Bengal.

References.—*Roxb., Fl. Ind., Ed. C.B.C., 429; Pharm. Ind., 46; Flück. & Hanb., Pharmacog., 129; U. S. Dispens., Ed. 15th, 280; Bentl. & Trim., Med. Pl., 55; Dymock, Mat. Med. W. Ind., 112; Moodeen Sheriff, Supp. Pharm. Ind., 33; U. C. Dutt, Mat. Med. Hind., 129; Stewart's Pb. Pl., 28; Brandis, For. Fl., 57; Kurz, Burm. Fl., 199; Balfour's Cycl., I., 33; Official Correspondence, Home Dept.'s Progs., 1880, 286.*

Habitat.—A tree, found in cultivation all over India, often curiously sending up off-shoots from the roots, which in time become trees. Wild in Sub-Himálayan forests from the Jhelum eastward, in Central and South India, and in Burma.

Botanic Diagnosis.—A small, deciduous, glabrous tree, with straight, strong, axillary spines. *Leaves* pale green, of three leaflets; lateral *leaflets* sessile, ovate-lanceolate, 3-5 inches long, terminal long petioled. *Flowers* an inch in diameter, greenish white, sweetly scented.

This tree has its nearest affinity to the elephant-apple or wood-apple, but the imparipinnate leaves, 1-celled fruit, and few stamens of the latter, at once remove it from the *Bél.* The *Flora of British India* remarks that there is a form in Burma with oblong fruits, of which no definite information exists, and it is not known whether this is a distinct species or only a local variety. In most bazars of India there are two kinds—the small or wild form, and the large or cultivated. In a correspondence with the Home Department, Government of India, communicated by **Honorary Surgeon Moodeen Sheriff, Khan Bahadur,** and forwarded by the Government of Madras, these two forms are carefully compared and contrasted.

Dr. Moodeen Sheriff says the cultivated form is generally free from spines: "The leaflets are broadly and abruptly acuminate, instead of oblong or broadly lanceolate, and, when bruised, have an agreeable and aromatic odour; fruit three or four times larger, edible, and very delicious when quite ripe." The officinal part is the full-grown fruit of both varieties just when it begins to ripen. It is green or yellowish green externally and yellow internally. The mucus contained within the cells of the fruit and around the seeds is "thick, very tenacious, transparent, and terebinthinate in smell and taste." "The pulp should be removed from the rind before the fruit is dry, cut into small pieces, and dried in the sun. When dry the pulp of the small variety retains its yellowness, while that of the large becomes brown or reddish brown. The pulp of both has an agreeable and aromatic odour and a terebinthinate and bitterish taste. The pulp of the small or common variety, however, is much stronger in these respects and preferable as a medicine. The dried pulp is not destroyed by keeping."

THE GUM.

GUM from the Stem. 535

THE STEM yields a good gum, occurring in tears like gum arabic, or in fragmentary pieces resembling coarse brown sugar.

from the Seeds. 536

From THE SEEDS a mucous fluid is secreted within the cells of the fruit which hardens into a transparent, tasteless, gummy substance. **Roxburgh**, who rarely overlooked any facts connected with the plants he had the opportunity of examining, clearly described this substance, but by modern authors it has apparently been confused with the opaque yellowish pulp. **Roxburgh** says: "Berry large, sub-spherical, smooth, with a hard shell, from 10 to 15-celled; the cells contain, besides the seeds, a large quantity of an exceedingly tenacious, tran sparent gluten, which, on drying, becomes very hard, but continues transparent; when fresh it may be drawn out into threads of one or two yards in length, and so fine as to be scarcely perceptible to the naked eye, before it breaks." "The *mucus* of the seed is for some purposes a very good cement." (*Roxb., Fl. Ind., Ed. C.B.C., p. 429.*)

§ "The seeds and the mucus are encased in a rough, opaque membrane, in the form of a white bean—the carpels or cells of the fruit. From 10 to 15 of these bean-shaped cells, with the gluten and seeds inside, are found in each fruit. They are embedded vertically in the yellowish opaque pulp of the fruit. As stated by **Dr. Roxburgh**, the mucus is transparent and very tenacious. It has the appearance of an exceedingly pure white gum, and is almost tasteless. I am not aware of any use to which it is put, but with lime it acts as a very good cement for mending porcelain-ware." (*L. Liotard.*)

THE DYE.

DYE. Rind. 537

A yellow dye is obtained from the rind of the fruit; the unripe rind is also used along with myrabolans in calico-printing.

THE MEDICINAL PROPERTIES OF BÉL.

No drug has been longer and better known nor more appreciated by the inhabitants of India than *bél;* but the descriptions given by English writers are very ambiguous. The unripe fruit acts as an astringent; the ripe fruit, taken in the *fresh* state, is laxative, but the *dried* ripe pulp is only mildly astringent. By some authors the astringency is denied; a few chemists maintain that the fruit contains tannin, while others assert that this is not the case. The drug used in India, for diarrhœa and dysentery, is the roasted or sun-dried unripe fruit cut up into slices.

Synopsis of the parts of the plant as used medicinally.

MEDICINE.

Unripe Fruit. 538

(*a*) The UNRIPE FRUIT is cut up and sun-dried, and in this form is sold in the bazars in whole or broken slices. It is regarded as astringent, digestive, and stomachic, and is prescribed in diarrhœa and dysentery with debility of the mucous membrane, often proving effectual in chronic cases after all other medicines have failed. It seems specially useful in chronic diarrhœa; a simple change of the hours of meals and an alteration in the ordinary diet, combined with *bél* fruit, will almost universally succeed.

Ripe Fruit. 539

(*b*) The RIPE FRUIT is sweet, aromatic, and cooling; made into a morning sherbet, cooled with ice, it is pleasantly laxative and a good simple cure for dyspepsia and is useful in febrile affections. The *dried* ripe pulp is mildly astringent and may be used in dysentery. A useful popular preparation, made in India, is the Bél-marmalade, which may be taken like jam at the breakfast table in convalescence from chronic dysentery or diarrhœa.

Root-bark. 540

(*c*) The ROOT (and sometimes the STEM) BARK is made into a decoction which is used in the treatment of intermittent fever. It constitutes an ingredient in the *dasamul* or ten roots. It is given in hypochondriasis and palpitation of the heart.

Leaves. 541

(*d*) The LEAVES are made into poultice, used in the treatment of ophthalmia. The fresh juice is bitter and pungent, and diluted with water is praised as a remedy in catarrhs and feverishness.

Rind. 542

(*e*) The ASTRINGENT RIND of the ripe fruit is employed in dyeing and tanning. It is also sometimes used medicinally.

§ "Sections of the dried rind are sometimes used as receptacles for medicines." (*Assistant Surgeon Bhagwan Das, Rawal Pindi.*)

Flowers. 543

(*f*) The FLOWERS are deemed fragrant by the Native physicians.

The difference between the appearance and properties of fresh ripe fruit, and of the dried slices of unripe fruit sold in the druggists' shops and exported to Europe, must not be overlooked. The ripe pulp is of a pale orange or flesh colour, is deliciously fragrant, and yields with water a pleasant orange-coloured sherbet, slightly laxative. The dried slices give a reddish solution, acid and astringent in its action (or by some authors considered stimulant to the mucous membrane, but not astringent), and not possessed of the characteristic fragrance of the ripe fruit. The dried slices are prepared from the unripe fruit before the pulp has either become flesh-coloured or acquired its characteristic odour. **Dr. Moodeen Sheriff** (*in the official correspondence quoted*) says it is "a tonic, stomachic febrifuge, nauseant, and a remedy in dysentery, scurvy, and apthæ. It is not astringent, and therefore not useful in all forms of dysentery and diarrhœa. Acute dysentery is the disease which is most benefited by it, particularly in its first stage. It seems to exercise a greater influence in altering the nature of the motions than in diminishing their frequency. Its usefulness is greatly enhanced by the combination of opium. (Pulv. Ipecac. Co.)" (Compare with medical opinions.)

Chemical Composition.

According to **Dr. Macnamara** and **Mr. Pollock**, *bél* contains tannic acid, a concrete volatile oil, a bitter principle, and a balsamic principle resembling Balsam of Peru. Speaking of this analysis, **Professor Flückiger** and **D. Hanbury**, in their *Pharmacographia*, say they are unable to confirm the conclusions arrived at; "Nor can we explain by any chemical examination upon what constituent the alleged medicinal efficacy of bael depends." With reference to the unripe dried fruit these learned authors say: "The pulp, moistened with cold water, yields a red liquid

Opinions of Medical Officers

MEDICINE.

containing chiefly mucilage and (probably) pectin, which separates if the liquid is concentrated by evaporation. The mucilage may be precipitated by neutral acetate of lead or by alcohol, but is not coloured by iodine. It may be separated by a filter into a portion truly soluble (as proved by the addition of alcohol or acetate of lead), and another, comprehending the larger bulk, which is only swollen like tragacanth, but is far more glutinous and completely transparent.

"Neither a per- nor a proto-salt of iron shows the infusion to contain any appreciable quantity of tannin, nor is the drug in any sense possessed of astringent properties." (*Pharmacographia, Ed. 1879, 131.*)

§ "A section of both the ripe and unripe fruit, when moistened with a solution of ferric chloride, gives a most marked tannic acid reaction strongest in those portions of the pulp nearest to the rind. **Flückiger** and **Hanbury's** statement (quoted above) that the drug does not contain any appreciable amount of tannin, and is therefore not possessed of astringent properties, requires to be modified. The clear mucilage which surrounds the seeds has an acid reaction, and is readily soluble in water. It gives no reaction with either ferrous or ferric salts, and does not possess any astringent properties. It contains lime." (*Surgeon Warden, Prof. of Chemistry, Calcutta.*)

Officinal Preparations.

PREPARATIONS. Extract. 544 Liquid Extract. 545

(1) Extract of Bél.—Made from the fresh unripe fruit. Dose from half a drachm to one drachm twice or thrice daily.

(2) Liquid extract of Bel.—Prepared from the dried slices of unripe fruit. This possesses in a much less degree the properties of the extract. Dose one to two fluid drachms.
Made officinal in *Pharmacopœia of India* in 1868.

(3) **Dr. Moodeen Sheriff** says that a powder of the dried pulp is the most convenient form of administration: it keeps well in tight bottles. Dose as a tonic 12 to 15 grains; as a febrifuge and remedy for scurvy and apthæ, 16 to 20 grains; and as a nauseant and remedy in dysentery, 20 grains to 2 drachms.

Opinions regarding the Unripe Fruit.

§ "The pulp of the unripe fruit is soaked in gingelly oil for a week, and this oil smeared over the body before bathing, to remove the peculiar burning sensation in the soles of the feet so common amongst natives." (*Surgeon-Major John Lancaster, M.B., Chittore.*)

"I have found this fruit very useful in cattarrh and diarrhœa, but of very little use in acute dysentery. I consider the taste unpleasant." (*Surgeon-Major H. J. Hazlitt, Ootacamund, Nilgiri Hills.*)

"The unripe fruit, in the form of decoction, is very effective in cases of chronic diarrhœa, such as occur in jails, where there is also probably a scorbutic element present. Its efficiency is increased by the addition of opium. I have not found it produce hæmorrhoids in any of my cases, although the drug has been continued for a long time." (*Surgeon S. H. Browne, M.D., Hoshangabad, Central Provinces.*)

"Pulp of green fruit, softened by roasting and sweetened with sugar-candy, is useful in chronic diarrhœa and dysentery. The 'sharbat' of ripe fruit is a pleasant, cooling drink, but heavy of digestion, often causing acidity and heartburn." (*Assistant Surgeon Shib Chunder Bhuttacharji, Chanda, Central Provinces.*)

"Very useful in chronic dysentery. In the acute form it is not so useful, owing to the rapidity with which inflammatory action proceeds and grave ulceration supervenes." (*Surgeon-Major C. R. G. Parker, Pallaveram, Madras.*) "Most useful in diarrhœa due to general relaxed state of health

MEDICINE.

(more particularly in summer.)" (*Surgeon H. D. Masani, Karachi, Bombay.*) "Astringent in chronic dysentery and diarrhœa, the unripe fruit made into decoction. The ripe fruit is eaten with sugar." (*Surgeon-Major A. F. Dobson, M.B., Bangalore.*) "It is also pickled, and in this state is of great benefit in chronic dysentery." (*Surgeon-Major A. S. G. Jayakar, Muskat, Arabia.*) "I always use this in chronic diarrhœa, especially after an attack of dysentery." (*Surgeon-Major H. D. Cook, Calicut, Malabar.*) "If used for any length of time, *bél* is apt to produce hæmorrhoids, but this is avoided by using a little sugar along with the *bél*." (*Surgeon G. A. Emerson, Calcutta.*)

"The powder is more useful in acute diseases, and the syrup in chronic affections. In acute dysentery, the powder should be administered in much larger doses than in any other disease. Its first good effect is the rapid disappearance of blood and increase of the fœculent matter in the motions. It seems to have, in fact, more power to alter the nature of the motions than to reduce their number. For the latter purpose, *i.e.*, to check the frequency of the motions, it requires to be combined with some preparation of opium. The powder of bael fruit is also useful in relieving the febrile symptoms in all forms of idiopathic fevers, including hectic and typhoid. The abnormal temperature, in febrile conditions, is reduced under its use, in a very remarkable manner. Doses of the powder, as a remedy in dysentery, from 20 grains to 1 drachm, 4, 5, or 6 times in the 24 hours; and for all other purposes, from 10 to 20 grains. Of the syrup, from 2 to 4 or 6 fluid drachms 3 or 4 times in the 24 hours." (*Hony. Surgeon Moodeen Sheriff, Madras.*)

"Used as an astringent in diarrhœa and dysentery. The following powder is found very efficacious in chronic diarrhœa and dysentery:—

Bél pulp.	Almond.
Bruised mango seeds.	Sugar."
Catechu.	(*Jaswant Rai, Mooltan.*)
Bruised seeds of Plantago Isphagula.	

"The ripe fruit, when fresh, is mucilaginous, astringent, and slightly acid; I have found it useful in diarrhœa. The powder or decoction of the dry fruit may be used in the place of fresh fruit." (*Assistant Surgeon Bhagwan Das, Rawal Pindi.*)

"The half-ripe fruit, freshly gathered, is very useful in cases of obstinate diarrhœa and dysentery, especially if scurvy be present. Made into a powder with arrowroot it is very useful in the bowel complaints of children." (*Brigade Surgeon J. H. Thornton, B.A., M.B., Monghyr.*) "A liquid extract is the best way of administering this drug; it is somewhat overrated. It is useful in dysentery with scorbutic taint." (*Surgeon G. Cumberland Ross, Delhi.*)

"I think the ripe pulp is of very little value as an astringent. The unripe fruit is decidedly astringent." (*Deputy Surgeon-General G. Bidie, C.I.E., Madras.*)

"I have used unripe *bél* fruit in two ways: (*a*) entire *bél* fruit, partially burnt, about half or one third of each fruit to a man once a day. The burning softens the pulp and makes it more digestible; (*b*) unripe *bél* cut into slices and sun-dried and boiled before eating with a little sugar. In both ways it has been found to be a *mild* astringent, stomachic, and nutritive, most useful in chronic dysentery and diarrhœa; slowly but steadily reducing the number of motions and the quantity of mucus. The ripe fruit made into a sherbet is a mild laxative and cooling drink; a little *dahi* or tamarind and sugar is added to give a subacid taste and to increase the cooling laxative property." (*Surgeon D. Basu, Faridpur.*) "Both the ripe and unripe fruits are useful in dysentery, especially after

ÆGLE
Marmelos.

The Bael Fruit.

MEDICINE.

the acute symptoms have been checked by ipecacuanha." (*Surgeon Price, Shahabad.*)

"I have used a strong decoction of the dried and sliced fruit in chronic dysentery and diarrhœa, also the sherbet and pulp of the ripe fruit and the different forms of *bél* powder and preserve." (*Surgeon Picachy, Purneah.*) "The unripe fruit is roasted in the fire, and the pulp eaten in chronic dysentery and diarrhœa." (*Surgeon Bensley, Rajshahye.*)

"Its astringent property is due to the presence of tannic acid." (*Surgeon C. M. Russell, Sarun.*) "Used in chronic gonorrhœa, when the pulp of a fresh fruit is mixed with milk and administered with cubeb powder. Supposed to act as a diuretic and astringent on the mucus membranes of the generative organs." (*Surgeon-Major J. T. Fitzpatrick, M.D., Coimbatore.*) "The unripe fruit is half roasted in hot ashes; the whole fruit, with the rind and all, is beaten into a pulp, mixed with a sufficient quantity of water, strained and taken in large draughts with a little *palmyra* sugarcandy in cases of chronic dysentery." (*Surgeon-Major D. R. Thompson, M.D., C.I.E., Madras.*)

"The unripe fruit, powdered and given in doses of gr. vi and kino Co. gr. i with sugar gr. ii in each powder, to a small child suffering from chronic diarrhœa, was most useful. Extract has been prepared from the same by powdering, mixing with water, and evaporating to proper strength." (*Surgeon G. F. Poynder, Roorkee.*)

"The unripe fruit is used as a pickle. It is also made into preserve and commonly used for cases of dysentery." (*Surgeon-Major Robb, Ahmedabad.*)

Regarding the Ripe Fruit and Sherbet,—"The use of ripe fruit, in the form of *sherbet*, is very valuable in seasons of prevalence of bowel complaints and cholera. The strained pulp of the half-roasted unripe fruit is more efficacious than the extracts sold by English druggists in diarrhœa and dysentery." (*Surgeon R. L.Dutt, M.D., Pubna.*) "Sherbet made from the ripe fruit is most useful in chronic dysentery and diarrhœa." (*Surgeon C. H. Joubert, Darjíling*).

"The ripe fruit is an excellent laxative. The sherbet should be made thick enough to be eaten with a spoon, and not, as many servants make it, so thin that it can be drunk. The quantity required to produce a laxative effect is a small tumblerful; a mixture of half milk and half sherbet is an agreeable drink." (*Surgeon Edw. Borill, Champarun.*)

"A very pleasant and extremely useful fruit. Thick sherbet made of the ripe fruit is the best and surest laxative I know; the quantity necessary to produce this effect being an ordinary tumblerful. Very useful in dyspepsia and habitual constipation." (*Surgeon G. Price, Shahabad.*) "The pulp of the fresh fruit in the form of a thick sherbet is much recommended in scurvy, acting at the same time as a purgative." (*G. W.*) "In subacute and chronic dysentery often invaluable, taken in the form of sherbet." (*Surgeon J. Maitland, M.B., Madras.*)

"Prescribed in diarrhœa and dysentery. I have used the liquid extract very successfully as an injection in cases of gonorrhœa. The ripe fruit is eaten with sugar by natives suffering from dysentery or diarrhœa." (*Surgeon W. Barren, Bhuj, Cutch, Bombay.*)

Regarding the Leaves.—"Leaves are very efficacious when pounded into a pulp without any admixture of water and applied cold in the form of a poultice to unhealthy ulcers." (*Assistant Surgeon Anund Chunder Mukerji, Noakhally.*) "The fresh juice of the leaves acts as a mild laxative in cases of fever and catarrh, and has probably the effect of remedying these conditions." (*Doyal Chunder Shome, Lecturer, Campbell Medical School, Sealdah.*) "The decoction of the leaves is used as a febrifuge and expectorant." (*Assistant Surgeon Nundo Lal Ghose, Bankipore.*) "The juice of the fresh leaves has a laxative action." (*Surgeon K. D. Ghose, Bankoora.*)

Regarding the Root.—"The root is said by the people here to be an antidote against poisonous snake-bite." (*Surgeon C. J. W. Meadows, Burrisal.*)

"For habitual constipation, root-bark 1 oz., boiling water 10 oz.; dose 1 to 2 oz." (*Apothecary Thomas Ward, Madanapalle, Cuddapa.*)

FOOD.

FOOD. 546

The fruit when ripe is sweetish, wholesome, nutritious, and very palatable, and much esteemed and eaten by all classes. The ripe fruit, diluted with water, forms, with the addition of a small quantity of tamarind and sugar, a delicious and cooling drink.

TIMBER.

TIMBER. 547

Structure of the Wood.—Yellowish white, hard, with a strong aromatic scent when fresh cut; no heart wood, not durable, readily eaten by insects. Weight 40 to 50 lbs. (*Brandis.*) **Wallich** gives 49; **Mr. Gamble's** specimens averaged 57 lbs.

Used in construction for the pestles of oil and sugar mills, naves and other parts of carts, and for agricultural implements. The wood is also valued for making charcoal, but is not often used. (*Stewart, Pb. Pl.*) The wood is used in the Panch Mahâls for oil-mills.

Dr. Warth gives the following analysis of the ash composition of the wood of **Ægle Marmelos** in the *Indian Forester, Vol. X., p. 63*:—

Soluble potassium and sodium compounds	0·16
Phosphates of iron, calcium, &c.	0·13
Calcium carbonate	2·16
Magnesium carbonate	0·19
Silica with sand and other impurities	0·01
Total ash	2·65

DOMESTIC USES.

DOMESTIC USES. 548

Domestic and Sacred Uses.—"The fruit is nutritious, warm, cathartic; in taste delicious, in fragrance exquisite; its aperient and detersive quality and its efficiency in removing habitual costiveness have been proved by constant experience. The mucus of the seed is for some purposes a very good cement. The fruit is called *Shriphula,* because it sprang, say the Indian poets, from the milk of *Shri,* the goddess of abundance, who bestowed it on mankind at the request of *Jowarra,* whence he alone wears a chaplet of *Bilva* flowers; to him only the Hindús offer them; and when they see any of them fallen on the ground, they take them up with reverence and carry them to his temple." (*Roxb., in As. Res., Vol. 2, 340; also quoted in his Flora of India.*)

"This is one of the most sacred of Indian trees, cultivated near temples and dedicated to *Siva,* whose worship cannot be completed without its leaves. It is incumbent upon all Hindús to cultivate and cherish this tree, and it is sacrilege to cut it down." (*U. C. Dutt, Civil Medical Officer, Serampore.*)

Birdwood, in his *Industrial Arts of India,* says it is sacred to the Trimurti, being a representative of Siva. It is also sacred to the Parvati, and is the *Vilva-rupra,* one of the Patricas, or nine forms of *Kálí.* It is one of the trees the planting of which by the waysides gives long life. "Leaves used in enchantments." (*Irvine.*)

Snuff-boxes. 549

"In Pesháwar, large numbers of snuff-boxes for Afghans are made from the shell of the fruit, which is prettily carved over, and fitted with a small bone plug for the opening in the end, which serves as entrance and exit for the snuff." (*Stewart, Pb. Pl.*)

Medicine dishes. 550

The young dry shell is also largely used for medicine dishes and bottles.

ÆRIDES, *Lour.; Gen. Pl., III., 576.*

Orchids. 551

Ærides, a large genus of tropical orchids, of which **Æ. odoratum** is the most common, and at the same time most handsome species, growing freely and perfuming the orchid house. The leaves in this genus are distichous, channelled, and unequally truncate, but sometimes round. The flowers are large and frequently scented. **Æ. tœniale,** a native of Sylhet and Mánipur, has flat green rootlets, closely embracing the twigs upon which it grows, somewhat like a tape-worm, hence the specific name. **Æ. affine** and **Æ. odontochilum** are also met with in Assam and Sylhet, while in Western India occurs the spotted species, **Æ. maculosum.** None are known to be of economic value, although all are much prized as cultivated plants.

ÆRUA, *Forsk.; Gen. Pl., III., 34.*

552

A small genus of shrubs or herbs (placed in the Natural Order AMARANTACEÆ), comprising in all some 10 species; they are inhabitants of tropical Asia and Africa.

Erect or scandent, closely covered with a short white tomentum. *Leaves* alternate, opposite or almost whorled, linear oblong or obovate, entire. *Flowers* small or minute, arranged on terminal or axillary simple or panicled spikes, white or rusty; hermaphrodite, polygamous or diœcious, with one large and two small bracts, concave and persistent. *Perianth* of 5 (rarely 4) short leaves, oblong-lanceolate, acute or acuminate and very hairy. *Stamens* 5 (rarely 4); *filaments* often unequal, subulate, united at the base into a short cup. *Ovary* sub-globose; *style* short; *stigmas* 2 or capitulate. *Ovary* one-celled with a single ovule suspended from a long funiculus. *Seed* inverted, ovoid or reniform, compressed; *testa* thin, coriaceous; *arillus* wanting; *albumen* farinaceous; *radicle* superior.

553

Ærua javanica, *Juss.; Gen. Pl., III., 34; Wight, Ic., t. 876.*

Syn.—ACHYRANTHES INCANA, *Roxb., Fl. Ind., Ed. C.B.C., 225;* CELOSIA LANATA, *Linn.;* ACHYRANTHES JAVANICA, *Wight's Ic., t. 876.*

Vern.—Probably same as Æ. LANATA.

References.—*Voigt's Hort. Calcut., 317; Dalz. and Gibs., Bomb. Fl., 216; DC. Prod. XIII., Part 2, 299.*

Habitat.—Common throughout the Peninsula, and in flower all the year.

Botanic Diagnosis.—An erect or ascending herbaceous plant, tomentose hoary. *Leaves* obovate-lanceolate, obtuse, shortly mucronate. *Spikes* solitary, sessile, ascending. *Calyx* a little longer than the acuminate one-nerved bracts.

554

Æ. lanata, *Juss.; Gen. Pl., III., 34; (Wight, Ic, t. 723?).*

Syn.—ACHYRANTHES LANATA, *Roxb., Fl. Ind., Ed. C.B.C., 227.*

Vern.—*Chaya,* BENG.; *Bhúi,* RAJ.; *Búi, jári,* SIND; *Búi-kallan* (flowers as sold in bazars), PB.; *Kul-ke-jar, khul,* DUK.; *Azmei, spirke, sassdi,* TRANS-INDUS; *Kapur-madhura,* MAR.; *Sirrú-púlay-vayr,* TAM.; *Pindie-conda,* TEL.; *Astmabayda,* SANS.

References.—*DC. Prod., XIII., Part 2, 303; Dymock, Mat. Med., W. Ind., 540; Murray's Pl. and Drugs, Sind, 101.*

Habitat.—Small, herbaceous weeds, common everywhere in the plains, ascending to 3,000 feet in altitude; from the Indus eastward to Bengal and Burma, and southward to the Madras Presidency.

Botanic Diagnosis.—A small, herbaceous plant, ash-coloured, and a little tomentose. *Leaves* ovate, obtuse, shortly mucronate, pubescent on both sides. *Spikes* solitary or in twos or threes, sessile, horizontal. *Calyx* twice as long as the bracts.

Properties and Uses—

MEDICINE. Flowering tops. 555

Medicine.—The flowering tops are officinal, and the roots are used in the treatment of headache, and by the natives of the Malabar coast are regarded as demulcent. The flowers are sweetly scented. (*Graham, Murray, &c.*)

Roots. 556

§ "I am not aware of its being used medicinally in South India." (*Deputy Surgeon-General G. Bidie, C.I.E., Madras.*)

DOMESTIC USES. Woolly spikes. 557

In Sind, **Stewart** says the woolly spikes are used for stuffing pillows; rats are fond of the seed. The stems are often covered with woody galls.

558

Ærua scandens, *Wall.; DC. Prod., XIII., pt. 2, 302.*

Syn.—ACHYRANTHES SCANDENS, *Roxb.; Fl. Ind. Ed. C.B.C., 227; Gamble's Trees and Shrubs, Darj., 63.*

Vern.—*Nuriya*, BENG.

Habitat.—A large climber, covering the tallest tree with its masses of handsome flowers and soft whitish leaves: common in the lower hills, ascending to 6,000 feet; Monghyr, the Terai, Kumaon, &c.

Botanic Diagnosis.—Stem sub-fruticose, climbing. *Leaves* elliptic-oblong, acuminate at both ends, mucronulate, pubescent, green.

559

ÆSCHYNOMENE, *Linn.; Gen. Pl., I., 515.*

A genus of herbs or shrubs (belonging to the Natural Order LEGUMINOSÆ), comprising some 30 species, distributed throughout the tropics.

Leaves odd pinnate with numerous sensitive leaflets. *Flowers* in sparse racemes. *Calyx* deeply two-lipped, the lips toothed. *Corolla* early caducous; *standard* orbicular; *keel* not beaked. *Stamens* in two equal bundles. *Ovary* stalked, linear, many-ovuled; *style* filiform, incurved; *stigma* terminal. *Pod* with a stalk longer than the calyx, composed of 4-8 joints, each one-seeded.

The name "Æschynomene" is derived from αἰσχύνομαι, to be ashamed. It was given to a sensitive plant (? a **Mimosa**) mentioned by Pliny, probably in allusion to its closing so readily on being touched. The genus to which the name is now applied belongs to the Sub-Order PAPILIONACEÆ, not to the MIMOSEÆ. There are two Indian species.

560

Æschynomene aspera, *Linn.; Fl. Br. Ind., II., 152; Wight, Ic., t. 299.*

Syn.—HEDYSARUM LAGENARIUM, *Roxb., Fl. Ind., Ed. C.B.C.*

Vern.—*Sola* or *shola, phul-shola*, BENG.; *Kuhilá*, ASS.; *Bhend*, MAR.; *Atunete, takke*, TAM.; *Nirjilúza, bend*, TEL.; *Paukpan, paukbyu*, BURM.

Habitat.—A small, sub-floating bush, frequenting marshes and growing mostly during the season of inundation in Bengal, Assam, Sylhet, Burma, and South India.

Botanic Diagnosis.—*Stems* robust, swollen (often 2 inches in diameter, full of white light pith, with a central channel and a thin yellowish-grey bark, not more than $\frac{1}{18}$ inch in thickness), simple, rarely if ever branched, erect. Peduncles, calyx, and large corolla hispid.

Properties and Uses—

FIBRE. 561

Fibre.—In Burma a fibre is obtained from the thin bark.

DOMESTIC USES. Floats. 562

Domestic Uses.—The so-called pith or *sola*, however, is the most valuable product of the plant; it is largely used by fishermen for floats; it is cleverly cut up into paper-like sheets and made into temporary decorations for idols during certain festivities. Europeans use it for making hats

Hats. 563 **Toys. 564**

(*sola topís*) which, while being perfect protectors from the sun, are extremely light. The pith models of bullock-carts and other articles of Indian interest, made chiefly at Tanjore in Madras, are both curious and artistic. Sola is also made into a multitude of highly-coloured toys.

§ "A common weed in tanks. The soft and spongy pith is used as a substitute for sponge for the preparation of surgical lints which are used for widening the narrow openings of sinuses and abscesses, and also for dilating the rigid os uteri. If a piece of the pith, about an inch long and shaped like a cone with a sharp knife, be pressed between the fingers, it becomes very thin, and if inserted into the narrow opening of an abscess or sinus, it absorbs the moisture and swells up to its original size, and thus enlarges the opening without cutting." (*Hony. Surgeon Moodeen Sheriff, Khan Bahadur, Madras.*) "The pith of the stem can be used as a substitute for corks in medicine bottles." (*Brigade Surgeon G. A. Watson, Allahabad.*)

Æschynomene cannabina, *Retz.*

Syn. for **Sesbania aculeata,** *Pers., var.* **cannabina,** which see.

Æ. grandiflora, *Linn.*

Syn. for **Sesbania grandiflora,** *Pers.,* which see.

565

Æ. indica, *Linn.; Fl. Br. Ind., II., 151; Wight, Ic., t. 405.*

Syn.—HEDYSARUM NELI-TALI, *Roxb., Fl. Ind., Ed. C.B.C.;* SMITHIA ASPERA, *Roxb., Fl. Ind., Ed. C.B.C.;* Æ. KASHMIRIANA, *Camb.;* NELI TALI, *Rheed, Mal., IX., t. 18.*

Vern.—*Tiga jilúga,* TEL.

Habitat.—From the plains of Bengal to the lower hills; ascending in Kashmír to 5,000 feet and to 4,000 in Kumaon. Distributed to Ceylon, Siam, Japan.

Botanic Diagnosis.—Stems slender, much branched. *Peduncles* viscid; *calyx* small; *corolla* glabrous.

Apparently not put to any economic purpose.

Æ. Sesban, *Linn.*

Syn. for **Sesbania ægyptiaca,** *Pers.,* which see.

566

ÆSCULUS, *Linn.; Gen. Pl., I., 398.*

A genus of trees (belonging to the Natural Order SAPINDACEÆ), comprising in all only 14 species, natives of the temperate parts of Asia and America.

Leaves exstipulate, opposite, digitate, deciduous. *Panicles* terminal, thyrsoid. *Flowers* large, polygamous, irregular. *Sepals* and *petals* 4-5, unequal. *Disk* annular or unilateral, lobed or entire. *Ovary* sessile, 3-celled. *Style* elongated; *stigma* simple; *ovules* 2 in each cell, superposed. *Fruit* capsular, 1-3 celled; *valves* loculicidal, coriaceous; *cells* 1-seeded. *Seeds* exalbuminous, with a broad *hilum*; *testa* coriaceous; *cotyledons* thick, corrugated, conferruminated.

The generic name is derived from the Latin word *Æsculus,* given by Virgil and Horace to a tree believed to have been a species of Oak.

567

Æsculus indica, *Colebr.; Fl. Br. Ind., I., 675; Bot. Mag., t. 5117.*

THE INDIAN HORSE-CHESTNUT.

Vern.—*Torjaga,* TRANS-INDUS; *Háne, hanudún,* KASHMIR; *Gún, kanor,* PB.; *Kishing,* N.-E. KUMAUN; *Bankhor, gugu, kanor, pánkar,* HIND.

Habitat.—A large tree, 60 to 70 feet in height, deciduous, found most abundantly in the North-West Himálaya, extending from the Indus to

Nepal, between 4,000 and 10,000 feet in altitude. It grows on any soil, and produces annually an abundant crop of nuts and elegant foliage.

Botanic Diagnosis.—*Leaflets* 7, acuminate and minutely serrate, distinctly petioled. *Panicles* oblong, nearly equalling or exceeding the leaves. *Flowers* secund. *Petals* 4, red and yellow, the place of the fifth vacant. *Capsule* ovoid, reddish brown, without spines, rough. *Seeds* dark. The bark peels off in long vertical strips, giving old trees a scaly appearance.

Properties and Uses—

FOOD. Fruits. 568

Fodder.—The fruits are in the Himálaya eaten greedily by cattle, and in times of scarcity by men after being steeped in water, and sometimes mixed with flour. The leaves are lopped for cattle fodder.

FODDER. 569

Medicine.—The fruit is given to horses during colic. It is also applied externally in rheumatism, but for this purpose the oil is generally extracted from the seed.

MEDICINE. The Oil. 570

Structure of the Wood.—White, with a pinkish tinge, soft, close-grained. Weight 34 lbs. per cubic foot.

TIMBER. 571

It is used for building, water-troughs, platters, packing-cases, and tea-boxes; it is easily worked.

DOMESTIC USES. Cups. 572

Domestic Uses.—Thibetan drinking-cups are sometimes made of it.

CULTIVATION.—The *Indian Forester*, February 1884, page 57, gives a useful and practical note regarding the rearing of this tree. It recommends that the seed should be collected in November or December, and sown in good rich soil in drills. In the following cold weather, when the young trees shed their leaves, they should be transplanted into lines, each seedling 18 inches apart. If kept free from weeds, by the following cold season they will be ready for re-transplantation into the forest.

Æsculus Hippocastanum, *Linn.; Fl. Br. Ind., I., 675.*

573

THE HORSE-CHESTNUT.

Vern.—*Pú*, PB. HIM. NAME (*Brandis*).

Habitat.—A well-known tree in Great Britain and in Europe generally. Is supposed to have been introduced most probably from Asia. It is found in Persia and the Caucasian region, in India only in a state of cultivation. The home of the common horse-chestnut is at present unknown.

Botanic Diagnosis.—*Leaflets* 7, digitate, the larger ones woolly when young, bi-serrate and with prominent lateral nerves. *Capsule* echinate.

As an ornamental tree for parks, pleasure-grounds, road-sides, and avenues in temperate countries, this is justly a great favourite. The avenue in Bushy Park, London, planted by William III., affords a fine example of the adaptability of the horse-chestnut for ornamental purposes. (*Smith's Dictionary.*) It is not particular as to soil.

DYE. 574

Dye.—An extract from the wood is said to be used in imparting to silk a black dye.

FOOD. Nuts. 575

Food.—The NUTS are variously utilised; in Turkey they are ground with other food and given to horses, hence the name; in France they are employed in the manufacture of starch; in Ireland they are macerated in water, and being saponaceous are used to whiten linen.

MEDICINE. 576

Medicine.—The fruit and bark have for long been regarded as useful in the treatment of fevers, as an antiperiodic. Esculine in doses of 15 grains is said to have been found most useful in malarial disorders.

Chemical Composition.—The bitter principle of the fruit has been termed *esculin* and may be obtained by precipitation with acetate of lead. It forms shining, white, prismatic crystals, inodorous, bitter, slightly soluble in cold water. Its formula according to **Schiff** is $C_{15}H_{16}O_9$. When treated with dilute sulphuric acid it is converted into grape-sugar. Tannin is also found in all parts, the leaves and bark more especially.

577 **Æsculus punduana,** *Wall.; Fl. Br. Ind., I., 675.*

Syn.—Æ. ASSAMICA, *Griff. (Kurz, 286).*

Vern.—*Cherinangri,* NEPAL; *Kunkirkola, ekuhea,* ASS.; *Dingri,* DUARS; *Bolnawak,* GARO.

Habitat.—A moderate-sized, deciduous tree, found in Northern Bengal, in the Khásia Hills, Assam, and Burma, ascending to 4,000 feet.

Botanic Diagnosis.—*Leaflets* 5-7, shortly petioled. *Panicles* narrowly lanceolate, nearly equalling the leaves, lower pedicels longer. *Petals* white and yellow.

TIMBER. 578

Structure of the Wood.—White, soft, close-grained. Weight 36 lbs. per cubic foot. Rarely used.

579

AFZELIA, *Sm.; Gen. Pl., II., 580.*

A small genus of erect unarmed trees (belonging to the LEGUMINOSÆ) comprising only 10 species; distributed throughout the tropics of the Old World. *Leaves* abruptly pinnate, with few pairs of opposite leaflets. *Disk* at the top of a long calyx-tube. *Sepals* 4, unequal; only one *petal* developed, the others absent or rudimentary. *Stamens* 3-8, free; *filaments* long, pilose. *Pod* large, oblong, flat.

There are four Indian species—all, however, confined to the Straits and the Andaman Islands.

580 **Afzelia bijuga,** *A. Gray.; Fl. Br. Ind., II., 274.*

Vern.—*Shoondul, hinga,* BENG.; *Pynkado,* BURM., IN THE ANDAMANS; *Pirijdá, dsagundá,* AND.

Habitat.—A moderate-sized, evergreen tree, found in the Sunderbans Bengal, Andaman Islands, and the Malay Archipelago.

TIMBER. 581

Structure of the Wood.—Sapwood white, moderately hard, relatively large in young trees. Heartwood reddish brown, hard, close-grained. Weight: young wood, 36 to 42 lbs.; old wood, 45 to 49 lbs. (*Brandis, Memorandum on Andaman Woods,* 1874, Nos. 12 and 13, gives 50 lbs.)

A valuable wood, used in the Andamans for bridges and house-building. It is, however, very little known in India.

Agallocha, see **Aquilaria Agallocha,** *Roxb.;* THYMELÆACÆ.

THE EAGLE-WOOD; ALOES-WOOD; CALAMBAC-WOOD; AGILA; AKYAW.

Agallocha, see also **Excæcaria Agallocha,** *Willd.*

582

AGANOSMA, *G. Don; Fl. Br. Ind., III., 663; Gen. Pl., II., 717, reduces this genus to* Ichnocarpus.

A genus of APOCYNACEÆ, belonging to the sub-tribe EUECHITIDEÆ, comprising some 5 species; confined to India and the Malaya. Nearly allied to ICHNOCARPUS, the æstivation of the corolla separating them.

Flowers in terminal tomentose cymes, large or middle-sized. *Sepals* narrow lanceolate-acuminate, with subulate glands at the base. *Corolla* salver-shaped, *tube* short, *throat* naked, except with longitudinal bands behind the anthers; *lobes* lanceolate to linear oblong or broad and rounded, in bud overlapping to the right and then straight. *Stamens* included; *anthers* sagitate, conniving over the stigma and adnate to it; *cells* spurred at the base. *Disk* 5-lobed. *Carpels* 2, distinct, hirsute, many-ovuled. *Follicles* short or long, terete, straight or curved, linear. *Seeds* ovate linear-oblong, flattened, glabrous, not beaked.

583 **Aganosma calycina,** *A. DC.; Fl. Br. Ind., III., 664.*

Syn.—ECHITES CARYOPHYLLATA, *Roxb.;* E. CALYCINA, *Wall.;* A. ROXBURGHII, *G. Don, ex Wight, Ic., t. 440.*

Habitat.—An evergreen, scandent shrub, met with in the forests of Tenasserim, flowering in September.

Botanic Diagnosis.—*Leaves* elliptic-oblong, acuminate, glabrous, 7-10 pairs of arching, slender, impressed nerves. *Cymes* terminal, lax-flowered, densely rusty-tomentose. *Sepals* $\frac{3}{4}$-1 inch long, eglandular. *Petals* ovate-acute. *Ovary* quite glabrous.

Aganosma caryophyllata, *G. Don; Fl. Br. Ind., III., 664.* 584

Syn.—ECHITES CARYOPHYLLATA, *Wall.; non Roxb.*

Vern.—*Málati*, HIND., BENG., and SANS. **Voigt** gives *Gandhomálati* as the Bengali name.

Habitat.—Lower Bengal (Monghyr, *Hamilton*); common on rocks at Risikund (*Wallich*), Deccan Peninsula (*Heyne*), &c.

Botanic Diagnosis.—*Leaves* ovate-acute, obtuse or acuminate, glabrous or tomentose, nerves red-coloured, 3 pairs, very oblique. *Cymes* lax-flowered, pubescent; *pedicels* shorter than the sepals; *sepals* glandular within. *Corolla-lobes* obliquely orbicular. *Ovary* hairy at the tip.

Properties and Uses—

MEDICINE. 585

Medicine.—The only mention I find of **Aganosma** being medicinal is in *U. C. Dutt's Materia Medica*, where it occurs in his Glossary of Indian medicinal plants mentioned by Sanskrit writers. He does not give its supposed properties, but states that the vernacular name for **A. caryophyllata,** *G. Don*, is *Málatí*, whereas all other writers give that as the vernacular name of **A. calycina,** *DC.* It seems, therefore, that the plant has not been carefully identified by the author of the *Hindú Materia Medica.*

§ "According to Sanskrit authors this plant is heating and tonic; useful in diseases caused by disordered bile and blood." (*U. C. Dutt, Civil Medical Officer, Serampore.*)

A. cymosa, *G. Don; Fl. Br. Ind., III., 665.* 586

Habitat.—A stout, rambling climber, met with in Sylhet, and also in the Western Peninsula from Bombay to Travancore.

Botanic Diagnosis.—*Leaves* acute or finely acuminate. *Cymes* dense-flowered, rounded, densely tomentose. *Sepals* $\frac{1}{4}$-$\frac{1}{2}$ inch long. *Corolla-tube* $\frac{1}{6}$-$\frac{1}{3}$ inch; *petals* ovate-acuminate.

A. marginata, *G. Don; Fl. Br. Ind., III., 663.* 587

Habitat.—A large, evergreen, scandent bush, met with in Sylhet, Chittagong, Tenasserim, Malacca, and distributed to Java, Sumatra, and the Philippine Islands.

Botanic Diagnosis.—*Leaves* oblong-acute, acuminate or caudate, nerves very strong beneath, accurately uniting towards the margin. *Cymes* lax *Corolla* glabrous, tube rather longer than the acute calyx-segments; *lobes* linear-obtuse.

TIMBER. 588

Structure of the Wood.—Light, coarsely fibrous, close-grained, soft, and pale coloured (*Kurz*).

Agar-agar OR Ceylon Moss, see **Gracilaria lichenoides,** *Greville*, LICHENES.

AGARICUS, *Linn.; Syst. Nat., 1735.* 589

A large and important genus of FUNGI, referred to five series, each containing a number of sub-genera. *Spores* of various colours. *Gills* membranaceous, persistent, with an acute edge. *Trama* (the layer of tissue which separates the gills at their union to the pileus) floccose, confluent with the inferior hymenium. Fleshy fungi, which, on being dried, putrefy and cannot again be revived.

Series I.—Leucospori—Spores white.
Series II.—Hyporhodii—Spores pink.

Series III.—Dermini—Spores brown.
Series IV.—Pratellæ—Spores purple.
Series V.—Coprinarii—Spores black.
Each of these is again sub-divided into—
* Hymenophore, distinct from the fleshy stem.
** Hymenophore, confluent and homorogenous with the fleshy stem.
*** Hymenophore, confluent with but heterogenous from the cartilaginous stem.

More than 1,000 species are known and referred to AGARICUS, the typical genus of the Order. (*Cooke, Hand-Book Br. Fungi.*)

590

Agaricus campestris, *Linn.*

THE MUSHROOM.

Vern.—*Kat phula*, Ass.; *Alombe, kalambe*, BOMB.; *Mánskhel* KASHMIR; *Moksha*, CHAMBA; *Kuti leubhá, khumha* SIND.; *Kagdana chhatra*, GUJ.; *Chattrak*, SANS.; *Khúmbah, khámbúr, chattri*, AFG. BAZAR NAMES; *Ot*, SANTH; *Kúmbh samarogh* (**Stewart**), *Herar*, Poisonous forms. *Kullalic-dív* (Fairies' cap), also *chatr-i-már* (Snake's umbrella) and *samárugh*, PERS.

Habitat.—There are in India several species of fungi eaten indiscriminately, but as these have not as yet been botanically determined, it is preferable to refer to all under the common name which in English they would doubtless receive, *viz.*, The Mushroom.

Botanic Diagnosis.—"*Pileus* fleshy, convexoplane, dry silky floccose or squamose, stem stuffed, even, white; *ring* medial, somewhat torn; *gills* free, approximate, ventricose, sub-deliquescent, flesh-coloured, thin brown." (*Cooke.*) This species belongs to the series PRATELLÆ and the sub-genus PSALLIOTA.

Chemical Composition.—Interesting information regarding the chemistry of certain species of AGARICUS will be found in the *Year-Book of Pharmacy, 1877, p. 142; 1881, p. 147.*

§ "**Dr. N. Chevers**, in his work on *Indian Medical Jurisprudence*, refers to a case in which symptoms closely resembling those of intoxication rapidly ensued after eating mushrooms, and the author, therefore, considers it probable that there exists in Bengal a fungus which closely resembles an edible variety in form and colour, but which contains *amanitine* or *muscarine*, the poisonous principle of the **Amanita Muscaria.** *Amanitine* is stated also to exist in **A. bulbosus** and **A. volvaceus**. This principle sometimes acts as an irritant, at other times as a narcotic or narcotous acrid, and the symptoms may be developed within a few minutes, or not for several hours. Muscarine is described as being soluble in water, and appears to be a somewhat stable compound. Its action upon the system is opposite to that of atropine. (*Jour. Ph. Soc.*) The **A. Muscarius** or fly mushroom is in England most frequently found in fir or beech woods. It has a rich vermillion pileus studded with white or slightly yellowish warts, white gills, and tall, white stem, swollen at the base into a bulb, and furnished with a ring a short distance below the pileus. This fungus is used in Siberia as an intoxicating agent, and one or two suffice to produce pleasant intoxication for a whole day. (*Spons' Encycl.*) Some persons through idiosyncracy are injuriously affected by ordinary edible varieties, the effects being usually confined to colic, purging, and vomiting. (*Parg.*) Edible mushrooms contain a non-saponifiable, buttery, fat Agaricin. (*Lefort.*) The recent researches of **Dupetit** indicate that all edible mushrooms contain a poisonous principle, which resembles the soluble ferments and not the known alkaloids. The poisonous principle is destroyed at a temperature of 100° C. and the mushrooms rendered innocuous. This author has also obtained two alkaloids from edible mushrooms." (*Surgeon C. J. H. Warden, Professor of Chemistry, Medical College, Calcutta.*)

Fungi.

AGARICUS campestris.

Properties and Uses—

FOOD. 591 592 593 594

Food.—In many parts of India, especially in the Panjáb, the true Mushroom is abundant in fields. It is universally eaten by the natives, fresh or dried in the sun. It is apparently a very common plant in Afghanistan. **Aitchison** mentions **A. Mitto,** *Pers.*, as met with in Kúrum district; he also mentions **Morchella esculenta, Helvella crispa,** and **Hydnum coralloides** as eaten by the Afghans, the last-mentioned being collected in August, and sun-dried.

The common mushroom, says **Dr. Stewart,** is abundant in cattle-fields in many parts of the central Panjáb after the rains, and is also frequent in the desert tracts of central and southern Panjáb. It is largely eaten by the natives, and is described as excellent and equal to the English mushroom by those Europeans who have eaten it. It is also extensively dried for future consumption, and is said to preserve its flavour tolerably well. Mushrooms are largely used in Europe in the manufacture of ketchup. A trade in Panjáb mushrooms might easily be established were they to be improved in quality by cultivation.

It may not be out of place to mention here a few of the characters by which a wholesome fungus may be recognised:—

1st—Wholesome fungi are found growing in fields or in open grassy places in forests.

2nd—They are scattered, each rising direct from the ground; never collected into clumps nor found growing upon trees.

3rd—The stem should break easily when touched; it should spring from the centre of the pileus. The cap should be thick relatively to the gills.

4th—They should not be acid in flavour nor smell. No fungus is so poisonous but that this test may be put into force; but it does not follow an acid fungus will be poisonous; indeed, **Hydnum repandum** and **Cantharellus cibarius** are both acid, yet are excellent articles of food. A hot burning taste or acid flavour should, however, as a rule, be avoided.

5th—The bright rosy or pink gills and the absence of any yellow stain when bruised are two good tests.

The natives of India seem to eat any fungus, and, indeed, if properly cooked, few are dangerously poisonous. If macerated in vinegar before being cooked, and if eaten with plenty of bread, there is almost no danger. It is a good practice with any acid or doubtful mushroom to slice it into hot water, and then to press the slices in a cloth before stewing. The Russians preserve mushrooms in salt, but this is far from destroying their poisonous property, witness the death of the wife of the Czar Alexis I. from eating mushrooms in Lent. The narcotic poison in certain fungi resembles, in its action, Indian Hemp. No antidote has as yet been discovered for this poison.

MEDICINE. 595

Medicine.—The small dried mushrooms are officinal in the Panjáb, and are sold as "*Mokshai,*" being regarded as alterative.

§ "The cultivation of several indigenous varieties of mushroom is of importance for supplying an excellent nutritive food for European convalescents. There are many valuable varieties procurable in the plains of India. Even Truffles—white and black—I have seen in abundance at Bankura. In taste and flavour they are not inferior to the French plant. They grow under the soil below 'Sal' trees, and are dug out by the Santals." (*Surgeon-Major R. L. Dutt, M.D., Pubna.*)

"Common in South India, and used by the natives." (*Deputy Surgeon-General G. Bidie, Madras.*)

The following species are also mentioned by Indian writers on Economic plants; but the subject is at present too imperfectly known to admit of more than the enumeration of the names used by authors. It is hoped that this confession of want of definite information, regarding the edible and officinal Asiatic fungi, may bring about an investigation of what, both economically and scientifically, cannot fail to prove valuable and instructive.

596

Agaricus igniarius and **A. albus** are referred to by **Dr. Stewart** in his *Panjáb Plants;* **Polyporus igniarius** by **Dr. Irvine,** and **P. officinalis** by **Dr. Dymock.**

Vern.—*Bulgar jangli,* KASHMÍR; *Búti-ka-mochka,* CHENAB; *Kiain,* PB. *Ghárikún,* a BAZAR NAME.

MEDICINE. 597

Medicine.—Stewart remarks: "This appears to come from the west, about 15 seers being annually imported *viâ* Peshâwar." It is officinal, being given for internal disorders. The tinder or ashes are also said by **Honigberger** to be used to stop hæmorrhage. **Dr. Irvine** in his *Short Account of the Native Materia Indica of Patna* describes what in all probability is the same as the above species under the name of **Polyporus igniarius.** He gives it the HINDI name *garigond* and the ARAB. *Agarikún,* and describes the drug as "used as a styptic externally, and internally as a bitter tonic and laxative." He gives the dose as one to five grains, and says that it fetches the extraordinary price of R26 a lb. This is most probably the same plant as **Dr. Dymock** describes under the name of **Polyporus officinalis,** the *Garikún* of Bombay, which he says is largely imported from the Red Sea and the Persian Gulf ports into Bombay, and is much used by hakims.

"This is the *Agaricon* of **Dioscorides** and the white *Agaric* of European medicine." "It is commonly kept by native druggists, being an important article in the Materia Medica of the Mohammedans, who prescribe it in a great number of disorders, but generally in combination with other drugs. According to their hakims it acts principally by expelling cold and bilious humors." (*Surgeon-Major Dymock, Mat. Med., W. Ind., 702.*)

Hanbury, in his *Science Papers,* p. 184, speaking of what appears to be the same as the above fungus, says: "During the Middle Ages it was exported from Asia Minor; and in the Paris Exhibition specimens from this region—that is to say, from the Gulf of Adalia—were exhibited. What is the tree from which this Asiatic Agaric is obtained? It is found upon the larch in Northern Russia."

§ "*Garikún* acts as a purgative, and is used in disorders of the liver and kidneys. Diuretic and emmenagogue properties have also been assigned to it. It has been employed with benefit in cases of gravel in the bladder. It is used as a gargle in affections of the throat, gums, and teeth. In combination with liquorice, it is employed in chronic bronchitis and asthma, and with syrup of vinegar, in jaundice and enlargement of the spleen. It is said to be of much repute as a safeguard against scorpion-bites. In large doses it acts as a poison. *Jund-i-bedastur* is said to be its antidote." (*Asst. Surgeon Gholam Nabi.*)

"**A. albus** furnishes a principle *agaricin* which was formerly renowned as a specific for lessening the night sweats of phthisis. It is again coming largely into use." (*Surgeon-Major E. G. Russell, Calcutta.*) "It is diuretic, laxative, and expectorant. In very small doses it is a nervine tonic." (*Assistant Surgeon J. N. Dey, Jeypore.*) "A large fungus is imported as a medicine from Central Asia to Leh and Kashmír, and is called *Gari-kún.*" (*Surgeon-Major J. E. T. Aitchison, Simla.*)

From the conflicting opinions received from most parts of India, it seems that a number of widely different species of fungi are imported into and sold in India as *Gari-kún*. Until this subject can be thoroughly and scientifically investigated, it has been thought advisable to publish in one place all notices regarding *gari-kún*. It seems probable that most of the forms of *gari-kún* belong to the Polyporus group of fungi and not to Agaricus. For further particulars regarding other Indian fungi, see **Polyporus, Morchella**, &c.

Agaricus ostreatus, *Jacq.* 598

Vern.—*Phanasa-alambé*, or vulgarly *phansámba*, CUTCH, BOMB.

"This is a dark, snuff-coloured fungus, which grows upon the stumps of old jack-trees (*Phanas*). It consists of a short, thick stalk, which supports a flat woody pileus, having a considerable resemblance to an oyster-shell, and consisting of a number of laminæ, upon the under-surface of which is situated the hymenium."

MEDICINE. 599

Medicine.—"*Phanas-alambé* is ground to a paste with water and applied to the gums in cases of excessive salivation. It appears to have much the same properties as *Amadou*, and to be a useful styptic." (*Surgeon-Major Dymock, Mat. Med., W. Ind., 704.*)

§ "Useful in stomatitis." (*Surgeon W. Barren, Bhuj, Cutch, Bombay.*)

FOOD. 600

Food.—The species is edible.

Agathotes Chirata, *D. Don*, see **Swertia Chirata,** *Ham.*; GENTIANACEÆ.

THE CHIRETTA.

AGATI, *Desv.; Gen. Pl., I., 502.* 601

A generic name, formerly applied to one or two species of plants, now reduced to the section AGATI under the genus SESBANIA (Natural Order LEGUMINOSÆ). They differ from the type of the other SESBANIÆ chiefly in having larger flowers with the flower-bud falcately recurved.

Agati grandiflora, *Desv.*, or **Æschynomene grandiflora,** *Linn.; Roxb.*; **Coronilla grandiflora,** *Willd.*; see **Sesbania grandiflora,** *Pers.*, LEGUMINOSÆ.

AGAVE, *Linn.; Gen. Pl., III., 738.* 602

The name of a large and important genus belonging to the Natural Order AMARYLLIDEÆ. There are several species, all originally natives of Central America, and chiefly of Mexico. They are now, however, widely acclimatised in most warm, temperate, or sub-tropical and tropical countries—in Spain, Italy, Africa, Western Asia, and India. They are first mentioned in Europe in 1561, and are supposed to have been introduced into India by the Portuguese. They are commonly, but erroneously, called American Aloes. From the Aloë proper they are botanically separated by the position of the ovary, which is inferior in the Agave but superior in the Aloë. Like the Aloes, however, they consist of a crowded whorl of thick, fleshy leaves, more or less spirally arranged on the top of a short stem, which, in the majority of species, rarely rises much above the level of the ground. Along the margins are arranged sharp prickles, and each leaf ends in a formidable apex, long, sharp, and spear-like. In most species so closely do the leaves in the bud embrace each other, that each impresses its outline upon the fleshy substance of the other, forming a graceful variation on the otherwise smooth, glaucous surface. In cultivation several species become variegated in colour, the most striking being one with golden bands along the leaves. They take several years to reach the flowering stage, and from the fact that in adverse circumstances their development may be retarded from 10 to 50 or even 100 years, they are popularly called the Century Plants. When about to flower, an axis is developed from the centre of the rosette of leaves, and rising at the rate of from 5 to 10 inches a day, it often attains the height of 20, 30, or even 40 feet. Thereafter it produces its flowers from subdivisions, and continues to do so for over a month. After flowering and seeding the plant dies, but from the ground daughter off-shoots spring up, and thus a hedgerow of Agave continues to flower year after year.

Generic Diagnosis.—*Leaves* thick, fleshy toothed and spiny, crowded on the apex of a short, erect, succulent stem. *Scape* rising from the centre of the leaves, erect often becoming very high and rapidly developed, branching towards the upper third into an immense thrysiform panicle. *Flowers* fascicled on the branches of the panicle, greenish white, erect. *Perianth tube* often very short, split into 6 sub-equal lobes. *Stamens* longer than the perianth upon which they are inserted; *filaments* tapering above, flattened below. *Ovary* inferior, globose-ovoid, often fleshy, 3-celled; *ovules* numerous on the central angle on each cell and 2-seriate; *style* filiform at the base, 3-celled like the ovary.

The genus Fourcroya is so nearly allied to that of Agave, that the various species are popularly viewed as mere varieties of **A. americana.** All that can be said regarding the fibre of the species of Agave is equally applicable to the fibres from the species of Fourcroya. Indeed, the reports of Aloe fibre cultivation are uniformly written in what may be briefly described as commercial language, and it is quite impossible to discover whether the good or bad varieties mentioned by writers on the subject are different species of Agave, or even of Fourcroya, or merely all local and accidental varieties of **A. americana.** It has been repeatedly observed that aloe fibre plants change their character when taken from one country to another, and at the present moment it is next to impossible to arrive at any definite knowledge regarding the species and varieties cultivated for their fibre. The foregoing diagnostic characters of Agave, if compared with Fourcroya, should help, however, to remove ambiguity regarding the forms of the so-called American Aloë fibre. An important step would be taken were the fibre-yielding plants to be carefully referred to their respective genera. The account of **A. americana** in the succeeding pages should be viewed more as aloë fibre plant, since it seems probable that the name **A. americana** is popularly given to a series of species and varieties yielding allied fibres.

The generic name is derived from the Gr. ἀγαυός, illustrious or admirable, in allusion to the stately form of the flowering stem.

603

Agave americana, *Linn.*; AMARYLLIDEÆ.

THE CENTURY PLANT, THE AMERICAN ALOE, THE CARATA, *Eng.*; PITA OR MAGUEY, *Sp.*

Vern.—*Rakas-pattah, banskeora, bara kanwar, kantala, hathi-sengar, rám kantá,* HIND.; *Jungli* or *bilati-ánanásh,* vulgarly (*ánarás*), *banskeora, bilatipát, koyan,* incorrectly called *murga murji,* BENG.; *Wilyati kaitalu,* PB.; *Kantala,* SANS.; *Seubbára,* ARAB.; *Párkánd,* MAR.; *Rakas-patta,* DUK.; *Jangli-kunvára,* GUJ.; *Anaik-kat rázhai, pitha kalabuntha,* TAM.; *Rákáshimatalu,* TEL.; *Panam-katrázha,* MAL.; *Bhuttále, budukattalenaru,* KAN.; *Kétgí,* HYDERABAD (a name applied in other parts of India to Pandanus).

References.—*Roxb., Fl. Ind., Ed. C.B.C., 296; Royle, Fib. Pl., 41-50; Christy, New Com. Pl., 45-47; Drury, Us. Pl., 21; Lindley and Moore's Treasury of Botany; Smith's Dict., Econ. Pl.; Spons' Encycl.; Official Correspondence.*

Habitat.—Originally a native of America; naturalised in many parts of India. While plentiful throughout the country, the Agave nowhere occurs gregariously. In the Madras Presidency several varieties are extensively used for hedges to protect the railways, and in many of the hotter portions of Central India they grow where scarcely anything else grows, and, as hedges, serve the valuable purpose of checking the translation of sand and surface soil by the hot winds, which for some months sweep across the dry arid tracts. As has been remarked, they "stand isolated in the midst of dreary solitude, and impart to the tropical landscape a peculiarly melancholy character." It is one of the most remarkble features of the Agave that, while thus luxuriating under warm, tropical influences, it is equally at home on the hills under widely different climatic conditions. This strikes the traveller forcibly when, after hurrying across the tropical plains of Madras, with their interminable hedges of Agave and Cactus, he finds, on the hills amid a temperate vegetation, equally luxuriant hedgerows of Agave.

If planted at regular distances, with a ditch on one side so as to carry off the excess of moisture, the plant will thrive under the greatest variations of temperature, ascending in India to the altitude of the tea plantation, where it is largely employed as a hedge.

Synopsis of the Properties and Uses of Agave.

The LEAVES and the ROOTS yield an excellent fibre, generally known as Pita, American Aloë Fibre, or Vegetable Silk. The word Pita is, by some authors, restricted to a form which appears to be a variety of this species. The large, moist, fleshy leaves are sometimes used as a poultice. They are occasionally used as fodder.

The ROOTS are diuretic and antisyphilitic, and are said to find their way to Europe mixed with Sarsaparilla. **Prescott**, in his *History of Mexico*, says that when properly cooked the root affords a "palatable and nutritious food."

The SAP.—If the central bud be lopped off at the flowering season, the cut stem discharges freely a sour liquid, which ferments rapidly and forms the Pulque Beer of the Spaniards, or, by distillation, a kind of brandy known as Mexical.

The EXPRESSED JUICE of the leaves is administered by American doctors as a resolvent and alterative, especially in syphilis. It may also be used as a substitute for soap.

The FLOWERING STEM, dried and cut into slices, may be used as natural razor-strops, or as a substitute for cork. Wall plaster, impregnated with the expressed juice, is said to be proof against the ravages of white-ants.

THE GUM.

GUM. 604

It is stated that in America a gum exudes from the leaves, which in some respects resembles gum arabic. It contains a much larger proportion of lime and is only partially soluble in water. The soluble portion resembles gum arabic, and the insoluble, gum bassora (*U. S. Dispens., 15th Ed.*). The exudation of gum has apparently not been observed in India.

THE SAP AND JUICE.

The Sap and the Spirit.—In their *Treasury of Botany*, **Lindley** and **Moore** state that "the most important product of the Agave, and especially of **A. americana**, is the sap," and that many varieties of this species are "common everywhere in equinoctial America, from the plains even up to elevations of 9,000 or 10,000 feet." The sap is further stated to have given a net revenue of £166,497 from three cities.

In the *Pharm. Journal, 3rd series, V., 461*, occurs an interesting article upon Agave by **Mr. J. R. Jackson**, an abstract of which will be found in the *Year-Book of Pharmacy for 1875*, from which the following passages have been extracted: "The juice forms a large article of internal trade in Mexico. The plant is known as the Maguey, or 'tree of wonders,' and even at the present time, in some parts of Mexico, it is considered one of the most important productions of the soil. The use of the juice of the plant as an intoxicating beverage, is said by some to date back to the days of the early inhabitants of the Mexican Continent.

"SEVERAL VARIETIES OF THE PLANT are cultivated in Mexico, each being known for the greater or lesser quantity of the juice it produces, its colour, whether yellow or greenish, its thickness or sweetness or bitter taste. These variations as to the properties or consistency of the juice depend a great deal upon the nature of the soil and of the range of temperature; thus it is the least mucilaginous in a somewhat clayey soil, and is cultivated with the greatest success at an elevation of about 9,000 feet.

"THE MODE OF PROPAGATION is by removing the young plants or suckers from the old ones, and after spreading them on the ground for two or three months to partially dry them, so that they may not rot instead of starting into growth; they are planted in rows, and barley sown between them, which is considered rather to assist their growth. In a good soil, the Agave plant requires a period of from 10 to 12 years before attaining maturity. The plant upon attaining its full growth, which is easily discernible by its height and the prodigious extension of its leaves, brings forth a tall stem crowned with yellow flowers, and then a certain amount of pruning becomes necessary, so as to form a kind of reservoir in the centre, and what is technically termed a *cara* or 'face' around it, so as to cause the juice to flow towards the same spot, and to facilitate the extraction of it by removing some of the interior leaves and thorns.

PULQUE. 605

"TO COLLECT THE JUICE or *pulque*, as it is called, as soon as the leaves begin to turn yellow, a small concave aperture is scooped in the core of the plant, and an elongated tube-like gourd, the air in which is exhausted by suction, is thrust into the aperture; each labourer carries with him, strapped to his back, an impervious sheepskin bag, into which the gourd tube is emptied as soon as it is filled. The juice is emptied into vats, and allowed to stand for about 36 hours, when fermentation ensues, and its yellow transparent colour changes into a milky white.

TRADE IN PULQUE.—"Not less than 20,000 mules and donkeys laden with the beverage enter the city of Mexico every month by the gate leading to the Maguey district. To the quantity paying duty must also be added a considerable quantity which is smuggled in, and including this it may be calculated that about 50,000,000 bottles are now annually introduced into the city of Mexico.

MEXICAL SPIRIT. 606
BRANDY. 607
SUGAR FROM AGAVE. 608
VINEGAR FROM AGAVE. 609

MEZCAL OR MEXICAL SPIRIT.—"Besides this *pulque*, which, as we have seen, is the chief product of the *Agave* in Mexico, a strong spirit is prepared from the sap, known as *mezcal*, also a kind of brandy of 80 degrees of strength, a sweet, thick substance resembling honey, a concentrated gum used in medicine, brown sugar, loaf-sugar, sugar-candy, and vinegar of very excellent quality, so that the *Agave*, the value of which to us is mostly for its fibre, is, in fact, one of the most important economic plants of Mexico." (*Year-Book of Pharm.*, *1875*, *232; also 1882*, *221.*)

FIBRE. 610

THE AMERICAN ALOE FIBRE.

The immense and growing demand for new and cheap textile materials has recently caused the fibre greatly to surpass the sap in importance. In fact, the latter may be said to be unknown to the natives of India. It is much to be regretted that the same might almost be said of the fibre, although various species of Agave are now widely cultivated, if not naturalised, in India.

The American Aloe Fibre.

Flourishing under the most diverse circumstances, the Agave is in India eminently suited for providing a supply of fibre, if not for trade and exportation, at least for domestic purposes. Every encouragement should therefore be given to develope an interest in this most useful plant. In America rapid progress is being made in this direction, and it is evident that, with the American inexhaustible supply, and with improved machinery, the Pita fibre must, sooner or later, affect certain branches of the jute and other textile industries. For cordage it is now held in high esteem, and the manufacturers assert that the fibre imported into England from America improves every year. It is composed of large filaments, white, brilliant, and readily separated by friction without danger to the fibre. It takes colour freely and easily. It is light, contracts under water rapidly, and becomes fixed, while it bears changes of humidity even more severe than can be resisted by the best hemp. In London the aloe fibre generally fetches from £35 to £40 a ton.

In Mauritius, Agave fibre has attracted considerable attention within the past few years, the sterilization of the soil, through constant cultivation of sugarcane, having forced upon the planter the necessity of adopting some new and less exhaustive crop. There are already some six companies in existence, with a capital of £150,000, working the fibre, which they appear to extract chiefly from **Fourcroya gigantea** ("The *green aloe*"), the fibre of which they export to Europe under the name of "Mauritius Hemp."

Cultivation.

The Agave is planted about 5 feet apart, and the furrows 5 feet distant from each other, so that an acre would contain from 1,600 to 2,000 plants. When about 7 to 8 years old the cutting may commence, but by planting with off-shoots this need be only some 3 or 4 years after the opening up of the plantation. They will continue to yield for four or five years. The Superintendent of the Hazaribagh Jail, **Dr. R. Cobb**, furnished the following report detailing the process adopted by him in the cultivation of the fibre, which is here published in the hope that it may prove useful to interested persons:—

"*Growth from seed.*—It may be grown from seed collected from the tall candelabra-like stems thrown up by the plant after it has reached the age of from five to seven years. The seed should be planted in a nursery in rows, 18 inches apart, and the seeds 12 inches from each other. The best time to plant them (in Hazaribagh) is during the rains; they will then rarely fail to germinate, and throw out leaves three or four inches long by the end of the year. If, however, they are put down in the dry season, they require watering at least twice a week. The young plants should be allowed to remain in the nursery till the following rains, when they may be transplanted to the hedge or plantation where they are intended to grow.

"*Growth from shoots.*—This is the best method, because there is no chance of failure of germination, the labour of sowing is saved and much time is gained. Young plants from one to two years old should be procured at the commencement of the rainy season, and put down where they are intended to grow permanently. If for hedgerows, a ditch should be dug, and the young plants put on the top of the earth thrown up. They should not be closer than two feet from each other. The holes in which they are placed should be eight inches in depth, and the earth should be well pressed round them. No further care is then required, and in about three or four years the plants will grow quite close together and make an excellent fence. If it is intended to make an aloe plantation, the young shoots should be planted in rows, ten feet apart, and five or six feet should be allowed between each plant in the rows.

The American Aloe Fibre.

"*Soil.*—A gravelly or laterite soil appears to be best suited for the growth of the aloe plant. If he plantation is made on high ground, it is not necessary to make ridges to plant on, and the plant is quite as prolific of young shoots, for experience has shown that they do equally well in the flat; but in low situations and hollows it is necessary to make ridges 12 to 18 inches high, the plant being very partial to a light dry soil, while a damp and water-logged soil is death to it. No manure is required, and it grows on the most stony ground, where apparently there is not sufficient soil to support life in the plant. In some places it may be seen growing in the clefts of the rocks. We have not found it necessary to hoe or dig up the land near the plants, and weeds, grass, &c., do not appear to interfere with its growth. From experiments which have been made here, the use of the expressed juice of the leaves, as manure, has appeared to accelerate the growth of the plants.

"*Cutting the leaves.*—The leaves should not be cut until the aloe is six or seven years old, after it has thrown up its tall candelabra-like stem; some of these grow to the height of 18 or 20 feet; they flower and produce seeds: before these are thrown up the fibre is weak and not fit for manufacture.

"*Protection of plantation.*—It is commonly supposed that cattle will not eat the aloe plant on account of its sharp pointed leaf and acrid sap, but our experience has shown this to be an error. Several growing plants have had their leaves eaten, and very young plants have been found cropped close to the ground. It is advisable, therefore, to keep off cattle by means of a ditch (outside) and a close aloe hedge round the plantation.

"*Value of crop per acre.*—After the plants are seven or eight years old, one acre of land may be expected to yield seven maunds of fibre per annum (it requires as much as forty maunds of leaves to make one maund of fibre). There is no doubt about this, as repeated experiments have been made in this jail. After the ground has been planted, no expenditure is required, and the cost of planting depends greatly on the distance from which the plants have to be brought.

"*Preparation of fibre.*—After the leaves are cut they are put through a crushing machine, invented by my jailor, Mr. Pimm, which breaks the hard bark of the leaf and crushes out the juice. It has been found that a great deal of manual labour is saved by this process. The machine is not unlike a sugar-crushing machine. This process should be carried on as near water as possible. Then the crushed leaves are pounded on a smooth stone by a wooden mallet until all the bark and woody matter are removed. The fibre is then washed until the whole of the sap and dirt is cleared out of it. It is dried in the sun and is then ready for use."

MANUFACTURE AND TRADE.

MACHINERY FOR AGAVE FIBRE.

M. Evenorde Chazal has recently published a most interesting account of the cultivation of the plant, and separation of the fibre, as practised in Mauritius, from which it would appear that the machinery used costs originally from £1,000 to £1,200, apart from the necessary buildings, roads, and purchase of land. From the outturn per acre, generally published at from 40 to 70 tons of green leaf, yielding about 1½ tons of fibre, valued at £40 per ton, it would take some four or five years before an Agave (or aloe) fibre-extracting factory could give any returns to the capitalist. It must not be overlooked, however, that a large part of the above expenditure is for motive power, and that if the industry were to be combined with that of the tea and coffee plantation or the indigo factory, waste lands in the vicinity would become productive, and periods when the factory is now silent would be fully and profitably occupied. A correspondent from Colombo in the *Tropical Agriculturist for 1882-83, p. 429*, strongly urges this view, and he calculates that a "gratteuse" or fibre-extracting machine could easily be constructed for R200 and worked by any available motive power. The subject seems well worthy of the attention of our planters.

In the spring of 1882, what appears an excellent machine for the separation of this fibre was patented in the United States and favourably noticed and illustrated in the *Scientific American*. It is exceedingly probable, however, that the Ekman or some other chemical process of extraction of the fibre will supersede completely the mechanical. The *Planters' Gazette* very justly remarks that a fibre would then be obtained fine enough for spinning and weaving purposes, the market value of which would be three or four times as much as that obtained for the coarse fibre extracted by mechanical contrivances. In India there should be no difficulty in introducing the Ekman process, since bi-sulphate of magnesia is easily and cheaply procurable. The Aloe fibre succeeds best where few, if any, trees will grow; the chemical process removes the difficulty regarding the fuel supply.

Two samples of Agave fibre from Mauritius were exhibited at the late Calcutta International Exhibition,—the one in the Mauritius Court, prepared by the usual process practised in that island; the other in the New South Wales Court, prepared by a patent process invented by **M. G. A. Lusignan** of Sydney.

ROPE.

In India, the Agave, as a fibre-producer, has attracted attention from time to time since 1798. In that year a **Mr. W. Webb**, who had a plantation of it near Madras, made ropes from it. He submitted a coil of this rope to the Military Board of Fort St. George, with a suggestion that he should be allowed to supply it in lieu of rope made in Europe. **Captain P. Malcolm**, of His Majesty's ship *Suffolk*, after trying it, reported it to be as strong as, if not stronger than, coir, and as having the advantage of pliability. A committee of the Military Board, after a trial with it, were of opinion that it was at least equal in point of strength to the best European rope of the same size. "As regards durability, it was mentioned that part of a coil which had been fixed to the anchor of a boat and kept constantly under water for six months, appeared to have undergone no other alteration than European rope would have done in the same situation."

At the beginning of the present century **Dr. Buchanan** found the villagers in Mysore employing the Agave for making strong hedges, and separating the fibre for cordage.

Agave Ropes. 611

In 1841, rope was made at the Alipore Jail (Calcutta) from Agave plants grown there by the convalescent insane. The rope was tested and gave the following results, as compared with other ropes:—

Agave rope broke with	2,519½
Coir rope (same dimensions)	2,175
Jute „ „ „	2,456½
Country hemp (sunn)	2,269½

In 1851, a varied assortment of string, cord, rope, and fibre, undyed and dyed,—orange, red, maroon, and green, as also paper made from the fibre,—was prepared at the suggestion of **Dr. Hunter** of Madras in his School of Arts, and by the prisoners in the jail at Madras. The assortment was shown and admired at the London International Exhibition of that year. There existed at the time a small export trade in Agave fibre from the Madras Presidency to the United Kingdom, Bombay, Cutch, Gujarát,

AGAVE americana.

Agave Fabrics.

Sind, and Bengal. The value of this trade was in the official year 1852-53 put down as R27,095 and in 1853-54 R21,506. The quantity exported continued to decline, until the trade practically died out. It may be mentioned, however, that this was in all probability mainly due to the rapid growth of the jute trade, which, at this period, began to develope into a large and important industry.

In 1852, **Dr. G. Tranter**, Surgeon in charge of the United Malwa Contingent (Central India), forwarded to the Agri-Horticultural Society of India some specimens of the fibre which he had extracted from Agave grown in Malwa, where it is plentiful; the specimens were reported to be quite equal in strength to the best Russian hemp.

Dr. Wight gives the following as the results of his experiments with the chief fibres in the Madras Presidency :—

	lbs.
Coir	224
Hibiscus Cannabinus	290
Sansevieria zeylanica	316
Gossypium herbaceum	346
Agave americana	362
Crotalaria juncea	407
Calotropis gigantea	552
The breaking strain of a rope is said to be 270 to 360 pounds as against Russian hemp at	160

All this was shown and urged by **Dr. Royle** nearly 30 years ago; but as far as India is concerned, no progress has been made, except that the plant has spread and taken a firm hold of immense scattered tracts. In concluding his notice of this fibre, **Dr. Royle** urges, what we cannot do better than repeat, that it is very desirable that experiments should be made with the view of discovering the climate best suited for the production of the strongest fibre, the age at which the leaves should be cut, how long they should be macerated, if at all, and the commercial value of the fibres yielded by the various species of **Agave** and **Fourcroya.** *Spons' Encyclopædia* contains an interesting account of the preparation of this fibre by mechanical means. The writer urges that, where cheap labour can be had, the hand-prepared fibres are preferable—an important consideration for India. He further states that the leaves should be cut before the flower appears, and that in fact they cannot be too young, as the older the leaf, the coarser the fibre. (Compare with **Dr. Cobb's** opinion based on his Hazáribágh experiments.)

FABRICS.

Agave Fabrics. 612

At the Calcutta International Exhibition, 1883-84, some excellent samples of Agave fibre were exhibited from Mysore in Madras and from Hazáribágh in Bengal. In the latter case large bundles of both dyed and undyed fibre were exhibited in the Economic Court, as also samples of small carpets woven from the fibre. So early as 1839, however, carpets are reported to have been experimentally made in Balasore (Bengal). Mention is also made of *sataranjis* (carpets) being made in the Bulundshar district, North-West Provinces, and in 1882 the Revenue and Agriculture Department received a sample of matting made of this fibre at Hoshiarpur in the Panjáb; but the preparation of Aloe fibre must be regarded as in an experimental condition only, as far as India is concerned. The chief difficulties are the want of a collective supply (the plant being scattered along roadsides and not cultivated as a crop) and a cheap and convenient appliance for the extraction of the fibre.

§ "The plant requires three years to come to perfection. In Mexico

5,000 to 6,000 plants go to an acre, the average number of leaves being 40, and yielding 6 to 10 per cent. of fibre. The leaves should be cut before they are over-ripe. It is better to cut them too soon than too late, as over-ripe leaves yield a coarse fibre of inferior colour. In Mexico the natives prepare the fibre in the following manner. The cut leaves are steeped in water, then beaten and scraped to remove non-fibrous portions, washed and bleached in the sun. Another plan is to deprive the leaves of about 6 inches of the pointed end, and after having been well beaten, they are tied in bundles, laid in heaps, and allowed to ferment. After the beating the bundles are macerated in water for 14 days, and then finally washed and dried. The process of retting having proved injurious to the fibres of all endogens, mechanical contrivances are now used for separating fibres from the leaves of the Agave, &c. The length of the fibre varies from 3 to 7 feet, and the commercial article is white to straw-white. The breaking strain of a rope has been stated at 270 ℔. to 362 ℔. as against Russian hemp at 160 ℔. In its native countries the fibre is used in the manufacture of ropes, hammocks, twine, &c. The short fibres have been carded and spun, while the waste is an excellent material for the manufacture of coarse paper. Slips of paper weighing 39 grains bore an average weight of 89 ℔. as against Bank of England pulp 47 ℔. (*Spons' Encycl.*). As the plant is exceedingly hardy, very prolific, and will grow in arid wastes, where scarcely any other plant can live, it might perhaps be advantageous to plant with Agave those districts in India in which the soil proves unfertile from the presence of alkaline salts, &c. The experiment would not be costly, and might give good results. **Bousingault** found the juice to contain over 6 per cent. of cane-sugar. The fermented juice, *pulque*, contained 35·4 grains of alcohol per litre." (*Surgeon C. J. H. Warden, Professor of Chemistry, Medical College, Calcutta.*)

AS A PAPER MATERIAL.

PAPER. 613

In India scarcely any Agave fibre is used in the manufacture of paper.

INDIAN EXPERIMENTS.

In October 1877 the Revenue and Agriculture Department of the Government of India issued a Resolution, requesting Provincial Governments to consider the utilization of the Agave fibre, and especially directed the attention of the Government of Bombay to the advisability of making an experiment with it, adding that there was a prospect of utilising large quantities of a material now almost altogether wasted. An experiment was accordingly made at the Girgaum Paper Mill, and, under the instructions of the Bombay Government, 300 maunds of leaves were, on 30th July, furnished by the Collector of Thána. The manager of the mill could not, however, conveniently use them, and they lay in a heap until the middle of August, when, without having been previously prepared, they were put into the steam-machine to be worked up direct into paper. The result was of course a failure.

In Bengal, at the Central Jail, Hazáribágh, an experiment was also undertaken. An area of 300 acres of waste land was put under Agave in 1878-79 to furnish stock to be experimented with at the jail paper-mill. Samples of this fibre were also supplied to the Bally Paper-Mills (Calcutta), where it was discovered that one of the greatest difficulties in the way of Agave fibre for paper manufacture was the fact that the young leaves yielded too fine a pulp. The best leaves were those three years old. A mixture was proved to be injurious, and therefore a difficulty exists in the necessity of getting leaves uniformly of the same age, and if possible leaves three years old.

For paper manufacture this fibre seems likely, however, to command a good market. "It is the most highly approved of all the paper fibres, making a strong, tough, smooth paper which feels like oiled paper, and, even while unsized, may be written upon, without the ink running." "Its price is governed by that of Manilla hemp, being generally £7 to £10 a ton less than the latter." "The fibre prepared in India is harsh and brittle, though of good colour; it is not met with in commerce." (*Spons' Encyclop.*) The character of the Indian-prepared fibre is chiefly due to the mode of its extraction.

SOAP. 614

SOAP SUBSTITUTE.

The juice is made into soap. For this purpose it is expressed and the watery part evaporated either by artificial heat or by simple exposure to the sun. On its reaching a thick consistence it is made up into balls along with lye-ash. This soap lathers with salt as well as with fresh water. A gallon of the sap yields about a pound of the soft extract. (*Treasury of Botany; U. S. Dispens., Ed. 15th; &c., &c.*)

MEDICINAL PROPERTIES OF AGAVE.

MEDICINE. The Leaves. 615 Roots. 616 Decoction. 617 Sap. 618 Leaves as a poultice. 619 Fresh juice. 620 Gum. 621

Medicine.—The leaves, as also the ROOT when cut, yield a saccharine juice which by evaporation may be converted into a syrup having a peculiar nauseous odour and acid taste; it reddens litmus paper. This is administered by American doctors, having attributed to it resolvent and alterative properties; it is regarded as specially useful in syphilis. **Dr. Ross** (*Pharm. Ind., 235*) is said to have employed the roots in secondary syphilis with great apparent benefit, in the form of a decoction, in the proportion of four ounces to one pint of water. The SAP is stated to be laxative, diuretic and emmenagogue, and in doses of two fluid ounces, three times a day, has been found very useful in scurvy. (*U. S. Dispens.*) General Sheridan is reported to have used the juice with great success amongst his men who were suffering from scurvy, in a small isolated post on the Texas border. The disagreeable smell of the juice, which has been compared to that of putrid meat, causes a person at first to turn from it in disgust, but after a while the odour is overcome, and a liking for it takes the place of the previous dislike (*Year-Book, Pharm., 1875, 232*). The large, moist, fleshy LEAVES are stated to have been used with much advantage as a poultice; the fresh juice is applied to bruises and contusions. The GUM found exuding from the leaves and lower part of the stem is used in Mexico as a cure for toothache.

§ "The pulp of the leaves placed between folds of muslin is applied to the eye in conjunctivitis; it is also used, mixed with sugar, in gonorrhœa twice a day, the dose being ℥ii (by weight). (*Hony. Surgeon P. Kinsley, Chicacole, Ganjam District, Madras.*) "Not used medicinally, I think, in South India; it is cultivated as a hedge, and for making fibre." (*Deputy Surgeon-General G. Bidie, C.I.E., Madras.*)

Roots. 622

"The roots are diuretic. (*Lindley.*)" (*Dr. H. W. Hill, Mânbhûm.*) "Used by the natives in chronic gonorrhœa." (*Surgeon-Major J. M. Zorab, Balasore.*)

DOMESTIC. Razor-strops. 623 A cork substitute. 624 Roofing. 625 Fuel. 626

DOMESTIC USES.

The dry flowering stem is cut up into useful razor-strops and used as a substitute for cork. "The leaves and stems are employed in roofing; the decayed leaves are also used as fuel when firewood is scarce; the terminal spines serve as pins and nails." (*Cleghorn.*) Wall plaster, impregnated with the expressed juice, is said to be proof against the ravages of white-ants. Sugar, vinegar, and a kind of beer and brandy are made from the sap.

Spines. 627 Wall plaster. 628

Agave angustifolia is cultivated as a source of fibre, but is not distinguished by the planter from **A. americana.** 629

A. saponaria is a powerful detergent; its roots are used as a substitute for soap. (*Lindley's Vegetable Kingdom, 1847, pp. 157, 158.*) Compare with remarks given under **A. americana** as to the detergent properties of the sap of that species. 630

A. sisalana.

THE HENEQUEN FIBRE OR SISAL HEMP OF AMERICA.

This fibre is rapidly gaining favour. It is said to be prepared without maceration. The leaf is laid upon a board and scraped with a wooden fork till all the pulp has been removed. The fibre is then bleached and dried in the sun. It is more easily dyed than any other fibre of this class, and is thus very useful for making fancy articles of different kinds. At the same time *Henequen* is now made into sacks and used in the grain trade. The following extract from *Christy's New Commercial Plants* will be found interesting as showing the progress made in the Sisal Hemp industry:—

FIBRE. Leaves. 631
Sacking. 632
Fancy articles. 633

"In Yucatan the two varieties of the fibre are distinguished as the *Yashqui henequen*, which produces the best quality, and the *Sacqui henequen*, which gives the greatest quantity. It is worked by machinery, and from July 1875 to June 1876, Yucatan produced 22,000,000 lbs. of Henequen fibre, 18,000,000 lbs. of which were sent to British ports. The remainder was sent to Cuba and Mexico. I am unable to give the figures as to the American importation in late years, but the amount must be considerable, as the fibre is now in high favour as a cordage material, manufacturers claiming that it has been growing better and better each year in quality. A few figures are given in the latter part of the flax and hemp report under the heading 'Other Fibres,' which will give some idea of the amount consumed at present in this country. A recent report, published in Yucatan, gives the following figures: Taking $1\frac{1}{4}$ lbs. of fibre for the yearly production of each Henequen plant, we come to the conclusion that at present there are more than 18,000,000 plants under cultivation. For this number of plants over 420 scraping-wheels are in operation, moved by 229 steam-engines, with a force of 1.732 horse-power, and 30 wheels moved by animal power. Each scraping-wheel cleans daily, on an average, 300 lbs. of fibre; so the 450 wheels in existence do not work at present 163 days in the year.

Yashqui fibre. 634
Sacqui fibre. 635

"It is estimated that in Yucatan alone a capital of over $5,000,000 is vested in this industry.

"A peculiarity of this fibre is that it resists the action of dampness for a greater length of time than hemp or similar fibres, which makes it very desirable in the manufacture of cable-ropes, &c., used in the rigging of ships."

CULTIVATION AND YIELD OF FIBRE.—As cultivated in Mexico, an acre generally contains 5,000 to 6,000 plants. A dry, stony soil is selected for its cultivation. Young plants, 2 to 3 feet high, are planted out 12 feet apart, and weeded twice a year. The yield commences by the lower leaves being cut off about the fifth year, and this is continued annually for ten years or more. Of the shoots that spring up at the 8th, 10th, or 12th year, one is left to replace the parent plant, which is then destroyed, and the other daughter off-shoots are transplanted. The annual yield of fibre is about a ton an acre.

A. vivipara, *L.* 636

THE BASTARD ALOE.

Syn.—AGAVE CANTALA, *Roxb.* This may prove but a variety of **A. americana.**

Vern.—*Khetki, háthi chingár*, OUDH; *Kathalai*, TAM.; *Petha-kalabantha erikatali* (BELLARY), TEL; *Kantala*, SANS.

Habitat.—Commoner in upper than in lower India, specially in the North-West Provinces; almost unknown in Bengal.

Botanic Diagnosis.—Some authors seem to think this should not be regarded as a variety of **A. americana**, but rather as a distinct species, differing, as it does, chiefly in the fact that it raises its cluster of leaves upon a short, erect stem, which produces viviparous buds. It is altogether a much less robust plant. The leaves are less fleshy and erect instead of reflexed. The flowering stem is not more than half the height of that of **A. americana**, and much thinner and red-coloured. When seen growing side by side they appear quite distinct.

Stewart identified this plant with **Roxburgh's A. Cantala,** and seemed to think it was indigenous to India. There is little doubt, however, that all the members of this group have been imported from America, and even **Roxburgh,** with the imperfect botanical literature at his disposal, arrived at the conclusion that the plant had been introduced by Europeans, from the name *bélati ánanásh* (European Pine Apple) applied to the species of Agave. **Royle** observed that on a rich soil this species becomes viviparous, while on a poor stony soil and under a dry climate, seeds alone are produced.

FIBRE. 637

Fibre.—The *Oudh Gazetteer* says it is chiefly grown as a hedge to keep back cattle, but in the jails good fibre is prepared from its leaves.

MEDICINE. 638

Medicine.—§ "The juice, along with the flower of **Eleusine coracana,** is applied to contusions of draught-cattle." (*Surgeon-Major W. Dymock, Bombay.*)

Agave Soap, see **A. saponaria** and **A. americana.**

Agila-wood, see **Aquilaria Agallocha,** *Roxb.*

639

AGLAIA, *Lour.; Gen. Pl., I., 334.*

A genus of trees or shrubs, glabrous, lepidote or stellately pubescent (belonging to the Natural Order MELIACEÆ and the Tribe TRICHILIEÆ), comprising some 50 species; inhabitants of China, India, the Malaya, and the Islands of the Pacific.

Leaves pinnate or trifoliolate, *leaflets* quite entire. *Flowers* polygamo-diœcious small, globose or turbinate (not oblong linear), numerous, paniculate. *Calyx* and *corolla* each 5-lobed. *Stamens* united into an urceolate tube; *anthers* 5, included or sub-exserted, erect. *Disk* small. *Ovary* 1-3-celled, with 2-1 ovules in each cell; *style* very short. *Berry* dry, 1-2-celled and seeded.

This genus is now made to include Nemedra, *Juss.*, and Milnea, *Roxb.*; the former used to be referred to Amoora, and the latter retained as a distinct genus. (See *Appendix to Gen. Pl., I.*) **Aglaia**, a Gr. proper name, Αγλαία, derived from ἀγλαία, beauty, splendour—the youngest of the Three Graces.

A species collected by **Kurz** in the Andaman Islands, **A. ? andamanica,** *Hiern*, is called in Burmese *Tau-ahnyeen.*

640

Aglaia edulis, *A. Gray; Fl. Br. Ind., I., 556.*

Vern.—*Late mahwa,* NEPAL; *Sinakadang,* LEPCHA; *Gumi,* GÁRO HILLS and SYLHET.

Habitat.—A middling-sized tree of Eastern Bengal, as also the Gáro Hills and Sylhet, flowering in June-July; fruit ripening two or three months later.

Botanic Diagnosis.—Shoots, leaves, and inflorescence with ferruginous scales, mixed with stellate hairs, leaflets 9-13, opposite or sub-opposite. Flowers shortly pedicelled, arranged in pyramidal panicles shorter than the leaves.

Properties and Uses—

FOOD. Fruit. 641

Food.—Fruit eaten by the natives. **Roxburgh** says the natives of the Gáro Hills and Sylhet "eat the large succulent aril which surrounds the seed under the cortex of the berry."

642

Aglaia minutiflora, *Bedd.; Fl. Br. Ind., I., 557.*

Habitat.—A handsome tree, 25 to 40 feet in height, with exceedingly hard wood, was collected in Courtallum by **Voigt**, and by **Beddome** at Travancore, altitude 2,500 feet. **Griffith** and **Maingay** found it in Tenasserim and Malacca.

Botanic Diagnosis.—*Leaflets* 7-15, pubescence ferruginous or rufous stellate, narrowly elliptic, acuminate. *Panicles* divaricately branched, many-flowered, half to as long as the leaves. *Fruit* sub-globose 1-2-seeded, $\frac{3}{8}$ to 1 by $\frac{1}{2}$-$\frac{7}{8}$ inch.

643

A. odorata, *Lour.; Fl. Br. Ind., I., 554; Wight, Ic., t. 511.*

Habitat.—An elegant shrub or small tree, met with in the Eastern Peninsula, often cultivated in gardens on account of its sweetly-scented flowers.

Botanic Diagnosis.—Extremities of the young shoots covered with stellate hairs, rapidly becoming glabrous. *Leaflets* 3-5, rarely 7, obtuse. *Panicles* lax-flowered; *ovary* hairy.

644

A. Roxburghiana, *Miq.; Fl. Br. Ind., I., 555; Wight, Ic., t. 166.*

Vern.—*Priyangu*, BENG., HIND., and SANS.

Habitat.—A large tree of the Western Peninsula; from the Konkan and Midnapore southwards; Ceylon, ascending to 6,000 feet; Singapore. *Distrib.*—Java, Sumatra, and other Malay Islands.

Botanic Diagnosis.—*Leaflets* 5, rarely 7 or 3, ellipti-cobtuse, glabrescent. *Panicles* dense-flowered, somewhat supra-axillary, pyramidal elongate; *flowers* on very short pedicels, $\frac{1}{12}$ in. diameter. *Calyx* yellow, often covered with stellate hairs. *Fruit* $\frac{3}{4}$ in. diameter, buff coloured, minutely pilose.

Properties and Uses—

FOOD. Fruit. 645

Food.—Fruit said to be edible.

MEDICINE. Fruit. 646

Medicine.—§" Is regarded by Sanskrit writers to be cooling and useful in burning of the body and painful micturition. The fruits are described as sweet, astringent, and tonic." (*U. C. Dutt, Civil Medical Officer, Serampore.*)

TIMBER. 647

AGRICULTURAL IMPLEMENTS AND MACHINERY, Timbers used for.

Acacia arabica.
A. Catechu.
A. ferruginea.
A. melanoxylon.
A. modesta.
A. planifrons.
Acer oblongum.
A. pictum (ploughs).
Adina cordifolia.
Ægle Marmelos.
Albizzia amara (ploughs).
A. procera.
Anogeissus pendula.
Bauhinia purpurea.
B. variegata.
Berrya Ammonilla.
Briedelia retusa (agricultural implements and cattle-yokes).
Buchanania latifolia (cattle-yokes).
Calophyllum inophyllum (machinery).
Capparis aphylla.
Caryota urens.
Cassia Fistula.
C. siamea (mallets).

Agricultural Implements and Machinery, Timbers used for—(continued.)

Castanopsis rufescens.
Chloroxylon Swietenia (ploughs).
Cordia Myxa.
C. Rothii.
Cratoxylon neriifolium.
Dalbergia cultrata.
D. latifolia.
D. Sissoo.
Daphnidium pulcherrimum (cattle-yokes).
Dolichandrone falcata.
Ehretia lævis.
Eugenia Jambolana.
E. operculata.
Feronia Elephantum.
Flacourtia Ramontchi.
Fraxinus floribunda (ploughs).
Har dwckia binata (machinery).
Heritiera Papilio.
Hymenodictyon excelsum.
Kydia calycina (ploughs).
Lagerstrœmia parviflora.
Melia Azadirachta.
Miliusa velutina.
Milletia pendula (harrows).
Morus serrata.
Murraya Konigii.
Odina Wodier (cattle-yokes).
Olea ferruginea.
Ougeinia dalbergioides.
Phyllanthus Emblica.
Pinus excelsa (spades).
Plectronia didyma.
Prosopis spicigera.
Pterocarpus Marsupium.
Quercus dilatata.
Q. fenestrata.
Q. Ilex.
Q. incana (ploughs).
Q. semecarpifolia (ploughs).
Randia dumetorum.
Salvadora oleoides.
Schleichera trijuga.
Securinega obovata.
Shorea robusta (Santal ploughs).
Soymida febrifuga (ploughs).
Stephegyne parvifolia.
Strychnos Nux-vomica.
S. potatorum.
Tamarix articulata (ploughs).
Tecoma undulata.
Tectona grandis.
Terminalia Arjuna.
T. belerica (ploughs).
T. Chebula.
T. paniculata (ploughs).
Tetranthera monopetala.
Trewia nudiflora.
Wendlandia exserta.
Xylia dolabriformis.
Zizyphus Jujuba.
Z. xylopyra.

648

AGRIMONIA, *Linn.; Gen. Pl., I., 622.*

A small genus of herbs (belonging to the Natural Order ROSACEÆ), containing in all some 8 species, 3 of which are met with in India.

Leaves interruptedly pinnate; *stipules* slightly adnate. *Flowers* small, yellow, in terminal spike-like racemes, 2 bracteolate; pedicels bracteate at the base. *Calyx* persistent; *tube* turbinate-spinous; *mouth* contracted; *lobes* 5, triangular, imbricate. *Petals* 5. *Stamens* 5-10 or more, inserted at the mouth of the calyx. *Disk* lining the calyx-tube, its margin thickened. *Carpels* 2, included in the calyx-tube; *styles* exserted; *stigmas* 2-lobed; *ovule* 1, pendulous. *Fruit* enclosed within the hardened spinous calyx.

649

Agrimonia Eupatorium, *Linn.; Fl. Br. Ind., II., 361.*

Habitat.—An herb of the temperate regions, frequenting hedgerows and thickets. It is common in England, America, and India; in the latter all along the Himálaya from Kashmír to Sikkim, altitude 3,000 to 10,000 feet, and to the Khásia, Naga, and Mishmi hills.

MEDICINE.

Medicine.—From the remotest times Agrimony has enjoyed a high reputation amongst the herbalists of Europe; it is strange that it should be apparently quite unknown to the native doctors of India. The root is a powerful astringent, a useful tonic, and a mild febrifuge. The whole plant also yields a dye, which seems to be unknown to the hill-people of India.

Root. 650
DYE. 651

Agrimony Hemp, see **Eupatorium cannabinum,** *Linn.*; COMPOSITÆ.

AGROSTIS, *Linn.*; *Gen. Pl., III., 1149.*

652

A genus of grasses, the type of the Tribe AGROSTIDEÆ (Natural Order GRAMINEÆ), comprising about 100 species, distributed through the colder temperate regions.

Creeping annual or perennial grasses. *Panicle* loose, *spikelets laterally compressed, 1-flowered. Glumes* membranous-acute, unarmed, *the upper being smaller* than the lower. *Flower with hairs at its base* and no rudiment. *Pales* unequal, scarious; dorsal awn falling short of the glumes, or wanting. *Stamens* generally 3. *Style* short, distinct; *stigma* feathered. **A. canina,** *Linn.*, does not possess the inner pale.

Mr. Duthie enumerates, amongst others, the following species as met with in the N.-W. Provinces. Very little is known regarding the economic uses of the Indian members of this genus.

Agrostis alba, *Linn.; Duthie's Grasses, 29.*

653

FIORIN OR WHITE BENT GRASS.

Syn.—A. STOLONIFERA, *Savi*; A. SYLVATICA, *Host.*

Habitat.—Northern India, and ascending the Himálaya up to 13,000 feet. Grows in all kinds of soils; delights in one that is rich and moist. In Europe it frequents fields and the sands on the sea-shore.

Botanic Diagnosis.—*Stem* procumbent, creeping, often with long stoles; *sheath* rough; *ligula* long, acute. *Panicle*-spreading in flower, afterwards becoming close. *Pedicels* very much toothed. *Florets* rarely awned; *glumes* nearly equal, lower toothed through its keel.

Fodder.—A most valuable fodder grass.

FODDER. 654

A. canina, *Linn.; Duthie's Grasses, 29.*

655

Syn.—A. RUBRA, *Linn.*; AGRANLUS CANINUS, *Beauv.*; TRICHODIUM CANINUM, *Schrad.*

Habitat.—Western Thibet, altitude 12,000 feet; in Europe common on heaths.

Botanic Diagnosis.—*Branches* and *pedicels* rough; *sheath* smooth; *ligule* oblong acute. *Glumes* unequal, acute, lower pale, jagged, at the top 4-ribbed, kneed and twisted, awn from below the middle of and exceeding the pale, lower setaceous and tufted.

A. ciliata *Trin.; Duthie's Grasses, 29.*

656

Syn.—LACHNAGROSTIS CILIATA, *Nees.*

Habitat.—North-West Himálaya from 8,000 to 15,000 feet.

A. diandra, *Linn.; Roxb., Fl. Ind., Ed. C.B.C., 106.*

THE *bena joni*, BENG. Syn. for **Sporobolus diander,** *Beauv.*, which see.

A. maxima, *Roxb.; Fl. Ind., Ed. C.B.C., 107.* Syn. for **Thysanolœna acarifera,** *Nees,* which see.

A. tenacissima, *Linn.; Roxb., Fl. Ind., Ed. C.B.C., 106.* Syn. for **Sporobolus tenacissimus,** *Beauv.*, which see.

Agar-wood, see **Aquilaria Agallocha,** *Roxb.*

657

AILANTHUS, *Desf.; Gen. Pl., I., 309.*

A small genus of lofty trees (belonging to the Natural Order SIMARUBEÆ), comprising 3-4 species, of which two are met with in India.

Leaves very large, unequally pinnate. *Flowers* small, polygamous, in terminal or axillary panicles. *Calyx* 5-fid; *lobes* equal, imbricate. *Petals* 5, valvate. *Disk* 10-lobed. *Stamens* 10 (in the hermaphrodite flowers 2-3). *Ovary* 2-5-partite; *ovules* one in each cell. *Fruit* a one-seeded samara.

The generic name is said to be derived from *Ailanto,* the vernacular name for a species met with in the Moluccas.

658

Ailanthus excelsa, *Roxb.; Fl. Br. Ind., I., 518; Wight, Ill., I., t. 67.*

Vern.—*Maha rukha, mahárukha, limbado,* HIND.; *mahánimb,* MAR.; *Mahanim, máhála, gormi-kawat,* URIYA; *Ghorkaram,* PALAMOW; *Moto-aduso,* GUJ.; *Varul, mahárúkh,* DUK.; *Arúa,* N.-W.P. and MEYWAR; *Peru, pee, perumaruttú,* TAM.; *Pedu, pey, pedda, peddámánu putta,* TEL.; *Perumarum,* MAL.; *Mádalá, aralu,* SANS.

Habitat.—A tree about 60 to 80 feet in height, somewhat resembling the ash; probably introduced into India; common in the North-West Provinces, Behar, the Western Peninsula, and the Carnatic. In the Bombay Presidency widely distributed over the Kaira, Panch Maháls, and Gujarát District; occasionally met with in Rájputana. (*Ráj. Gaz.,* 27.) Common on the Coromandel Coast and in Ceylon.

Botanic Diagnosis.—*Leaves* 1-2 feet long, glandularly hairy; *leaflets* very coarsely toothed. *Stamens* with the filaments about half the length of the anther. *Samara* 2 inches by ½ inch, red, twisted.

GUM. 659

Gum.—A red gum, sent from Madras to the Panjáb Exhibition, is said to have been prepared from this plant at Chingleput. It resembles Moringa gum, and consists of large rounded tears of a deep vinous red.

MEDICINE. Bark. 660

Medicine.—THE BARK is aromatic and used for dyspeptic complaints; it is also regarded as tonic and febrifuge in cases of debility. Expectorant and antispasmodic, given in chronic bronchitis and asthma (*Bomb. Gaz., VI., p. 15*). The leaves and the bark are used as a medicine (*Bomb. Gaz., VII., 42*). "This bark has a pleasant and somewhat aromatic taste, and is prescribed by the native practitioners in infusion, in dyspeptic complaints, to the extent of three ounces twice daily" (*Ainslie*). **Dr. Dymock** says this description is scarcely correct; the bark is intensely bitter, like quassia. "In Bombay the bark and the leaves are in great repute as a tonic, especially in debility after child-birth. The name *Maharúk* is also applied to a species of cinnamon by the Konkanist gardeners of Bombay." (*Dr. Dymock, Mat. Med., W. Ind., 116.*) The Surgeon-General, Madras, in forwarding, through the local Government to the Supreme Government, certain proposals regarding a future edition of the *Pharmacopœia of India* suggested "that the bitter principle of the bark" of this plant should be made officinal. A powder made from the resin mixed with milk is given in small doses in dysentery and bronchitis.

§ "Used also as an astringent in diarrhœa and dysentery. Natives mix it with curds." (*Surgeon-Major W. D. Stewart, Cuttack.*)

Chemical Composition.—This substance has not been carefully examined, but **Dr. Dymock** informs me that "**Mr. N. Daji** separated an acid principle which he named *Ailantic acid.* It is reddish brown, very bitter, very easily soluble in water, less in alcohol and ether, and insoluble in chloroform and benzol." **Mr. N. Daji** also found a bitter non-crystallizable principle, but he attributes the medicinal virtue to Ailantic acid.

Ailantic Acid. 661

"Ailantic acid may be given in doses of 1 to 3 grains, and is said to be tonic and alterative. In large doses it causes nausea and vomiting, and is purgative. He recommends its use in dyspepsia with constipation. Mr. Narayan Daji's paper is of a high class." (*Deputy Surgeon-General G. Bidie, Madras.*)

TIMBER. 662

Structure of the Wood.—Soft, white; similar to that of **A. malabarica.** Weight 28 lbs. per cubic foot. **Dr. Dymock** says that the microscopic structure "of the bark shows large stony cells collected together in groups. There are also many conglomerate raphides."

Used to make floats for fishing nets and lines, sword-handles, spear-sheaths, and catamarans. (*Ainslie; Roxburgh.*) The wood is used in making drums and sword-sheaths. (*Bomb. Gaz., VII., p. 42.*)

663

Ailanthus glandulosa, *Desf.; Fl. Br. Ind., I., 518; Brandis. For. Fl., 58.*

Incorrectly called the JAPAN VARNISH TREE, *Eng.;* GÖT-TERBAUM (*Tree of the Gods*), *Ger.;* VERNIS DU JAPAN, *Fr.*

Habitat.—A lofty tree, met with in North India, most probably introduced from Japan. Extensively cultivated on the Continent as an avenue tree along with the tulip-tree, the horse-chestnut, the plane, &c. The leaves are not liable to be attacked by insects, and therefore, until the first frosts of November, the tree remains covered with its large leaves, affording a grateful shade. "It grows rapidly, throwing up abundant root-suckers, and has for that reason been employed in plantations made to clothe barren stony hills in the south of France." (*Gamble.*)

Botanic Diagnosis.—*Leaves* often exceeding 1 foot, pubescent or sub-glabrous; *leaflets* very numerous, coarsely toothed at the base. *Stamens* exserted; *filaments* several times the length of the anthers. *Samara* 1 inch by ⅓ inch, membranous, linear-oblong.

SERICULTURE. 664

Sericulture.—Upon the leaves of this tree the wild silk-worm Attacus Cynthia, *Drury*, is reared in Europe, and it is perhaps the most successful tree for the experimental rearing of different species of silk-worms. It grows freely even in England, and the insects thrive upon it. It is anticipated that the rearing of Attacus Cynthia upon this tree may become an established industry in Europe.

In connection with the subject of the value of **Ailanthus** as a food for silk-worms, it seems highly desirable that experiments be performed in India, with the object of producing a *reelable hybrid-eri* cocoon, which would still preserve the valuable property of feeding upon an annual plant such as the castor-oil. The **Ailanthus glandulosa** has proved a most convenient plant for experimenting with Indian wild insects in Europe, but, both for experiments in India and in Europe (as far as the respective climates will permit), the following plants are those which would most probably afford the means of prosecuting the investigations necessary for the production of hybrids of Indian indigenous silk-worms. The plants have been grouped in a way which brings out the overlappings in habit, as also some of the structural affinities of the more important species of silk-worm.

665

TEMPERATE PLANTS.

* **Actias and Caligula Series.**

1st.—Rosaceous plants such as **Prunus Cerasus** (the wild cherry), **Pyrus communis** (the wild pear), and **Cydonia vulgaris** (the wild quince); small trees met with in India on the Himálaya and the hills of the Eastern Peninsula, at altitudes of from 5,000 to 10,000 feet.

The following species of silk-worms feed upon these plants in their wild state: Actias selene, Caligula simla, C. thibeta.

A Classified List of the Plants

2nd.—**Pieris ovalifolia**, an exceedingly plentiful, Ericaceous, small tree, coming into fresh green foliage just before, and flowering during, the rains, on the Himálaya and mountains of the Eastern Peninsula, at altitudes of from 4,000 to 8,000 feet.

The following insects feed upon it: Actias selene, Caligula thibeta.

** **Actias and Attacus Series.**

3rd.—**Ailanthus excelsa** and **A. glandulosa** (the former would most probably not succeed in England).

The following are the insects regularly found feeding on these trees in India: Attacus ricini (the *eri silk-worm*) and A. cynthia.

4th.—**Coriaria nepalensis**, a small leafy bush, belonging to the Natural Order CORIARIEÆ (allied to MORINGEÆ, and LEGUMINOSÆ), plentiful on the Himálaya and mountains of the Eastern Peninsula, Burma, and the Straits; altitude from 5,000 to 10,000 feet. Should grow freely in England. This is one of the most curious plants enumerated in this list, and for the purpose of rearing hybrids seems the most hopeful.

The following insects feed upon it in their wild state: Actias selene, Attacus canningi, and A. ricini.

666

WARM TEMPERATE PLANTS.

*** **Attacus, Antherœa, and Cricula, or the Eri, Munga, Cricula. and Tusser Series.**

5th.—**Symplocos cratægoides, S. grandiflora,** and **S. ramosissima,** Small trees or shrubs on the Himálaya and lower hills of India, ascending to 7,000 feet in altitude.

The following insects are known to feed upon these plants, or are actually fed upon them, in their semi-domesticated condition: Attacus atlas, A. ricini (*small red form of Eri*), and Antheræa assama (*the Munga silk-worm*).

6th.—**Ricinus communis** (the common castor-oil plant), cultivated in the plains of India and on the hills up to an altitude of 7,000 to 8,000 feet. Grown as an annual in England, ornamental forms having been produced by the gardeners.

The following are the insects which feed on this plant: Attacus ricini (*the Eri silk-worm*). This is its principal food in domestication and also in its wild state. A. cynthia and Antheræa myletta (*the Tusser silk-worm*).

7th.—Species of Laurels, in India chiefly **Machilus odoratissima** (up to altitude 8,000 feet, the principal food of the *Munga*) and **Tetranthera polyantha.**

The following are the insects which feed on these plants, as also on one or two allied species of laurels: Antheræa assama (*the Munga silk*) and Cricula trifenestrata (the common wild, yellow, reticulated cocoon of Burma and of the South and West of India). Upon the former tree these insects chiefly feed, both in their wild and semi-domesticated conditions.

667

TROPICAL PLANTS.

It is necessary to add to the experimental plantation one or two other trees with the view of admitting of a more thorough investigation of the forms and possible hybridisation of the *Tusser silk-worm.*

**** **The Tusser Series.**

8th.—**Zizyphus Jujuba.**

9th.—**Lagerstrœmia indica.**

Two small trees or bushes which experience has shown to be perhaps upon the whole the best plants for the cultivation of the tusser worm.

10th.—**Terminalia tomentosa** and one or two allied species (the myrobolan or wild almond family). These are the trees which the *tusser* worm seems to prefer most in its wild condition.

In the above brief indication of the food materials of certain silk-worms, only the more important species or the genera most likely to afford useful hybrids have been mentioned.

It is interesting to observe that this climato-botanical classification brings the indigenous silk-worms of India into groups closely corresponding to those formed upon a more scientific principle. It would almost seem that hybridisation to be successful must pass through these natural affinities. It is remarkable that none of the Indian SATURNIDÆ (the family to which the foregoing silk-worms belong) show the slightest tendency to feed on the plants upon which the mulberry silk-worms (the BOMBYCIDÆ) are reared,—a fact which gives some weight to the idea that the latter are not truly indigenous to India. (*For further information consult the account given under "Silk."*)

MEDICINAL PROPERTIES AND USES.

MEDICINE. The Bark. 668

According to **Prof. Hetet** the bark of **Ailanthus glandulosa** is an active vermifuge; in powder it has a strong, narcotic, nauseating odour. It exercises a powerful depressing influence on the nervous system similar to that of tobacco. Various preparations of the bark administered by **Prof. Hetet** to dogs had a purgative effect with the discharge of worms. The powdered bark has been given in one or two cases of tape-worm in the human subject and proved remarkably successful in expelling the worm and at the same time operating on the bowels.

It was found that the depressing effects on the nervous system were due to the presence of the volatile oil, the resin having no such influence. The oleo-resin produces the same effects as the powdered bark and has the advantage that it keeps better. The dose of the powdered bark, sufficient for the expulsion of tape-worm, was found to be from 8 to 30 grains, the oleo-resin somewhat smaller. (*U. S. Dispens., 15th Ed., 1564.*)

TIMBER. 669

Structure of the Wood.—"Extremely durable, pale yellow, of silvery lustre when planed, and therefore valued for joiners' work; it is tougher than oak or elm, easily worked, and not liable to split or warp." (*Baron F. Von Mueller.*)

It grows exceedingly rapidly, sending out numerous suckers from the roots, and as it is not particular about soil, it is admirably suited for the reclamation of waste lands. **Professor Meehan** states that it interposes the spread of the rose-bug, to which the tree is destructive. (*Extra-Trop. Plants, Baron F. Von Mueller.*)

670

Ailanthus malabarica, *DC.; Fl. Br. Ind., I., 518; Wight, Ic., t. 1604.*

Vern.—*Peru, peru-marattup-pattai, maddi-pál,* TAM.; *Perumarum, pedda-mánu-patta, maddi-pálu,* TEL.; *Peru-marat-toli, mattip-pál,* MAL.; *Guggula-dhúp, ud,* MAR.; *Dhúp, baga-dhúp, gogul-dhúp,* KAN.; *Manda-dúpa,* HASSAN; *Máttipal,* ANAMALAIS; No Burmese name; *Kambalu, walbiling, koombaloo-gass, wal-biling-gass,* CINGH.

Habitat.—A large, deciduous tree, of the evergreen tropical forests abundant in the Western Gháts; rare in Pegu, but met with on the eastern slopes, and in the valley of the Tsit-toung. Often planted in South India for ornamental purposes.

Botanic Diagnosis.—*Leaves* very large; *leaflets* distant, almost entire, nearly glabrous. *Stamens* exserted, upon filaments many times longer than the anthers. *Samara* large, rounded at both ends, not twisted.

RESIN. 671

Resin.—On incision the bark yields a dark-coloured soft resin known as *Mattipál*, which, in time, hardens into a brittle resin with a strong balsamic odour.

§ "Exudes a reddish gum." (*J. E. Hardinge, Rangoon*).

Mr. Broughton, Quinologist to Government of Madras, reported upon the resin as follows: "This resin, as commonly met with, is dark brown or grey in colour, is plastic, opaque, and has an agreeable smell. It contains much impurity. The pure resin is very soft, having the consistence of thick treacle; and this is doubtless the reason why it is always mixed with fragments of wood and earth, which make it more easy to handle. The sample which I examined contained but 77 per cent. of resin, the remainder being adulterations. Alcohol readily dissolves the resin, and on evaporation leaves it as a very viscous, transparent, light-brown semi-liquid, which does not solidify by many days' exposure to a steam heat; when burned it gives out a fragrance, and hence it is sometimes used for incense. Its perfume is, however, inferior to that produced by many other resins employed in the concoction of the incense employed in Christian and heathen worship. The peculiar consistency of the resin would enable it to substitute Venice turpentine for many purposes, though its price (R6 for 25 lbs. in the crude state) forbids an extensive employment." "Resin burnt as an incense in Hindú temples." (*Bomb. Gaz., XV., Pt. I., 61.*)

MEDICINE. Resin. 672

Medicine.—The RESIN called *muti-pál* was first discovered by **Dr. Buchanan.** It is used medicinally, especially in dysentery. "**Dr. Gibson** regards it as a good stimulant in bronchitic affections." (*Pharm. Ind., 50.*)

Fruit. 673

The FRUIT is considered useful in cases of ophthalmia. (*Aitchison, Kuram Valley Plants, in Jour. Linn. Soc.*) "The fruit, triturated with mango and mixed with rice, is reckoned useful in cases of ophthalmia." (*Surgeon-Major Dymock, Bombay.*)

Bark. 674

The BARK is bitter and given in the treatment of dyspepsia. **Wight** describes this bark as rough and very thick, studded with bright garnet-looking grains, apparently of a resinous nature, which do not dissolve either in spirit or water. "A further knowledge of this bark and its exudation is desirable." (*Pharm. Ind.*)

§ "A valuable substitute for Ipecacuanha in the treatment of dysentery. Fresh juice of bark (1 oz.), with equal quantity of curd, morning and evening, proves highly useful. Commonly used by natives of all classes." (*Surgeon E. W. Savinge, Rajamundry, Godavery, Madras.*) "This tree is very common in the Vizagapatam District. I have frequently used a decoction of the bark in chronic dysentery with the very best effect." (*Hony. Surgeon Easton Alfred Morris, Negapatam.*) "The root-bark, coarsely bruised and kept soaking in gingelly oil, when given internally, is said to be an antidote for cobra-poisoning." (*Surgeon-Major D. R. Thompson, Madras.*)

TIMBER. 675

Structure of the Wood.—White, very soft and spongy. Weight 23 lbs. per cubic foot. Useless.

Ajowan, see **Carum copticum**, *Benth.*; UMBELLIFERÆ.

676

AJUGA, *Linn.; Gen. Pl., II., 1222.*

A small genus of herbaceous plants (belonging to the Natural Order LABIATÆ), containing some 30 species. *Corolla* with upper lip very short,

2-lobed, lower 3-lobed and much longer than the upper. *Calyx* ovate, bell-shaped, nearly equally 5-toothed. *Stamens* parallel, protruding beyond the upper lip of the corolla; the lower pair the longest. One of the most marked features of the genus is the prevalence of leaves or bracts in the spike-like inflorescence, causing them to appear more like an ACANTHACEÆ than LABIATÆ.

677

Ajuga bracteosa, *Wall.; DC. Prod., XII., 589.*

Vern.—*Ratpathá,* KUMAON; *Kauri bóti,* JHELUM; *Karkú, nilkantihi,* SUTLEJ; *Khurbanri,* TRANS-INDUS PANJABI NAME.—The bazar names are *Janiadam, mukund babri, nílkantti.* Mr. Baden Powell gives *jan-i-adam* as the vernacular of **Ajuga reptans,** a European species, and **Stewart** further gives that name to **Salvia lanata.**

Habitat.—A small, herbaceous plant, met with on the Himálaya, altitude 2,000 to 3,000 feet, extending from Afghánistan to Nepal.

MEDICINE. 678

Medicine.—"*Ján-i-adam* is described as a bitter astringent, nearly inodorous; sometimes substituted for cinchona in the treatment of fevers" (*Baden Powell*). "*Múkand babrí.*—On the Salt Range it is used to kill lice, and is regarded as depurative." (*Stewart.*) "An aromatic tonic, specially useful in ague." (*Baden Powell.*)

There appears to be some confusion as to the identification of the medicinal products sold in the bazars of the Panjáb and North-West Provinces under the names of *Ján-i-adam* and *Múkand babri.* The leaves of the species of **Ajuga** have a peculiar resinous, not disagreeable odour, and a bitter, balsamic taste. They are said to be stimulant, diuretic, and aperient. They have been given in rheumatism, gout, palsy, and amenorrhœa in doses of from 1-2 drachms. (*U. S. Dispens.*)

A. fruticosa, *Roxb.; Fl. Ind., Ed. C.B.C., 458.* Syn. for **Anisomeles malabarica,** *R. Br.*

Akakia.—This is an extract prepared from a species of Acacia,—see **A. arabica.**

679

Akakiya, a redstone said to be used medicinally. **Dr. Irvine,** in his *Medical Topography of Ajmere,* mentions this drug, and says it contains iron. It is used as a tonic.

680

ALANGIUM, *Lam.; Gen. Pl., I., 949.*

A genus of shrubs or small trees containing only 2 species (belonging to the Natural Order CORNACEÆ). ***Leaves*** **alternate, petioled, entire, 3-nerved at the base, persistent.** ***Flowers*** **hermaphrodite, fascicled upon the naked twigs, silky white, jointed on the pedicel:** ***bracts*** **absent.** ***Petals*** **narrow, much elongated.** ***Stamens*** **twice or thrice the petals.** ***Ovary*** **inferior, 1-celled, surmounted by a disk;** ***style*** **long;** ***stigma*** **capitate;** ***ovule*** **pendulous.** ***Fruit*** **a berry crowned by the disk and the enlarged calyx.** ***Seed*** **with crumpled** ***cotyledons*** **and ruminated** ***albumen.***

The generic name appears to be the Tamil name *Alangi* Latinised.

681

Alangium Lamarckii, *Thwaites; Fl. Br. Ind., II., 741; Wight, Ic., t. 194.*

Syn.—A. HEXAPETALUM, *Roxb.; Fl. Ind., Ed. C.B.C., 404;* A. DECAPETALUM, *Lam.* (*Kurz, I., 543*).

Vern.—*Akola, thaila-ankúl, dhera,* HIND.; *Kalá-akolá, ankola,* BOMB.; *Onkla,* GUJ.; *Ankol,* MAR.; *Akar-kanta, bagh-ankurá, dhalákura* (*U. C. Dutt*), BENG.; *Dela,* SANTAL; *Kimri,* MAL. (*S. P.*); *Ankol,* KOL.; *Ankula, -dolanku,* URIYA; *Azinghi-maram, azhinji, alangi,* TAM.; *Urgu, údugachettu, woodiya-chettoo* (in Godavari Dist.),

ALANGIUM Lamarckii.

The Alangium.

kudagu, amkolam-chettu, TEL.; *Ankola, anisaruli, udagina-gida, ansaroli,* KAN.; *Uru, ankola,* GOND; *Ankota,* SANS.; *Eepatta,* CINGH.

Habitat.—A deciduous shrub or small tree met with throughout India and Burma in tropical forests.

MEDICINE. Root-bark. 682

Medicine.—The ROOT-BARK is used in native medicine, being regarded as anthelmintic and purgative. It is mentioned by Sanskrit writers as the *Ankota,* and has a reputation in leprosy and skin diseases. **Dr. Moodeen Sheriff**, in his valuable *Supplement to the Pharmacopœia of India,* says: "It has proved itself an efficient and safe emetic in doses of fifty grains; in smaller doses it is nauseant and febrifuge. The bark is very bitter, and its repute in skin diseases is not without foundation." In an official correspondence, forwarded to the Supreme Government regarding the *Pharmacopœia of India,* **Dr. Moodeen Sheriff** says further of this drug: "It possesses the emetic and nauseant properties of ipecacuanha. Is used by natives in cases of leprosy and syphilitic and other skin diseases, and appears to be valuable in this respect." "It is useful in simple continued fever." **Drury** says: "it is also employed in dropsical cases; and pulverised, is a reputed antidote in snake-bites." The Malays believe the fruit to be a hydragogue purgative.

Assistant Surgeon S. Arjun (*Bombay Drugs, p. 70*) states that "the leaves are used as a poultice to relieve rheumatic pains." (*Dymock's Mat. Med., W. Ind., 332.*)

§ "My experience of the root-bark of the white-flowered variety of **Alangium Lamarckii (A. decapetalum)** is much greater than before, and I am now able not only to confirm my former opinion as to its efficiency as an emetic in 45 or 50 grain doses, but also to speak of some of its other medicinal properties in more favourable terms. In the early stage of leprosy, psoriasis, secondary syphilis, and some other skin diseases, its benefit is satisfactory if it is used sufficiently long according to the nature of each disease and individual case. It is a good substitute for ipecacuanha and proves useful in all the diseases in which the latter is indicated, except dysentery. As a diaphoretic and antipyretic it has been found useful in relieving pyrexia in many cases of simple, slight, continued, and idiopathic fevers. It is very frequently resorted to as an alexiteric by natives of this country, especially in cases of bites from rabid animals. Powder is the most convenient form of using the root-bark. "*Doses.*—As an emetic, from 45 to 50 grains; as a nauseant, diaphoretic and febrifuge, from 6 to 10 grains; and as an alterative tonic, from 2 to 5 grains." (*Hony. Surgeon Moodeen Sheriff, Madras.*)

"The root is described by Sanskrit writers as heating, pungent, and acrid. It is laxative and useful in worms, colic, inflammations, and poisonous bites. The fruit is said to be cooling, tonic, nutritive, useful in burning of the body, consumption, and in hæmorrhages." (*U. C. Dutt, Civil Medical Officer, Serampore.*)

"There are two or three sorts, with flowers dark, white, and red; the bark of the latter is used as an antidote in snake-poisoning." (*Surgeon-Major J. J. L. Ratton, Salem.*) The root-bark, pulverised and mixed with nutmeg, mace, and cloves, of each grs. 20, is given to check the progress of leprosy; 40 grains of the powder of this bark made into a bolus is given in cases of cobra-poisoning. It is well worth trying in such cases." (*Surgeon Lee, Mangalore.*) The oil of the root-bark is said to be a useful external application in acute rheumatism. No personal experience." (*Surgeon Joseph Parker, M.D., Poona.*)

FOOD. Fruit. 683

Food.—The fruit, a fleshy one-seeded drupe, is eaten, though astringent and acid. (*Bom. Gaz., XV.*)

Structure of the Wood.—Sapwood light yellow; heartwood brown, hard, close and even-grained, tough and strong, easily worked, with a beautiful glossy surface. "The wood is beautiful." (*Roxb.*) **Wight** found it to sustain a weight of 310 lbs. Weight 49 to 56 lbs. According to the *Mysore Gazetteer* the wood is strong and beautiful.

TIMBER. 684

It is used as pestles for oil-mills, wooden bells for cattle and other purposes, and is valuable as fuel.

ALBIZZIA, *Duraz.; Gen. Pl., I., 596.*

685

A genus of unarmed trees (belonging to the Natural Order LEGUMINOSÆ and the Sub-Order MIMOSÆ), comprising some 30 species, distributed through the tropics of the Old World.

Leaves bipinnate, often glandular at the base of the petiole, or between certain pinnæ. *Flowers* in globose heads, sessile or pedicellate, usually pentamerous and all hermaphrodite. *Calyx* campanulate or funnel-shaped, distinctly toothed. *Corolla* funnel-shaped; *petals* firmly united below the middle. *Stamens* indefinite, monodelphous at the base (free in Acacia); *filaments* several times the length of the corolla; *anthers* minute, not gland-crested. *Ovary* sessile or only shortly stalked; *stigma* minute. *Pod* large, thin, flat, strap-shaped, straight, sutures not thickened.

The Flora of British India refers the Indian species to two sections:—

* LEAFLETS OBLONG, AT LEAST $\frac{1}{4}$-$\frac{1}{3}$ INCH BROAD.

A. Lebbek, A. pedicellata, A. odoratissima, A. procera, A. lucida, and A. glomeriflora.

** LEAFLETS NARROW, DIMIDIATE-LANCEOLATE, WITH THE MIDRIB CLOSE TO THE UPPER EDGE.

A. Julibrissin, A. stipulata, A. myriophylla, and A. amara.

Albizzia amara, *Boivin.; Fl. Br. Ind., II., 301.*

686

Syn.—A. AMARA and A. WIGHTII, *Grah.* (*Beddome, t. 61, xcvi.*); MIMOSA AMARA and M. PULCHELLA, *Roxb., Fl. Ind., Ed. C.B.C., 418.*

Vern.—*Lulai* or *láli*, MAR.; *Moto sarsio*, GUJ.; *Thuringi, wúnja, suranji, shekram*, TAM.; *Nallarenga, shekrani, sikkai, narlingi*, TEL.; *Wusel*, MADURA (MADRAS); *Bil-kambi*, KAN.; *Kadsige*, COORG; *Oosulay*, MAL.; *Krishna sirish*, SANS.

Habitat.—A moderate-sized, deciduous tree, met with in South India and the Deccan, also Ceylon and distributed to Abyssinia and Kordofan.

Botanic Diagnosis.—*Pinnæ* 8-20 and 1-3 inch long; *rachis* densely pubescent; *leaflets* 30-60, $\frac{1}{4}$-$\frac{1}{3}$ inch long, sessile, caducous, finely pubescent; *stipules* minute, caducous. *Heads of flowers* crowded in the axils of much-reduced leaves. *Pods* distinctly stalked, 6-9 inches by $\frac{3}{4}$-1 inch and 6-10-seeded.

Properties and Uses—

Gum.—It yields a good gum, not very much known.

GUM. 687

Medicine.—§ "Described by Sanskrit writers as cooling and useful in erysipelas, eye disease, inflammation, and ulcers." (*U. C. Dutt, Civil Medical Officer, Serampore.*)

MEDICINE. 688

Structure of the Wood.—Sapwood large; heartwood purplish brown, beautifully mottled, extremely hard, with alternate, concentric, light and dark bands.

TIMBER. 689

Skinner gives the weight at 70 lbs.; **Gamble's** specimens weighed 61 to 62 lbs. **Skinner** also says: "The wood is strong, fibrous, and stiff, close-grained, hard, and durable, superior to *sal* and teak in transverse strength and direct cohesive power; also that it is used for the beams of native houses and carts; the wood of the crooked branches for ploughs." **Beddome** states that it is a good fuel, and is extensively used for the locomotives at Salem and Bangalore. (*Gamble.*)

In the *Bombay Gazetteer, XV., I., p. 61,* occurs the following (in the account of the Konkan): "The tree is common and yields dark-brown, close-grained, and very strong and durable timber, one of the most favourite woods in Kánara. A seasoned cubic foot weighs about 70 pounds."

DOMESTIC USES. 690

Domestic Uses.—The natives use the leaves as a detergent for washing the hair.

691

Albizzia anthelmintica.

A native of Abyssinia, has recently attracted considerable attention as an anthelmintic. It is known as *Musenna* or *bisenna*. The Abyssinians employ the powdered bark to expel the tape-worm, to which they are subject owing to the habit of eating raw flesh. About two ounces of the powdered bark are taken in the morning either suspended in water or made into a confection with honey. It produces no pain and does not purge. The same day portions of the worm are expelled, the remainder next morning.

This tree might with advantage be introduced into India.

692

A. Julibrissin, *Durazz.; Fl. Br. Ind., II., 300.*

THE PINK SIRIS.

Syn.—MIMOSA KALKORA, *Roxb.; Fl. Ind., Ed. C.B.C., 418.;* (?) A. JULIBRISSIN, *Willd.*

Vern.—*Sirín, kurmru, surangru, shirsh, shishi, búna, tandái, mathirshi, brind,* PB.; *Lal siris, baraulia, barau, bhokra,* HIND.; *Kalkora* (?), BENG.

Habitat.—A moderate-sized, deciduous, ornamental tree, with fragrant blossoms, met with in the Himálaya, from the Indus to Sikkim, ascending to 7,000 feet.

Botanic Diagnosis.—*Pinnæ* 8-24; *rachis* downy with a gland between the upper pinnæ; *leaflets* 20-50, sessile, sensitive, ¼ inch long, cuspidate, with the mid-rib close to the straight upper edge; *stipules* and *bracts* caducous. *Heads of flowers* not panicled but crowded in the leafless upper nodes. *Corolla* 3 times as long as the calyx; *pods* glabrous, 5-6 inches long by ¾ to 1 inch, 8-12 seeded, narrowed to the beak and to the short stalk.

MEDICINE. 693

Medicine.—Used like **A. Lebbek. Stewart** says the word *Julibrissin* is derived from *Gul abresham;* in Egypt "J" being pronounced as "G."

TIMBER. 694

Structure of the Wood.—Sapwood large; heartwood dark brown, almost black in old trees; beautifully mottled, shining. Annual rings distinctly marked by a sharp line. Weight 43 to 52 lbs. per cubic foot.

Used to make furniture.

695

A. Lebbek, *Benth.; Fl. Br. Ind., II., 298.*

THE SIRIS TREE.

Syn.—ACACIA SIRISSA, *Ham.*; A. LEBBEK, *Willd.* (*DC. Prod.*); A. SPECIOSA, *Willd.*; MIMOSA SPECIOSA, *Jacq.*; M. SIRISSA, *Roxb.*; ALBIZZIA LATIFOLIA, *Boivin.*

Vern.—*Siris, sirín, sirár, siras, sirai, mathirshi, lasrin, kalsis, tantia, garso,* HIND.; *Sirisha, siris,* BENG.; *Pit shirish,* SANS.; *Chapot-siris,* SANTAL; *Tinia,* URIYA; *Vaghe, kot vaghe,* TAM.; *Dirasan, darshana, kat vaghe, pedda duchirram,* TEL.; *Kal baghi, bengha, dirisana, godda hunshe,* KAN.; *Chichola, mothá siras,* MAR.; *Doli, saras,* PANCH MAHALS; *Pilo sarshio,* GUJ.; *Sirasa, shirrus, surí,* SIND; *Kokko* or *kúk-ko,* BURM.; *Beymadá, gachodá,* AND.

Habitat.—A large, deciduous, spreading tree, found wild or cultivated in most parts of India; grows in the evergreen mixed forests in the Sub-

Himálayan tract from the Indus eastward, in Bengal, Burma, Central and South India, ascending to 5,000 feet in altitude.

§ "A handsome and common tree in the town of Madras, grows best when self-sown; yields a gum." (*Deputy Surgeon-General G. Bidie, C.I.E., Madras.*)

Botanic Diagnosis.—*Pinnæ* 4-8; *leaflets* 8-18, short-stalked, obtuse, oblique, 1-1½ inch long and ¾ broad. *Heads of flowers* not panicled, 3-4 together. *Corolla* greenish yellow, twice the length of the calyx. *Pod* strap-shaped, yellow-brown, ½-1 foot long by ¾-1½ inch broad, 6-10-seeded.

An ornamental tree with light-coloured bark, exceedingly good for avenues. Its roots do not penetrate very deep; it may be propagated readily by cuttings.

Properties and Uses—

GUM. 696

Gum.—It yields a gum, which is said not to be soluble in water, but merely to form a jelly. It resembles gum arabic. **Roxburgh** states that he has often seen large masses of pure gum upon this plant, while other authors give conflicting opinions regarding its properties. **Mr. Baden Powell** says that, under the name of *lera*, it is used as an adulterant for pure gum arabic in calico-printing and in the preparation of gold and silver leaf cloths. The *Mysore Gazetteer* remarks that the tree yields a good gum. (*Vol. I., p. 47.*) "A dark gum oozes from wounds in the bark." (*Bombay Gazetteer, XV., Pt. I., 61.*)

TAN. 697

Tan.—The bark is said to be used in tanning leather.

OIL. 698

Oil.—An oil extracted from the seeds is considered useful in leprosy.

MEDICINE. Seeds. 699 Bark. 700

Medicine.—The SEEDS are officinal, forming part of an *anjan* used for ophthalmic diseases (*Stewart*). They are astringent, and are given in piles, diarrhœa, gonorrhœa, &c. The oil extracted from them is considered useful in leprosy. The bark is applied to injuries to the eye (*Madden.*)

MEDICINE. Flowers. 701 Leaves. 702

The FLOWERS are considered by the natives a cooling medicine, and are externally applied to boils, eruptions, and swellings; they are regarded as an antidote to poisons. The LEAVES are regarded as useful in ophthalmia. (*Baden Powell's Panjáb Prod., s. v.* **Acacia speciosa**, *p. 345.*)

§ "Powdered seeds in doses of 6 mashas=ʒ 1½ have been successfully administered in cases of scrofulous enlargement of the glands. A paste of pounded seeds and water is useful as a local application at the same time." (*Asst. Surgeon Gholam Nabi, Pesháwar.*) "This is sometimes used in ophthalmia, but my experience of it is too limited to enable me to give any opinion." (*Surgeon-Major C. J. McKenna.*) "The powder of the root-bark is used to strengthen the gums when they are spongy and ulcerated." (*Native Surgeon Ruthnam Moodelliar, Chingleput, Madras.*)

Bark. 703

FODDER. 704

Fodder.—The leaves are used for camel fodder. It is often cultivated as a fodder plant in Mysore. The tree grows rapidly and flourishes on almost any soil, especially on canal embankments and roadsides, affording both fodder and fuel where these are otherwise scarce. Deserves to be cultivated to a much greater extent than at present.

TIMBER. 705

Structure of the Wood.—Sapwood large, white; heartwood dark brown, hard, shining, mottled, with deeper coloured longitudinal streaks. The annual rings in trees grown in the Panjáb are marked by a distinct line. Weight 40 to 60 lbs. per cubic foot. It seasons, works, and polishes well, and is fairly durable. The value of the tree may be inferred from the fact that the Burmese Government fixed a higher tax upon the felling of *kúk-ko* than for teak or any other tree. (*Burma Gazetteer, I., 128*).

It is used for picture-frames, sugarcane-crushers, oil-mills, furniture, well-curbs, canoes (Burma), and wheel-work; in South India for boats. In the Andamans, where trees of large size are procurable, it is utilised for building, but more usually for house-posts. Used for furniture and picture-frames in Mysore. In the Deccan the wood is regarded as of excellent quality. In Northern India it is considered unlucky to employ the timber in house-building. (*Drury; Roxb., &c.*) It is a common practice to pollard the tree, the cuttings being used as firewood.

706

Albizzia lophantha, *Benth.; Fl. Br. Ind., II., 298.*

Habitat.—An Australian small tree or bush, now largely grown in India; naturalised on the Nilgiri Hills.

Botanic Diagnosis.—Closely resembles **A. amara,** only that the flowers are in spikes. One of the most rapidly growing trees for copses, affording temporary shade in exposed localities.

TAN. Bark. 707

Pods. 708

Tan.—The bark may be used in tanning. It contains about 8 per cent. of *mimosa tannin;* "but **Mr. Rummel** found in the root about 10 per cent. of *saponin,* valuable in silk and wool factories. Saponin also occurs in **Xylia dolabriformis** of South Asia." (*Baron F. Von Mueller, Extra-Trop. Pl.*) The pods when crushed give out a peculiar smell, said to be due to some allied compound to that met with in mustard. (*Flück. and Hanb., Pharm.*)

709

A. lucida, *Benth.; Fl. Br. Ind., II., 299.*

Syn.—MIMOSA LUCIDA, *Roxb.; Fl. Ind., Ed. C.B.C., 417.*

Vern.—*Sil koroi,* BENG.; *Sil-karai,* ASS.; *Tapria-siris,* NEPAL; *Ngraem* LEPCHA; *Messguch,* ASS.; *Gunhi,* MAGH.; *Thanthat,* BURM.

Habitat.—A large, deciduous tree, met with in Eastern Bengal, Assam, and Burma.

Botanic Diagnosis.—*Pinnæ* unijugate; *leaflets* 2 to 4, rarely 6, oblong-acute, 2 to 4 inches long, glabrous, bright green. *Panicles of heads* umbellate or corymbose.

TIMBER. 710

Structure of the Wood.—Heartwood hard, brown, with dark streaks and alternating dark and light-coloured concentric bands. Average weight 50 lbs. per cubic foot.

It is hard and good, and used for pots in Assam.

711

A. odoratissima, *Benth.; Fl. Br. Ind., II., 299.*

Syn.—MIMOSA ODORATISSIMA, *Roxb.; Fl. Ind., Ed. C.B.C., 418.*

Vern.—*Siris, sira, bhandir, bersa, bás, bassein, bansa,* HIND.; *Jang siris,* SANTAL; *Lasrín, karmbru, polach,* PB.; *Chichwa, chichola, yerjoohetta,* GOND.; *Chichora,* KURKU; *Jati-koroi, siris,* ASS.; *Sisoo,* GARO; *Koroi,* CACHAR; *Tedong,* LEPCHA; *Kalthuringi, kar vaghe, karu-vengé, bilwara, solomanim, sela vanjai, karuvaga,* TAM.; *Shinduga, chindu, telsu, yerruchinta, karu vage,* TEL.; *Pullibaghi, billawar, bilvara,* KAN.; *Siris, chichna, chicháda, siras, shiras,* KONKAN, BOMB., DUK., MAR.; *Kalo-sarasio,* GUJ.; *Kali saras* or *harreri,* PANCH MAHALS; *Húre márá,* CINGH.; *Thitmagyí,* BURM.

Habitat.—A large, deciduous tree, met with in the Sub-Himálayan tract from the Indus eastward, ascending to 3,000 feet in altitude; in Bengal, Assam, Burma, and Central and South India.

Botanic Diagnosis.—*Pinnæ* 6-8, with a gland between the 1-2 pairs; *leaflets* 16-40, sessile, obtuse, very oblique, glaucose beneath, strongly veined, with the mid-rib parallel to and at a little distance from the upper edge, $\frac{3}{4}$-1 inch by $\frac{1}{4}$-$\frac{1}{3}$ inch. *Heads of flowers* copiously panicled, each few-flowered, apricot-scented, pale-greenish white.

Properties and Uses—

GUM. 712

Gum.—It yields a dark-brown gum in rounded tears, tasteless but soluble in water.

DYE. 713

Dye.—§ "The bark is boiled by the Gáro people, together with the leaves of the *dúgál*, **Sarcochlamys pulcherrima,** and the yarn for their cloth, to give the latter a brownish colour." *(Mr. G. Mann, Conservator of Forests, Assam.)*

MEDICINE. Bark. 714

Medicine.—The bark, applied externally, is considered efficacious in leprosy and in inveterate ulcers.

§ "The leaves boiled in *ghí* are used by the Santals as a remedy for coughs." *(Rev. A. Campbell, Pachumba.)*

FODDER. 715

Fodder.—The leaves and twigs are used for fodder, and in the Konkan they are regarded as of excellent quality. *(Bomb. Gaz.)*

TIMBER. 716

Structure of the Wood.—Sapwood large, white; heartwood dark brown, with darker streaks; very hard. Dark, narrow, concentric bands (annual rings?), alternating with bands of lighter colour. "Grain ornamental, but rather open." *(Cleghorn.)* It seasons, works, and polishes well, and is fairly durable. Weight 42 to 60 lbs. per cubic foot.

It is used for wheels, oil-mills, and furniture. The timber is excellent for all purposes requiring strength and durability. One of the most valuable of jungle timbers found around the villages of the Indian peasant. It is the principal wood used for cart-wheels in Gujarát.

717

Albizzia procera, *Benth.; Fl. Br. Ind., II., 299.*

Syn.—MIMOSA ELATA, *Roxb.; Fl. Ind.; Ed. C.B.C., 418.*

Vern.—*Safed siris, gurar, karra, karo, karanji, gurbári, gurkur, baro, karolu, garso,* HIND.; *Koroi* or *kori,* BENG., ASS.; *Kili* or *khili,* GÁRO; *Pandrai,* KÓL.; *Garso,* KHARWAR; *Laokri,* MECH.; *Sitto siris,* NEPAL; *Takmur,* LEPCHA; *Passerginni,* GOND.; *Sarapatri, tinia,* URIYA; *Karallu, kinai tihiri,* BOMB.; *Kinni,* BHÍL; *Kanalu,* DUK.; *Kinai,* MAR.; *Kínhai* (THANA), KONKAN; *Konda vaghe,* TAM.; *Pedda pattseru, tella sopara, tella chindagu,* TEL.; *Chikul,* KAN.; *Choi,* MAGH.; *Seet* or *sit,* BURM.; *Búrdá,* AND.

Habitat.—A large, deciduous, fast-growing tree, found in the Sub-Himálayan tract from the Jumna eastward; in Bengal and Behar, in the Satpura Range, in the Central Provinces, in Gujarát, and South India and Burma.

Properties and Uses—

GUM. 718

Gum.—This tree yields large quantities of gum.

TAN. 719

Tan.—The bark is sometimes used as a tan.

TIMBER. 720

Structure of the Wood.—Sapwood large, yellowish white, not durable; heartwood hard, brown, shining, with alternate belts of darker and lighter colour. The wood is straight and even-grained, seasons well, and the heartwood is durable. Weight 26 to 60 lbs. per cubic foot. Yields excellent timber and is in great request. *(Bomb. Gaz., XV., p. 61.)*

It is used for sugarcane-crushers, rice-pounders, wheels, agricultural implements, bridges, and house-posts. It is used by tea-planters for stakes for laying out tea-gardens, as it is found to split well, and occasionally it is also used for tea-boxes; it is found to be very good for charcoal.

A. (Pithecolobium) Saman, *F. v. Mueller.*

THE RAIN TREE OR GUANGO.

Habitat.—A native of Mexico, Brazil, and Peru; it is experimentally cultivated in most warm-temperate countries, and would succeed well in many parts of India, especially in the vicinity of the sea or salt-lakes. Is one of the best trees for roadsides.

TIMBER. 721

"The wood is hard and ornamental, but the principal utility of the tree lies in its pulpy pods, which are produced in great abundance, and constitute a very fattening fodder for all kinds of pastoral animals, which eat them with relish." (*Mueller, Extra-Tropical Plants.*)

722

Albizzia stipulata, *Boivin; Fl. Br. Ind., II., 300.*

Syn.—Mimosa stipulacea, *Roxb.; Fl. Ind., Ed. C.B.C., 418.*

Vern.—*Siran, kanujera, pattia, samsundra,* Hind.; *Chakua, amluki,* Beng.; *Oë, oi, sirín, shirsha, kasír,* Pb.; *Chapún, kera serum,* Kol; *Bunsobri,* Mechi; *Kala siris,* Nepal; *Singriang,* Lepcha; *Sow, sau,* Ass.; *Selcho,* Gáro; *Kat turanjí,* Tam.; *Konda chiragu, chindaga,* Tel.; *Udala,* Bomb.; Mar.; *Phalári* (Thana), Konkan; *Shembar,* Panch Mahals; *Kal baghi, hote baghi, bagana,* Kan.; *Kabal,* Cingh.; *Pokoh, bhúm-mai-za,* Magh.; *Cabal-mará-gass, búmaiza, bnuméza,* Burm.

Habitat.—A large, deciduous, fast-growing tree, met with in the Sub-Himálayan tract from the Indus eastward, ascending to 4,000 feet; in Oudh, Bengal, Burma, and South India.

Botanic Diagnosis.—*Pinnæ* 12-40, with many glands on the rachis; *leaflets* 40-80, $\frac{1}{12}$ inch or less in breadth, sessile, finely downy with a slightly recurved acute point; *stipules* and *bracts* large, membranous, downy, cordate-acute, persistent. *Heads of flowers* panicled, terminal racemes densely pubescent. *Pod* 5-6 inches by $\frac{3}{4}$-1 inch, pale brown, thin indehiscent, sub-sessile, 8-10-seeded.

This tree is attracting considerable attention in Assam. It has been found that tea flourishes better under it than when exposed to the sun. The most favourable explanation of this fact is that the leaves manure the soil; the roots, which do not penetrate deep, tend to open up the soil; while the shade is not to severe as to injure the tea, the leaves closing at night and during early morning.

Properties and Uses—

GUM. 723

Gum.—It yields a gum, which exudes copiously from the stem, and is used by the Nepalese for sizing their "Daphne" paper.

FODDER. 724

Fodder.—The branches are lopped for cattle fodder. (*Gamble.*)

TIMBER. 725

Structure of the Wood.—Sapwood large, white; heartwood brown, generally not durable, soft, shining. Weight 25 to 45 lbs. per cubic foot.

Fuel. 726

It is also used as fuel.

It is said by **Beddome**, probably quoting **Skinner**, to be used for building and for naves of wheels. **Kurz** says it is good for cabinet-work, furniture, and similar purposes. *Brandis' Burma List, 1862,* No. 27, says it is prized for cart-wheels and for wooden bells. In Bengal it has been tried for tea-boxes, for which purpose it will probably suit well; also for charcoal. Said to be much used in South Kanara.

ALBUMEN.

727

Albumen.

A term which, in chemistry, means a compound containing nitrogen in addition to the carbon, hydrogen, and oxygen of the starches. It is readily known by its coagulating with heat. The white of an egg is a good example of this compound in animal matter, but it is also largely present in vegetable substances, and especially so in the sap of plants.

In botanical science, however, the term "An Albumen" has come to have a widely different meaning. It is a layer of albuminous matter (albumen, fibrine, and casein, together with starches), surrounding the embryo and within the seed-coats. In the pea albuminous matter is stored within the embryo itself, filling its seed-leaves (the halves of the pea), and such a seed is therefore *exalbuminous* in botanical terminology. In the castor-oil seed, on the other hand, the albuminous matter forms a distinct and complete layer around the embryo (or infant plant), and such is therefore regarded as an *albuminous* seed. An exalbuminous seed does not imply the absence of albuminous matter (chemically), but the absence of a peculiar layer of such matter around the infant plant and within the seed-coats.

MEDICINE. 728

Medicine.—Albumen is described in the *Indian Pharmacopœia* as emollient, demulcent, and nutritive. It acts as an antidote to the soluble salts of copper and zinc, and corrosive sublimate or creosote.

Chemical Note.—§ "Albuminoid or proteid is a generic term given to the chief mass of nitrogenous material of plants and animals. All proteids contain nitrogen, carbon, hydrogen, oxygen, (sulphur and phosphorus). The white of egg is an example of an animal proteid, while in vegetable juices which are coagulated by heat, a substance exists which is either identical with or closely resembles egg albumen. Proteids have been divided into classes by **Hopper-Seyler**, and comprise egg-albumen, serum-albumen, myosin, globulin, fibrinogen, vitelleri-filiere. Besides these there are devoid albumens, obtained by the reaction of reagents on an albumen. Thus an albuminoid, which has been digested or dissolved by the gastric juice, is called a peptoric." (*Watts.*) "A peptoric differs in a most marked manner from the proteid from which it has been obtained. It is very soluble in water, and is not precipitated by heat. It is also soluble in dilute alcohol. It is uncrystallizable and devoid of odour and almost tasteless." (*Surgeon Warden, Prof. of Chemistry, Medical College, Calcutta.*)

§ "Eggs are very useful in cases of anæmia resulting from loss of blood or chronic discharges. They also act as an aphrodisiac. In combination with *kundur* it is employed in chronic bronchitis, and with *kuttan* in asthma. A mixture composed of eggs, *kobroba* and *tahashir* is said to act as a powerful astringent, and is used in hæmorrhage and chronic diarrhœa. Yolk of egg is often applied locally to the part bitten by a snake.

"A liniment composed of eggs, *rogungul,* and *babuna,* is said to be a very useful local application in ophthalmia and orchitis. An ointment made with *mom rogun* has been employed with benefit in cases of severe neuralgia and other painful affections. Yolk of egg, mixed with *zira kirmani,* and spread over a piece of paper and applied while warm over the loin, is said to remove the pain. Eggs are also used as a local application in cases of burns and scalds. Eggs burnt to ashes and mixed with honey are said to be very efficacious in removing the opacity of the cornea." (*Assistant Surgeon Gholam Nabi, Pesháwar.*)

ALCOHOL.

729

Alcohol.

The product of vinous fermentation. Through the agency of the fungus—Yeast—sweet liquids have their chemical constituents rearranged. They are then said to be fermented, and the spirit or pure alcohol formed may be separated from admixture by distillation.

Chemical Note.—§ "Chemically, alcohol means a neutral compound of oxygen, carbon, and hydrogen, from which an ether can be obtained. Usually, however, the term is restricted to ethylic alcohol—spirits of wine. Alcohol is a product of the fermentation of saccharine matter by the action of a fungus, the **Saccharomyas cererisiæ**, a constituent of yeast. In commerce three varieties of alcohol of different strengths are recognised—Absolute Alcohol, Rectified Spirits, and Proof Spirits. Absolute Alcohol is alcohol which has been deprived of water; Rectified Spirit is Absolute Alcohol mixed with 16 per cent. of water by weight; and Proof Spirit,

Absolute Alcohol wlth 50·76 per cent. of water. The strength, therefore, of an alcoholic liquid may be expressed in terms of one of these three varieties of alcohol. For excise purposes, "Proof," "under Proof," and "over Proof" are terms which are constantly employed. Formerly the strength of spirit was ascertained by pouring some of it over gunpowder and igniting the spirit. If the powder inflamed, the spirit was "Proof," but if weaker the gunpowder was too much moistened by the water, and would not explode, and the spirit was "under Proof." The composition of Proof spirit has been defined by Act of Parliament, and is of the strength already stated. If the spirit be stronger than Proof spirit, it is said to be so many degrees or per cent. over proof, or O. P., and if weaker, so much per cent. under Proof, U. P. A liquor described as being 20 degrees U. P. means that 100 parts of the spirit contain 80 parts of Proof spirit and 20 parts of water, while a liquor 20 O. P. means that if 100 parts of the spirit were diluted with water till the mixture measured 120 parts, the product would be Proof spirit. For purposes of manufacture, &c., Rectified spirit is issued duty-free, after admixture with a certain percentage of commercial wood, *naphtha*. This addition renders the spirit unfit for potable purposes, and the spirit so treated is known as Methylated Spirit of Wine. In India, for certain trade purposes, the Excise Department permit the addition of caoutchouc in lieu of wood spirit." (*Surgeon Warden, Prof. of Chemistry, Medical College, Calcutta.*)

Proof spirit. 730

Rectified spirit. 731

Methylated spirit. 732

MEDICINE. 733

Medicine.—It is chiefly used for chemical purposes and in the preparation of tinctures. Rectified spirit is a powerful diffusible stimulant, useful as an evaporating lotion, but not administered internally *per se.*

Alder, see **Alnus glutinosa** and **A. nepalensis.**

734

ALEURITES, *Forst.; Gen. Pl., III., 292.*

A small genus of EUPHORBIACEÆ, containing trees with long-petiolate, simple or lobed leaves with 2 glands at the base. *Calyx* 2-3-partite, valvate in bud. *Petals* 5, twisted, longer than the sepals. *Disk* present in both sexes, often minute or reduced to glands. *Stamens* indefinite, on a conical naked torus, erect in bud; *anthers* 2, parallel, dehiscing longitudinally. *Ovary* 2-5-celled, with a solitary ovule in each cell; *styles* as many as the cells, deeply bifid. *Fruit* drupaceous, of 2-5 cocci; endocarp crustaceous, exocarp succulent. *Seeds* compressed-globular, with a spurious white aril; *albumen* oily; *cotyledons* large, the radicle minute.

The generic name is derived from **ἀλευρίτης**, made of wheaten flour, because of the mealiness of the plant.

735

Aleurites cordata, *Müll.*

Syn.—ALEURITIES VERNICIA, *Hassk.;* A. CORDATA, *R. Br.;* ALÆOCOCCA VERNICIA, *Spreng.*

Vern.—*Tung*, CHINESE. Sometimes called the WOOD-OIL of CHINA.

Habitat.—This exceedingly interesting tree is said to have been found by **Wallich** in Nepal (*Wall., Cat. N., 7958*), but apparently it must be very rare, or its valuable properties are quite unknown to the natives of India.

VARNISH. Chinese wood-oil. 736

Varnish.—In the Kew Report for 1880, p. 11, this is said to be the plant which yields the Chinese varnish (formerly supposed to be the same as the Japanese varnish, **Rhus vernicifera**). Samples of variously coloured lacquers were exhibited at the Calcutta International Exhibition from Tonquin, which were most probably obtained from this plant. **Mueller** (*Extra-Tropical Plants*) says: "This tree, for its beauty and durable wood, deserves cultivation in our plantations in humid districts." "The oil is an article of enormous consumption amongst the Chinese, who use it in the caulking and painting of junks and boats, for preserving wood-work, varnishing furniture, and also in medicine." (*Flück. and*

Hanb., Pharm., 91.) **Lindley** (*Veg. Kingdom, 1847, pp. 278, 280*) states that the Ceylon gum-lac is made from **A. laccifera.**

Aleurites moluccana, *Willd.; DC. Prod., XV., pt. 2, 723.* 737

THE BELGAUM OR INDIAN WALNUT; THE CANDLE-NUT.

Syn.—A. TRILOBA, *Forst.*

Vern.—*Akrót, akola, jangli-ákrót,* HIND., BENG.; *Akshota,* SANS. (*Sakháràm Arjún, Bombay*); *Khasife-hindí, jouzebarri,* ARAB.; *Girdagáne-hindí, chahár-maghze-hindí,* PERS.; *Jangli eranda, jelapa, jangli ákhróta* or *ákrót, jábhal,* BOMB.; *Akhoda,* GUJ.; *Jáphala, akhod,* MAR.; *Akrota,* CUTCH; *Náttu-akrótu-kottai,* TAM.; *Nátu-akrótu-vittu,* TEL.; *Nát-akródu,* KAN.; *Kakkuna,* SINGH.; *To-sikya-sí,* BURM.; *Shih leih,* CHINA. The names given, in most parts of India, to this plant are those which more properly belong to the Walnut, the *ákrót.* It is therefore advisable to add the word "wild"=*Janglí-ákrót.*

Habitat.—A handsome tree, introduced from the Malay Archipelago, and now found in cultivation or run wild in many parts of South India. **Roxburgh** says of it: "A large tree, now pretty common in gardens about Calcutta." "Flowering time the hot season; seeds ripen in August." **Cleghorn** remarks that it thrives well in Madras.

Botanic Diagnosis.—Leaves and twigs covered with a brownish, stellate, scaly, minute tomentum. *Leaves* ovate, base truncate-obtuse, having two glands, acute or acuminate, often 3-lobed. *Panicles* on the extremities of the branches, covered with scaly tomentum and crowded with white flowers.

Properties and Uses—

GUM. 738

Gum.—Bark smooth, olive-green, a gum often naturally exuding from the stem and found also upon the fruit. This gummy substance is said to be chewed by the Tahitians, especially that from the fruit.

DYE. 739

Dye.—The *Treasury of Botany* says the root of the tree affords a brown dye, which is used by the Sandwich Islanders for dyeing their native cloths. This may be the brown dye of Tonkin, of which samples were exhibited at the Calcutta International Exhibition of 1883-84.

OIL. 740

Oil.—NUT OIL OR ARTIST'S OIL.—The nuts of this plant contain 50 per cent. of oil, which is extracted and used as food and for burning. It is known as *Kekuna* in South India and Ceylon. The nuts when strung upon a thin strip of bamboo and lighted are said to burn like a candle. Strung upon strips of the wood from the palm leaf they are regularly used by the inhabitants of the Sandwich Islands, where the plant is called *Kukni,* and the torches are reported to burn for hours, giving a clear and steady light. The yearly production of the *kukni* oil in the Sandwich Islands is said to be 10,000 gallons. It is now exported to Europe for candle-making, and is reported to be equal to gingelly (Sesame) or rape oil. **Simmonds** reports that 31½ gallons of the nut yield 10 gallons of oil, which bears a good price in the home market. It may be obtained either by boiling the bruised seeds or by expression.

"The oil is very fluid, of an amber colour, without smell, congealing at 32° F., insoluble in alcohol, readily saponifiable, and very strongly drying. (*U. S. Dispens., 15th Ed.*) "The cake, after expression of the oil, is a good food for cattle, and useful as manure." (*Drury's Us. Pl.*) "The cake, left after the expression of the oil, given to a dog in the dose of about half an ounce, produced no vomiting, but acted strongly as a purgative." (*U. S. Dispens.*) These opinions would seem to be rather conflicting.

§ "The oil makes a capital dressing for ulcers." (*Surgeon W. Barren, Bhuj, Cutch, Bombay.*)

MEDICINE. Kernels. 741

Medicine.—The kernels "yield on expression a large proportion of a fixed oil, which has been pronounced by the Madras Drug Committee

(1855, p. 428) to be superior to linseed oil for purposes connected with the arts." Medicinally, a dose of about two ounces has been found to act in from three to six hours as a mild purgative, its action being unattended with either nausea, colic, or other ill effects. It approaches castor oil, and has been found quite as certain in its action, with the advantage of possessing a nutty flavour; dose ½ to 1 oz. (*Pharmacopœia of India.*) **Dr. Irvine** says the nut is a stimulant and sudorific: dose ʒii to ℥i.

Dr. Calixo Oxamendi (*Anales de Medicine de la Habana*) performed a series of experiments by which he arrived at the conclusion that the oil must be administered in much smaller doses than is commonly stated. He found that half an ounce was quite sufficient to move the bowels of an adult. He recommends that it be used as a substitute for other aperients on account of its having a pleasant nut-like taste and acting freely in three hours without giving pain or griping. **Dr. Oxamendi** attributes this property not only to the oil itself but to a peculiar resin which irritates the intestinal mucous membrane. He recommends gum arabic to be combined with it, and for external application in obstruent constipation, he suggests that it should be combined with Tinct. of Cantharidis and Ammonium Carbonate: ℞. Ol. Nucis Aleurites Trilobæ, ½ oz., Tinct. Cantharid. and Ammon. Carb. a. a. ʒiii. (*M. Linam.*)

FOOD. 742

Food.—It is cultivated for the sake of its fruit, which is generally 2 inches in diameter. **Roxburgh** says: "The kernels taste very much like fresh walnuts, and are reckoned wholesome."

Algarobilla and Algaroba.

TAN. 743

A tan obtained, chiefly in America, from certain members of the genus **Prosopis**, of which **P. pallida**, *Kunth*, **P. glandulosa**, **P. dulcis**, and **P. spicigera**, are the most important species. See **Prosopis**.

By some authors **Algaroba** is restricted to the Carob tree, **Ceratonia Siliqua**, which see. The *U. S. Dispensatory, 15th Ed.*, says that **Algarobilla** is the pod of **Balsamocarpon brevifolium**, a drug containing 60 to 68 per cent. of tannin and a large quantity of ellagic acid, but none in the seeds. It is obtained from Chili. **Dr. R. Godeffroy** (*in Archiv. der Pharm., XIV., p. 449*) regards this as a good source from which to prepare tannin. (*Year-Book, Pharm., 1879, 215; and 1882, 208.*)

The word Algaroba is said to be derived either from Algarobo, a town in Andalusa, or from the Arabic *Al*,=the, and *Kharroub*, the Carob-tree. It seems probable that the name is applied to a number of plants the pods of which contain a sweet mucilage—the pods which are alluded to in the Scriptures as the husks or beans.

744

ALHAGI, *Desv.; Gen. Pl., I., 512.*

A low shrub, armed with hard spines ½-1 inch long, belonging to the Natural Order LEGUMINOSÆ. *Leaves* simple, drooping from the base of the spines or branches, oblong-obtuse, coriaceous, glabrous. *Flowers* 1-6, axillary to a spine, on short pedicels. *Calyx* campanulate-glabrous, $\frac{1}{12}$ to ⅛ inch; teeth 5, minute. *Corolla* reddish, 3 times the size of the calyx; *standard* broad; *keel* obtuse. *Stamens* 10, diadelphous; *anthers* uniform. *Ovary* linear, sub-continuous; *joints* small, turgid, smooth.

The generic name is the Arabic for the plant *Al-hagu*, pronounced by the Egyptian Arabs *el-hagu*.

745

Alhagi maurorum, *Desv.; Fl. Br. Ind., II., 145.*

THE CAMEL THORN; THE PERSIAN MANNA PLANT.

Syn.—HEDYSARUM ALHAGI, *Willd.; as in Roxb., Fl. Ind., Ed. C.B.C., p. 574.*

The Persian Manna Plant.

ALHAGI maurorum.

Vern.—*Juwásá, jawása,* or *junvásá,* or *yavásá,* or *javásá,* or *javánsá,* HIND., BOMB.; *Zuwasha,* CUTCH; *Dulal-labhá, javáshá,* BENG.; *Duralabha, girikarnika-yavása,* SANS.; *Shutar-khár,* or *ushtar-khár, khár-i-shutr,* PERS.; *Alhaju, háj, aáqúl, shoukul jamal,* ARAB.; *Girikarmika, tella, giniya-chettu,* TEL.

The names *Unt-katárá* and *únt-katyah,* **Moodeen Sheriff** says, are sometimes, but incorrectly, applied to this plant. The *Manna* is known as *Taranjabin.*

Habitat.—A widely-spread shrub of the Ganges valley and of the arid and northern zones. A native of the deserts of South Africa, Egypt, Arabia, Asia Minor, Greece, to Beluchistan and Central India, the Konkan, and the plains of the Upper Ganges and North-West Provinces. Very common near Delhi.

Properties and Uses—

Medicine.—The HERB is cooling and bitter and has antibilious properties. The twigs are often resorted to as a poultice or fumigation for piles; the FLOWERS are also sometimes used for this purpose. The thorny TWIGS are sold as the medicinal product, and the preparation generally used is the extract by evaporation of a decoction of these. This is called *Yávasarkará.* It is sweetish-bitter, and is a favourite remedy for the coughs of children. By the Hindús the FRESH JUICE is used as a diuretic in combination with laxatives and aromatics. The "expressed juice is applied to opacities of the cornea, and is directed to be snuffed up the nose as a remedy for megrim." (*Dymock, Mat. Med., W. Ind., 179.*)

MEDICINE. Flowers and Herb. 746 **Poultice. 747** **Twigs. 748** **Extract. 749** **Fresh Juice. 750**

§ "The infusion has a diaphoretic action." (*Surgeon W. Barren, Bhuj, Cutch, Bombay.*)

Oil.—The oil, prepared with the leaves, is used as an external application in rheumatism.

OIL. 751

Manna.—The Sanskrit writers do not appear to refer to the MANNA or sweet sugary excretion obtained naturally from the plant by shaking its twigs over a cloth. This is chiefly collected in Khorasan, Kurdistan, and Hamadan, and imported into Bombay from November to January. It is called *Taranjabin.* It occurs in small, round, unequal grains, of the size of coriander seeds, caking together and forming an opaque mass. **Royle** states that the Indian plant does not yield the manna, and that the *Taranjabin* of the bazars is imported into India from Persia and Bokhara. (*O'Shaughnessy.*)

MANNA. 752

§ "I have never observed any manna or sweet sugary excretion on this plant, although I have seen it in every stage of its growth in large quantities in all parts of the Panjáb and North-West Provinces." (*Brigade Surgeon G. A. Watson, Allahabad.*) "The *Juwásá* trees in the districts of Muzaffernagar, Meerut, &c., on the banks of the Jumna yield *Taranjabin,* but only in small quantities. My assistant has seen it growing and has collected the manna in these districts." (*Surgeon-Major C. W. Calthrop, M.D., Morar.*)

"The editors of the *Pharmacographia* state that Alhagi MANNA is collected near Kandahar and Herat, where it is found on the plant at time of flowering. Specimens sent them by **Dr. E. Benton-Brown** and **Mr. T. W. H. Talbot** had the form of roundish, hard, dry tears, varying in size from a mustard seed to that of a hemp seed, of a light-brown colour, and agreeable, saccharine, senna-like smell. The leaflets, spines, and pods, mixed with the grains of manna, are characteristic. It is imported into India from Kábul and Kandahar to the extent of 2,000 lbs. annually, and is valued at 30 shillings per lb. According to **Ludwig**, it contains cane-sugar, dextrine, a sweetish mucilaginous substance, and a little starch." (*Surgeon Warden, Prof. of Chemistry, Calcutta.*)

Chemical Composition.—"According to **Villiers**, Alhagi Manna, after being boiled with animal charcoal and evaporated to a syrup, crystallized after some months in small brilliant crystals, which, on crystallization from alcohol, formed large white crystals of the formula $C_{12}H_{22}O_{11} + H_2O$. It is dextrorotatory, its power being + 94° 48', or for the sodium flame, + 88° 51'. On boiling with an acid, it is converted into glucose, and its rotatory power is reduced to that of glucose, *viz.*, + 53. It then reduces Fehlig's solution; nitric acid oxidizes it to mucic and oxalic acids. Its melting point is 140°. It is thus seen to be identical with Berthelot's melezitose. It crystallizes in monoclinic (clinorhombic) prisms. The mannite of Alhagi also contains cane-sugar, which may be isolated by treating the mother liquor of the melezitose with alcohol, and adding ether till a slight precipitate is formed. Crystals of cane-sugar are then deposited. The mother liquor acts like a solution of cane-sugar containing dextrorotatory foreign substances which are not fermentable with beer-yeast. (*Vide Jour. Chem. Soc., April, 1877.*)" (*Dymock's Mat. Med., W. Ind., 180.*)

FODDER. (Camel.) 753

Fodder.—In the hot season, when almost all the smaller plants die, this puts forth its leaves and flowers, which are used as a camel fodder. Just about this time the leaves and branches exude a gummy-looking liquid which soon thickens into solid grains; these are gathered by shaking the branches, and constitute the edible substance known as manna. This secretion, however, is apparently not found on the Indian plant, but is collected at Kandahar and Herat, whence small quantities of the manna are imported into Pesháwar.

DOMESTIC USES. Tatties. 754

Domestic Uses.—The twigs are much used for making the tatties (cooling mats) used in Upper India in the hot season.

ALISMACEÆ.

755

Alismaceæ; *Gen. Pl., III., 1003; Mono. Phanerg., DC., III., 29.*

A Natural Order of aquatic monocotyledons, with radicle sheathing, strongly marked leaves. *Flowers* hermaphrodite or monœcious. *Perianth* 6-merous, 2-seriate—a distinct calyx and corolla. *Stamens* hypogynous or perigynous, equal to or double the number of the perianth leaflets. *Ovaries* more or less numerous, whorled or capitate, distinct, 1-celled and 1- to 2-ovuled. *Ovules* campylotropous. *Fruit* a follicle. *Seeds* recurved exalbuminous; *embryo* hooked.

This Natural Order has its chief affinity to JUNCAGINEÆ (in the NAIADACEÆ), which only differ in their extrorse anthers, anatropous ovules, and straight embryo. The BUTOMEÆ are so closely related that they have been reduced to a tribe of the ALISMACEÆ, being only separated by their placentation and the number of the ovules.

FOOD. 756

Economic Properties.—There are 12 genera in this Order, ALISMA and SAGITTARIA being the largest and most abundant, both of which have representatives in India, found in tanks and marshes. Their economic uses are apparently unknown to the natives of India. For some time they enjoyed in Europe the reputation of being useful in the treatment of hydrophobia, having been pitched upon by empirics. The rhizomes are, however, largely eaten in many parts of the world. In China **Sagittaria chinensis** is cultivated as an article of food, and so also, in North America, is **S. obtusifolia.** In India **S. sagittæfolia** is found in every tank throughout the plains, and by desiccation the rhizomes of this species lose their acridity; in this condition they are eaten by the Tartar Kalmucks. Apparently the natives of India are ignorant of this property, and it would appear that great advantage might be taken of the edible rhizomes of **Sagittaria** in times of famine. (See **Sagittaria.**)

Alisma Plantago, *Linn.* 757

Common in tanks in Bengal; also in marshes and lakes; it extends throughout the Himálaya to Kashmír.

ALKALINE ASHES. 758

Alkalis, or **Alkaline Ashes,** or **Pearl-Ash.** 759

The ash produced by the incineration of plants may be referred to many classes, each characterised by the prevailing constituent present. Amongst these may be mentioned pearl-ash or alkaline earths; these contain potash. Barilla is a vegetable ash containing soda salts; Kelp, bromine, and iodine ash. Silicon is also frequently present, especially in the ash of graminaceous plants, and so also is lime in others. The first three are those of commercial importance. The following are the chief plants which yield pearl-ash in India:—

Abrus precatorius.
Achyranthes aspera.
Adhatoda Vasica.
Alstonia scholaris.
Amarantus spinosa.
Anthrocnemum indicum.
Bamboo ash.
Borassus flabelliformis.
Butea frondosa.
Cæsalpinia Bonducella.
Caroxylon fœtidum.
C. Griffithii.
Calotropis gigantea.
Cassia Fistula.
Cedrus Deodara.
Euphorbia neriifolia.
E. Tirucalli.
Erythrina indica.
Gmelina arborea.
Holarrhena antidysenterica.
Indigofera tinctoria.
Luffa ægyptiaca.
Musa sapientum.
Nerium odorum.
Penicillaria spicata.
Plumbago zeylanica.
Pongamia glabra.
Shorea robusta.
Stereospermum suaveolens.
Suœda indica.
S. nudiflora.
Symplocos racemosa.
Vallaris dichotoma.
Vitex Negundo.

These salts are largely used in India as mordants, but rarely in a pure form.

Of minerals alum and *sajjí-mátí* (an impure carbonate of soda, found as a natural earth) are those most used. (See **Auxiliaries, Dye.**)

ALKANET. 760

Alkanet, said to be derived from *al-kanna,* a dye supposed originally to mean the *henna* dye or **Lawsonia alba.** It is now restricted to the root of **Anchusa tinctoria** of China, a red dye, much used in colouring liquids. The Alkanet of Sikkim is obtained from **Onosma Hookeri,** *Clarke* (which see). **Dr. Dymock** informs me that a root is imported from Afghanistan as an alkanet which he thinks may prove a species of ARNEBIA. **Alkanna tinctoria,** *Tausch,* grows on sandy places on the Mediterranean coast.

ALLAMANDA, *Linn.; Gen. Pl., II., 690.* 761

A handsome genus of climbing APOCYNACEÆ; there are 12 species, chiefly inhabitants of Brazil and other parts of South America. They have been introduced and form much-prized additions to the flower-gardens of India.

762

Allamanda cathartica, *Linn.*

Syns. & References.—A. AUBLETII, *Rohl.; DC. Prod., VIII., 318; Dymock, Mat. Med., W. India, p. 421.*

Vern.—*Jaharí sontakká, pivli kanher, píli-kaner,* BOMB.

Habitat.—A large yellow-flowered shrub from America, much cultivated in India and run wild in the tidal back-waters of the western coast. (*Beddome.*)

Botanic Diagnosis.—This is the species most frequently seen in Indian gardens. The leaves are in fours, oblong-lanceolate; the flowers at the extremities of long trailing branches, tube 1 inch long, and the bell-shaped portion 2 inches long. This is a native of Guiana; it flowers freely in Calcutta gardens during the hot and rainy seasons.

MEDICINE. Leaves. 763

Medicine.—**Dr. Dymock** remarks: "Though not used in India, it has a medicinal reputation, the leaves being considered a valuable cathartic in moderate doses." **Ainslie** (*Mat. Ind., II., 9*) says that the Dutch consider an infusion of the leaves as a valuable cathartic.

A. Schottii, a native of Brazil, is even a still more showy species, having much larger flowers, with an extra tooth between the petals. **A. neriifolia**, another Brazilian species, is much more compact and shrubby, with broader leaves. The flowers are deep yellow streaked with orange, occurring in dense panicles. **A. violacea** has reddish-violet flowers. (*Treasury of Botany.*) **Firminger** says that he has never seen the Allamandas produce seed in India, but that they are all easily propagated by cuttings.

764

ALLÆANTHUS, *Thwaites; Gen. Pl., III., 361.*

Allæanthus Zeylanicus, *Thw.;* URTICACEÆ.

Vern.—*Allandoo-gass,* CINGH.

Habitat.—A tree met with in the central province of Ceylon, altitude 1,000 to 2,000 feet.

FIBRE. 765

Fibre.—A very tough fibre is obtained from the inner bark of this tree, which is used by the Cinghalese for a variety of purposes. (*Thwaites, Enumaratio Plantarum Zeylanicæ, p. 263.*)

Allmania, see **Amarantaceæ.**

766

ALLIUM, *Linn.; Gen. Pl., III., 802.*

A genus of bulbous, herbaceous plants, belonging to the Natural Order LILIACEÆ, containing some 250 species, confined to Europe, and the temperate and extra-tropical regions of Africa, Asia, and America.

Bulb tunicated. *Spathe* many-flowered. *Umbels* crowded. *Flowers* regular, 6-merous; *segments* distinct or only slightly united below. *Stamens* 6; *anthers* oblong, attached by the middle and on the back. *Ovary* superior, sessile, 3-celled; *stigma* 3-fid; *ovules* mostly 2 in each cell.

Allium is the classical name for the garlic, leek, &c.

767

Allium ascalonicum, *Linn.; Roxb. Fl. Ind., C.B.C. Ed., 288.*

THE SHALLOT.

Vern.—*Gandhan, gandana,* PB.; *Gandana,* SAHARANPUR, N.-W. P.; *Gandana,* AFG.; *Gundhun,* BENG.

Roxburgh gives this species the vernacular name of *peeaj*, but this would seem to be a mistake. The specific name is in honour of the ancient city of Ascalon, where Richard the First, King of England, defeated Saladin's army in 1192.

Habitat.—A hardy, bulbous perennial, native of Ascalon in Palestine. Has been cultivated from the remotest times by all the nations of

The Onion.

the East, entering largely into their diet. It is regarded as much milder than garlic. It was most probably introduced into England about the middle of the sixteenth century. (*Treasury of Botany.*) Flowers greenish white or purplish white; bulbs about the size of a nut, white.

FOOD. The Cloves. 768

Food.—The bulbs separate into what are termed cloves, like those of garlic, and are used for culinary purposes, being of milder flavour than the onion. They also make excellent pickle. It is cultivated apparently in Afghánistan for the sake of the leaves, which may be cut two or three times a year for 25 or 30 years. **Firminger** says that it is little known to Europeans in India, but that the cloves or small bulbs should be planted out in October about 6 inches asunder, and that by the beginning of the hot season the crop will be ready for use. **Balfour** recommends that it should be sown in the commencement of the rains in beds and propagated by dividing the roots: it will yield a crop in the cold season.

Allium Cepa, *Linn.; Roxb. Fl. Ind., C.B.C. Ed., 287.* 769

THE ONION, *Eng.*; OIGNON, *Fr.*; ZWIEBEL, *Ger.*

Vern.—*Piyás,* HIND.; *Piyáj, palandu,* BENG.; *Piyás,* ASS.; *Piaj,* SANT.; *Palándu,* SANS.; *Basl,* ARAB.; *Píyás,* PERS.; *Dúngari,* GUJ., SIND.; *Kanda, piyaj,* BOMB.; *Kándá,* MAR.; *Kándá,* CUTCH; *Vella-vengáyam, irulli, ira-vengáy-am,* TAM.; *Vulli-gaddalu, nirulli,* TEL.; *Vengáyam, nirulli, kunbali,* KAN.; *Bawang,* MAL.; *Lúnú,* SINGH.; *Ky-et-thwon-ni, kesun-ni,* BURM.

References.—*Bentley and Trimen, Med. Pl., p. 280; Moodeen Sheriff's Supp. to Pharm. of India, p. 37; Baden Powell, Panjáb Products, I., p. 381.*

Dr. **Moodeen Sheriff** says that in some Indian languages the same names are applied to the onion as to the garlic, the latter being called the white onion—a name very easily confused with the names applied to the white forms of the true onion. *Kándá* is the HIND. for squill; it very much resembles the MAR. *kándé* for this plant.

Habitat.—Cultivated all over India. There are, in Bengal, two forms, known as the Patna and the Bombay; the onions of Janjira, Bombay, are much prized, being small and white. (*Bomb. Gaz., XI., 425.*) English seed does not as a rule succeed so well as country, because, before it can come to India in time for the Indian season, it is two years old. Onion seed will not keep for certain more than one year. **Firminger** recommends that selected bulbs be planted, and seed obtained from these. If planted in the cold season, they will seed about the beginning of the hot season; and if carefully preserved, after being well ripened and dry, the seed obtained in this way will be found to yield a good crop in the following cold season, from October to February.

Chemical Composition.—**Fourcroy** and **Vauquelin** obtained from the onion a volatile oil containing sulphur, albumen, much uncrystallizable sugar and mucilage, phosphoric acid, both free and combined with lime, citrate of lime, and lignin. The expressed juice is susceptible of vinous fermentation. The oil is essentially the same as that from **A. sativum,** consisting chiefly of allyl-sulphide $(C_3 H_5)_2 S$. (*U. S. Dispens., 15th Ed.*)

MEDICINE. The Bulb. 770

Medicine.—The bulbs contain an acrid volatile oil, which acts as a stimulant, diuretic, and expectorant. Onions are occasionally used in fever, dropsy, catarrh, and chronic bronchitis; in colic and scurvy; externally as rubefacients, and when roasted, as a poultice. Considered by the natives hot and pungent, useful in flatulency. Said to prevent the approach of snakes and venomous reptiles. (*Baden Powell.*)

They are also described as aphrodisiac and carminative. Eaten raw they are emmenagogue. The juice rubbed on insect-bites is said to allay irritation; the centre portion of a bulb, heated and put into the ear, is a good remedy for earache. The warm juice of the fresh bulb is also used for this purpose. In addition to the oil obtained from the bulbs, the seeds yield a colourless clear oil used in medicines.

Opinions of Medical Officers.—§ "The bulb is crushed and the acrid smell emitted is utilised like smelling-salts for fainting and hysterical fits." (*Surgeon-Major Robb, Ahmedabad.*) "Said to increase the peristaltic action of the intestines, and is prescribed in obstruction. Used by natives in jaundice, hæmorrhoids, and prolapsus ani, also in hydrophobia. As an external application, onions are used in scorpion-bite and to allay irritation in skin diseases. They have antiperiodic properties attributed to them, and are said to mitigate cough in phthisis, and mixed with vinegar are used in sore-throat." (*Surgeon J. McConaghey, Shahjahanpore.*) "Used as decoction in cough." (*Surgeon G. C. Ross, Delhi.*)

Juice.
771

"Onion juice, mixed with mustard oil in equal proportions, is used as a liniment to allay rheumatic pains. The bulbs, made into a necklace, are worn as a charm to ward off the attack of cholera, and frequently kept suspended in front of the entrance to houses." (*Asst. Surgeon Anund Chunder Mukerji, Noakhally.*) "The onion promotes appetite and sexual desire; it acts also as a deodoriser, and is employed to correct the ill effects of the atmosphere, when cholera or any other epidemic disease is prevailing. Eaten raw, it acts as a diuretic and emmenagogue. Cooked with vinegar, it has been employed with benefit in cases of jaundice, enlargement of the spleen, and dyspepsia. The fresh juice is said to be a useful local application in cases of the bite of mad dogs; its internal exhibition at the same time accelerates the recovery. In scorpion-bite it has attributed to it the same properties." (*Asst. Surgeon Gholam Nabi, Peshâwar.*)

"I have found the onion very useful in preserving natives from scurvy." (*Surgeon L. Cameron, Nuddea.*) "The juice of the bulb in ʒiv to ʒviii doses, mixed with about ʒii of sugar, is a capital remedy for bleeding piles; one dose a day." (*Asst. Surgeon Nundo Lal Ghose, Bankipur.*) "A medium-sized onion is eaten twice a day with two or three black peppercorns as a favourite remedy in malarial fevers; a decoction of onion is used in cases of strangury." (*Surgeon-Major John North, Bangalore.*) "Upon the cut surface of a large onion a little slaked lime is placed; this rubbed over the part stung by a scorpion gives immediate relief." (*Surgeon-Major D. R. Thompson, Madras.*)

"Soporific when eaten raw. The juice is an excellent stimulant in cases of faintness; it should be applied freely to the nostrils. Is also used locally for the cure of scorpion-bites. It is said that the aphrodisiac properties of onions are enhanced by preserving them in a well-stoppered pot and then permitting the latter to remain in 'a cowdung yard' for a period of four months. One onion treated after this method is said to produce strong aphrodisiac effects." (*Surgeon W. Barren, Bhuj, Cutch, Bombay.*)

"The natives use this largely in cases of dysentery. It is prescribed thus: a grain of opium is buried in a bulb, and this is roasted under hot ashes, and is then administered to a patient suffering from acute dysentery. Good success follows this mode of treatment. Three ordinary-sized bulbs, with a handful of the leaves of the **Tamarindus indicus**, is made into a paste and used as a purgative." (*Surgeon Lee, Mangalore.*)

"Fresh juice of the bulbs rubbed on the body in case of sunstroke is attended with apparent benefit. A popular embrocation with the natives of Upper India, where, in the hot season, parents hang a number of onions on the chests of their children as a safeguard against hot winds. Roasted, they are commonly given to children as a stomachic." (*Asst. Surgeon Shib Chunder Bhuttacharji, Chanda, Central Provinces.*)

FOOD. 772

Food.—The onion is cultivated very extensively all over India, especially in the neighbourhood of large towns, and is consumed both by Europeans and natives. The Mussulmans of India never cook curry without onions, but the strict Hindús of Bengal regard them as objectionable, and rarely if ever eat them. The Patna onion is of a superior kind, and is much sold in the Calcutta markets. The onions of the northern provinces are larger and more succulent than those of Bengal and the southern provinces. Deprived of its essential oil by boiling, the onion becomes a mild esculent.

Onions, leeks, and garlic were cultivated in Egypt in the time of Moses, and Herodotus (B.C. 413) mentions an inscription stating that 1,600 talents, equal to £428,800, were paid for the onions and garlic eaten by the workmen engaged upon the erection of the great pyramid.

§ "When pressure of work or any other cause prevents the cooking of curry, the natives frequently eat onions with their daily meal, which, in the case of the poorer Bengalis, is stale rice and water with salt, and with the natives of Upper India coarse bread. The onion in these cases is eaten raw, for the purpose, apparently, of flavouring the meal." (*Mr. L. Liotard.*)

Allium fistulosum, *Linn.*

773

THE WELSH ONION; ROCK ONION; STONE LEEK.

A native of Siberia, said to have been introduced into Europe in 1629. Cultivated in gardens, but not admired as a culinary vegetable. It is a strong-rooted perennial plant, with sharp-pointed leaves, a foot or more in length. It never forms a bulb like the true onion, but has long tapering roots. From being very hardy it is generally sown to supply early onions for salad. (*Smith's Dictionary.*)

A. leptophyllum, *Wall.*

774

THE HIMÁLAYAN ONION.

The bulbs are regarded as sudorific: they are said to have a stronger pungency than ordinary onions. The leaves form a good condiment. Is this the species said to be exported from Lahoul?

A. Porum, *Willd.; Roxb., Fl. Ind., C.B.C. Ed., 287.*

775

THE LEEK.

Vern.—*Kiráth or Kirás*, ARAB.; *Parú*, BENG.; *Tan kyet thoon*, BURM. (*Balfour.*)

This esculent plant has been known from time immemorial. According to some authors it was originally a native of Switzerland, but more probably, like the onion, it came from the East. It is mentioned in the sacred writings, and was cultivated by the Egyptians in the time of Pharoah. Pliny says leeks were brought into notice by the Emperor Nero. The leek has been the badge of Welshmen ever since the sixth century, and is worn on St. David's day in commemoration of a victory they had over the Saxons, when they were instructed to wear the leek as a distinguishing badge during the battle. (*Treasury of Botany.*)

FOOD. 776

Firminger says leeks are best propagated in India by sowing the seed broadcast on a small bed immediately the rains stop. When the seedlings are about six inches high they should be carefully transplanted, taking care not to injure the roots. They should then be planted in rows six inches apart. They require plenty of water and should be earthed up once or twice.

777

Allium Rubelium, *Bieb.*

Vern.—*Jangli piaz, barani piaz, chiri piazi*, HIND.

Habitat.—Slender-leaved species, common in North-West Himálaya, extending into Lahoul.

FOOD. 778

Food.—The root is eaten raw or cooked.

779

A. sativum, *Linn.; Roxb., Fl. Ind., C.B.C. Ed., 287.*

THE GARLIC.

Vern.—*Lasan*, or *lahsan*, HIND.; *Rasún*, or *lasún*, or *lashan*, BENG.; *Naharu*, ASS.; *Rasun*, SANT.; *Maha-ushadha, lasuna*, SANS.; *Sum*, ARAB.; *Sir*, PERS.; *Lasunas*, MAR.; *Lasan*, GUJ.; *Shunam*, DUK.; *Vallai-púndu*, TAM.; *Vellulli tella-gadda*, TEL.; *Belluli*, KAN.; *Gokpas*, BHOTE; *Samsak*, TURKI; *Kyat-thou-bega, kesúm-phiu, kyet-thwunbya*, BURM.; *Sudu-lúnú*, CINGH.

References.—*Bentley & Trimen, Med. Pl., 280; U. S. Dispens., 15th Ed.; Supp. to Pharm. of India, p. 38; Treasury of Botany, Pt. I, p. 39; Watt's Dictionary of Chemistry, Vol. I., p. 143; Sharp, in Proc. Amer. Pharm. Association for 1864.*

Habitat.—Cultivated all over India; **Firminger** says it is very much cultivated by the natives in most parts of India, and can be had cheap at any bazar.

It is propagated by planting out the cloves singly, in October, in drills, about 7 inches apart and 2 or 3 inches deep. The crop is taken up in the hot weather, and after being dried in the sun the BULBS are stored for future use.

Botanic Diagnosis.—*Bulbs* numerous; enclosed in a common membranous covering. *Stem* simple, about 2 feet in height. *Leaves* long, flat, acute, sheathing the lower half of the stem. *Scape* smooth and shining, solid, terminated by a membranous pointed spathe enclosing a mass of flowers and solid bulbils and prolonged into leafy points. *Flowers* small, white.

Chemical Composition.—§ "Allylic sulphide is the chief constituent of the oil obtained by the distillation of garlic with water; it also occurs, but to a smaller extent, in oil of onions. From the herb and seeds of the **Thlaspi arvense**, it can also be obtained together with sulphocyanide of allyl, and oil of mustard. The leaves of the **Sisymbrium Alliaria** yield oil of garlic, and the seeds oil of mustard. A mixture of these two oils is likewise yielded by the following: **Capsella Bursa-pastoris, Raphanus Raphanistrum,** and **Nasturtium.** In some cases the oils do not exist ready formed; for example, the seeds of **Thlaspi arvense** emit no odour when bruised, and they must be macerated in water some time before distillation." (*Watts.*) "Allylic sulphide is a colourless oil of sharp unpleasant odour, lighter than water. The crude oil has a most intense odour of garlic." (*Surgeon Warden, Prof. of Chemistry, Medical College, Calcutta.*)

Properties and Uses—

OIL. 780

Oil.—The seeds yield a medicinal oil, clear, colourless, and limpid. **Dr. Ainslie** remarks that an expressed oil is prepared from the garlic, which is called *Vallay pundu unnay;* it is of a stimulating nature, and the Vytians prescribe it internally to prevent the recurrence of the cold fits of intermittent fever; externally, it is used in paralytic and rheumatic affections. (*Cooke.*)

MEDICINE. 781

Medicine.—"As a medicine garlic was held in great repute by the ancient physicians, and was also formerly much used in modern practice; but in this country it is now rarely prescribed by the regular practitioner.

MEDICINE.

although it is still employed to some extent in the United States. Garlic is stimulant, diaphoretic, expectorant, diuretic, and tonic, when exhibited internally; and rubefacient when applied externally. It is also regarded by some as anthelmintic and emmenagogue." (*Bent. & Trim., Med. Plants, 280.*)

In India, Garlic is considered hot and aperient; given in fevers, coughs, piles, leprosy, being regarded as carminative, diuretic, stomachic, alterative, emmenagogue, and tonic, and much used by the natives in nervous affections. It is esteemed by the Hindús as a remedy in intermittent fevers. The bulb is given in confection for rheumatism. Externally, the juice is applied to the ears for deafness and pain. Garlic is, in fact, chiefly employed at the present day as an external remedy; it is resolvent in indolent tumours. Is largely used as a liniment in infantile convulsions and other nervous and spasmodic affections. It is also frequently used as a poultice, as, for example, in retention of the urine from debility of the bladder.

The properties of garlic depend upon a volatile oil which may readily be obtained by distilling the bruised bulbs. When purified this oil is colourless, and may be distilled without decomposition. When garlic has been eaten, the odour of the oil may be detected in the various secretions of the body.

Opinions of Medical Officers.—§ "Mixed with vinegar garlic is used as an astringent in relaxed sore-throat and relaxation of the vocal chords. It is also used in asthma, general paralysis, facial paralysis, gout, and sciatica; is much thought of in the treatment of flatulent colic. Supposed to prevent the hair turning grey when applied externally." (*Surgeon G. A. Emerson, Calcutta.*) "Eaten in its green condition by persons in the cold season, from an idea that it wards off attacks of rheumatism and neuralgia." (*Surgeon-Major J, Robb, Ahmedabad.*) "Sometimes used for blistering purposes, but takes a long time before having any effect." (*Surgeon-Major C. J. McKenna, Cawnpore.*) "Garlic is an excellent medicine in several forms of atonic dyspepsia. It appears, like onions, to be useful in keeping up the temperature of the body. It is a good antispasmodic. In bronchial and asthmatic complaints it is decidedly beneficial." (*Surgeon-Major R. L. Dutt, M.D., Pubna.*) "Mustard oil, in which garlic has been fried, is an excellent application for scabies, and for maggots infesting ulcers." (*Assistant Surgeon Nobin Chunder Dutt, Durbhunga.*) "The juice or the whole bulb is used with salt as a poultice in bruises and sprains, also in neuralgia, rheumatism, gout, and rheumatoid arthiotis, and to relieve earache." (*Brigade Surgeon J. H. Thornton, Monghyr.*) "The smell of garlic is said to kill snakes; they never come where it is kept. Garlic poultice is used for rheumatic pains and also in neuralgia; if kept long it is rubefacient. Garlic oil is stimulant and rubefacient, largely used in the bronchitis of children." (*Asst. Surgeon J. N. Dey, Jeypore.*)

"A clove or two of garlic, boiled in half an ounce of gingelly oil, (Sesamum) and used as an ear-drop in atonic deafness, has proved very successful in my practice. The juice in elongated uvula is used with the same effect as that of nitrate of silver." (*Honorary Surgeon Easton Alfred Morris, Negapatam.*) A necklet of the bulbs is worn by children suffering from whooping cough. The juice is sometimes given with hot water for asthma. (*Surgeon James McConaghey, Poona.*)

"The expressed juice is a common application as a rubefacient." (*Native Surgeon Ruthnam T. Moodelliar, Chingleput, Madras.*) "Expressed oil used for elongated uvula, is said to act better than nit. arg." (*Surgeon-Major J. J. L. Ratton, M.D., Salem.*)

Chives and Rocambole.

MEDICINE.

"Habitually eaten by many persons subject to rheumatism. Cocoanut or mustard oil, in which a few pieces of garlic have been boiled, is useful in scabies and other parasitic skin diseases." (*Assistant Surgeon Shib Chunder Bhuttacharji, Chanda, Central Provinces.*) "The juice is used by the natives to destroy lice. It also acts as a blister, and as such is frequently used by native practitioners." (*Surgeon S. H. Browne, Hoshangabad, Central Provinces.*)

"The bulb is washed and applied to the temples, and, acting as a counter-irritant, has been known to relieve severe hemicrania and other forms of headache." (*Surgeon-Major A. S. G. Jayakar, Muskat, Arabia.*)

FOOD. 782

Food.—Used as a condiment in native curries throughout the country. § "The bulbs of garlic are eaten almost daily by the natives." (*Brigade Surgeon G. A. Watson, Allahabad.*)

783

Allium Schœnoprasum, *Linn.*

CHIVES OR CIVES.

A cultivated pot-herb, allied to garlic, with purple flowers; a native of North Europe. Its hollow grass-like leaves, forming clustered tufts, are commonly seen in kitchen gardens in Scotland. It is indigenous to Great Britain, and is accordingly very hardy, standing repeated cutting off close to the ground; the leaves are used in salad and to flavour soup. **Firminger** says it is little known in India, but is propagated by division of the roots in October.

784

A. scorodosprasmum, *Linn.*

THE ROCAMBOLE.

A perennial, esculent lily, closely allied to garlic, but regarded as much milder in flavour. It is a native of Denmark and other parts of Europe; it is used in the same way as garlic and the shallot, but its small cloves are considered more delicately flavoured than either.

785

A. sphærocephalum, *Linn.*

Grows wild in Lahoul.

The root and dried leaves are eaten (**Stewart**).

786

A. tuberosum, *Roxb.; Fl. Ind., Ed. C.B.C., 287.*

Vern.—*Bunga-ghundeena*, BENG.

Roxburgh says of this plant: "The natives use it as an article of diet, as leeks are used in Europe." **Royle** simply refers to this as a plant collected by **Roxburgh.** There seems to be some mistake regarding it, however, for it appears not to have been collected since **Roxburgh's** time, and even **Voigt** in his *Hortus Suburb. Calc.* says this is "unknown to us, as well as to our oldest native gardeners, who have hitherto been unsuccessful in their endeavours to procure it from the neighbourhood" of the Calcutta Botanic Gardens, where **Roxburgh** most probably collected the plant.

The greatest possible confusion exists in India regarding the cultivated forms, and, indeed, regarding even the wild forms of the genus **Allium**; and it is probable that, in addition to all the preceding forms, many others are regularly known to the natives of India, and even cultivated and sold in our bazars. **Stewart** enumerates the following unknown species as met with on the Panjáb Himálaya: **A. sp. (? odorum)**—vern. in Jhelum valley *bhúk;* in Ladák, *skodze.* **A. sp.**—vern. Jhelum, *khan;* Spiti, *phúndú.* **A. sp**—vern. *kiúr* in the Ravi Valley, and *kosse gokpa* in Ladák. **Stewart** also says that an unknown species is exported from Lahoul to Kúllú, to be eaten as a condiment by Hindús. (See **A. leptophyllum.**)

ALLOPHYLLUS, *Linn.; According to Gen. Pl., I., 396, reduced to* Schmidelia.

787

Allophyllus Cobbe, *Bl.; Fl. Br. Ind., I., 673;* SAPINDACEÆ.

Syn.—ORNITROPHE COBBE, *Willd.; Roxb., Fl. Ind., Ed. C.B.C., 328;* SCHMIDELIA COBBE, *Beddome, lxxiii.*

Vern.—*Thaukjot,* BURM.

Habitat.—A deciduous shrub of East Bengal, South India, Burma, and the Andaman Islands.

TIMBER. 788

Structure of the Wood.—Grey, soft.

ALLSPICE.

789

Allspice or **Pimento.**—A small bush or tree. **Pimenta acris,** *Wight,* and **P. officinalis,** *Linn.,* MYRTACEÆ.

Habitat.—Native of the West Indies. **Cleghorn** reports that several trees are in Madras, but that the climate of the Carnatic does not seem to suit them. **Mason** states that this large tree is repeatedly met with in Tavoy, but it does not flower; he is probably mistaken, as the plant is a small tree.

It is much cultivated in the West Indies for the sake of its aromatic leaves and berries. They partake of the smell and flavour of the cinnamon, clove, and nutmeg. Largely cultivated in Jamaica in what are known as pimento-walks. The berries are highly spoken of as a substitute for tobacco, and are said to be very pleasant, but require to be smoked in a long pipe. They are also used as a spice to flavour food. An oil is obtained by distillation, equal to nutmeg oil; reputed to allay tooth-ache. The bruised berries are carminative, stimulating the stomach, and promoting digestion; they also relieve flatulency. **P. acris** is regarded as inferior to **P. officinalis.**

Almonds, Bitter and **Sweet,** see **Prunus Amygdalus,** *Bail.,* ROSACEÆ.

Almonds, Country, see **Terminalia Catappa,** *Linn.,* COMBRETACEÆ.

ALNUS, *Gærtn.; Gen Pl., III., 404.*

790

A genus of trees belonging to the tribe BETULEÆ, of the Natural Order CUPULIFERÆ, a tribe which formerly was viewed as a Natural Order. The genus contains some 14 species, inhabitants of Europe, temperate Asia, and America, chiefly delighting in a moist soil, and most of them preferring the northern or alpine regions to the warm southern tracts of the temperate zones.

Leaves alternate, deciduous, rounded, blunt, serrate-penninerved, and furnished with tufts of whitish down in the angles of the veins beneath. *Flowers* monœcious; *male catkins* long, pendulous appearing, in autumn; *stamens* 4. *Female-spikes* ovate cone-like, appearing in spring; after fructification the thickened scales of the cone open and allow the seeds to escape, the cone-like bodies remaining attached to the tree until next year. Compare with the 2-stamened condition of BETULA with its caducous cones.

Alnus dioica, *Roxb.; Fl. Ind., Ed. C.B.C., 658.*

A Euphorbiaceous plant. Syn. for **Aporosa dioica,** *Müll.-Arg.,* which see.

The Nepal Alder.

791

Alnus glutinosa, *Linn.; Brandis, For. Fl., 461; Hooker's Students' Fl., 346.*

THE ENGLISH ALDER; SCHWARZERLE, *Ger.;* AUNE GLUTINEUX, *Fr.;* ONTANONERO, *It.*

Habitat.—The English alder has apparently not been introduced into India.

Properties and Uses—

DYE & TAN. The Bark. 792

Dye and Tan.—The bark is used in dyeing and tanning. It contains about 20 per cent. of a peculiar tannin.

MEDICINE. The Bark. 793 The Leaves. 794 Cones. 795

Medicine.—The bark and the leaves are very astringent and somewhat bitter. The former has been used in intermittent fever, and the latter as an external remedy in the treatment of wounds and ulcers. Bruised leaves are also sometimes applied to the breasts with the object of arresting the milk. A decoction of the cones is used as a gargle.

TIMBER. 796

Structure of the Wood.—White, soft, and light, on exposure to air turning pale red; will decompose in a year if exposed alternately to wet and dry weather, but if buried in the ground or submerged in water no wood is more durable. There is no heartwood. Knotty trees often yield beautifully mottled wood. The alder is the badge of the Clan Chisholm. The wood furnishes the best charcoal for gunpowder. Used extensively in Europe for herring-barrels.

The following are indigenous species :—

797

A. nepalensis, *D. Don; Brandis, For. Fl., 460; Wall., Pl. As. Rar., t. 131.*

THE NEPAL ALDER.

Vern.—*Kohi, Kœ,* PB.; *Udish,* KUMAON; *Udis, udish, wústa,* N.-W. P.; *Udis, utis (Gamble), boshi, swa (Brandis),* NEPAL; *Kowal,* LEPCHA.

Habitat.—A tall, sparsely-branched, deciduous tree, whose leaves soon become completely perforated by insects. It occurs from the Ravi eastward, between 3,000 and 9,000 feet in altitude, extending eastward to the Khásia and Nagá Hills (*Watt*), and to the Kakhyen hills in Ava (*Kurz*).

Botanic Diagnosis.—The tree is easily recognised by its dark-green coloured bark in young trees, becoming brown and fissured with age. *Leaves* oblong to elliptic on a slender petiole, obtuse at the base, shortly acuminate. *Catkins* sessile; *cones* ovoid, shortly stalked. *Nuts* irregular, winged, often broadest at the apex. *Fruit* ripe in March.

Properties and Uses—

DYE. 798

Dye.—The bark is used for dyeing and tanning. By the Nagás and Mánipuris it is used in combination with **Rubia sikkimensis** and **R. cordifolia** to deepen the colour. See **Rubia.** "The bark is used in tanning and dyeing. It is also said to enter into the composition of native red inks." (*Madden.*)

OIL. 799

Oil.—Said to yield an oil resembling birch oil.

TIMBER. 800

Structure of the Wood.—Similar to that of **A. nitida,** but the pores are fewer and somewhat larger, and the medullary rays are broad and very numerous. Weight 27 to 28 lbs. per cubic foot.

It is used for tea-boxes in Darjeeling.

This is perhaps one of the commonest plants in North Mánipur and the Nagá Hills, extending into the mountains of North Burma. It is rare in Mánipur proper, owing to the prevalence of dry red clay. Every moist soil and river-bank in the region indicated from 2,000 to 8,000 feet in altitude is, however, full of it, and so also are some portions of the country to the south and west of the valley of Mánipur, extending into the

Northern Cachar hills. It might be propagated to an unlimited extent, and not only supply timber and fuel, but its bark might easily form an article of internal trade. How far it might prove practicable to extend the cultivation of this useful tree into the tea districts of Cachar, Sylhet, and Assam, remains to be proved, but if not already attempted, it seems worthy of a trial. It grows rapidly, stands being pollarded freely, and if not propagated for its light soft wood for tea-boxes, would form a valuable source of fuel, luxuriating in the damp, uncultivatable banks of rocky streams and river-beds.

Alnus nitida, *Endl.; Brandis, For. Fl., 460; Gamble, Man. Timb., 373.* 801

Vern.—*Shrol, saroli, sawáli, sílein, rikúnra, cháp, chámb, tsápú,* or *chápu, piák, kúnsh, kúnsa, kúnich, niú, kosh, koe, rajáin, kúndash,* PB.; *Paya, udesh,* KUMAON; *Gira, ghuzhbe,* AFG.

Habitat.—A large tree, 80 to 100 feet in height, met with in the Pánjab Himálaya, ascending from 1,000 feet to 9,000 feet in altitude. Brandis says the largest trees are seen in the basins of the Jhelum and Chenab.

Properties and Uses—

Dye.—The bark is used for dyeing and tanning. DYE AND TAN. The Bark. 802

Fibre.—The young twigs are used for tying loads, rope-bridges, &c., and in the construction of baskets. FIBRE. 803

Fodder.—Leaves are sometimes used as fodder. FODDER. 804

Structure of the Wood.—Reddish-white, soft, close, and even-grained tough to cut; annual rings distinctly marked by harder wood near the inner edge of each ring.

Used for bedsteads and for the hooked sticks of rope-bridges. TIMBER. 805

The Indian Alders do not seem to possess, in the same degree at least, the properties of the English and American species. It is probable, however, that their properties are only unknown, and that they require to be made known in order to take an important place amongst the indigenous products of the country. "The bark of several alders is of great medicinal value, and a decoction will give to cloth saturated with lye an indelible orange colour." (*Porcher.*) "It contains a peculiar tannic principle. American alder has come into use for tanning; it renders skins particularly firm, mellow, and well coloured." (*Easton.*) "The bark contains 36 per cent. of tannin (*Muspratt*)." (*Baron F. Von Mueller, Extra-Trop. Plants.*)

ALOCASIA, *Schott.; Gen. Pl., III., 975; Mono. Phaner., DC., II., 497.* 806

A genus of herbs, belonging to the AROIDEÆ in the tribe COLOCASIEÆ, containing about 20 species, inhabitants of the tropics in Asia and the Malayan Archipelago. Tall herbs, with a succulent, sub-erect stem, marked with scars. *Leaves,* the younger all peltate, the older sagitato-cordate, the petiole elongated and possessed of a well-developed sheath. *Peduncles* often numerous and pointing upwards. *Spathe-tube* naked, ovoid or oblong, convolute, acrescent, and persistent. *Spadix* inappendiculate. *Flowers,* the upper male, the lower female. *Perianth* absent. *Ovary* 1-locular, or at the apex 3-4-locular; *ovules* few, orthotropous, erect from a basilar placenta; *micropyle* on the apex.

Alocasia cucullatum, *Schott.;* **Syn.**—ARUM CUCULLATUM, *Lour.* 807

A native of Bengal.

A. fornicatum, *Schott.;* **Syn.**—ARUM FORNICATUM, *Roxb.* 808

Used medicinally (*Roxb., Fl. Ind., Ed. C.B.C., 626*).

809

Alocasia indica, *Schott.*

Syn.—ARUM INDICUM, *as in Roxb., Fl. Ind., Ed. C.B.C., 625;* A. INDICUM, *Roxb., in Wight, Ic., III., t. 794.*

Vern.—*Mánkanda,* HIND.; *Mánkachú,* BENG., ASS.; *Mánaka,* SANS.; *Alú,* MAR.

Habitat.—Generally cultivated around the huts of the poorer classes in Bengal, its large leaves forming a striking feature of such localities.

Properties and Uses—

FOOD. 810

Food.—Its esculent stems and root-stocks are eaten in curries by people of all ranks. It is, in fact, an important article of food in Bengal; the edible stems, often 2 or 3 feet in length, may be preserved for months.

MEDICINE. 811

Medicine.—Said to be useful in anasarca; the flour obtained by pounding the dried stems is boiled along with rice-flour until all the water has evaporated, and is given to the patient, no other food being allowed. (Compare with **Colocasia antiquorum.**)

Opinions of Medical Officers.—§ "I have never used it solely as a medicine; but as food taken frequently, it seems to act as a mild laxative and diuretic. In piles and habitual constipation it is useful. Plants grown in loose soils and ash-pits are best. The tough portions should be rejected, and the stems and root-stocks boiled and the water thrown away, otherwise they are likely to irritate the throat and palate." (*Surgeon D. Basu, Faridpur.*) "The flour of old dried stems is a valuable article of food for invalids. It is an excellent substitute for arrowroot and sago, in place of which I have used it in many instances." (*Assistant Surgeon Shib Chunder Bhuttacharji, Chanda, Central Provinces.*)

"The starch contained in the edible stem seems much more easily digestible than rice; and in the milk treatment of cases of malarious saturation with anasarca, I have found this root useful to accustom the patient to return to starchy food. It has no medicinal virtue. I have used it extensively in the Rungpur Jail." (*Surgeon K. D. Ghose, Khulna.*) "*Mánkuchú* is a very agreeable vegetable in the convalescence of natives from bowel complaints. It is light and nutritious and somewhat mucilaginous. I prescribe it often in such cases. The root-stock is decorticated and cut in small pieces and boiled in water. It is then mixed with *brinjal* or some other vegetable and made into a curry with a little turmeric and salt." (*Surgeon-Major R. L. Dutt, M.D., Pubna.*) "The root-stocks are largely used for patients; they are juicy and easily digestible." (*Surgeon-Major E. C. Bensley, Rajshahye.*) "The ash of the root-stocks mixed with honey is used in cases of aphthæ." (*Assistant Surgeon Anund Chunder Mukerji, Noahkally.*)

812

A. montana, *Schott.*

Syn.—ARUM MONTANUM, *Roxb.*

The natives of the Northern Circars use the roots to poison tigers. (*Roxb., Ed. C.B.C., 652.*)

813

A. odora, *C. Kock.*

Syn.—ARUM ODORUM, *Roxb.*

Flowers are fragrant.

814

A. rapiformis, *Schott.*

Syn.—ARUM RAPIFORME, *Roxb.*

A native of Pegu.

815

ALOE, *Linn.; Gen. Pl., III., 776.*

A genus of plants with thick, succulent, and spiny leaves, belonging to the Natural Order LILIACEÆ, and comprising some 80 species, chiefly inhabitants of Africa, more particularly of South Africa and the Island of Socotra; now cultivated in all tropical and extra-tropical countries.

Stem absent or short erect, or arborescent and sometimes even branched. *Leaves* succulent, forming a rosette on the extremity of the stem, often spinous at the apex and sparsely spinously serrate on the margin. *Flowers* forming spikes, axillary in the uppermost leaves or terminal, nodding, cylindrical, connivent by the short teeth. *Stamens* 6, hypogynous, as long as the perianth or longer; *filaments* subundulate; *anthers* oblong dorsifixed. *Ovary* sessile, 3-celled, many-ovuled. *Fruit* membranous, 3-celled, bursting by loculicidal dehiscence. *Seeds* compressed.

The word Aloë is said to be derived from the Syriac, *Alwai*, and the Greek derivative ἀλόη, but in the 10th century the drug was known as Succotrina. In India it is known by the generic name of *Elwa, Elia*, or *Mushabbar*. Several species yield a bitter juice which, when inspissated, forms a drug of varying commercial value, according to the care with which it has been prepared, and the specific peculiarities of the plant or plants from which obtained. Indeed, it would seem probable that several species of Aloë afford each of the important commercial forms,—*viz.*, Barbados, Socotrine, Cape Aloes, and Natal Aloes. This being so, it has been deemed the most satisfactory course to treat first of aloes as met with in the druggist's shop, and then to endeavour to discuss the principal facts known regarding the plants from which the aloes met with in Indian commerce are obtained.

History of the Drug Aloes.—The following abstract, taken chiefly from the *Pharmacographia*, will be found to contain the more important facts connected with the early history of this drug: Aloes appear to have been known to the Greeks in the 4th century B. C., for the Arabian historian **Edrisi**, accounting for the Greek occupation of Socotra, attributes this to Aristotle having persuaded Alexander to take possession of the island that produces *aloes*. The original inhabitants were removed and Ionians placed in charge of the island, and of its aloe plantations. In the 10th century aloes were produced only in Socotra, and it is reported that they had greatly improved under the Greek management. Aloes were known to **Celsus, Dioscorides**, and **Pliny**, as well as to the later Greek authors and Arabian physicians. Aloes were recommended to Alfred the Great by the Patriarch of Jerusalem, and it may therefore be inferred that the drug was not unknown in Britain as early as the 10th century. In 1516, **Pires**, a Portuguese apothecary, the first ambassador sent to China, reports to Manuel, King of Portugal, that aloes are found in the Island of Socotra, Aden, Cambaya, and other parts, the most esteemed being that of Socotra. In the 17th century a direct trade was established between Socotra and England. **Wellstead**, who travelled in Socotra in 1835, reported that it must once upon a time have been far more extensively cultivated than at present. He describes it as growing abundantly in parched and barren places on the sides and summits of limestone mountains, at an altitude of 500 to 3,000 feet. **Mr. Wykeham Perry**, in 1878, brought specimens of the Socotra aloe to Kew, London, when it was discovered that it was quite distinct from the plant which, by botanists, had come to bear the name of Socotra. This new species was, by **Mr. Baker**, named in honour of its discoverer, and this interesting fact regarding aloes has since been confirmed by **Prof. J. B. Balfour**, who has clearly shown that the plant **A. Perryi** is the true source of the Socotrine Aloes of commerce.

From about the 10th century it seems the cultivation of the aloe became in all probability distributed over the tropical portions of the globe.

ALOE.

Medicinal Properties of the Aloe.

In the 17th century, **Ligon**, who visited the Island of Barbados, speaks of the Aloë as if it were indigenous. This was about 20 years after the first settlers had taken up their residence. He mentions a number of useful plants which they had introduced into the island. Barbados aloes first appeared in the London drug warehouses in 1693, Cape aloes in 1780, and Natal aloes in 1870. (*Flückiger and Hanbury's Pharmacog.*)

For further particulars regarding the history of Indian aloes the reader is referred to **A. vera.** (See page 186.)

816

MEDICINAL PROPERTIES AND USES.

Cultivation and Manufacture.—In Barbados, where the plant is systematically cultivated, the plants are set 6 inches apart, in rows 1 to 1½ feet wide, the ground having been carefully prepared and manured. The plants are kept dwarf and free from weeds. The leaves are 1-2 feet long; they are cut annually. The cut leaves are rapidly placed in a V-shaped trough with the cut end downwards, and so arranged that the juice may drain from all the leaves into a jar below. These troughs are placed all over the plantation so as to be easily accessible to the cutters. By the time five troughs have been filled, the first is exhausted. The leaves are neither boiled nor pressed, and no use is made of them after the juice has drained off. The juice is next inspissated by evaporation, but it does not injure by being left for some time in the jar. (*Pharmacographia.*)

Description and Properties of the Commercial Varieties of Aloes.—In an interesting paper of much practical value, **Dr. Squibb** (*Pharmacist, 1873, p. 33; Year-Book of Pharm., 1874, p. 38*) discusses the merits of the commercial forms of Socotrine and Barbados aloes, describing the former in its therapeutic effect as comparatively mild and gentle and unirritating, with tonic and aromatic qualities, while the latter is harsh and drastic, producing more irritation, and much more liable to over-action. He recommends that only the better qualities of the so-called Socotrine aloes be prescribed to human beings, while he regards the various forms of Barbados as better adapted to the medication of animals. While this therapeutic distinction holds good—clearly separating the two classes—there are a whole series of forms of the drug, which, in their external and physical properties, blend into each other so completely that it requires almost professional skill to distinguish them. The former class is lighter in colour, with a pleasant but feeble aromatic odour. It is mainly by the odour that the forms of aloes are valued by the dealers, and it is therefore impossible to describe this test: the inferior kinds of aloes are harsh, strong, often fœtid, while the better qualities are faint and aromatic. Recently, the numerous forms of Socotrine aloes have, in the trade, been referred to two classes—red Socotrine and yellow Socotrine, the former being held in the highest esteem. **Dr. Squibb** made the curious observation that the red form is at first yellow, becoming red as it dries, while the yellow only deepens in colour, but never becomes red. The term Hepatic Aloes is exceedingly vague, and "appears to us unworthy to be retained." (*Flück. and Hanb.*) It is applied to any sort of liver-coloured aloes, its opacity having formerly been attributed to the presence of crystals, whereas it is now believed to be due to the presence of a feculent matter, the nature of which is unknown, although it is most probably produced through the fermentation caused by impurities such as pieces of sheepskin. **Dr. W. Craig** recommends that aloes should be administered only in the form of aloïn. He bases his opinion upon one or two important considerations:

"*1st.*—Aloïn may, by exposure to the air, undergo considerable chemical change without losing its physiological action as an active aperient.

"*2nd.*—The resin of aloes, when thoroughly exhausted of aloïn, possesses no purgative properties, and therefore cannot be the active principle of aloes.

"*3rd.*—The resin of aloes is not the cause of the griping which sometimes follows the administration of the drug; it is a perfectly inert and harmless substance.

"*4th.*—Aloïn is an activea perient, and in all probability is the only active principle of aloes.

"*5th.*—Being uniform in strength, its dose can be more accurately determined.

"*6th.*—Its dose being only half a grain to one grain, it can easily be introduced into tonic pills without making these too large.

"*7th.*—By using the active principle all impurities are excluded which are apt to cause griping." (*Year-Book of Pharm., 1875, p. 289.*)

The *Makhzan-ul-Adwiya* mentions four kinds of aloes as met with in India, *viz.*, Socotrine, Bokhára or Persian, Indian, and Arabian. The latter is said to be often adulterated with akakia and with gum arabic. The Bokhárian is pronounced the worst kind, being full of stones.

Chemical Composition.

References.—*Flück. & Hanb., Pharmacog., 1879, 686; U. S. Dispens., 15th Ed., 155; Tilden, Year-Book of Pharm., 1875, 540; Schmidt, Archiv. der Pharm., V., No. 6, 1876; Pharm. Journ., 3rd Series, VII., 70; Year-Book of Pharm., 1877, 38.*

All varieties of aloes have an odour very much of the same character, and a bitter, disagreeable taste. The odour is due to a volatile oil which the drug contains in minute proportions. The most interesting constituents of aloes, however, are the substances known under the generic term of aloïn. Chemically, these principles appear to be complex phenols. The name aloïn was originally used to designate the substance found in Barbados aloes, but this aloïn is now named Barbaloïn, and the aloïns contained in Natal, Socotrine, and Zanzibar aloes are known respectively as Nataloïn and Socaloïn or Zanaloïn. The three varieties of aloïn are crystalline, and by chemical tests can be readily distinguished. Aloes also contain resins, certain of which are soluble and the others insoluble in water. By the action of reagents aloes afford a large number of derivatives. (*Pharmacographia.*)

The aloïn which is met with in commerce is prepared chiefly, if not entirely, from Barbados aloes. **Dr. Tilden** recommends the following method for the separation of this substance. Crushed aloes is dissolved in nine or ten times its weight of boiling water acidified with sulphuric acid. After cooling and standing for a few hours the clear liquid is decanted from the resin and evaporated. The concentrated solution will be found to deposit a mass of yellow crystals which can be purified by washing, pressure, and by recrystallization from hot spirit. After repeated crystallization the aloïn is obtained in the form of beautiful yellow needles, which are fairly soluble in water and in alcohol, but difficultly so in ether. **Dr. Tilden** recommends for the isolation of *zanaloïn* **Mr. Histed's** process, which, though troublesome, is the only process at present known. Powdered aloes should be macerated in proof spirit to make a paste, and the liquid gradually expressed from the mass. The yellow cake remaining is purified by crystallization from water and afterwards from rectified spirit.

E. von Sommaruga and **Egger** and also **Rochleder** consider the aloïns to form a homologous series, for which they have assigned the formula $C_{17}H_{20}O_7$ barbaloïn, $C_{16}H_{18}O_7$ nataloïn, and $C_{15}H_{16}O_7$ socaloïn—compounds derived from anthracene $C_{14}H_{10}$. This opinion has not, however, met with the support of subsequent experimenters, and **Tilden** is of opinion that, on the contrary, barbaloïn and zanaloïn, and in

CHEMISTRY.

all probability socaloïn, are chemically identical and must be expressed by the formula $C_{16}H_{18}O_7$. **Tilden**, speaking of barbaloïn and zanaloïn, says "the two bodies resemble each other in appearance and in taste":—

(*a*) *Zanaloïn* is slightly paler in colour; more soluble; it contains more water of crystallization than barbaloin. Moistened with nitric acid, it gives no immediate coloration, but on the application of heat an intense orange-red is developed.

(*b*) *Barbaloïn* gives with nitric acid an instant coloration, which fades quickly to orange.

Both zanaloïn and barbaloïn, under the prolonged action of nitric acid, give chrysammic acid, and both yield crystallizable chloro- and bromo-substitution derivatives which resemble each other very closely. Socaloïn is believed to be identical with zanaloïn.

(*c*) *Nataloïn* is widely distinct from these crystalline principles, but chiefly in its not forming chrysammic acid nor chloro- or bromo- derivatives. It is also much less soluble than either of the preceding.

On the main features of these opinions **Dr. Schmidt** supports **Dr. Tilden**, confirming the formula given for barbaloïn and zanaloïn.

Histed published a most beautiful and convenient test for the three principal forms of aloïn. A drop of nitric acid, placed on a slab, gives with barbaloïn and nataloïn a bright crimson, but produces little or no effect on socaloïn. Barbaloïn is further distinguished from nataloïn by adding a minute quantity to a drop of oil of vitriol, and allowing the vapour from a rod moistened with nitric acid to pass over the surface of the solution. Barbaloïn (and also socaloin) will undergo no change, but nataloïn will assume a fine blue colour. The nitric acid colour produced with barbaloïn rapidly fades, but that with nataloin is permanent unless heat be applied. These reactions may sometimes be produced with the crude drugs. "Aloes yields its active matter to cold water, and when good is almost wholly dissolved by boiling water; but the inert portion, or apothême of Berzelins, is deposited as the solution cools. It is also soluble in alcohol, rectified or diluted. Long boiling impairs its purgative properties by oxidising the aloïn and rendering it insoluble. The alkalies, their carbonates, and soap, alter in some measure its chemical nature, and render it of easier solution. It is ïnflammable, swelling up and decrepitating when it burns, and giving out a thick smoke which has the odour of the drug." (*U. S. Dispens.*, *157*.) The fact that heat affects the properties of the drug must be clearly borne in mind; since the heat used in melting, straining, and drying which, on account of the presence of impurities, often becomes necessary before it can be used medicinally, will, unless carefully performed, impair considerably the action of the drug.

817

Trade Returns of Aloes.

Aloes (Foreign Trade by Sea).

Years.	Imports.		Exports and Re-exports.	
	Quantity.	Value.	Quantity.	Value.
	Cwt.	R	Cwt.	R
1879-80	792	21,330	752	20,684
1880-81	1,029	24,721	840	19,561
1881-82	1,023	23,780	469	16,489
1882-83	1,345	29,514	783	24,113
1883-84	1,611	31,639	610	21,676

TRADE RETURNS.

Jafferabad Aloes.

Detail of Imports, 1883-84.

Provinces into which imported.	Quantity.	Value.	Countries whence imported.	Quantity.	Value.
	Cwt.	R		Cwt.	R
Bombay . .	1,593	30,709	Aden . . .	1,102	15,591
Sindh . .	18	930	Arabia . . .	489	15,768
			China—Hong-Kong	20	280
TOTAL .	1,611	31,639	TOTAL .	1,611	31,639

Detail of Exports, 1883-84.

Provinces from which exported.	Quantity.	Value.	Countries to which exported.	Quantity.	Value.
	Cwt.	R		Cwt.	R
Bombay . .	520	20,798	United Kingdom .	341	18,097
Madras . .	90	878	Straits Settlements	229	2,746
			Other Countries .	40	833
TOTAL .	610	21,676	TOTAL .	610	21,676

The following are the principal forms of Aloë met with either in cultivation in India, or in the drug of which a large import and internal trade exists:—

Aloë abyssinica, *Lam.; Baker, Linnæan Jour., XVIII., 174.* 818

JAFFERABAD ALOES.

Syn.—A. MACULATA, *Forsk.;* A. VULGARIS, *var.* ABYSSINICA, *DC. Plantes Grasses sub. t. 27.*

Habitat.—Dr. Dymock informs me that this plant is common on the coast of Bombay and Gujarát, and that it furnishes the round cakes known as Jaferabad Aloes. It is a native of Abyssinia and Central Africa.

Botanic Diagnosis.—*Stem* simple, 1-2 feet in height, 2-3 inches in diameter. *Leaves* ensiform, $1\frac{1}{2}$-$2\frac{1}{2}$ feet long, broad-acuminate, green, often white spotted, base rounded. *Raceme* dense oblong; *bracts* lanceolate-acuminate, 3-4 lines long. *Flowers* yellow, base green, tube short, teeth long. *Stamens* often exserted. Introduced into Europe in 1777.

MEDICINE. Aloes. 819

Medicine.—It seems probable that this species may contribute, along with the two following, to the so-called *Moka* Aloes imported from the Red Sea coast into Bombay, or by way of Zanzibar from Socotra, and re-exported to Europe.

Leaves. 820

Under the heading **Aloe abyssinica,** *Baker,* **Dr. Dymock,** in his *Materia Medica of Western India,* p. 667, gives an interesting history of the Indian Aloes. The Mohammedans, he informs us, learned the preparation and uses of the drug from the Greeks. Its manufacture spread from Africa and Arabia, ultimately reaching India. He further states that the Hindús, though unaware of the method of preparing the inspissated sap, have long been in the habit of using the plant medicinally under the name *Ghirta kumarí.* In Eastern and Southern India, the *Ghirta kumarí* is one of the forms of **A. vera,** probably an indigenous plant to India, while **A. abyssinica** seems to have been introduced. It may be the case, however, that all wild or cultivated Aloes of India go by the name of *Ghirta kumari*. (See **A. vera**, *var.* **officinalis.**)

Pickled. 821

Pickled Aloes.—Dr. Dymock says that the leaves and flower-stalks of this aloe (? **A. abyssinica**) are pickled by the natives of Gujarát after having been soaked in salt and water.

§ "The aloes cultivated in the Kanara district are more succulent, yielding larger quantities of viscid juice. The fresh leaves are used as emollients in lieu of linseed poultices, meal, flour, &c., for abscesses or whitlow. The leaf is first roasted and then laid open on its inner side, and applied whilst warm. It hastens suppuration." (*Surgeon-Major W. Nolan, M.D., Bombay.*)

822

Aloë ferox, *Miller*, is one of the principal plants from which Cape Aloes are obtained.

823

A. spicata, *Thunb.*, is also one of the Cape Aloes plants.

824

A. succotrina, *Lam.*, and **A. Perryi,** *Baker.*

THE SOCOTRINE ALOES OF COMMERCE; THE YAMANI OR MOKA ALOES OF BOMBAY.

In the present state of our knowledge of the subject it has been deemed advisable to discuss the various forms of the drug, commonly known as Socotrine aloes, under this head, without attempting to separate the two species mentioned above. Recent discoveries in the Island of Socotra itself, made by **Mr. Perry**, and later by **Professor Bayley Balfour**, have shown **A. Perryi** to be the species from which the pure Socotrine aloes is derived; but much of what is sold commercially as Socotrine aloes is doubtless got from other species, and chiefly from **A. succotrina**. This is specially true of the Socotrine aloes not obtained from the Island of Socotra.

Syn.—A. PERFOLIATA, *var.* SUCCOTRINA, *Curt.*; A. VERA, *Miller*. See BAKER *in Linnæan Journal, XVIII., 173.*)

Vern.—In the chemist's shop Aloes go by the same names as are given to the next species, *viz.*, *Sibr*, or *Sabir*, ARAB.; *Bóle-síyáh*, PERS.; *Musabbar, ilvá, yalvá*, HIND.; *Móshabbar*, BENG.; *Élvá, eliyá, musanbar*, DUK.; *Musambarból*, MAR.; *Kariya pólam*, TAM.; *Múshámbaram*, TEL.; *Chennanáyakam*, MAL.; *Eliyá*, CUTCH; *Yelíyo*, GUJ.; *Mou*, or *mo*, BURM.; *Kalu-bólam, karibólam*, SINGH. The name of the Socotrine aloes is *Sibr-sagótari*, ARAB. Met with in all the bazars of India.

References.—*Flück. and Hanb., Pharmacog., 679; Med. Plants, Bentl. & Trim., IV., 283; U. S. Dispens., 15th Ed., p. 151; Pharm. Ind., 236; Royle, Mat. Med., Ed. Harley, 398; Pereira in Pharm. Jour., ser. I., vol. XI., p. 439; Forskal, Fl. Ægypt. Arab, p. 73; Wellstead in Journ. R. Geogr. Soc., V., p. 197; Kew Report, 1880, pp. 21 and 54; Year-Book of Pharm., 1874 to 1884; Pharm. Jour., vols. III., IV., V., VI., VII., VIII.; American Pharm. Soc. Proc., vol. XXV.*

Habitat.—Dr. Trimen writes me that he regards **A. succotrina** as indigenous to South Africa and not Socotra or the Red Sea districts; this is the modern opinion, although, as the name implies, it was formerly viewed as a native of the Island of Socotra. **A. Perryi,** as far as has been discovered hitherto, is peculiar to Socotra, and the presumption is that it is the species from which the Socotrine aloes is obtained in the Island of Socotra itself.

Botanic Diagnosis.—A. succotrina, *Lam.*—*Stem* woody, often 6 feet in height, strongly ridged with scars of the fallen leaves, often becoming dichotomously branched. *Leaves* crowded, the rosette 2 to 3 feet in diameter, ensiform, falcate, sessile, amplexicaul, 15 to 20 inches long, tapering to an acute point. *Scape* exceeding the leaves, angled, purplish green; *flowers* numerous, forming a narrow, erect, spike-like raceme; *pedicels* erect, curved so that the flowers are pendulous; *bracts* shorter than the pedicels. *Flowers* about 1⅛ inch long, red below, orange-red or pinkish in the middle, with greenish white tips, deeply cut into 6 obtuse segments. *Stamens* 6, 3 sometimes exceeding the perianth.

A. Perryi, *Baker.—Stem* simple, 1 inch in diameter, scarcely rising above ground. *Leaves* crowded, much shorter than in the preceding and rounded at the base. *Racemes* dense; *bracts* lanceolate, sub-equal to the pedicels. *Flowers* red, 9-10 lines long; *segments* oblong, three times the length of the tube. *Stamens* included.

MEDICINE. 824a

Medicine.—In small doses, the drug aloes, prepared from the juice of the leaves, is stomachic, tonic; in larger doses, purgative, and indirectly emmenagogue. It is a remedy of great value in constipation caused by hysteria and atony of the intestinal muscular coat. It is also very useful in atonic dyspepsia, jaundice, amenorrhœa, and chlorosis. Locally applied, dissolved in glycerine, it is valued as a stimulant application in skin diseases. (*Pharm. Ind.*)

Socotrine Aloes. Pure. 825

Impure. 826

Dr. Dymock informs me that a mixture of aloes and myrrh is known in the Deccan as *mussabar*, and that Socotrine aloes is largely imported into Bombay. It appears that **Dr. Dymock** (*pp. 669 and 670*) regards the Socotrine aloes as distinct from the *moka*. He may be correct in this opinion, but it seems doubtful if any Socotrine aloes can be said to be the product of but one species even when imported direct from Socotra, and it is more than probable that the purer forms of African aloes are regularly sold under the time-immemorial reputation of the Island of Socotra, the more so since all the imports from Socotra, preserving the old trade route, come by way of Zanzibar to Bombay and England. The distinction which it is possible to establish is that pure Socotrine aloes is the product of **A. Perryi**, while the aloes to which the name *moka* or Arabian aloes may in the future be restricted, is the product chiefly of **A. succotrina** and one or two allied species inhabiting the Red Sea coast and the Arabian coast of the Persian Gulf. It may accordingly be found convenient to refer Socotrine aloes of commerce to two great sections:—

827

(A) The Pure Forms of Socotrine Aloes (contain Socaloïn).

Speaking of these separate forms **Dr. Dymock** says of Socotrine aloes: "This drug is imported into Bombay *viâ* Zanzibar and the Red Sea ports. It is packed in skins, the packages varying much in size and shape, and often containing a large proportion of rubbish, such as pieces of hide, stones, &c. In Bombay the skins are opened, and the aloes repacked in boxes for exportation to Europe. The best Socotrine aloes is of a reddish-brown colour, hard externally, soft internally; the odour is aromatic and peculiar; when powdered or in thin fragments it is orange-brown; sometimes it is almost fluid."

Flückiger and **Hanbury**, in their *Pharmacographia* (*1879, p. 684*), say of this form: "The SOCOTRINE, so-called BOMBAY, EAST INDIAN, or ZANZIBAR ALOES, which when opaque and liver-coloured is also known as HEPATIC ALOES, is imported from Bombay into England in kegs and tin-lined boxes. When moistened with spirit of wine, and examined in a thin stratum under the microscope, good Socotrine aloes is seen to contain an abundance of crystals. As imported it is usually soft, at least in the interior of the mass, but it speedily dries and hardens by keeping," losing about 14 per cent. in the process. "Some fine aloes from Zanzibar, of which a very small quantity was offered for sale in 1867, was contained in skin." When it is fluid it is known as *Liquid Socotrine Aloes.* This was at one time supposed to be different from *Hepatic*

Aloes, and that the latter owed its opacity to crystals. But it has been shown that the opacity is due to some feculent matter, and that therefore opaque aloes, from whatever plant derived, equally deserves the commercial name of hepatic aloes.

Bentley and **Trimen** say the colour of Socotrine aloes varies: "the reddish tint is also liable to great variation; thus sometimes the masses are garnet-red, at other times they are much paler, and when quite dry, are golden-red, and yield a golden-yellow powder. By exposure to air the colour is deepened. The fracture is usually smooth and resinous, but sometimes rough and irregular."

Impure Socotrine. 828

(B) The Impure Forms, or Yamani or Moka Aloes (contain Zanaloïn ?).

This, **Dr. Dymock** informs us, is imported from Arabia into Bombay. It is the kind of aloes most in use by the natives of India. "It varies much in quality. It is of a black colour in mass, and somewhat porous, but thin fragments are translucent and yellowish brown; the odour is powerfully alœtic, without the aroma of Socotrine or Jafferabad aloes; medicinally it appears to be sufficiently active. With nitric acid it gives a deep-red colour, like Barbados; the solution in sulphuric acid is not affected by nitric acid fumes." (*Mat. Med., W. Ind., 670.*) The *Pharmacographia* says: "A very bad, dark, fœtid sort of aloes is brought to Aden from the interior. It seems to be the *Moka Aloes* of some writers."

Special Opinions.—§ "Useful in combination with sulphate of iron in cases of irregular or suspended menstruation, also in hysteria, headache, constipation, and flatulence." (*Brigade Surgeon J. H. Thornton, B.A., Monghyr.*) "In sprains and inflammations, applications of aloes and opium are found to be very beneficial in allaying pain, &c." (*Assistant Surgeon Doyal Chunder Shome, Campbell Medical School, Sealdah, Calcutta.*) "Purgative, and emmenagogue when applied externally over the abdomen in puffiness of the abdomen. The tincture, in combination with simple soap liniment, when applied over the abdomen of children who cannot tolerate aperients by the mouth, acts freely on the bowels." (*Surgeon W. Barren, Bhuj, Cutch, Bombay.*)

"Given internally by the hakims in bronchial catarrh and jaundice; externally applied with lime-juice in contusions and sprains." (*Surgeon-Major J. T. Fitzpatrick, M.D., Coimbatore.*) "Formed into a paste with hot water, it is useful when applied to severe sprains and contusions." (*Surgeon-Major J. J. L. Ratton, M.D., M.C., Salem.*) "Aloes rubbed up with opium, myrrh, and white of egg, applied to any swelling, causes absorption, soothes and relieves pain." (*Surgeon-Major Henry David Cook, Calicut, Malabar.*)

829

Aloe vera, *Linn.*

BARBADOS ALOES, INDIAN ALOES, *Eng.*; ALOES, *Fr.*; ALOE, *Germ.*

Syn.—A. BARBADENSIS, *Miller*; A. VULGARIS (Bauhin), *Lam.*

Vern.—*Ghi-kavár, Ghi-kanvar*, or *ghigvár, kumári*, HIND.; *Ghirta-kumári, girta-kunvár*, BENG.; *Ghirta-kumári, kanyá*, SANS.; *Sabbárá, nubátussibi*, ARAB.; *Darakhte-sibr*, PERS.; *Eliyá* (the resin), *kor-kand* (the plant), *kumári, ghi-kanvar, kanvar-patha*, DUK.; *Koraphad*, MAR.; *Kunvar*, GUJ.; *Kanvár, kora kanda, kora-phad, lephee*, SIND.; *Katrazh-ai*, or *Kattalai, shóttu-katrazhai*, or *shóttu-katrazhai*, or *kattalai*, TAM.; *Kalabanda*, TEL.; *Kattruvázha*, or *kattála*, MAL.; *Lola-sorá*, KAN.; *Komarika*, SINGH.; *Tazávon-le-pá*, or *shazaon-le-pá*, BURM.

The resinous extract is generally known as *Sibr*, PERS. (See also names given under **A. succotrina.**)

Mr. J. G. Baker, in the *Linnæan Society's Journal, Vol. XVIII., p.*

176, has established the synonyms above given, and formed under this species two varieties. **Bentley** and **Trimen**, in their *Medicinal Plants*, reduced all the names for the forms of this species to mere synonyms, under the name of **A. vulgaris**, *Lam.* **Mr. Baker** seems correct, and the varieties formed by him are well known to the natives of India, and their individual properties have been recognised in native practice from almost time immemorial.

Habitat.—There are many sub-varieties of this plant met with in cultivation throughout India, some of which have run wild, as, for example, on the coast of South India. All the forms of this species must, however, be described as natives of Northern Africa, from Morocco eastward; of the Canary Islands and of Southern Spain. They have long been cultivated in the West Indian Islands, Jamaica, Antigua, and Barbados, where they were probably introduced at an early date from the Canary Islands.

Botanic Diagnosis.—*Stem* short, 1-2 feet, 2-3 inches diameter. *Leaves* ensiform, densely crowded, 1½ to 2 feet long, broad at the base, attenuated at the apex to a blunt point, pale green, glaucose, distantly dentate. *Scape* 2-3 feet long, simple or branched. *Racemes* dense, ½ to 1 foot long; *bracts* lanceolate-acute, 3-4 lines long. *Perianth-tube* yellow, cylindrical, 9-12 lines long; *segments* three times as long as the tube; *stamens* and *style* distinctly exserted.

Properties and Cultivation of Barbados Aloes.

830

The plant is readily cultivated, growing in the driest situations and poorest soils. The bitter juice of the aloe is contained in vessels placed just below the epidermis. It escapes when the leaves are cut off close to the stem: it is at first colourless, but quickly acquires a brownish tinge on exposure to the air. Its activity seems to vary with the age of the leaves from which it is drained, and the season of the year. In Barbados, where this species is systematically cultivated, the leaves are cut annually in March and April, during the heat of the day. The better quality of aloes is that obtained by allowing the leaves to drain naturally, for if the leaves be artificially pressed, the juices of the leaf are mixed with the lateciferous fluid and the quality of the drug greatly impaired. The natural heat of the sun is also the best means of drying the inspissated sap, for if artificial heat be used, the active property of the drug is weakened. (*Bentley & Trimen.*)

General Character.—In addition to the forms of **A. vera**, the following are also cultivated in Barbados: **A. succotrina, A. purpurescens**, and **A. arborescens**, all of which and many hybrids between them yielding the Barbados Aloes of commerce.

In colour Barbados aloes is not uniform; it varies from a deep reddish brown or chocolate to almost black. It has usually a dull waxy fracture, and is almost perfectly opaque even at the margins. When it presents a smooth glassy fracture it is known as "Capey Barbados." Its odour is disagreeable and even nauseous; the powder of Barbados aloes is of a dull olive-yellow. It is much more powerful than Socotrine aloes, but more subject to produce griping pains. It is almost entirely soluble in proof spirit, and under the microscope the solution exhibits numerous crystals. It is said to give in aqueous solution a fine rose colour with chloride of gold or with tincture of iodine, a reaction which does not take place with other aloes.

DYE. 831

Dye.—In *Spons' Encyclopædia* there occurs an account of the preparation of the dye "Chrysammic Acid." It is prepared by heating 8 parts of nitric acid with 1 part of aloes. After the violent action has subsided, a second proportion of aloes is added to the mixture until the fumes of hypo-nitric acid subside. The mass is then poured into water, when chrysammic flakes settle in the bottom of the vessel. These are washed several times in water. The crystals change their colour under varying circumstances, giving a purple colour to silk, black to wool, and pink to linen. A French firm has recently used it to give a beautiful brown known as vegetable brown, which is produced through the agency of sulphuric acid. This dye is bright; it resists strong alkaline action; it combines with most of the anilines and other dyes, economising them and rendering them thoroughly fast; and it is not expensive.

It would be exceedingly interesting to know if the existence of this dye or dye auxiliary be known to the cultivators of Indian aloes, and if it has ever been extracted in India. The uses of the dye are likely to be greatly developed, and it therefore seems desirable that it should receive the attention of the Indian authorities.

FIBRE. 832

Fibre.—The leaves yield a good fibre. It seems highly desirable that the idea of combining the preparation of the drug aloes with the separation of the fibre should be brought before the public. In the account of the preparation of the drug practised in Barbados (given in the preceding page), it is stated that after the sap has drained off the leaves are rejected and no further use made of them. This seems an unnecessary waste of material, since from these rejected leaves a most useful fibre could be prepared.

MEDICINE. Juice. 833

Medicine.—As a medicine the *inspissated juice* from the forms of this species is in India regarded as but little inferior to the imported Socotrine aloes. It is an aperient, and deemed highly beneficial to persons predisposed to apoplexy. The FRESH JUICE from the leaves is said to be cathartic, cooling, and useful in fevers, spleen, and liver disease, enlarged lymphatic glands, and as an external applicant in certain eye diseases. The PULP of the leaves is, in native practice, applied to boils, and is regarded as acting powerfully on the uterus, and useful as an emmenagogue. It is also largely used in veterinary medicine. The ROOT is supposed to be efficacious in colic.

Pulp. 834 Root. 835

836 **Var. littoralis**, sp., *Koening.*

Vern.—*Chhótá-ghi-kanvar, chhótá-kanvár,* HIND., DUK.; *Chhótá-jangli ánanash,* BENG.; *Shiru-katrázh-ai,* or *shiru-kattálai,* TAM.; *Chinna-kalabanda,* TEL.; *Cheru-katru-vazha,* MAL.; *Shime-kattáli,* KAN.; *Dhákutá kunvára,* BOMB.; *Naháni-kunvar,* GUJ.; *Laháni-kumári,* MAR. Ainslie gives the plant the Sanskrit name of *kúmari. Koyangali* is the Burmese name for a species of Crinum, but it is also sometimes applied to this plant.

Syn.—§ "This, in my opinion, is a stunted variey of **A. indica,** *Royle.*" (*Deputy Surgeon-General Bidie, C.I.E., Madras.*)

Habitat.—This is altogether a much smaller form than the typical condition of the species, having yellow flowers in simple spikes, with the bases of the leaves not half so broad as in the preceding, and always of a pale green colour. It has become quite naturalised on the southern coast of the Madras Presidency.

Botanic Diagnosis.—*Leaves* 15-18 inches long, 1½ broad; *scape* simple, 2 feet long.

MEDICINE. 837

Medicine.—Ainslie says: "The pulp of the leaves of this small and very succulent plant, when well washed in cold water, is prescribed as a refrigerant medicine in conjunction with a small quantity of sugar-candy. The same pulp so purified, and with the addition of a little burnt alum, the native practitioners consider as a valuable remedy in cases of ophthalmia." The opinion of Madras officers as to this local form of aloes would be most acceptable. **Dr. Waring,** in the *Pharmacopœia of India,* says: "By inspissating the viscid juice of the leaves of **A. littoralis** collected at Cape Comorin, where the plant is in great abundance, the editor in 1853 prepared several ounces of excellent aloes, which proved actively purgative in the same doses that the officinal aloe is usually prescribed in. **Dr. W. Dymock**, of Bombay, corroborates the statement that this plant yields very good aloes, adding that he has tried it both in the fresh and dried state. It appears certain that, with a little care, aloes of good quality might be obtained from this source, in considerable quantities, at a cost far less than that of the imported article. The aloes procurable in the bazars (chiefly imported) is generally of a very inferior description. "The freshly-expressed juice is in almost universal use as an external refrigerant application to all external or local inflammations."

Dr. Dymock (*Mat. Med., W. Ind., 668*) gives an interesting account of Indian aloes from the pen of **Garcia de Orta**, a Portuguese physician, who, in 1534, accompanied **Admiral Martin Alfonso de Souza** to Goa. From this it would appear that the juice and fresh plant were at that time used in South India, and the species was most probably the **var. littoralis**. In **Clusius'** translation of **de Orta's** work occurs a prescription for the use of the fresh plant, *viz.*, aloe leaves sliced, 3 ozs., salt 3 drms., heat to boiling, strain, add 1 oz. of sugar: to be taken cold early in the morning.

§ "Laxative, tonic; useful in diseases of the spleen. The decoction of the root is prescribed as a febrifuge." (*Surgeon W. Barren, Bhuj, Cutch.*)

"Very largely used in Mysore as an aperient and emmenagogue." (*Surgeon-Major John North, Bangalore.*)

Var. officinalis, sp., *Forsk.*

838

Syn.—A. RUBESCENS, *DC.*; A. INDICA, *Royle.*

Vern.—*Lal-kumári, lal-ghígavár, kanvár*, HIND.; *Ghikawár*, N.-W. P.; *Ghirta-kamári*, or *ghirta-kanvár*, BENG.; *Ghí-kanvár, lal-ghi-kanvar, kanvár phod, kalbandá*, DUK.; *Kumár*, GUJ.; *Korphad*, MAR.; *Nabátussibr, e-ahmar*, ARAB.; *Darakhte-sibre-surkh*, PERS.; *Sirrúghá kuttalay* (see **Ainslie**), *shivappu katrázh-ai*, or *shivapp-kuttalai, shivappushhóttu-katrazh-ai, kumári*, TAM.; *Ena-kalabanda*, TEL.; *Chovannakatru-vazha*, MAL.; *Kempu-lóla-sará*, KAN.; *Ruta-komárika*, SINGH.; *Ava-tazávon-le-pa, shazávn-le-pa*, BURM.

Habitat and Botanic Diagnosis.—This is the form met with in a semi-wild condition in Bengal and the North-West Provinces. It has beautiful reddish and orange flowers, with the bases of the leaves purple-coloured and so dilated as to have in all probability suggested the name **A. perfoliata**, given by popular writers to this and many other species of aloe.

MEDICINE. 839

Medicine.—§ "In cases of chronic fissures and ulcers about the rectum, indigenous aloes have been largely used by the natives both internally and externally. It acts also as an emmenagogue and anthelmintic. It is a favourite medicine for intestinal worms in children. As an aperient it is generally given in combination with *turkud* or scammony. Dissolved in uttar of roses it is used in various affections of the eye. Mixed with *bartung* it is said to be very useful in chronic discharges from the nose or ears. Dissolved in spirit it is used as a *hair dye*, and it is said that it also stimulates the hair to grow. Dissolved in warm water and spread over a betle-leaf and applied while hot to the belly of a child, it is said to act as an aperient." (*Asst. Surgeon Gholam Nabi.*) "A sweetmeat, *halwa*, is prepared from the pulp of the leaves and given in cases of piles, and apparently with very good effect." (*Surgeon-Major C. W. Calthrop, M.D., Morar.*)

"The resinous extract obtained from this plant is applied to swellings in the form of a paste to cause absorption. It is used internally by native practitioners in melancholia and brain diseases, complicated with gastric symptoms. It produces griping, to correct which is added confection of roses and mastich. Given as a night pill in hæmorrhoids. A paste of fresh aloes and turmeric relieves the pain of contusions." (*Surgeon G. A. Emerson, Calcutta.*) "The pulp with a solution of alum is very extensively used by native practitioners in every form of ophthalmia, but especially in catarrhal and purulent ophthalmia." (*Asst. Surgeon Jaswant Rai, Mooltan.*)

"The inspissated juice, in combination with gum asafœtida, is applied as a warm plaster in colic and the pneumonia of infants. It is also given internally in these cases in doses of 1 grain with borax in the same quantity with the mother's milk." (*Lal Mahomed, Hoshangabad, Central Provinces.*) "It is applied over the abdomen for constipation and tympanitis." (*Surgeon-Major Robb, Ahmedabad.*) "I have seen the juice administered with powdered turmeric by village native practitioners in enlarged spleen." (*Assistant Surgeon Shib Chunder Bhuttacharji, Chanda, Central Provinces.*) "Aloes have been found useful in piles, mixed in small quantities with sulphur. It is applied by natives externally in the form of *lep*—paste—in pleurisy." (*Assistant Surgeon Bhugwan Das, Rawal Pindi.*) "A sort of pickle, prepared with aloe, salt, and ajowan, is very useful in colic and dyspepsia." (*Surgeon J. C. Penny, M.D., Amritsur.*)

"Inspissated juice, mixed with sugar, frequently given in gonorrhœa with great advantage." (*Brigade Surgeon S. M. Shircore, Moorshedabad.*) "The fresh juice of the leaves is taken with milk and water as a remedy for gonorrhœa and methritis. It acts as a mild purgative, emollient, and demulcent." (*Brigade Surgeon J. H. Thornton, Monghyr.*) "Hospital Assistant Gopal Chunder Gangooly, of the Noakhally Dispensary, reports that he has used the fresh pulp of the leaves, mixed with sugar, in cases of gonorrhœa, with good results; it acts as a demulcent." (*Surgeon Anund Chunder Mukerji, Noakhally.*) "The fresh juice from the leaves is cooling, diuretic; largely used by the natives in gonorrhœa. The tender pulp is eaten in rheumatism." (*Assistant Surgeon J. N. Dey, Jeypore.*)

"I have used it as a stomachic purgative in veterinary practice with much effect. It makes a good adjunct to sulphur, for internal use, in bad cases of mange. In the human subject, in cases of chronic cough due to dyspepsia, and in cases of foul evacuations, I have given it in 5-grain doses with *ghi*, in the former with sulphate of iron, in the latter two or three times a day, with much benefit." (*Surgeon K. D. Ghose, Khulna.*) "The indigenous drug, known in the bazars as *Musubbar*, has all the properties of the Socotrine or Barbados aloes." (*Surgeon R. D. Murray, M.D., Burdwan.*)

"A piece of the fleshy pulp (peeled), about two inches square, with 4 grains of turmeric, and 10 grains of burnt borax, is a favourite remedy for enlargement of the spleen associated with constipation of the bowels." (*Surgeon-Major E. C. Bensley, Rajshahye.*)

"One grain of bazar aloes, with 1 grain of bazar sulphate of iron and 1 grain of asafœtida, is often used by natives in the form of a pill for spleen enlargement." (*Surgeon K. D. Ghose, Bankura.*)

FOOD. 840

Food.—The pulp of the leaves is eaten by the poorer people in times of famine; the seeds also are eaten.

841

ALOPECURUS, *Linn.; Gen. Pl., III., 1140.*

A genus of grasses belonging to the Tribe PHALARIDEÆ, comprising in all some 20 species, inhabitants of Europe and temperate Asia.

Spike compressed, one-flowered, peduncled in the sheaths of the upper leaves; *glumes* 3-4, the 2 exterior empty, compressed, connate below, membranous and awnless. *Pale* 1, scarious, 5-veined, awned on the back; *stamens* 3; *style* long; *stigma* filiform, elongated, shortly hairy.

Alopecurus agrestis, *Linn.; Duthie's List of Grasses, 25;* GRAMINEÆ. 842

SLENDER FOX-TAIL GRASS; BLACK GRASS.

Habitat.—Found in the Panjáb in cultivated ground.

Botanic Diagnosis.—*Stem* erect, 1 to 2 feet high, roundish above; *panicle* tapering, slender. *Glume* glabrous, but with a row of short cilia on the back, acute, connected below; awn from near the base of the pale and projecting half its length beyond. A troublesome weed.

FODDER. 843

Fodder.—**Duthie**, quoting **Parlatore**, describes it as a good fodder grass, fresh or dry.

A. geniculatus, *Linn.; Duthie's List of Grasses, 25.* 844

FOX TAIL-GRASS.

Vern.—*Pumila*, N.-W. P.

Syn.—A. FULVUS, *Sm.*

Habitat.—Inhabits the plains of Northern India, in wet places ascending the Himálaya to Kumaon and Kashmír valley.

Botanic Diagnosis.—*Stems* ascending, smooth, kneed and swollen at the joints, about a foot long, branching below, knots generally fleshy. *Panicle* cylindrical, 1-2 in. long. *Glume* blunt, connected below, ciliate, exceeding the pale; *awn* from near the base of the pale and projecting half its length beyond it. *Pale* when opened out oblong, blunt, slightly notched. *Anther* ultimately violet-yellow. *Styles* mostly combined.

FODDER. 845

Fodder.—**Mueller** describes it as a good fodder grass for swampy land. A variety, **A. pumila**, was found by **Royle** on the banks of the Jumna.

A. pratensis, *Linn.; Duthie's List of Grasses, p. 26.* 846

MEADOW FOX-TAIL GRASS.

Habitat.—Inhabits the North-West Himálaya, 5,000 to 8,000 feet, ascending in Lahoul to 13,000 feet; also found in Kashmír and on the Panjáb plains. Is fond of rich pasture lands.

FODDER. 847

Fodder.—A perennial pasture grass, considered one of the best of its class. Sheep thrive well on it. **Loudon** mentions it as an excellent fodder grass in England.

Since it requires two or three years to attain perfection, it is disqualified from becoming part of a rotation of crops. For fallow and waste lands it is, however, very valuable, especially in damp soils. It has been ascertained that if mixed with white clover, this grass, after the second year, will support five ewes and five lambs on an acre of sandy loam, especally if the soil contains lime. For permanent pastures in warm temperate climates this grass is one of the best.

ALPINIA, *Linn.; Gen. Pl., III., 648.* 848

A genus of SCITAMINEÆ belonging to the tribe ZINGIBEREÆ, containing some 40 species, inhabiting the tropical and sub-tropical regions of Asia, Australia, and Pacific Islands.

Rhizome thick, often aromatic, horizontal, creeping. *Inflorescence* a thyrsate, dense-flowered raceme, rarely a lax panicle. *Calyx* superior, forming a loose tube cut into 3 lobes. *Corolla* with the tube nearly as long as the calyx or sometimes a little longer. *Andrœcium* of six staminodes, in two rows of 3; *outer row* inserted at the mouth of the corolla, the two *posterior* or abortive stamens small, forming thickened, glandular, horn-like bodies or absent, the anterior forming the labellum or inner and fourth petal; *inner row* with the two *anterior* staminodes reduced to glands, inserted upon the apex of the ovary, and the *posterior* developed into the solitary fertile stamen. *Stamen* equal or nearly equal to or only half the length of the corolla; *filament* generally flattened, concave, embracing the style; *connective* flattened, not prolonged beyond the anthers, or if prolonged forming a short, broad, rounded, entire or bifid appendix. *Ovary* inferior, 3-celled, many-seeded; *style* filiform, passing between and behind the anthers, within the staminal sheath, often compressed below by the glandular staminodes; *stigma* capitate. *Fruit* globose, an indehiscent berry or rarely bursting into 3 valves.

The genus is named in honour of the Italian botanist **Prosper Alpinus.**

849

Alpinia Allughas, *Roscoe.*

Vern.—*Taro, taruko,* BENG.

Habitat.—A native of Bengal, Assam, Burma, and Celon; also of the Konkan in Western India.

MEDICINE. 850

Medicine.—The aromatic rhizomes are used by the natives medicinally.

851

A. calcarata, *Roxb.; Fl. Ind., Ed. C.B.C., 23.*

Habitat.—A native of China, cultivated in gardens in India; introduced in 1799.

Botanic Diagnosis.—*Leaves* linear-lanceolate, polished. *Spikes* compound, erect. *Flowers* large, in pairs or more, expanding at different times. Outer *petals* 3, linear, equal; *labellum* ovate-oblong, apex curved and bifid.

MEDICINE. 852

Medicine.—Dr. Moodeen Sheriff says that this is sold as a substitute for galangal in Haidarabad and other parts of India. The rhizomes, however, possess no pungency.

A. Cardamomum, *Roxb.*, see Elettaria Cardamomum, *Maton.*

853

A. Galanga, *Willd.*

The Rhizome.—THE GREATER GALANGAL, OR JAVA GALANGAL, *Eng.;* GALANGA, *Port.*

The Fruit.—THE GALANGAL CARDAMOM.

Syn.—AMOMUM GALANGA, *Lour.*

Vern.—*Kúlanján, kúlinján, bará-kaliján,* or *bará-kúlanján,* HIND., BENG.; *Kolinjan,* GUJ.; *Bari-pán-ki-jar, malabari-pánki-jar,* BOMB.; *Kosht-kulinján,* MAR.; *Bará-khúlanjan, bari-pán-ki-jar,* or *sufédpán-ki-jor,* DUK.; *Kunjar, kathi,* SIND; *Dumparástma, kúlin-jána,* SANS.; *Khúlanjáne-qasbi, khúlanján-e-kabír,* ARAB.; *Khusravedúrúé-kalán,* PERS.; *Péra-rattai,* TAM.; *Pedda-dumpa-rásh-trakam,* TEL.; *Péraratta,* MAL.; *Dumpa-rásmi,* KAN.; *Padagoji,* BURM.

Many of the names for this root-stock indicate an erroneous idea that it is the root of the betel-leaf plant, *viz., pan-ki-jar.* **Hanbury** suggests that the Arabic word *khalanjan* may have been derived from the Chinese *liang-kiang* (wild ginger), which in Europe became further corrupted into galangal, garingal, and (in German) galgant.

Habitat.—A perennial plant, native of Java and Sumatra, now cultivated in East Bengal and South India.

Botanic Diagnosis.—A perennial with broad, lanceolate, sessile, sheathing *leaves,* having a short, rounded, ciliate ligule, from 12-24 inches long by

The Greater Galangal and Galangal Cardamom.

4-6 broad; *stem*, when in flower, 6 feet high, the lower half embraced by the smooth leaf-sheaths. *Panicle* terminal, erect, oblong, composed of numerous spreading, simple, dichotomous branches, each supporting 2-3-6 pale-greenish white, faintly fragrant flowers. *Calyx* scarcely the length of the corolla-tube. *Labellum* oblong, stalked, arching towards the stamen, lip bifid. *Capsule* the size of a cherry, deep orange-red; *seeds* often only one in each cell. (*Roxb.; Hance, Linn. Journ., XIII.*)

Description of Bazar Products.—THE RHIZOME.—*The Greater Galangal.*—Recognised from the *Lesser Galangal* by its larger size, feebler odour and taste, and by its deep orange-brown skin, contrasting prominently with the pale buff hue of the internal structure.

THE FRUIT.—*The Galangal Cardamom.*—About half an inch long, oblong, somewhat constricted in the middle, or at times even pear-shaped, obscurely three-sided. Often shrivelled on one side from being collected when immature. In colour from pale to deep reddish brown; externally and internally whitish. Seeds united in a three-lobed mass, invested by a white integument, each mass consisting of two seeds, one above the other. Seeds ash-coloured, three-cornered, finely striated towards the hilum; connected to the axillary placenta by a long, broad funiculus. Aril tough, nearly surrounding the seed; seeds pungent, burning, with an aroma resembling that of the rhizome. (*D. Hanbury, Science Papers, 107.*)

History.—**Garcia de Orta,** physician to the Portuguese Viceroy of India at Goa in 1563, was the first writer who pointed out that in India there were two forms of Galangal, the lesser and more powerful root-stock imported from China, and the larger a native of Java. The former alone is that met with in Europe, a rhizome partaking of little medicinal virtues that are not possessed by ginger, but which, nevertheless, enters into many ancient prescriptions still in use.

DYE.
854

Dye.—**Mr. Buck** says that this root-stock is imported into the North-West Provinces from the Panjáb, and is used in calico-printing along with myrabolans.

MEDICINE.
Rhizome.
855

Galangal
Cardamoms.
856

Medicine.—The rhizomes of this species are aromatic, pungent, and bitter, and are used in the form of an infusion in fever, rheumatism, and catarrhal affections. As a drug they are supposed to improve the voice. The aromatic tubers are sometimes used as carminative or fragrant adjunct in complex prescriptions, but they have nothing peculiar in their properties or action. (*U. C. Dutt.*) How far these properties may have been intended to be attributed to this root-stock or should have rather been given to **A. officinarum** cannot be accurately determined. The statements of Indian authors have to be accepted for the present, but it seems probable that future enquiry may show that, while both the greater and the lesser galangals are regularly imported into India, as far as their medicinal properties are concerned, the former is only used as a substitute for the latter, being commercially less valuable and less active in its therapeutic properties. It is, however, difficult to determine in many cases to which species authors refer. **Dr. Irvine,** in his *Medical Topography of Ajmere,* says: "Root of this plant is hot and stimulating; used in *mesalihs,* has a sweet scent; is put into bazar spirits to make it more intoxicating." This habit of flavouring spirits with galangal also prevails in Russia,—see under **A. officinarum.** The seeds also possess similar medicinal properties.

Seeds.
857

§ "Hakims use it in impotence, bronchitis, and dyspepsia. It is disinfectant, used to destroy bad smells in the mouth or any other part of the body. It is also advocated in diabetes mellitus." (*Assistant Surgeon J. N. Dey, Jeypore.*) "In Mysore a domestic medicine, much used by old people with bronchial catarrh." (*Surgeon-Major John North,*

Bangalore.) "I have known natives who think this drug improves the voice." (*Surgeon-Major C. J. McKenna, Cawnpore.*)

858

Alpinia Khulanjan, *Moodeen Sheriff, Suppl. Ind. Pharm., 268.*

Habitat.—§ "This plant is found growing in several gardens at Madras, and its rhizome, when dried, bears the greatest resemblance to the Lesser Galangal **(A. chinensis).** The root is not sold in the bazar, but when sent there, it was recognised by the same native names as those of the Lesser Galangal.

"A few years ago, when I first found the plant, I thought it to be **A. chinensis,** but on examining it several times when in flower, I found it to be a new species of **Alpinia,** not hitherto described by anybody, as far as my knowledge extends. I have therefore named it **Alpinia Khulanjan,** after its native appellation *khúlanján,* and have described it as minutely as I could in the *Supplement to the Pharm. of India,* pp. 268 and 269."

Description.—"If the root of this plant is cut into pieces and dried, it presents the following characters: Tuberous, about the thickness of the little finger, somewhat thicker at one end than at the other, from one and a half to three inches long, often knotty and forked, reddish brown externally and greyish internally, annulated or marked with white rings, slightly wrinkled; smell warm and aromatic, and taste strongly pungent and peppery. This root is somewhat smaller and lighter in colour than the Lesser Galangal of the shops, but slightly stronger in smell and taste."

MEDICINE. 859

Medicine.—"With regard to the medicinal properties of the root of **A. Khulanjan,** it is not only stimulant, carminative, stomachic, and expectorant like ginger, but also a very good stimulant tonic. In addition to all the diseases in which ginger is indicated, it is very useful in some nervous disorders, as neuralgia, functional impotence, nervous debility, &c. It has also proved useful in several cases of incontinence of urine. Its preparations and doses are the same as those of ginger, to which it is also preferable in another respect, *viz.,* that it is neither attacked by insects, nor destroyed by any length of time. It is best administered in powder and tincture, the latter being prepared exactly in the same manner as the tincture of ginger, except the quantity of the root, which is to be four ounces instead of two and a half. Doses of the powder, from 10 to 30 grains, and of the tincture, from 30 minims to two drachms." (*Honorary Surgeon Moodeen Sheriff, Khan Bahadur, Triplicane Dispensary, Madras.*)

I have not seen a specimen of the plant referred to above by **Dr. Moodeen Sheriff,** but from his description (which was published in the *Supplement to the Indian Pharmacopœia, 1869*), it would seem to be the same plant which **Dr. Hance** described in the *Linnæan Journal, Vol. XIII., p. 6, 1871,* under the name of **A. officinarum.** If this conjecture proves correct, according to the rule of priority, the information given under **A. Khulanjan,** and that which, further on, has been given under **A. officinarum,** it would seem, should be reduced to one place under the name of **A. Khulanjan,** *Moodeen Sheriff.* The difference between the rhizomes of the Madras plant and the imported Chinese specimens might easily be due to the former being cultivated in India. **Dr. Moodeen Sheriff's** description is scarcely a botanical one, but the honour of associating his name with this plant would be a deserved recognition of his distinguished labours in the field of economic science. I have left the information given, however, under the two names until it can be ascertained whether the Madras garden plant is in reality the Chinese species described by **Hance** or not. (*Compare with Hanbury's "Science Papers," p. 373.*)

Alpinia nutans, *Roscoe.*

860

LIGHT GALANGAL.

Vern.—*Punag-champa*, BENG.; *Kasta-serambet*, PERS.; *Pa-ga-gyis*, BURM.

Habitat.—A native of the Eastern Archipelago, found also in Burma, Sylhet, and on the Coromandel Coast; much cultivated in Indian gardens.

Botanic Diagnosis.—"*Leaves* lanceolar, short-petioled, smooth. *Racemes* compound, by the lower pedicels being two or three-flowered, drooping. *Lip* (the labellum) broad three-lobed, the lateral lobes incurved into a tube, the exterior curled and bifid. *Capsule* spherical, opening on the sides. *Seeds* few." (*Roxb., Fl. Ind., Ed. C.B.C., p. 22.*)

MEDICINE. Rhizome. 861

Medicine.—The rhizome is often used as a substitute for **A. Galanga,** and even as a substitute for ginger. It is much larger than the large galangal and not so pungent.

A. officinarum, *Hance.*

862

THE LESSER GALANGAL; ALPINIA CHINENSIS of Chemists.

Vern.—This is the article which is most frequently sold in the bazars under the names of *kulinján* and *kolijána*, or *pán-ki-jer* or *chandápushpi. Chhota-pan-ki-jar*, *chota* or *choti-kulijan*, HIND., BENG., and BOMB.; *Shitta-rattai*, TAM.; *Chhóté-pán-ki-jor*, or *kálé-pán-ki-jor*, DUK.; *Sanna-elumparásh trákam*, TEL.; *Khúlanján*, ARAB.; *Khusro-dáru*, PERS.

Habitat—The root-stock is a native of China, and is largely exported to Europe and India.

Botanic Diagnosis.—*Leaves* 9-14 inches long, narrowly lanceolate, much attenuated at the apex, leathery, bright green; *ligule* oblong, subacute, decurrent at the base, and along the margin of the sheath. *Flowers* sessile, closely packed in an erect, dense, terminal spike; *bracts* three, longer than the flowers, the outer green, the inner white; calyx and corolla finely pubescent. *Labellum* about ¾ inch long and broad-ovate, entire, acute or bi-lobed, crispid and denticulate, white striated with dark-red veins which coalesce into a distinct fan-shaped spot near the apex.

MEDICINE. The Rhizome. 863

Medicine.—This is the Galangal of the European shops. In India it is generally known as the *Pan-ki-jar*. It is stomachic, tonic, used by native practitioners to reduce the quantity of urine in diabetes. Is said to correct foul breath when chewed, and the juice swallowed is reputed to arrest irritation in the throat. It is considered a nervine tonic and an aphrodisiac.

The botanical source of this plant—the true or officinal Galangal—was determined in 1870 by **Dr. Hance,** who published an account of it in the *Journal Linn. Society*, *1873, Vol. XIII., 6.* (Compare with **A. Khulanjan.**) Although a native of China, it has been imported into India and used by the Hindú and Mohammedan physicians from time immemorial. **Meer Muhammad Hussain** says that if given to infants it makes them talk early, and that a paste of the powdered drug made with oil or water will remove freckles. "Galangal is one of the ingredients of *Warburg's Tincture*. It is not used in English medicine, but there is a considerable demand for it in Russia." (*Dymock, Mat. Med., W. Ind., 637.*)

D. Hanbury (*Linn. Soc. Journ., 1871, and in his "Science Papers," p. 373*) says: "As a medicine, the manifold virtues formerly ascribed to it (the lesser Galangal) must be ignored; the drug is an aromatic stimulant, and might take the place of ginger, as, indeed, it does in some countries. That it is still in use in Europe is evident from the exports from China, and from the considerable parcels offered in the public drug sales of London. The chief consumption, however, is not in

ALPINIA officinarum.

Trade Returns of Galangal.

Flavouring the Liqueur Nastoilla. 864
Spice. 865
Cattle medicine. 866

England but in Russia. It is there used for a variety of purposes, such as for flavouring the liqueur called Nastoilla. The drug is also employed by brewers, and to impart a pungent flavour to vinegar." "As a popular medicine and spice it is much sold in Livonia, Esthonia, and in Central Russia; and by the Tartars it is taken with tea. It is also in requisition in Russia as a cattle medicine; and all over Europe there is a small consumption of it in regular medicine."

In concluding his interesting paper upon this drug, **Hanbury** says: "According to **Roudot**, writing in 1848, the trade in this drug is on the decline; and the statistics which I have examined tend strongly to show that this is the fact." The foregoing notes may thus be summarized:—

"1. Galangal was noticed by the Arab geographer, **Ibu Khurdadbah**, in the ninth century, as a production of the region which exports musk, camphor, and aloes-wood.

"2. It was used by the Arabians and later Greek physicians, and was known in Northern Europe in the twelfth century.

"3. It was imported during the thirteenth century with other eastern spices by way of Aden, the Red Sea, and Egypt, to Akka, in Syria, whence it was carried to other ports of the Mediterranean.

"4. Two forms of the drug were noticed by **Garcia de Orta** in 1563; these are still found in commerce, and are derived respectively from **Alpinia Galanga**, *Willd.*, and **A. officinarum**, *Hance.*

"5. Galangal is still used throughout Europe, but is consumed most largely in Russia. It is also used in India, and is shipped to ports in the Persian Gulf and Red Sea."

867

Galangal (Foreign Trade by Sea).

Years.	Imports.		Exports and Re-exports.	
	Quantity.	Value.	Quantity.	Value.
	Cwt.	R	Cwt.	R
1879-80	3,129	25,503	1,817	17,668
1880-81	2,289	19,603	815	7,960
1881-82	3,813	29,625	1,971	15,986
1882-83	3,354	30,952	1,164	10,146
1883-84	3,870	35,982	1,670	13,306

Detail of Imports, 1883-48.

Province into which imported.	Quantity.	Value.	Countries whence imported.	Quantity.	Value.
	Cwt.	R		Cwt.	R
Bengal	686	7,831	China—Hong-Kong	1,230	10,741
Bombay	1,750	14,897	Straits Settlements	2,540	24,446
Madras	1,434	13,254	Other Countries	100	795
Total	3,870	35,982	Total	3,870	35,982

The Dita Bark.

Detail of Exports, 1883-84.

Province whence exported.	Quantity.	Value.	Countries to which exported.	Quantity.	Value.
	Cwt.	R		Cwt.	R
Bengal . . .	51	339	United Kingdom .	480	3,840
Bombay . .	1,544	12,596	Arabia . . .	397	3,736
Madras . . .	75	371	Persia . . .	249	1,471
			Other Countries .	544	4,259
TOTAL .	1,670	13,306	TOTAL .	1,670	13,306

Note.—It is impossible to say how far these trade returns refer to the Greater and Lesser Galangal respectively.

ALSEODAPHNE, *Nees ;* PERSEA, *Gærtn., in Gen. Pl., III., 157.*

868

Alseodaphne, ? sp.; LAURINEÆ.

Vern.—*Dowki poma,* Ass.

Habitat.—A tree met with in Assam.

Structure of the Wood.—Soft, red, even-grained.

Used for boats, furniture, and building.

TIMBER. 869

ALSTONIA, *R. Br.; Gen. Pl., II., 705.*

870

A genus of trees or shrubs belonging to the APOCYNACEÆ (the Dogbane family), in the tribe CERBEREÆ. There are about 30 species in the genus, inhabitants of tropical Asia, the Malaya, and Australia.

Leaves 3-4-nately whorled, rarely opposite. *Calyx* short, 5-lobed or partite, glandular within. *Corolla* salver-shaped; *throat* naked, annulate or with reflexed hairs; *lobes* overlapping to right or left. *Stamens* near the top of the tube included; *anthers* subacute. *Carpels* two, distinct. *Follicles* 2-linear; seeds many and many-seriate in each carpel, oblong flattened, peltately attached, often ciliate (or comose) on both ends; *cotyledons* oblong flat; *radicle* superior.

The generic name is in honour of **Alston**, once Professor of Botany, Edinburgh.

Alstonia scholaris, *R. Br.; Fl. Br. Ind., III., 642.*

871

Commercially known as DITA BARK.

Vern.—*Chatwan, chhatin, chatiun,* BENG.; *Satiún, chatiún, satwin, satni,* HIND.; *Sapta-parna,* SANS.; *Chhatnia,* URIYA, SANTAL, and MAL. (S. P.); *Chatin, bomudu,* KÓL.; *Chatiwan,* NEPAL; *Purbo,* LEPCHA; *Satiana, chatian,* ASS.; *Satvin,* BOMB., MAR.; *Sattni,* CACHAR; *Ezhilaip-pálai, wodrase,* TAM.; *Edakula-pála pala-garuda, édákula-ariti, édá-kula-ponna,* TEL.; *Mukampala, pála,* MAL.; *Janthalla,* KAN.; *Rúk-attana,* SINGH.; *Chaile, chalain,* MAGH.; *Let-top, toungmayobeng,* BURM.

Dr. Rice thinks that *Sapta chhada* and *Sapta parna* (seven-leaf) are ancient Sanskrit names for this] tree. **Dr. Moodeen Sheriff** says the Singhalese name *rúk-attana* is sometimes given to it; *attana* is, however, the name for Datura. **Dr. Trimen** informs me, however, that *rúk*-attana is the correct Singhalese name for this tree.

Habitat.—A tall, evergreen tree, widely cultivated throughout India, and found in the Sub-Himálayan tract from the Jumna eastward, ascending to 3,000 feet; in Bengal, Burma, and South India. Distributed to Java, Tropical Africa, and Eastern Australia. An exceedingly useful as it is a highly ornamental tree.

Botanic Diagnosis.—*Leaves* 4-7 in a whorl, ovate or elliptic oblong, white beneath, 4 to 8 by 1 to 2½ inches, nerves 30-60 pairs joining an intramarginal one, base acute, tip obtuse or obtusely acuminate; *cymes* peduncled or sessile, umbellately branched or capitate; *corolla* white, pubescent, throat villous, lobes rounded. *Follicles* very long and slender.

Properties and Uses—

Caoutchouc. 872

Caoutchouc.—It yields an inferior quality of Caoutchouc or Gutta-percha, the *Gutta-pulei* of Singapore, which see under **Caoutchouc.**

MEDICINE. The Bark. 873

Medicine.—The bark is used medicinally as an astringent tonic, anthelmintic, alterative, and antiperiodic. It is a valuable remedy in chronic diarrhœa and the advanced stages of dysentery. It is also useful in catarrhal fever. The milky juice is applied to ulcers, and, mixed with oil, is put into the ear in earache. It has also been found most useful in restoring the tone of the stomach in debility or after fever. *Ditain*, the uncrystallizable substance obtained from the bark, is reported to be equal in efficacy to the best sulphate of quinine, while being free from the disagreeable secondary symptoms of that drug.

Description of the Bark.—The drug consists of irregular fragments of bark ⅛ to ⅓ inch thick, easily breaking with a short fracture. The external layer fissured, dark grey or brownish, sometimes with black spots: it readily separates when handled. Inner substance of a bright buff. It has no smell, is bitter but not disagreeable when chewed. (*Official Report of the Bombay Committee regarding a future edition of the Pharmacopœia of India.*)

Officinal Preparations and Dose.—*Preparations.*—An infusion, a tincture and the dry powdered bark, also the active principle *Ditain.*

Doses.—Of the powder 3-5 grains combined with ipecacuanha and with the infusion of gentian used for bowel complaints. Of the infusion 1-2 ounces; of the tincture 1-2 drachms. The active principle may be given in from 5-10 grain doses, repeated every 3-4 hours, not exceeding 1 drachm daily.

DITAIN. 874

Chemical Composition.—§ "The bark of **A. scholaris** has been frequently examined. **Gruppe**, an apothecary of Manilla, separated an uncrystallizable bitter principle which he called *Ditain*, and to which he ascribed the febrifuge properties of the drug. **Sorup-Bosanerj** obtained from ditain a crystallizable substance, which possessed all the properties of an alkaloid. The bark was next examined by **J. Jobst** and **O. Hesse**, who isolated the following: an alkaloid, *Ditamia*, another substance the nature of which as an alkaloid was not clearly established, a crystallizable acid, as well as a fatty acid, and fatty resinous substances. The fatty resins have been named *Echicaoutchin, Echicerin, Echitin*, or *Echiteïn* and *Echiretin*, and a fifth resin not fully investigated. **Hesse**, continuing his investigations, subsequently discovered two other alkaloids, in addition to *Ditamine*, i.e., *Echitamine* and *Echitenine*.

"The **A. constricta**, an Australian species, and the **A. spectabilis**, or *poclé* bark of Java, have also been examined by **Hesse**. From the bark of the first-mentioned variety he isolated *Alstonine* or *Chlorogenine*, *Porphyrine, Porphyrosine*, and *Alstonidine*. The alkaloids contained in *poclé* bark are thus present in the **A. scholaris**, together with a fourth, *Alstonidine*. In a report on the Centennial Exhibitions, presented to the American Pharmaceutical Association, 1877, it is stated that equal doses of ditain from the **A. scholaris** and sulphate of quinine have the same medicinal effects, while the disagreeable secondary symptoms which so frequently follow the administration of a large dose of quinine are absent.

"The results arrived at in the Manilla hospitals and in private practice with ditain are described as simply marvellous. The report further adds that in military hospitals and in penitentiary practice (*Manilla*) ditain has perfectly superseded quinine, and is now being largely employed with most satisfactory results in the Island of Mindanso, where malignant fevers are prevalent." (*Surgeon Warden, Professor of Chemistry, Calcutta.*)

Special Opinions.—§ "The tender leaves, roasted and pulverised and made into poultices, act as a useful local stimulant to unhealthy ulcers with foul discharges." (*Surgeon-Major D. R. Thompson, Madras.*)

TIMBER. 875

Structure of the Wood.—White, soft, even-grained; seasons badly, and soon gets mouldy and discoloured. It is not durable, but is easily worked. Weight about 28 lbs. per cubic foot.

Boxes and Furniture. 876

It is used for boxes, furniture, scabbards, coffins, &c. In Burma it is made into blackboards, and in Darjeeling, Assam, and Cachar is occasionally used for tea-boxes.

ALTERNANTHERA.

877

Alternanthera sessilis, *R. Br.;* AMARANTACEÆ (which see).

Vern.—*Mokú-nú-wanna,* SINGH.

FOOD. 878

Food.—Dr. Trimen writes me that this is largely eaten in Ceylon as a vegetable, especially by mothers to increase the flow of milk; also used as a wash for the eyes.

ALTHÆA, *Linn.; Gen. Pl., I., 200.*

879

A genus of sub-bushy, herbaceous, erect, or procumbent plants, belonging to the Natural Order MALVACEÆ, inhabitants of the temperate regions: 12 known species.

Althæa, classical Latin name for the Marsh Mallow; ἀλθαία, Gr.

Althæa officinalis, *Linn.; Fl. Br. Ind., I., 9.;* MALVACEÆ.

880

THE MARSH MALLOW, *Eng.;* GUIMAUVE, *Fr.;* ALTHIEWUREL EIBISCHWURZEL, EIBISCH, *Germ.;* ALTEA, *It.;* ALTEA MALVA VISCO, *Sp.*

Vern.—*Gul-khairo, Khitmi-ká-jhár, khaira-ka-jhor,* PERS., HIND., DUK., and BOMB.; *Gulkhair,* MAR., GUJ., and CUTCH; *Shémai-tutti,* TAM. The fruits are *Tukm-i-khitmi,* PERS. and BOMB.; the roots *Résha-i-khitmi,* PERS. and BOMB.

Habitat.—A native of Kashmír and the Panjáb Himálaya.

DYE. 881

Dye.—Often cultivated in Indian gardens for its flowers, rarely for its dye,—a rich blue, obtained from the leaves. **A. rosea,** *Linn,* the HOLLYHOCK, yields the dye even more freely than **A. officinalis,** *L.;* it is met with plentifully in Kashmír.

Information as to whether this dye is actually prepared in India would be exceedingly interesting.

MEDICINE. Leaves. 882 Flowers. 883 Root. 884

Medicine.—This plant was held in great esteem by the Greeks and Latins for its healing properties. The Mohammedans also describe it as a suppurative and emollient; they use the leaves in the form of poultice. The leaves and flowers mixed with oil form an application to burns and venomous bites. A decoction of the root with sugar is given in cough and irritation of the intestines and bladder. (*Dymock.*)

§ "The juice of the leaves, boiled to a proper consistence with castor oil in equal parts, is given internally in parasitic affections of the skin." (*Surgeon-Major R. Thompson, M.D., Madras.*) "The boiled leaves are in common use in Ceylon as a local application to sprains, bruises, and other injuries." (*Surgeon W. H. Morgan, Cochin.*) "The boiled leaves are used as an emollient and suppurative by native hakims." (*Honorary Surgeon Easton Alfred Morris, Negapatam.*) "An excellent application for ulcers." (*Surgeon W. Barren, Bhuj, Cutch.*) "Is not known to the inhabitants of Kashmír as medicinal, but I have there used the leaves in a poultice. Very common in the Sind Valley near streams. (*Surgeon George Cumberland Ross, Delhi.*) "The leaves are useful as a fomentation to relieve pain and itching." (*Surgeon-Major Henry David Cook, Calicut, Malabar.*)

"The powdered seeds are employed in cases of gravel from the kidneys. Chewing the leaves is said to allay thirst. Root-bark acts as a purgative and is used in colic. An infusion of root-barks has also been used as an eye-wash with success. Seeds dissolved in vinegar are employed generally to remove toothache." (*Assistant Surgeon Gholam Nabi.*)

FOOD. 885

Food.—Is used as a green vegetable.

886

Althæa rosea, *L.; Fl. Br. Ind., I., 319.*

THE ENGLISH HOLLY-HOCK, *Eng.;* GUIMAUVE, *Fr.*

Habitat.—Largely cultivated in Indian gardens, flourishing freely at all hill stations; it probably bears the same vernacular names as have been given above for the MALLOW.

MEDICINE. Seeds. 887 Flowers. 888 Roots. 889 Leaves. 890

Medicine.—The SEEDS of this plant are demulcent, diuretic, and febrifuge. The FLOWERS have cooling and diuretic properties. The ROOTS are supposed to be astringent and demulcent, and are much used in France to form demulcent drinks. Boiled with sugar they yield a decoction largely used in India in the treatment of coughs and irritable conditions of the intestines and bladder. (*Dymock.*) The LEAVES are used as a poultice or fomentation, and, mixed with oil, are applied to burns or sores caused by snake-bites.

§ "**Althæa officinalis** and **A. rosea** are cultivated in many gardens in Madras for medicinal and ornamental purposes. The fresh petioles, stems, and roots of both plants yield a mucilage when bruised, broken and shaken in water. The mucilage is cooling, demulcent, and a very useful adjunct to other medicines in dysentery. In mild cases it is sufficient by itself to relieve some dysenteric symptoms, as tormina and tenesmus. Dose of the mucilage, from one ounce to two ounces. If used alone it should be repeated frequently. (*Honorary Surgeon Moodeen Sheriff, Khan Bahadur, Madras.*)

891

ALTINGIA, *Noronha; Gen. Pl., I., 669.*

A genus of trees, containing only 2 species (belonging to the Natural Order HAMAMELIDEÆ).

Leaves alternate, persistent petioled, ovate or oblong, glandularly serrate; *stipules* deciduous or persistent. *Flowers* in dense heads: heads enclosed by a large bract, males racemose, females solitary. *Male* heads a mass of stamens with very short filaments; *anthers* obverse-pyramidal, the valves when young thrown inwards so as to become pseudo-4-celled, dehiscing longitudinally. *Female* heads of 12-20 flowers. *Calyx* confluent, teeth absent, but with rudimentary anthers inserted on the rim. *Ovary* $\frac{1}{2}$ inferior, 2-celled; *carpels* prolonged into 2 distinct deciduous styles; *ovules* numerous on an axile placenta. *Fruiting-head* globose, harsh, many-capsuled; *seeds* imperfect, the lowest winged and fertile, the upper without wings and sterile.

The genus is named after the botanist **Altingia**; there is only one species met with in India:—

Altingia excelsa, *Noronha; Fl. Br. Ind., II., 429.* 892

Syn.—Liquidamber Altingia, *Bl.*

Vern.—*Siláras,* Hind. and Duk.; *Jutili,* Ass.; *Méaahe-sáyelah,* Arab.; *Aslelubní,* Pers.; *Neriuriship-pál,* Tam.; *Shilá-rasam,* Tel.; *Rasa-mála,* Mal.; *Siláras,* Guj.; *Shiláras,* Mar.; *Nan-ta-yok* or *nan-ta-yu,* Burm.

Habitat.—A magnificent tree of the tropical evergreen forests of the Indian Archipelago, Burma, Assam, and Bhután; abundant in the Tenasserim Province of Burma.

Gum.—In Java it yields in small quantity an odorous resin, known in Europe under the name *Storax,* which is obtained by incisions in the trunk; the tree is not regularly cultivated. In Burma, it is said (in the *Pharmacographia*) to afford a fragrant balsam of two varieties: one pellucid and of a light-yellowish colour, obtained by simple incision; and the other, dark, opaque, and of a terebinthinous odour, procured by boring the stem and applying fire around the trunk.

RESIN. Storax. Pellucid form. 893

Opaque form. 894

Medicine.—Yields a form of the resin known in Europe under the name "Storax." For medicinal properties see **Liquidamber orientalis,** *Miller.*

MEDICINE. Storax. 895

§ "In orchitis a very thin layer of storax is laid on a tobacco leaf and applied to the inflamed part." (*Surgeon Joseph Parker, Poona.*)

Structure of the Wood.—Soft, reddish-grey, with lighter streaks. Annual rings marked by a narrow belt of firm wood without pores. Weight 46 lbs. per cubic foot.

TIMBER. 896

Used in Assam for building and ordinary domestic purposes.

ALUMEN.

Alumen or **Alum.** 897

Vern.—*Phitkari,* Hind.; *Phatkirí,* Beng.; *Sphatikari,* Sans.; *Shib, záj,* Arab.; *Zák, záke-safed,* Pers.; *Phatki, turti, patakri,* Mar.; *Pati-káram,* Tam.; *Pati-káram,* Tel.; *Patik-káram,* Mal.; *Keo-khin* or *kyankchin,* Burm.

Preparation of Indian Alum.—Alum is prepared from alum shale in Behar, in Cutch, and in the Panjáb. It is often met with in different shades of colour—white, yellow, red, and black, depending upon impurities.

In the *Bombay Gazetteer, Vol. V., pp. 19-20,* is given an interesting account of the manufacture of alum in Cutch. It is said to have been carried on for the past two or three centuries. During certain months of the year a large quantity of alum is made at Madh. The material used is pyritous dark-grey or black shale, closely associated with a soft aluminous pseudo-breccia of the sub-nummulitic group. This appears to overlie or enclose the shale or to have invaded it. The native burrowings give a poor chance of studying the relations of the rocks; the air in them is exceedingly bad and it is difficult to obtain light, and much of the ground may have been disturbed by "*old men's*" workings, which, according to **Colonel Grant,** fall in every year. Each work is entered by a narrow passage, the sides cut vertically, the floor sloping. About 20 feet below the surface the open air passage stops, and an underground gallery, about 6 feet high and from 3 to 4 feet wide, slopes down to the alum bed, through which, owing to the accumulation of water, no passage has ever been driven.

Cutch Shale. 898

ALUMEN. **Alum.**

Alum Seed. 899

The alum earth is dug out and exposed for months in heaps. It is then spread in squares and sprinkled with water. After about 12 days it consolidates into efflorescing and mamillated crystalline plates of sulphate of alumina, called alum seed, *phatakarí-ká-bín* or *turi*. These plates are boiled in water mixed with salt-potash in the proportion of 15 parts of the sulphate of alumina to six of the salt-potash. Before the salt-potash has time to dissolve, the fluid is ladled into small earthen vessels, crystallization taking place in less than two days. These crystals are again boiled to concentrate the solution, which is finally ladled into large bladder-shaped earthen jars, *matkás*, sunk in the ground to prevent breaking. The alum in each jar forms a solid crystal in about four days.

Alum Matka. 900

Cutch Alum. 901

In 1867 the yearly outturn of Cutch alum was estimated at about 294 tons. But after 1867, owing partly to an idea that Cutch alum tinges cloth, and partly because the working of the mines was a mismanaged monopoly, the demand for Cutch alum almost entirely ceased. The Bombay Chemical Examiner analysed Cutch alum in 1878, and according to him it is better than either English or Chinese alum, as the Cutch alum contains only 13 per cent. of impurities and 10·73 per cent. of alumina, being 0·12 per cent. less than the theoretic quantity. The Cutch State has lately discontinued the monopoly of the mines and begun to sell the alum on its own account.

Sind Alum. 902

Irvine, in his *General and Medical Topography of Ajmere*, published in 1841, says that alum "comes from Sind, where it is made: about 300 camel-loads annually arrive: red alum is brought from Lahore: used in medicine as an astringent; but chiefly employed in dyeing: one maund for R10."

Panjab Alum Kalabagh. 903

Baden Powell, in his *Panjáb Products*, says that "European alum is white and pure, and is on that account preferred to Indian alum in medicine. Bituminous shale, yielding more or less alum, is abundant, all through the Salt-range in the Panjáb, although the manufacture is confined to two places—Kálábagh and Kutki. The alum made at Kálábagh is always of a pinkish colour, due to the presence of chloride of iron. It is remarkable that the alkaline base of Kálábagh alum is soda, while that of English alum is potash. The shale strata at Kálábagh are nearly 200 feet thick. The shale is very soft, and contains a large amount of iron pyrites in crystalline nodules. The red mound-like alum kilns form a striking feature at Kálábagh. In making an alum kiln, layers of brushwood, generally "jhau" or "pilchi" (*Tamarix*) are spread on the ground; then a layer of alum shale is laid upon them, then more brushwood, and so on, the half-formed pile is lighted first, and subsequently more layers of shale and brushwood are added, till the pile reaches a height of 20 to 60 feet. A pile takes 6 or 8 months to burn. The calcined shale is next lixiviated with water in large tanks of baked earth about 12 feet square and 18 inches deep. The liquid is after some time allowed to flow off into a similar tank at a lower level, where it deposits by subsidence its mud and impurities, and is again drawn off into a third vat. It is then poured into iron evaporating pans and mixed with a dirty-looking salt, called "*jamsan*," which appears to be similar to the saline efflorescence of *reh* lands, and consists of sulphate of soda with a little common salt, and a very little carbonate of soda. The alkali contained in *jamsan* converts the solution into the alum of commerce. When the mixture has settled, the solution appears as a clear brown fluid, and is drawn off to be evaporated in vats under a shade, where the alum is formed in crystals of a pink colour. These crystals are next washed slightly with cold water on strainers of "*sirki*" grass, after which they are liquified by heat in iron pans. The liquid is poured into earthen jars, where it again crystallizes. The

Jamsan Salt. 904

contents of the jar, broken into lumps, form the alum of commerce. The manufacture appears to have been carried on at Kálábagh for many generations.

Kutki. 905

"The alum works at Kutki across the Chichalli range is of much more recent date than at Kálábagh. The cost of manufacture at Kutki is less than at Kálábagh, owing to the shale being cheaper and the fees lower at the former place.

"There is no difference in the quality of the alum produced at Kálábagh and at Kutki, but the value of Kálábagh alum is R3-4 a maund on the spot, while Kutki alum sells at R2-8. About 12,000 maunds of alum are made annually at Kálábagh, and 10,000 maunds at Kutki. It is exported to all parts of India." (*Baden Powell, Panjáb Products, Vol. I., p. 84.*)

Burma Shale. 906

§ "According to **Dr. Brandis,** alum can be obtained from shale which exists in abundance in the Shwegyin district in Burma." (*J. C. Hardinge, Esq., Secy., Agri.-Horticultural Society, Rangoon.*)

Burma Alum. 907

MORDANT. 908

Mordant.—It has been found difficult to obtain any very definite information regarding the Indian trade in this valuable salt. **Mr. E.C. Buck** says it is imported into the North-West Provinces from Calcutta, and is much used as a mordant in dyeing, especially with madder and turmeric. Potash-alum is largely imported into Bombay from Europe. (*Dymock.*)

MEDICINE. 909

Medicine.—In the *Indian Pharmacopœia* this substance has been described as astringent, styptic, and antiseptic. Used internally in passive hæmorrhages, atonic diarrhœa, infantile cholera, catarrhal affections of the stomach, colica pictonum, whooping-cough, and bronchorrhœa; in the form of lotion or powder, as a local application for catarrhal ophthalmia, granular eyelids, and many other diseases of the eye, in leucorrhœa, gonorrhœa, menorrhagia, prolapsus of the uterus and rectum, and ulcerations. Burnt powdered alum is used as a snuff to stop bleeding from the nose.

Chemical Note.—"The officinal alum of the *Indian Pharmacopœia* is a double sulphate of alumina and ammonia. Several other alums are known, of different bases, such as potash, soda, iron, &c., replacing ammonia in the salt. It is probable that the burnt alum which so frequently enters into native doctors' nostrums is often useless from too high a temperature having been employed in its preparation. Above 400° Farh., alum is decomposed—inert and insoluble alumina remaining." (*Surgeon C. J. H. Warden, Professor of Chemistry, Calcutta.*)

Special Opinions.—§ "It is useful in aphthæ, ulcerated sore-throat, spongy gums, salivation, chronic ulcers." (*Brigade Surgeon J. H. Thornton, B.A., M.B., Monghyr.*) "It has been recommended as useful in relieving the pain caused by a carious tooth." (*Surgeon G. F. Poynder, Roorkee.*) "The domestic use of alum is to clear water. It is a strong cement when liquified by boiling." (*Assistant Surgeon T. N. Ghose, Meerut.*) "It is used as a gargle in relaxed and inflamed throat." (*Brigade Surgeon G. A. Watson, Allahabad.*) "Burnt alum is sometimes very effective as an external application to scorpion-bite." (*Surgeon Joseph Parker, M.D., Poona.*) "If, after the umbilical cord has dropped off, there be any ulceration of the navel, a little burnt alum sprinkled over the part will effect a speedy cure." (*Surgeon W. Wilson, Bogra.*)

"Found beneficial in early abortions, where a difficulty exists to extract the debris; it also lessens hœmorrhage. Should be used in the following manner: finely-powdered alum placed in a muslin bag (the size of a walnut or large-sized marble), with a long thread attached to hang out of the passage. This is introduced into the vagina as far as the os uteri and left there for 24 hours. Should no irritation be felt, the bag can

MEDICINE.

be left for another 24 hours, after which, in its removal, the debris will be found lying in the passage and can easily be removed." (*Honorary Surgeon Peter Anderson, Guntur, Madras Presidency.*) "This and the decoction of babúl bark is useful in dysentery, used as injection." (*Surgeon-Major P. N. Mukerji, Cuttack, Orissa.*) "A piece of alum burnt and applied to the part stung by a scorpion allays the pain rapidly. Burnt alum mixed with lime-juice is a useful remedy in ophthalmia." (*Surgeon-Major John Lancaster, Chittore.*) "Alum is one of the best remedies for whooping-cough, but it seems to have gone out of use lately. For this purpose, it should be given in doses of from 10 to 20 grains three times a day." (*Surgeon-General William Robert Cornish, F.R.L.S., Madras.*) "Useful in doses of 5 grains in diarrhœa and latter stage of dysentery combined with opium. Useful astringent lotion in conjunctivitis and purulent ophthalmia (alum 4 grains, rose-water one ounce); also in gonorrhœa and gleet." (*Assistant Surgeon Shib Chunder Bhuttacharji, Chanda, Central Provinces.*)

910

ALYSICARPUS, *Neck.; Gen. Pl., I., 522.*

A genus of annual or biennial spreading or erect herbs, comprising about 15 species (belonging to the Natural Order LEGUMINOSÆ).

Leaves simple, rarely 3-foliolate, stipulate, subcoriaceous. *Flowers* in copious axillary racemes. *Calyx* glumaceous; *teeth* deep, often imbricated, the two upper frequently connate. *Corolla*, not exserted; *standard* broad; *keel* obtuse, adhering to the wings. *Stamens* diadelphous; *anthers* uniform. *Ovary* nearly or quite sessile, many-ovuled; *style* incurved; *stigma* capitate. *Pod* terete or turgid, composed of several indehiscent 1-seeded joints.

911

Alysicarpus vaginalis, *DC.*, *var.* **ummularifolius;** *Fl. Br. Ind., II., 158.*

Vern.—*Nág bala* (*Stewart's Pb. Pl.* 57). **Sakhárám Arjun** says that in Bombay *Nág bal* is the name for SIDA ALBA.

Habitat.—Himálaya to Malacca and Ceylon, ascending to 4,000 feet in the North-West Provinces.

MEDICINE. 912

Medicine.—This may be the officinal plant referred to by some authors; information very imperfect.

913

AMARANTACEÆ.

A Natural Order of herbaceous or suffruticose. glabrous, pubescent, or woolly plants; erect, sparely branched or scandent. *Leaves* opposite or alternate, simple, usually entire, membranous or fleshy, exstipulate. *Flowers* small, scarious, diclinous or hermaphrodite, rarely polygamous, sessile, solitary or in glomerulate heads or spikes, the lateral ones sometimes arrested or developed only into crests, awns, or hooked hairs. *Bracts* three, the lateral smaller, often keeled, the central and lowest large, sometimes leafy. *Calyx* 4-5, sepals (in Mengea) distinct or coherent at the base, erect, equal or sub-equal (2-3 interior sepals smaller), green, scarious, rarely petaloid, persistent, imbricate in æstivation. *Corolla* wanting. *Stamens* hypogynous, inserted at the base of the sepals, 1-5, fertile, included or early exserted, opposite the sepals (rarely fewer), with or without alternate staminodes, all free or united below, forming a cup or tube; *filaments* filiform, subulate or dilated, sometimes 3-fid; *staminodes* entire or fringed flat or rarely concave, sometimes very small and toothed; anthers introrse 1-2 celled, dorsifixed, dehiscing longitudinally. *Ovary* free, ovoid or globose compressed, rarely depressed, 1-carpelled and 1-celled; *style* terminal, simple, sometimes obsolete; *stigma* capitate-emarginate, or two or more lobed or 2-3-fid. *Ovule* solitary, erect or suspended from an ascending funiculus. *Fruit* usually enveloped in the calyx, sometimes membranous, a 2- or more-seeded utricle, bursting by circumsciss dehiscence or irregularly, rarely a berry. Seeds usually compressed, reniform, testa crustaceous, black, shining; *hilum* naked or early arillate, albumen abundant. *Embryo* peripheric, annular or curved; *cotyledons* incumbent, radicle near the hilum, inferior or sub-ascending.

Affinities of the Order.—The Amarantaceæ have their closest affinity with the Chenopodiaceæ, the latter differing chiefly in habit, and in having distinct styles and a herbaceous calyx.

Its Habitat.—There are 480 species, in the whole world, referred to this Natural Order, and they are mostly tropical or extra-tropical plants, taking the place in the tropics of the Chenopodiaceæ, which extend into the temperate regions. In India there are a little over 80 species, chiefly extra-tropical, or if met with in the tropical regions they are annuals, which appear, or are cultivated in the plains, during the cold season only. A few species are strictly tropical, and these, compensating for the sparsity of forms, make up in abundance of individuals, since they are perhaps the most plentiful weeds on roadsides and waste places met with in the plains of India. The Indian species are referred to 15 genera; of these the genus Amarantus contains 27 species, and 17 occur in Ærua and Achyranthes,—two genera, the species of which are undoubtedly the most typical and prevalent Indian representatives of the order. It is not far from correct to say that the Amarantaceæ attain their maximum development in the tropical regions of the New World, the greatest number of species perhaps occurring in Mexico. There are few species in the temperate zones, and none in cold countries, some 5 or 6 species only being met with in Europe.

DeCandolle (in *L'Orig. des Pl. Cult.*) very truly remarks that all the species of Amarantus spread themselves on cultivated lands, among rubbish-heaps and on roadsides, and have thus naturalised themselves in most warm countries as well as in Europe; hence great difficulty exists in distinguishing the species, and above all in guessing or proving their origin.

The following brief classification of the Indian genera may be found useful; it will at least serve to direct attention to the respective alphabetical positions where fuller details will be found regarding the more important members of this order. It is necessary to explain, however, that the information given is of the most meagre kind, since there is perhaps no family of Indian economics regarding which greater confusion exists. Most writers give the information published by them under vernacular names only, and the few authors who do associate these vernacular names with the scientific names for the plants referred to, are unfortunately most conflicting in their statements, so that it has been found next to impossible to arrive at any satisfactory conclusion. It is hoped that this confession, which must of necessity run through the greater portion of the present work, but which is specially true with regard to the Amarantaceæ, may call forth new material, based upon the present attempt at grouping scientifically the available economic information. For museum purposes it is absolutely necessary that all collections of Amarantaceous food-stuffs or drugs be accompanied with dried specimens of the plants from which they are obtained, together with the various vernacular names given to these plants. Were such collections to be made by the local authorities there would then be no difficulty in having the present confusion regarding the Amarantaceæ completely removed, and this much-to-be-desired result would, without doubt, prove most convenient and valuable both to the cultivator of the soil and to the administrator, since, in times of scarcity and famine, few sources of food are more valuable than the various forms of Amarantaceous grains. They reach maturity in little more than two months, and require scarcely any rain, so that they often succeed when other crops fail. That confusion and ambiguity should exist in official correspondence regarding plants of such importance to India is much to be regretted. The conflicting opinions in vernacular and botanic names applied to the species of economic interest render it impossible to do more than indicate the probable species to which the Indian forms belong, and it is difficult, from available literature, to arrive at any conclusion regarding the vernacular which should be associated with the scientific names. For example, it may be mentioned that in the Introduction to Part I. of **Duthie** and **Fuller's** *Field and Garden Crops, p. v.*, in a list of crops are found the words "*ramdana* (**Amarantus frutescens**)." What is **Amarantus frutescens?** **DeCandolle** gives it as a synonym for **Iresine amarantoides,** a plant apparently not met with in India, while **Moquin** places it under "*Species non satis notæ.*" According to **Duthie and Fuller,** it is one of the *rabi* crops of the North-West Provinces.

914

CLASSIFICATION OF THE INDIAN AMARANTACEÆ.

[NOTE.—*Fuller details of the genera marked * will be found in their respective alphabetical positions.*]

TRIBE I.—CELOSIEÆ.

Anthers 2-locular. *Ovary* 2, many-ovuled. *Leaves* alternate.

SECTION 1ST.—*Fruit a berry, perianth spreading, stalked.*

1. **Deeringia.**—Flowers racemose.

This genus is now made to include the species formerly referred to **Deeringia** and **Cladostachys.** The following are the commoner species:—

D. baccata, *Moquin; Wight, Ic., t. 728.*

Syn.—D. INDICA, *Spreng.*; D. CELOSIOIDES, *R.Br.*; *Roxb., Fl. Ind., Ed. C.B.C., 229.*

Vern.—*Latman*, HIND.; *Gola-mohaní*, BENG.

Habitat.—An extensive climber, very common in Bengal, covering the *babúl* trees with its racemes of small scarlet berries, which ripen from December to January. Apparently not put to any economic use.

D. muricata, formerly **Cladostachys muricata,** *Moquin,* and **Achyranthes muricata,** *Linn.*

D. tetragyna, *Roxb.; Wight, Ic., t. 729.*

SECTION 2ND.—*Fruit membranous, perianth erect.*

2.* **Celosia.**—Filaments connate at the base, fruit bursting by circumsciss dehiscence.

TRIBE II.—AMARANTEÆ.

Anthers 2-locular. *Ovary* 1-ovuled.

Sub-tribe 1. EUAMARANTEÆ.—Ovule erect, funicle short, radicle inferior. Leaves alternate.

SECTION 1ST.—*Flowers hermaphrodite or diœcious. Perianth segments spreading in fruit.*

3. **Rodetia.**—Flowers 2-4—bracteolate.

SECTION 2ND.—*Flowers hermaphrodite. Perianth segments erect in fruit.*

4. Banalia.—Flowers in panicled spikes. Stigma 2. Fruit membranous indehiscent. Arillus absent.

B. thrysiflora, *Moquin,* a native of the Nilgiri Hills. **Achyranthes thrysiflora,** *Wall.* A herbaceous climber.

5. Allmania.—Herbaceous plants. Flowers capitate. Stigma capitate. Fruit bursting by circumsciss dehiscence. Seed arillate. The following are the principal Indian species :—

A. esculenta, *R.Br.*, Singapore.

A. nodiflora, *R.Br.*, Coromandel.

6.* Digera.—Herbs. Flowers spicate. Stigma 2-fid. Nut crustaceous.

SECTION 3RD.—*Flowers unisexual. Perianth segments erect in fruit.*

7.* Amarantus.—Flowers most frequently monœcious. Fruit various.

This genus is now made to include the species formerly referred to the following genera : **Amarantus, Amblogyne, Mengea,** and **Euxolus.**

Sub-tribe II. ACHYRANTHEÆ.—Ovule suspended from the apex of an elongated funiculus. Fruit indehiscent. Seed inverted ; radicle ascending or superior.

SECTION 4TH.—*Flowers hermaphrodite, 1-3 perfect and bracteate, the others imperfect and stipulate.*

8. Cyathula.—Flowers forming fascicled or capitate spikes. Staminodes or teeth ascending from the connate base of the filaments. Leaves opposite. The following are the principal Indian species :—

C. capitata, *Moquin.*

C. prostrata, *Blume ; Wight, Ic., t. 733 ; Roxb., Fl. Ind., Ed. C.B.C., 226.*

C. tomentosa, *Moquin.*

9. Pupalia.—Flowers forming simple or panicled spikes perfect flowers solitary amongst the imperfect ones. Staminodes absent. Leaves opposite. The following are the Indian species :—

P. atropurpurea, *Moquin ; Wight, Ic., t., 731.*

Vern.—*Duiya-khuiya,* BENG.

P lappacea, *Moquin.*

P. velutina, *Moquin.*

SECTION 5TH.—*Flowers hermaphrodite, solitary, 2-bracteolate. Interfilamentary teeth absent.*

10. Psilotrichum.—Exterior segments of the perianth thick, 3-costate. Herbs or under-shrubs with opposite leaves and dense-flowered axillary spikes. The following are the Indian species :—

P. ferrugineum, *Moquin (? Endl.) ; Wight, Ic., t. 721 ;* Achyranthes ferruginea, *Roxb., Ind. Ed., Wall.*

P. nudum, *Moquin.*

P. trichotomum, *Blume.*

11. Ptilotus.—Perianth segments free, thin, transparent. Leaves alternate.

12.* Ærua.—Perianth segments hairy. Teeth short. Herbs or under-shrubs (lanate), leaves opposite or alternate.

13.* Achyranthes.—Perianth, after flowering, deflexed, segments and bracts spinescent. Herbs with opposite leaves, flowers in long lax spikes. This genus is now made to include **Achyranthes** and **Centrostachys.**

TRIBE III.—GOMPHRENEÆ.

Anthers 1-locular. *Ovule* suspended from the apex of a prolonged funiculus which ascends from the base of 1-celled ovary. *Leaves* generally opposite.

SECTION 1ST.—*Stigma simple, capitate.*

14. Alternanthera.—Flowers hermaphrodite, capitate, axillary, rarely terminal, solitary or 2-5. Staminal tube nearly as long as the ovary; stamens 3 (rarely 5); staminodia nearly as long as the filaments. Ovary orbicular compressed. (*See page 199.*)

The above definition has been restricted to the characters of the Indian representatives of the genus, which form a small section or sub-genus by themselves.

A. denticulata, *R. Br.*

A. nodiflora, *R. Br.; DC. Prod., XIII., 2nd, 356; Thw. En. Cyl. Pl., 350.*

A. sessilis, *R. Br.; Wight, Ic., t. 727.*

Syn.—GOMPHRENA SESSILIS, *Linn.*; ALTERNANTHERA TRIANDRA, *Lam.*; ACHYRANTHES TRIANDRA, *Roxb.*; *Fl. Ind., Ed. C.B.C., 227.*

Vern.—*Shanchi,* BENG.; *Mokú-nú-wanna,* SINGH.

SECTION 2ND.—*Stigma 2-subulate or filiform.*

15. Gomphrena.—*Perianth* often softly hairy or lanate, segments free or united at the base. *Staminal tube* elongated, antheriferous teeth 5. *Inflorescence* capitulate or hemispheric; *bracts* lateral, concave, fleshy, crested on the back. Under-shrubs with opposite generally semi-amplexicaul leaves.

G. globosa, *Linn.*

THE GLOBE AMARANT.

Common in Indian gardens. There are two varieties—*Lal-gûl-makmal,* BENG., or the crimson form; and *Safed-gúl-makmal,* the yellowish-white form.

The Globe Amaranth is one of the most prolific and ornamental flowers in the Indian flower garden, largely cultivated by natives. Flowering time the rainy season.

915

AMARANTUS, *Linn.; Gen. Pl., III., 28.*

A genus of tropical plants, belonging to the Natural Order AMARANTACEÆ, comprising some 45 species, of which 27 are most probably natives of India.

Leaves alternate, contracted at the base, ovate-lanceolate or linear, entire or rarely sinuate-dentate, apex often mucronate. *Flowers* minute, monœcious or polygamous, bracteate, arranged in dense axillary or terminal panicled spikes. *Perianth* segments 5, rarely 1-3, membranous, equal or sub-equal, in the male ovate-lanceolate, in the female oblong, white or coloured, generally purplish red, thickened at the base, erect in fruit. *Stamens* 5, rarely 1-3, filaments subulate free. *Ovary* ovoid, compressed; *style* short or wanting; *stigmas* 2-3, subulate or filiform and papillose. *Ovary* 1, sub-sessile, erect;

fruit often included by the persistent perianth, orbicular or ovoid, compressed, indehiscent or opening by a circumsciss, membranous, or coriaceous apex, simple or 2-3 dentate.

This genus, according to the *Genera Plantarum*, includes the species formerly referred to **Euxolus, Mengea,** and **Amblogyne,** as well as to **Amarantus** proper.

The generic name is derived from the poetic flower, the Amarant, supposed never to fade. *Amaranth,* GER.; *Amarante,* FR.; *Amaranto,* IT., SP., and PORT.; *Amarantus,* LAT., and *ἀμάραντος,* GR., non-fading (from *ἀ* and *μαραίνω,* to quench).

O'Shaughnessy says that nearly all the species of Amarantus "may be used as emollients for enemata, cataplasms, diluents and drinks." These properties doubtless depend upon the amount of nitre which they contain. **Boutin** found that **A. Blitum** yielded for 100 parts of the plant 11·68 grains of nitrate of potash (*Journ. Pharm. and Chem., 4th series*).

The following brief classification of the more important forms of Amarantus met with in India may assist the reader to recognise the species as defined by botanists:—

(A.) *Spikes branched, terminal and axillary. Stamens 5.*

Amarantus Anardana.—Erect-branched. *Leaves* oblong. *Spikes* erect, cylindrical obtuse. *Calyx* shorter than the bracts; sepals oblong-elliptic, mucro-nulate.

A. frumentaceus.—Stems and branches erect. *Leaves* broad lanceolate. *Spikes* adpressed, crowded. *Calyx* longer than the stamens. *Capsule* wrinkled; seeds pellucid with a white margin. (This may prove to be but a form of a **A. paniculatus.**)

A. tristis.—Erect, very much branched near the ground. *Leaves* rhomboid-ovate, obtuse. *Spikes* long erect, sparsely branched, green.

A. spinosus.—Erect, much-branched, with round spikes in the axils. *Spikes* terminal, almost simple, with sessile axillary glomeruli.

A. paniculatus.—Erect-branched. *Panicle* 1-2 feet long, decompoundly branched, crimson. *Leaves* long petioled, broad lanceolate, concave. *Sepals* obtuse, shorter than the capsule.

(B.) *Spikes simple and terminal; axillary ones very short and distant. Stamens 3.*

A. gangeticus.—Erect-branched above the middle. *Leaves* rhomboid-ovate. *Glomerules* axillary or spicate; terminal spikes very often ovate obtuse rigid, axillary glomerules ovate. *Calyx* longer than the slightly rugose capsule and shorter than the bracts.

A. lanceolatus.—Straight, erect. *Leaves* long petioled, lanceolate, tapering at both extremities. *Glomeruli* axillary scarcely spiked. *Calyx* 3-membranous, with green keel. *Anthers* sagittate.

A. oleraceus.—Erect, sparsely branched. *Leaves* broad, rhom-boid-ovate lanceolate. *Calyx* cuspidate, longer than the rugose capsule.

A. mangostanus.—Terminal *spike* oblong or sub-globose, very obtuse sub-flexuose, axillary glomeruli rotund. *Fruit* shorter than the calyx.

A. lividus.—Erect, smooth. *Leaves* long petioled, sub-ovate retuse. *Calyx* 3-5, shorter than the compressed capsule.

A. viridis.—*Leaves* elliptic-emarginate. *Glomeruli* on the ends of axillary twigs. *Sepals* obtuse, much shorter than the rugose capsule.

A. fasciatus.—*Leaves* rhomboid-ovate. *Panicles* terminal, composed of a few cylindrical branches. *Bracts* minute, shorter than the obtuse sepals, which are shorter than the rugose capsule.

Forms of Amarantus in India.

Amarantus polygamus.—Diffuse. *Leaves* petioled, rhomboid-emarginate with a bristle. *Glomeruli* rarely spicate. *Calyx* twice the length of the capsule.

A. atropurpureus.—Erect-branched. *Leaves* lanceolate. *Glomeruli* axillary and forming terminal spikes. *Calyx* 3-5, cuspidate, longer than the capsule.

(C.) *Flowers in axillary glomerules, but never forming terminal spikes; stamens 3 or 2.*

A. polygonoides.—A small, diffuse plant. *Leaves* obovate, sub-sessile or petioled. *Glomeruli* axillary, two parted but never spiked. *Capsule* equal or longer than the calyx. *Seeds* black, shining.

A. melancholicus.—Much-branched, 6-12 feet high. *Leaves* rhomboid-ovate. *Calyx* cuspidate, longer than the capsules. *Glomeruli* almost surrounding the stem.

A. tenuifolius.—A diffuse annual with shortly petiolate deltoid leaves. *Male flowers* diandrous, female irregular. *Capsule* with 6 longitudinal furrows.

916

Amarantus Anardana, *Hamilt.*; AMARANTACEÆ.

Vern.—*Chúa*, HIND.; *Chuko*, GUJ.; *Ganhár, tawal, chauláí, síl* (seed), PB.

With the exception of *Chúa*, the vernacular names given above are also associated with **A. mangostanus.** It seems probable that the word *Anárdáná* may be applied to the grain obtained from most AMARANTACEÆ.

References.—*DC. Prod. XIII., II., 256; Stewart's Pb. Pl., 181.*

Habitat.—Cultivated in the mountain tracts of Bengal and the Upper Provinces.

From the information available in works on Indian Economic Botany, it is almost impossible to arrive at any conclusion as to the plant which generally bears this name. According to some authors it is only a variety of **A. frumentaceus** (or of a **A. paniculatus**); by others it is viewed as a distinct species. The similarity in vernacular names with those given for **A. mangostanus,** *Linn.*, would lead one to the conclusion that it is nearly related to that plant. **Stewart** combines the names **A. Anardana** and **A. gangeticus** in a common paragraph describing their properties, and it is therefore impossible to separate the vernacular names, or to understand how far his description applies to the one or the other. In the *Prodromus DeCandolle,* **Moquin** seems to regard **A. Anardana** as a better known species than **A. frumentaceus**; but his description of the latter agrees accurately with the plant cultivated on the Himálaya (the *bathú* of Simla), so that it would seem the Himálayan plant should be known as **A. frumentaceus.**

MEDICINE. 917

Medicine.—They are used in scrofula and as a local application for scrofulous sores; administered in the form of a liquid.

§ "Astringent in diarrhœa." (*Surgeon Barren, Bhuj, Cutch.*)

FOOD 918

Food.—The leaves are eaten as a pot-herb. The seeds, after being parched, are used in some places as a food-grain, but are considered heating.

919

A. atropurpureus, *Roxb.; Fl. Ind., Ed. C.B.C., 662; DC. Prod., XIII., 2, 264.*

Said to be cultivated as a pot-herb in Bengal and to be known as *Lál-natí, kunka-natí,* or *bansh-pata-lál-natí.*

920

A. Blitum, *Linn.*

Said to be cultivated in India.

Amarantus caudatus, *Linn.; DC. Prod., XIII., pt. 2, 255.* 921

LOVE LIES BLEEDING, *Eng.*; QUEUE DE RENARD, DISCIPLINE DE RELIGIEUSE, *Fr.*

Vern.—*Kedari chúa,* HIM. NAME (according to **Atkinson**).

Habitat.—Cultivated in gardens throughout India. A well-known plant with drooping tail-like spikes of flowers.

Atkinson says it is cultivated in the hills for local consumption. The seed is sown in May and June, and the crop is ripe in October. This is the only mention of this plant being cultivated as a food-crop. I found one or two plants the other day growing on the border of a field a few miles from Simla. On enquiry it was found the people had no name for it, and they stated that it was an introduced form of *bathú,* which latter they regard as indigenous. This species does not seem to have been introduced into India during **Roxburgh's** time; at all events he does not mention it.

A. farinaceus, *Roxb.; DC. Prod., XIII., pt. 2, 266.*

MEDICINE. 922

Medicine.—The plant is said to possess diuretic and purifying properties.

A. fasciatus, *Roxb.; Fl. Ind., Ed. C.B.C., 663; Wight, Ic., t. 717.* 923

Vern.—*Tún-túni-nati, ban-nati,* BENG.; *Chilaka-tóta-kúra, chilaka-kúra,* TEL.; *Hilamóchaka,* SANS.

A common green weed (*Roxb.*).

A. flavus, *Linn.; DC. Prod., XIII., 2, 258.* 924

Said to have been collected by **Wallich** in Nepal, where it was cultivated.

A. frumentaceus, *Buch.; Wight, Ic., t. 720.* 925

Syn.—It seems probable that this may be reduced to A. PANICULATUS, *Miq.*; A. SPECIOSUS, *Sims; Bot. Mag., t. 2227.*

Vern.—*Bathu, báthu, batu,* or *bathú,* PB.; *Betu,* KUMAON; *Larka baha,* SANTAL; *Pungi-kírai,* TAM.

References.—*Roxb., Fl. Ind., Ed. C.B.C., 663; DC. Prod., XIII., 2, 265; DC. L'Orig. Pl. Cult.; Atkinson's Him. Districts, 697.*

It seems likely that this is the plant which most generally receives the names *chúa, chúa-mársa, rámdána, anardána.* The name *báthú* is in the plains of India generally applied to **Chenopodium album.**

Habitat.—Cultivated on the Himálaya from Kashmír to Sikkim. **Roxburgh** says it was first discovered by **Dr. Buchanan** on the hills between Mysore and Coimbatore. (Compare with the remarks under **A. Anardana** in explanation of the above region over which this species is cultivated.) **Atkinson** says: "*Chúa* is largely grown in the northern *parganas* up to 9,500 feet, where it forms the staple food of the poorer classes, and is a favourite crop in newly-cleared jungle, as it is not easily injured by bears and deer."

FOOD 926

Food.—One of the mosti mportant sources of food to the hill tribes of India. There are two varieties,—a golden-yellow and a red. The former seems to be preferred, since it is more cultivated than the latter; but most fields contain, as a rule, a few red plants here and there amongst the golden-coloured crop. Although, no doubt, the young tops are to a certain extent eaten as a vegetable, most of the hill-men speak of this as only an occasional thing,—the small seed is the product for which it is cultivated.

FOOD.

Indeed, from the fact that it rarely branches, it seems probable that the plant would be injured were the young tops or leaves to be lopped as a vegetable. The unbranched habit is, however, the result of thick, broadcast sowings. When grown singly it seems to branch. It is perhaps one of the most elegant crops cultivated by the hill people. When young, the large leaves (seen at a distance) remind the traveller of a turnip-field, but when the terminal golden-yellow or red crowded spikes of flowers appear in the centre of each terminal rosette of leaves, it becomes truly lovely.

Dr. Roxburgh records the following facts: "In the Botanic Garden 40 square yards of ground sown with this plant in June, yielded 21 lbs. weight of the clear ripe seed in September. It also grows well during the cold season, *viz.*, from October till February, inclusive." My friend **Mr. Campbell**, of the Santal Mission, informs me that the plant is sparingly cultivated hy the Santals and eaten as a pot-herb.

Atkinson says: "It is sown in May and June in first and second class unirrigated land, and yields about twenty loads to the acre. The produce of an acre is worth about R16, and the estimated outlay is about half that sum." The yield and the shortness of the period required for the production of this food-supply, seem to justify the opinion already indicated, that this, as also several other species of Amarantus, might with advantage be resorted to, when, through want of rain, scarcity or even famine is threatened.

927

Amarantus gangeticus, *Linn.*

Vern.—*Lál-ság*, HIND.; *Ranga-shák, lál-shák, dengua*, BENG.; *Arak, gandhari*, SANTAL; DeCandolle (*in L' Orig. Pl. Cult.*) says that the forms of this plant are called in Telugu *Tóta-kúra*, with an adjective to denote the special form. These are the names that Roxburgh gives to the plant he calls **A. oleraceus**, and it is quite probable that the South Indian forms of *lál-ság* belong to **A. gangeticus.** Until, however, this has been clearly established, it has been thought advisable to leave the Madras plants in the position assigned them by Roxburgh. *Wight, Ic., t. 715*, gives a figure of this species under the name of **A. oleraceus,** *Linn.* See **A. oleraceus.**

In India cultivated as a vegetable. There are a large series of forms of this species varying as to colour and shape of leaf, but referable to two sections—those which may be placed under **A. gangeticus** proper, the *lál-ság*, and those which would more naturally fall under—

928

Var. angustifolius.—A. ANGUSTIFOLIUS, *Roxb. Hert.; and* A. LANCEOLATUS, *Roxb., Fl. Ind., Ed. C.B.C., 662; Wight, Ic., n. 716, Ic., t. 713.*

Vern.—*Báns-patá-natíya*, BENG.

References.—*Roxb., Fl. Ind., Ed. C.B.C., 662; DC. Prod.; DC., L'Orig. Pl. Cult.; Stewart, Pb. Pl., 181; U. C. Dutt, Hindú Mat. Med.*

Habitat.—A small annual plant common in Bengal and Assam. **Dr. Roxburgh** says: "The varieties of this useful species cultivated in Bengal alone are endless." "They are in more general use among the natives of Bengal than any other species or variety. The varieties are tolerably permanent, and differ in colour chiefly, which varies from green with the slightest tinge of red, to rufous, liver-coloured, and bright red. One variety has particularly broad leaves, with the margins green, and the centre dark purple." Most probably this species was originally indigenous to India, but it is now extensively cultivated in many parts of the world, and even claimed as a native of Egypt and Abyssinia. From the fact that all the species allied to **A. gangeticus** are indigenous to Asia, it may be presumed that it is a native of India.

FOOD. 929

Food.—This and **A. frumentaceus** are perhaps the two most important species of Amarantus met with in India. But while the latter is cultivated entirely for its seed, **A. gangeticus** is grown as a green vegetable only. It is extensively cultivated by the natives of Bengal, sown broadcast, under what is commonly known as garden cultivation, by professional vegetable-producers. The plants are pulled up when young and sold in the bazars entire; the leaves and tender stalks are the parts of the *lál-shák* chiefly used; they are made into curry by all classes of natives. Largely cultivated in Chutia Nagpur. **De Candolle** (*L'Orig. Pl. Cult.*) says the young stems are sometimes used as a substitute for asparagus on the English table. He also states that several species of annual Amarantus are cultivated in Mauritius, Bourbon, and Seychelles under the name of Bréde de Malabar, of which **A. gangeticus** seems to be the chief species. The Japanese cultivate as a vegetable the variety **melancholicus** amongst many others, such as **A. polystachyus,** *Blume.*

MEDICINE. 930

Medicine.—Used in India in the form of an emollient poultice.

Amarantus hypochondriacus, *Linn.; DC. Prod., XIII., 2, 256.* 931

THE PRINCE'S FEATHER.

An exceedingly handsome annual, common in Indian gardens, the leaves as well as the spikes being of a rich crimson.

"The leaves are said to be astringent, and to be used internally and topically in the complaints to which astringents generally are applicable." (*U. S. Dispens., 15th Ed., 1568.*)

A. lanceolatus, *Roxb.* 932

See **A. gangeticus,** *var.* **angustifolius.** There seems quite as much ground for this being kept up as a distinct species as for any other species; but most authors seem to place it under **gangeticus,** and it has therefore been deemed advisable to adhere for the present to that view.

A. lividus, *Linn.* 933

Syn.—EUXOLUS LIVIDUS, *Moquin; DC. Prod., XIII., 2, 273.*

Vern.—*Gobura-nati,* BENG.

A native of America: cultivated in India.

O'Shaughnessy says it is "held in great esteem by the natives."

A. mangostanus, *L.; DC. Prod., XIII., 2, 261.* 934

Vern.—*Chaulai, ganhar,* UPPER INDIA; *Ság,* BENG. "Ság is a generic name for pot-herbs: *Chouláí* is in South India applied to **Portulaca quadrifolia.**" (*Moodeen Sheriff.*)

Habitat.—Occasionally cultivated in the plains.

FOOD. 935

Food.—The leaves are used as a pot-herb.

MEDICINE. 936

Medicine.—**Mr. Baden Powell** gives **Amarantus,** sp., *Chuláí* (*p. 425*), amongst his rare medicinal oils. No other mention can be found of this or any other oil made from Amarantus, and the fact is therefore of considerable interest. It would be exceedingly interesting to have fuller particulars, and also samples of the oil, and of the plant from which it is made. The name *chauláí* is applied to this species, as also to at least a half of all the known Amarantuses, and it may be quite wrong to refer this curious oil to **A. mangostanus.**

A. melancholicus, *Linn.; Moquin, DC. Prod., XIII., 2, 262.* 937

Syn.—A. TRICOLOR, *Linn.*

There are numerous cultivated forms of this species, which have received gardeners' names, many of which are highly ornamental. In *L'Orig. Pl. Cult.*, **DeCandolle** says that the forms of this species should all be referred to **A. gangeticus**. This may be correct, but if so the definition of the species will have to be enlarged, since in inflorescence and number of stamens they do not agree. See classification of species of Amarantus.

938

Amarantus oleraceus, *Linn.*

Syn.—EUXOLUS OLERACEUS, *Moquin; DC. Prod., XIII., 2, 273.*

Vern.—*Sada-natia, natiyá-ság,* BENG.; *Bhájí* and *támbadámáth,* MAR.; *Tótakúra, erra-tóta-kúra, tella-tóta-kúra* and *pedda-tóta-kúra,* TEL.; *Tand-kirai, kirai-tand,* TAM.; *Dat, dant-ké-bhájí,* DUK.; *Dant,* GUJ. AND MAR.; *Márisha,* SANS. Compare with the note under vernacular names of **A. gangeticus.**

This plant, if it really exists in India, must have been introduced, and the various forms attributed to it seem to take the place in Southern India of **A. gangeticus** in Bengal. The descriptions of these two species as published by **Roxburgh** hardly differ, however, and it seems exceedingly probable that the plants which in India have been called **A. oleraceus** should be referred to **A. gangeticus.** Until this opinion can be clearly established, however, it has been deemed advisable to leave them in the position assigned to them by **Roxburgh.** The following are the forms described by the father of Indian botany:—

"There are several varieties cultivated as pot-herbs over India, particularly on the coast of Coromandel. The most conspicuous after the common green sort are—

"*1st.—Erra-tóta kúra* of the Telingas, a very beautiful variety, with a clear, bright red stem; branches, petioles, nerves, and veins, and the leaves themselves, rather ferruginous. (*Lál-daṭ,* Duk.)

"*2nd.—Tella-tóta kúra* of the Telingas; here all the parts that are red in the last variety are of a clear, shining, white colour. *Saféd-daṭ,* DUK.)

"*3rd.—Rosa* or *Pedda-tóta kúra* of the Telingas, is a very large variety, which **Konig** called **A. giganteus.** In a rich soil it grows to from 5 to 8 feet high, with a stem as thick as a man's wrist. The tender succulent tops of the stem and branches are sometimes served up on our tables as a substitute for asparagus. (*Gulábí-dat,* DUK.)

"The other varieties are more changeable and not so well marked. I will not therefore take notice of any more of them."

939

A. paniculatus, *Miq., var.* cruentus; *DC. Prod., XIII., 2, 257.*

Vern.—*Rajagaro,* GUJ.; *Táj-e-khurús,* PERS., PESHAWAR; *Bustán-afróz,* PERS., KASHMIR; *Rájgirá,* DUK.

Habitat.—Most probably a native of China, cultivated in India.

MEDICINE. 940

Medicine.—Used medicinally for purifying the blood and in piles, and as a diuretic in strangury. (*Baden Powell, Pb. Prod., I., 373.*) (See **A. frumentaceus.**)

941

A. polygamus, *Linn.*

Syn.—EUXOLUS POLYGAMUS, *Moquin; DC. Prod., XIII., 2, 272*; A. POLYGAMUS, *Willd.; Roxb., Fl. Ind., Ed. C.B.C., 661.*

Vern.—*Chámpá-natíya,* or *lál-chámpá-natiya* (or *nutí*), BENG.; *Chumli-ság, chowlai-ka-bhají* (according to **Murray**), (see note under **A. mangostanus**), HIND.; *Doggali-kúra,* TEL.; *Tanduliya,* SANS. (according to **U. C. Dutt**).

Cultivated throughout the peninsula of India as a pot-herb. It admits of being freely lopped. **Roxburgh** says both the green and red "sorts are extensively cultivated all over the southern parts of Asia." **Atkinson** says it is a common species grown as a pot-herb along the edges of fields in the sub-montane tracts.

Amarantus polygonoides, *Linn.; Wight, Ic., t. 512.*

942

Syn.—A. POLYGONOIDES, *Willd.; Roxb., Fl. Ind., Ed. C.B.C., 661*; AMBLOGYNA POLYGONOIDES, *Rafin; DC. Prod., XIII., 2, 270? Wight, Ic., t. 719.*

Vern.—*Cherú-natia, chelú-natipa*, BENG.; *Chira kura*, HIND.

Though not cultivated, the natives use it as a pot-herb, as it is considered very wholesome, especially for convalescents. (*Roxburgh; O'Shaughnessy; &c.*)

A. spinosus, *Willd.; DC. Prod., III., 2, 260.*

943

PRICKLY AMARANTH.

Vern.—*Kántá naté* or *kánta nutia, kánta-maris*, BENG.; *Cholái*, ? HIND.; *Janum arak'*, SANTAL; *Kánte-mát*, MAR.; *Mulluk-kírai*, TAM.; *Kántá-nu-dánt*, GUJ.; *Mundla-tóta-kura, nalla-doggali, erra-mulu-góranta*, TEL.; *Mullan-chíra*, MAL.; *Mulla-dantu, mulharavesoppu*, KAN.; *Tanduliya*, SANS.; *Hinkanoe-súba* or *hinnoe-súbá*, BURM.

Habitat.—Frequent in the plains of India, chiefly in Bengal and Malabar.

Dye.—**Dr. McCann**, in his *Report on the Dye-stuffs of Bengal*, states that in Cuttack the ashes of this plant are used in dyeing with **Mallotus philippinensis.**

DYE. 944

Food.—The leaves make a good spinach and pot-herb, though the sharp spines in their axles are troublesome to pick. The poor among the natives use the leaves as pot-herbs, especially in times of scarcity.

FOOD. 945

Medicine.—The whole plant is used as an antidote for snake-poison, and the root as a specific for colic. The root has been found useful in the treatment of gonorrhœa: it is said to arrest the discharge.

MEDICINE. Root. 946

"The Hindú physicians prescribe the root in combination with other drugs in menorrhagia. A poultice of the leaves was officinal in the Bengal Pharmacopœia." (*Dymock's Mat. Med., W. Ind.*) It is also considered a lactagogue, and, boiled with pulse, is given to cows.

Leaves. 947

§ "The root has lately been introduced into European practice as a remedy for gonorrhœa, and is advertised by some of the London druggists." (*Surgeon-Major Dymock.*) "Roots made into poultice are applied to buboes and abscesses for hastening suppuration. (*Surgeon Anund Chunder Mukerji, Noakhally*). "Supposed to be an excellent remedy for gonorrhœa. Dose of the decoction of the root one to two oz." (*Surgeon W. Barren, Bhuj, Cutch.*) "*Kántá nutia* is a cooling diuretic. An infusion in hot water I have used in some cases of gonorrhœa. It lessens burning and relieves pain." (*Surgeon R. L. Dutt, M.D., Pubna.*)

Decoction. 948

"Used frequently for colic pain and for scorpion-bite." (*Surgeon C. J. W. Meadows, Burrisal.*) "Emollient and used in the form of poultice." (*Deputy Surgeon-General G. Bidie, Madras.*) "Given to cows as a lactagogue." (*Asst. Surgeon Shib Chunder Bhuttacharji, Chanda, Central Provinces.*)

In a recent correspondence, however, with the Government of India in regard to the proposed issue of a revised edition of the *Pharm. Ind.*, the Surgeon-General of Madras expressed the opinion that **A. spinosus** should be excluded from the future edition of that work.

949

Amarantus tenuifolius, *Willd.; Wight, Ic., t. 718.*

Syn.—MENGEA TENUIFOLIA, *Moquin; DC. Prod., XIII., 2, 271.*
Vern.—*Ghénti-natí, jélchumlí,* BENG.

Cultivated in Bengal in the neighbourhood of Calcutta—a stemless spreading plant (*O'Shaughnessy*).

950

A. tristis, *Linn.; Wight, Ic., t. 514 & t. 713.*

Syn.—? A. CAMPESTRIS, *Willd, sp., 382.*
Vern.—*Pond-gandhari,* SANTAL; *Mekanada, ganna,* SANS.; *Churí-ki-bhájí,* DUK.; *Sirru-kirai, kappi-kirri* (**Roxburgh**), *shiru-kírai, kurví-kírai,* TAM.; *Sirru-kúra, koyya-tóta-kúra,* TEL.
References.—*DC. Prod., XIII., 2, 260; Roxb., Fl. Ind., Ed. C.B.C., 661.*

Habitat.—An erect herb, branching freely even from the ground. There are probably many sub-species met with in cultivation in the plains of India (especially of the south and west portions of the peninsula), of which **A. campestris** is by some authors regarded as a distinct species.

MEDICINE. 951

Medicine.—The roots have attributed to them demulcent properties.

FOOD. 952

Food.—**Roxburgh** says of this plant: "It is held in great esteem by all ranks of the natives, and is much cultivated by them; it grows readily all the year round if watered." It has "always terminal spikes; besides, it may be cut down several times without destroying the plants, for they soon shoot out vigorously again. This renders it much more useful to the poorer natives, who are possessed of but a very small spot of ground and little time to spare for its culture; besides, it is in higher esteem than **A. oleraceus,** which yields but one crop."

§ "Used also as diuretic in form of a decoction combined with some other medicines." (*Surgeon-Major J. T. Fitzpatrick, M.B., Coimbatore.*)

953

A. viridis, *Linn.*

Syn.—EUXOLUS VIRIDIS, *Moquin; DC. Prod., XIII., 2, 273;* A. BLITUM of most authors but not of *Linn.*

Roxburgh says of this plant: "A native of various parts of India, appearing most frequently as a weed in gardens during the rainy and cold seasons. The tender tops are eaten by the natives, though not so much esteemed as the cultivated sorts." (*Fl. Ind., III., 605.*)

AMARYLLIS.

954

Amaryllis grandiflora, AMARILLYDEÆ, the *sukhdarsan,* PB., is alluded to by both **Stewart** and **Baden Powell.** It is said to yield a medicinal oil. If this plant has been correctly identified, it can only be cultivated in gardens; its modern name is **Brunsvigia grandiflora. Stewart** says: "A correspondent of the Agri.-Horti. Society states that the strained juice of two drams, reduced to a pulp with water, is a good emetic, and that one drop into the ear will generally cure earache." From the existence of a vernacular name for this plant, however, it seems probable that the authors mentioned above have given incorrectly the name **Amaryllis grandiflora** to some other Amaryllidaceous plant. "*Sukhadarsan* is a well-known Hind. name for a species of **Crinum.**" (*Moodeen Sheriff.*)

955

AMBER.

Vern.—*Kahrubá,* HIND., DUK., PERS.; *Inqitriyún, qarnul bahr,* ARAB.; *Ambeng,* BURM.

A fossilised resin, yielded by trees, chiefly pines, (?) which grew during the cretaceous period of geologists, usually found in connection with tertiary lignites. It is hard, brittle, easily cut, of various shades of yellow, and semi-transparent. It is very useful to the physicist, becoming negatively electric by friction. The amber supply is chiefly from the Baltic region, Samland being the great centre. Crude amber occurs in commerce in irregular pieces. When ground or heated it emits a pleasant odour. It is completely soluble in alkaline solutions containing camphor. On being boiled for 20 hours in rape or linseed oil, it becomes transparent and ductile, and may then be moulded into any desired form. It is chiefly used for ornamental purposes such as necklaces, bracelets, and brooches, for mouth-pieces of pipes and cigar-holders, for the preparation of a varnish, and for the manufacture of amber-oil and succinic acid. See **Varnish** and also **Gum Copal.**

Irvine, in his *Materia Medica of Patna,* says that it is used as an aphrodisiac in native medical practice, and in his *General and Medical Topography of Ajmere* (published in 1841) he says that the natives of Ajmere do not know real amber, but that they use a mixed sort of scent called amber, which is of the consistence of plaster, and seems an imitation of ambergris. It is chiefly used as an aphrodisiac, and costs R5 to R6 a tolah.

AMBERGRIS.

956

Abr-i-amber, Aanbar, anber, or *araba* of the Indian bazars, is produced from ambergris. "Ambergris is found in pieces floating in the sea near the coasts of India, Africa, and Brazil; it is of an ash-grey colour, spotted like marble with black spots; but it appears to vary considerably in colour, some pieces being white, some black, and some grey with yellow spots. It is very light and easily takes fire. It is most probably a concretion formed in the stomach or intestines of the spermaceti whale, *Physeter macrocephalus.* Several specimens have been found full of the embedded beaks of a species of sepia which is the food of the *Physeter:* it is supposed by some to be formed only during disease, as the specimens of the whales in the stomach of which ambergris was found were sickly." (*Baden Powell, Panjáb Products, Vol. I., p. 190.*) **Dr. Irvine** says that ambergris is brought from Singapore. It is used as an aphrodisiac. It costs R80 a ℔. (*Mat. Med., Patna, p. 10.*)

"Ambergris has a peculiar aromatic, agreeable odour, is almost completely volatilizable by heat, and is inflammable. It is insoluble in water, but is readily dissolved, with the aid of heat, by alcohol, ether, and the volatile and fixed oils. It consists chiefly of a peculiar fatty matter analogous to cholesterin and denominated by **Pelletier** and **Caventou** *ambrein.* This may be obtained by treating ambergris with heated alcohol; filtering the solution and allowing it to stand, crystals of amberin are deposited. It is incapable of forming soaps with alkalies. When pure it has little or no odour." (*U. S. Disp., 15th Ed., 1568.*)

Amblogina polygonoides, *Rafin.;* Syn. for **Amarantus polygonoides,** AMARANTACEÆ, which see.

AMMANNIA, *Linn.; Gen. Pl., I., 776.*

957

A genus of annual glabrous herbs, belonging to the Natural Order LYTHRACEÆ. *Stems* square; *leaves* opposite and alternate, sometimes whorled, entire; *stipules* wanting. *Flowers* small, axillary, solitary and sub-sessile, or in small trichotomous cymes; bracteoles usually 2. *Calyx* membranous, campanulate or tubular-campanulate 3-5-toothed, often with minute teeth or folds. *Petals* 3-5 or 0, small, inserted between the calyx-teeth. *Stamens* 2-8, inserted on the calyx-tube. *Ovary* enclosed in the calyx-tube, 1-5-celled, the septa often absorbed ; *style* filiform or short ; *stigma* capitate ; *ovules* many ; *placentas* axile. *Capsule* membranous, dehiscing irregularly or by a circumsciss. *Seeds* many.

A genus of sub-aquatic herbs, named after **John Ammann,** Professor of Botany, St. Petersburgh.

958

Ammannia baccifera, *Linn.; Fl. Br. Ind., II., 569.*

BLISTERING AMMANIA.

Syn.—A. VESICATORIA, *Roxb.*

Vern.—*Dád-mári*, HIND. ; *Ban-marach, dádmári*, BENG. ; *Banmarich, agin-búti, guren, bhár jambol, agiya*, BOMB., DUK.; *Kallu rivi, nírumél-neruppu*, TAM. ; *Agnivenda-páku*, TEL. ; *Kallúrvanchi*, MAL.

Dr. Sakhárám Arjun says that in Bombay the *dád-mári* is **Cassia Tora.**

Habitat.—A small, herbaceous plant, generally met with on wet places throughout India, and extending to Afghánistan and China.

Botanic Diagnosis.—Cauline *leaves* opposite or alternate, oblong or narrow-elliptic, tapering to the base. *Flowers* in dense clusters forming knots on the stem, or in looser but very short axillary *cymes*. *Capsule* globose.

MEDICINE. 959

Medicine.—Dr. Roxburgh says: "It has a strong muriatic but not disagreeable smell. Its leaves are exceedingly acrid; they are used universally by the natives to raise blisters in rheumatic pains, fevers, &c. The fresh leaves, bruised and applied to the part intended to be blistered, perform their office in the course of half an hour or a little more, and most effectually." **O'Shaughnessy** says: "We made trial of this article in eight instances; blisters were not produced in less than 12 hours in any, and in three individuals not for 24 hours. The bruised leaves had been removed from all after half an hour. The pain occasioned was absolutely agonising until the blister rose. We should not be justified in recommending these leaves for further trial; they cause more pain than cantharides, and are far inferior to the plumbago (*lál chitrá*) in celerity and certainty of action." "The juice of the plant is given internally in spleen; but it causes great pain, and the result is not certain." (*Amster. Desc. Cat.*) In a recent correspondence with the Government of India about the revision of the *Pharmacopœia of India*, the Surgeon-General of Madras recommended that the **Ammannia baccifera** should be excluded from a future edition of the work.

§ "I find that an etherial tincture blisters well, without more pain than Liquor Lythæ." (*Surgeon-Major Dymock, Bombay.*) "As a blister it is much more painful than cantharides." (*Deputy Surgeon-General G. Bidie, Madras.*) "The bruised fresh leaves are applied as a poultice to abscesses, and the juice of the root is used in rheumatic affections as a topical application." (*Honorary Surgeon P. Kinsley, Ganjam, Madras.*) "The leaves are used as a vesicant." (*Assistant Surgeon Shib Chunder Bhuttacharji, Chanda, Central Provinces.*)

960

A. senegalensis, *Lamk.; Fl. Br. Ind., II., 570.*

Vern.—*Faugli mehndí, dádmárí*, PB.

Habitat.—Grows in wet places in the plains of the Panjáb and of the North-West Provinces, ascending to 5,000 feet in altitude.

Botanic Diagnosis.—*Leaves* opposite, elongate-oblong, sessile, sub-auriculate at the base. *Cymes* peduncled compound. *Capsule* $\frac{1}{12}$ in. diameter, globose.

MEDICINE. 961

Medicine.—Used as a blistering agent.

A. vescicatoria, *Roxb.;* Syn. for **A. baccifera,** *Linn.*, which see.

Ammoniacum, see **Dorema Ammoniacum,** *Don ;* UMBELLIFERÆ.

AMMONIUM.

Ammonium chloride.

962

SAL AMMONIAC, *Eng. ;* HYDROCHLORATE D'AMMONIAC, *Fr. ;* SALMIAK, *Germ.*

Vern.—*Nousádar,* HIND. ; *Nishedal, nóshágar,* BENG. ; *Navsár, navaságar,* GUJ. ; *Navságar,* MAR. ; *Navá-sagara,* KAN. ; *Nava-charum, navách-chárum,* TAM. ; *Navá-ságaram, navá-chárum,* TEL. ; *Navasáram,* MAL. ; *Milhunnár, arminá,* ARAB. ; *Nóshádar,* PERS. ; *Giatsah,* BHOTE ; *Naváchàram,* CINGH. ; *Zarasa, dza-wet-tha,* BURM.

This substance is largely manufactured in the Panjáb and used in tinning and forging metals, in the formation of freezing mixtures, and also in the separation of ammonia.

Baden Powell gives an interesting account of the manufacture of sal-ammoniac in the Panjáb. He states that sal ammoniac has been, for ages, largely manufactured by the potters (*Kúmhárs*) of the Kurnál district, chiefly in the village of Gumtallah. The process of the manufacture, which is similar to the Egyptian method, is as follows: From 15,000 to 20,000 bricks, made of the dirty clay to be found in certain ponds, are put all round the outside of a brick-kiln, which is then heated. When these bricks are half burnt, there exudes from them a substance of a greyish colour which resembles the bark of a tree. This substance is of two sorts: (*a*) an inferior kind, called the *mittíkhám* of *naushádar,* produced at the rate of 20 to 30 maunds for each kiln, and sells at 8 annas a maund; (*b*) the superior kind called *papri,* of which not more than 1 or 2 maunds is obtained from each kiln, and sells at R2 to 2¼ a maund. Merchants who deal in sal ammoniac buy both sorts. The *khám mittí* is passed through a sieve, and then dissolved in water and allowed to crystallize, the solution being repeated four times to clear away all impurities. The pure substance that remains is then boiled for nine hours, to allow the liquid to evaporate, and the resulting salt has the appearance of raw sugar. The *pápri* is next taken and pounded, after which it is mixed with the first preparation. The whole is put in a large pear-shaped vessel made of thin black-coloured glass, having a neck 2½ feet long and 9 inches round. The vessel is closed at the mouth, or, more properly speaking, the vessel has no mouth,—the composition being inserted by breaking a hole in the body of the vessel near the neck. This hole is eventually closed by placing a piece of glass over it. The vessel is then coated over with seven successive coatings of clay. It is placed in a large earthen pan filled with *nausádar* refuse to keep it firm. The neck of the vessel is further enveloped in a glass cover and plastered with fourteen different coatings of clay to exclude all air. When thus arranged it is placed over a furnace kept burning for three days and three nights, the cover being removed once every twelve hours to insert fresh *nausádar* so as to supply the place of what has been sublimed. After three days and three nights the vessel is taken off the furnace, and when cool the neck is broken off, and the rest of the vessel calcined. A substance called *phálí* is produced by the sublimation of the salt from the body of the vessel into the hollow neck. There are two kinds of *phálí;* the superior kind is that produced after the *nausádar* has been on the fire for only two days and two nights, in which case the neck is only partially filled with the substance and the yield is about 5 or 6 seers. This is sold at the rate of R16 a maund. The inferior kind is produced by the *nausádar* being kept on the fire three days and three nights; the neck of the vessel is completely filled with *phálí* when it yields 10 or 12 seers, and the salt is sold at R13 a maund.

"That portion of the sublimed *nausádar* which is formed in the mouth, and not in the neck of the vessel, is distinctively called *phúl*, and not *phálí*; it is used in the preparation of *surma*, and is esteemed of great value, selling at R40 a maund. The production of *nausádar* in brick-kilns is probably owing to the decomposition of watery vapours by the red-hot bricks in presence of the nitrogen of air and of common salt. The amount of sal ammoniac manufactured in the Kurnál district is estimated at 2,300 maunds, valued at R34,500. The merchants buy it on the spot from the manufacturers on an average at R8 a maund, who export it to Bhawani, Dehli, Farakabad, Mirzapur in the North-Western Provinces, and to Firozpur and Amritsar in the Panjáb, and who also sell it at R15 a maund. "It is also occasionally extracted from brick-kilns in other districts of the Panjáb than Kurnál, but in small quantities. It is found in Europe near burning beds of coal, in England and Scotland, and also near the volcanoes of Vesuvius, Etna, &c."

"It is used as a freezing mixture with nitre and water, and in the arts in tinning and soldering metals, and in the operation of forging the compound iron used for making gun-barrels by native smiths. (*Baden Powell, Panjáb Products, I., pp. 89, 90.*)

MEDICINE. 963

Medicine.—In medicine it is prescribed in inflammation of the liver and spleen. According to **Dr. Irvine**, it is not used internally in native practice. (*Mat. Med., Patna, p. 74.*)

§ "Dissolved in oil or yolk of eggs, it is used as a local application in cases of leucoderma." (*Assistant Surgeon Gholam Nabi.*) "Used largely in congested liver, in bronchitis, and in glandular enlargement as an external application." (*Assistant Surgeon Nehal Sing, Saharunpore.*) "Useful in guinea-worm, both internally and externally." (*Surgeon-Major G. Y. Hunter, Karachi.*) "Largely imported into Bombay from Europe." (*Surgeon-Major Dymock, Bombay.*) "An excellent remedy in affections of the bronchial tubes. Relieves hemicrania if given in 10 or 20 grain doses." (*Surgeon-Major W. Barren, Bhuj, Cutch, Bombay.*) "Useful in neuralgic headache in doses of 20 grains. I have used it largely in chronic diseases of the liver with benefit. As obtained in the bazar it is very impure and should be recrystallized." (*Assistant Surgeon Shib Chunder Bhuttacharji, Chanda, Central Provinces.*) "It is useful when ammonia is not procurable. This and quicklime can be procured in any bazar, and in cases of sudden fainting or hysterical fits mix the two in a phial; with gentle heat ammonia will be given off." (*Surgeon K. D. Ghose, Khulna.*) "Invaluable in neuralgia, 20-grain doses every 3 hours; relief after 3 doses or not at all; also in laryngeal cough or spray. In catarrh of Bhyondic and of the urinary tract, in whooping-cough, migraine." (*By a Surgeon who has not signed his contributions.*) "Mixed with *kalakootkee* (**Helleborus niger**) and softened with water, applied to the temple and forehead in the form of a paste in cases of hemicrania." (*Assistant Surgeon Anund Chunder Mukerji, Noakhally.*) "In 10-grain doses 3 or 4 times a day, I have found the medicine to be of the greatest use as an alterative in different affections of the liver." (*Assistant Surgeon Doyal Chunder Shome, Calcutta.*)

964

AMOMUM, *Linn.; Gen. Pl., III., 644.*

A genus of herbaceous plants belonging to the Natural Order SCITAMINEÆ, and the Tribe ZINGIBEREÆ, comprising some 50 species, chiefly inhabiting the tropical regions of Asia and Africa, a few extending to Australia and the Pacific Islands.

Root-stock horizontal, thick or elongated, rooting. *Leafy branches* ascending from the ground, destitute of flowers. *Leaves* lanceolate, spreading distichously, sessile sheathing. *Scape* short-oblong, crowded spike with the flowers expanding spirally (strobuliferous), or elongated, leafless, with a few scales, ascending in spring from the rhizome, very rarely terminating the leafy branches (compare with inflorescence in **Alpinia**); *bracts* imbricate, solitary or 2-3 flowered. *Calyx* tubular, dilated upwards (spathaceous) obliquely 3-fid, the posterior sepal very much larger than the others. *Corolla-tube* most frequently exceeding the calyx; limb 3-lobed, equal and prominent or the posterior very much larger, erect and hooded, the lateral ones long, narrow, spreading. *Stamens* theoretically 6, in two series; *the outer* petaloid, of which the two posterior are reduced to two small awn-shaped teeth, inserted upon the mouth of the corolla-tube, and the anterior one developed into a large labillum or spreading lip, entire and undulated on the margin or more or less trilobed, convolute at the sheathing base. *Filament* of the fertile stamen short, ascending from the mouth of the corolla, posterior and within the erect-hooded petal; *connective* more or less dilated, concave, upon which the two large diverging anthers are inserted, prolonged beyond into a small appendage which may be entire or variously cut or produced into a crest, entire or trilobed and often highly coloured. *Ovary* inferior 3-locular, many-ovuled; *style* thin, prolonged behind the anther-cells; *stigma* subglobose, fitting into the space formed through the anthers diverging upward. *Fruit* globose or oblong, embraced by the fleshy receptacle, pericarp fleshy rough or echinate, indehiscent or bursting irregularly or into 3 valves. *Seeds* many, globose or obovoid-truncate.

A genus closely allied to **Alpinia**, differing chiefly in the habit and inflorescence, and in the diverging anther-cells, and the prolonged or crested connective.

Amomum aromaticum, *Roxb.*; SCITAMINEÆ.

965

THE AROMATIC CARDAMOM PLANT.

Vern.—*Mórang-ilách*í, BENG. and HIND.; *Veldode*, MAR.

Habitat.—"A native of the villages on the eastern frontier of Bengal." (*Roxb.*)

MEDICINE. 966

Medicine.—The *Pharmacopœia of India* refers the Greater Cardamom to this plant, following apparently an error which exists in all the earlier works on Indian Economic Science. **Mr. Hanbury** made the same mistake. See **A. subulatum.**

"The fruit ripens in September; the capsules are then carefully gathered by the natives, and sold to the druggists, who dispose of them for medicinal and other purposes, where such spices are wanted, under the name of *mórang-ilách*í or *cardamom,* though the seed-vessels of this species differs in form from all hitherto-described sorts of this drug; however, the seeds are similar in their shape and spicy flavour." (*Roxburgh.*)

Apparently this fruit is not now used, or there was some mistake on the part of **Dr. Roxburgh** as to this being the Greater Cardamom of Bengal. He does not call it by the name of Greater Cardamom; but the plant which is sold and used at the present day as the Greater Cardamom of Bengal, and presumably the *mórang-ilách*í of **Roxburgh,** has been identified as the fruit of **A. subulatum.**

§ "Astringent and tonic, used as a tooth-powder; and said to be a good dentifrice." (*Surgeon John McConaghey, M.D., Shahjehanpore.*)

A. dealbatum, *Roxb.*

967

Habitat.—A native of Eastern Bengal and the adjoining frontier; a stately species, flowering in March and April, and ripening its insipid seed in September and October.

FOOD. 968

Food.—According to **Mr. Baden Powell** (*Pb. Prod.*, *380*), this is the species which yields the Cardamom, the *Ilách*í *bari* or *kalán* of the

Panjáb bazars. He says of it: "Said to be more powerful than the smaller kind, but to resemble it in other respects. An agreeable aromatic stimulant." It seems probable that this is a pure case of mistaken identity, and that the above quotation should be referred to **A. subulatum.**

969

A. masticatorium, *Thw.; En. Cy. Pl., 317.*

Habitat.—Common in the forests of the central provinces of Ceylon up to an elevation of 4,000 feet.

FOOD. 970

Food.—The Singhalese chew the rhizomes of this plant with their betel.

971

A. maximum, *Roxb.*

Habitat.—A native of Java. This was supposed by **Dr. Pareira** to be the Greater Cardamom of Bengal. **Dr. Roxburgh** says it was introduced into Bengal from the Malay Islands by the late **Colonel Kyd.**

FOOD. 972

Food.—The flowering time is the hot season, and the seeds ripen three or four months afterwards; they possess a warm pungent taste somewhat like that of Cardamoms, but by no means so grateful. (*Roxburgh.*)

973

A. Melegueta, *Roscoe.*

Habitat.—Cultivated to a small extent in Indian gardens. A native of, and widely distributed in, West Tropical Africa, extending from Sierra Léon to the Congo.

FOOD. 974

Food.—Grains of Paradise or Melegueta Pepper are the produce of this species. They are carminative, aromatic, and are used to flavour cordials, and to give false strength to beer and other liquors.

MEDICINE. 975

Medicine.—They are also used in cattle medicines. (*Smith; Pharmacographia; Bent. & Trim.*) About 1,000 cwt. are annually exported to Great Britain from the Gold Coast, which is chiefly consumed in the preparation of cattle medicine.

§ "Used commonly as carminative." (*Nehal Sing, Saharunpore.*)

976

Amomum subulatum, *Roxb.*

THE GREATER CARDAMOM.

Vern.—*Bari-iláchi*, HIND.; *Bara-eláchi*, BENG.; *Elachi, elcho, moto-iláchi*, GUJ.; *Bari-iláyechi*, DUK.; *Moté-veldode*, MAR.; *Periya-yélak-káy, káttu-yelak-káy*, TAM.; *Pedda-yéla-káyalu, adaviyéla-káya*, TEL.; *Doddá-yalakki*, KAN.; *Pérélam, periya-elattari*, MAL.; *Brihat-upakun-chiká, ela*, SANS.; *Qákilhahe-kibár, hél-zakar*, ARAB.; *Qákilahe-kalán, qáqilahe-zakar*, PERS.; *Ben, pala*, BURM.

The meaning of almost all the vernacular synonyms is, according to Moodeen Sheriff, the *Larger Cardamom.* The Greater Cardamom is most readily obtained in Calcutta, Hyderabad, Bombay, and other places under the Arabic name *Qákilahe-kibár;* in Madras it is to be had under the names *Jangli-ilichi*, DUK.; *Kattu-elakkáy*, TAM.; and *Adavi-éla-káya*, TEL.;—all signifying the wild Cardamom.

Habitat.—A native of Nepal.

The Greater Cardamom has a fruit about the size of a nutmeg, irregularly obcordate, flattened antero-posteriorly, having 15 to 20 irregular dentate-undulate wings, which extend from the apex downwards for two-thirds of the length of the Cardamom. **Dr. King**, in the *Linnæan Journal, Vol. XVII., p. 3* (reproduced in *Kew Report, 1877, p. 27*), clearly showed that the larger Cardamoms were the produce of this species and not of **A. aromaticum**, *Roxb.*, to which plant **Dr. Roxburgh** attributed them, but he presumes that it may be possible the latter plant was used in **Roxburgh's** time, though out of use now.

MEDICINE. Oil. 977 Seeds. 978 FOOD. 979

Medicine.—The seeds yield a medicinal oil. It is an agreeable, aromatic stimulant, pale yellow in colour, having the odour and flavour of the seeds. The seeds are aromatic and camphoraceous. Medicinal properties will be found under the true cardamom, *i.e.*, **Elettaria Cardamomum**, for which the greater cardamom is used as a cheap substitute.

Food.—The greater cardamoms are much used in the preparation of sweetmeats on account of their cheapness.

The **Opinions** of medical officers received, appear chiefly to refer to the true cardamom, although communicated under this species.

§ "Used as a carminative and stomachic." (*Assistant Surgeon Jaswant Rai, Mooltan.*) "It acts as a stomachic, and is said to allay irritability of the stomach produced either by cholera or some other affections. The decoction of cardamom is used as a gargle in affection of the teeth and gums. In combination with the seeds of melons it is used as a diuretic in cases of gravel of the kidneys." (*Assistant Surgeon Gholam Nabi.*) "Invaluable in certain disorders of the digestive system, marked by scanty and viscid secretion from the intestines, promotes elimination of bile, and is useful in congestion of the liver." (*Surgeon J. Maitland, M.B., Madras.*) "Very useful in liver affections, especially where abscess threatens; dose x grains. (*Surgeon-Major C. R. G. Parker, Pallaveram, Madras.*) "I have found it most useful in neuralgia in large doses, 30 grains, in conjunction with quinine. It is also a useful carminative in dyspepsia and diarrhœa." (*Surgeon-Major Henry David Cook, Calicut, Malabar.*) "Used in gonorrhœa as an aphrodisiac." (*Surgeon-Major J. J. L. Ratton, M.D., Salem.*)

Amomum xanthioides, *Wall.*

980

Vern.—§ "The seeds, *Ilâyechi-dâné*, HIND. and DUK.; *Elam*, TAM.; *Elakulu*, TEL.

Habitat.—§ "The seeds (not entire capsules) are imported from China and Singapore, and met with in every large bazar of South India.

Description.—"They are angular and very irregular seeds, generally inclining to be triangular, and sometimes compressed or flat; smaller in size than the common cardamom seeds, colour pale brown; odour strongly aromatic and agreeable, and taste aromatic and slightly pungent. Although the smell and taste of these seeds are stronger than those of the common or Malabar cardamom (**Elettaria Cardamomum**), yet they are more agreeable; and there is the same difference between the tinctures prepared from these drugs.

MEDICINE. 981

Medicine.—"The seeds are stimulant and carminative, and are useful in all the affections in which the common cardamoms are indicated. They are also of great service in relieving tormina and tenesmus, and even frequency of motions, in some cases of dysentery, and for this purpose they must always be used in powder with butter. They are administered in simple powder and compound tincture, the latter being prepared in the same way as the Tincture Cardamom Co. of the Pharmacopœia of India. Dose of the powder, from 20 to 40 grains, and of the tincture, from ʒi to ʒii." (*Moodeen Sheriff, Khan Bahudur, Madras.*)

AMOORA, *Roxb.; Gen. Pl., I., 334.*

982

A genus of trees belonging to the Natural Order MELIACEÆ, comprising some 15 species, inhabitants of the tropical and extra-tropical regions of Asia and Australia; 12 occurring in India, and 1 being endemic to Australia.

Leaves usually unequally pinnate; *leaflets* oblique, quite entire. *Inflorescence* subdiœcious, paniculate, female spicate or racemose. *Calyx* 3-5-partite or -fid. *Petals* 3-5, thick, concave, imbricated, rarely slightly combined at the base. *Staminal* tube sub-globose or campanulate, inconspicuously 6-10 crenate; *anthers* 6-10 included. *Disk* obsolete. *Ovary* sessile, short 3-5-celled; cells 1-2-ovuled. *Stigma* sessile or style elongated. *Capsule* sub-globose, coriaceous, 3-4-celled and seeded, loculicidally 3-5-valved. *Seeds* with a fleshy aril; *hilum* ventral.

The generic name is derived from the Bengali vernacular name *Amúr.*

983

Amoora cucullata, *Roxb.; Fl. Br. Ind., I., 560.*

Syn.—ANDERSONIA CUCULLATA, *Roxb.; Fl. Ind., Ed. C.B.C., 310.*

Vern.—*Amúr, latmí, natmi,* BENG.; *Thitnee,* BURM.

Habitat.—A moderate-sized, evergreen tree, met with on the coasts of Bengal and Burma: in Nepal and in the Andaman Islands.

Botanic Diagnosis.—A large tree of slow growth with cinereous bark; sub-glabrous, *leaflets* 3-13, opposite or sub-opposite, obliquely-oblong, obtuse at both ends, terminal one often hooded at the apex. *Flowers* panicled, not spicate; males drooping, about as long as the leaves, with numerous diverging branches, sparingly lepidote; female racemes few-flowered. *Petals* 3. *Anthers* 6-8. *Style* short, ovary 3-celled, cells 2-ovuled. *Fruit* sub-globose, 3-lobed, 3-celled, and 3-valved.

TIMBER. 984

Structure of the Wood.—Red, hard, close-grained, but apt to split. Weight 44 lbs. per cubic foot.

DOMESTIC. 985

Used for posts and other purposes in Lower Bengal, and for firewood in the Sundarbans.

986

A. decandra, *Hiern.; Fl. Br. Ind., I., 562.*

Vern.—*Tangarúk,* LEPCHA.

Habitat.—A large, spreading tree, found in the Eastern Himálaya, Nepal, Sikkim, from 2,000 to 4,000 feet.

Botanic Diagnosis.—*Leaflets* 7-13, opposite, oblong-acuminate, base somewhat cuneate or nearly rounded, sub-glabrescent and sub-membranous. Male panicles equalling the leaves. *Flowers* fragrant, on slender pedicels. *Sepals* very short. *Petals* 5. *Anthers* 10. *Ovary* 3-5-celled; cells 1-ovuled. *Fruit* globose, 5-furrowed, umbilicate, 5-celled and seeded.

TIMBER. 987

Structure of the Wood.—Pinkish white, hard.

988

A. Rohituka, *W. & A.; Fl. Br. Ind., I., 559.*

Syn.—ANDERSONIA ROHITUKA, *Roxb.; Fl. Ind., Ed. C.B.C., 311.*

Vern.—*Harin hara, harin khana,* HIND.; *Tikta-raj, pitraj,* BENG.; *Sikru,* KOL.; *Sohága,* OUDH; *Bandriphal,* NEP.; *Tangarúk,* LEPCHA; *Lota amari, amora amari,* ASS.; *Okhioungza, okhyang,* MAGH.; *Shem maram (the red-wood plant),* TAM.; *Chaw-a-manu, rohitakah,* TEL.; *Shem-maram,* MAL.; *Rohituka,* SANS.; *Hingal gass,* SINGH.; *Thitni, chayan-ka-you,* BURM.

Habitat.—An evergreen tree with large crown of branches, met with in Oudh, Assam, Sylhet and Cachar, Northern and Eastern Bengal, Western Ghâts, and Burma, the Andaman Islands, and Malacca.

Botanic Diagnosis.—*Leaves* 1-3 feet long, leaflets 9-15, in size 3 to 9 by $1\frac{1}{3}$ to 4 inches; young parts tawny, closely pubescent, early glabrescent. *Flowers* white, bracteate, sub-sessile; male spikes panicled, female simple. *Calyx* 5-partite, petals 3. *Anthers* 6 *Ovary* 3-celled with 2 superposed ovules in each cell.

OIL. 989

Oil.—In Bengal an oil is expressed from the seeds. The natives, where the tree grows plentifully, extract this oil, which they use for various economic purposes. (*Roxburgh.*)

MEDICINE. 990

Medicine.—The bark is used as an astringent.

§ "The ripe seeds yield an oil which is burnt by the poorer classes and is used as a stimulating liniment in rheumatism. The seeds are fried and bruised, then boiled with water, when the oil floats on the top. (*Surgeon D. Basu, Faridpur, Bengal.*)

Structure of the Wood.—Reddish, close and even-grained, hard. The concentric bands in this species are remarkable, since they are absent from the two other species. Average weight 40·5 lbs. per cubic foot.

TIMBER. 991

The timber is of good quality, but is little used. In Chittagong canoes are sometimes made of it.

Amoora spectabilis, *Miq.; Fl. Br. Ind., I., 561.* 992

Vern.—*Amari,* Ass.

Habitat.—An evergreen tree, found in the eastern moist zones of Assam and Burma.

Botanic Diagnosis.—*Leaflets* 11-13, opposite or sub-opposite, oblong, acutely acuminate base obtuse, glabrescent, shining, glaucescent beneath, petiolate. *Male panicles* peduncculate, with alternate unequal branches. *Calyx* stellately puberulent, obtusely 3-lobed, short. *Petals* 3, imbricate, sub-stellate, velutinous along the back. *Staminal tube* urceolate, glabrous, shortly and obtusely 8-dentate. *Anthers* 8. *Fruit* obovoid-pyriform, 1¼ to 1⅜ by 1 to 1⅛ inch. Some doubt exists in botanical works regarding this species.

Structure of the Wood.—Red, hard, close-grained, durable, and takes a good polish. Weight 48 lbs. per cubic foot.

TIMBER. 993

Used for boat-building and furniture in Assam.

AMORPHOPHALLUS, *Blume; Gen. Pl., III., 970; Monogr. Phanerog., DC., II., 308.* 994

A genus of tuberous-rooted herbs belonging to the Natural Order AROIDEÆ, Tribe PYTHNONIEÆ. There are in all some 25 species, inhabitants of tropical Asia and Africa, of which 7 are met with in India and Ceylon.

Leaves generally solitary, ascending from the flattened corm after the spathe has faded; *petiole* erect, variously spotted; blade large, primarily 3-sected and bulbiferous; segments pinnatifid or bipinnatifid or dichotomous, ultimate divisions oblong-acute. *Spathe* broad-ovate, base infundibuliform or campanulato-convolute, spreading above and exposing the spadix. *Spadix* erect, fleshy, as long as the spathe, appendix dilated fungus-like. *Male flowers* crowded, forming a fusiform section placed immediately above the *female* cylindrical section, *neuter* flowers none. *Perianth* none. *Female flowers,* ovary globose, 1-4 celled; *style* short or elongated; *stigma* entire, 2-4-lobed; *ovules* solitary in the cells, anatropous or half anatropous, decurved, funiculus short or sub-elongated, placenta basilar, micropyle inferior.

The word Amorphophallus is derived from ἀμορφος and φαλλός in allusion to the shapeless form of the plant, or rather to the barren appendix of the spadix, which is not only devoid of flowers, but assumes an irregularly crumpled form.

Amorphophallus bulbifer, *Blume;* AROIDEÆ. 995

Syn.—ARUM BULBIFERUM, *Roxb.; Fl. Ind., Ed. C.B.C., 310.*

Vern.—*Umla bela,* BENG.

Habitat.—A native of Bengal, plentiful in the neighbourhood of Calcutta, where it blossoms in May, the leaves appearing in the rainy season.

A. campanulatus, *Blume; Wight, Ic., t. 785.* 996

Syn.—ARUM CAMPANULATUM, *Roxb.*

Vern.—The tuber, *Zamin-kand,* PERS. and HIND.; *Ol,* BENG.; *Kanda, arsaghna,* SANS.; *Jangli-súran,* BOMB., CUTCH; *Súran,* MAR.; *Karu-*

naik-kizhangu or *karuna-kalang, nalle-karuna-karang* (a variety), TAM.; *Kanda-godda, kanda, poti-kanda, durada-kanda-godda, manchik-anda* or *ghensi-kanda* (variety), TEL.; *Karuna-kizhanna* or *karuna-karang*, MAL.; *Kandá*, DUK., HIND.; *Wa*, BURM.

It seems probable that one of the forms of this plant affords the *madan-mast* of Bombay druggists, described by **Dr. Dymock** (*Mat. Med., W. Ind., 664*) under the name of **Amorphophallus sylvaticus.** A note which **Dr. Dymock** has kindly supplied would seem to justify this inference, while, on the other hand, it is possible that the tubers of **Arum sylvaticum,** *Roxb.* (now known as **Synantherias silvatica,** *Schott.*,—see *Engler, Mono. Phaner. DC., 320*) affords the drug referred to by **Dr. Dymock,** the more so since that species is a native of Bombay. The name *madan-mast* appears, however, to be also given to **Amorphophallus campanulatus,** *Blume.*, "and in Madras to **Artabotrys odoratissima,** *R. Br.*" (*Moodeen Sheriff.*)

Habitat.—A native of India and Ceylon; cultivated throughout the peninsula, in rich moist soils.

MEDICINE. Corms. 997

Medicine.—The corm (or tuber) and the seeds are used as irritants and relieve the pain of rheumatic swellings when applied externally. It is considered a hot carminative in the form of a pickle.

Mr. Baden Powell says: "The roots contain a large quantity of farinaceous matter, mixed with acrid poisonous juice, which may be extracted by washing or heat. When fresh it acts as an acrid stimulant and expectorant, and is used in acute rheumatism." **U. C. Dutt** says: "The tubers contain an acrid juice, which should be got rid of by thorough boiling and washing, otherwise the vegetable is apt to cause troublesome irritation in the mouth and fauces. Medicinally *súrana* is considered serviceable in hœmorrhoids; in fact, one of its Sanskrit synonyms is *arsoghna*, or the curer of piles. It is administered in this disease in a variety of forms. The tuber is covered with a layer of earth and roasted in a fire; the roasted vegetable is given with the addition of oil and salt."

§ "The dried corm sliced is sold in Bombay, under the name of *madan-mast*, as a restorative, tonic, and carminative." (*Surgeon-Major W. Dymock, Bombay.*) "The tubers, first boiled with tamarind leaves and paddy husk, and then made into a curry with the usual condiments, is efficacious in bleeding piles. It produces intense itching of the tongue when tasted, and it is to remove this irritating quality that tamarind is largely used when cooking it." (*Honorary Surgeon P. Kinsley, Ganjam, Madras.*) "The cultivated or pinkish white variety is used as food. The tuber is cut into small pieces, boiled in water to get rid of the irritation, and then used as a '*bhurta*,' or in curry, or is fried with cocoanut pulp. Medicinally I have seen its benefit in bleeding piles. It should be used in the form of '*bhurta*.'" (*Surgeon-Major R. L. Dutt, M.D., Pubna, Bengal.*) "Recommended by native physicians in piles. I tried it in various forms without success." (*Assistant Surgeon Shib Chunder Bhuttacharji, Chanda, Central Provinces.*) "It is used externally in the form of poultice in the bites of insects, scorpions, &c. Internally in the form of a pickle, it is used as a laxative in hœmorrhoids." (*Brigade Surgeon J. H. Thornton, B.A., M.B., Monghyr.*) "Used as a stimulating poultice." (*Surgeon W. Barren, Bhuj, Cutch, Bombay.*) "Used in boils and ophthalmia." (*Surgeon H. W. Hill, Mánbhúm, Bengal.*)

Speaking of the *jangli súran*, **Dr. Dymock** says the tuber is peeled and cut into segments, and in that condition is sold in the Bombay native druggists' shops as *madan-mast*. "The segments are usually threaded upon a string, and are about as large as those of an orange, of a reddish-brown colour, shrunken and wrinkled, brittle and hard in dry weather; the surface is mammilated. When soaked in water they swell up and become very soft and friable, developing a sickly smell." "*Madan-mast* has a mucilaginous taste, and is faintly bitter and acrid; it is supposed to have restorative powers, and is in much request." The above extract may probably be describing the properties of a plant quite distinct from **Amorphophallus campanulatus** (see remarks under vernacular names and also under **Synantherias silvatica,** *Schott.*).

FOOD. Corms. 998

Food.—The corms or solid bulbs are considered nutritious and wholesome when cooked, and are accordingly in common use as an article of food. They are boiled like potatoes and eaten with mustard; they are cooked in curries; they are cut into slices, boiled with tamarind leaves, and made into pickles; and they are also cooked in syrup and made into preserves.

The larger corms have small lateral tuberosities; these are separated and form cuttings for propagation. They are planted immediately after the first rains (say May and June) in loose, rich soil, repeatedly ploughed. In twelve months they are fit to be taken up for use. If cultivated under favourable circumstances, each corm will weigh from 4 to 8 lbs., and they may be preserved for some time if kept dry. The average outturn is about 200 to 400 maunds per acre, and the price is about R2½ a maund.

§ "When cultivated the tuber becomes large and loses much of its irritant properties, and when boiled or otherwise cooked makes a substantial starchy vegetable. It is sold largely in the Calcutta bazars." (*Surgeon K. D. Ghose, Khulna.*) "Used as food, possessing most of the properties of **Alocasia indica.** If not properly cultivated in loose soil it becomes irritant in its action." (*Surgeon D. Basu, Faridpur, Bengal.*) "When cooked the tubers are wholesome and nutritious." (*Deputy Surgeon-General G. Bidie, Madras.*)

Amorphophallus dubuis, *Blume.*

999

A native of Ceylon and the Malabar Coast of India.

A. giganteus, *Blume.*

1000

Syn. for **Conophallus giganteus,** *Schott.*

A native of Malabar, Ceylon, Java, &c.

A. lyratus (Arum lyratum, *Roxb.*)

1001

Imperfectly known; said to be a native of the Circars, Madras.

A. margaritifer, *Kunth,* and Arum margaritiferum. *Roxb.* (*Dymock, Mat. Med., Western India, 664*); Syn. for **Plesmonium margariti ferum,** *Schott.*, which see.

A. tuberculiger, *Schot.*

1002

A native of the Khásia Hills and of Sikkim.

A. zeylanicus, *Blume.*

1003

A native of Ceylon and Java.

AMPELIDEÆ.

1004

"Small trees or shrubs, usually climbing by means of tendrils, more rarely radicant (sometimes herbaceous in *Leea*); juice copious, watery. *Stems* angled, compressed or cylindric, with numerous very large proper vessels. *Leaves* alternate, usually petioled, simple or digitately or pedately 3-9 foliolate, rarely pinnate or decompound. *Flowers* umbellately-paniculately- or spicately-cymose. *Peduncles* often transformed into simple or compound tendrils or adhering to rocks or trees by viscid pads terminating the ultimate segments, or expanded into a broad floriferous membrane (**Pterisanthes**). *Flowers* regular, hermaphrodite, rarely unisexual. *Calyx* small, entire or 4-5-toothed or -lobed. *Petals* 4-5, distinct, or cohering, valvate, caducous. *Stamens* 4-5, opposite the petals, inserted at the base of the disk or between its lobes, filaments short-subulate; anthers free or connate, 2-celled, introrse. *Disk* free or connate with the petals, stamens or ovary, annular or variously expanded. *Ovary* 2-6-celled; *style* short, slender, conical or 0; stigma minute or large and flat, sublobed; ovules 1-2 in each cell, ascending, anatropal, raphe ventral. *Berry* 1-6-celled; cells 1-2-seeded. *Seed* erect, often rugulose, albumen cartilaginous; *embryo* short basal, *cotyledons* ovate. *Distrib.*—Species about 250, inhabiting the tropical and temperate regions of the whole world.

Scandent shrubs, usually bearing tendrils. Flowers racemose or cymose. Ovary 2-celled, cells 2-ovuled . . . . 1. **Vitis.**

Flowers sessile on the dilated membranous peduncle . 2. **Pterisanthes.**

Erect shrubs destitute of tendrils, petals and stamens connate with the disk. Ovary 3-6-celled, cells 1-ovuled . . . . 3. **Leea.**"

(*Flora of British India, I., 645.*)

The above extract has been published here with the object of suggesting the names of the genera of this family; the economic information will be found under these in their respective alphabetical positions.

Distribution of the Ampelideæ.—There are in all some 250 species belonging to this order, chiefly met with in the tropics, extending to the temperate regions. They are rare in America, and exceedingly rare in the Pacific Islands; none are indigenous to Europe. The vine-grape appears to have been originally a native of Georgia and Mingrelia, but it is now cultivated in all countries with a mean summer temperature not below 66° Fh. Where the temperature falls below 66° Fh. the grapes never become sweet; where it is much above that temperature they do not mature, although the plant may flourish, as in Indian gardens.

In India there are in all 94 species, grouped in three genera. Of these 52 or 55·3 per cent. are confined to the plains, 34 or 35·1 per cent. are found up to an altitude of 5,000 feet, and 8 or 8·5 per cent. up to an altitude of 10,000 feet. Geographically, 50 or 53·1 per cent. are confined to East India, 10 or 10·6 per cent. to West India, 8 or 8·5 per cent. to South India, and 3 or 3·2 per cent. to North India, all three being in the Upper Gangetic Sub-Division. Of the remaining 23 species, 8 or 8·5 per cent. are found in two or more regions not including North India, and 15 or 15·9 are found in North India as well as in one or more of the other divisions. All these 15 species are found in the Upper Gangetic Sub-Division; 5 of them are also found in the West Panjáb, and 2 in the dry tracts of the Panjáb and Sind Sub-Division of North India.

1005

Amygdalus communis, *Linn.*, see **Prunus Amygdalus,** *Baill.*, ROSACEÆ.

Amyris commiphora, *Roxb.*, see **Balsamodendron Roxburghii,** *Arn.*, BURSERACEÆ.

ANABASIS, *Linn.; Gen. Pl., III., 72.*

Anabasis multiflora, *Moq.*; CHENOPODIACEÆ; *DC. Prod., XIII., 2, 212.*

Vern.—*Ghalme, lána, metra láne, gora lane, dána, shor lana, búi, choti,* PB.

Habitat.—Met with in the Panjáb, a short distance east of the Sutlej.

MEDICINE. 1006

Medicine.—Mr. Baden Powell mentions this plant amongst his drugs, but says nothing of its medicinal property.

FODDER. 1007

Fodder.—Camels are fond of the plant.

1008

ANACARDIACEÆ.

"Trees or shrubs; juice often milky and acrid. *Leaves* alternate, opposite in **Bouea**, exstipulate, simple or compound. *Inflorescence* various; flowers small, regular, unisexual, polygamous, or bisexual. *Calyx* 3-5-partite, sometimes accrescent, spathaceous in **Gluta**. *Petals* 3-5, alternate with the sepals, free, rarely O, imbricate or valvate in bud, sometimes accrescent. *Disk* flat, cup-shaped or annular, entire or lobed, rarely obsolete. *Stamens* as many as the petals, rarely more, inserted under, rarely on, the disk; filaments usually subulate; anthers 2-celled, basi-or dorsi-fixed. *Ovary* superior, half inferior in **Holigarna**, 1- or 2-6-celled, rudimentary or 2-3-fid in the ♂; of 5-6 free carpels in **Buchanania**; styles 1-4 or stigma sub-sessile; ovules solitary in the cells, pendulous from the top or wall or from an ascending basal funicle. *Fruit* usually a 1-5-celled 1-5-seeded drupe; stone sometimes dehiscent. *Seed* exalbuminous; embryo straight or curved, cotyledons plano-convex, radicle short. *Distrib.*—Chiefly tropical; genera about 45; speciesabout 450.

"**Sorindeia madagascariensis**, *DC.* (*Wall. Cat., 8491*), is cultivated in gardens in India.

"Tribe I. Anacardieæ. *Ovary* 1-celled, or if 2-celled, with one cell early suppressed.

(A.) OVULES PENDULOUS FROM A BASAL FANICLE.

* *Sepals and petals not accrescent.*

Calyx 4-5-partite. Petals 4-6. Stamens 4-10. Leaves alternate, usually compound . . 1. **Rhus.**

Calyx 5-partite. Petals O. Stamens 3-4. Leaves alternate, compound . . . 2. **Pistacia.**

Calyx 4-5 partite. Petals 4-5. Stamens 1-5. Style filiform. Leaves alternate, simple . 3. **Mangifera.**

Calyx 5-partite. Petals 5. Stamens 8-10, all or a few only perfect. Torus stipulate. Style filiform. Leaves alternate, simple . . *3. **Anacardium.**

Calyx 3-5 partite, valvate. Petals 3-5. Stamens 3-5, all perfect. Style short. Leaves opposite, simple . . . . . 4. **Bouea.**

Calyx spathaceous. Petals 4-6. Stamens 4-6. Torus stipulate. Style filiform. Leaves alternate, simple . . . . . 5. **Gluta.**

Calyx 3-5 lobed. Petals 3-5. Stamens 10. Carpels 5-6, one only perfect. Styles short. Leaves alternate, simple . . . . 6. **Buchanania.**

** *Sepals or petals accrescent. Leaves simple.*

Calyx spathaceous. Stamens 5 or numerous . 7. **Melanorrhœa.**

Calyx 5-partite. Stamens 5 . . . . . 8. **Swintonia.**

(B.) OVULES PENDULOUS FROM THE TOP OF THE CELL OR FROM THE WALLS OF THE OVARY ABOVE THE MIDDLE.

* *Leaves 3-foliolate or pinnate.*

Calyx not accrescent. Petals valvate. Stamens 10. Style 1 . . . . 9. **Solenocarpus.**

Calyx not accrescent. Petals imbricate. Stamens 10. Style 1 . . . 10. **Tapiria.**
Calyx not accrescent. Petals imbricate. Stamens 5, with 5 staminodes. Style very short 11. **Pentaspadon.**
Calyx not accrescent. Petals imbricate. Stamens 8-10. Styles 3-4 . . . . 12. **Odina.**
Calyx accrescent. Petals 4. Stamens 4. Style 3-fid . . . . 13. **Parishia.**

** *Leaves simple.*

Petals imbricate. Stamens 5. Styles 3. Drupe on a much-enlarged peduncle . . . . 14. **Semecarpus.**
Petals imbricate. Stamens 5. Style 1. Drupe superior . . . . 15. **Drimycarpus.**
Petals valvate. Stamens 5. Styles 3. Drupe inferior . . . . 16. **Holigarna.**
Petals valvate. Stamens 5. Style 1. Drupe superior . . . . 17. **Melanochyla.**
Petals imbricate. Stamens 4. Style 1, short. Drupe superior . . . . . . 18. **Nothopegia.**
Petals imbricate. Stamens 6-10. Style 1. Drupe superior . . . . 19. **Campnosperma.**

Tribe II. Spondieæ. *Ovary* 2-5-celled; ovules pendulous. Leaves pinnate.

Flowers polygamous. Stamens 8-10. Styles 4-5, free above . . . . . . 20. **Spondias.**
Flowers bisexual. Stamens 10. Style 5, thick, connate at the lips . . . . . . 21. **Dracontomelum.**

Doubtful genus.

Calyx 3-fid. Stamens 3. Ovary 3-celled. Leaves entire . . . . 22. **? Rumphia.**"

(*Fl. Br. Ind., Vol. II., pp. 7-8.*)

The above analysis of the genera of Anacardiaceæ will be found useful, in enabling the reader to recognise the plants of economic interest which belong to this family; for fuller details consult their respective alphabetical positions in this work.

1009

Distribution of the Anacardiaceæ.—There are in all some 450 species belonging to the Anacardiaceæ as defined by the *Genera Plantarum.* They are chiefly inhabitants of the tropical regions of the Old World, but are fairly represented in tropical America, and less frequent in Australia. Only a few species (but these abundant in individuals) reach South Europe, South Africa, or North America. There are in India 116 species, referred to 23 genera. Of these 83 or 71·5 per cent. are peculiar to the plains, 28 or 24·8 per cent. ascend to 5,000 feet in altitude, and 5 or 4·3 per cent. reach higher altitudes. In their distribution over the peninsula of India they show a corresponding preference for the moist and extra-tropical regions. Sixty-eight species or 58·6 per cent. are peculiar to the eastern division of India, 17 or 14·6 per cent. to South India and Ceylon; 7 are peculiar to North India (3 of these in the Upper Gangetic basin, 1 in the South-Eastern Panjáb and 3 diffused over both these sub-divisions of North India, but none of the endemic North Indian species seem to pass into the drier and desert tracts of the Eastern Panjáb and Sind). The remaining 12 Indian species are less local, being diffused through at least two or more of the divisions of India, 11 of them passing to Upper India, of which 3 occur in the East Panjáb and Sind, and one species follows the coast of India, appearing to require the sea atmosphere.

1010

Affinities of the Anacardiaceæ.—They are placed in most works on systematic botany before the Leguminosæ and after Rutaceæ, Zygophylleæ, Simarubeæ, Burseraceæ; Rhamneæ and Sapindaceæ, to which they bear their closest relations. Through Leguminosæ they have many features of resemblance to the Amygdaleæ in Rosaceæ, especially in habit, woody stems, alternate leaves, perigynous stamens, and polypetalous corolla, the solitary carpel, drupaceous fruit and exalbuminous seed. They have a strong affinity to the Juglandeæ; indeed, certain authors (**Kunth, Endlicher,** &c.) have combined the latter with the Anacardiaceæ (or Terebinthaceæ). **De Candolle,** while retaining Burseraceæ, excludes Juglandeæ from them. They are also closely related to Connaraceæ and Burseraceæ, the latter, by **Baillon** and other authors, being viewed as a tribe; they differ chiefly in the two-ovuled condition of Burseraceæ, the ovule having also a superior micropyle. **Baillon,** in addition to the above tribes, places the Mappieæ and Phytocreneæ as tribes under this family. Most modern authors, including **Sir J. D. Hooker,** exclude Burseraceæ, Sabiaceæ, and Juglandeæ, these forming respectively independent natural orders, while Anacardiaceæ has been restricted to the tribes Anacardieæ and Spondieæ.

1011

Properties and Uses of the Anacardiaceæ.—They yield food, medicine, oil, gum, and resin, turpentine, varnish, dye, tan, and useful woods. The Pistachio nut, the Mango, the Cashew-nut, the Spanish Plum (Spondias), and the nut of **Semecarpus Anacardium** are regularly eaten and prized as amongst the best of Indian fruits. The barks, leaves, young fruits, seeds, and oils obtained from these plants, as also many others, are regarded as possessing remedial properties. The resin is often very valuable. **Pistacia Lentiscus** yields the resin *mastic,* much used in the East to perfume the breath, strengthen the gums, as also to flavour wines and confectionery. In England it is used for varnishing pictures and in dentistry. **P. Terebinthus,** a Mediterranean tree, yields Cyprus turpentine. **Melanorrhœa usitatissima** yields the celebrated black varnish of Burma. **Rhus succedanea,** Japanese vegetable wax; **R. Vernix,** Japanese varnish; the Indian **Holigarna longifolia** also yields a good varnish. A large number of Indian species yield gum at certain seasons of the year: **Odina Wodier** is simply covered with its brown gum streaking down the stem and ultimately becoming black.

1012

Rhus coriaria and **R. cotinus,** the Sumach, are much-prized tans, the wood of the last species yielding a good orange dye. In Europe a tincture is chiefly used for this purpose; with Cochineal or Prussian blue it gives chamois or green tones. The juice of the pericarp of **Semecarpus** gives an indelible black ink, used for marking linen.

1013

ANACARDIUM, *Rottb.; Gen. Pl., III., 420.*

A genus of shrubs or trees belonging to the Natural Order ANACARDIACEÆ, comprising 6 species, natives of America, one of which has been naturalised in India.

Leaves alternate, simple, quite entire. *Panicles* terminal, bracteate. *Flowers* small, polygamous; *sepals* and *petals* not accrescent. *Calyx* 5-partite; *sepals* erect, deciduous; *disk* erect, filling the base of the calyx. *Petals* 5-linear-lanceolate recurved imbricate. *Stamens* 8-10, all or only a few fertile; *filaments* connate and adnate to the disk. *Ovary* obovoid or obcordate; *style* filiform, excentric *stigma* minute; ovule 1, ascending from a lateral funicle. *Nut* kidney-shaped, seated on a large pyriform fleshy body formed of the enlarged disk and top of peduncle; pericarp cellular and full of oil. *Seed*

kidney-shaped, ascending; *testa* membranous, adherent; *cotyledons* semi-lunar; *radicle* short-hooked.

The generic name is derived from **ἀνά**, resemblance, and **καρδία**, heart, in allusion to the form of the nut.

1014

Anacardium occidentale, *Linn.; Fl. Br. Ind., II., 20.*

THE CASHEW-NUT.

Vern.—*Kájú*, HIND., GUJ., DEC.; *Hijlí-bádám, káju*, BENG.; *Kájú, káju-kalíyá*, BOMB.; *Kájúcha-bi, kájú*, MAR.; *Mundiri, kottai, kottai-mundiri*, TAM.; *Jídi-mámidi vittu, muntamámidi-vittu, jídi-ánti, jiedí pundú* (fruit), TEL.; *Jídi váte, kempu géru bíja, gera-poppu geru váte, gerabíja*, KAN.; *Paranki-máva kuru, kappal-chérun-kuru, kappa-mávakuru*, MALA.; *Káju* or *kaju-atta*, SINGH.; *Thee-noh, thayet, thee-hot, síhosayesi* or *tihotiya-si*, BURM.

Habitat.—A tree, 30 to 40 feet; originally introduced from South America, now established in the coast forests of India, Chittagong, Tenasserim, and the Andaman Islands, and over South India.

"The local name *káju* appears to be restricted to the Konkan. The tree is indigenous to the West Indies. It is probable that the Portuguese on its introduction to the west coast of India called it *káju* as a rendering of the Brazilian name *acajau*. The French, by a similar transliteration, called it *Cashew*." (*Bomb. Gaz., X., 38.*)

Properties and Uses—

GUM. 1015

Gum.—Rai Kanai Lal De, Bahadur, in his *Indigenous Drugs of India*, mentions that the bark of this plant yields a gum.

§ "This gum occurs in large stalactitic pieces; it is yellow or reddish, and only slightly soluble in water. It is obnoxious to insects." (*Surgeon-Major Dymock, Bombay.*)

Sap. 1016

The sap.—The juice issuing from incisions in the bark is in demand as an indelible marking ink. (*Br. Burm. Gaz., I., 136.*) The astringent juice is used by native workmen as a flux for soldering metals. (*Bomb. Gaz., X., p. 38.*)

DYE. 1017

Dye.—The bark may be used for tanning. The pericarp gives an oil, called *Cardol*, which is very astringent, and is used by the Andamanese to tan or colour fishing-nets, so as to preserve them. **Dr. Dymock** informs me that this oil is called *Dík* in Goa, where it is much used as a tar for boats and nets.

OIL from the nut. 1018

Oil.—From this plant two distinct oils are obtained:—

1st.—The kernels, when pressed, yield a light-yellow bland oil, very nutritious; the finest quality in every respect equal to almond oil, and considered superior to olive oil. The yield is about 40 per cent. The kernels are so extensively eaten in India, however, that it is almost impossible that a trade could at present be done in this oil. Samples of this fixed oil, and information as to methods of preparation and extent of trade, are much required. The kernels have been once or twice exported to Europe under the name of "Cassia Nuts."

Cardole or oil from the shell. 1019

2nd.—Cardole or *Cashew-apple-oil.*—This is prepared from the pericarp or shell of the nut. It is black, acrid, and powerfully vesicating. In the Andamans, it is used to colour and preserve fishing lines. It is an effective preventive against white-ants in carved wood-work, books, &c. The yield is 29½ per cent. (The *British Burma Gazetteer, I., 131*, says: "The pericarp of the nuts produces a black acrid oil, *Cardole*.")

MEDICINE Acrid oil. 1020

Medicine.—The medicinal uses of this plant are many. The acrid oil is used as an anæsthetic in leprosy, and as a blister in warts, corns, and ulcers. Between the laminæ of the shell of the kernel there is a black caustic fluid, which contains an acrid, oily principle, *Cardol*, and a peculiar acid, *Anacardic Acid.* It possesses powerful rubefacient and vesicant properties. The spirit distilled from the expressed juice of the fruit may be used as a stimulant.

Spirit. 1021

§ "Fruit eaten as a remedy for scurvy. The juice of the nut is used as a substitute for iodine, locally." (*Surgeon W. Barren, Bhuj, Cutch, Bombay.*) "The oil I have used with benefit in the anæsthetic variety of leprosy." (*Assistant Surgeon Bolly Chand Sen, Calcutta.*) "The oil obtained from the shell by maceration in spirit is the very best application for cracks of the feet, so common with natives." (*Brigade Surgeon C. Joynt, M.D., Poona.*) "It is locally applied to the sole of the foot as a remedy for cracking of the cuticle." (*Surgeon-Major Henry David Cook, Calicut, Malabar.*) "The oil is efficacious when faintly brushed as a local stimulant in psoriasis." (*Assistant Surgeon Devendro Nath Roy, Calcutta.*)

Food.—Produces a small fruit, within which is the nut known as the Cashew nut, commonly eaten roasted,—a process which improves the flavour. § "The ripe fleshy stalk or torus of this plant is eaten as a fruit. In Purí they used to call it *lanká ám*. The kernels are fried and eaten; they are also made into confectionery with sugar. They are sold in the markets in Purí under the name of *Hidjlí badam.*" (*U. C. Dutt, Serampore.*) "The seeds deprived of their shell are eaten." (*Deputy Surgeon-General G. Bidie, Madras.*)

FOOD. Nuts. 1022 Receplace. 1023

Structure of the Wood.—Red, moderately hard, close-grained. Weight 38 lbs. per cubic foot.

Used in Burma for packing-cases, for boat-building, and charcoal.

TIMBER. 1024

ANACHARIS, *L. C. Rich.; Gen. Pl., III., 450.*

1025

Anacharis or American duckweed. A delicate, much-branched, aquatic plant, belonging to the Natural Order HYDROCHARIDEÆ. By the *Genera Plantarum* this genus has been reduced to **Elodea,** *Mich.*

This curious plant made its appearance simultaneously in various parts of Great Britain, about 30 years ago, and its spread from lake to lake has attracted much attention. How it was introduced is unknown, and it is equally difficult to know how it is propagated with such rapidity; for the plant is diœcious, and only female flowers have been discovered in Great Britain; it cannot therefore produce fertile seeds. The date of its introduction into India seems equally difficult to determine; but at the present moment most of the tanks and lakes of the plains have become almost impossible of navigation from the immense masses of this plant which choke up every piece of water under 8 to 10 feet in depth. It affords rare feeding-ground for aquatic birds, both from the tender leaves, which are greedily eaten, and from the multitude of insects and snails which live amongst the portions of the plant which reach above the surface of the water. I have failed to discover any vernacular name for it, other than the generic appellation to all aquatic weeds. It is sometimes used along with **Vallisneria** in the native process of refining sugar.

ANACYCLUS, *Linn.; Gen. Pl., II., 419.*

1026

Anacyclus Pyrethrum, *DC.*; COMPOSITÆ; *DC. Prod., VI., 15.*

THE PELLITORY OF SPAIN.

Vern.—*Akarkará,* HIND., BENG., BOMB.; *Akkalkará, akkirakáram,* TAM. and TEL.; *Akkalkádhá,* MAR.; *Akkala-karé,* KAN.; *Akorkaro,* GUJ.; *Agal-górú,* DUK.; *Akki-karuká, akkilá-káram,* MUL.; *Aáqarqarhá* or *áquarqarhá, aúdul-qarhá, údal-qarha,* ARAB.; *Akara karava, akarákarabha,* SANS.

Habitat.—Indigenous to North Africa, whence it has been introduced into South Europe. "The root is collected chiefly in Algeria and is exported from Oran, and to a smaller extent from Algiers." A large amount is also shipped from Tunis to Leghorn and Egypt. (*Dymock.*)

MEDICINE. Root. 1027

Medicine.—THE ROOT of this plant has stimulant properties, and when locally applied, acts as an irritant and rubefacient. It is also used as a sialagogue. In India it is often given to parrots, with the idea of helping to make them talk. It is imported into India, chiefly from Algeria. **Ainslie** (*Mat. Ind., I., 300*) gives a long account of this medicine. He informs us that vegetarians prescribe an infusion of it, in conjunction with the lesser galangal and ginger, as a cordial and stimulant in lethargic cases, in palsy, and in certain stages of typhus fever, and that they also order it to be chewed as a masticatory for toothache. It certainly possesses powerful stimulant properties, but is scarcely ever employed in Europe as an internal remedy; though it has been found useful as a sialagogue, and as such, **Dr. Thomson** says, has been given with success in some kinds of headache, apoplexy, chronic ophthalmia, and rheumatic affections of the face.

Infusion. 1028

Special Opinions.—§ "The root is used by natives as a nervine tonic in cases of facial palsy, paralysis, hemiplegia, epilepsy, and cholera. It is also employed in rheumatism, sciatica, and dropsy. As a sialagogue it is used to allay toothache. Aphrodisiac, emmenagogue, and diuretic properties have also been assigned to it. Its local application to the forehead is said to remove headache. A gargle is reputed to be very efficacious in affections of the teeth, throat, and tonsils. In a drachm-and-a-half doses it is said to act as a purgative. Dissolved in olive oil and rubbed over the skin, it is reported to produce profuse perspiration, and thus to cut short an attack of fever. As an expectorant it has been employed with benefit in cases of chronic bronchitis." (*Assistant Surgeon Gholam Nabi.*) "It is used in toothache, in which it sometimes gives instantaneous relief." (*Surgeon J. C. Penny, M.D., Amritsar.*) "It is expectorant." (*Surgeon-Major J. T. Fitzpatrick, M.D., Coimbatore.*) "Decoction used in bronchitis as an expectorant." (*Surgeon-Major J. J. L. Ratton, M.D., Salem.*) "Decoction of it is used by native practitioners as a gargle in sore-throat." (*Brigade Surgeon S. M. Shircore, Moorshedabad.*) "Is a powerful sialagogue, and I have seen it give relief in rheumatic pains in the face." (*Surgeon-Major John North, Bangalore.*) "Frequently given to infants in the Deccan and the Konkan from the idea of its assisting to make them talk. As a stimulant the dose is one to five grains." (*Surgeon W. Barren, Bhuj, Cutch, Bombay.*) "Is used by the natives of India as an aphrodisiac. Applied in the form of a powder to a carious tooth, it is said to remove toothache; useful in flatulent dyspepsia as a carminative." (*Assistant Surgeon Jaswant Rai, Mooltan.*)

Gargle. 1029

Decoction. 1030

Powder. 1031

1032

ANADENDRON, *Schott.; Gen. Pl., III., 991.*

Anadendron, sp.; AROIDEÆ.

Vern.—*Yolba*, AND.

FIBRE. 1033

Fibre.—In the Andaman Islands, bow-strings are made from the fibre of the bark of this plant, to which, to increase strength, a coating of black bee's-wax (*Tobul-pid*) is frequently applied. Netted reticules are also prepared from this fibre, which are used by women for carrying small objects. (*Mr. Mann's Andaman and Nicobar Islands Catalogue, Calcutta Exhibition.*)

ANAGALLIS, *Linn.; Gen. Pl., II., 637.*

1034

A genus of slender herbs belonging to the Natural Order PRIMULACEÆ, comprising some 12 species, inhabitants of the north temperate zones, 1 occurring on the temperate Himálaya.

The generic name is derived from *ἀνά*, again, and *ἀγάλλω*, to make glorious, or to cause mirth, from its fabled virtue to remove sadness. This name was most probably suggested from the beauty of the flowers, or from the fact that as the sun rises and sets, so the sparkling Anagallis opens and closes, hence the popular name Poorman's weather-glass.

Anagallis arvensis, *Linn., var.* **cœrulea**; *Fl. Br. Ind., III., 506.*

1035

Vern.—*Jonkhmári, jainghani*, N.-W. P.

Habitat.—Found on the mountains of Bengal and of the North-West India, and the Himálaya generally, from Nepal westward, ascending to 8,000 feet; common in the neighbourhood of Simla, on rubbish-heaps and walls around fields; Central India, the Nilgiri Hills, and Ceylon (perhaps introduced).

MEDICINE. The Plant. 1036

Medicine.—Used to intoxicate fish and to expel leeches from the nostrils. (For this purpose the juice of the various species of **Begonia** would also seem admirably suited,—see **Leech.**) It is used in cerebral affections, leprosy, hydrophobia, dropsy, epilepsy, and mania. Formerly it was used in Europe in epilepsy, mania, hysteria, delirium, enlargement of the liver, spleen, dropsy, emaciation, stone, the plague, bites of serpents and mad animals, and in numerous other diseases.

Said to be poisonous to dogs, producing inflammation of the stomach. (*Baden Powell, Panjáb Products, I., 368.*)

ANAMIRTA, *Colebr.; Gen. Pl., I., 35.*

1037

A climbing shrub (belonging to the Natural Order MENISPERMACEÆ), a native of Eastern Bengal, the Khásia hills, and Assam; and from the Konkan to Orissa and Ceylon.

Flowers panicled. *Sepals* 6, with 2 adpressed bracts. *Petals* O. *Male flowers:* anthers sessile, many, arranged upon a vertical column, 2-celled, bursting transversely. *Female flowers:* staminodes 9, clavate, 1-seriate. *Ovaries* 3 on a short gynophore; *stigma* subcapitate reflexed. *Drupes* on a 3-fid gynophore obliquely ovoid, dorsally gibbous; *style* scar sub-basal; *endocarp* woody. *Seed* globose, embracing the sub-globose, hollow, intruded endocarp; *albumen* dense, composed of horny granules; *embryo* curved; *cotyledons* narrow-oblong, thin-spreading.

Anamirta Cocculus, *W. & A.; Fl. Br. Ind., I., 98.*

1038

COCCULUS INDICUS of Pharmacy.

Vern.—*Kákmári, kákmári-ké-binj*, HIND., DUK.; *Kákamári*, SANS., BENG.; *Káka-phala, vátoli, kákphal*, BOMB.; *Kakkáy-kolli-virai* or *káká-koli-viari*, TAM.; *Káki-champa, káká-mári, vittu*, TEL.; *Kákamári-bija*, KAN.; *Karanta-kattin-káya, polluk-kaya*, MALA.; *Tittaval*, SINGH.; *Tuba bidji*, MALAY.

Habitat.—A climbing shrub of South and East India, Burma, and Oudh forests.

OIL. 1039 Fat from Seeds.

Oil.—The fruit contains a large quantity of fixed oil. The fat expressed from the seeds, which amounts to about half their weight, is used in India for industrial purposes.

1040 MEDICINE. Berries. 1041 Ointment. 1042

Medicine.—The bitter berries of this plant are used in India to poison fish and crows. In medical practice they are never administered internally, but are sometimes used in the form of an ointment. This ointment is employed as an insecticide, to destroy pediculi, &c., and in some obstinate forms of chronic skin diseases. (*Bentley & Trimen.*)

The Pine Apple.

The Surgeon-General of Madras proposed to exclude this drug from the new edition of the *Pharmacopœia of India.* (*Home Department Official Correspondence.*)

Dr. Dymock (*Mat. Med., W. India, 20*) says: "This plant, which is a large climbing shrub with rough bark, abounds on the western coast of India. Its properties have been known to the Hindús from an early date, and the fruit appears to have been long in use as a remedy in certain skin affections, possibly of parasitic origin. The Arabs were probably also acquainted with it, but there is no satisfactory evidence upon this point to be gathered from their writers upon materia medica."

Oily solution. 1043

§ "The berries of **Anamirta cocculus** are an active poison in large doses, but not in small ones. I have taken the drug myself up to five grains, three times a day, without any effect, good or bad. The cheapest and most convenient way of using it externally is in the form of oily solution, with cocoanut oil, in the proportion of one drachm of the former to one ounce of the latter." (*Honorary Surgeon Moodeen Sheriff, Khan Bahadur, Madras.*)

Chemical Properties.—§" The poisonous properties of the seeds are due to Picrotoxin $C_{12}H_{14}O_5$. The neutral principle readily crystallises, and has an intensely bitter taste; in consequence of this fact, **Cocculus Indicus** has been employed as a substitute for hops in the manufacture of beer. In addition to Picrotoxin two other crystalline principles have been described as existing in the berries, *Menispermine* and *Paramenispermine.*" (*Surgeon C. J. H. Warden, Calcutta.*)

1044

ANANAS, *Adans.; Gen. Pl., III., 662.*

A genus of almost stemless plants, with a rosette of lanceolate leaves, belonging to the Natural Order BROMELIACEÆ, and comprising some 5 or 6 species, inhabitants of tropical America.

Inflorescence densely crowded and spirally arranged into a strobiliform head. *Bracts* and *ovary* with the receptacle, developing into a succulent compound fruit. *Sepals* and *petals* distinct on the apex of each ovary.

The generic name is supposed to be a Latinised form of the Guiana or South American name *Nanas.*

1045

Ananas sativa, *Linn.*

THE PINE APPLE.

Vern.—*Anánas, anannas,* HIND.; *Anánash* (vulgarly *anáras*), BENG.; *Anánas,* GUJ., MAR.; *Anáshap-pasham,* TAM.; *Anása-pandu,* TEL.; *Anánasu-hannu,* KAN.; *Annanas, kaita-chakka,* MALA.; *Aainunnás,* ARAB. and PERS.; *Annási,* SINGH.; *Nanna-ti,* BURM.

Habitat.—This and all the other members of the same order now met with in India have been introduced from America. From the vernacular names of this species, one would suppose it had reached India through Persia.

History.—A perennial universally cultivated in all tropical and subtropical countries. The pine apple was unknown to Europe, Africa, and Asia, prior to the discovery of the Western Continent. It is apparently a native of Brazil, and it was first made known to Europe by **Goncatlo Hernandez** in 1513; it was introduced by the Portuguese into Bengal in 1594. "Its introduction is expressly mentioned by Indian authors such as **Abul Fuzl** in the *Ayeen Akbarí,* and again by the author of *Dhara Shekoih* (*Royle*). The rapidity with which it spread through Europe, Asia, and Africa is unparalleled in the history of any other fruit. It seems to have met with universal acceptance; hence, apparently, the purity with which its American name *Anasi* or *Nanas* has passed through so many languages. The Asiatic recipient of a living plant seems to have carried off, and adopted as his own, the name by which so valuable a treasure was made known to him. The first pine apples which appear to have reached England were those presented to Cromwell. The next notice is of the "Queen pine" presented to Charles II., on the 19th July 1688, having been sent from Barbados, and the first pine apple grown in England seems to have been reared from the rejected crowns of these. It was first systematically cultivated in Europe by M. Le Cour, a Dutch merchant near Leyden. It was first fruited in England in the year 1712; since then its cultivation may be said to have become universal all over Southern Europe. The largest pine apple on record was reared in England, and it weighed over 14 lbs.

Properties and Uses—

FIBRE. 1046

Fibre.—The leaves, which require to be steeped in water for 18 days, yield a beautiful fibre, which, but for the difficulty of extraction, would be largely used. This fibre is in request in India for threading necklaces, as it does not rot, and is very strong. Both the wild and cultivated pine apple yield fibres which, when spun, surpass in strength, fineness, and lustre those obtained from flax. It can be employed as a substitute for silk, and as a material for mixing with wool or cotton. For sewing-thread, twist, trimmings, laces, curtains, and the like, its particular qualities render it specially applicable. (*Chambers' Journal.*) In 1839, **Miss Davey,** in answer to an advertisement published by the Agri.-Horticultural Society of India, submitted some thread made from pine apple leaves, of which she remarked that it was equal to the finest flax thread manufactured in Europe, and considered it comparable to the best cambric thread. This lady, with some difficulty, owing to the conservative objections of the Dacca weavers, whom she tried to induce to make some cloth from this fibre, manufactured handkerchiefs, cuffs, and some cloth which are alluded to in the Proceedings of the Society as "elegant specimens." Some thread was sent home, but the English spinners seem to have been as prejudiced against this fibre as the Dacca weavers were.

Thread. 1047

Silk substitute. 1048

In the Agri.-Horticultural Society's Journal for 1853, some trials of various fibres made by Harton & Co., Calcutta, are published: a ¾-inch in circumference rope made of pine apple fibre easily bore a weight of 42 cwt., and broke only with a weight of 57 cwt. (*Tropical Agriculturist, III., 522.*) In *Royle's Fibrous Plants of India* will be found some interesting information regarding the pine apple, of which the following may be given as an abstract of the more important facts. It has become quite naturalised in some parts of this country. It flourishes in Assam and is very abundant on the Khásia hills. **Captain Turner** found it plentiful at the foot of the Himálaya. According to **Dr. Helfer,** the pine apple is so abundant in Tenasserim as to be sold in Amherst Town, in June and July, at the rate of one rupee for a boat-load. The natives know it only by the American name, which they transform into *Nana thi* or *Nanna* fruit. The pine apple of the Phillipine Islands is much valued for its fine hair-like fibres, of which the famous pine apple cloth is manufactured. **M. Perottet** considers this to be a distinct species and has named it **Bromelia Pigna.** **Mr. Bennett** visited a plantation near Singapore made by a Chinaman, who prepared the fibre for export to China, where it is used "in the manufacture of linens." The leaves, in the green state, were laid upon a board and the epidermis removed with a knife. The fibres were then easily detached by the hand on being raised with a broad knife. The separation of the pine apple fibre is practised in many parts of the East Indies. The natives of Burma, however, do not seem to have been acquainted with the *Ananas* fibre, although the plant is very abundant there. From some experiments

Pine Apple Cloth. 1049

which were made by **Dr. Royle**, it appears that a certain quantity of the fibre prepared at Madras bore 260 lbs., and a similar quantity prepared in Singapore bore 350 lbs., while New Zealand flax bore only 260 lbs. **Mr. Zincke** took out a patent for the manufacture of thread from this fibre. Bleaching destroys the adhesion between the bundles of fibres, and renders it fit to be spun in the same way as flax. Twine, cord, and fishing-lines are also sometimes made of it. Pine apple ropes are said to bear constant immersion in water, a property which is increased by the natives in some places by tanning the fibre. (*Fibrous Plants of India, pp. 37-41.*)

Ropes. 1050

Mr. Thomas Christy, in his *New Commercial Plants* (*No. 6, page 40*), says that in Rungpore shoemakers largely use the fibre for twine, but that the plant is chiefly cultivated there for its fruit, the fibre being but little appreciated. He also says: "The filaments of pine apple are very fine and flexible and also resistant. They are easily divided after treatment in the alkaline bath, and after being submitted to trituration. The isolated fibres are very fine, of a tolerably regular diameter from one end to the other, but of very different size. The inferior canal, which is very perceptible in the largest, is not so in the smaller ones. They are very flexible, curling and crisping readily under mechanism. The points are rarely sharp, and taper gradually to the extremities. They are rounded, or rather blunt, at the end."

Twine. 1051

MEDICINE. Juice. 1052

Medicine.—In India the fresh juice of the leaves is regarded as a powerful anthelmintic, and that of the fruit as an antiscorbutic. A friend informs me that the natives regard the fresh juice of the fruit as poisonous if hypodermically injected.

Special Opinions.—§"The pure juice of the ripe fruit is said to allay gastric irritability in fever. The fresh juice of the leaves given with sugar is said to relieve hiccup." (*Surgeon-Major Bankabehari Gupta, Púrí.*) "Raw pine apple is used to produce criminal abortion." (*Surgeon-Major Henry David Cook, Calicut, Malabar.*) "It is antiscorbutic, cholagogue. The green fruit is emmenagogue, produces abortion." (*Assistant Surgeon Devendro Nath Roy, Calcutta.*) "The fresh juice of the white portions of the leaves, mixed with sugar, is used as a purgative and anthelmintic. The juice of the ripe fruit is diuretic, diaphoretic, and refrigerant. In large quantities it is believed to have the property of causing strong uterine contractions." (*Brigade Surgeon J. H. Thornton, B.A., M.D., Monghyr.*) "The fruit is antiscorbutic, and the fresh juice of leaves anthelmintic." (*Surgeon C. J. W. Meadows, Burrisal.*) "The juice of the ripe fruit is useful in jaundice." (*Assistant Surgeon K. N. Acharji, Dacca.*) "Used in animal abortion, fruit eaten." (*Surgeon-Major J. J. L. Ratton, Salem.*)

Leaf juice. 1053

Fruit. 1054

"In the Straits of Malacca the juice of the leaves is used to produce abortion, also as an emmenagogue. The ripe fruit eaten freely has the same effect in a less degree." (*Honorary Surgeon P. Kinsley, Ganjam, Madras.*) "Besides its value as an antiscorbutic, it seems to have an irritant action on the uterus, as it is reported to have caused abortion in weakly or predisposed women." (*Surgeon-Major R. L. Dutt, M.D., Pubna.*) "Its preserve is much employed by the hakims as an excellent nutritive and tonic." (*Surgeon Mokund Lal, Agra.*)

FOOD. Fruit. 1055

Food.—The pine apple is generally regarded as one of the most delicious fruits met with in tropical regions. To avoid the dangerous consequences attributed to it, however, many persons will only eat it when stewed, while others prefer to eat it fresh, with a little sugar or even salt. During the season in Calcutta good pine apples can be purchased for a pice each (*i. e.*, less than a halfpenny.) The pine apples of Burma and of the Straits are, of the Indian forms, those most prized.

§ "On the Malabar coast near Mahé, and in British Burma, near Myanoung, the pine apple is remarkably abundant. In the former tract the natives have a prejudice against eating the fruit, from an idea that it is poisonous, and they consequently destroy it, or give it away. In Myanoung, **Monsieur d'Avera** is trying to make use of the large quantities that grow there to manufacture champagne. I am in correspondence with him on the subject, and he seems hopeful of success. Should the experiment succeed, it could be repeated on the Malabar coast." (*Mr. L. Liotard.*)

Pine apple champagne. 1056

Chemical Note.—§" The oil or essence of pine apple, used for flavouring purposes in confectionery, is a solution of ethybutyrate in alcohol. This compound has also been employed to give the pine apple flavour to Jamaica rum." (*Surgeon Warden, Prof. of Chemistry, Calcutta.*)

Chemically prepared Pine apple juice. 1057

Anatherum muricatum, *Retz.*, see **Andropogon muricatus,** *Retz.*

ANCHUSA, *Linn.; Gen. Pl., II., 2, 855.*

Anchusa tinctoria, *Linn.*; BORAGINEÆ.

1058

Syn. for **Alkanna tinctoria,** *Tausch.*

Mr. Baden Powell mentions an oil as obtained from this plant. Other references to the root of this plant occur in works on Indian Economic Science; it is incorrectly described as yielding the *Ratanjote* (see **Onosma echioides,** *Linn.*). **Anchusa** is not indigenous to India, and could only occur in gardens at hill stations.

ANCISTROCLADUS, *Wall.*

1059

Ancistrocladus Vahlii, *Arn.; Fl. Br. Ind., I., 299;* DIPTEROCARPEÆ.

Vern.—*Coná-wel*, SINGH.

Habitat.—Central and Southern parts of Ceylon up to an altitude of 2,000 feet.

Fibre.—**Dr. Trimen** informs me that the long tough stems are used as jungle ropes.

Fibre. 1060

ANDRACHNE, *Linn.; Gen. Pl., III., 270.*

1061

Andrachne cordifolia, *Müll.-Arg.*; EUPHORBIACEÆ.

Vern.—*Kurkni, gurguli, kurkuli*, PB.; *Kúrkni, gúrgúli*, JHELAM; *Bersu*, CHENAB; *Barotri, madare*, RAVI; *Mútkar, chirmútti, pín*, BEAS; *Tsátin*, SUTLEJ.

Habitat.—A small shrub, met with in North-West Himálaya, from the Indus to Nepal, ascending to 8,000 feet.

Poison.—"The twigs and leaves are said to kill cattle when browsed in the early morning on an empty stomach." (*Dr. Stewart.*)

POISON. 1062

Structure of the Wood.—White, moderately hard, close-grained. Weight 45 lbs. per cubic foot.

TIMBER. 1063

A. trifoliata, *Roxb.; Fl. Ind., Ed. C.B.C., 703.*

Syn. for **Bischoffia javanica,** *Bl.* (*Gamble's Man. Timb., 335*), which see.

ANDROGRAPHIS, *Wall.; Gen. Pl., I., 1099.*

An Indian genus of annual herbaceous or shrubby plants, erect or procumbent, belonging to the Natural Order ACANTHACEÆ, and comprising some 19 species.

Leaves entire. *Corolla* small, tubular, 2-lipped white or pink, with dark purple lower lip, pubescent. *Ovary* 6-12-ovuled, thinly hairy; *style* slender, tip minutely bifid. *Capsule* linear-oblong or elliptic, compressed contrary to the septum, 6-12 seeded. *Seeds* osseous, sub-quadrate or oblong, not compressed, rugose-pitted, glabrous.

The generic name is derived from ἀνήρ, a stamen, γραφίς, a writing style, in allusion to the form of the filaments.

1064

Andrographis paniculata, *Nees; Fl. Br. Ind., III., 501; Wight, Ic., t. 518;* ACANTHACEÆ.

THE CREAT.

Syn.—JUSTICIA PANICULATA, *Roxb. Fl. Ind., Ed. C.B.C., 40.*

Vern.—*Kiryát, charáyetah, mahátitá,* HIND.; *Kálmegh, mahátita,* BENG.; *Olenkiráyat,* MAR.; *Kiryáta, olikiryát, kiryáto, kariyátu,* GUJ.; *Charayetah, kalafnáth,* DUK.; *Nila-vémbu, shirat-kuchchi,* TAM.; *Nela-vému,* TEL.; *Nila-veppu, kiriyattu,* MALA.; *Nela-bevinágidá, kreata,* KAN.; *Kirata, bhunimba,* SANS.; *Qasabuzzarírah, qasabhuvá,* ARAB.; *Nainehavandi,* PERS.; *Nin-bin-kohomba,* SINGH.

Dr. **Moodeen Sheriff** says that *kara-kanniram* or "*cara caniram*" is the Malayan name found in *Hortus Malabaricus,* which means the ack *Strychnos Nux-vomica,* and which he considers as neither correct nor safe to be applied to this plant.

Habitat.—An annual common in hedgerows throughout the plains of India, from Lucknow to Assam and Ceylon. Cultivated in gardens in some parts of India.

Botanic Diagnosis.—*Leaves* lanceolate glabrous. *Racemes* lax paniculate-divaricate, pedicels manifest. *Capsules* thrice as long as broad, nearly glabrous.

MEDICINE. Alui. 1065

Medicine.—This bitter shrub is well known under the name of *Kálmeg,* and forms the principal ingredient of a household medicine called *Aluí,* extensively used in Bengal. The expressed juice of the leaves, together with certain spices, such as cardamoms, cloves, cinnamon, &c., dried in the sun, is made into little globules, which are prescribed for infants to relieve griping, irregular stools, and loss of appetite. The medicinal properties of this plant are many. The ROOTS and the LEAVES are febrifuge, stomachic, tonic, alterative, and anthelmintic. According to **Murray**, the plant is very useful in general debility, dysentery, and certain forms of dyspepsia. **U. C. Dutt** says that there is some doubt as to the Sanskrit name of this plant. The name *Yavatikta* with its synonyms *Mahátikta, sankhini,* are by some supposed to refer to this plant, but **Dutt** is of opinion that it was not used in Sanskrit medicine. Great confusion exists between this plant and chiretta, and samples of the latter are frequently adulterated with *Kálmeg.* **Flückiger** and **Hanbury** point out that this plant has been wrongly supposed to be an ingredient in the famous bitter tonic called by the Portuguese of India *Droga amara.*

Roots. 1066

Leaves. 1067

Drs. **Carter, Dymock,** and **Sakhárám Arjun** reported on this drug as follows: "A bitter tonic and stomachic. It is used in general debility, in convalescence after fevers, and in advanced stages of dysentery. It is also used as a tonic, stimulant, and gentle aperient in the treatment of several forms of dyspepsia, and in the torpidity of the alimentary canal. The expressed juice of the leaves is a common domestic remedy in the bowel complaints of children. Dose: 1 to 2 ounces of the infusion, and 1 to 4 drachms of the tincture." (*Home Department Official Correspondence.*)

Expressed Juice. 1068

Infusion. 1069

Tincture. 1070

It is officinal in the *Pharmacopœia of India,* where directions will be found as to the preparation of a compound infusion and compound tincture. **Irvine**, in his *Mat. Med. of Patna,* says that the root is used as a stomachic bitter; and in the "drogue amere." The dose is from ʒss to ʒi in infusion.

Chemical Composition.—The intensely bitter taste appears to be due to an indifferent non-basic principle, since the usual reagents fail to indicate the presence of an alkaloid. Tannic acid produces an abundant precipitate. The infusion is but little altered by salts of iron. It contains a considerable quantity of chloride of sodium. (*Flück. & Hanb., Pharmacog.; Bent. & Trim.*, *195*.)

Pills. 1071

§ "The Yanadees, a wandering gipsy tribe in the Madras Presidency, constantly carry a supply of pills made of Creat fresh leaves, and the pulp of the ripe tamarind, which they consider antidotal to the venom of the cobra. A pill made into a paste with water is applied to the bitten part, and some of it is put into the eyes; two pills are given for a dose every hour or two internally." (*Honorary Surgeon P. Kinsley, Chicacole, Ganjam, Madras.*) "Creat leaves with the leaves of Indian birthwort **(Aristolochia indica)** and the fresh inner root-bark of country sarsaparilla, made into an electuary, is used by native hakims as a tonic and alterative in syphilitic cachixia and foul syphilitic ulcers. I have seen many cases successfully treated by this electuary." (*Honorary Surgeon Easton Alfred Morris, Negapatam.*) "The green leaves are given with aniseed (4 to 20) as stomachic and anthelmintic." (*Assistant Surgeon Devendro Nath Roy, Calcutta.*) "This is called Indian Chiretta, and is used as a tonic." (*Surgeon-Major Lionel Beech, Cocanada.*) "Decoction of all parts of the plant acts as a mild antiperiodic." (*Surgeon-Major John Lancaster, M.B., Chittore.*) "It is efficacious in certain forms of skin diseases, especially in eczema." (*Assistant Surgeon J. N. Dey, Jeypore.*) "Febrifuge, used in infusion." (*Surgeon-Major J. J. L. Ratton, Salem.*)

ANDROPOGON, *Linn.; Gen. Pl., III., 1133.*

1072

A genus of grasses (GRAMINEÆ) belonging to the tribe ANDROPOGONEÆ, of which about 25 species are met with in India.

Spikes polygamous, arranged in pairs (or many) on a common slender peduncle, at the bent basal node of which occurs a large leafy bract which in bud encloses the pair of spikes; *peduncles* arising from the zig-zag flattened branches of a panicle. *Panicles* pendulous, single or clustered from the axils of the upper leaves. *Spikelets* articulated to the rachis of the spike in pairs, the one sessile and hermaphrodite (or rarely feminine), the other pedicillate, exaristate and masculine, rarely hermaphrodite, sterile, or reduced to empty glumes. Terminal spikelets often in threes, the middle fertile, one lateral male and the other neuter. *Glumes* sub-equal, often longer than the hermaphrodite flowers; the lowest largest; those of the stalked spikes many-veined, often sessile; the lower glume is flattened on the back against the rachis and veinless. *Pales* very small, the lowest deeply bifid, from the sinus of which arises a long awn.

The species of economic interest met with in India belong chiefly to the section which corresponds to the sub-genus CYMBOPOGON, characterised by their large bracts, and by the veination of the glumes. The generic name is derived from ἀνήρ, a stamen, and πώγων, a beard, in allusion to the bearded appearance of the stamen.

The greatest confusion exists in the identification of the plants yielding the essential oils from this genus. In all collections intended for museum specimens, the plants (in flower) should, if possible, accompany the oils, so as to secure accurate identification. In fact, until such collections have been made it will be impossible to remove the unavoidable errors which must creep into all pure compilations of the literature of a subject so difficult as that of the economic uses of the Indian species of **Andropogon.**

1073 **Andropogon aciculatus,** *Roxb., in Fl. Ind., Ed. C.B.C., 88 (given as Linn. Sp. Pl. Ed. Willd., IV., 906).*

Syn. for **Chrysopogon acicularis,** *Retz.; Duthie's List, p.* 22 (the *chora-kanta* of Bengal).

1074 **A. ampliflorus,** *Stend.* Met with in the North-West Himálaya.

1075 **A. ariani,** *Edgew.* Met with in the sandy deserts of the Panjáb.

1076 **A. Bladhii,** *Retz.*

Syn.—Lepeoceris Bladhii, *Nees.*

Vern.—*Loari,* Beng.; *Donda* or *dhunda, nilon, janewar* (Captain Wingate's Report for 1876 on the grass farms of Allahabad and Cawnpore), N.-W. P.

Habitat.—Described by **Roxburgh** as a native of hedges and roadsides, but chiefly of old pasture grounds. **Duthie** says it is found in the plains of the North-West Provinces and the Panjab.

FODDER. 1077

Fodder.—**Captain Wingate,** in the report quoted above, seems to speak highly of this as a fodder grass. He says: "At Allahabad the indigenous grass is not *dhoob,* but *janewar,* **Andropogon Bladhii,** and *unjan,* **Pennisetum anchroides.** These two grasses may be seen in their perfection in the Alfred Park of this station in the early part of October." **General Sir Herbert Macpherson,** in a report of a silo experiment in the Allahabad fort with this grass, says: "Both *janewar* and *unjan* are first-rate fodder grasses."

1078 **A. brevifolius,** *Sw.*

Syn.—A. exertus, *Stend.;* Pollinia brevifolia and vaginata, *Spreng.*

Collected at Hazaribagh (altitude 2,000 feet) by **Mr. C. B. Clarke.**

1079 **A. citratus,** *DC.*

The Lemon Grass.

Syn.—Cymbopogon citratum, *DC.;* A. Schœnanthus, *Wall.; Plant. Asiat. Rariores, III., t. 280;* A. Schœnanthus, *Roxb., in Dalz. and Gib., Bomb. Fl., Sup., 99; Rheed. Hort., Mal., 12, t. 72;* A. Schœnanthus. *Linn., as in Roxb., Fl. Ind., Ed. C.B.C., 92; Voigt, Hort. Sub. Calc., 706;* and *U. C. Dutt, Mat. Med., Hindús, 271.*

Vern.—*Gandha bená,* Beng.; *Gandha trina,* Hind.; *Hirvacha* or *olancha,* Mar.; *Lilichá, lilacha,* Guj.; *Hazár-ma-sálah,* Duk.; *Vashanup-pulla, kurpura-pulla,* Tam.; *Nimma-gaddi, chippa-gaddi,* Tel.; *Vasanap-pulla, shambhára-pulla,* Mala.; *Púr-hali-hulla,* Kan.; *Cháe-kashmiri,* Pers.; *Bhústrina,* Sans.; *Penquin,* Singh.

The vernacular names *Gandha-bená,* Beng., and *Malutrinukung, bhústrinung,* Sanskrit, are by **Roxburgh** given to a plant he describes as **A. Schœnanthus,** *Linn.* This may probably be **A. citratus,** *DC.,* but it seems to agree equally well in certain respects with **A. laniger,** *Desf.*

Habitat.—A large, coarse, glaucous grass, found under cultivation in various islands of the Eastern Archipelago, and in gardens over an extensive tract of country in India and Ceylon; it rarely or never bears flowers. It is also largely cultivated in Ceylon and in Singapore for its odoriferous oil. "I have seen it in flower more than once." (*Dr. Dymock, Bombay.*)

OIL. 1080

Oil.—The lemon grass yields lemon-grass oil, verbena oil, or Indian *molissa* oil.

This oil is chiefly employed in Europe in adulterating true verbena oil. It is largely employed to perfume soaps and greases. The annual production of otto of lemon grass in Ceylon is above 1,500 lbs., valued at 1*s.* 4*d.* per ounce. There is a large consumption of this otto in the manufacture of Eau de Cologne. This oil is said to be more costly and less extensively produced than citronella; it is chiefly manufactured in Ceylon and Singapore. More than half the annual exports go to America. In 1875 Ceylon exported 13,515 ounces of this oil. In India it is used in native perfumery.

PERFUMERY. Verbena oil. 1081 Otto of Lemon Grass. 1082 Soaps and Greases. 1083

Food.—The leaves are often resorted to to flavour tea, and the centre of the stems are cooked in curries.

FOOD. Leaves. 1084 Stems. 1085

Medicine.—In the Indian Pharmacopœia this oil is regarded as officinal. When pure it is of a pale sherry colour, transparent, with an extremely pungent taste, and a peculiar fragrant lemon-like odour. The properties attributed to it are stimulant, carminative, antispasmodic, and diaphoretic; locally applied it is a rubefacient. It is recommended to be administered in flatulent and spasmodic affections of the bowels and in gastric irritability. In cholera it has been spoken highly of as a remedy of great value, allaying and arresting the vomiting, and aiding the process of reaction. **Dr. Waring**, in the appendix to the Indian Pharmacopœia, records a high testimony in its favour both as an external application in rheumatism and in other painful affections, and as a stimulant and diaphoretic internally. He states that amongst the Indo-Britons of South India, it is one of their most highly esteemed remedies in cholera. **Dr. Ross**, in the same notice, reports very favourably of a warm infusion prepared by macerating about four ounces of the leaves in a pint of hot water. This he has used very successfully as a diaphoretic in febrile affections, especially in weakly subjects, or when the fever is of a typhoid type. (*Pharm. Ind., 464.*)

MEDICINE. Oil. 1086

It seems probable that in different parts of India the various species of grass oil are used indiscriminately; indeed, considerable doubt in many cases exists as to the correct species to which the facts contained in the writings of Indian authors should be referred. Externally, they are nearly all used in rheumatism, and are administered internally as carminatives in colic, and an infusion of the leaves is a popular diaphoretic stimulant and antiperiodic, very frequently given in simple catarrh. "An infusion of it as a fever drink has great effect in inducing a remission or intermission by bringing on sweat." (*Dalz. and Gibs., Bomb. Fl.*)

§ **Special Opinions.**—"Infusion of the leaves (tea) is largely used as an agreeable sudorific in mild cases of fever, and as a medicinal vapour bath for the same purpose. It is often combined with **Mentha arvensis**, when used with the above object." (*Assistant Surgeon Sakhárám Arjun Rávat, Bombay.*) "Taken internally, in some parts of India, in the form of an infusion like tea or with milk, it is said to be a stimulant and diaphoretic. The vapour of a hot infusion is inhaled by fever patients to produce diaphoresis." (*Surgeon W. Barren, Bhuj, Cutch, Bombay.*) "An infusion of the leaves (known as "Lemon tea") is very refreshing." (*Honorary Surgeon P. Kinsley, Chicacole, Ganjam, Madras.*) "The roots and tender levees are sometimes given with black pepper in cases of disordered menstruation and in the congestive and neuralgic forms of dysmenorrhœa. The oil is useful in flatulent colic and other spasmodic affections of the bowels, and as an application in chronic rheumatism, &c." (*Brigade Surgeon J. H. Thornton, B.A., M.B., Monghyr.*) "Carminative and tonic to the intestinal mucus membrane, useful in vomiting and diarrhœa, externally it forms a useful liniment." (*Surgeon-Major Henry David Cook, Malabar.*) "Lemon grass oil, applied with prolonged friction, is a pleasant and useful application in lumbago." (*H. DeTatham,*

Infusion. 1087

Roots and leaves. 1088

Liniment. 1089

M.D., M.R.C.P. Lond., Ahmednagar.) "It is used as a stomachic for children, and also as a diaphoretic. Externally it is used for ringworm." (*Surgeon H. W. Hill, Mánbhúm.*) "The oil is used to conceal the odour of iodoform in the Dispensary of the Royal Infirmary, Edinburgh." (*Dr. Forsyth, Dinajpore.*)

1090

Andropogon contortus, *Linn.*

THE SPEAR GRASS.

Vern.—*Sarwála, sariála, sarári,* PB.; *Parba parbi, parva, bandar pnucha, sarwar, musel, lap,* N.-W. P.; *Panri-pullu,* TAM.; *Eddi, eddigaddi,* TEL.

Habitat.—Grows on pasture grounds; a very troublesome weed; reduced to **Heteropogon contortus,** *R. & S.*, which see.

1091

A. Hookeri, *Munro.*

Habitat.—A native of the Panjáb (at Pathankote), altitude 1,500 feet. (*C. B. Clarke.*)

1092

A. Ischæmum, *Linn.*

Syn.—A. ANGUSTIFOLIUM, *Sibth. & Sm.*

Vern.—*Palwal jarga,* N.-W. P.

Habitat.—"Grows (at Aligarh) on barren wet soil and is eaten by cattle and horses." (*Lang.*) "Excellent for hay; the seeds are nutritious. This is considered (at Muttra) one of the best fodder grasses." (*Crooke; Duthie's List of Grasses, 20.*)

1093

A. laniger, *Desf.*

THE JUNCUS ODORATUS and HERBA SCHŒNANTHI of Pharmacists.

Syn.—A. IWARANCUSA, *Roxb.* (*in part*); CYMBOPOGON LANIGER, *Ders.*; A. OLIVIERI, *Boiss.*

Vern.—*Lámjak, búr, kháwi,* or *kháví, khoi, panni, solára, san, ibharankusha, karan kusha, ghat-yári,* HIND. and PB.; *Káránkusa, ibharankusha,* BENG.; *Miriya ban, ganguli, bad, piriya,* N.-W. P.; *Lámajjaka,* SANS.; *Izkhir,* ARAB. and BOMB.

The *Makhzan-ul-Adwiya* describes two forms of the plant called *Izkhir* and gives their synonyms, from which it would appear that under the first form it confuses **A. Schœnanthus** and **A. laniger** as one. The second form is apparently **A. muricatus,** the *khas-khas* of India. Compare with note under the last species.

Habitat.—Native of the Lower Himálayan tract (said to occur in Thibet at an altitude of 11,000 feet), extending through the plains of the North-West Provinces and Panjáb, to Sind. An inhabitant of dry desert tracts.

History.—"It is particularly mentioned by **Arrian** in his account of Alexander's journey through the Panjáb and Sind, and was gathered by the Phœnician followers of the army in Lus, who called it spikenard. It is common about Kurrachee, and is used as a scent by the natives." (*Dalz. & Gibs., Bomb. Fl., 302, under* **A. Iwarancusa,** *Roxb.*) "This plant has a wide distribution, extending from North Africa, through Arabia and North India to Thibet; it is the σχοῖνος ἀρωματικὸς of **Dioscorides,** and the Herba Schœnanthi and Juncus odoratus of Latin writers on Materia Medica. It has also been named **Fænum camelorum** from its use as a forage for camels. The Arabic name *Izkhir* is given in the best lexicons as derived from the root ذخر, the same root furnishes the derivative *Zákhirah*, a common term in India for stored-up forage, &c." "Western India is supplied from the Persian Gulf ports." (*Dymock, Mat. Med., W. Ind., 691.*) "**Mr. Tolbort** has sent us specimens under the name of *Khávi* gathered by himself, in 1869, between Multan and Kot Sultán, and quite agreeing with the drug of pharmacy." (*Flück. and Hanb., Pharm., 728.*)

MEDICINE. 1094

Medicine.—Used to purify the blood, and in coughs, chronic rheumatism, and cholera. It is recommended as a valuable aromatic tonic in dyspepsia, especially that of children; it is also used as a stimulant and diaphoretic both by natives and Europeans, in gout, rheumatism, and fever. (*Baden Powell.*) "The grass has an aromatic, pungent taste, which is retained in very old specimens. We are not aware that it is distilled for essential oil." (*Flück. and Hanb., Pharmacog., 728.*) **U. C. Dutt** says of this species: "Its virtues seem to reside in the larger roots marked with annular cicatrices."

FODDER. 1095

Fodder.—**Roxburgh** says it grows in large tufts, each tuft composed of a number of plants adhering together by their roots. The roots are aromatic. Cattle are said to be very fond of the grass. The plant has been called **Fænum camelorum** from its use in dry desert tracts as a forage for camels.

§ "A jungle grass, does not grow about cultivation. Is grazed when tender, but not when full grown; may be stacked, and is then useful in times of scarcity; will last 10-12 years in stack. When cattle eat much of this grass the milk becomes scented." (*Mr. Coldstream, Commissioner, Hissar.*)

Andropogon miliaceus, *Roxb.*

1096

THE HILL GRASS.

Syn.—A. MILIFORMIS, *Stend.*

Habitat.—An erect grass, from 6 to 10 feet in height, inhabiting the mountains north of Oudh.

Roxburgh writes: "The seeds of this most beautiful stately grass were sent me from Lucknow by the late **Gen. Claude Martin**, under the name of Hill Grass." It blossoms during the latter part of the rains.

A. muricatus, *Retz.*

1097

CUSCUS, KHUS-KHUS, or KOOSA.

Syn.—A. SQUAROSUS, *Linn.*; VETIVERIA ODORATA, *Virey*; ANATHERUM MURICATUM, *Retz.*; RHAPHIS MURICATA, *Nees*; PHALARIS ZIZANOIDES, *Linn.*

Vern.—(The plant) *Khas, bená, panni, senth, ganrar, oneí, balah,* or *bala*, HIND.; *Panni*, PB.; *Bálé-ka-gháns*, DUK. (the root); *Khas-khas, shanadér jhar, válá bálá* (the root), BENG., HIND., and DUK.; *Sirom*, SANTAL; *Tin*, OUDH; *Válo*, GUJ.; *Válá*, MAR.; *Válá, khasakhasa*, BOMB.; *Válá*, CUTCH; *Vetti-vér, vizhalvér, ilámich-chamvér, viranam*, TAM.; *Vattí-véru, avvuru-gaddi-véru, lamaj-jakamu-véru, vidavali-véru, ouru-véru*, TEL.; *Lávanchá*, KAN.; *Vettivér, ramachcham-vér*, MAL.; *Usira, vir-ana*, SANS.; *Usir*, ARAB.; *Khas*, PERS.; *Savandra-múl*, SINGH.; *Miya-móe*, BURM.

According to **Moodeen Sheriff**, the lowest parts of the culms of this grass, with or without a portion of its roots, are sold under the Arab name *Izkhir* in South India, while in Haidarabad, Calcutta, &c., *Izkhir* is used for **A. Schœnanthus**. The true *Izkhir* of Arabia does not, acccrding to him, exist in India. He seems also to regard the name *khas-khas* as being doubtfully a true Bengal synonym. It is the name by which the roots are universally known in Bengal, and there seems no possible chance of confusing them with *khas-khas*, the poppy seed. **Dr. Dymock** (*Mat. Med., W. Ind.*) gives *Izkhir* as the Arab and Bombay name for **A. laniger**, which see.

Habitat.—A perennial, tufted grass, very common in every part of the coast (Coromandel, Mysore), and in Bengal and Burma, where it meets with a low, moist, rich soil, especially on the banks of water-courses, &c. (*Roxb.*) It covers large tracts of waste land in Cuttack. It inhabits the plains of the Panjáb and North-West Provinces, and ascends into Kumaon, 1,000 to 2,000 feet in altitude. (*Duthie.*) Cultivated in Rájputana and Chutia Nagpur (Gobindpur). This plant is alluded to on some copper-plate inscriptions discovered near Etawah (dated A. D. 1103 and 1174) as being one of the articles on which the kings of Kanauj levied imports. (*Proc., As. Soc., Beng., Aug. 1873, p. 161.*)

RESIN. 1098

Resin.—The roots contain a resin and volatile oil, which is rather difficult to extract. (*Dr. Bidie's Paris Exhibition Catalogue, p. 47.*)

FIBRE. 1099

Fibre.—The roots are extensively made into aromatic scented mats, hung over doors, and kept wet to cool the atmosphere during the hot season, and are also in great demand for making fans, ornamental baskets, and other small articles, &c. The grass is suited for the manufacture of paper, and it is estimated that from 60,000 to 70,000 maunds are annually available in the Hissar district of the Panjáb alone. In the *Gazetteer of the Central Provinces* the grass is described as a "nuisance to the cultivators, as it grows on the rich soils, and is very difficult to eradicate." **Mr. Baden Powell** says the fibre is much used as a packing material.

OIL. Perfumery. 1100

Oil.—The roots, when distilled with water, yield a fragrant oil, which is used as a perfume, and as such it deserves the extended attention of European perfumers. **Dr. Irvine**, in his *Medical Topography of Ajmere*, mentions the preparation of attar from the roots of this plant, which he says is used in sherbet.

MEDICINE. Roots. 1101

Medicine.—An infusion of the roots is given as a febrifuge and a powder in bilious complaints. *Khas-khas* is regarded as stimulant, diaphoretic, stomachic, and refrigerant. The essence (or otto) is used as a tonic. The roots are regarded as a cooling medicine, and are given mixed with other medicines having similar properties. A paste of the pulverised roots in water is also used as a cooling external application in fevers.

The *Pharmacopœia of India* says of this plant: "Antispasmodic, diaphoretic, diuretic, and emmenagogue properties have been assigned to it; but beyond being a gentle, stimulant diaphoretic, it seems to have no just claims to notice as a medicine. An account of the uses to which it has been put in Europe is given by **Pereira** (*Mat. Med., Vol. II., Pt. I., p. 132*). Its uses in native medicine are detailed in the *Taleef Shereef, p. 14, No. 47*. According to the analysis of **Gager**, it contains a resin, a bitter extractive, and a volatile oil. The dose of the powdered root is about twenty grains, or it may be given in infusion (two drachms of the bruised root to ten ounces of boiling water) in doses of an ounce or more. As a medicine, as far as is at present known, it is an article of minor importance." **U. C. Dutt** says: "It is described as cooling, refrigerant, stomachic, and useful in pyrexia, thirst, inflammation, irritability of the stomach, &c. It enters into the composition of several cooling medicines, as, for example, the preparations called *Shadanga pániya*. A weak infusion of the root is sometimes used as a febrifuge drink. Externally it is used in a variety of ways. A paste of the root is rubbed on the skin to remove oppressive heat or burning of the body. The use of this drug appears to have been popular with the ancients." "An aromatic cooling bath is prepared by adding to a tub of water the following substances in fine powder, namely, root of **Andropogon muricatus, Pavonia odorata** (*bálá*), red sandal-wood, and a fragrant wood called *padma kashtha.*" (*Hindú Mat. Med.*, *271.*)

Chemical Composition.—" *Khas-khas* has been analysed by Vanquilin, who has obtained from it, *1st*, a resinous substance of a deep red-brown colour having an acrid taste and an odour like myrrh; *2nd*, a colouring matter soluble in water; *3rd*, a free acid; *4th*, a salt of lime; *5th*, a considerable quantity of oxide of iron; *6th*, a large quantity of woody matter. (*Vanquilin, Annales de Chimie, t. LXXII., p. 302.*)" (*Dymock's Mat. Med., W. Ind., 692.*)

§" The grass with its roots is boiled in water and used as a steam bath in fevers of a continued type, producing speedy and profuse diaphoresis. The otto is given in two minim doses to check the vomiting of cholera." (*Surgeon-Major J. M. Honston, Travandrum.*)

" Used in the form of cigarettes with benzoin, it relieves headache; a cold infusion is refrigerant." (*Surgeon-Major John Lancaster, Chittore.*)

" Infusion of the root is used as a febrifuge." (*Surgeon-Major J. J. L. Ratton, Salem.*)

" Refrigerant." (*Surgeon C. M. Russell, Sarun.*)

FODDER. 1102

Fodder.—The grass when young affords good fodder. After the rains are over it is cut as bedding for horses at Saharanpur.

§" Principally used for thatching; not grazed upon except in times of excessive drought. Cattle will eat the young leaves after the stems have been burned down." (*Mr. Coldstream, Commissioner, Hissar.*)

DOMESTIC. Thatching grass. 1103 Tatties. 1104 Fans. 1105 Boxes. 1106 1107

Domestic Uses.—*Tin*, a grass in universal use for thatching purposes, the reeds being made into brooms. The roots of it supply the "*khas*" with which our hot-weather tatties are made. It grows on the banks of rivers and marshes, and is generally strictly preserved, as it takes time to spread. Proprietors are averse to its being dug up for the *khas.* (*Oudh Gazetteer, III., 176.*) The fibre is made into fans, ornamental work-boxes, and other small objects.

Andropogon Nardus, *Linn.*

THE CITRONELLA.

Syn.—A. FLEXUOSUS, *Nees;* A. COLORATUS, *Nees;* A. MARTINI, *Thw.* (*En. Ceylon Plants, 361, not of others*); A. IWARANCUSA, *Roxb.* (*in part*); CYMBOPOGON FLEXUOSUS, *Nees;* C. NARDUS, *Linn.* (*in Pharm. Ind.*).

Vern.—*Ganjni, or ganjni-ká-ghás, pust-burn*, HIND.; *Kamá-khér*, BENG.; *Ganjni*, DUK.; *Usadhan*, MAR.; *Kámákshi-pullu, mándap-pullu, kávattam-pullu, shunnárip-pullu*, TAM.; *Kámákshi-kasuvu, kámanchi-gaddi*, TEL.; *Kámákshi-pulla, chóra-pulla*, MALA.; *Ganda-hanchi-khaddi*, KAN.; *Maana*, SINGH.; *Sing-ou-miá*, BURM.

Habitat.—A grass common in the plains and lower hills of the North-West Provinces and Panjáb; extensively cultivated in Ceylon and Singapore for the production of oil of citronella. Abundant about Travancore. As cultivated in Ceylon it often rises to the height of 6 or 8 feet. It is most readily recognised from all the other species by its rufous colour, short spikes, and narrow leaves.

OIL. 1108

Oil.—The leaves are distilled with water, and yield over 3 ozs. of essential oil from 1 cwt. The pure oil is thin and colourless, with a strong aromatic odour, and an acrid, citron-like flavour.

The average exportation of citronella from Colombo is about 40,000 lbs., valued at £8,000, valued at about 4*s.* 1*d.* per lb. It is largely used to give the peculiar flavour to what is known as " Honey-soap."

The extract published below, giving a mode of detection of admixture of other oils with citronella, taken from the *Year-Book of Pharmacy for*

1875, will be found useful, since doubtless the same process would be applicable to all the grass oils. It may be remarked, however, that the author is incorrect in attributing citronella to **A. Schœnanthus,** but the adulteration of the Indian *rusa* oil (the oil from **A. Schœnanthus**) is daily becoming more and more serious, and an easy mode of detecting the adulterations is thus very important. This same mistake is made by **Dr. Wright** (*Year-Book of Pharm., 1874, 631*), who gives some interesting chemical information regarding citronella oil, which he appears to regard as obtained from **A. Schœnanthus.**

Chemical Composition.—Dr. Wright informs us that the essential oil of citronella mainly consists of an oxidized substance boiling near 210°, which becomes to a certain extent resinized, losing partially the elements of water, by continued heat. On analysis it gave quantities which would agree with the formula $C_{10}H_{18}O$, a formula corroborated by its behaviour with bromine, zinc, chlorine, &c. **Prof. Gladstone,** who experimented with citronella oil some time previous to **Dr. Wright** (1872), arrived at the conclusion that it owed its peculiar property to an oxidised oil which he called *Citronellol.* This he separated by fractional distillation into two portions, the one boiling at 200-205° C. and the other at 199-202° C. The composition of each portion as arrived at by **Gladstone** would be represented by the formula $C_{10}H_{16}O$.

The chemistry of Citronella oil would thus seem to require further investigation to remove the slight disparity in these opinions.

1109

"**Estimation of Fixed Oil in adulterated Citronelle** (*Chem. News, XXX., 293*).—The following method yields constant results when managed with care, and when taken in conjunction with the specific gravity of the sample, may give a good approximation as to the quantity and the class of the adulterating oil:—

"(*a.*) Dissolve about one ounce of caustic potash in five ounces of alcohol in a flask; put on a sand bath, and leave to boil.

"(*b.*) Take an eight-ounce beaker, and weigh into it 400 to 500 grains of the citronella; add two volumes of alcohol; boil on a sand bath.

"(*c.*) When (*a*) and (*b*) are both boiling add one volume of the alcoholic solution of potash to the three volumes alcohol and citronella. Boil for a minute or so, and then fill to within an inch of the top with distilled water. Stir gently, and let boil for half an hour, or until the upper layer is perfectly clear, and the under-fluid semi-transparent. Then allow to cool.

"(*d.*) When quite cold, siphon off the under-fluid [containing water, alcohol, potash, and *soap*, if any fixed oil was in the sample] very carefully into another beaker, and boil gently. Acidify with dilute H_2SO_4. Add 50 or 100 grains of wax, continue gently boiling till the oily layer is perfectly clear, and then allow to cool gradually.

"(*e.*) When cold remove the cake of fat, dry and weigh. The weight, less 50 or 100 grains of wax, is the amount of fatty acid contained in the fixed oil. A simple calculation will show the amount per cent. of the adulterant in the citronella." (*Year-Book of Pharmacy, 1875, p. 302.*)

1110

Cultivation of the grass and distillation of Citronella Oil.—In Ceylon the citronella grass is raised from seed and planted like guinea grass, and will give two or three crops a year. When fit to cut, the grass is carried to a large boiler and the oil is distilled. It is estimated to give about three dozen bottles of oil to the acre, but the demand is limited, and price fluctuates from 2*s.* 6*d.* to 4*s.* 6*d.* a bottle. At the latter price it pays handsomely, while at the former it little more than covers expenditure. A still, capable of turning out a dozen bottles a day, costs £300. (*Report by Major Wimberley, Officiating Deputy Superintenden, Port Blair, from the Tropical Agriculturist, Vol. III., p. 58.*)

Citronella Grass-stuff as a Paper Fibre.—A correspondent, writing in the *Ceylon Observer*, suggested the use of citronella grass as paper material. In extracting oil from the grass, it is boiled or subjected to steam, under pressure, and as this is one of the first operations to which the raw material is subjected in paper manufacture, grass which has been thus treated should be much more easily utilised than material not previously boiled. Citronella grass, like esparto, can be supplied entirely free from knots, which is a great advantage in paper manufacture. At present about 3,500 tons of citronella are available for export in Ceylon. (*Tropical Agriculturist, Vol. III., p. 831.*)

FIBRE. 1111

Medicine.—This is regarded as officinal by the Indian Pharmacopœia; the essential oil of citronella used medicinally being imported from Ceylon. In its properties it closely approaches that from **A. citratus.**

MEDICINE. Oil. 1112

Dr. Irvine, in his *Materia Medica of Patna,* says that the infusion of the leaves in doses of ½ to 2 ounces is used as a stomachic.

Infusion. 1113

Domestic Uses.—Dr. Trimen writes me that this grass is in Ceylon largely used for thatching.

Andropogon pertusus, *Willd.*

1114

Syn.—A. PUNCTALUS, *Roxb.*; HOLEUS PERTUSUS, *Linn.*; A. ANNULATUS, *Forsk.*; LEPEOCERCIS PERTUSUS, *Hassk.*

Vern.—*Pulwal, pulúah, rukar,* N.-W. P.; *Palwán,* or *palwa minyár,* PB.

Habitat.—Found on old pasture grounds, generally shaded by trees, in the plains of the Panjáb and North-West Provinces, and at lower elevations of the Himálaya. Abundant in Hissar.

Fodder.—Dr. Stewart, writing under **A. annulatus,** *Forsk.*, says: "It is considered excellent fodder for bullocks, &c., and for horses, when green." In Australia it is regarded as one of the best grasses to withstand long droughts, while it will bear any amount of feeding. (*Baron Von Mueller.*)

FODDER. 1115

§ "Good for stacking, will remain for 12 or 13 years: much stacked at the Hissar farm. Is especially grazed by buffaloes." (*Mr. Coldstream, Commissioner, Hissar.*)

A. scandens, *Roxb.*

Habitat.—Found in the Panjáb, in Kashmír, and Bundelkhand. It is a coarse grass growing commonly in hedges. It flowers during the rains.

Fodder.—Cattle are apparently not fond of it.

FODDER. 1116

A. Schœnanthus, *Linn.*; *Royle, Ill. Him., t. 97.*

1117

THE GERANIUM GRASS; RUSA OIL GRASS; OIL OF GINGER GRASS.

Syn.—A. MARTINI, *Roxb.*; A. NAEDOIDES, *Nees*; A. CALAMUS AROMATICUS, *Royle*; A. PACHNODES, *Trin.*; CYMBOPOGON MARTINI, *Munro.*

Vern.—*Rúsá ghás, rousá-ghás, rousá-kú-ghás, musel, mirchia gand,* HIND.; *Agyá-ghás, gandha bena,* BENG.; *Bujina, pálá-khari,* N.-W. P.; *Raúns,* PB.; *Mirchia-gard,* SIWALIKS; *Rosegavat, rohisha,* BOMB., MAR.; *Rusa-ka-tel, roshel* (the oil).

In **Moodeen Sheriff's** Supplement to the Pharmacopœia of India the vernacular names for the Lemon grass are given under **A. Schœnanthus** instead of **A. citratus.**

Habitat.—This grass is wild in Central India, the North-West Provinces, and the Panjáb. **Dr. Roxburgh** first saw the plant from seeds, forwarded to him by **General Martin,** collected in Balaghat during the last war with Tippu Sultan. This is the roussa paper grass, abundant everywhere in the Deccan.

Geranium Grass.

OIL. 1118

Oil.—The oil obtained from this plant has come to bear a number of names, which appear for the most part to be of modern origin, and to indicate the use to which it is put. Perhaps the name by which it is most generally known is rusa oil, *roshel* (*rusa-ka-tel*). As pointed out by the distinguished authors of the *Pharmacographia*, these names look exceedingly like a corruption from rose oil, the more so since the principal consumption is as an adulterant for attar of roses. It is curious, however, that, as stated by **Dr. Dymock** (*Mat. Med., W. Ind., 690*), the Indian distillers and dealers know nothing of this use. About 40,000 lbs. are annually exported from Bombay to the Red Sea ports, chiefly to Jedda; a small amount finds its way to Europe, but the great bulk is sent to European Turkey. In Arabia and Turkey it appears under the name of *idris yaghi*, and in the attar-producing districts of the Balkan, it is known, at least to Europeans, as geranium oil or Palmarosa oil. The name Geranium oil, which has caused much confusion with the true geranium oil or oil from the geranium plant, has apparently come into existence from the fact that the so-called geranium grass oil is used to adulterate the true geranium oil, which in its turn is used as an adulterant for attar of roses. **Piesse** says that true geranium oil is worth 3*s.* an ounce, whereas geranium grass oil is not worth more than that sum per lb.

The *Pharmacographia* gives some interesting information regarding the mode by which *rusha* oil is refined so as to prepare it for admixture with the attar of roses. The *rusha* oil is shaken with water, acidulated with lemon-juice, and then exposed to the sun and air. By this process it loses its penetrating after-smell and acquires a pale-straw colour. "The optical and chemical differences between grass oil thus refined and attar of roses are slight, and do not indicate a small admixture of the former. If grass oil is added largely to attar, it will prevent its congealing" (*p. 728*). "It was formerly added to the attar only in Constantinople, but now the mixing takes place at the seat of the manufacture. It is said that in many places the roses are sprinkled with it before being placed in the still. As grass oil does not solidify by cold, its admixture with rose oil renders the latter less disposed to crystallize" (*p. 267*). The degree of admixture with grass oil is thus determined by the crystallizing of the attar.

From the fact of one of the largest supplies of Indian grass or *rusha* oil being in Nimar district, Khándesh, Bombay Presidency, the oil has come to bear the commercial name of Nimar (or, as it is commercially written, Namar) oil. **Dr. Dymock** gives some interesting information regarding this industry. The first mention apparently of the oil is by **Maxwell** in 1825 (*in the Calcutta Med. Phys. Trans., Vol. I., p. 367*). It was afterwards described by **Forsyth**, 1827 (*Ibid., Vol. III., p. 213*). It is only within comparatively recent times that the oil has become an article of commercial value. **Dr. Dymock** thus describes its manufacture: "It is now chiefly distilled in the Nimar district, an iron still being used and a very small quantity of water; when the still is carelessly worked, the grass burns and communicates a dark colour to the oil, which should be of a pale sherry colour. I am assured by the Bombay dealers that all the oil of commerce is more or less adulterated; a comparison of the commercial article with some oil distilled by myself supports this statement; the adulteration is said to be practised by the distillers, who, I am informed, are regularly supplied with oil of turpentine from Bombay. The grass flowers in October and November, and is then fit for cutting: 373 lbs. of grass received from Khándesh, and submitted to distillation under my own superintendence in Bombay, yielded 1 lb. 5½ ozs. of oil." It appears that the chief substances used to adulterate the Khándesh *rusha* oil are turpentine, ground nut, rape and linseed oils. With the two first the turbidity passes off in a day or two, hence they are preferred, and turpentine is chiefly used, because it cannot be detected by the evaporation test. **Dr. Dymock** adds: "The genuine oil is dextrogyre; the ray is rotated 39° to the right by 100 min.: 200 min. rotate it 78°. Some samples of the commercial article rotate the ray about 13° to the right; some have little or no effect upon it."

MEDICINE. Oil. 1119

Medicine.—The oil is used as a liniment in chronic rheumatism and neuralgia, and is believed to have the property of curing baldness. It resembles, in quality and appearance, the lemon-grass oil. The oil is seldom taken internally by the natives, but is considered a powerful stimulant when applied externally.

Ainslie calls **A. Nardus** (?) ginger grass or spice grass, and says that an infusion of it is used as a stomachic, and that occasionally an essential oil is prepared from it which is useful in rheumatism; but the plant he refers to is probably **A. Schœnanthus.** Similar confusion exists in the writings of most Indian authors. § "There is a grass about Bombay which smells like fresh ginger. I think it must have been introduced, as it is only found about fields and gardens." (*Surgeon-Major W. Dymock.*) It seems probable this is **Ainslie's** plant, which may prove a distinct species from **A. Schœnanthus.**

The *Bombay Gazetteer* gives an interesting account of the manner in which *rusa* oil used to be prepared at the Panch Mahâls. Paper, soap, and grass-oil were formerly prepared, but these industries have almost entirely disappeared: "The grass-oil made from the large-bladed aromatic grass known as *roisa,* which used to grow over large stretches of waste land, was, at the rate of 4*s.* (R2) a pound, bought in considerable quantities, and used partly as a remedy for rheumatism, partly to mix with attar of roses. The oil was extracted by distillation. A rough stone oven was built by the side of a stream, and in it a large metal caldron was placed and filled with bundles of grass and water. When full, a wooden lid was put on, and sealed with a plaster of ground pulse, *adad.* Through a hole in the lid one end of a hollow bamboo was thrust and the other end passed into a smaller metal vessel securely fixed under water in the bed of the stream. The oven was then heated, and the vapour passing through the hollow bamboo was, by the coldness of the smaller vessel, precipitated as oil." (*Bombay Gazetteer, III., 251.*)

Mr. Baden Powell appears to be referring to this plant, under **Cymbopogon aromaticus** (**Vern.**—*Khas, usar, balam*). He says: "Considered by natives cool and astringent, useful in skin diseases, bilious affections, and special diseases. It is an aromatic stimulant, useful in fever, and to make *tatties.* The roots are dug up in March and April. Used as an aromatic in fever. Gives a fragrant oil." (*Panjáb Products, Vol. I., 383.*) This quotation might, as far as its facts are concerned, be equally applicable to **A. muricatus**; but since in another part of his work **Mr. Baden Powell** alludes to the *khas-khas* under its correct botanical synonym, and the more so since **A. Schœnanthus** is a native of the Panjáb, it seems a just inference that by **Cymbopogon aromaticus** is meant **A. Schœnanthus.** If this be correct, the above vernacular names become of interest as those in use in the Panjáb for this species. **Dr. Stewart** describes two species as common in the plains of the Panjáb, but he gives their characters and vernacular names jointly, and it is therefore impossible to separate them. The scientific names which he gives to these Panjáb forms are equally confusing—**Cymbopogon Iwarancusa,** *Schult.*, and **C. laniger,** *Desf.* Regarding the latter there seems little or no doubt it is **Andropogon laniger,** *Desf.*, the species which inhabits the dry, sandy, and desert tracts of the Panjáb, Sind, &c. The former seems to be **A. Schœnanthus,** *Linn.*, which by **Stewart** has been incorrectly named **Cymbopogon Iwarancusa,** *Schult.* The oil which **Stewart**, on the authority of **Vigne**, says is prepared near Hassan Abdál would, according to this view, be a form of rusa or Nimar grass oil. **Stewart** adds the exceedingly interesting fact, which does not appear to have been observed by other authors: "A spirit" (*arak*), he says, is "also distilled from the grass with spices, &c., and is said to be useful in indigestion and fever." **Madden** mentions that the roots are sometimes luminous.

§ "The decoction of it is a febrifuge, and I have used it in cases of cold and feverishness with benefit." (*Assistant Surgeon Bolly Chand Sen, Calcutta.*) "Excellent external application in rheumatism." (*Assistant Surgeon Nehal Singh, Saharunpore.*) "Mixed with tea in equal proportion and infused, is sometimes used as a diaphoretic in fevers." (*Surgeon-Major C. J. McKenna, Cawnpore.*)

FODDER. 1120

Fodder.—**Duthie** writes: "The grass is a favourite fodder for cattle, and **Mr. Miller** tells me that at Banda (North-West Provinces) it is grown in meadows kept for the purpose and sold in the bazar."

General **Martin** collected seed of this grass in the high-lands of Balaghat, while there with the army during the war with Tippu Sultan; and after growing it in Lucknow sent specimens to **Dr. Roxburgh,** with the remark that he had noticed the cattle were voraciously fond of it, but that it had so strong an aromatic and pungent taste that the flesh of the animals fed upon it, as also the milk and butter, were strongly scented with the plant. There seems some mistake in this account of General Martin having collected **A. Schœnanthus** on high-lands: it frequents swamps. **A. laniger,** *Desf.*, is found on dry places.

§ "Not very good for grazing. Grows in swamps, but not abundant in Hissar. A tall grass, too coarse to stack but used for thatching and screens." (*Mr. Coldstream, Commissioner, Hissar.*)

1121

ANEILEMA, *R. Br.; C. B. Clarke, in DC. Monogr. Phaner., III., 195; Gen. Pl., III., 849.*

A genus of herbaceous perennials, belonging to the Natural Order COMMELINACEÆ, and the Tribe COMMELINEÆ, comprising 60 species, chiefly Asiatic and African, only 5 or 6 occurring in America. In India there are 28 species met with in the plains, one or two ascending to 2,500 feet in altitude.

Stem erect, delicate rarely robust, simple or branched. *Leaves* scattered or crowded and radical. *Peduncles* often terminal or in the axils of the upper leaves, simple or branched. *Inflorescence* a many-flowered panicle with the middle branches the longest, or corymbose, or few-flowered, rarely 1-flowered; *bracts* few, rarely herbaceous and sheathing; in a few species they resemble the spathaceous condition of Commelina. *Stamens* 6, of which 3-2 are fertile, free filaments, tapering, hairy or glabrous; of these 2 are opposite the lateral and interior sepals and have large anthers, the cells of which are parallel and united throughout their length; 1 opposite the anterior petal generally fertile, but with the anthers diverging; the other 3 stamens have sterile anthers. *Ovary* sessile, 3-locular. *Seeds* few or solitary, arranged in 1 or 2 vertical series.

1122

Aneilema scapiflorum, *Wight, Ic., t. 2073.*

Syn.—COMMELINA SCAPIFLORA, *Roxb., Fl. Ind., Ed. C.B.C., 59;* A. TUBEROSUM, *Buch., Ham. in Wall., Cat.;* MURDANNIA TUBEROSA, *Royle, Him. Ill., t. 95.*

Vern.—*Kureli,* BENG.; *Siyah músli,* or *músli-e-siyah,* PERS., HIND.; *Sismuliá,* GUJ.

The above (Hindustani) vernacular names are by authors said to be applied to the tubers of this plant, but it would seem that this can only be the case when used as substitutes for the true *Siyah músli*—the roots of **Curculigo orchioides,** *Gærtn.;* it is difficult to discover whether it possesses any properties of its own or not.

References.—*DeCandolle, Monog. Phaner., by C. B. Clarke, III., 200; Clarke's Commel et Cyrt., Beng., t. 14; Stewart's Pb. Pl., 236.*

Habitat.—A native of the Himálaya from the Jumna to the Khásia Hills; Tenasserim, and south to Cape Comorin, ascending to 1,000 feet in altitude.

Botanic Diagnosis.—Roots perennial fascicled, tuberous-fusiform. *Scape* composed of dichotomous elongated pikes, branches angled; bracts large, sheathing or spathaceous, often ochreate. *Petals* equal, rounded, concave, blue. *Stamens* 3 perfect, with 3 sterile alternating ones; filaments all hairy; anthers blue. *Capsule* 3-angled ellipsoid, apex acute or mucronate. *Seeds* white, minutely reticulated and very minutely glandular.

Properties and Uses—

MEDICINE. Tuberous Root. 1123

Medicine.—Said to have astringent and tonic properties, and considered by natives to be hot and dry; useful in headache, giddiness, fever, jaundice, and deafness. It is also regarded as an antidote to poisons, and a cure for snake-bite.

§ "The roots of this plant bear no resemblance to the *Siyah Múslí* of the bazars." (*Surgeon-Major Dymock, Bombay.*) "The dried powder, mixed with sugar, is used as an aphrodisiac. With the juice of the *túlsí* leaves it is administered for pains in the kidneys, and is one of the chief remedies used by hakims for spermatorrhea." (*Surgeon Emerson, Calcutta.*) "Root-bark dried in the shade is said to have been employed with benefit in cases of asthma. It is a remedy of great repute for impotence and spermatorrhœa. It is used also in colic, piles, and infantile convulsions. In combination with *Dárfilfil* it is employed in bites of mad dogs, both internally and externally." (*Asst. Surgeon Gholam Nabi.*) "It is used for incontinence of urine." (*Surgeon-Major C. W. Calthrop, Morar.*)

ANEMONE, *Linn.; Gen. Pl., I., 4.*

1124

A genus of perennial herbs, belonging to the Natural Order RANUNCULACEÆ. *Leaves* radical, lobed or divided. *Flowers* occurring in 1 or more flowered simple or branched scapes; involucre 3-partite, bracts free or connate. *Petals* O. *Stamens* many, outer deformed or petaloid. *Carpels* many; *ovule* 1, pendulous. *Fruit* a head of sessile achenes, with naked or bearded styles.

There are some 80 species, natives of the cold and temperate regions, of which 15 occur in India.

Anemone obtusiloba, *Don; Fl. Br. Ind., I., 8.*

1125

Vern.—*Rattanjog, padar,* PB.

Habitat.—Temperate and Alpine Himálaya, from Kashmír to Sikkim; altitude 9,000 to 15,000 feet.

Chemical Composition.—Several varieties of Anemone contain a principle, *anemone* or *anemone camphor,* which is stated to be an acrid poison. *Anemonic acid* is also contained in the fresh herbs.

MEDICINE. Root. 1126

Medicine.—In Hazára the pounded root, which is acrid, is mixed with milk and given internally for contusions. In Bissahir it is said to be used as a blister, but to be apt to produce sores and scars. (*J. L. Stewart, Pb. Pl.*)

§ "The seeds (which are as sweet as almonds), if given internally, produce vomiting and purging. The oil extracted from them is used in rheumatism." (*Bappoojee Joyaram Bhoslay, Junagarh.*)

Anethum Sowa, *Kurz,* and A. graveolens, *Linn.*

See **Peucedanum graveolens,** *Benth.;* UMBELLIFERÆ.

1127

ANGELICA, *Linn.; Gen. Pl., I., 916.*

Angelica glauca, *Edgw.; Fl. Br. Ind., II., 706;* UMBELLIFERÆ.

Vern.—*Chora* or *Churá,* PB.

Habitat.—From Kashmír to Simla, altitude 8,000 to 10,000 feet; found also in the Dhanla Dhar Range, above the Kangra Valley.

MEDICINE. 1128

Medicine.—A cordial and stimulant remedy, formerly used in the cure of flatulence and dyspepsia. It is also used in obstinate constipation, and in bilious complaints.

FOOD. 1129

Food.—Its aromatic root is added to food to give it a flavour like that of celery.

Angustura Bark, see **Galipea Cusparia,** *St. Hil.;* RUTACEÆ.

Anise-seed, see **Pimpinella Anisum,** *Linn.;* UMBELLIFERÆ.

Anise, Star, see **Illicium Anisatum,** *Linn.;* MAGNOLIACEÆ.

1130

ANISOCHILUS, *Wall.; Gen. Pl., II., 1177.*

Anisochilus carnosus, *Wall.; DC. Prod., XII., 81;* LABIATÆ.

Vern.—*Panjiri-ká-pát, sitá-ki-panjiri,* HIND.; *Kápurli, chora-onva, kapúrli,* BOMB., MAR.; *Ajmánu-pátru, ajamá,* GUJ.; *Ajván-ka-patta, panjiri-ká-pattá,* DUK.; *Karppúra-valli,* TAM.; *Karpúra-valli, ǀómamu-áku, róga-chettu,* TEL.; *Chómara, kattu-kúrkká, kúrkká, patu-kúrkká,* MAL.; *Dodda-patri,* KAN.

Habitat.—Found in the North Circars and Malabar.

MEDICINE. Leaves. 1131

Medicine.—**Ainslie** says that the fresh juice of the leaves mixed with sugar-candy is given by the Tamil doctors in cynanche, and, mixed with sugar and gingelly oil, is used as a cooling liniment for the head. The leaves and stems are given in infusion in coughs and colds as a mild expectorant, especially for children. The plant yields a volatile oil which is said to be stimulant, diaphoretic, and expectorant. (*Dymock; Pharm. Ind.; Bidie, Madras Quart. Med. Journ., 1862.*)

§ "The juice of the leaves mixed with sugar and human milk is in Mysore a popular domestic remedy for coughs in children." (*Surgeon-Major North, Bangalore.*)

"Juice used in catarrh." (*Surgeon-Major J. J. L. Ratton, M.D., Salem.*)

Anisodus luridans, *Linth. & Otto.,* see **Scopolia lurida,** *Dunal.;* SOLANACEÆ.

1132

ANISOMELES, *R. Br.; Gen. Pl., II., 1207.*

Anisomeles malabarica, *R. Br.; DC. Prod., XII., 456; Wight, Ic., t. 864;* LABIATÆ.

Vern.—*Chodhará,* MAR., BOMB.; *Mogbíre-ka-pattá,* DUK.; *Peymarutti, péya-verutti, irattai-péy, marutti,* TAM.; *Moga-bírákú, moga-bíra, maga-bíra, mogbíra, mabheri-china-ranabheri,* TEL.; *Péyi-meratti, peruntumba, karintúmba,* MAL.; *Bútan-kúshum,* SANS.

§ "This plant, in Bombay, is known as *Chodhara* (four-angled). It is aromatic, but is not the *Gule-gao-zubán,* which is mucilaginous." (*Sakhárám Arjun Rávat, Bombay.*)

Dr. Birdwood mentions this as one of the plants which in Bombay are sold as ***gao-zubán***, but this seems to be a mistake. I have, however, received that as the name by which it is known in Cutch; the true *gao-zubán* is imported into India from Persia,—see **Echium**; but doubtless many other plants are sold under that name.

Habitat.—Found in South India. **Dr. Dymock** says it is pretty common on the Ghâts in the Bombay Presidency, but appears to be better known in South India.

Medicine.—"Few plants are held in higher esteem, or are more frequently employed in native practice than this. An infusion of the aromatic bitter leaves is in common use in affections of the stomach and bowels, catarrhal affections, and intermittent fevers. According to **Dr. Wight**, in addition to its internal use in the cure of fevers, patients are made to inhale the vapour of a hot infusion so as to induce copious diaphoresis." (*Pharm. Ind.*) **Ainslie** tells us that an infusion of the leaves is given to children in colic, dyspepsia, and fever arising from teething. A decoction of the plant, or the essential oil distilled from the leaves, is used externally in rheumatism. (*Dymock.*)

MEDICINE. Leaves. 1133 Infusion. 1134 Decoction. 1135

§ "A handful of the leaves is used in Mysore with a vapour bath when profuse diaphoresis is required." (*Surgeon-Major John North, Bangalore.*) "Diaphoretic, dose of infusion 1 to 2 oz." (*Surgeon W. Barren, Bhuj, Cutch.*)

Anisomeles ovata, *R. Br.*

1136

Vern.—*Gobura.*

Habitat.—Found in Ceylon, Coromandel, Bombay, Bengal, and Nepal.

Medicine.—The whole plant has a strong camphoraceous smell In Ceylon a distilled oil is prepared from it, and found useful in uterine affections. It has also carminative, astringent, and tonic properties.

MEDICINE. 1137

ANISOPHYLLEA, *Brown; Gen. Pl., I., 683.*

1138

A genus of trees or shrubs, belonging to the Natural Order RHIZOPHOREÆ.

Leaves **alternate, ex-stipulate.** ***Flowers*** **minute, in axillary simpled or fascicled spikes, ebracteate or minutely bracteolate, bi- or uni-sexual.** ***Petals*** **4, small, involute.** ***Stamens*** **8;** ***filaments*** **short, subulate;** ***anthers*** **small, didyamous.** ***Ovary*** **inferior 4-celled;** ***styles*** **4.** ***Embryo*** **exalbuminous.**

Anisophyllea zeylanica, *Benth.; Fl. Br. Ind., II., 442.*

1139

Vern.—*Wellipyanne*, SINGH.

Habitat.—A tree of the southern and central parts of Ceylon, ascending to 1,500 feet.

Structure of the Wood.—Greyish brown, moderately hard.

TIMBER. 1140

ANODENDRON, *A.DC.; Gen. Pl., II., 719.*

1141

Anodendron paniculatum, *A.DC.; Fl. Br. Ind.*, III., *668;* APOCYNACEÆ.

Vern.—*Dúl*, SINGH.

Habitat.—An immense climber, occurring from Sylhet to Martaban; the Deccan Peninsula; Western Ghâts from the Konkan southward; Ceylon; altitude from the plains up to 2,000 feet.

Fibre.—Tbe stem yields a fine and very strong fibre, much used by the Singhalese (*Thwaites; Trimen.*)

FIBRE. 1142

1143

ANŒCTOCHILUS, *Bl.; Gen. Pl., III., 598.*

Anœctochilus setaceus, *Blume; Wight, Ic., t. 1731;* ORCHIDEÆ.

Vern.—*Wanna-rajah,* (vulg.) SINGH.

Habitat.—Damp forests of Ceylon.

MEDICINE. 1144

Medicine.—Dr. **Trimen** informs me that this is considered a valuable medicine in Ceylon.

1145

ANOGEISSUS, *Wall.; Gen. Pl., I., 687.*

A genus of trees or shrubs, belonging to the Natural Order COMBRETACEÆ, comprising some 4 or 5 species, natives of Africa and India.

Leaves alternate or falsely opposite, petioled entire. Flowers in dense globose heads, on axillary peduncles, much shorter than the leaves. *Calyx-tube* long, attenuated above the ovary, sub-persistent, with 5 deciduous lobes. *Petals* wanting. *Stamens* 10, in two series, and without glands or staminodes at their base (the character of the tribe COMBRETEÆ as distinguished from GYROCARPEÆ). *Ovary* inferior 1-celled; *style* filiform simple; *ovules* 2-pendulous from the top of the cell. *Fruits* small, coriaceous, compressed, 2-winged, packed horizontally into dense heads. *Seed* 1.

1146

Anogeissus acuminata, *Wall.; Fl. Br. Ind., II., 450; Bedd., Fl. Syl., t. 16.*

Syn.—CONOCARPUS ACUMINATA, *Roxb.; Fl. Ind., Ed. C.B.C., 384.*

Vern.—*Chakwa,* BENG.; *Gara hesel, pandri, pansi,* KOL; *Panchi, pasi,* URIYA; *Numma,* TAM.; *Páchi mánu, panchman, paunchinan, bucha karum,* TEL.; *Phás,* or *phassi,* MAR.; *Saikamehhia, thekri, napay,* MAGH.; *Yung, sehoong,* ARACAN; *Yungben, yón,* BURM.

Habitat.—A large, deciduous tree, met with in some districts of Bengal, Orissa, South India, Chittagong, and Burma.

Botanic Diagnosis.—*Leaves* elliptic or oblong, acute at both ends, villous or pubescent beneath, peduncles solitary (rarely clustered), very rarely divided, ripe fruits shining, glabrous.

There are two distinct varieties of this plant:—

Var. 1st, **typica.**—Leaves broad lanceolate, fulvous beneath, peduncles with obovate bracts, often large and leaf-like, fruit broadly winged with a deflexed pubescent beak much longer than the nucleus.

Habitat.—Northern edge of the Deccan, ascending to altitude 3,000 feet; Bundelkhand, the Circars, Godáveri, and North-West India.

Var. 2nd, **lanceolata.**—Leaves narrow lanceolate, grey beneath, bracteoles on the peduncles small, linear, very deciduous, fruit winged, subquadrate, with an erect beak shorter than the nucleus.

Habitat.—Pegu, Tenasserim.

Properties and Uses—

TAN. Leaves. 1147

Tan.—The leaves are used in Gamsur for tanning. (*Gamble.*)

TIMBER. 1148

Structure of the Wood.—Grey, sometimes yellowish grey, with a greenish tinge, shining, in structure moderately hard, resembling that of **Anogeissus latifolia.** It warps and cracks in seasoning, and is not very durable, especially where exposed to water. Weight 57 lbs. per cubic foot.

Used in Burma and in Madras for building. **Roxburgh** says it is durable if kept dry, but soon decays if exposed to wet.

1149

Anogeissus latifolia, *Wall.; Fl. Br. Ind., II., 450; Wight, Ic., t. 994; Bedd., Fl. Sylv., t. 15.*

Syn.—CONOCARPUS LATIFOLIA, *Roxb.; Fl. Ind., Ed. C.B.C., 384.*

Vern.—*Dhává, dháurá, dháuri, dhau, dhauta, dohu, bakla, bakli,* HIND.

Gólra, goldia, dhaukra, dhokri, dau, RAJ.; *Dhauk,* ULWAR; *Dháurá,* C. P.; *Khardháwa,* BANDA; *Vellay naga, namme, veckali,* TAM.; *Chirimánu, sherimán* or *siri-mánu, yettama, tirman, yella maddi,* TEL.; *Dham,* OUDH; *Dohu, dhobu,* URIYA; *Hesel,* KOL. and SANTAL; *Dhau,* MAL. (S. P.); *Dábriá, dhávdo,* GUJ.; *Dhámorá, dhávdá,* MAR.; *Dhává,* DUK.; *Dinduga, dindiga, dindlu, bejalu, dindal,* KAN.; *Arma, yerma,* GOND.; *Dhawa,* BAIGAO; *Dhaundak,* BHIL; *Dhaura,* KURKU; *Daawú,* SINGH.

Habitat.—A large, handsome tree, met with in the Sub-Himálayan tract from the Rávi eastward, ascending to 3,000 feet in Central and South India. Very plentiful in Melghát; common in the Upper Godáveri; not met with in the trans-Gangetic Peninsula.

Botanic Diagnosis.—*Leaves* broad elliptic-obtuse at both ends; peduncles 1 or more from the same axil, often branched, bracteoles inconspicuous; ripe fruits shining, glabrous, the beak as long as the nucleus or longer.

Properties and Uses—

GUM. 1150

Gum.—It yields a gum, which is extensively sold for use in calico-printing. It occurs in clear straw-coloured, elongated tears, adhering into masses, sometimes honey-coloured or even brown from impurities. As an adhesive gum it is inferior in strength to gum arabic, in consequence of which it commands a much lower price in Europe, the more so since it is nearly always mixed with the bark of the tree, sand, and other impurities, and adulterated with the brown tears which are probably derived from some other plant than **Anogeissus.** In India the reputation of this gum stands high with the calico-printers, especially of Lucknow, and it probably possesses some specific peculiarity justifying this preference, since it is used with certain dye-stuffs, such as with *haldí* **(Curcuma longa),** while gum arabic or *babúl* is used with *madder* **(Rubia cordifolia).** The *Dhává* or *bakli* gum is generally collected in April.

According to the **Rev. A. Campbell,** Pachumba, the gum of this tree is used by the Santals in the treatment of cholera. It is much eaten by the people in Thána, and in the Central Provinces it is regarded as more adhesive than *babúl* gum. (See *Gazetteers.*)

DYE & TAN. Leaves. 1151

Dye and Tan.—In addition to the gum being used by calico-printers, **Dr. Dymock** informs me that the leaves are in Bombay used as a tan. They were analysed by **Dr. Lyon** and were found to contain as much tannin as those of the *Sumach* tree. The leaves yield a black dye and are very useful in tanning. (*Bomb. Gaz., XIII., Part I., 24.*) **Mr. Duthie** reports that they are also used as a tan in the N.-W. Provinces.

TIMBER. 1152

Structure of the Wood.—Grey, hard, shining, smooth, with a small, purplish-brown, irregularly shaped, extremely hard heartwood. Sapwood in young trees and branches yellow. Annual rings marked by darker lines. Weight about 65 lbs. per cubic foot.

It is highly valued on account of its great strength and toughness, but it splits in seasoning, and unless kept dry is not very durable. It is used for axe-handles, poles for carrying loads, cart-axles, in the construction of furniture, agricultural implements, and in ship-building. It has been recommended for sleepers. Out of 18 sleepers which had lain seven to eight years on the Mysore State Railway, there were found, when taken up, four good, ten still serviceable, and four bad. It gives an excellent charcoal. A very valuable firewood; it is strong and tough and much used for cart-axles, poles, and in calico-printing. *The Bombay Gazetteer* (III, p. 199) says that although it does not rank as a timber tree it makes good fuel, and is used for ploughs in the Panch Maháls, Gujarát. In the Deccan it is regarded as one of the commonest and most useful of trees.

1153

Anogeissus pendula, *Edgw.; Fl. Br. Ind., II., 451.*

Syn.—CONOCARPUS MYRTIFOLIA, *Wall.*

Vern.—*Dháu, dháukra, kala dháukra,* MEYWAR; The lesser *dhauk,* ULWAR; *Kardahi,* HIND.

Habitat.—A small, gregarious tree, with pendulous branches, found in the arid and northern dry zones of Rájputana-Malwa plateau, as far as the Nerbudda, in Nimar, and in the Mandla District.

Botanic Diagnosis.—*Leaves* small, elliptic or obovate-acute or obtuse, always narrowed at the base; peduncles, solitary simple; fruit subquadrate, ultimately glabrous; beak much less than half the height of the nucleus.

TIMBER. 1154

Structure of the Wood.—Hard, yellowish white, with a small, irregular, blackish-purple heartwood. Weight 59 lbs. per cubic foot.

It coppices well, but the wood is not in general use.

1155

ANONA, *Linn.; Gen. Pl., I., 27.*

A genus of trees or shrubs, belonging to the Natural Order ANONACEÆ, comprising some 50 species, inhabitants of America and Africa, a few being naturalised in Asia. Three species are met with in India, of which two may be described as naturalised.

Flowers solitary or fascicled, terminal or leaf-opposed. *Sepals* 3, small, valvate. *Petals* 3-6, valvate, 2-seriate, the inner series sometimes wanting, outer triquetrous, base concave. *Stamens* numerous; anther-cells narrow, dorsal, contiguous, top of connective ovoid. *Ovaries* many, subconnate; style oblong, ovule, 1 erect. *Ripe carpels* confluent into a many-celled ovoid or globose, many-seeded fruit.

The word Anona is said to be derived from the Malay name *Manoa,* pronounced in Banda Islands *Menona.* Some authorities derive the word from *Annona,* Lat., the year's produce or grain; *annus,* a year. It is difficult to see what could have suggested the name if this derivation of the word be correct.

FOOD. Fruit. 1156

Anona Cherimolia, *Miller.* A native of America, and of Jamaica and the other West Indian Islands, nearly allied to **A. squamosa,** is regarded by the Creoles as the most delicious fruit in the world—an opinion not generally confirmed.

A. muricata, *Linn.; DC. Prod., I., 84.*

THE SOUR SOP.

FOOD. Fruit. 1157

The fruit of this tree is sometimes to be seen in India, for, although not so common as either of the following species, it is occasionally seen in cultivation. It often attains a weight of upwards of two pounds. It is greenish and covered with prickles; the pulp is white, and has an agreeable, slightly acid flavour. Drury says this species is sparingly cultivated in Madras; "the fruit is muricated with soft prickles."

1158

A. reticulata, *Linn.; Fl. Br. Ind., I., 78.*

THE BULLOCK'S HEART, OR TRUE CUSTARD APPLE of the West Indies.

Vern.—*Louná, rám-phal,* HIND.; *Nóna,* BENG.; *Góm,* SANTAL; *Rámphal,* BOMB., DUK., MAR., GUJ., KAN.; *Rámsitá* or *rámsítu-plam,* TAM.; *Rámá-pandu, rámá-phalam* or *rámá-chandar-pandu,* TEL.

Habitat.—A small tree, supposed by some authors to be a native of Asia, naturalised in some parts of India, extensively cultivated; occurring everywhere in Bengal, Burma, and South India.

DYE. Fruit. 1159

Dye.—The dry unripe fruit yields a black dye, and the fresh leaves a fairly good quality of indigo.

TAN. Leaves. 1160

Tan.—§ "The leaves and young twigs are largely used for tanning." (*E. A. Fraser, Rájputana.*)

FIBRE. 1161

Fibre.—A good fibre is prepared from the bark of the young twigs.

FOOD. Fruit. 1162

Food.—The fruit, which resembles a bullock's heart, ripens during the latter part of the rainy season, and is eaten by natives, but only rarely by Europeans.

MEDICINE. Bark. 1163; Fruit. 1164

Medicine.—The bark is said to be a powerful astringent and to be much used as a tonic by the Malays and Chinese. The fruit is reported to be used in the West Indies, and by the natives of America, as an anti-dysenteric and vermifuge.

TIMBER. 1165

Timber.—Very little has been written regarding the timber obtained from this small tree. **Skinner** gives 40 lbs. as the weight per cubic foot.

1166

Anona squamosa, *Linn.; Fl. Br. Ind., I., 78; Roxb., Fl. Ind., Ed. C.B.C., 453.*

CUSTARD APPLE of Europeans in India; SWEET-SOP or SUGAR APPLE of the West Indies and America.

Vern.—*Sharífah, át* or *átá sitáphal,* or *sitáfal,* HIND., DUK., GUJ., MAR., ARAB.; *Sharífa, behli,* N.-W. P.; *Áta, lúna,* BENG.; *Átá, kátál,* ASS.; *Mandar gôm,* SANTAL; *Sirpha, átta,* MAL.; *Sita-palam,* or *sítá-pázham,* TAM.; *Sitapandu,* TEL.; *Atta,* CINGH.; *Auza,* or *awza,* BURM. The fruit is called *Sharífah* and *Káj* in PERSIAN.

Habitat.—A small tree, naturalised in Bengal and the North-West Provinces. It is common in Rájputana (*Ráj. Gaz., p. 27*). Abundant in many places in Burma, but principally at Prome, where the neighbouring hills are covered with orchards of this fruit (*British Burma Gaz., Vol. I., 430*). "Cultivated as far north as Gurdáspur in the Panjáb. Almost wild in the Central Provinces and Bundelkhand (near old forts) and in swamps near Burmdeo in the Kumaon Bhábar." (*Brandis.*) "Grows wild in the Deccan up to Sholapore. In Sind it is also cultivated, but the fruit is inferior." (*Murray.*) "Grows in a semi-wild condition in Kaira, and in the Panch Mahâls, Bombay." (*Gazetteer.*)

1167

HISTORY.—Custard apples have been identified among the sculptures of the Ajanta caves as well as of the Bhárhut Stupa. This identification is opposed to the theory that the custard apple is an introduced tree. On this subject **General Cunningham** remarks: "My identification of this fruit amongst the Máthura sculptures has been contested on the ground that the tree was introduced into India by the Portuguese. I do not dispute the fact that the Portuguese brought the custard into India, as I am aware that the East India Company imported hundreds of grindstones into the fort of Chunár, as if to illustrate the proverb about carrying coals to Newcastle. I have now travelled over a great part of India, and I have found such extensive and such widely distant tracts covered with the wild custard apple, that I cannot help suspecting the tree to be indigenous. I can now appeal to one of the Bhárhut sculptures for a very exact representation of the fruit and leaves of the custard apple." (*Bhárhut Stupa, 55.*) "The names of the two varieties of custard apple, *Rámphal* and *Sítáphal*, are in themselves almost enough to show that from very early times the trees have been grown and honoured by the Hindús." (*Bomb. Gaz., XII., 490.*)

Botanical evidence is opposed to **General Cunningham's** opinion on this subject; indeed, there seems hardly any doubt as to **Anona squamosa,** *Linn.,* being an introduced plant (see account given under ANONACEÆ); the date of its introduction is, however, very obscure. **Anona reticulata,**

ANONA squamosa.

The Custard Apple.

HISTORY.

Linn., is wild in Cuba, Jamaica, and the West Indies, and, together with **A. squamosa,** has been naturalized in India. It is impossible, however, to point to any forest where either species shows the slightest indication of being indigenous. The representations referred to by **General Cunningham** might be associated with a large number of plants; they may prove to be conventional representations of the jack-fruit tree or some other allied plant: they are not unlike the flower-heads of the sacred *kadamba* or **Anthocephalus.** It may be remarked that the Bengali names *nona* and *ata* are as much opposed to the custard apple tree being indigenous to India as are the names referred to by **General Cunningham** in favour of that idea. We know that the natives of India have adopted pre-existing names for introduced plants (an indefinite series of examples of this might easily be produced), and there is no evidence to show that this is not the case with the vernacular names now given to the custard apple. (See **Argemone.**)

Properties and Uses—

FIBRE. 1168

Fibre.—An inferior quality of fibre may be prepared from this species.

MEDICINE. Ripe fruit. 1169 Seed & leaves. 1170 Unripe fruit. 1171 Infusion. 1172 Root. 1173

Medicine.—The RIPE FRUIT is medicinally considered a maturant, and when bruised and mixed with salt, is applied to malignant tumours to hasten suppuration. The SEEDS, LEAVES, and immature fruits contain an acid principle fatal to insects, and the dried UNRIPE FRUIT, powdered and mixed with gram flour, is used to destroy vermin. An infusion of the leaves is considered efficacious in prolapsus ani of children. The root is regarded as a drastic purgative; natives administer it in acute dysentery. It is also employed internally in depression of spirits, and spinal diseases.

§"The *Makhzan* notices the poisonous action of the seed upon lice, and says that when applied to the os uteri, they cause abortion." (*Surgeon-Major Dymock.*) "The leaves are applied for the extraction of guinea-worm in Secunderabad." (*Surgeon-Major John North, Bangalore.*) "Green leaves pounded and made into a paste, and without admixture of water, are applied to unhealthy ulcers." (*Surgeon Anund Chunder Mukerji, Noakhally.*) "Leaf used for anthelmintic." (*Surgeon-Major C. M. Russell, Sarun.*) "The leaves pounded with tobacco leaves and a little quicklime are frequently used by the natives in ill-conditioned ulcers to destroy maggots in the inferior animals." (*Honorary Surgeon Easton Alfred Morris, Negapatam.*) "The leaves are applied to sores infested with maggots." (*Brigade Surgeon G. A. Watson, Allahabad.*) "Fruit good for digestion." (*Surgeon-Major J. J. L. Ratton, M.D., Salem.*) "The seeds are used to destroy lice. The bark is employed internally in depression of spirits, in asthma and fever." (*Surgeon W. Barren, Bhuj, Cutch.*) "The fresh leaves made into a pulp are applied to abscesses to hasten suppuration." (*Surgeon-Major C. J. W. Meadows, Burrisal.*)

Chemical Composition.—"The seeds yield an oil and resin; the latter appears to be the acrid principle." (*Dymock, Mat. Med., W. Ind., 18.*)

FOOD. Fruit. 1174

Fruit.—The plant may be described as completely domesticated in Indian gardens. The fruit ripens in summer, is of a more delicate flavour than the fruit of **A. reticulata,** and is eaten with relish by both natives and Europeans. The opinion formed of this fruit in India is much more favourable than in the West Indies, a fact which would seem to indicate that it is much superior in quality in India than in the country where it is supposed to be indigenous. In the West Indies an agreeable fermented drink, like cider, is made from the expressed juice; in India the juice is chiefly used to flavour ice-puddings. The fruit has, in times of famine, literally proved the staff of life to the natives of some parts of India. Its extended cultivation should be encouraged.

Juice. 1175 Beverage. 1176

TIMBER. 1177

Structure of the Wood.—Soft, close-grained. Weight 46 ℔ per cubic foot.

1178

ANONACEÆ.

A Natural Order of shrubby or arborescent Thalamiflorals, comprising some 400 species, inhabiting the tropical moist regions of Asia, America, and Africa, rare in the extra-tropical zones, and almost absent from dry regions. They extend over the whole world for about 40° on each side of the equator; in Africa to about 20° north latitude. None are indigenous to Europe. In India there are in all 192 species grouped into 26 genera, of which 109 are referred to the following genera: 19 species in **Uvaria**, 18 in **Unona**, 25 in **Polyalthia**, 17 in **Goniothalamus**, 16 in **Melodorum**, and 14 in **Xylopia**. Of the whole Indian species only 14. or 7·3 per cent. ascend above the level of the plains. The large majority are confined to the warm moist plains and lower hills of the eastern side of India, extending to the Straits, *viz.*, 124 species or 64·5 per cent. South India, including Ceylon, naturally stands next in importance, containing 43 species or 22·4 per cent.; 16 are met with in the Western Peninsula, or 8·3 per cent., while 9 occur, chiefly in cultivation, over the greater part of India. None are met with, however, as endemic to the dry tracts of India, such as Sind, Rájputana, Central India, or the Panjáb, and, except as stunted plants under garden cultivation, they do not exist even in these regions. This remarkable isolation in the type vegetation of the eastern and southern from the western and northern divisions of India, is borne out by many other Natural Orders, suggesting forcibly a strong resemblance to the moist tropical regions of America. Not only in the tropical and extra-tropical regions is this idea borne out, but also in the assemblages of temperate and even alpine plants which occur on the mountain tracts of the eastern division of India. **Sir J. D. Hooker**, in his *Himálayan Journals* (Vol. II., 39), says: "At first sight it appears incredible that such a limited area" (Lachen valley, Sikkim), "buried in the depths of the Himálaya, should present nearly all the types of the flora of the north temperate zone." After enumerating the trees which are common to Europe and North America, which occur also in the Lachen valley, **Sir Joseph Hooker** adds: "Of North American genera not found in Europe were **Buddleia, Podophyllum, Magnolia, Sassafras?, Tetranthera, Hydrangea, Diclytra, Aralia, Panax, Symplocos, Trillium,** and **Clitonia.** The absence of heaths is equally a feature of the flora of North America."

This idea of the similarity of the flora of the eastern division of India to many of the features of the new world has a remarkable confirmation in the distribution of the species belonging to the Magnoliaceous genus ILLICUM, the Star Anise. Commencing in China and Japan, they extend through the Straits to the mountains of northern Mánipur and the Khásia Hills, and passing almost round the globe they re-appear again in the basin of the Mississippi. The ease with which tropical American plants become naturalised in India, until they almost appear as indigenous to the Gangetic plains, must largely depend on the causes which have given birth to the similarities indicated. No better example of this could possibly be given than the distribution of the species of **Anona** now met with in India. The pine apple has, also in eastern India and the Straits, become a troublesome weed. **Anona** belongs to the series **Duguetia, Rollinia, Cymbopetalum, Bocagea,** and **Anona,** a group of genera characteristic of the eastern side of America from the United States to the south of Brazil; only 1 or 2 species of **Anona** are supposed to be indigenous to Africa or occur outside the region indicated. **Dr. Martius** (*Fl. Bras. Anonac.*, *51*) in his history of the Anonas is of opinion that the cultivated species are natives of the Antilles, from whence they were introduced into South America and the Old World. According to **DeCandolle**, the custard apple is a native of the West Indies; no **Anonaceæ** with a united ovary occurring in Asia. (See account of **Anona squamosa.**)

For a definition of the Natural Order ANONACEÆ the reader is referred to the *Flora of British India* (Vol. I., 45), where will also be found the following analysis of the genera :—

"**Tribe I. Uvarieæ.**—*Petals* 2-seriate, one or both series imbricate in bud. *Stamens* many, close-packed; their anther cells concealed by the overlapping connectives. *Ovaries* indefinite.

Flowers 1-sexual; ovules many; torus conical . 1. **Stelechocarpus.**
Flowers 2-sexual; ovules many, rarely few; torus almost flat . . . . . 2. **Uvaria.**
Flowers 1-2 sexual; ovule solitary . . . 3. **Elipeia.**

Tribe II. Unoneæ.—*Petals* valvate or open in bud, spreading in flower, flat or concave at the base only, inner subsimilar or 0. *Stamens* many, close-packed; their anther-cells concealed by the overlapping connectives. *Ovaries* indefinite.

* *Petals conniving at the concave base and covering the stamens and ovaries.*

Ovaries 1-3, many-ovuled; peduncles not hooked 4. **Cyathocalyx.**
Ovaries many, 2-ovuled; peduncles hooked . 5. **Artabotrys.**
Ovaries many, ovules 4 or more; peduncles straight . . . . 6. **Drepananthus.**

** *Petals flat, spreading from the base.*

Ripe carpels indehiscent.
Ovules many, 2-seriate; petals lanceolate . . 7. **Cananga.**
Ovules many, 2-seriate; petals broad ovate . 8. **Cyathostemma.**
Ovules 2-6, 1-seriate on the ventral suture . . 9. **Unona.**
Ovules 1-2, basal or sub-basal . . . . 10. **Polyalthia.**
Ripe carpels follicular . . . . . 11. **Anaxagorea.**

*** *Inner petals valvate tip incurved* . . . 12. **Popowia.**

Tribe III. Mitrephoreæ.—*Petals* valvate in bud, outer spreading; inner dissimilar, concave, connivent, arching over the stamens and pistil. *Stamens* many, close-packed; anther-cells concealed by the overlapping connectives. *Ovaries* indefinite.

Inner petals not clawed.

Inner petals smaller than the outer . . . 13. **Oxymitra.**
Inner petals much larger than the outer . . 14. **Phæanthus.**

* *Inner petals clawed, usually smaller than the outer.*

Ovules 1-2, near the base of the ovary . . 15. **Goniothalamus.**
Ovules many . . . . 16. **Mitrephora.**

Tribe IV. Xylopieæ.—*Petals* valvate in bud, thick and rigid, connivent, inner similar but smaller, rarely 0. *Stamens* many, close-packed; anther-cells concealed by the produced connectives. *Ovaries* indefinite.

Ovules solitary; fruit fleshy, of many connate carpels . . . . 16.***Anona.**
Ovules 2—many; outer petals broad; torus convex 17. **Melodorum.**
Ovules 2—many; outer petals narrow; torus flat or concave . . . . . . 18. **Xylopia.**

Tribe V. Miliuseæ.—*Petals* imbricate or valvate in bud. *Stamens* often definite, loosely imbricate; anther-cells not concealed by the overlapping connectives. *Ovaries* solitary or indefinite.

* *Ovaries indefinite.*

Petals valvate, inner largest; ovules definite . 19. **Miliusa.**
Petals valvate, inner largest; ovules indefinite . 20 **Saccopetalum.**
Petals valvate, subequal; ovules 4-8 . . . 21. **Alphonsea.**
Petals valvate, inner shortest; ovules 2-4 . . 22. **Orophea.**
Petals imbricate, subequal; ovules 2-8 . . . 23. **Bocagea.**

** *Ovaries solitary.*

Outer petals valvate, inner imbricate . . . 24. **Kingstonia.**
All the petals valvate . . . . . . 25. **Lonchomera."**

(*Fl. Br. Ind., I., 46.*)

1179

Affinities in Structure of the Anonaceæ.—They resemble the Myristiceæ in the 3-partite valvate perianth, extrorse anthers, solitary erect unatropous ovule, copious ruminate albumen, minute basilar embryo with inferior radicle, woody aromatic stem and alternate more or less distichous leaves, folded lengthwise in bud. They are separated by the fact that the Anonaceæ have usually hermaphrodite flowers, petaloid, with indefinite stamens, and generally numerous carpels, with exarillate or only imperfectly arillate seeds. The ternary arrangement of the parts of the flower bring them also very near to the Menispermaceæ and to the Magnoliaceæ. From the former they differ in habit, size of the flower, the inflorescence, the structure of the stamens, and the fruit; from the latter by the leaves being stipulate, the testa of the seed fleshy and the albumen not ruminate. They also approach very near to the Dilleniaceæ, but in the latter the leaves are often stipulate, the flowers terminal and quinary, and the albumen not ruminate.

1180

Properties and Economic value of the Order.—The bark is usually aromatic, stimulant, and astringent, the inner layer affording a useful bast fibre. The leaves also possess the aromatic properties of the bark, being often acid and sometimes even nauseous. The non-aromatic fruits are generally edible. The flowers are sometimes strongly perfumed, some species being used in European perfumery, such as the Ilang Ilang **(Cananga odorata)**, which is also used in India to perfume cocoanut oil, and by rubbing over the skin to bring back the heat of the body in the cold stage of fever. The species of **Xylopia** are most useful. **X. æthiopica** supplied the ancients with the so-called Ethiopian pepper. The wood of the species of **Xylopia** is exceedingly bitter, and from the Brazilian species strong ropes are made from the bark.

The fruit of some of the American Uvarias is edible; **U. triloba** affording the Papaw of the United States, from which an alcoholic drink is prepared. Other species are used as stimulant drugs. Blume has shown that the barks of certain species are efficacious in affections arising from obstruction of the portal vein, but they require to be used with caution.

The leaves of **Artabotrys suaveolens** afford an aromatic medicine, said to be efficacious in inducing reaction during the cold stage of cholera. The flowers of **Artabotrys odoratissima** are used in native perfumery, while the roots of **Polyalthia macrophylla** are strongly aromatic, an infusion in Java being used in eruptive fevers. Many species of Anona yield edible fruits. The timber is often extremely light, as, for example, in the genus **Polyalthia**, such as the lance-wood, a timber extensively used by coach-builders. **Polyalthia longifolia** is valued in India as an avenue tree, because of the rapidity of its growth and the dense mass of graceful undulated leaves; in spring this tree produces a flush of pale delicate green leaves, becoming at this season one of the most elegant of Indian roadside trees.

ANT-GREASE.

GREASE. 1181 OIL. 1182

Ant-grease is prepared by boiling white-ants and skimming off the oil which floats on the surface. An oily substance is also obtained by expression. Ant-grease is reported to be an article of food.

§ "Red-ants are eaten by the Santals." (*Rev. A. Campbell.*)

1183

ANTHEMIS, *Linn.; Gen. Pl., II., 420.*

A genus of annual or perennial COMPOSITÆ, belonging to the Tribe ANTHEMIDEÆ, and comprising some 80 species, chiefly inhabitants of the temperate regions of Europe, Asia, and Africa.

Heads radiant, receptacle convex or conical, scaly throughout. *Flowers of the ray* female or neuter, ligulate, in one row; *flowers of the disk* perfect, tubular. *Bracts of the involucre* of few rows. *Fruit* terete or bluntly tetragonal, without pappus, but with a more or less prominent margin.

1184

Anthemis nobilis, *Linn.; DC. Prod., VI., 11.*

COMMON OR TRUE CHAMOMILE.

Vern.—The flowers: *Babúní-phúl, bábúne-ké-phúl*, or *bábúnah*, HIND., DUK.; *Shímai-chamantipú*, TAM.; *Síma-chámanti-pushpam*, or *síma-chámanti-purvu*, TEL.; *Shíma-jevanti-pushpam*, MAL.; *Shíme-shyá-mantige, chamundi-huvvu*, KAN.; *Bábúnaj*, ARAB.; *Bábúnah, gule-bábúnah*, PERS. The *Anthemis* of **Dioscorides.**

Dr. Dymock says: "The Persian name *Bábúna his* said to be derived from the name of a village in Irak-Arabi; this the Arabians convert into *Bábunaj.*"

References.—*Ind. Pharm., 121; Flück. and Hanb., Pharmacog. (1879), 384; Medical Plants by Bentley and Trimen, 154; U. S. Dispens., 15th Ed., 195; Dymock's Mat. Med., W. Ind., 370.*

Habitat.—A perennial herb, indigenous in England, but is also plentiful in France, Spain, Germany, and Russia. Imported into India and also cultivated in the gardens of the rich.

Chemical Composition.—§ "The flowers yield about 15 per cent. of essential oil, first of a pale blue colour, becoming yellowish brown in the course of a few months. The researches of **Demarcay** show this oil to be a mixture of butylic and amylic angelate acid and valerate. A bitter crystalline acid has been isolated from double chamomile, which is regarded as identical with anthemic acid from **Anthemis arvensis.** According to **Flückiger** and **Hanbury,** the bitter acid principle is apparently a glucoside; they also state that no alkaloidal principle is contained in the flowers. According to **Bley,** the seeds contain a bitter substance, which appears to be an alkaloid." (*Surgeon C. J. H. Warden, Professor of Chemistry, Calcutta.*)

According to **Kopp** (*Liebig's Annalen., CXCV., 81-92*), the oil saponified by boiling with alcoholic potash and after fractional distillation was found to contain angelic ($C_5H_5O_2$) and tiglic ($C_5H_8O_2$) acids in about equal proportions, isobutyric acid in much smaller amounts, and a fourth acid, most probably methacrylic. He also confirms the statement made by **Demarcay** that by gentle heat angelic acid is transformed into isomeric tiglic acid.

MEDICINE. Flower-heads. 1185

Medicine.—The medicinal properties of this imported drug are too well known to require a detailed account in a work on the Economic Products of India. The dried flower-heads are officinal in the *Indian Pharmacopœia.* They are described as stimulant, tonic, and carminative, and useful in constitutional debility, hysteria, and dyspepsia. Formerly they were employed in intermittent fevers, but this usage has been superseded by the introduction of cinchona. In large doses they act as emetic. Bags containing hot moist chamomile flower-heads are sometimes used externally on account of their retention of heat. Chamomile oil possesses stimulant and antispasmodic properties, and is considered a valuable remedy for flatulence. Persian chamomile is obtained from **Matricaria Chamomilla,** *Linn.* (**M. suaveolens,** *Linn.*), which see.

§ "Flower-tops by boiling in hot water are used by natives as an external application and fomentation in rheumatic and other painful affections." (*Assistant Surgeon Bhagwan Das, Rawal Pindi.*)

"I have found a warm infusion of dried chamomile flower-heads a most soothing application in cases of conjunctival irrigation; a poppy capsule or two is a useful addition." (*Surgeon Joseph Parker, Poona.*)

Substitutes and Adulterations.—In Europe the following plants are chiefly used for this purpose: **Matricaria Parthenium,** *Linn.;* **Matricaria parthenodies,** *Desf.;* **M. Chamomilla,** *Linn.* The single chamomile flowers are said to be yield more oil than the double. (*Bent. & Trim., Med. Pl.*)

Anthemis Pyrethrum, *Linn.*, see **Anacyclus Pyrethrum,** *DC.;* COMPOSITÆ.

ANTHISTIRIA, *Linn.; Gen. Pl., III., 1136.*

1186

A genus of grasses belonging to the Tribe ANDROPOGONEÆ; GRAMINEÆ.

Anthistiria arundinacea, *Roxb.; Fl. Ind., Ed. C.B.C., 84.*

Vern.—*Bharua, askhun,* PANJABI NAMES (according to **Duthie**); *Ulú, úllah, kangar, khandura,* N.-W. P.? (**Atkinson**).

Habitat.—A grass met with chiefly in the North-West Provinces.

FIBRE. 1187

Fibre.—The culms yield a fibre used for cordage, and for the sacrificial, strings used by the Hindús where **Saccharum Munja** is not available. The leaves are also used for thatching.

A. ciliata, *Linn. f.*

1188

THE KANGAROO GRASS of Australia.

Syn.—ANTHISTIRIA CILIATA, *Linn.*, and A. SCANDENS, *Roxb.;* as in *Fl. Ind., Ed. C.B.C., 83.*

Vern.—*Musel.*

Habitat.—A widely-spread grass stretching through Africa and South Asia to Australia. In India it is reported at Banda, Saharanpore, Garhwál, and Kumaon, ascending to 7,000 feet in altitude.

FODDER. 1189

Fodder.—This is one of the most useful fodder grasses in India, and its cultivation should be extended. In Australia it is much valued, being one of the chief grasses upon which Australian cattle are fed. It luxuriates in a warm temperate or tropical climate.

A. polystachia, *Roxb.; Fl. Ind., I., 248.*

FODDER. 1190

Fodder.—Grass given as fodder.

ANTHOCEPHALUS, *A. Rich.; Gen. Pl., II., 29.*

1191

A genus of glabrous trees comprising 3 or 4 species, belonging to the Natural Order RUBIACEÆ, one of which occurs wild or cultivated throughout the greater part of the plains of India.

Leaves petioled; *stipules* lanceolate, caducous. *Flowers* in terminal globose heads without bracteoles, united by their confluent calyx-tubes. *Corolla* funnel-shaped. *Stigma* simple. *Ovaries* confluent, 2-celled, with a solitary ovule in each cell.

1192 **Anthocephalus Cadamba,** *Bth. & Hook.f.; Fl. Br. Ind., III., 23.*

Syn.—NAUCLEA CADAMBA, *Roxb., Fl. Ind., Ed. C. B. C., 1723; Beddome, t. 35;* SARCOCEPHALUS CADAMBA, *Kurz, II., 63.*

Vern.—*Kadam, kadamb,* HIND.; *Kadam,* BENG.; *Bol-kadam,* CHITTAGONG; *Sanko,* KÓL.; *Pandúr,* LEPCHA; *Kodum,* MECHI; *Roghu,* ASS.; *Kadambo,* URIYA; *Kadamba, nhyú,* BOMB.; *Kadam, kadamb, nhiv,* MAR.; *Kalam, nhio* or *nhiu,* PANCH MAHALS; *Kadamb,* GUJ.; *Vellai cadamba,* TAM.; *Kadamba, rudraksha-kamba,* TEL.; *Heltega, arsanatega,* MYSORE; *Kaada vailu, kadaga, kadwal,* KAN.; *Kadamba, nipa,* SANS.; *Kadamba,* SINGH.; *Ma-ú, sanyepang,* MAGH.; *Maú, maúkadún,* BURM.

Habitat.—A large, deciduous tree, wild in Northern and Eastern Bengal, Pegu, and the Western Coast; cultivated in Northern India.

It grows very fast. During the first two or three years, the rate is about 10 feet a year, the girth increasing at the rate of 1 inch a month. After 10 or 12 years, however, the growth becomes very slow. (*Ind. For., X., 246.*) It grows to a large size in the forest of the Panch Mahâls, Bombay.

MEDICINE. Bark. 1193 Leaves. 1194

Medicine.—The bark is used medicinally as a febrifuge and tonic.

§ "Decoction of the leaves is used as a gargle in cases of aphthæ and stomatitis." (*Assistant Surgeon Anund Chunder Mukerji, Noakhally.*)

FOOD. Fruit. 1195

Food & Fodder.—The fruit is eaten, and the foliage sometimes used as fodder for cattle.

TIMBER. 1196

Structure of the Wood.—White, with a yellowish tinge, soft, even-grained. Weight about 40 lbs. per cubic foot.

It is used for building; in Assam, Cachar, and occasionally in Darjíling, for tea-boxes. **Cunningham** (1854) says that it is used for beams and rafters, on account of its cheapness and lightness, and that it is good for joiner's work, but is a brittle wood. If a little less heavy, it would be much valued for gun-stocks. (*Bomb. Gaz., XI., 24.*)

DOMESTIC. Flowers. 1197

Domestic Uses.—The flowers are offered at Hindú shrines; they are sacred to *Siva.* Often cultivated for ornament and for the grateful shade its large close foliage affords. (*Brandis.*)

1198 **ANTIARIS,** *Lesch.; Gen. l., III., 371.*

A small genus of trees containing 5 or 6 species, belonging to the Natural Order URTICACEÆ and the Tribe ARTOCARPEÆ.

Leaves elliptic, oblong, rough, on short petioles; sap milky; leaves distichous; *stipules* in pairs, axillary. *Flower-heads* axillary, clustered, monœcious, the males crowded within an imbricated involucre opening into a convex receptacle; female flowers solitary within a many-bracteated involucre and devoid of a perianth. *Male perianth* segments and stamina 4, rarely 3; segments spathulate and imbricate with the stamens placed opposite. *Ovary* connate to the involucre with a solitary pendulous ovule; *style* short 2-cleft, the *stigma* filiform recurved. *Fruit* a fleshy purple drupe, the pericarp formed of the enlarged fleshy involucre. *Seed* pendulous; *testa* leathery; *albumen* wanting.

1199 **Antiaris innoxia,** *Bl.;* Syn. A. SACCIDORA, *Dalz.*

THE TRAVANCORE SACKING TREE.

1200 **A. toxicaria,** *Lesch.; Wight, Ic., t. 1958.*

THE UPAS TREE.

According to some authors, these are kept up as distinct species, and by others reduced to one. The former occurs in Burma and the Indian Archipelago, and the latter on the Western Ghâts and in Ceylon.

Vern.—*Chándla, chánd kudá, charvár mádá, karvat,* or *kharvat,* BOMB., MAR.; *Karwat,* KONKAN; *Alli, netávil, nettávil maram,* TAM.; *Jazúgri,*

Ajjanapatte, jagúri, KAN.; *Araya-angely, nettávil,* MALA.; *Riti,* SINGH.; *Hmyaseik* (formerly *Myeh-seik*), BURM.

References.—*Brandis, For. Fl., 427; Kurz, Fl. Burm., II., 462; Gamble, Man. Timb., 332;* A. SACCIDORA, *Dalz., in Hook. Jour. Bot., III., 232, and Dalz. and Gib.'s Fl., Bombay, 244; Dymock's Mat. Med., W. Ind., 615; U. S. Dispens., 15th Ed., 1772; Drury's Us. Pl., 45; Treasury of Botany; Smith's Economic Dict.*

Habitat.—A large evergreen tree of Burma, the Western Ghâts, and Ceylon. § "The 'Upas tree' of Java grows all along the eastern slopes of Pegu Yoma, Martaban, and down to Tenasserim." (*J. C. Hardinge, Rangoon.*) **Beddome** says it is the largest tree of the western forests, attaining a height of 250 feet.

HISTORY. 1201

History.—The most absurd accounts of the properties of this tree have become current. While it cannot by any means be regarded as harmless (the juice forming a deadly arrow-poison), still it has been freed from the superstitious ideas which lingered around its very name. The following extract from the *Treasury of Botany* will be found both interesting and instructive: "The Upas tree, when pierced, exudes a milky juice, which contains an acrid virulent poison, called *antiarin*. Most exaggerated statements respecting this plant were circulated by a Dutch surgeon about the close of the last century. The tree was described as growing in a desert tract, with no other plant near it for the distance of 10 or 12 miles. Criminals condemned to die were offered the chance of life if they would go to the Upas tree and collect some of the poison. They were furnished with proper directions, and armed with due precaution, but not more than two out of every twenty ever returned. The Dutch surgeon, **Foersch**, states that he had derived his information from some of those who had been lucky enough to escape, albeit the ground around was strewn with the bones of their predecessors; and such was the virulence of the poison, that 'there are no fish in the waters, nor has any rat, mouse, or any other vermin been seen there; and when any birds fly so near this tree that the effluvia reaches them, they fall a sacrifice to the effects of the poison.' Out of a population of 1,600 persons, who were compelled, on account of civil dissensions, to reside within 12 or 14 miles of the tree, not more than three hundred remained in less than two months. **Foersch** states that he conversed with some of the survivors, and proceeds to give an account of some experiments that he witnessed with the gum of this tree, these experiments consisting principally in the execution of several women, by direction of the Emperor! Now, as specimens of this tree are cultivated in botanic gardens, the tree cannot have such virulent properties as it was stated to have; moreover, it is now known to grow in woods with other trees, and birds and lizards have been observed on its branches. It occasionally grows in certain low valleys in Java, rendered unwholesome by an escape of carbonic acid gas from crevices in the ground, and which is given off in such abundance as to be fatal to animals that approach too closely. These pestiferous valleys are connected with the numerous volcanoes in the island. The craters of some of these emit, according to **Reinwardt**, sulphureous vapours in such abundance as to cause the death of great numbers of tigers, birds, and insects; while the rivers and lakes are in some cases so charged with sulphuric acid, that no fish can live in them. So that doubtless the Upas-tree has had to bear the opprobrium really due to the volcanoes and their products: not that the Upas is by any means innocent, for severe effects have been felt by those who have climbed the tree for the purpose of bringing down the branches and flowers. The inner bark of the young trees, which is fabricated into a coarse garment, excites the most horrible itching. It clings to the

The Poisonous Upas or Sacking Tree.

skin, if exposed to the wet before being properly prepared. The dried juice, mixed with other ingredients, forms a most venomous poison, in which the natives dip their arrows."

RESIN. 1202

Resin.—It exudes a white resin, used for poisoning arrows. On cutting the stem or unripe fruit, a white milky viscid fluid exudes in large quantities, which shortly hardens, becoming of a black and shining colour. This inspissated sap is probably the so-called resin of authors (*Br. Burm. Gaz., I., 135.*)

FIBRE. 1203

Fibre.—The natives strip the bark of this tree into large pieces, soak them in water, and beat them well, when a good white fibre is obtained—a natural cloth worn by the natives. It is in Western India well known as the *sacking tree*, on account of the tough, inner, fibrous, felted bark, being removed entire, thus forming natural sacks. Small branches are made into legs of trousers and arms of coats, the larger ones forming the bodies of the garments. In this way felt costumes are made which require no more sewing than is necessary to connect the parts together. If passed through rollers, and at the same time dyed and tanned, these natural cloths or felts are very interesting. The samples exhibited at the late Calcutta International Exhibition (contributed by the Bombay Committee) were very much admired, and proved very attractive. In making sacks sometimes a disk of the wood is left attached to the fibre so as to form the bottom of the sack. At other times a vertical incision is made on the tree and a transverse cut around the stem at the top and bottom of this vertical one. The bark is then peeled off, and after being beaten in water and dried, the top and bottom are sewed up (forming the sides of the sack). These sacks are extensively used for storing rice. In Ceylon ropes are made of the bark. "The bark yields strong fibre suited for cordage, matting, and sacking. In making sacks a branch or trunk is cut to the required length, soaked in water, and beaten till the fibre separates from the wood. It is then turned inside out and the wood sawn off, except a small piece at the bottom." (*Bombay Gazetteer, XV., Part 1., 62, Konkan District.*) There seems every likelihood that the bark of this tree may come into use as a paper fibre.

Natural Felts. 1204

Sacks. 1205

Ropes. 1206

Paper material. 1207

MEDICINE. Seeds. 1208

Medicine.—The bitter seeds contain a peculiar principle, which may prove an active medicinal agent. Information regarding the properties of these seeds would be desirable. (*Pharm. Ind.*)

§ "The seeds are used as a febrifuge and in dysentery: dose $\frac{1}{3}$ to $\frac{1}{2}$ seed three times a day." (*Surgeon-Major W. Dymock, Bombay.*)

1209

CHEMICAL COMPOSITION.

"**Upas Antiar and Upas Tieute.**—Under these names, two poisons have long been used by the natives of Java and other East India islands for poisoning their arrow-heads; and very exaggerated notions have prevailed among the people of the Western World in relation to the tremendously destructive power over animal life of the Upas tree in Java, from which it was supposed that the poison was derived. The tale was told that birds and animals perished when within the influence of its exhalations, and that man came into its near vicinity at the peril of life. All such accounts have proved to be fabulous; but there is no doubt as to the exceedingly poisonous character of the arrow poison to which reference has been made. It seems now to be pretty well determined that the active ingredient of the *Upas Antiar* is a gum-resinous exudation proceeding from incisions in the trunk of the **Antiaris toxicaria,** a large tree belonging to the URTICACEÆ, growing in Java, Celebes, and the neighbouring islands, and described in *Lindley's Flora Medica* (p. 301). Like certain species of Rhus, this plant exhales an æriform matter, which

CHEMISTRY.

very unpleasantly affects many of those who approach it, causing eruptions upon the skin and exterior swelling, while others seem altogether insensible to its influence. The juice is mixed with various substances, which probably have little other effect than to give a due consistence to the poison. This, whether taken internally or introduced into the system through a wound, acts with extreme violence, producing vomiting, with great prostration, a feeble irregular pulse, involuntary evacuations, and convulsive movements, which are soon followed by death, which **Brodie** ascertained to be due to cardiac paralysis. From a chemical examination by **Pelletier** and **Caventou** it appears that the Antiar owes its activity to a peculiar principle *antiarin*, crystallizable, soluble in water and alcohol, but scarcely so in ether, and consisting of carbon, hydrogen, and oxygen $C_{14}H_{20}O_5$ (see *A. J. P., 1863, p. 474*). Antiarin appears to act directly as a paralyzant on the cardiac muscle, diminishes the irritability of the peripheral vagus, and stimulates the vaso-motor centres (*N. R., 1875, p. 308*").

[§ "I failed to obtain *antiarin* from the seeds of this plant; they contain a bitter principle." (*Surgeon-Major W. Dymock, Bombay.*)]

"The *upas tieute* is even more poisonous than the *antiar*. This is said to be obtained from a climbing woody plant growing exclusively in Java, and belonging to the genus STRYCHNOS especially designated by **Leschenault** as *Strychnos Tieute*. It is from the bark of the root, according to this author, that the poison is prepared. A decoction of the bark is concentrated to the consistence of syrup, then mixed with onions, garlic, pepper, &c., and allowed to stand till it becomes clear. **Leschenault**, having dipped the point of an arrow in the poison, and allowed it to dry, pricked a chicken with it, which died in a minute or two in violent convulsions. **MM. Delille** and **Magendie** found that the poison had not lost its strength in four years." (*Hammond, Am. Journ. of Med. Sci., 1860, p. 366.*) Three grains have produced very violent symptoms resembling those caused by strychnine, which alkaloid has indeed been found in the poison. (*Chemist and Druggist, May 15, 1863.*)

"**Dr. Wm. A. Hammond** made some experiments with a poisonous substance brought by **Dr. Ruschenberger** from Singapore, which proved to have the combined effects of the two poisons above mentioned, both diminishing directly the power of the heart, and causing tetanic spasms of the muscles, suggesting that it might be a mixture of the antiar and tieute; but **Dr. Hammond** seems, from other considerations, to have been led to the opinion that it had a different origin from either. (*Am. Journ. of Med. Sci., 1860, p. 371.*)" (*U. S. Dispensatory, 15th Ed., 1771.*)

TIMBER. 1210

Structure of the Wood.—Pale brown, very coarse, fibrous (*Kurz*). "White, soft, even-grained, annual rings faintly marked."

1211

ANTIDESMA, *Burm.; Gen. Pl., III., 284.*

A genus of trees or shrubs belonging to the Natural Order EUPHORBIACEÆ, comprising some 60 species, inhabitants of the warm temperate regions of the world.

Leaves alternate, entire, stipulate, penniveined. *Flowers* diœcious, numerous, small, the male flowers in deciduous spikes, the female racemose. *Calyx* 3-5, imbricate. *Petals none*. *Stamens* opposite the calyx-lobes, inserted round a rudimentary ovary; *filaments* free. *Disk* of distinct glands, alternating with the filaments and calyx segments. *Ovary* 1-celled, with 2 pendulous ovules; styles 3-4, short, united at the base. *Fruit* indehiscent, generally a 1-seeded drupe. (*Wight's Ic., tt. 766, 768, 819, and 821.*)

1212

Antidesma Bunias, *Müll.-Arg.; DC. Prod., XV., 2, 262.*

Vern.—*Himal cheri*, NEPAL; *Kantjer*, LEPCHA; *Nolai-tali*, TAM.; *Nuli-tali*, MAL.

Habitat.—A small tree of North and East Bengal, South India, and Tenasserim.

Botanic Diagnosis.—Branchlets and buds tawny pubescent; *stipules* ovate-cordate. *Flowers* small, green, sessile, forming robust and branched spines in the axils of the leaves. *Calyx* obsoletely 3-toothed. *Stamens* 3. *Drupes* elliptic, red, becoming black; stone compressed.

FOOD. Leaves. 1213 Fruit. 1214

Food.—According to **Mr. Gamble** *(List of Darjeeling Shrubs, Trees, &c., p. 69)*, the leaves and fruit are eaten. **Dr. Trimen** informs me that the fruits are also eaten in Ceylon.

TIMBER. 1215

Structure of the Wood.—Reddish, hard; weight 46 lbs. per cubic foot.

Antidesma diandrum, *Tulasne; DC. Prod., XV., 2, 266.*

Syn.—STILAGO DIANDRA, *Roxb., Fl. Ind. Ed. C.B.C., 714.*

Vern.—*Aamári, sarshoti, gúr-mussureya, ban-mussureya, dhakki,* HIND.; *Mutta,* BENG.; *Nuniári,* URIYA; *Kantjer,* LEPCHA; *Pella, gumudu masúrbauri,* GOND.; *Pella-gumudu* (?), TEL.; *Matha,* SANTAL; *Amtua sag,* MAL. (S. P.); *Patimil,* NEPAL; *Kimpalin,* BURM.

Habitat.—A small tree found in Garhwál, Kumaon, Oudh, Bengal, South India, and Burma, common in the hill forests of the Santal Pergunnahs.

Botanic Diagnosis.—Branchlets, petioles, and under-side of leaves along the midrib with scattered rust-coloured hairs. *Flowers* pedicillate. *Calyx* cup-shaped. *Stamens* 2-3.

FOOD. Leaves. 1216 Fruit. 1217

Food.—The leaves are acid; they are eaten. They resemble sorrell and are made into chutney; the fruit is also eaten. (*Gamble.*)

TIMBER. 1218

Structure of the Wood.—Pinkish grey, hard, close-grained. Weight 41 lbs. per cubic foot.

A. Ghæsembilla, *Müll.-Arg.; DC. Prod., XV., 2, 251.*

Syn.—A. PUBESCENS, *Willd.,* and A. PANICULATUM, *Roxb., Fl. Ind., Ed. C.B.C., 717.*

Vern.—*Khúdi jamb, limtoá,* BENG.; *Umtoá,* HAZARIBAGH; *Mata sure,* KOL.; *Pulsur, polári, jána-pa-laseru, pollai,* TEL.; *Jondri,* MAR.; *Búambilla,* SINGH.; *Byaitsin,* BURM.

Habitat.—A small, deciduous tree, met with in Nepal, Oudh, Bengal Burma, Chanda, and South India.

Botanic Diagnosis.—Branchlets, young leaves, and inflorescence soft-tomentose. *Flowers* sessile. *Calyx* deeply 5-cleft. *Stamens* 5.

FOOD. Leaves. 1219 Berry. 1220

Food.—The leaves are eaten in Bengal. The berry is dark purple with a pleasant sub-acid flavour.

TIMBER. 1221

Structure of the Wood.—Red, with darker-coloured heartwood, smooth, hard, close and even-grained. Weight 49 lbs. per cubic foot.

A. Menasu, *Müll.-Arg.*

Vern.—*Kumbyúng, tungcher,* LEPCHA; *Kin-pa-lin,* BURM.

Habitat.—A small tree, found in Sikkim, Khásia Hills, Burma, and the Andaman Islands.

FOOD. Fruit. 1222

Food.—The fruit is eaten; it is of a red colour.

TIMBER. 1223

Structure of the Wood.—Darkish red, similar to that of **A. Ghæsembilla,** but the pores smaller and the medullary rays finer. Weight 52 lbs. per cubic foot.

ANTIMONIUM.

1224

Antimonium or Antimony, Black.

Vern.—*Surmé-ká-patthar, surmah,* HIND.; *Surmá* or *shurmá,* BENG.; *Surmah-i-Isfahani,* BOMB.; *Surmo, surmá-no-pahro,* GUJ.; *Anjan,*

anjan-ká-patthar, DUK.; *Anjanak-kallu*, TAM.; *Anjana-ráyi*, TEL.; *Annanak-kalla*, MAL.; *Anjanam*, SANS.; *Ismad, kohal*, ARAB.; *Surmah, sange surmah*, PERS.

A black ore of antimony, a tersulphide, and called *surma*, occurs in various parts of the Panjáb. The ore is imported from Kandahár and Ispahan, but is also obtained in great abundance in the Himálayan range. This tersulphide is often confused by natives with galena, as both can be reduced to a black powder. Iceland spar (carbonate of lime) is also called *surmá*, but is distinguished as *surma-sufaid* or white antimony. Natives do not seem to be acquainted with the use of this metal as an alloy, or even as a pure metal, for scientific purposes. It is in fact only used as a cosmetic for the eye. It is also supposed to act as a tonic to the nerves of the eye and to strengthen the sight. But much of the antimony sold by druggists is really galena and is imported from Kábul and Bokhára. (*Baden Powell, Panjáb Products, Vol. I., p. 10.*)

Antimony. 1225

§ "It is called *Surmai Ispahani* in Bombay to distinguish it from galena, but it is by no means common in Bombay: a few Persians only import it when there is a demand." (*Surgeon-Major W. Dymock, Bombay.*) "Largely used by women in India as an application to the edges of the eyelids to improve personal appearance." (*Surgeon-Major G. A. Emerson, Calcutta.*) "It is also used to prevent the injurious effects of the glare of light on the eyes, which it does by absorbing the rays." (*Brigade Surgeon G. A. Watson, Allahabad.*) "It is supposed to have a cooling effect on the eyes, protecting them from the glare of the sun." (*Surgeon K. D. Ghose, Bankura.*) "Very good tonic for horses." (*Surgeon H. D. Masani, Karachi.*)

Antirrhinum glaucum, *Linn.*, see **Linaria glauca,** *Spreng.*; SCROPHULARIACÆ.

APIUM, *Linn.*; *Gen. Pl., I., 888.*

1226

A genus of glabrous herbs (belonging to the UMBELLIFERÆ), comprising some 14 species, scattered over the world; one met with in India.

Leaves pinnate, 3-partite or compound. *Umbels* compound, often leaf-opposed. *Bracts* and bracteoles absent (in the Indian species). *Flowers* white. *Calyx-teeth* obsolete. *Petals* ovate-acute, tip inflexed. *Fruit* orbicular or elliptic, slightly longer than broad, laterally sub-compressed; *carpels* semi-terete, subpentagonal, plain on the inner surface; *primary ridges* distinct, filiform; secondary absent; *furrows* 1-vittate; *carpophore* undivided or shortly 2-fid. *Seed* semi-terete, dorsally sub-compressed.

A genus very nearly allied to Carum. The word Apium was probably the Latin name for parsley; it literally means water plant, the words *ap, ab,* and *av,* in various languages, meaning water,—*e. g.*, Panjáb, or five waters.

Apium graveolens, *Linn.*; *Fl. Br. Ind., II., 679.*

1227

WILD AND CULTIVATED CELERY.

Vern.—*Ajmúd, bori-ajmúd karafs*, HIND.; *Chanú, rándhuni*, BENG.; *Bori ajamoda*, or *ajmúd*, BOMB.; *Ajwankaputa, budiajiwan*, CUTCH; *Karafs*, ARAB.; *Karasb*, PERS.; *Bhút jhata*, PB. Vulg. *Saleri* in the bazars of India. "The Hindi, Arabic, and Persian names here given are more generally applied to the fruit of **Carum Roxburghianum.**" (*Moodeen Sheriff.*)

Habitat.—A native of England and other parts of Europe. Cultivated in different parts of India during the cold weather, chiefly as a garden crop in the vicinity of towns, for the use of the European population, by whom it is eaten as a salad and pot-herb or made into soup. It is also cultivated sometimes in Bengal for its seed, and in the Panjáb for its root.

Botanic Diagnosis.—Peduncle short, leaf-opposed

MEDICINE. 1228

Medicine.—The officinal root is considered alterative and diuretic, and given in anasarca and colic. The seeds are also given as stimulant and cordial.

§ "As an antispasmodic celery is used in bronchitis and asthma, and to some extent, by natives, for liver and spleen diseases; it is regarded as emmenagogue." (*Surgeon G. A. Emerson, Calcutta.*)

"Celery is emmenagogue, and is used by hakims for the expulsion of stone." (*Asst. Surgeon J. N. Dey, Jaipur.*)

"Carminative, useful in colic; dose 1 to 2 leaves." (*Surgeon W. Barren, Bhuj, Cutch, Bombay.*)

FOOD. 1229

Food.—The seed is eaten as a spice by the natives, and the blanched stems and leaf-stalks by Europeans. In the wild state it is to a certain degree poisonous, but under cultivation it becomes a wholesome salad and pot-herb. In the Levant it is not blanched; the green leaves and stalks are used as an ingredient in soups. A form met with in France and Germany (and occasionally in India) is eaten as a vegetable after being boiled; this is known as the turnip-rooted celery. The seeds tied into a piece of cloth are sometimes used to flavour soup.

Apium involucratum, *Roxb.*, see **Carum Roxburghianum,** *Benth;* UMBELLIFERÆ.

1230

A. petroselinum or **Parsley,** see **Petroselinum sativum,** *L.*

A variety of this plant called **fusiformis** is grown chiefly as a vegetable; the tap-roots being boiled and eaten like parsnip.

1231

APLOTAXIS, *DC., Gen. Pl., II., 472.*

A genus of COMPOSITÆ chiefly found on the temperate Himálaya,; by modern botanists reduced to **Saussurea.**

Aplotaxis auriculata, *DC.*, see **Saussurea hypoleuca,** *Spreng.*

A. candicans, *DC.*, see **Saussurea candicans,** *C.B.C.*

A. gossypina, *DC.*, see **Saussurea gossypifera,** *Don.*

A. Lappa, *Decne.*, see **Saussurea Lappa,** *C.B.C.* (THE COSTUS).

1232

APLUDA, *Linn.; Gen. Pl., III., 1137.*

Apluda aristata, *Linn.;* GRAMINEÆ.

Syn.—A. ROSTRATA, *Nees; Roxb., Fl. Ind., Ed. C.B.C., 109.*

Vern.—*Bhanjuri, bhanjra, send,* BUNDLEKHAND; *Baru,* SAHARANPUR, *Goroma,* BENG.; *Putstrangali* (?), TEL.

Habitat.—A creeping, perennial grass, commonly found in hedges or other shady places, in the plains of northern India, and in the Himálayas, ascending to 7,000 feet in altitude.

FODDER. 1233

Fodder.—Used for fodder in the Banda district. (*Duthie's List of Grasses, p. 24.*)

1234

A. communis, *Nees.*

Habitat.—Plains of the Panjáb and North-West Provinces, ascending to low altitudes on the Himálaya.

Apluda geniculata, *Roxb.; Fl. Ind., Ed. C.B.C., 109.* 1235

Vern.—*Tachla,* MUSSOORIE.

Habitat.—Panjáb and the Himálayas up to 9,000 feet; Simla, Pesháwar.

APOCYNACEÆ—The Dogbanes. 1236

"Erect or twining shrubs, rarely herbs or trees. *Leaves* opposite or whorled (scattered in *Cerbera* and *Plumeria*), quite entire, exstipulate. *Flowers* in terminal or axillary cymes, hermaphrodite, regular. *Calyx* inferior; lobes 5, rarely 4, imbricate, often glandular within at the base. *Corolla* rotate or salver-shaped; lobes 5, rarely 4, spreading, contorted and often twisted in bud, very rarely valvate. *Stamens* 5, rarely 4, on the tube throat or mouth of the corolla, filaments usually short; anthers oblong, linear or sagittate, conniving, connective, sometimes adhering to the stigma; cells 2, dehiscing lengthwise, sometimes produced downwards into an empty spur; pollen granular. *Disk* annular, cupular or lobed, or of glands, or 0, sometimes concealing the ovary. *Ovary* 1-celled with 2 parietal placentas, or 2-celled with axile placentas, or of 2 distinct or partially connate carpels; style simple or divided at the base only; top thickened; stigma 2-fid, acute or obtuse. *Ovules* in each cell 2, or few or many and 2-8-seriate, rarely solitary. *Fruit* a dry or fleshy drupe, berry or samara, or of 2 drupes, berries or follicles. *Seeds* various, often winged, or with a terminal pencil of long silky hairs (*Coma*); albumen hard, fleshy, or scanty or 0; embryo straight, cotyledons flat, concave, convolute or contorted, radicle usually superior. *Distrib.*—Species about 900, chiefly tropical.

Tribe I. Carrissæ. *Anthers* included, free from the stigma; cells rounded at the base. *Ovary* of 2 wholly combined carpels, 1-2-celled. *Fruit* large, usually fleshy or pulpy within. *Seeds* without wing or pencil of hairs. *Corolla-lobes* overlapping to the left in all.

* *Ovary 1-celled, with parietal ovules.*

Fruit indehiscent. Albumen 0 . . . 1. **Willoughbeia.**
Fruit 2-valved. Albumen horny . . 2. **Chilocarpus.**

** *Ovary 2-celled, with axile ovules.*

Flowers 4-merous. Erect shrubs. Seeds exalbuminous . . . . . . 3. **Leuconotis.**
Climbing shrubs. Corolla-mouth with lobed scales . . . . . . 4. **Melodinus.**
Climbing slender unarmed shrub. Corolla-mouth naked . . . . . 5. **Winchia.**
Erect or stout climbing armed shrubs . 6. **Carissa.**

Tribe II. Plumerieæ. *Anthers* included, free from the stigma, cells rounded at the base. *Ovary* of two distinct carpels united by the style. *Fruit* various. *Seeds* peltate. *Corolla-lobes* overlapping to the left except in **Ochrosia.**

Subtribe 1. Rauwolfieæ. *Calyx* eglandular within. *Carpels* 1-2, rarely 4-6-ovuled. *Fruit* of 2 1-seeded drupes or berries, rarely moniliform (of superposed drupes).

Leaves usually whorled. Disc present. Albumenwen . . . . . . 7. **Rauwolfia.**
Leaves usually whorled. Disc 0. Albumen ruminate . . . . . 8. **Alyxia.**
Leaves opposite. Disc 0. Albumen smooth . . . . 9. **Hunteria.**

Subtribe 2. Cerbereæ. *Calyx* glandular within. *Carpels* 2-, rarely 4-ovuled; ovules on opposite sides of a thick placenta. *Drupes* or berries 1-seeded or 2-seeded, the seeds separated by the enlarged placenta.

Leaves scattered, alternate. Corolla funnel-shaped . . . . . . 10. **Cerbera.**
Leaves opposite Corolla salver-shaped, lobes overlapping to the right . . 11. **Ochrosia.**
Leaves opposite. Corolla salver-shaped . 12. **Kopsia.**

Subtribe 3. Euplumerieæ. *Calyx* glandular within. *Carpels* 6-∞-ovuled. *Fruit* (in the Indian genera) of 2 follicles.

* *Ovules 2-seriate.*

Disc annular or obscure. Seeds winged. Leafless shrub . . . . . 13. **Rhazya.**
Disc of 2 scales. Seeds truncate at both ends . . . . 14. **Vinca.**

** *Ovules ∞-seriate.*

Erect trees. Leaves scattered, alternate. Seeds winged . . . . . . 14. ***Plumeria.**
A climber. Leaves opposite or whorled. Seeds winged . . . . . . 15. **Ellertonia.**
Erect trees or shrubs. Leaves whorled. Seeds comose. Style distinct . . 16. **Alstonia.**
Erect trees. Leaves whorled. Seeds winged. Style 0 . . . . . 17. **Dyera.**
Erect trees. Leaves opposite. Seeds comose. Style short . . . 18. **Holarrhena.**

Subtribe 4. Tabernæmontaneæ. *Calyx* glandular within. *Carpels* ∞-ovuled. *Fruit* fleshy or coriaceous, dehiscent or not.

Erect trees or shrubs . . . . . 19. **Tabernæmontana.**

Tribe III. Echitideæ. *Anthers* included or exserted, conniving in a cone around the top of the style, and adherent to it by a point on the connective; cells produced downwards into a subulate empty spur. *Ovary* of 2 distinct carpels united by the style. *Fruit* of 2 follicles. *Seeds* comose at one or both ends. Exceptions, see **Parsonsia.**

Subtribe 1. Parsonsiæ. *Corolla* rotate or salver-shaped, throat naked, except **Wrightia.** *Anthers* more or less exserted.

Corolla-lobes valvate. Carpels *connate in flower* . . . . . . 20. **Parsonsia.**
Corolla rotate, mouth naked. Connective thickened at the back . . 21. **Vallaris.**
Corolla salver-shaped, mouth naked . 22. **Pottsia.**
Corolla rotate or salver-shaped, mouth with scales . . . . . 23. **Wrightia.**

Subtribe 2. Nerieæ. *Corolla-throat* broad, with 5-10 scales. *Anthers* included.

Shrubby, erect. Leaves whorled. Corolla-lobes short. Follicles erect . 24. **Nerium.**
Shrubby or twining. Leaves opposite. Corolla-lobes long or tailed. Follicles spreading . . . . . . 25. **Strophanthus.**
Herbaceous. Leaves opposite. Corolla-lobes short. Follicles slender . . 26. **Apocynum.**

Subtribe 3. Euechitideæ. *Corolla* various, mouth naked. *Anthers* included.

* *Corolla-lobes valvate, overlapping to the left.*

Flowers small or minute. Corolla urceolate, lobes valvate . . . 27. **Urceola.**
Flowers small. Corolla subcampanulate, lobes overlapping . . . 28. **Parameria.**

** *Corolla-lobes overlapping to the right.*

α *Corolla very large.*
Immense climbers, corolla bell- or funnel-shaped . . . 29. **Beaumontia.**
Lofty climbers. Corolla salver-shaped . 30. **Chonemorpha.**

β *Corolla minute, urceolate, lobes very short.*
Ovary exserted from the disc. Seeds beaked . . . . 31. **Ecdysanthera.**

γ *Corolla small or medium-sized, salver-shaped, lobes nearly straight or slightly twisted to the left in bud.*
Ovary hidden in the disc. Seeds slender . . . . . 32. **Baissea.**
Ovary hidden or not in the disc. Seeds ovate or oblong . . . 33. **Aganosma.**

δ *Corolla small, salver-shaped, lobes sharply twisted to the left in bud, tips not deflected.*
Ovary hidden in the disc. Seeds slender . . . . 34. **Epigynum.**
Ovary exserted from the disc. Seeds beaked . . . . . 35. **Rhynchodia.**
Ovary exserted from the disc. Seeds not beaked . . . . . 36. **Trachelospermum.**
Ovary hidden in the disc. Seeds ovate, beaked . . . . . 37. **Anodendron.**

ε *Corolla small, salver-shaped, lobes sharply twisted to the left in bud, with the tips deflected.*
Seeds ovate, beaked . . . . 38. **Ichnocarpus.**
Seeds ovate, not beaked . . . 39. **Micrechites.**"

(*Flora of British India, III., 621.*)

The above extract from the *Flora of British India* has been published here in the hope that it may be found useful in suggesting the genera belonging to APOCYNACEÆ, many of which will be found described in greater detail in their respective alphabetical positions.

1237

Distribution.—The Natural Order APOCYNACEÆ contains in all some 900 species, chiefly inhabiting the intertropical zones of the Old and New Worlds. Few species are indigenous to the extra-tropical or warm temperate zones, and still fewer are truly temperate.

In India there are 147 species, (some 15 of which are doubtfully distinct or are introduced), arranged in 40 gener Of these 99 or 67·3 per cent. are confined to the plains; 44 or 27·9 per cent. occur on the lower hills, and 7 or 4·7 per cent. ascend above 5,000 feet in altitude. In the Eastern Peninsula 97 or 66 per cent. occur; South India and Ceylon have 14 or 9·5 per cent.; in the Western Peninsula only 8 species are met with or 5·4 per cent., and in North or Upper India only 2 species appear to be peculiar to the drier tracts or 1·3 per cent.; while in cultivation all over India or wild in two or more of the four preceding divisions some 26 species are met with, equal to 17·7 per cent. of the entire Indian Apocynaceæ. Of these less local species the Eastern Peninsula and South India take the greater number. South India and the Western Peninsula are next in importance; few only extend from the other divisions into North India.

From this analysis it appears that in the Eastern Peninsula (and on the plains of that division) of India the vast majority of the Apocynaceæ occur; the eastern region thus preserving its feature as the home of Indian tropical plants, it is consequently the division of India most suited for introduced Apocynaceæ. (Compare with the account given under **Anonaceæ, Asclepiadeæ,** &c.)

1238

Affinities of the Apocynaceæ.—They have their closest resemblances to ASCLEPIADEÆ, LOGANIACEÆ, and GENTIANEÆ on the one side, and to OLEACEÆ and JASMININEÆ on the other. From ASCLEPIADEÆ they are at once separated by the condition of the anthers, generally united together in the dogbanes, with free filaments and free from, although embracing, the stigma—the anthers and filaments in ASCLEPIADEÆ being completely united to the style and stigma. From LOGANIACEÆ they are separated by the absence of intrapetiolar stipules; in most APOCYNACEÆ the carpels are two, free from each other, except by the united and common stigma. When the fruit of the APOCYNACEÆ, like that of the LOGANIACEÆ, is a solitary capsule, berry or drupe, they are distinguished by the milky sap, always isostemonous corolla and united anthers, and by the leaves being opposite or whorled and exstipulate. The LOGANIACEÆ have winged but never comose seeds.

Through LOGANIACEÆ they have an affinity to RUBIACEÆ from which the position of the ovary at once separates them, superior in APOCYNACEÆ but inferior in RUBIACEÆ. From GENTIANEÆ they are at once distinguished by their woody stems or more or less arborescent habit with milky juice, the GENTIANEÆ being herbs with watery juice, free anthers and a one-celled ovary having two parietal placentas with many seeds.

From OLEACEÆ and JASMINEÆ they are at once separated by the anisostemonous corolla of these orders.

1239

Properties and Uses of the Apocynaceæ.—Most of the species possess a milky sap, often rich in India-rubber and Gutta-percha. In India **Alstonia scholaris, Chonemorpha macrophylla, Parameria glandulifera, Urceola elastica, U. esculenta, Willoughbeia edulis,** and other species yield India-rubber. A form of **Alstonia scholaris** at Singapore is said to afford part of the *Gutta-pulei*, and two species of **Dyera** met with in the forests of Malacca, Singapore, and Sumatra yield the *Gutta-jelutong* of commerce. Other India-rubber genera, belonging to this order, are **Vahea gummifera,** *Lam.* in Madagascar, **Hancornia** in Brazil, and **Landolphia** in West Africa. In other species the milky sap, while containing less caoutchouc, has often other properties. It is purgative and febrifuge in **Allamanda, Carissa,** and **Plumeira.** It is mildly bitter and laxative in **Cerbera,** and in the fruits of certain species it is acid sweet; they are accordingly eaten as edible fruits, such as **Carissa Carandas, Willoughbeia edulis, Urceola elastica,** and **Tabernæmontana utilis.** At other times the sap is acrid and very poisonous, as in the Madagascar ordeal-plant, **Tanghinia venenifera,** a seed of which is sufficient to poison 20 persons. The wood, flowers, and leaves of the Oleander—**Nerium odorum** and **N. Oleander**—are very poisonous. So also are the nuts of **Thevetia neriifolia.**

Some are medicinal; the CONESSI BARK or the bark of **Holarrhena antidysenterica** and the bark of **Thevetia neriifolia** and of **Alstonia scholaris** are most valuable antiperiodics, useful in malarious fevers.

TIMBER.
1240

Structure of the Wood.—Most of the APOCYNACEÆ are arborescent, but the timber is, as a rule, of poor quality. It is white, soft, without heart-wood, the pores are small and the medullary rays fine and numerous **Alstonia** is perhaps the only exception to this character, the pores being of moderate size and the rays distant.

Many are handsome, ornamental trees, bushes or climbers, much cultivated in Indian gardens. Amongst the most important may be mentioned the species of **Allamanda** and **Alstonia**—the latter are large shrubs or trees; **Beaumontia grandiflora,** a truly superb climber, with pendulous white flowers, 6-8 inches long, is also common. **Kopsia fruticosa,** with its rose-coloured flowers; **Echites caryophyllata,** the clove-scented Echites (*mâlati*) and other species are elegant chimbers, frequently seen rambling over trees. **Nerium odorum,** or Indian Oleander, with its pink, dark red, or white flowers, is a constant companion with the mango tree in the gardens of the natives, and justly deserves its great popularity. **Plumeria acutifolia** (*gúl-i-chín*) is less frequent in native gardens; it is a very ornamental small tree, when covered with its clusters of whitish pink flowers and large spreading leaves. Its long naked branches and gouty stem are perhaps its chief drawback; it is exceedingly plentiful in gardens in the neighbourhood of Calcutta. **Parsonsia corymbosa (spiralis,** *Wall.*) is a beautiful scandent shrub, its bright crimson flowers appearing in the hot season. **Rauwolfia serpentia,** of which **Sir W. Jones** says: "Few shrubs in the world are more elegant, especially when the vivid carmine of the perianth is contrasted not only with the milky-white corolla, but with the rich green berries which at the same time embellish the fascicles." **Thevetia neriifolia** (*zard kunél*) has made far more progress probably than any other introduced dogbane. Every garden wall in Bengal, one might almost say, is decorated with a few plants of this elegant, small, spreading tree or bush, which throughout the year is covered with its large, yellow, sweetly-scented flowers. The odour of these flowers, at first much too strong for most Europeans, becomes more delicate after a time, and, indeed, exercises such an influence over the olfactory nerves that they lose the power of smelling it. **Tabernæmontana coronaria** (*chándui*) is probably after **Nerium** the next most popular plant in Indian gardens. Flowers large, single or double, and pure white, delicately scented at night. **Vinca alba** and **rosea** must not be omitted from this list of garden Apocynaceæ. Indeed, two or three varieties of the periwinkle are, perhaps, the most constant herbaceous favourites, and, associated with the balsam and the Indian yellow marigold (*géndha*), abound in every native garden.

A few species of Apocynaceæ yield dyes. **Wrightia tinctoria** is used in some parts of India for the manufacture of indigo, and **W. tomentosa** yields a yellow juice used as a dye.

APONOGETON, *Thunb.; Gen. Pl., III., 1013.*

1241

A genus of submerged aquatic herbs belonging to the Natural Order NAIADACEÆ.

Leaves on long petioles floating on the surface, very much like a Potamogeton, only green coloured. *Spike* solitary, generally bifurcating into two recurved portions. *Flowers* hermaphrodite, situate within two highly coloured bracts; *perianth* absent. *Stamens* 6-many, hypogamous; *filaments* unequal subulate, *anther* small. *Ovary* of 3-6 distinct *carpels,* oblique, sessile; *stigma* oblique, disciform. *Ovules* 2-many, erect, anatropous basilar; *carpels* 3, mature, or more.

Aponogeton monostachyum, *Linn.*

1242

Vern.—*Ghechu,* HIND.; *Khaangi,* SANS.; *Kotti-katang* or *kotti-kishangu,* TAM.; *Kotti-gadda nama,* TEL.; *Kotti-kang,* DUK.

Habitat.—A native of shallow, standing, sweet water, in Bengal, appearing during the rains.

FOOD. Roots. 1243

Food.—The natives are fond of the roots, which are said to be nearly as good as potatoes. (*Roxb.*) It is remarkable that this property should have been detected, while the tubers of its associate, **Sagittaria**, have escaped discovery.

1244

APOROSA, *Blume; Gen. Pl., III., 282.*

A genus of trees or shrubs (belonging to the Natural Order EUPHORBIACEÆ), comprising some 20 species, inhabiting Asia, and chiefly the Malayan Peninsula.

Leaves alternate, simple, petiolate, entire, penninerved. *Flowers* diœcious, minute, enclosed by the bracts; forming spikes or racemes. *Males* forming clusters; *females* short, few-flowered. *Calyx* 3-6, often unequal, membranous, imbricate. *Petals*and *disk* absent. *Stamens* 2-5, long, free, inserted around rudiment of the pistil; *anther* small; *cells* subglobose, distinct, united throughout their length to a more or less thickened connective, dehiscense by a longitudinal slit on each cell. *Female flowers* solitary within the involucre, sessile. *Ovary* 3-, rarely 2-4-celled, with 3 ovules in each cell. *Capsule* fleshy indehiscent, endocarp referred to 3 one-seeded pyrenes.

1245

Aporosa dioica, *Müll-Arg.; DC. Prod., XV., 2, 472;* EUPHORBIACEÆ.

Syn.—A. ROXBURGHII, *Baill.*; ALNUS DIOICA, *Roxb.*; LEPIDOSTACHYS ROXBURGHII, *Wall.*

Vern.—*Kokra*, BENG.; *Sanpau*, GARO; *Tanprengjan*, MAGH.

Habitat.—A tree of North and East Bengal and Burma.

Botanic Diagnosis.—"*Style-lobes* simple, short. *Ovary* thinly appressed pubescent, glabrescent." (*Kurz.*)

TIMBER. 1246

Structure of the Wood.—Dark brown, very hard, close-grained with white sapwood, weighing 79 ℔ per cubic foot.

This has by botanists been identified as the tree which yields in the West Indies the Coco-wood of commerce. The Indian plant should be carefully examined to ascertain if the wood obtained from it is of equally good quality with that obtained from the West Indies.

1247

A. villosa, *Baill.; DC. Prod., XV., 2, 471.*

Vern.—*Ya-mein*, BURM.

Habitat.—A tree frequent in the *Eng* forests of Burma from Pegu to Martaban. (*Kurz.*)

Botanic Diagnosis.—"*Leaves* shortly and softly pubescent beneath. *Ovary* villous, tomentose, or pubescent. *Berries* densely velvety tomentose." (*Kurz.*)

RESIN. 1248

Resin.—Yields a red resin.

DYE. 1249

Dye.—The bark is used as a red dye.

Apple, The, see **Pyrus Malus,** *Linn.;* ROSACEÆ.

Apricot, The, see **Prunus armeniaca,** *Linn.;* ROSACEÆ.

1250

AQUILARIA, *Lam.; Gen. Pl., III., 200.*

A genus of trees belonging to the Natural Order THYMELÆACEÆ, comprising only 2 or 3 species, inhabitants of tropical South-West Asia—the Malaya and Borneo.

Leaves alternate, entire or nearly so; *petiolate* exstipulate, penninerved, nerves close, parallel. *Flowers* pedunculate in subsessile umbels, axillary or terminal. *Bracts* absent. *Flowers* hermaphrodite. *Perianth* forming a distinct campanulate tube (sometimes described as the campanulate flora receptacle); (calyx) *teeth* 5 broad ovate-acute or obtuse, imbricate in bud. *Squamules* (or corona-like scales) equal in number to the stamens and alternate with them, inserted on the mouth of the tube, erect, very hairy. *Stamens* 10, inserted below the squamules, and with the sepals therefore perigynous; *filaments* short, very rarely elongated; *anthers* basifixed, ovate or oblong, introrse. *Ovary* subsessile in the bottom of the tube, free, perfect or imperfect, hairy, 2-locular or 1-locular, the placentas being along the middle of the valves. *Fruit* drupaceous, becoming capsular, surrounded below by the persistent perianth tube. *Seeds* 1-3, often 2, ovoid, raphe-ventral, produced in a more or less spongy cone; *cotelydons* fleshy plano-convex; *radicle* short, inferior.

Aquilaria is derived from the Lat. *Aquila,* the eagle; hence the name Eagle-wood.

Aquilaria Agallocha, *Roxb.; DC. Prod., XIV., 601.*

1251

CALAMBAC, AGALLOCHUM OR ALOE-WOOD, OR EAGLE-WOOD; LIGNUM-ALOES, AGLIA, AKYAW.

Vern.—*Agar,* HIND.; *Agaru, ugar,* BENG.; *Agaru,* SANS.; *Agare-hindi, úd,* or *aúd,* or *aúde-hindi,* or *úde-hindi,* ARAB.; *Agre-hindi, agar,* PERS.; *Hindiagara,* BOMB.; *Agar,* GUJ., *Agar, aggalichandana,* TAM.; *Agru,* TEL.; *Sási,* ASS.; *Akyau,* BURM.; *Kihay, sinnah,* SINGH.; *Kayu, garu,* MALAY; *Nwahmi,* SIAM; *Nyaw-chah,* CHINESE.

Habitat.—A large evergreen tree of Sylhet and Tenasserim; distributed to the Malay Peninsula and Archipelago.

Botanic Diagnosis.—"*Capsules* wrinkled, softly and densely tomentose." (*Kurz.*)

HISTORY, 1252

History.—Since the time **Dr. Roxburgh** described this plant scarcely any further information has been obtained. The conclusion he arrived at seems correct, namely, that the much-prized wood is obtained from eastern India, and from the forests to the east and south-east of Sylhet, extending through Mánipur, Chittagong, Arakan, to Mergui and Sumatra. From India it finds its way to China, and from Cochin China it was first re-exported to Europe; hence, in all probability, the association of the plant with that country. **Loureiro** described a plant under the name of **Alœxylon Agallochum,** said by him to be a native of Cochin China, and to yield the true Calambac-wood or Agallocha. His description is incomplete, and his genus has therefore been set aside by **Bentham** and **Hooker** in their *Genera Plantarum,* while the plant has never since been identified or re-collected. **DeCandolle,** in the *Prodromus,* refers it to Leguminosæ.

Dr. Royle regards the Aloes-wood of the Scriptures as the *Ahila* or *Ahalim* of the East, so famed for its fragrance, and that it is yielded by **Aquilaria Agallocha. Gamble** says that "*Akyau* (the Burmese name for Agallocha) is the most important produce of the forests of South Tenasserim and the Mergui Archipelago. It is found in fragments of various shapes and sizes in the centre of the tree, and usually, if not always, where some former injury has been received."

An enquiry into the history of Aloes-wood shows that an odoriferous wood bearing the name of *Ahalot* was known to the ancient Jews; the same substance appears to have been called agallochon by the Greeks and Romans. The early Arabs corrupted this term into *Aghalúkhí,* but subsequently adopted the terms *Ood* (or *aúd*), meaning wood, and *Ood-hindi* (Indian wood), as the technical names for Aloes-wood. In Sanskrit it is called *Agaru,* from which is derived the modern Hindi name *Agar.* Upon the subject of *Ood,* **Mir Muhammad Hosein** has the following remarks: "*Ood,* in Hindi *Agar,* is the wood of a tree which grows in the Jaintiya hills near Sylhet, a dependency of the Suba of Bengal, situated towards the north-east of Bengal proper. The tree is also found in the islands to the south of Bengal, situated north of the equator, and in the Chatian islands belonging to the town of Nawaka, near the boundaries of China. The tree is very large, the stem and

HISTORY.

branches generally crooked, the wood soft. From the wood are manufactured walking-sticks, cups, and other vessels; it is liable to decay, and the diseased part then becomes infiltrated with an odoriferous secretion. In order to expedite this change it is often buried in wet ground. Parts which have undergone the change above mentioned become oily, heavy, and black. They are cut out and tested by being thrown into water; those which sink are called *Gharki*, those which partly sink *Neem Gharki*, or *Samaleh-i-aala*, and those which float *Samaleh;* the last kind is much the most common. *Gharki* is of a black colour, and the other qualities dark and light brown. According to other and older authorities, *Ood* is classified as *Hindi*, *Samandooree*, *Kamari*, and *Samandalee*. *Hindi* is described as black, *Samandooree* as more oily than *Hindi*, *Kamari* as pale-coloured, *Samandalee* as very odoriferous. Elsewhere it is described as *Barree* and *Jabali*, the latter having black lines in it, the former white; others again described *Barree* as having black lines, and *Jabali* white."

"*Samandooree Ood* is said to be called after the place whence it is obtained, also *Kamari*."

"The best kind for medicinal use is *Gkarki Ood* from Sylhet; it should be bitter, odoriferous, oily, and a little astringent; other kinds are considered inferior. In most receipts raw Ood (*Ood-i-kham*) is enjoined to be used to prevent the use of wood from which the oil has been abstracted by crushing and maceration in water, or by crushing and admixture with almonds, which are afterwards expressed. This precaution is the more necessary as Ood shavings are an article of commerce in India under the name of *Choora agar;* they are often adulterated with chips of Sandalwood, or Tagger, an odoriferous wood much like Aloes and common in India. I have also heard mention made of the kind of Ood which is described by the author of the *Ikhtiárát-i-badiæ* as coming from Bunder Cheeta, ten days' sail from Java, and esteemed equal in value to its weight of gold; this kind is said to have no smell until warmed. When taken in the hand it diffuses a delightful odour (possibly Bunder Cheeta has been written in mistake for Chatiyan). There is another kind of wood common in India, which resembles Aloes in appearance, consistence, and oiliness; it is called Tagger, and is often sold for Ood. (*Makhzan*, article [Ood or *Aúd*].)"

"**Rumphius, Kœmpfer,** and others have written at some length on Aloes-wood, but have not thrown much light upon the subject (*Confer. Guibourt Hist. Nat., Tom. III., p. 336*). **Guibourt** describes five kinds of Aloes wood, from the examination of specimens which he has met with in Europe. The first, a specimen in the *Ecole dePharmacie*, he attributes to **Alœxylon Agallochum**; the second, which he considers to be the produce of an Aquilaria, is the ordinary Aloes-wood of European commerce. The fifth, which is very heavy, oily, and resinous, he thinks must be produced by **Excæcaria Agallocha**; in this wood there are vities filled with a reddish resin. **Guibourt's** first and second kinds are more minutely described by **Planchon.** The varieties of *Agar* found in the Bombay market are three: Siam or Mawurdhee, Singapore, and Gágulee. Besides these we have Tagger from Zanzibar and a false Aloes-wood." (*Dymock, Mat. Med., W. Ind., pp. 239-241.*)

No further evidence having come to light of the existence of Agallocha wood in Cochin China, it is probable that the odoriferous wood was not the product of the tree described by **Loureiro,** but was an importation obtained from India. There are many plants, however, which resemble the Agallocha in the odour of their wood, resin or sap, and it is therefore probable that Cochin China may possess one of these. The saps of **Excæcaria Agallocha,** *L.*, a small tree found along the coast of

Burma from Chittagong to Tenasserim, is supposed to resemble Agallocha, hence the specific name (see **Excæcaria**). So also the resinous excretions from various members of the Myrrh family have been erroneously associated with the *Agar*. This, in all probability, is the explanation of **Balsamodendron Agallocha,** *W. & A.*, as in **Drury**, the description of which most probably contains a compilation of the characters attributed to **B. Mukul, B. Roxburghii,** and **Aquilaria Agallocha. Smith,** in his *Dictionary of Economic Plants,* seems to lay stress upon Agallocha being the vernacular name for **Excæcaria Agallocha,** but the name Agallocha does not appear to be of Indian origin.

RESIN. 1253

Resin.—The wood of this tree is impregnated with a resinous principle, often found collected in masses here and there throughout the stem. This curious fact is in all probability due to some diseased condition, which might be artificially produced in order to increase the formation or collection of the resin. To obtain this sweetly-scented resin the trees are hewn down and cut to pieces while searching for the masses of resin.

PERFUME. 1254

THE WOOD CHIPS (*chúrá-agar*) are largely sold in the bazars, and used, either by themselves or associated with *Bdellium,* as incense burned at Hindu temples. They are also boiled, and the water thereafter distilled, in order to prepare *Agar-atar* (or *agar-ká-itr*), a perfume much admired by the people of India.

Fibre.—"The bark is used in Assam for covers of unbound books." (*Mr. H. Z. Darrah, Director of Agriculture, Assam.*)

MEDICINE. Resinous substance. 1255

Medicine.—The fragrant resinous substance is considered cordial by some Asiatic nations. It has been prescribed in gout and rheumatism (*Ainslie, ex Voigt's Hortus Suburb. Calcut.*). **Lourerio** observes that the Calambac is a delightful perfume, serviceable in vertigo and palsy, and the powder is useful as a restrainer of the fluxes and vomiting. In decoction it is useful to allay thirst in fever.

Baden Powell says that aloe-wood is supposed to owe its fragrance to the rotting of the wood, and the best specimens are therefore buried in earth for some time. It was formerly much used in Europe in gout, rheumatism, diarrhœa, vomiting, and palsy. The name aloe-wood has nothing to do with aloes, but is a corruption of the Arabic term *Al-'úd* (or *al-aúd*). (*Baden Powell, Panjáb Products, I., 337.*) An essential oil prepared from the wood is also used medicinally.

OIL. 1256

§ "Internally in fevers, externally in colic." (*Surgeon-Major D. R. Thompson, Madras.*) "The otto of *ood* is considered cooling, and is an ingredient in many eastern perfumes." (*Surgeon-Major A. S. G. Jayakar, Muskat, Arabia.*)

TIMBER. Ordinary. 1257

Structure of the Wood.—White soft, even-grained, scented when fresh cut. Weight about 25 lbs. per cubic foot. In the interior of old trees are found irregular masses of harder and darker-coloured wood, which constitute the famous Eagle-wood of commerce, called *Kaya garu* by the Malays, and *Akyau* by the Burmese (*agarú*, Sans.).

Kurz says of this wood: "very light, yellowish white, coarse, fibrous, but closely grained, takes a pale-brown polish. Used by the Karens for bows." The fragrant wood *Ood* is also largely used for making jewel-cases, and, indeed, precious stones are very frequently set in it. Aloes-wood is also largely used for making ornaments and rosary beads.

Selected piece of perfumed. 1258

Eagle-wood is stated to bring about £30 per cwt. for 1st quality (Sumatran); £20, 2nd quality (Malaccan); and £2-10, 3rd quality (Malaccan and Indian). It should melt like wax and emit an agreeable odour. There seems considerable confusion in the use of the word Eaglewood. It apparently is applied to the masses highly impregnated with the gummy substance as well as to the timber.

1259

Aquilaria malaccensis, *Lamk.; DC. Prod., XIV., 602.*

GARODE MALACCA OF MALACCA EAGLE-WOOD.

Habitat.—Said to be met with in Tenasserim.

Botanic Diagnosis.—" An evergreen tree, the young shoots of which are covered with adpressed hairs. *Capsules* smooth and glabrous." (*Kurz.*)

Arabic Gum, see Acacia arabica and A. Senegal.

1260

ARACHIS, *Linn.; Gen. Pl., I., 518.*

A genus of Brazilian herbaceous prostrate annuals (belonging to the LEGUMINOSÆ, Sub-order PAPILIONACEÆ), comprising some 6 or 7 species, one of which is now cultivated throughout the tropical and extra-tropical regions of the globe.

Leaves abruptly pinnate, 2-jugate, or rarely 3-foliolate; *leaflets* exstipulate; *stipules* adnate to the base of the petiole. *Flowers* in dense axillary spikes. *Receptacle* more or less concave lined by the disk. *Calyx* gamosepalous, either tubular or sacciform at the base or else 2-partite, anterior sepal free to the base, 4 superior connate to a considerable height and membranous; *teeth* imbricate. *Petals* very unequal; *standard* suborbicular, scarcely tapering at the base, thickened gibbose at the back; *wings* oblong, free; *keel* curved, beaked and tapering for a considerable distance at the apex. *Stamens* 9 or 10, 1-adelphous; *tube* more or less thickened and fleshy at the base; *anthers* of 2 forms, 5 oppositipetalous shorter sub-globose, versatile, 5 alternipetalous elongated, basifixed. *Legume* sub-sessile, few-ovuled, oblong, thick, reticulated, sub-torulose.

The generic name is derived from the Greek name for a leguminous plant, ἄρακος or αραχος, referred to by **Pliny**, which had neither stem nor leaves. The specific name of the plant met with in India **(A. hypogæa)** is derived from ὑπόγειος = subterranean.

1261

Arachis hypogæa, *Linn.; Fl. Br. Ind., II., 161;* LEGUMINOSÆ.

THE GROUND NUT OF EARTH NUT OF PEA NUT.

Vern.—*Buchanaka,* SANS.; *Mát-kalái, chiner-bádám, biláti-mung,* BENG.; *Múngphali, viláyeti-múng,* HIND.; *Bhoni-mug,* SIND.; *Bhóya-chená, bhæ-mag,* GUJ.; *Bhúi-chane, bhui-múga,* or *bhui-sheng, viláyati múg,* BOMB., MAR.; *Vérk-kadalai, nilak-kadalai,* TAM.; *Verushanaga-káya, verushanaga,* TEL.; *Nelak-katalá, vérk-kalá,* MALA.; *Nelgale-káyi, kadale-kayi,* KAN.; *Rata-kaju,* SINGH.; *Mibé, myépe, maibai,* BURM.

Habitat.—An annual of South America, now generally cultivated throughout India, but chiefly in South India and Bombay; also in certain parts of Bengal, and more rarely in Upper India.

This plant was not known in the Old World before the discovery of America. **Dr. Dymock** thinks the ground-nut reached India through China. It does not appear to have been cultivated for more than 50 years It may have come to Western India from Africa.

Botanic Diagnosis.—After the flowers wither, the torus (which supports the ovary) becoming elongated in the form of a thick rigid stalk, and curving downwards, by alternately bending upon itself from one side to another, forces the pod underground, and in this position the peas are ripened. In India this curious plant often attracts to itself a large number of red-ants, which, in gardens in Bengal, seem regularly to soften and pulverize the soil as if to facilitate the movement of the pod. It would be interesting to know whether the same fact has been observed in other parts of the world, and if so, to discover whether the plant feeds these useful insects in return for their assistance. They do not appear to eat the nuts or peas.

Cultivation.—Ground-nuts do not appear to require much care in cultivation. They grow best on dry, sandy soil. Watering is not needed, nor any particular observance as to the time when supplies of seed are put down, nor when they are reaped. With the close of the grain-reaping season, extensive plots of land, especially in the Chingleput, South Arcot, parts of the Tanjore and Trichinopoly districts, are sown broadcast with ground-nuts. (*Trop. Agri., III., 774.*) An average good crop will yield as much as 50 bushels from the acre. In 1879 the total area under the cultivation of ground-nuts in India was ascertained to be 112,000 acres, and almost exclusively confined to Madras, Bombay, Berar, and Mysore. In hard soils the crop proves objectionable, from the difficulty to remove the nuts from the soil, many remaining and the plant thus becoming a troublesome weed.

OIL. 1262

Oil.—The seeds of this plant afford on expression a clear straw-coloured, non-drying oil, which resembles olive oil in taste, and in India is now being used as a substitute for it in medicinal preparations. As a lamp oil it has for some time been used, and has the reputation of burning longer, but giving a less luminous flame. It has, however, one important advantage over other lamp oils—it will keep for a longer time without becoming rancid. In North Arcot it is reported to be used for adulterating gingelly oil, and in Pondicherry it is said to be mixed with cocoa-nut oil.

In Europe it is now extensively used as a substitute for olive or salad oil, both medicinally and for alimentary purposes. It has taken a distinct place in the soap manufacture, and it is largely consumed for lubricating machinery, as a lamp oil, and for dressing cloth. According to **Dumas**, a Marseilles merchant was the first who experimented with the ground-nut oil as a substitute for olive oil. The suggestion to do so is said to have been given by a French colonist, **Joubert**, in 1840. Little more than 40 years ago the oil was unknown to European commerce, and now the annual consumption is perhaps little short of 100,000 tons prepared oil a year. The chief emporiums of the European trade are Barcelona, Marseilles, and Genoa.

The yield of oil is often as much as 50 per cent., the average Pondicherry about 37 per cent., and the Madras 43 per cent. The quality of the oil from cold expression is much finer than when heat is employed, but in the latter case the volume is much increased. Formerly this oil was more extensively expressed in India than at the present date. In an official communication from the Board of Revenue, Madras, to the Government of India, it is stated that, in 1877, 7,130 cwt. of nuts and 20,387 cwt. of oil were exported from Madras. With the increased trade in the nuts the preparation of the oil seems to have declined in India. The Suez Canal having lessened the time occupied on the voyage to Europe, the nuts can now be exported in a good condition; and this fact, together with the improved machinery used on the Continent, have combined to render the oil the least important part of the trade in this product. The nuts, either shelled or not, constitute the chief export trade, and within the past few years this trade has rapidly developed, so that at the present moment it must be viewed as a most important item in the exports of South India. An effort is, however, being made to open out a company, with European machinery, to express the oil in India; there would seem to be a good future for such an industry.

Trade Returns.

1263

In a Resolution of the Government of India, Revenue and Agriculture Department, dated November 1877, will be found some interesting facts regarding the trade in ground nuts. The following extract shows the condition of the French trade in 1875: "Although the exports to foreign

ARACHIS hypogæa.

Trade Returns of Ground Nuts.

countries from British India are trifling, considerable quantities are sent from Pondicherry to France, as will be seen from the following figures. which have been extracted from the French trade returns of 1875:—

	Kilogrammes.	Francs.
Imports from British India	1,231,803	406,494
" " French "	6,404,899	2,113,616

The total imports into France in that year from all countries were 101,524,468 kilogrammes, or nearly 100,000 tons, worth 33,503,000 francs. Thus out of 33½ millions of francs, only 2½ millions stand against India. Nearly all the rest is imported from the West Coast of Africa."

Exports of Earth-nuts from British India.

Years.	Weight.	Value.
	Cwt.	R
1878-79	25,472	1,68,420
1879-80	48,435	2,85,519
1880-81	188,381	10,07,818
1881-82	373,317	17,35,269
1882-83	265,743	13,13,918
1883-84	712,954	37,65,462

Analysis of the Exports for 1883-84.

Province from which exported.	Weight.	Value.	Countries to which exported.	Weight.	Value.
	Cwt.	R		Cwt.	R
Bombay	595,822	32,04,357	United Kingdom	24,211	1,10,576
			Belgium	290,450	14,80,459
Sind	15	85	France	332,602	18,58,141
			Italy	15,913	82,113
Madras	117,117	5,61,020	Egypt	32,750	1,73,116
			Straits Settlements	12,769	42,634
			Other Countries	4,259	18,423
Total	712,954	37,65,462	Total	712,954	37,65,462

Exports of Earth-nuts from the French Ports in India.

Years.	Quantity in bags.	Value.
	Nos.	R
1880-81	233,533	14,39,340
1881-82	355,121	14,70,972
1882-83	412,415	16,18,659
1883-84	453,366	29,68,698

A. 1263

ARACHIS hypogæa.

Ground-nut Oil.

Analysis of the Exports for 1883-84.

Ports from which exported.	Quantity in bags.	Value.	Countries to which exported.	Quantity in bags.	Value.
	Nos.	R		Nos.	R
Pondicherry	450,170	29,58,176	United Kingdom	43,906	2,88,949
Karikal	3,196	10,522	France	403,964	26,59,845
			Réunion	50	329
			Straits Settlements	5,446	19,575
TOTAL	453,366	29,68,698	TOTAL	453,366	29,68,698

Accepting a mean between the discrepancies of the published figures of exports from British India, and of those of the imports from British India into other countries, we learn that the trade in ground-nuts has developed from 20,000 cwts. in 1875-76 to 700,000 cwts. in 1883-84. The above tables show that the United Kingdom is mainly supplied from French India, while the bulk of the British India exports are consigned to France and Belgium. Both in British and in French India the ground-nut trade has thus developed with marvellous rapidity, and this is doubtless due, in a large measure, to the action taken by the Government of India in the Resolution quoted above.

The following extract reporting the condition of the Pondicherry trade during May 1884 will be found interesting: "The ground-nut trade between Pondicherry and France is in full swing, and has been so since the month of February. The South Indian Railway Company has been running special trains with nuts from Punrooty to Pondicherry every day during the past nine weeks, and will probably continue to do so for two or three months longer. The ground-nut trade is the most important in the chief town of the French Settlements in India. Three ships are loading nuts in the Pondicherry Roads and more are expected. The European and Native merchants are fully engaged in this trade for at least six months in the year, and to facilitate shipments of nuts, the South India Railway has laid down a railway line from the Pondicherry railway station to the pier, so that the bags are shipped off as fast as they come in from the interior. Coolies find ample employment during the present season, and the price of labour is high. The European merchants in the port have entered heart and soul into the enterprise, and it is surprising how the South Arcot districts can produce such an immense quantity of nuts." (*Madras Standard,* quoted in the *Tropical Agriculturist, III., 830.*)

The oil known as *gora-tél* (*góră-tíl*) or sweet oil of the Indian bazars is obtained from a mixture of safflower, sesamum, and ground-nut seed.

Chemical Composition of Ground-nut Oil.

1264

"The oil consists of the glycerides of four different fatty acids, the common *Oleic acid* $C_{18}H_{34}O_2$,—that is to say, its glycerin compound is the chief constituent of Arachis oil. *Hypogæic acid* $C_{16}H_{30}O_2$ has been pointed out by **Gossmann** and **Scheven** (1854) as a new acid, whereas it is thought by other chemists to agree with one of the fatty acids obtained from whale oil. The melting point of this acid from Arachis oil is 34° to 35° C. The third acid afforded by the oil is ordinary *Palmitic acid* $C_{20}H_{40}O_2$. The fourth constituent has also been met with among the fatty acids of butter and olive oil, and according to **Oudemans** (1866), in the tallow of **Nephelium lappaceum,** *L.*, an Indian plant of the Order Sapindaceæ.

When ground-nut oil is treated with hyponitric acid, which may be most conveniently evolved by heating nitric acid with a little starch, a solid mass is obtained, which yields by crystallization from alcohol *Elaïdid* and *Gæidinic* acids, the former isomeric with oleic, the latter with hypogæic acid." (*Pharm., by Flück. & Hanb., p. 187.*)

MEDICINE. Oil. 1265 Ointment and Plasters. 1266

Medicine.—Arachis oil forms a good substitute in pharmacy for olive oil. It has now almost entirely superseded that of olive oil in India, both for pharmaceutical and other purposes. It is well adapted for the preparation of ointments and plasters. **Dr. Dymock** says that at the Government Medical Store Depôt of Bombay the oil is expressed for pharmaceutical purposes to the extent of about 6,000 lbs. a year. It is used as a substitute for olive oil. For a good plaster 90 lbs. of the oil take 41 lbs. of oxide of lead.

§ "The oil is hardly known in South India. We use gingelly (Sesamum) oil in our dispensaries for olive oil. The ground-nuts are largely eaten by natives in the Madras Presidency." (*Deputy Surgeon-General G. Bidie, C.I.E., Madras.*)

"Oil of Arachis appears to me to be a very efficient substitute for olive oil in pharmacy, and is fully as useful." (*Surgeon-Major H. W. E. Catham, Ahmadnagar.*) "Useful in catarrh of the bladder." (*Surgeon-Major Joseph Parker, M.D., Poona.*) "Ground-nuts roasted are eaten freely by natives of South India. They are said to be very bilious." (*Honorary Surgeon P. Kinsley, Chicacole, Madras.*) "The pod is largely grown in the sandy soil of the South Arcot District and is exported to France." (*Surgeon-General Willam Robert Cornish, C.I.E., Madras.*)

1267

Indian Prices.—The price of ground nuts in South India is generally R15 to 19 per *candy* (=5 cwts.), but during the season of 1883, when every available bag was bought and shipped to France, it rose to R24 (*Trop. Agri., III., 774, quoting Madras Mail*). "In the Bombay market the price of the seeds ranges from R28 to 30 per candy, according as the supply is abundant or otherwise." (*Dymock, Mat. Med., W. Ind., 202.*)

African Nuts. 1268

African Trade.—Ground-nuts are also grown on the west coast of Africa, and a large trade exists between Senegal and the Mediterranean ports. The African trade has one very important advantage over the Indian trade, in the fact that Genoa and Barcelona are only from fifteen to twenty days distant by steamer from Senegal. The African nuts can accordingly be landed in a far better condition than the Indian. The yield of oil is stated to be from 42 to 50 per cent. The *Pharmacographia* (p. 187) says: "The pods are exported on an immense and ever-increasing scale from the west coast of Africa. From this region not less than 66 millions of kilogrammes, value 26 millions of francs (£1,040,000), were imported in 1867 almost exclusively into Marseilles. From the French possessions on the Senegal, 24 millions of kilogrammes were exported in 1876." One of the learned authors of the *Pharmacographia* regards Africa as the probable home of the ground nut, but most botanists are of opinion, as already stated, that it is more likely to have been originally a native of Brazil. It is nowhere met with in a wild condition at the present day.

FOOD. 1269

Food.—It produces the well-known ground nut, so called because the pod attains maturity underground.

In India the nuts are sold in the bazars or by the street hawkers either parched, with the shell on and put up in paper packets, or shelled and roasted in oil. They are eaten by natives of all classes, especially in South India. In Bombay they are a favourite food of the Hindús during certain fasts. They are occasionally seen roasted in shell as a dessert on the European table, and are eaten with salt. Hand-shelled nuts are also sometimes made into confectionery. The roasted seeds may be used as a substitute for chocolate. "According to Dr. Davey, they abound with starch, as well as oil, a large proportion of albuminous matter, and in no other instance had he found so great a quantity of starch mixed with oil."

Ground-nut Meal. 1270

Chemical Composition of the Meal.—"Dr. Muter, after giving the following analysis of ground-nut meal, urges its more general use as an important article of food:—

Moisture	9·6
Fatty matter	11·8
Nitrogenous compounds (flesh-formers)	31·9
Sugar, starch, &c.	37·8
Fibre	4·3
Ash	4·6
TOTAL	100·0

"From this analysis it is evident," he observes, "that the residue from them, after the expression of the oil, far exceeds that of peas, and is even richer than lentils in flesh-forming constituents, while it contains more fat and more phosphoric acid than either of them. On these grounds we are justified in urging the adoption of the ground-nut meal as a source of food, it being superior in richness of all important constituents to any other vegetable products of a similar nature. Although in the raw state it possesses a somewhat harsh odour, similar to that of lentils, this flavour entirely passes off in cooking, and when properly prepared it has a very agreeable flavour.

"This seed is held in such estimation for eating in the United States (where it is known as the 'pea nut'), that flourishing sale-stands are seen at almost every street corner of New York. They are not much appreciated in England, except by children.

"There are fully 550,000 bushels sold annually in the city of New York alone. Previous to 1860 the product in the United States did not amount to more than 150,000 bushels, and of this total nearly five sixths were from North Carolina. Formerly it was largely imported into America, now they are supplied by the home crops raised in Virginia and the Carolinas. (*Tropical Agriculturist, P. L. Simmonds, pp. 403-4.*)

CATTLE FOOD. Cake. 1271 FODDER. 1272

Cattle Food and Fodder.—The leaves and branches of the plant are greedily eaten by cattle, and form an excellent fodder. The hay is very nutritious, much increasing the milk of cows. The cake holds a high reputation as a food upon which cattle rapidly fatten.

ARALIA, *Linn.; Gen. Pl., I., 936.*

1273

Aralia cachemirica, *Dcne.; Fl. Br. Ind., II., 722;* ARALIACEÆ.

Vern.—*Banakhor, churial,* PB.

Habitat.—A rank plant growing in the basins of the Jhelum and the Chenab.

FODDER. 1274

Fodder.—Eaten by goats.

ARALIACEÆ.

1275

"Trees or shrubs, very rarely herbs, sometimes scandent or scandent when young, and finally self-supporting, not rarely prickly. *Leaves* alternate the uppermost rarely sub-opposite, long-petioled, large, simple or compound; *stipules* adnate to the petiole, sometimes in conspicuous or o. *Flowers* regular, small, sometimes polygamous, in umbels racemes or panicled heads; bracts and bracteoles small or conspicuous; pedicels continuous with the base of the calyx or there jointed. *Calyx-tube* adnate to the ovary; limb truncate, obsolete or with smallteeth. *Petals* 5, rarely 6-7 or many, valvate or subimbricate, expanding or deciduous in a cap. *Stamens* as many as and alternate with the petals (very many in l*Tupidanthus*), inserted round an epigynous disc. *Ovary* inferior, 2-cel ed, or cells as many as the stamens (in *Arthrophyllum* 1-celled); styles as many as the cells, distinct or united; ovules solitary and pendulous in each cell. *Fruit* coriaceous or drupaceous, usually small, one or more cells sometimes suppressed. *Seed* pendulous, albumen uniform or ruminated; embryo minute, radicle next the hilum. *Distrib.*—Species 340, chiefly tropical and sub-tropical, a few in the cool temperate zones."

SECTION I. Araliеæ. *Petals* imbricated (but only lightly). *Pedicels* jointed.

Styles 2-5, free. Leaves compound . . 1. **Aralia.**
Styles 5, combined . . . . . . 2. **Pentapanax.**
Styles 4-3, free, leaves pinnatifid . . . 3. **Aralidium.**

SECTION II. Panaceæ. *Petals* valvate. *Albumen* uniform.

* *Ovary 2-celled.*

Pedicels jointed. Leaves decompound . . 3. * **Panax.**
Pedicels continuous. Leaves digitate . . 4. **Acanthopanax.**

** *Ovary 4-10 celled.*

† *Umbels sessile on the back of the leaf.*

Leaves simple . . . . 5. **Helwingia.**

†† *Pedicels jointed.*

Leaves once pinnate . . . . . . 6. **Polyscias.**

††† *Pedicels continuous. Leaves not pinnate.*

Fruit angular the size of a pea . . . . 7. **Heptapleurum.**
Fruit more than ½ in. long . . . . . 8. **Trevesia.**
Flower sessile embraced by 4 bracteoles . . 9. **Brassaia.**
Leaves simple (except the lowermost) . . 10. **Dendropanax.**

SECTION III. Hedereæ. *Petals* valvate. *Albumen* ruminated.

* *Ovary 1-celled.*

Leaves pinnate or undivided . . . 11. **Arthrophyllum.**

** *Ovary 2-celled.*

Pedicels continuous. Styles distinct . . 12. **Heteropanax.**
Pedicels continuous. Styles combined . . 13. **Brassaiopsis.**
Pedicels jointed. Styles combined . . . 14. **Macropanax.**

*** *Ovary 5-4-celled; styles combined.*

† *Leaves simple lobed or pinnate.*

Pedicels continuous . . . . . . 15. **Hedera.**
Pedicels jointed . . . . 16. **Hederopsis.**

†† *Leaves digitate.*

Tree. Leaflets ciliate . . . . . . 17. **Gamblea.**

SECTION IV. **Plerandreæ.** *Petals* valvate. *Stamens* 20-50.

Petals united, falling off in a cap . . . 18. **Tupidanthus.**"

(*Flora of British India by Sir J. D. Hooker, Vol. II.*, 720-21.)

1276

Distribution.—In the Natural Order ARALIACEÆ there are in all 340 species referred to 38 genera; they abound in both hemispheres, but not beyond latitude 52°. They are particularly plentiful in the mountains of Mexico and New Grenada. They are rare in the parallel region of Europe and Asia. There are 54 species met with in India belonging to 19 genera. In the plains of India there are 12 species or 22·2 per cent.; ascending to 5,000 feet, 20 species or 37 per cent.; up to 10,000 feet, 18 species or 33·3 per cent.; and above 10,000 feet, 4 species or 7·4 per cent. These are distributed over India as follows: in Eastern India, 36 or 66·6 per cent.; in South India, 7 or 13 per cent.; in West India, *nil;* and peculiar to North India, 1 or 1·8 per cent. Besides these the following are met with: in two or more of the four regions or divisions of India, not including North India, 3, or 5·5 per cent.; and including North India, 7, or 13 per cent.: thus in North India there are in all 8 species, one being endemic and the others of a wider distribution. From these figures the Indian ARALIACEÆ are shown to have their head-quarters in the warm temperate and temperate regions of the eastern division, while their entire absence from the western division is exceedingly remarkable. **Dalzell** and **Gibson**, in their *Flora of Bombay*, mention a plant which they call **Hedera Wallichiana** as met with "at Moolus, foot of the Rám Gháut, and other similar places," but the *Flora of British India* takes no notice of this species. The authors of the *Bombay Flora* mention, in addition, 7 species of introduced Araliaceæ which do not seem to have succeeded very well, most of them occurring in one garden only, and others appear to have died out.

1277

The Affinities of the Araliaceæ are with UMBELLIFERÆ (with which they have in fact been combined by **Baillon**), with AMPELIDEÆ, and with CAPRIFOLIACEÆ. They are also closely connected to CORNEÆ, which in fact only differ in their drupaceous fruit and opposite leaves.

1278

Properties and Uses.—The family contains few species of any great importance to man. The small tree, **Fatsia papyrifera**, a native of the Island of Formosa, yields the rice-paper of China. The young shoots of a species of **Helwingia** are eaten in Japan, while **Panax Ginseng** affords the celebrated medicine *Ginseng*, famed as a tonic and aphrodisiac in the east. This is regarded by the Chinese as the most potent of restoratives, but to the European practitioner its remedial value is regarded as entirely overrated. The leaves of the Iv (**Hedera Helix**) have, from remote antiquity, enjoyed the reputation of possessing remedial virtues, especially as a dressing for ulcers and to destroy vermin on the body. For this purpose an oil medicated with the Iv was used.

ARAUCARIA, *Juss.; Gen. Pl., III., 437.*

1279

Araucaria Cunninghamii, *Ait.;* CONIFERÆ.

Habitat.—A large and handsome evergreen tree of Australia (Queensland), occasionally planted for ornamental purposes in Calcutta.

TIMBER. 1280

Structure of the Wood.—Soft, light yellow, perishable.

Several other species are also occasionally met with in Indian gardens.

ARCTOSTAPHYLOS, *Adans.; Gen. Pl., II., 581.*

1281

Arctostaphylos Uva Ursi, *Spreng.;* ERICACEÆ.

Syn.—ARBUTUS UVA URSI, *Linn.*

The Ardisia.

Habitat.—A native of North America, Europe, and Asia.

MEDICINE. Leaves. 1282

Medicine.—The leaves are astringent and diuretic. They are imported and sold by druggists.

1283

ARDISIA, *Swartz.; Gen. Pl., II., 645.*

A genus of shrubs or small trees belonging to the Natural Order MYRSINEÆ, comprising some 200 species, native of tropical regions—45 being met with in India.

Leaves petioled. *Flowers* hermaphrodite, in axillary or terminal, simple or compound umbels or racemes; *bracts* small, deciduous; *calyx* 5- (rarely 4- or 3-) lobed, persistent, often somewhat enlarged in fruit. *Corolla* red, white or spotted, 5-partite; *segments* acute, twisted to right in bud. *Stamens* 5; *filaments* generally short; *anthers* free, ovate-lanceolate, acute; *Ovary* globose, *narrowed* upwards; *style* cylindrical, often much longer than the corolla lobes; *stigma* punctiform; *ovules* few. *Fruit* globose or sub-globose. *Seeds* solitary, globose; *albumen* pitted or ruminated; *embryo* horizontal.

A genus in which many of the characters are exceptionally variable in certain species. The generic name is derived from, ἄρδις, a point, in allusion to he acute petals. The economic properties of the Indian plants are very imperfectly known.

1284

Ardisia colorata, *Roxb.; Fl. Br. Ind., III., 520; Fl. Ind., Ed. Carey and Wall., II., 271;* A. anceps, *Wall; Kurz, For. Fl., II., 107.*

Habitat.—A shrub frequent in Cachar, Assam to (?) Malacca.

MEDICINE. Bark. 1285

Medicine.—Said to be the *dan* of Ceylon, the bark of which is used as a febrifuge in fever and in diarrhœa, and also applied externally to ulcers.

1286

A. crenata, *Roxb.; Fl. Br. Ind. III., 524.*

Syn.—A. CRENULATA, *Lodd.;* A. CRISPA, *A. DC.; Kurz, For. Fl.; and Gamble, Man. Timb.;* A. GLANDULOSA, *Blume.*

Vern.—*Chamlani,* NEPAL; *Denyok,* LEPCHA.

Habitat.—"A small, erect shrub, met with in Eastern Himálaya, from 4,000 to 8,000 feet, and at Martaban at similar elevations." (*Gamble.*)

There seems to be some mistake regarding this species. Gamble gives the above locality, while the *Flora of British India* says it is a native of "Penang, Malacca, and Singapore, frequently distributed to Malaya, China, Japan." It seems probable that Gamble is alluding to another species, probably **A. macrocarpa**, *Wall.*

TIMBER. 1287

Structure of the Wood.—White, moderately hard. Very common under-growth in the hill forests.

1288

A. humilis, *Vahl.; Fl. Br. Ind., III., 530.*

Syn.—A. SOLANACEA, *Roxb., Fl. Ind., I., 580; Wight, Ic., t. 1212; Brandis, For. Fl., 287; Kurz, For. Fl., II., 110;* A. UMBELLATA, *Roth.; Roxb., Fl. Ind., I., 582;* A. LITORALIS, *Adr.; Kurz, For. Fl., II., 110;* A. POLYCEPHULA, *Wight, Ill., t. 145;* CLIMACANDRA LITTORALIS, *Kurz, Jour., As. Soc., 1871, Pt. II., 68.*

Vern.—*Banjám,* BENG.; *Bisi,* MAL. (S.P.); *Kudna,* URIYA; *Conda-mayúru-káki-nérédu,* TEL.; *Kantena, mayarawa,* C. P.; *Bodina gidda,* MYSORE; *Gyengmaope,* BURM.; *Balu-dan,* SINGH.

Habitat.—A small shrub, met with throughout India, ascending to altitude 5,000 feet. Not met with in North India and Ceylon. *Distrib.*—Singapore, Malaya, and China.

Dye.—The red juice of the berries yields a good though unknown yellow dye.

DYE. Yellow. 1289

Structure of the Wood.—Grey, moderately hard, used as fuel.

TIMBER. 1290

Ardisia involucrata, *Kurz; Fl. Br. Ind. III., 528.*

Vern.—*Denyok,* LEPCHA.

Habitat.—A small shrub, 3 to 6 feet, with yellow corky bark, altitude 2,000 to 5,000 feet, in Sikkim.

Structure of the Wood.—Pinkish white, with small, scanty pores, and broad, white, wavy, medullary rays.

TIMBER. 1291

A. paniculata, *Roxb.; Fl. Br. Ind., III., 519.*

1292

A small tree of the Khásia Hills and of Chittagong, with handsome pink flowers. The bark is thin, greyish brown, and the wood pinkish white, with small pores radially disposed between the short, broad, wavy, medullary rays.

ARECA, *Linn.; Gen. Pl., III., 883.*

1293

A genus of PALMÆ belonging to the Tribe ARECEÆ, comprising some 24 species, inhabitants of tropical Asia, the Malaya, and Australia.

Stem tall, slender, attaining 80 feet or more with a diameter 12-15 inches; *leaves* terminal, equally pinnatisect; *petiole* on a long smooth green sheath; *segments* lanceolate, acuminate, plicate, with the margin recurved; base broad, with numerous parallel nerves; *raches* angled, and convex below with an acute margin above. *Spadix* much-branched, pendulous, appearing from the axil of the lowest leaf. *Spathes* 3 or more, the lower enclosing the spadix, the upper generally bractiferous. *Flowers* monœcious, male and female on the same inflorescence, female flowers solitary, surrounded by numerous slender spikes of white fragrant male flowers. *Male flowers* compressed, small. *Sepals* small, imbricate, free or connate. *Petals* much larger, obliquely lanceolate, acute or acuminate-valvate. *Stamens* 3 or 6; *filaments* short or obsolete; *anthers* sagittate, basifixed, erect, surrounding a minute rudiment of the ovary. *Female flowers* much larger than the male. *Sepals* orbicular, concave, broadly imbricate. *Petals* much larger than the sepals, valvate. *Staminodia* minute or obsolete. *Ovary* ovoid, 1-locular; stigma 3, sessile, subulate, erect or recurved; *ovule* basilar, erect. Fruit orange-coloured, ovoid, surrounded by the persistent coriæceous perianth. *Seed* ovoid or sub-hemispherical, truncated at the base; *albumen* ruminated; *embryo* basal.

Areca is said to be the Latinised form of the Malayan name.

Areca Catechu, *Linn.;* PALMÆ.

1294

THE ARECA- or BETEL-NUT PALM; NOIX D'AREC, *Fr.;* AREKANÜSSE, BETELNÜSSE, *Germ.*

Vern.—The nut: *Supári, supyári,* HIND. and DUK.; *Gua, supári,* BENG.; *Tambul,* ASS.; *Póka-vakka, vakka,* TEL.; *Kamugu, pákku, kottai-pákku,* TAM.; *Adike,* KAN.; *Sopári, hopári, phophal,* GUJ.; *Supári,* MAR.; *Adaka, kavugu, atakka,* MAL.; *Púga-phalam, gubak,* SANS.; *Fófal,* or *foufal,* ARAB.; *Gird-chób, pópal,* PERS.; *Puwak, puvakka,* SINGH.; *Kwam-thee-beng, kúnsi, kun, kun theè-bin,* BURM.; *Áh-búd-dah, ah-pur-rud-dah,* AND.

Habitat.—A native of Cochin China, Malayan Peninsula and Islands. Cultivated throughout tropical India; in Bengal, Assam, Sylhet, but will not grow in Mánipur, and only indifferently in Cachar, Burma, Siam. In Western India, below and above the Ghâts. Does not grow at any distance from the sea, and will not succeed above 3,000 feet in altitude. It flourishes, however, in the dry plateau of Mysore, Kanara, and Malabar. Most villages in Burma, Bengal, and South India have their clumps or avenues of betel palms. The betel-palm groves and pepper

ARECA Catechu.

Betel-nut Palm.

CULTIVATION.

betel-leaf houses are perhaps the most characteristic features of the river-banks in Sylhet, and from these plantations the inhabitants of Cachar and Mánipur obtain their supplies.

1295

Cultivation and Yield.—*In Mysore.*—There are two varieties of the Areca in Mysore, the one bearing large and the other small nuts, the produce of both kinds being nearly equal in value and quantity.

The manner of Areca-nut cultivation is different in different districts of Mysore. The method followed in Channapatna is as follows. The seed is ripe about the middle of January to February, and is first planted in a nursery. Trenches are dug and half-filled up with sand, on the surface of which is placed a row of the ripe nuts. These are again covered with sand and rich black mould, and are watered once in three days for four months. The young palms are then transplanted to the garden, which had been previously planted with rows of plantain trees at the distance of about four feet. Two young Arecas are set in one hole between every two plantain trees. When there is no rain, the plants are watered every third day. In the rainy season, a trench is dug between every third row of trees to carry off superfluous water, and to bring a supply from the reservoir when wanted. At the end of three years the original plantain trees are removed and a row planted in the middle of each bed and kept up ever afterwards in order to preserve a coolness at the roots of the Areca. The trees are five feet high in five years, and begin to produce fruit. The plantation requires no more watering except twice a month during the dry weather.

The methods followed in other parts of Mysore differ in some respects from the one above, but they agree in the essential point, namely, plantain trees are planted with the Areca palms, and in most districts trenches are dug to carry off superfluous water. The seedlings, except in one district, are first raised in a nursery and then transplanted. Manure is used in some districts, but watering is resorted to everywhere. A rich black mould or a black soil containing calcareous nodules is preferred for Areca nut cultivation.

The areca plantations in Mysore are interspersed with cocoanut, lime, jack, and other trees, which add to the shade and to the freshness of the soil. (*Mysore Gazetteer, Vol. I., pp. 125-131.*)

In Kolába.—In Kolába, the betel palm is grown in large numbers in cocoa-nut plantations along the Alibág coast. The nuts are buried two inches deep "in loosened and levelled soil. When the seedlings are a year old, they are planted out in July and buried about two feet deep. The soil is then enriched by a mixture of salt and ***náchni***, sometimes with the addition of cow-dung." No watering is required at first, but after four months the plant is watered either daily or at an interval of one or two days. If water is not stinted, the betel palm yields nuts in its fifth or sixth year. "The tree yields twice or thrice a year, about 250 nuts being an average yearly yield." (*Bomb. Gaz., Vol. XI., pp. 97-98.*)

In Janjira or Shivardha.—In Janjira, the betel palm is the most important of garden crops. "*Shrivardhan* betel-nuts are known over the whole of the Bombay Presidency. The seed-nut is sown in February or March about half a foot deep and is carefully watered. After about four months the plant appears and is watered every second day. When it is four years old it is planted out about two feet and a half below the surface, a foot and a quarter of the seedling being buried under the ground, while a round trench of the same depth is left for the water." When the tree is nine or ten years old, it begins to bear fruit, the yearly yield varying from 25 to 400 nuts. (*Bomb. Gazetteer, Vol. XI., 425.*) This variety fetches, relatively, a much higher price in the market than any of the others.

ARECA Catechu.

Betel-nut Palm.

CULTIVATION.

In Thána.—The betel-nuts are grown largely in Thána, Bombay. The best nuts are carefully selected in October and dried in the sun; unhusked nuts are considered the best for seed. They are planted in a well-ploughed plot of land in pits three inches wide and three inches deep, and at a distance apart of from six inches to a foot. "For the first three months the young palm is watered at least every fourth day, and afterwards every third day" When the plants are a year or a year and a half old, they are fit for planting out. "The selling price of young plants varies from 6 pies to 1 anna."

"The betel palm usually grows in red soil, but it flourishes best in sandy soil that remains moist for some time after the rains. Before planting the young palm, the ground is ploughed, levelled, and weeded, and a water-channel is dug six inches deep and a foot and a half wide. Then pits nine inches deep and two feet wide are dug at least four feet apart, nearly full of earth, but not quite full, so that water may lie in them. Where the soil allows, plantains are grown in the beds to shade the young palms. Except during the rainy season, when water is not wanted, the young trees are watered every second day for the first five years, and after that every third or fourth day. During the rains, manure is sometimes given." (*Bombay Gazetteer, XIII., pt. I., 298-299.*)

The cost of betel-nut cultivation in Thána is calculated as follows: "An acre entirely given to betel palms would, it is estimated, hold 1,000 trees. The total cost of rearing 1,000 betel palms for five years—that is, until they begin to yield—is about £127 13*s*., including compound interest at 9 per cent. After five years a thousand trees are estimated to yield about £50 a year, from which, after taking £18 14*s*. for watering, assessment and wages, and £11 9*s*. 11½*d*. as interest at the rate of 9 per cent. on £127 13*s*, there remains a net estimated profit of £19 16*s*. 3½*d*., or 15·52 per cent." (*Bombay Gazetteer, Vol. XIII., pt. I., 301.*)

In Ceylon.—There are several varieties cultivated in Ceylon, but they are not so good as the Indian ones, and fetch the lowest price in the Bombay market. In poor soil, the plant at first grows slowly. It can thrive among an undergrowth of weeds;—in fact, clean weeding is not probably beneficial to its growth. Neither tree nor fruit is liable to attacks by enemies of any kind. Young trees are continuously produced from nuts that have been allowed to drop and supply the place of those that have become worn out or unfruitful. Areca-nut trees can be planted very closely, 1,200 per acre being not at all too high an estimate; 12,000 cured nuts make on an average one cwt. At 300 per tree, the average yearly yield would be about 30 cwt. per acre.

The wholesale value at Galle or Colombo is usually R8 per cwt., giving R240 per acre, or leaving out R100 for expenditure, R140 per acre as net profit.

At Madras and Bombay, Ceylon nuts fetch about R15 a cwt. The demand for areca-nut is practically unlimited, as hundreds of millions of people in China, India, &c., use it. (*Notes by a Ceylon planter in Trop. Agri, II., 791.*)

In Bengal.—"The *supárí* or betel nut is common in Eastern Bengal, especially in Tipperah, Backergunge, and Dacca; and its cultivation is very profitable to proprietors of land. It bears fruit in the eighth year, and is most productive from that time to the sixteenth year, when the produce falls off. The nuts are gathered in November." (*Administration Report of Bengal, 1882-83, p. 14.*)

"The betel-nut cultivation is very extensive, especially in the police circles of Tubkibágará and Hájiganj. A considerable trade in this article is carried on with Dacca, Náráinganj, and Calcutta. The cultivators of the betel-nut palm or *supárí* (**Areca Catechu**) usually own a large

Betel-nut Palm.

piece of ground, slightly raised above the level of the surrounding country, and surrounded by ditches. In the centre of this they build their dwellings, and all around them they plant betel-nut trees. An acre of land will obtain about 3,000 trees. When first planted, the betel-nut requires to be protected from the sun; for this purpose rows of *mádár* trees are planted between the lines of betel-nut trees, and the growth of jungle is encouraged. When the betel-nut trees have grown strong, and no longer require the shade, the cultivators are too lazy and thoughtless to remove the jungle; and the result is that 'whole parganas which were once fully cultivated are now covered with dense jungle,' in which even the betel-nut trees cannot grow; while 'thousands of the inhabitants have been swept away by cholera and malarious fever of a very virulent type.' The unhealthiness of the neighbourhood of betel-nut plantations is variously attributed to the dense jungle and undergrowth above mentioned, to the exhalations from the trees, and to the malarious gases generated by decomposing vegetable matter in the ditches surrounding the plantations. The betel-nut trees grow to a height of about 60 feet; and in some parganas they are cultivated to such an extent as to almost entirely exclude rice cultivation." (*Dr. W. W. Hunter's Statistical Account of Bengal, Vol. VI., pp. 391-92.*)

1296

Commerce.—The betel-nut palm is very largely cultivated on account of its seed (popularly called its nut). As stated in the previous pages, the average yield per tree is about 300 fruits, each of which contains one large seed, about the size of a small hen's egg. The chief trade seems to centre around Bombay. Ceylon and Madras export their nuts to the Western Capital, from which they are re-exported to the principal Asiatic centres, and diffused by land all over India. Sumatra and Singapore also export large quantities of nuts to Bombay. The following extract from **Dr. Dymock's** *Materia Medica of Western India* will be found useful us indicating the chief trade classes of areca-nuts met with in Bombay:—

"The kinds of betel-nut met with in Bombay are—

Gowai, from Goa, value R40 to R50 per candy of 5¼ cwt.

Mangalore, value R70 to R110 per candy.

Rupasai, from Alpai, R60 to R80 per candy.

Calcutta, value R60 to R65 per candy.

Asigree, from Singapore, value R60 to R70 per candy.

Kanarese, value R80 to R100 per candy of 5¼ cwt.

Severdani (*Shrivardhan*, Ed.) value R4½ to R4¾ per ¼ cwt.

All these are known as white betel-nut. The following kinds of red betel-nut are met with:—

Malabari, value,	R70 to	R80	per candy of	5¼ cwt.	
Kúmpta	,,	60	90	,,	,,
Marorkadi	,,	80	85	,,	,,
Goa	,,	65	90	,,	,,

Wasai from Bassein, value R6 to R8 per ¼ cwt.

Sewali value R5 per maund of ¼ cwt.

Malwan, value	R60 to	R65	per candy of	5¼ cwt.	
Vingorla	,,	60	65	,,	,,
Calcutta	,,	50	60	,,	,,

AREGA Catechu.

1297

Betel-nut Trade.

The following are the imports and exports of Areca-nut :—

Foreign Trade by Sea.

Years.	Imports.		Exports.	
	Quantity.	Value.	Quantity.	Value.
	℔	R	℔	R
1879-80	21,585,601	20,57,059	2,287,900	2,57,508
1880-81	27,973,690	23,29,395	967,005	96,930
1881-82	24,918,002	20,19,918	605,484	53,545
1882-83	23,693,555	21,68,061	948,165	77,199
1883-84	30,390,994	34,06,458	466,722	52,417

Details of Imports—1883-84.

Province into which imported.	Quantity.	Value.	Countries whence imported.	Quantity.	Value.
	℔	R		℔	R
Bengal . .	6,393,388	4,64,858	Ceylon . .	9,900,147	14,39,963
Bombay . .	1,006,734	1,23,227	Straits Settlements	16,821,823	13,61,827
Madras . .	20,343,284	26,48,425	Sumatra . .	3,650,869	6,03,532
British Burma .	2,647,588	1,69,948	Other Countries .	18,155	1,136
Total .	30,390,994	34,06,458	Total .	30,390,994	34,06,458

Details of Exports—1883-84.

Province whence exported.	Quantity.	Value.	Countries to which exported.	Quantity.	Value.
	℔	R		℔	R
Bengal . .	158,172	13,171	Mozambique .	47,964	5,878
Bombay . .	290,559	35,715	Zanzibar . .	171,419	20,156
Madras . .	17,991	3,531	Mauritius . .	114,061	11,931
			South America .	15,269	1,431
			Aden . . .	14,644	2,853
			Arabia . . .	9,147	1,320
			China—Hongkong	29,488	2,643
			,, Treaty Ports	5	1
			Maldives . .	36,257	3,462
			Other Countries .	28,468	2,742
Total .	466,722	52,417	Total .	466,722	52,417

Note.—Of the betel-nuts imported about 60,000 ℔, valued at R6,000, are re-exported annually to foreign countries.

AREKA Catechu.

Betel-nut Trade.

Imports of Betel-nuts in 1883-84. (Coasting Trade.)

Ports from which imported.	Into Bengal.		Into Bombay.		Into Sind.		Into Madras.		Into British Burma.		Total.	
	Quantity.	Value.	Quantity.	Value.	Quantity.	Value.	Quantity.	Value.	Quantity.	Value.	Quantity.	Value.
	lb	R	lb	R	lb	R	lb	R	lb	R	lb	R
British Ports in other Presidencies.												
From Bengal . .	...	...	96,068	9,541	3,696	500	52,852	6,499	17,078,376	27,22,139	17,230,992	27,38,679
,, Bombay . .	131,068	15,648	...	...	451,764	67,650	42,685	2,577	38	13	625,555	85,888
,, Sind . .	...	...	6,664	646	...	...	...	...	...	...	6,664	646
,, Madras . .	716,147	69,633	5,984,703	5,55,889	128,550	15,771	...	...	24,853	3,813	6,854,253	6,45,106
,, British Burma .	58,025	5,758	16,800	900	...	...	...	...	...	...	74,825	6,658
Total .	905,240	91,039	6,104,235	5,66,976	584,010	83,921	95,537	9,076	17,103,267	27,25,965	24,792,289	34,76,977
	3,225,877	3,43,865	10,530,609	12,05,655	63,929	7,295	1,211,327	1,16,675	648,188	77,727	15,679,930	17,51,217
British Ports within the Presidency.												
Indian Ports not British—												
Portuguese—												
From Diu . .	...	...	560	45	...	...	...	...	...	...	560	45
,, Goa . .	...	...	2,678,703	2,40,444	...	...	...	...	...	...	2,678,703	2,40,444
Native—												
From Cutch .	...	...	182	7	2,698	262	...	...	...	...	2,880	269
,, Kattywar .	...	...	31,665	2,826	...	...	...	...	...	...	31,665	2,826
,, Konkan .	...	...	252,791	44,146	...	...	...	...	...	...	252,791	44,146
,, Travancore	1,680	153	31,444	5,454	...	...	812	65	...	...	33,936	5,672
,, Cochin .	...	...	...	...	...	...	196	18	...	...	196	18
Total .	1,680	153	2,995,345	2,92,922	2,698	262	1,008	83	...	...	3,000,731	2,93,420
Total of all Ports	4,132,797	4,35,057	19,630,189	20,65,553	650,637	91,478	1,307,872	1,25,834	17,751,455	28,03,692	43,472,950	55,21,614

A. 1297

AREKA
Catechu.

Betel-nut Trade.

Frontier Trade by Land.

YEARS.	IMPORTS.		EXPORTS.	
	Quantity.	Value.	Quantity.	Value.
	lb	R	lb	R
1881-82	7,280	1,408	5,442,976	6,58,552
1882-83	560	125	6,876,128	6,79,010
1883-84	1,456	141	6,078,240	7,52,441

Details of Imports—1883-84.

Province into which imported.	Quantity.	Value.	Country whence imported.	Quantity.	Value.
	lb	R		lb	R
Bengal . .	1,456	141	Nepál . . .	1,456	141

Details of Exports—1883-84.

Province whence exported.	Quantity.	Value.	Countries to which exported.	Quantity.	Value.
	lb	R		lb	R
Panjáb . .	5,824	873	Nepál . . .	676,368	61,573
N.-W. P. and Oudh.	26,432	3,759	Bhutan . .	246,624	21,216
Bengal . .	893,424	78,944	Manipur . .	363,104	38,512
Assam . .	394,016	39,607	Upper Burma .	4,438,672	5,77,349
British Burma .	4,758,544	6,29,258	N. Shan States .	2,81,456	41,379
			Karennee .	30,464	7,875
			Other countries .	41,552	4,537
TOTAL .	6,078,240	7,52,441	TOTAL .	6,078,240	7,52,441

Decoction.
1298

Extract.—A decoction of the nut yields an inferior resinous extract, known sometimes as "Areca Catechu."

The water in which areca-nuts are boiled becomes discoloured and thick; this on being inspissated "forms *Kossa*, or the catechu of the greatest astringency; but the best catechu of a red or brown colour" is obtained by boiling in fresh water nuts which have been previously boiled. (*Baden Powell, Panjáb Products, I, 302.*)

The ripe fruit is boiled for some hours in an earthen or a tinned copper vessel, and the nuts, together with the boiling water, are poured over a basket. The boiled water is caught in a tinned copper vessel and is allowed to thicken of itself or is thickened, by boiling, into a black, very astringent, catechu. Sometimes these nuts are boiled a second time in

ARECA Catechu.

Betel-nut Palm.

fresh water, when the boiled water gives a yellowish-brown catechu. The refuse after boiling is sticky and is used for varnishing wood and for healing wounds. (*Bombay Gazetteer, XIII., pt. I, 300.*)

No definite information can be obtained as to the extent of the manufacture of this form of catechu. It is apparently rarely if ever exported from India.

DYE. 1299

Dye.—The preparation of *pán*, acting chemically upon the saliva, colours it red. A decoction of the *nut* is used in dyeing, and a kind of inferior catechu is prepared from it. With *tún* (**Cedrela Toona**) it is said to give a red dye. *Pán* is also used in Dinajpur as a subsidiary in red dyeing with **Morinda tinctoria.** (*M'Cann.*)

TAN. 1300

Tan.—*Spons' Encyclopædia* says: "An astringent extract, prepared from **Areca Catechu,** is said to contribute to commercial cutch; if so, it is a totally distinct product" from the true catechu.

FIBRE. 1301

Paper Material.—The spathe which covers the flowering axis may be used for paper-making, and so also might the fibrous pericarp which is removed from the nut. The spathes are largely used in India for packing and in the preparation of small articles for personal use. (See **Domestic Uses.**)

MEDICINE. Nuts. 1302

Dentifrice. 1303

Medicine.—Young nut is said to possess astringent properties, and is prescribed in bowel complaints and bad ulcers. It contains a large proportion of tannic and gallic acids, and hence its astringent property. The burnt nuts when powdered form an excellent dentifrice. According to **Dr. J. Shortt,** the powdered nut, in doses of 10 or 15 grains every three or four hours, is useful in checking diarrhœa arising from debility. It has also been found very useful in urinary disorders, and is reported to possess aphrodisiac properties. The dried nuts when chewed produce stimulant and exhilarant effects on the system.

Powder. 1304

"The powdered seeds have also long been held in some reputation as an anthelmintic for dogs, and Areca has now been introduced into the *British Pharmacopœia* on account of its supposed efficacy in promoting the expulsion of the tape-worm in the human subject. It is also reputed to be efficacious against round worm (*Ascaris lumbricoides*). **Dr. Barclay,** who appears to have been the first practitioner who called attention to the remedial value of the areca-nut in the expulsion of tape-worm, administered it, in powder, in doses of from four to six drachms, stirred up with milk." (*Bentl. & Trim. Med. Pl.*) **Dr. Waring** says: "Anthelmintic virtues have been assigned to the nut, but it can hardly have any claim to this character, as amongst the Hindús and Burmese, who use it habitually as a masticatory, intestinal worms (*lumbrici*) are almost universally met with."

Juice. 1305

Petioles as splints. 1306

Tincture. 1307

The nut is regarded as a nervine tonic and emmenagogue, and is used as an astringent lotion for the eyes. The juice of the young leaves mixed with oil is said to be used externally in lumbago. The dry expanded petioles may be used as ready-made splints.

§ "Is useful in checking the pyrosis of pregnancy. 'Control experiments' made with tincture of catechu showed the superiority of the nut, and would seem to demonstrate that this is not merely due to astringent action; possibly its property as a nervine stimulant enhances its utility." (*Surgeon G. King, Madras.*) "Used as an astringent for bleeding gums; native women employ it both internally and locally for stopping watery discharges from the vagina." (*Assistant Surgeon Jaswant Rai, Mooltan.*) "Is very useful as a vermifuge in dogs. I have given half a nut powdered, mixed with butter, to terriers with remarkable effect." (*Surgeon K. D. Ghose, Khulna.*) "There are various kinds; some are stimulant when chewed and their juice swallowed, causing an agreeable sense of warmth generally felt in the ears, but sometimes a disagreeable

CHEMISTRY.

sensation of constriction in the throat and chest with profuse flow of mucus. Powder of roasted nuts forms a good tooth-powder." (*Assistant Surgeon Shib Chunder Bhuttacharji, Chanda, Central Provinces.*)

"The powdered young bark is anthelmintic, used for tape-worm; useful in animals; supposed to be the principal ingredient in Naldire's worm tablets." (*Surgeon W. D. Stewart, Cuttack.*) "It is a good anthelmintic, and expels thread-worms. I have often given half a nut to a dog mixed up in butter with very good effect. The worms are expelled after one or two doses." (*Surgeon K. D. Ghose, Bankura, Bengal.*) "Is a good vermifuge for dogs in ʒi doses (powdered)." (*Surgeon-Major J. Byers Thomas, Waltair, Vizagapatam.*) "Nut cut small and soaked in milk is a good vermifuge for dogs." (*Surgeon-Major P. N. Mukerji, Cuttack, Orissa.*)

"Very useful in worms in dogs and other domestic animals. A piece kept in the mouth allays thirst on long marches in sandy deserts where water is scarce." (*Surgeon H. D. Masani, Kurrachee.*) "Most useful in the preparation of tooth-powder. The burnt nuts to be reduced to a fine powder, and mixed with powdered chalk, in the proportion of 3 of former to 1 of latter." (*Dr. S. M. Sircar, Moorshedabad.*) "The young and undried nut is distinctly astringent; when well dried under the sun the astringency becomes less, and the softer portions become slightly sweetish in taste. The young undried nuts possess something which when chewed in excess gives rise to temporary giddiness." (*Surgeon D. Basu, Faridpur, Bengal.*)

"Is a valuable vermifuge for dogs, especially for round-worms." (*Surgeon George Cumberland Ross, Delhi.*)

1308

Chemical Composition.—"We have exhausted the powder of the seeds, previously dried at 100° C., with ether, and thereby obtained a *colourless* solution which after evaporation left an oily liquid, concreting on cooling. This fatty matter, representing 14 per cent. of the seed, was thoroughly crystalline, and melted at 39° C. By saponification we obtained from it a crystalline fatty acid fusing at 41° C., which may consequently be a mixture of lauric and myristic acids. Some of the fatty matter was boiled with water: the water on evaporation afforded an extremely small trace of tannin but no crystals, which had catechin been present should have been left.

"The powdered seeds which had been treated with ether were then exhausted by cold spirit of wine (·832), which afforded 14·77 per cent. (reckoned on the original seeds) of a red amorphous *tannic matter*, which, after drying, proved to be but little soluble in water, whether cold or boiling. Submitting to destructive distillation, it afforded *Pyrocatechin*. Its aqueous solution is not altered by ferrous sulphate, unless an alkali is added, when it assumes a violet hue, with separation of a copious dark purplish precipitate. On addition of a ferric salt in minute quantity to the aqueous solution of the tannic matter, a fine green tint is produced, quickly turning brown by a further addition of the test, and violet by an alkali. An abundant dark precipitate is also formed.

"The seeds having been exhausted by both ether and spirit of wine, were treated with water, which removed from them chiefly mucilage precipitable by alcohol. The alcohol thus used afforded on filtration traces of an acid, the examination of which was not pursued. After exhaustion with ether, spirit of wine and water, a dark brown solution is got by digesting the residue in ammonia: from this solution, an acid throws down an abundant brown precipitate, not soluble even in boiling alcohol. We have not been able to obtain crystals from an aqueous decoction of the seeds, nor by exhausting them directly with boiling spirit of wine. We have come therefore to the conclusion that *Catechin* (p. 243) is not a constituent of areca nuts, and that any extract, if ever made from them, must be essentially different to the *Catechu* of *Acacia* or of *Nauclea*, and rather to be considered a kind of tannic matter of the nature of *Ratanhia-red* or *Cinchona-red*.

"By incinerating the powdered seeds, 2·26 per cent. were obtained of a brown ash, which, besides peroxide of iron, contained phosphate of magnesium." (*Flück. & Hanb., Pharmacographia, pp. 670-71.*)

FOOD. 1309

Food.—The nut is one of the indispensible ingredients which enter into the preparation of the *pán* or betel leaf, which is chewed so universally by natives of all classes. The betel nut is often chewed by itself in small pieces, and is sold in every bazar throughout India. It is said to stimulate digestion. Small pieces of the prepared betel nut are rolled up with a little lime, catechu, cardamoms, cloves and even rose water within the betel-pepper leaf. This combination forms the *pán* which gives to the lips and teeth the red hue which the natives admire. In the course of time it has the effect, however, of colouring the teeth black, at least along the edges, thus destroying the appearance of the teeth. The chewing of *pán* is supposed to prevent dysentery. "It is said to dispel nausea, excite appetite, and strengthen the stomach. Besides being used as an article of luxury, it is a kind of ceremonial which regulates the intercourse of the more polished classes of the East. When any person of consideration visits another, after the first salutations, betel is presented: to omit it on the one part would be considered neglect, and its rejection would be judged an affront on the other." (*McCulloch's Dictionary of Commerce and Commercial Navigation.*)

Panjab nuts. 1310

Assam. 1311

Manipur. 1312

Bengal. 1313

Bombay. 1314

Trade forms of Supari Phulbari. 1315

Preparation of the Nut.—After the nuts are husked, they are boiled till soft, and taken out and sliced; the slices are rubbed with the inspissated water in which the nuts were boiled, which became impregnated with the astringent principle contained in the nuts; the slices are then dried in the sun, and in this condition sent to market. Instead of being sliced and boiled, the nuts are also largely sold entire (*Baden Powell*). In Mánipur they are sold in the streets with the husk neatly opened up like a fringe to show that the nuts are fresh. In the Bombay and Mysore *Gazetteers* interesting details are given regarding the methods of preparing the nuts for the market. In Thána the growers sell the fruit wholesale to a tribe called Vánis, who, by different treatment, prepare six classes of nuts. To prepare *phulbardi supári*, or those with flower-like fissures, the nuts are gathered when yellow but not quite ripe. The husk is stripped off, and the kernels are boiled in milk or water, in an earthen or tinned copper vessel. When the nut grows red, and the water or milk thickens like starch, the boiled nuts are removed and dried in the sun for seven or eight days. The red, *támbdi*, betel-nuts are prepared by boiling ripe fruits stripped of its husk, in milk or water, with a small quantity of pounded *kath*, lime, and betel leaves. "As soon as the boiling is over, the nuts and boiling milk or water are removed in a basket with a copper vessel under it to catch the droppings." To make *chikni* or tough betel-nut, "the nuts are gathered when they are beginning to ripen," and when the boiling is over, the catechu-like substance left on boiling is rubbed on the nuts, when they are dried in the sun. "This process is repeated until the nuts grow dark red." To make *lavangachuri* or clove-like betel nuts, the kernels of tender fruits are cut into clove-like bits, and after boiling in water are dried in the sun. *Pandhri* or white betel nut is made by boiling the ripe fruit with its husk, and afterwards drying in the sun till the husks are easily removed. To prepare *dagdi* or strong nut, the fruit is gathered when ripened into hardness, and after stripping it of the husk, it is boiled

Tarahdi. 1316

Chikni. 1317

Lavanga-churi. 1318

Pandhri. 1319

Dagdi. 1320

and dried in the sun. To make *kápkadi*, or cut betel-nut, the kernels are cut out of the nut when tender, and dried in the sun without being boiled or soaked in water. (*Bombay Gazetteer, Vol. XIII., Pt. I., pp. 299-300.*)

Kapkadi. 1321

In Mysore, after removing the husk, the nuts are boiled in water, then cut into pieces and dried in the sun; or they are first cut into pieces, then boiled in water with cutch, leaves of **Piper betel**, and afterwards dried in the sun, when they are fit for sale. (*Mysore Gazetteer, Vol. I., pp. 126, 127.*)

Mysore nuts. 1322

Structure of the Wood.—The areca nut is one of the most elegant of Indian palms, with thin straight stem and crown of leaves looking like an arrow stuck in the ground. It often attains 100 feet in height, with a slender, cylindrical, annulate stem, the inner part of which is generally hollow. The vascular bundles are brown, forming a hard rind on the outside of the stem. Weight 57 lbs. per cubic foot.

TIMBER. 1323

Used for furniture, trenails, bows, spear, handles, and for scaffolding poles in Ceylon. In the Bombay Presidency "The trunk of the betel palm is used as roof rafters for the poorer class of houses, and for building marriage-booths; it is slit into slight sticks for wattle-and-daub partition-walls, and it is hollowed into water-channels. In some places it is used for spear-handles." (*Bombay Gazetteer, Vol. XV., Pt. I., 300.*)

Domestic Uses.—"The soft, white, fibrous flower-sheath, called *kácholi* or *poy*, is made into skull-caps, small umbrellas and dishes; and the coarser leaf-sheath, called *viri* or *virhati*, is made into cups, plates, and bags for holding plantains, sweet-meats, and fish." (*Bombay Gazetteer, Vol. XV., Pt. I., p. 300.*)

DOMESTIC. Caps. 1324 Umbrellas. 1325 Dishes. 1326 Cups. 1327 Bags. 1328

The nut is used in many religious ceremonies, and forms one of the chief articles of trade in Kanára. (*Bomb. Gaz., Vol. XV., Pt. I., p. 62.*)

Areca concinna, *Thw.; En. Ceyl. Pl., 328.*

Vern.—*Laina-terri*, SINGH.

Habitat.—A small palm indigenous to Ceylon.

Food.—The natives eat the nuts as a substitute for the ordinary betel nuts: it is never cultivated. (*Dr. Trimen.*)

FOOD. 1329

A. gracilis, *Roxb.*

1330

Syn.—PINANGA GRACILIS, *Kurz.*

Vern.—*Gua supari, ramgua*, BENG.; *Khur*, LEPCHA; *Ranga*, ASS.

Habitat.—A slender-stemmed palm, often gregarious, found in undergrowths of damp forests in Sikkim, Assam, Eastern Bengal, and Burma.

Structure of the Wood.—It is used for native huts and roofing in Assam. The outer portion is hard and closely packed with fibro-vascular bundles; the inner is soft, as the cane shrinks in drying.

TIMBER. 1331

ARENARIA, *Linn.; Gen. Pl., I., 149.*

1332

Arenaria holosteoides, *Edge.; Fl. Br. Ind., I., 241;* CARYOPHYLLEÆ.

Vern.—*Kakua, gandial*, PB.; *Chiki*, LADAK.

Habitat.—A herb found in the Western Himálaya and Western Thibet, from Kumaon to Kashmír, altitude 7,000 to 12,000 feet, and distributed into Afghánistan.

Food.—Used as a vegetable in Chumba and Ladák.

FOOD. 1333

1334

ARENGA, *Labill.; Gen. Pl., III., 917.*

An erect palm, with simple stem, often 40 feet in height. *Leaves* terminal, and seen at a distance, somewhat resembling the crown of leaves of the date-palm, except that they are longer; petiole thick, leaflets sub-opposite, 3-5 feet long, ensiform, the base dilated into 1 or 2 ears, upper half dentate serrate, apex somewhat obliquely cut, white beneath, green above. *Flowers* unisexual, monœcious, numerous, sessile, bracts 2 or more to each flower imbricate in bud; spadices several, 6-10 feet long, coming from among the leaves and developing downwards, the tree dying when the last and lowest spadix is ripe. *Male flowers* with 3 *sepals*, concave, rounded, fleshy; *petals* 3, longer than the sepals, valvate, purple outside, yellow within. *Stamens* numerous; *filaments* shorter than thea nthers; no trace of pistil. *Female flowers* with *petals* not much longer than sepals; *stamens* none; *ovary* large, 3-lobed, smooth, 3-celled, with a single erect *ovule*; *stigma* sessile conical. *Fruit* the size of an apple, depressed at the top, 3-celled, with a single *seed* in each cell.

1335

Arenga saccharifera, *Labill.; Kurz, For. Fl. Burm., II., 533; Brandis, For. Fl., 550;* PALMÆ.

THE SAGO PALM of Malacca and the Malaya.

Syn.—SAGUERUS RUMPHII, *Roxb., Fl. Ind., Ed. C.B.C., 669;* BORASSUS GOMUTUS, *Lour.;* GOMUTUS SACCHARIFERA, *Spr.*

Vern.—*Taung-ong, toung-ong,* BURM.; *Ejú* (fibre), MALAYA; *Gumúti* (tree), *kobong,* MALACCA.

Habitat.—A Malayan tree, generally cultivated in India, but said by **Kurz** to be wild in Burma, also mentioned by **Hooker** and **Thomson** as found wild in Orissa. One or two trees were observed growing along with **Caryota urens** on the mountains of North Mánipur, apparently wild (especially in the Kabú valley).

Properties and Uses—

FIBRE. 1336

Fibre.—At the base of the petiole is found a beautiful black, horse-hair-like fibre, known as the *Ejú* or Gomuta Fibre. Within the sheaths is also found a layer of reticulated fibres, which is said to be in great demand in China, being applied, like oakum, in caulking the seams of ships. It is also largely used as tinder for kindling fires. The Mánipuris value very much this reticulated fibre, which they use, as also that obtained from **Caryota urens,** for making mechanical filters. A bundle of these black reticulated fibres, tied firmly together, is placed in the bottom of a perforated vessel; the water, percolating through, is cleansed of mechanical impurities. The fibre has a high reputation for lasting under water. **Mueller** (*Extra-Tropical Plants*) says: "The black fibres of the leaf-stalks are adapted for cables and ropes intended to resist wet very long." **Roxburgh** (*Fl. Ind., Ed. C.B.C., 669*) remarks: "I cannot avoid recommending to every one who possesses lands, particularly such as are low, and near the coasts of India, to extend the cultivation" of this plant "as much as possible. The palm wine itself, and the sugar it yields, the black fibres for cables and cordage, and the pith for sago, independent of many other uses, are objects of very great importance, particularly to the first maritime power in the world, which is in a great measure dependent on foreign states for hemp, the chief material of which cordage is made in Europe." **Simmonds,** writing of this palm, says: "It furnishes a highly valuable black fibrous subtance, Ejoo fibre, superior in quality, cheapness, and durability to that obtained from the husk of the cocoanut, and renowned for its power of resisting wet."

FOOD. Sago. 1337

Food.—THE SAGO, from the interior of the stem, although inferior in flavour to that obtained from the true sago palm, is nevertheless an important article of food. It is the source of the Java sago, and although chiefly cultivated for its sap, from which a wine, and also sugar and vinegar, are prepared, the sago is an important article of food throughout the Malaya. After the tree ceases to yield sap or toddy, the stem furnishes the starchy substance. It is said that a single tree will often yield 150 to 200 pounds; **Arenga** is doubtless the source of a good deal of the sago of commerce (*Bentl. & Trim.*). It is generally stated that the trees which produce female spadices yield the best sago and scarcely any sap, whereas the male spadix gives a copious flow of the sap, from which toddy, wine, sugar, and vinegar are made.

Leaves. 1338 **Kernels. 1339** **Preserves. 1340**

The Mánipuris eat the young and blanched leaf-stalks as a pickle. "The young kernels are made, with syrup, into preserves." (*Mueller.*)

SAP. Wine. 1341 **Sugar. 1342** **Vinegar. 1343**

The Sap.—The following interesting account of the process of extraction of the sap is taken from **Simmonds'** *Tropical Agriculture* (*p. 248*): "One of the spadices is, on the first appearance of fruit, beaten on three successive days with a small stick, with the view of determining the sap to the wounded part. The spadix is then cut a little way from its root (base), and the liquor which pours out is received in pots of earthenware, in bamboos, or other vessels. The Gomuti palm is fit to yield toddy when nine or ten years old, and continues to yield it for two years, at the average rate of three quarts a day.

"When newly drawn the liquor is clear, and in taste resembles fresh *must*. In a very short time it becomes turbid, whitish, and somewhat acid, and quickly runs into the vinous fermentation, acquiring an intoxicating quality. In this state great quantities are consumed; a still larger quantity is applied to the purpose of yielding sugar. With this view the liquor is boiled to a syrup, and thrown out to cool in small vessels, the form of which it takes, and in this shape it is sold in the markets. This sugar is of a dark colour and greasy consistence, with a peculiar flavour; it is the only sugar used by the native population. The wine of this palm is also used by the Chinese residing in the Indian islands in the preparation of the celebrated Batavian arrack.

"In Malacca, the Gomuti, there termed Kabong, is principally cultivated for the juice which it yields for the manufacture of sugar. Like the cocoanut palm it comes into bearing after the seventh year. It produces two kinds of 'mayams,' or spadices, male and female. The female spadix yields fruit, but no juice, and the male *vice versâ*. Some trees will produce five or six female spadices before they yield a single male one, and such trees are considered unprofitable by the toddy collectors, but it is said that in this case they yield sago equal in quality, though not in quantity, to the **Cycas circinnlais,** though it is not always put to such a requisition by the natives; others will produce only one or two female spadices, and the rest male, from each of which the quantity of juice extracted is the same as that obtained from the cocoanut spadices. A single tree will yield in one day sufficient juice for the manufacture of five bundles of jaggery, valued at two cents each. The number of *mayams* shooting out at any one time may be averaged at two, although three is not an uncommon case. When other occupation or sickness prevents the owner from manufacturing jaggery, the juice is put into a jar, where in a few days it is converted into excellent vinegar, equal in strength to that produced by the vinous fermentation of Europe. Each *mayam* will yield toddy for at least three months, often for five, and fresh mayams make their appearance before the old ones are exhausted; in this way a tree is kept in a state of productiveness for a number of years, the first *mayam* opening at the top of the stem, the next lower down, and so on, until at last it yields one at the bottom of the trunk, with which the tree terminates its existence.

"**Dr. J. E. de Vry** states that this palm contains a great proportion

Arenga Sugar.

SAP.

of cane-sugar, although the natives in Java extract it by a very rude and entirely primitive mode. He thus describes the process, which differs little from that pursued for obtaining sap and sugar from other palms :—

"As soon as the palm begins to blossom, they cut off the part of the stem that bears the flower; there flows from the cut a sap containing sugar, which they collect in tubes made of bamboo cane, previously exposed to smoke, in order to prevent the fermentation of the juice, which, without this precaution, would take place very quickly under the double influence of the heat of the climate and the presence of a nitrogenous matter.

"The juice thus obtained is immediately poured into shallow iron basins, heated by fire, and is thickened by evaporation, till a drop falling on a cold surface solidifies; this degree of concentration attained, the contents of the kettle are put in forms of great prismatic lozenges. Several thousand pounds of sugar are thus obtained yearly. I have collected some of the sap in a clean glass bottle, and I found that the unaltered juice does not contain any glucose, but a nitrogenous matter, which, by the heat of the climate, quickly converts a part of the cane-sugar into glucose. In order to prove, without employing any artificial means, that the juice exuding from the tree contains pure cane-sugar, I collected a sample directly in alcohol; the nitrogenous principle is thus eliminated by coagulation; a mixture of equal parts of juice and alcohol has been, after filtration, evaporated on the sand-bath to the consistence of syrup. I brought this syrup with me on returning from Java; and during the voyage it became solid, presenting very fine and well-defined crystals of cane-sugar, immediately recognised as such by all the experts. At the Congress of Giessen, I spoke of the preparation of sugar from palms as the only rational mode of obtaining sugar in the future, basing my opinion on the following grounds: Sugar, by itself, being only composed, in a state of purity, of carbon, hydrogen, and oxygen, does not take anything from the soil; but the plants now mainly cultivated for extracting sugar, *viz.*, the **Beta vulgaris** and the **Saccharum officinerum** require for their development a great amount of substances from the soil in which they grow, whence it follows that their culture exhausts the soil. But this is not the only evil; what is worse is, that the space now occupied by beet-roots in Europe, and by sugar-cane between the tropics, might and ought to serve for the culture of wheat or of forage in Europe, and for rice under the tropics; and it is my opinion that, considering the increase of population, the time is not far distant when it will be absolutely necessary to devote to the culture of wheat or rice the lands now employed for beet-root or cane. While the cane and beet-root require a soil fit for cereals, the Arenga palm prospers on soils entirely unfit for their culture,—so unfit, indeed, that one might try in vain to grow on them rice or cereals; the Arenga palm thrives in the profound valleys of Java and in some parts of the island extends from the shores of the sea to the interior, where the tree is found in groups, and it is very possible to make rich plantations of that fine tree. There is one drawback, but not a very serious one; the tree must be eleven or twelve years old before it will yield sugar. When, however, it commences, the operation can be repeated during several years, and the preparation of the sugar becomes a continuous industry, and not an interrupted one, as it is now. According to my average, a field of thirty ares (¾ acre) planted with those trees should produce yearly 2,400 kilogrammes of sugar in a soil quite unfit for any other kind of culture."

TIMBER. 1344

Structure of the Wood.—"The trunk of the dead palm becomes soon hollow, and furnishes very durable underground water-pipes; also good for troughs or channels for water." (*Kurz.*)

ARGANIA, *Röem. et Schult.; Gen. Pl., II., 656.*

1345

A genus closely allied to SIDEROXYLON, containing only one species, the Argan tree of Morocco, sometimes attaining a height of 70 feet, but generally much lower, with wide-spreading branches, often covering a space of 220 feet.

Argania Sideroxylon, *R. & S.; Linn. Jour., Vol. XVI., 563;* SAPOTACEÆ.

1346

Habitat.—This is the Argan tree of Morocco, which is found growing gregariously in forests in the Atlas Mountains; in its wild state over but a very limited area.

Properties and Uses—

OIL. 1347

Oil.—An oil, resembling olive oil, is extracted from the seeds. It has a clear, light-brown colour, and a rancid odour and flavour. It is an important domestic oil among the Moors, and to a certain extent finds its way to India.

FOOD. Fruits. 1348 FODDER. Leaves. 1349

Food and Fodder.—The fruit, of the size of a small plum, is used for feeding cattle, the skin and pulp being much relished. The leaves are also given as fodder.

There seems no reason why this exceedingly valuable tree might not be successfully introduced into India. The attempt to do so appears, however, to have failed. In the *Kew Report* for 1879, page 12, an interesting account of this tree is given, from which the following has been extracted :—

"The husk of the fruit is greatly valued for cattle food, while the seed-kernel is the source from which an excellent oil is extracted.

"At different times the seed has been procured and distributed to various colonies, where, however, its slow growth has led to disappointment. In 1879 a supply was obtained through the kindness of Mr. C. F. Carstensen, H. B. M. Vice-Consul, Mogador.

"Amongst other places the Botanic Garden at Saharanpore was supplied, where, however, the plant, though probably well suited for North-West India, does not appear to have survived." (*See also Kew Report 1882, p. 17.*)

ARGEMONE, *Linn.; Gen. Pl., I.,* 2.

1350

A small American genus (six species) belonging to the Natural Order PAPAVERACEÆ; one species naturalized in India.

An erect, prickly annual; juice yellow. *Flowers* bright yellow; *sepals* 2-3; *petals* 4-6. *Stamens* indefinite. *Ovary* 1-celled; *style* very short; *stigma* 4-7-lobed; *ovules* many, on 4-7 parietal placentas. *Capsule* short, dehiscing at the top by short valves that alternate with the stigmas and placentas. *Seeds* many.

The Indian representative of this genus has now passed completely over the plains of India, ascending the hills to about 2,000 feet in altitude, and, but for its known history, no one could hesitate in pronouncing it wild and indigenous. It is, however, one of the numerous introduced plants which have made India their home; it has even received by adaptation vernacular names known to oriental literature before the introduction of the plant. There are many illustrations of this nature, *i.e.*, names being given by modern usage to plants which only very fancifully resemble the originals. Thus the names for the species of Tamarix are universally given to the introduced **Casuarina.**

The name Argemone is derived from the name of a small ulcer in the eye, for which this was supposed to be a specific.

1351

Argemone mexicana, *Linn.; Fl. Br. Ind., I., 117;* PAPAVERACEÆ.

THE MEXICAN OR PRICKLY POPPY.

Vern.—*Baro-shiálkánta, siál-kántá,* BENG.; *Gokhula janum,* SANTAL; *Bharbhand, piládhutúrá, farangi-dhutúrá, ujar-kántá, shiál-kántá,* HIND.; *Bharbhurwa, karwah, kantela,* N.-W. P.; *Kandiári, sialkántá, bhatmil, satya nasa, bherband, katci, bhat kateya,* PB.; *Srigála kantá, brahmadandi,* SANS.; *Farangi dhatura, bharamdandi, dárúri, pílá-dhaturá,* DUK.; *Dárudi,* GUJ.; *Firangi dhotra, dáruri, pinvalá-dhotrá, kánte-dhotrá,* MAR.; *Birama-dandu, kurukkum-ckedi,* TAM.; *Brahma-dandi-chettu,* TEL.; *Datturi, datturi-gidda,* KAN.; *Brahma-danti,* MAL.; *Kantá-kusham,* URIYA; *Khyáa,* BURM.

Datturí or *Datheri gida* is the Kanarese name under which this plant is generally known in Mysore, Bangalore, and Bellary; but this name is liable to be confounded with that of **Datura alba** in several other languages, as has been pointed out by **Moodeen Sheriff.**

Habitat.—A spiny, herbaceous annual, introduced into India within historic times, common everywhere from Bengal to the Panjáb on roadsides and waste places, self-sown, and appearing in the cold season. The *Bombay Gazetteer* (Vol. III., 206) says that in the Panch Maháls this plant "is as common here as elsewhere, and not the least like a foreigner."

SAP. 1352

THE SAP.

The milky sap, on drying, forms a substance resembling opium.

OIL. 1353

THE OIL.

The seeds yield a pale yellow, clear, limpid oil, used in lamps and medicinally in ulcers and eruptions. In Bengal, and more or less throughout India, the seed is collected and pressed for the oil, which is yielded as copiously sa that from mustard seed. The drawn oil is allowed to stand for a few days to deposit a whitish matter, after which it remains clear and bright. (*Spons' Encycl.*) According to **M. Lepine,** this oil might with advantage be used in the arts (*Journ. de Pharm., Juillet 1861, p. 16*). **Charbonnier** describes it as of a light-yellow colour, limpid, transparent, retaining its fluidity at 5° C., of a nauseous odour and slightly acrid taste, which, however, is not very disagreeable. It dries on exposure to the air, but is entirely soluble in 5 or 6 measures of alcohol at 32·2° C. **Flückiger** has not found this statement to be correct, however. He says it has the specific gravity of ·919 at 16·5° C., and remains clear at —6° C., but on exposure dries slowly and completely. **Dr. Dymock** informs me that the oil changes to a deep red colour. It may be readily separated by means of carbon disulphide. It is thought that it is likely to come into great demand as an oil for painting; if so, India could supply a practically unlimited amount, as in many parts of the country the plant is so abundant as to have become a source of anxiety to the cultivator.

MEDICINE. Juice. 1354

MEDICINAL PROPERTIES.

The YELLOW JUICE of this plant is used as a medicine for dropsy, jaundice, and cutaneous affections. In the West Indies it is reported to be used as a substitute for ipecacuanha. It is also diuretic, relieves blisters, and heals excoriations and indolent ulcers. The native practice of applying the juice of this plant to the eye in ophthalmia is dangerous, although interesting historically, the same practice having in all probability suggested the name Argemone (see generic description). The SEEDS have narcotic properties. They yield on expression a fixed OIL, which has long been in use amongst West Indian practitioners as an aperient. It exercises a soothing influence when applied externally in headache, and also to herpetic eruptions and other forms of skin

Seeds. 1355 Oil. 1356

MEDICINE.

disease. **Mr. Baden Powell** says that this is supposed to be the *fico del inferno* of the Spaniards, who consider the seeds more narcotic than opium. An INFUSION of the plant is regarded as diuretic.

Infusion. 1357

The following extracts from the writings of Indian authors will show the diversified opinions which are held with regard to this drug:—

"The juice of the plant in infusion is diuretic, relieves strangury from blisters, and heals excoriations. The seeds are very narcotic, and said to be stronger than opium. **Simmonds** says: 'The seeds possess an emetic quality. In stomach complaints, the usual dose of the oil is thirty drops on a lump of sugar, and its effect is perfectly magical, relieving the pain instantaneously, throwing the patient into a profound refreshing sleep. and relieving the bowels.' This valuable but neglected plant has been strongly recommended as an aperient anodyne, and hypuotic by **Dr. Hamilton** and other experienced practitioners in the West Indies. Samples of the oil were produced at the Madras Exhibition. It is cheap and procurable in the bazars, being used chiefly for lamps. (*Ainslie, Lindley, Simmonds, &c.*)" (*Baden Powell's Panjáb Products, I., 326.*)

"The seeds and seed oil have been used by European physicians in India, and there has been much difference of opinion regarding their properties, some considering them inert, and others asserting that the oil in doses of from 30 to 60 minims is a valuable aperient in dysentery and other affections of the intestinal canal. The evidence collected in India for the preparation of the *Indian Pharmacopœia* strongly supports the latter opinion; my experience is also in favour of it; and **Charbonnier**, who examined the oil in 1868, found it aperient in doses of from 15 to 30 minims. Possibly those who have used the oil unsuccessfully, purchased it in the bazar, and were supplied with a mixed article: no bazar-made oils can be relied upon. An extract made from the whole plant has been found to have an aperient action, and the milky juice to promote the healing of indolent ulcers. I have not noticed any bad effects from its application to the eyes. . . . . Recently (1878) a case has occurred in Bombay in which a number of people suffered from vomiting and purging after using sweet oil which had been adulterated with Argemone oil. The adulteration may be detected by the rich orange-red colour developed when strong nitric acid is added to Argemone oil, or to mixtures containing it." (*The Vegetable Mat. Med. of Western India, by W. Dymock, 40.*)

"The seeds are used in Jamaica as an emetic, a thimbleful being bruised with water. **Barton** again describes them as being more powerfully narcotic than opium.

"The Editor has subjected the seeds to numerous experiments, and has never found them to show any emetic or narcotic influence; they contain a bland oil resembling that of the poppy, and which can be used in ounce-doses without producing any purgative effect.

"The juice, which exudes on wounding or bruising this plant, is of a bright yellow colour, and is used by the natives as an application to indolent ulcers, and to remove specks on the cornea. It has by some writers been described as possessing the activity of gamboge. If this expressed juice of the plant be rendered alkaline by ammonia, a precipitate falls which is partially soluble in hot alcohol, giving a rich golden tincture; on cooling and spontaneous evaporation, silky crystals of an alkaline principle are deposited, which we propose to term *argemonine*.

"We have given this argemonine in considerable quantities to dogs, and did not find it induce any acrid or narcotic effects. It has not been as yet tried in hospital practice." (*The Bengal Dispensatory and Pharmacopœia by W. B. O'Shaughnessy, M.D., p. 183.*)

1358

Special Opinions.—§ "I have used the seeds of **Argemone mexicana** in many cases and found them to be laxative, emetic, nauseant, expect-

ARGEMONE mexicana.

The Mexican Poppy.

MEDICINE.

orant, and demulcent, and the oil obtained from them is a drastic purgative, nauseant, and expectorant. The seeds and oil have also a beneficial control over asthma. The largest dose of the seeds I have used is two drachms and a half, and even in so large a dose as this there was nothing in their action to lead to the suspicion of their being a narcotic as is generally supposed. It is difficult to account for such a supposition without suspecting that some other seeds were confounded with those under immediate consideration. I shall, therefore, describe the latter as minutely as I can, and if due attention is paid to this description, there will be no difficulty in distinguishing them from all other seeds. The seeds are small, round, hard, striated, dark brown, and about the size of a small mustard-seed. They are full of oil, and if one of them is placed on paper with a hard substance underneath and pressed with the nail of the finger, it breaks with a noise and leaves an oily stain on the paper; the kernel is white, minute, and albuminous. Although the doses of the seeds I have mentioned above are very large, yet this is no disadvantage, because they are always used in emulsion, which is tasteless and can be sweetened, if necessary. The emulsion is much liked by the patient.

"There is also a great difference of opinion as to the action and dose of the oil of **Argemone mexicana**. Some say that thirty minims of it act as an efficient cathartic, while others consider it to be quite inert and incapable of producing any purgative effect in ounce-doses. I got this oil prepared three or four times in my own presence and tried it in many cases. The former opinion is quite correct, and with regard to the latter, it is necessary to say that, so far from being inert in 'ounce-doses,' it is unsafe to administer the oil in more than forty minims, and produces a dangerous hypercatharsis when its dose is increased to one drachm. If the oil is fresh, its average dose is twenty-five minims, and if old thirty-five. It is a good drastic or hydrogogue cathartic in such doses, and generally produces from five to ten or twelve motions. Its advantage over jalap, rhubarb, castor oil, &c., is the smallness of its dose; and over the croton oil, its freeness from unpleasant, nauseous, and acrid taste. Its disadvantages as a purgative are, *firstly*, that its action is not uniform even in its average dose, which produces more than 15 or 16 motions at one time, and only 3 or 4 at another; and *secondly*, that it is generally accompanied by vomiting at the commencement of its operation. Though the latter is not severe, it is undesired, and an unpleasant effect in a purgative medicine. Hypercatharsis, however, from the use of this oil is not generally attended with great debility, nor with the other dangerous symptoms frequently observed under a similar condition from croton oil and some other purgatives." (*Honorary Surgeon Moodeen Sheriff, Madras.*)

"Very common all over South India and the Deccan, and the juice is in vogue as a native remedy." (*Deputy Surgeon-General G. Bidie, C.I.E., Madras.*) "The yellow juice mixed with *ghi* is given internally in gonorrhœa." (*Surgeon-Major D. R. Thompson, M.D., Madras.*) "Oil aperient, sedative in colic, dose 30 minims; noticeable effect when applied externally to skin diseases." (*Apothecary Thomas Ward, Madanapalle, Cuddapah.*) "The juice of this plant is much used by the inhabitants of Mysore for indolent and syphilitic ulcers and for itch." (*Surgeon-Major John North, Bangalore.*) "The yellow juice is often used by natives in simple conjunctivitis." (*Honorary Surgeon Easton Alfred Morris, Negapatam.*)

"I found the juice very useful in scabies. Asst. Gowry Coomar Mukerji found the powdered root in drachm doses useful in tapeworm." (*Surgeon R. L. Dutt, M.D., Pubna.*) "In Cuttack the seed of this plant is mixed with mustard seed as an adulteration." (*Surgeon-Major P. N. Mukerji, Cuttack, Orissa.*) "The yellow juice of the plant and

MEDICINE.

the cold-drawn oil of the seed is useful in scabies. I have seen the insect killed (under the microscope) on the application of ether." (*Surgeon K. D. Ghose, Khulna.*) "Oil obtained from the seed is largely used by the Santals for the purpose of burning. A valuable remedy for itch." (*Brigade Surgeon S. M. Shircore, Moorshedabad.*) "The juice of the plant is used as a detergent in chronic ulcers and sinuses with good effect." (*Assistant Surgeon Nundo Lal Ghose, Bankipore.*) "Useful in scabies." (*Surgeon-Major C. J. W. Meadows, Burrisal.*) "The fresh juice is used in scabies and indolent ulcers." (*Brigade Surgeon J. H. Thornton, B.A., M.B., Monghyr.*) "Known as *Karwah* in Oudh. The oil is used in the West Indies on sugar for colic. Has been tried in Oudh for the same ailment with advantage in several cases." (*Surgeon-Major Bonavia, Etawah.*) "The seeds contain an alkaloid which gives reactions similar to morphia (*Dragendorff*)." (*Surgeon-Major W. Dymock, Bombay.*) "The juice is efficacious in scabies. I saw a case of dangerous inflammation of the eye caused by the application of the juice in ordinary conjunctivitis." (*Assistant Surgeon Shib Chunder Bhuttacharji, Chanda, Central Provinces.*)

ARGENTUM.

1359

Argentum or Silver.

Vern.—The leaf: *Chándi-ká-varaq*, HIND.; *Rúpali*, BENG.; *Taka*, SANTAL; *Rupyáchá-varkh*, MAR.; *Ruperivarakh, chándi, rupu*, GUJ.; *Rupéritagat*, DUK.; *Velli-rékku*, TAM.; *Vendi-réku*, TEL.; *Vellit-takita*, MAL.; *Belli rekhu*, KAN.; *Varqul-fizah*, ARAB.; *Varqe-sim, varqe-nuqrah*, PERS.; *Ridi-tahadu, ridi-tagadu*, SINGH.; *Noye-saku*, BURM.

The metal: *Chándi, rupá, ruppá*, HIND., BENG. BHOTE, MAR., DUK., GUJ.; *Rupo*, SIND; *Velli*, TAM.; *Vendi*, TEL.; *Velli*, MAL.; *Belli*, KAN.; *Fizah*, ARAB.; *Sin, nuqrah*, PERS.; *Roupya, rajata*, SANS.; *Ridi*, SINGH.; *Noye*, BURM.

Silver is too well known to require to be dealt with here in detail. The following special opinions and notes regarding the Indian medicinal uses may, however, be found interesting.

MEDICINE. 1360

Medicine.—§ "Silver-leaf is used medicinally combined with other metals, chiefly with gold or iron, for nervous diseases." (*Surgeon G. A. Emerson, Calcutta.*)

"Argentum Nitras gr.x admixture most useful collyrium in acute conjunctivitis. Its action can be modified by a solution of common salt, applied to the eye after the application of the argt. nit." (*Surgeon Joseph Parker, M.D., Poona.*)

"Silver-leaf is daily prescribed by the hakims, along with the different preserves, particularly that of 'Anvala' fruit (**Phyllanthus Emblica**) in nervous palpitation, dyspepsia, and general debility." (*Surgeon Mokund Lall, Agra.*)

"Nervine tonic. The ash of silver (*raupya bhasma*), which is administered internally as a nervine tonic, is prepared by mixing together 1 part of arsenic with lemon-juice gr. ½ and ⅓ part of silver leaves in a mortar, and then enveloping the mixture in mud and clean cloth, burn freely until reduced to ashes. The ashes should be again covered over with mud and cloth, and burnt (14 times altogether)." (*Surgeon W. Barren, Bhuj, Cutch.*)

"It is used as an amalgam in stopping teeth." (*Brigade Surgeon G. A. Watson, Allahabad.*)

ARGYREIA, *Lour.; Gen. Pl., II., 869.*

1361

A genus of scandent shrubs, containing some 30 species, belonging to the Natural Order CONVOLVULACEÆ. The species are chiefly Indian (25 species); but one occurs in Africa, and a few others in China and the Malaya.

Leaves from cordate-ovate to narrow-lanceolate, silky hirsute or pubescent.

The Elephant Creeper.

Cymes sessile or peduncled, capitate or corymbose. *Flowers* showy purple or rose, rarely white. *Sepals* from orbicular to lanceolate, sub-equal or the inner smaller, adpressed to the fruit, often somewhat enlarged.

The generic name is derived from ἀργύρειος=silvery, in allusion to the silvery tomentum of the under-surface of the leaves.

1362

Argyreia speciosa, *Sweet.; Fl. Br. Ind., IV., 195; Wight, Ic., t. 851.*

THE ELEPHANT CREEPER.

Syn.—LETTSOMIA NERVOSA, *Roxb.; Fl. Ind., Ed. C.B.C., 164;* CONVOLVULUS SPECIOSUS, *Linn.*

Vern.—*Samandar-ká-pát, samandar-sóf, samandar-sokh, samandar-phaind,* HIND.; *Bichtárak, guguli,* BENG.; *Kedok' arak',* SANTAL; *Samudra-palaka, vriddhadáraka,* SANS.; *Samudra soka,* or *shokh,* MAR.; *Samandar-ká-pattá,* DUK.; *Shamuddirap-pach-chai,* TAM.; *Samudra-pála, chandra-poda, kokkita, pála-samudra,* TEL.; *Samudra-pach-cha, samudra-yógam, samudra-pala,* MAL.; *Mahadoomooda,* SINGH.

Moodeen Sheriff points out that the Hindustani and Deccan names of this plant are to be distinguished from *Samandar-phal,* the Hindustani and Deccan name of **Barringtonia acutangula.**

Habitat.—A twining, perennial plant, found all over India, ascending to 1,000 feet in altitude from Assam to Belgaum and Mysore, frequent in Bengal; cultivated in China and the Mauritius. Extremely common in Western India. **Dr. Bidie** informs me that it also extends to the extreme south of the peninsula of India.

Botanic Diagnosis.—*Leaves* large, ovate-cordate, acute, glabrous above, persistently white tomentose beneath; *peduncles* long; *flowers* sub-capitate; *bracts* large, ovate-lanceolate, acute, thin, softly woolly, deciduous; *corolla-tube* woolly; *fruit* brown-yellow, stout, nearly dry; *stem* white, tomentose, almost woody. (*Fl. Br. Ind.*)

Properties and Uses—

OIL. 1363

Oil.—Reported to yield oil; but no definite information regarding this fact can be discovered.

MEDICINE. Leaves. 1364 Root. 1365 Powder. 1366

Medicine.—The LEAVES are maturative and absorptive, and are used as emollient poultices for wounds, and externally in skin diseases, and by some authors they are even said to have rubefacient and vesiccant properties. The ROOT is regarded as alterative, tonic, useful in rheumatic affections and diseases of the nervous system. In synovitis the powdered root is given with milk.

"The large leaves, which have the under-surface covered by a thick layer of silky hairs, afford a kind of natural impermeable piline, and are used as a maturant by the natives. With regard to the alleged blistering properties of the upper surface of the leaf there must be some mistake, as I find it has no effect when applied to the skin." (*The Vegetable Materia Medica of Western India by Dr. W. Dymock, 474.*)

1367

Special Opinions.—§ "Mixed with vinegar the sap is rubbed over the body to reduce obesity." (*Surgeon G. A. Emerson, Calcutta.*) "Used externally in chronic eczema and as emollient poultices. Internally the root is given to rheumatic patients, dose 5 to 20 grains." (*Surgeon W. Barren, Bhuj, Cutch.*) "Leaves are used as a poultice in guinea-worm." (*Surgeon Joseph Parker, M.D., Poona.*) "Useful when applied to foul ulcers." (*Assistant Surgeon Shib Chunder Bhuttacharji, Chanda, Central Provinces.*) "The juice of this plant, mixed with an equal quantity of gingelly oil, and a little powdered dill seed, is used as an external application in scabies and other cutaneous diseases of children." (*Surgeon W. A. Lee, Mangalore.*) "In cases of unhealthy ulcers and sinus, the white surfaces of the young leaves are applied, the hairs causing irritation and promoting the secretion of healthy pus. When the sores are progressing favourably the smooth surfaces of the leaves are applied." (*Brigade Surgeon J. H. Thornton, B.A., M.B., Monghyr.*) "The leaves are both maturative and absorptive; when the under-part is applied to the inflammation it hastens suppuration, and the upper part resolution. It is also efficacious in skin diseases." (*T. N. Mukarji, Revenue and Agriculture Department, Calcutta.*)

ARISÆMA, *Martius; Gen. Pl., III., 965; Engler in DC. Mono. Phan., II., 533; N. E. Brown in Linn. Soc. Jour., XVIII., 246.* 1368

A genus of herbaceous plants with tuberous (often edible) corms, belonging to the Natural Order ARACEÆ. There are about 50 species belonging to the genus, inhabitants of temperate and extra-tropical Asia, with a few in America and Abyssinia. In India there are some 22 species.

Leaves 1 to 3, each 3-sected or pedate or verticillately lobed, with 5 to many segments, each broad, acute or acuminate; *margin* entire or crenulate; *petiole* sheathing at the base. *Spathe* deciduous, tube oblong, convolute at the base, not infrequently many-veined; *mouth* contracted; *blade* large, acuminate or caudate. *Peduncles* solitary. *Spadix* with an appendix included within the spathe or exserted. *Flowers* diœcious, rarely monœcious, male scattered, female crowded with neuter subulate flowers above. *Perianth* none. *Male flowers* with 2-5 stamens, sub-sessile. *Female-flowers* with the ovary ovoid-oblong or globose, 1-locular, style short or O; *ovules* 1-many, orthotropous, erect; *panicles* short, attached to a basilar placenta. *Fruit* an obconic berry, 1 or few-seeded.

Arisæma curvatum, *Kunth.; Engler in DC. Mono. Phan., II., 544.* 1369

Syn.—ARUM CURVATUM, *Roxb.; Fl. Ind., Ed. C.B.C., 628; Wight, Ic., t. 788.*

Vern.—*Bir-banka,* NEP.; *Gúrin, dor, kirkichálú, kirakal, jangúsh,* PB.

Habitat.—This plant grows at many places in the Panjáb Himálaya, from 4,000 to 6,500 feet.

Medicine.—It is stated to have poisonous qualities. In Kúlú the seeds are said to be given with salt for colic in sheep.

MEDICINE. Root. 1370 Seed. 1371

A. cuspidatum, *Engl.; Mon. Phan., DC., II., 536.* 1372

Syn.—ARUM CUSPIDATUM, *Roxb.; Fl. Ind., Ed. C.B.C., 628; Wight, Ic., t. 784.*

A. erubescens, *Schott.* 1373

Syn.—ARUM ERUBESCENS, *Wall.; Pl. As. Rar., II., 30, t. 135.*

Habitat.—The Himálaya and Western Ghâts.

A. intermedium, *Blume.* 1374

Habitat.—The Western Himálaya (Simla, 2,600 feet).

A. Jacquemontii, *Blume.* 1375

Habitat.—The Himálaya, 2,000 to 4,000 feet.

A. Leschenaultii, *Blume; Mono. Phan., DC. II., 552.* 1376

Syn.—A. PAPILLOSUM, *Stend.*

Vern.—*Wal-kidáran,* SINGH.

Habitat.—A native of the Himálaya (Nepal), Khásia Hills, the Nilgiri Hills, and Ceylon.

Medicine.—"The roots are employed as a medicine by the Singhalese." (*Thwaites, En. Plant Zeyl., 335.*)

MEDICINE. 1377

1378 **Arisæma Murrayi,** *Graham ; Cat. Pl., Bombay.*

THE SNAKE LILY OF THE KONKAN.

A. papillosum, *Stend.*, see **A. Leschenaultii,** *Blume.*

1379 **A. speciosum,** *Mart. ; Stewart, Pb. Pl., 247.*

Syn.—ARUM SPECIOSUM, *Wall.*

Vern.—*Samp-ki-khumb, kiri-ki-kukri, kiralu,* PB.

Habitat.—Found in the Panjáb Himálaya, from 6,000 to 8,500 feet.

MEDICINE. Root. 1380

Medicine.—In Hazára the root is stated to be poisonous; in Chumba it is applied pounded to snake-bites. In Kúlú, where the root is given to sheep for colic, the fruit is said to have deleterious effects on the mouth when eaten by children.

381 **A. tortuosum,** *Schott. ; Engler, DC. Mono. Phan., II., 545.*

Vern.—*Kiri-ki-kukri,* PB.

Habitat.—Found in Chumba at about 7,000 feet, also eastward to Nepal.

EDICINE. Root. 1382

Medicine.—The root of the plant is used to kill the worms which infest cattle in the rains.

ARISTIDA, *Linn. ; Gen. Pl., III., 1140.*

1383 **Aristida depressa,** *Retz ; Duthie's List of Grasses, 26 ;* GRAMINEÆ.

Vern.—*Spin-khalak, spin-wege, jandar lamba, lamp,* PB.; *Nalla-putiki,* TEL.

Habitat.—Inhabits the plains in North India; also found in the Southern Provinces. Grows in a dry, barren, binding soil.

FODDER. 1384

Fodder.—**Roxburgh** did not find that it was put to any use; but **Stewart** says it is a favourite food for cattle in North India. "Cannot be cut with a scythe, as it is too fine. Particularly relished by cattle, and is nutritious. It is too short and light to stack." (*Mr. Coldstream, Commissioner, Hissar.*)

1385 **A. setacea,** *Retz.*

Vern.—*Shipur-gaddi,* TEL.; *Thodapga-pullu,* TAM.

Habitat.—Common in dry parts of the Panjáb and North-West Provinces; also in South India, where it grows in dry, barren, binding soil.

FODDER. 1386

Fodder.—"Cattle do not eat it, yet it is very useful." (*Roxburgh.*) As to the remark that cattle do not eat this grass, **Roxburgh** was apparently mistaken, for **Bidie** says it is eaten by bullocks.

DOMESTIC. Culms. 1387 Tooth-picks. 1388 Tatties. 1389

Domestic Uses.—The Telinga paper-makers construct their frames of the culms; it also serves to make brooms and tooth-picks. It is employed in preference to other grasses for making the screens called *tatties ;* for this purpose it is spread thin in bamboo frames and tied down : these placed on the weather side of the house during the hot land-winds, and kept constantly watered during the heat of the day, renders the temperature of the air in the house exceedingly pleasant, compared to what it is without." It is used, in fact, like the *khas-khas* roots in Northern India.

1390

ARISTOLOCHIACEÆ.

A Natural Order of herbaceous plants with creeping rhizomes and creeping or twining stems; wood, when present, scented, composed chiefly of parallel plates held loosely together by soft medullary processes; no concentric zones nor liber fibres. There are in all some 200 species in the world, referred to 6 genera. They are inhabitants chiefly of tropical America, are rare in the north temperate zones, occasional in tropical Asia, and somewhat frequent in the Mediterranean region. In India there are in all some 6 or 7 indigenous species belonging to **Aristolochia** and **Bragantia**, with as many more introduced species, chiefly seen under garden cultivation.

The genera are referred to three tribes, the diagnosis of which, if taken collectively, constitute the characters of the order—an order which must be admitted as exceedingly artificial, since it includes tribes dissimilar in vital characteristics. The affinities of the family are, accordingly, very obscure.

Tribe I.—Asareæ.

Herbs with perennial rhizomes, having the lower leaves scab-like, the upper reniform. *Flowers* terminal, solitary. *Calyx* persistent; *limb* regular, 3-lobed. *Stamens* 12, all free, the outer and shorter whorl opposite the styles; *anthers* introrse or extrorse. *Ovary* more or less inferior, 6-celled, short and broad; a *capsule*, opening, when ripe, irregularly.

Asarum, Heteropa.

For the former of the two genera in this tribe see **Asarum.**

Tribe II.—Bragantieæ.

Shrubs or under-shrubs. *Leaves* reniform, oval or oblong, lanceolate, reticulate. *Flowers* in spikes or racemes, small **(Bragantia)** or large and campanulate **(Thottea).** *Calyx* deciduous, closely appressed to the top of the ovary and 3-lobed. *Stamens* 6-36, equal and free. *Ovary* completely inferior, elongated, slender, stipitate, 4 gonous, 4-celled; *ovules* numerous, 2-seriate on the middle of the septa. *Capsule* siliquose 4-valved.

Bragantia and Thottea.

For the former genus see its place in this Dictionary.

Tribe III.—Aristolochiæ.

Twining herbs (rarely scandent). *Calyx* deciduous, constricted above the top of the ovary, irregular, tubular, limb various. *Stamens* 6 (rarely 5); *anthers* sessile, extrorse, adnate by their whole dorsal surface to the column or style. *Ovary* completely inferior, elongated, slender, stipitate, 6-gonous, 6 (rarely 5) celled; *ovules* numerous, inserted at the central angle of the cells and 2-seriate. *Capsule* oblong or globose, 6-angled, 6-valved, opening at the top or bottom.

1391

Affinities of the Aristolochiaceæ are very obscure; the gynandrous condition of the stamens and twining habit bring Aristolochiæ near to Asclepiadeæ, but the opposite leaves and superior ovary of the latter at once separates them. By some botanists, an affinity to Cucurbitaceæ is seen in the twining stem, alternate leaves, inferior ovary, and extrorse stamens; but Cucurbitaceæ differ in their didymous double-perianth, imbricated æstivation, in the number and condition of the stamens, the mode of placentation and ex-albuminous seeds. It seems almost impossible to assign a natural position for this family. The *Genera Plantarum* places them after Nepenthaceæ, Cytinaceæ, and before Piperaceæ, Monimiaceæ, Laurineæ, Santalaceæ, Balanophoreæ, &c., and this would seem their most natural position. Like the Cytinaceæ they have a mono-perianthed flower, inferior, often 1-celled ovary, but Cytinaceæ are aphyllous and parasitic. With Nepentheæ they have many affinities, and although it is only fanciful, the pitcher-like glandular development on the leaves of Nepentheæ is exceedingly like the flower of the Aristolochias.

1392

Properties and Uses.—Most of the Aristolochieæ contain in their roots a volatile oil, a bitter resin, and an extractable acid, from which they derive their virtue as stimulants of the glandular organs and of the functions of the skin. The name Aristolochia is derived from ἄριστος, best, and λοχεία, child-birth, or herbs which promote child-birth, in allusion to their reputation as emmenagogues. They are also administered as anti-hysterics. As medicines they may be described as aromatic, stimulating, tonic, and useful in the latter stages of low fever. They are bitter and acrid with a disagreeable odour. They are also described as purgative, and are in India chiefly taken advantage of as mild aperients for children. They are all attributed with the property of being antidotes to snake-bite, the two best known examples being the Virginian snake-root and Guaco roots. The roots of the **Asarum** are emetic.

1393

ARISTOLOCHIA, *Linn.; Gen. Pl., III., 123.*

As this is the only Indian genus belonging to the Tribe ARISTOLOCHIÆ, it is scarcely necessary to add other characters to what have been already given. Aristolochia chiefly differs, however, from Holostylis in the pitcher-like form of the flower and more numerous stamens.

1394

Aristolochia acuminata, *Willd.*

Botanic Diagnosis.—An extensive twining plant. *Leaves* cordate, entire, acuminate, from 4-6 inches long by 2 to 4 broad. *Flowers* large and pendulous.

1395

A. bracteata, *Retz.*

THE BRACTEATED BIRTHWORT.

Vern.—*Kirámár, gandán, gandati,* HIND., DUK.; *Pattra bunga, katrabungá,* SANS.; *Paniri,* URIYA; *Gandháti, kidámári,* BOMB.; *Ádu-tina-pálai,* TAM.; *Gádide gada-para-áku, kadapara,* TEL.; *Átutintáp-pála,* MAL.

Habitat.—Found on the banks of the Jumna and Ganges and in the Deccan. Seems to luxuriate on the black soils of Western India.

Botanic Diagnosis.—*Leaves* reniform, glaucous; *flowers* axillary, solitary-peduncled; *peduncles* furnished at the base with sessile, reniform bracts.

Properties and Uses—

MEDICINE. 1396

Medicine.—Every part of this plant is nauseously bitter, and is much used by the Hindú physicians on account of its anthelmintic and purgative properties. Two fresh leaves rubbed up in a little water, and given to an adult for a dose, once in 24 hours, are considered a cure for purging with gripes. (*Roxburgh.*) The leaves are applied to the navel to move the bowels of children, and are also given internally along with castor oil as a remedy for colic. The natives squeeze the juice of this plant into wounds to kill worms. (*Dr. Gibson.*) It is spoken of by **Dalzell** as possessing "a merited reputation as an antiperiodic in intermittent fevers." Other authors affirm that it holds a high reputation as an antiperiodic in the treatment of fevers. For this purpose it is often made into a paste along with the seeds of **Barringtonia actulangula, Celastrus paniculata,** and Black Pepper, the whole body being rubbed with this paste in malarial fevers. (*Dymock; Dalzell; Gibson's Flora of Bombay.*) It is also supposed to be an emmenagogue. **Dr. J. Newton** reports that in Sind the dried root, in doses of about a drachm and a half, is administered during labour to increase uterine contractions.

A committee, consisting of **Drs. Carter, Dymock,** and **Sakhárám Arjun,** reported on this drug as follows: "The drug consists of the whole plant in fruit; it is nauseously and persistently bitter." "Anthelmintic, antiperiodic, and emmenagogue. Used in the bowel complaints of children, when depending on worms, in intermittent fevers, and to increase uterine contractions during labour. The juice of the leaves is applied to foul and neglected ulcers to destroy maggots. Dose—1½ drachms of the dried root is given in powder or infusion in cases of labour. The juice of the fresh plant is chiefly used by native practitioners." According to the same authorities it is found on banks of rivers and water-courses, and in the black soils of Gujarát and Deccan. The price is R3-8 per maund. (*Home Department Correspondence, 1880, p. 323.*) It was recommended by the Surgeon-General of Madras to be excluded from the proposed new edition of the *Indian Pharmacopœia* (*page 240*).

§ "This species, or one resembling it, is considered a powerful abortive, acting similarly in animals. The root is given mixed with round pepper." (*Surgeon W. D. Stewart, Cuttack.*) "The leaves bruised and applied as a poultice remove maggots from ulcers." (*Surgeon-Major John Lancaster, M.B., Chittore.*) "Antiperiodic, anthelmintic, also similar action to ergot on the uterus, produces violent contractions of the womb during labour. Dose one to two drachms." (*Surgeon W. Barren, Bhuj, Cutch.*) "Antiperiodic (slight), tonic. Infusion of whole plant, dried ½ oz., boiling water 10 oz., dose 1 to 2 oz." (*Apothecary Thomas Ward, Madanapalle, Cuddapah.*) "Common in Madras, regarded as anthelmintic and emmenagogue by natives." (*Deputy Surgeon-General G. Bidie, C.I.E., Madras.*)

Aristolochia hastata, *Nuttal.*

Habitat.—A species met with on the banks of the Mississippi.

Medicine.—Used medicinally in America.

MEDICINE. 1397

A. indica, *Linn.*

1398

THE INDIAN BIRTHWORT.

Vern.—The root: *Isharmul, isharmúl-ki-jar*, HIND.; *Isarmul*, BENG.; *Bhedi janetet*, SANTAL; *Sápasan*, BOMB., MAR.; *Isharmúl, issharmúl-ki-jar*, DUK.; *Arkmula, ruhimula*, CUTCH, GUJ.; *Sápús*, GOA; *Sunandá hari, jovari, arkamulá*, SANS.; *Zarávande-hindi*, ARAB., PERS.; *Ich-chura-múli*, or *ích-chura-múli-vér, peru-marindu, perum-kizhangu*, TAM.; *Ishvara-véru, dúla-góvela, govila*, TEL.; *Karalekam, haruhakpulla, karal-vekam, ishvaramúri*, MAL.; *Ishveri-vérú*, KAN.

Habitat.—A twining perennial, found all over India,—Bengal, Konkan, Travancore, and Coromandel.

Botanic Diagnosis.—*Leaves* cordate, wedge-shaped, three-nerved, with an undulated margin from 2-4 inches long by 1-2 broad; *flowers* small, erect.

Properties and Uses—

MEDICINE. 1399

Medicine.—The root possesses emmenagogue and antiarthritic properties. It enjoys, like all members of this genus, the reputation of being a valuable antidote for snake-bite, and is said to be used to effect abortion. It is also held in much esteem by the natives as a stimulant and tonic, and is used by them in intermittent fevers and other affections. The early Portuguese settlers called it *Raiz de Cobra*, owing to its supposed efficacy against the bite of the cobra, being both taken internally and a powder of the root applied externally to the injured part.

A committee in Bombay, consisting of **Drs. Carter, Dymock,** and **Sakhárám Arjun,** reported on this drug as follows: "The drug, as found in the shops, consists of the root and stem; the latter is by far the largest portion. In many parcels the stem only is to be found. It is either in short pieces, or the whole stem may be twisted into a kind of circular bundle. The thickest portion of the stem is from ¼ to ½ an inch in diameter, and has a central woody column made up of about ten wedge-shaped woody portions. The bark is thick and corky, marked with longitudinal ridges and numerous small warty projections; it is of a yellowish-brown colour. The taste is bitter, camphoraceous." "A stimulant, tonic and antiperiodic. It is chiefly used in the bowel complaints of children and in intermittent fevers. The juice of the leaves is believed to be efficacious in cases of snake-bites. Emmenagogue properties have also been attributed to it. Dose—of the decoction 1 to 2 ounces. Price six annas per lb. The drug can scarcely be called an article of commerce." It is common in the jungles of West India. (*Home Department Official Correspondence, 1880, 323.*)

§ "Tonic and stimulant, excellent antidote for scorpion-bite, used externally and internally. Dose—one to two drachms. Produces abortion, used as a cathartic in dropsy. Dose of the expressed juice, half to two drachms." (*Surgeon W. Barren, Bhuj, Cutch.*)

"It is undoubtedly used to produce abortion." (*Brigade Surgeon S. M. Shircore, Moorshedabad.*)

1400

Aristolochia longa, *Linn.*

LONG-ROOTED BIRTHWORT.

Vern.—The root: *Zarávande-tavíl, zarávand,* ARAB.; *Zarávand-dardz,* PERS.

Habitat.—Indigenous to South Europe; imported into India.

MEDICINE. 1401

Medicine.—The leaves are said to be useful in the cure of snake-bite, especially cobra-bites. The root is bitter, and used as an emmenagogue and in diseases of the womb and affections of the gums or ulcers; also in indigestion and bowel complaints of children. It is said to act as a tonic and febrifuge.

§ "Used by natives in apoplexy, jaundice, paralysis, gout, and chronic rheumatism." (*Surgeon G. A. Emerson, Calcutta.*)

"Much used by native hakims in these Provinces in cases of ulcer, &c. Suppositories prepared from this drug are supposed to produce abortion. It is also applied locally in cases of scorpion-sting." (*Surgeon J. Anderson, M.B., Bijnor, N.-W. P.*)

1402

A. reticulata, *Nuttal.*

An American species used medicinally, but only as a substitute for the true **Serpentaria**; it is a coarse species.

1403

A. rotunda, *Linn.*

ROUND-ROOTED BIRTHWORT.

Vern.—*Zarávand-emudahraj,* ARAB.; *Zarávnde-gird,* PERS.

Habitat.—Indigenous in South Europe; imported into Bombay.

MEDICINE. 1404

Medicine.—Used in coughs. The root is hot and aromatic. It is used by natives in the treatment of itch, lice, and intestinal worms; also in leprosy and ulcers, and to promote secretion of urine. It is also known as an antidote for poisons. **Dr. Dymock**, in his *Materia Medica of Western India*, says it is difficult to get this drug pure, corms of an aroid being often substituted for it.

§ "It is also used in rheumatism, fever, emphesema, chronic bronchitis, caries of tooth, enlargement of spleen." (*Assistant Surgeon J. N. Dey, Jeypore.*)

Aristolochia serpentaria, *L.*

1405

THE VIRGINIAN SNAKE-ROOT.

Habitat.—A native of North America.

Medicine.—The root of this species is given in the *Pharmacopœia of India* as the officinal form of **Aristolochia.**

MEDICINE. 1406

Medicinal Properties and Uses.—"As its common and specific names of Snake-root and Serpentaria imply, Serpentary had formerly a high reputation for the cure of the bites of venomous serpents; indeed, it was first introduced into regular medical practice as a remedy in such cases, but like all the so-called specifics of vegetable origin which have been introduced for destroying the effects caused by venomous reptiles, it is no longer regarded as of any remedial value. As a stimulant tonic, diaphoretic, and diuretic, it is, however, a medicine of some repute, but in too large doses it causes nausea, flatulency, griping pains in the bowels, and tendency to diarrhœa. It has been extensively employed in typhus and typhoid fevers; and has also been highly recommended in intermittent fevers, but in the latter it is commonly given as an adjunct to bark or sulphate of quinia, whose effects it is said to increase in a marked degree. It has likewise been employed as an antidote against the bite of a mad dog, but it has no more value in destroying the effects in such a case than as a remedy in the bites of venomous reptiles. It is, however, used with good results in diphtheria, chronic rheumatism, atonic dyspepsia, and in exanthematous diseases to promote eruption. A strong infusion is also reputed to be serviceable as a gargle in malignant sore-throat. **Garrod** states that, from observations made during many years, he "is inclined to think that serpentary is a remedy of some considerable power, acting in a manner not unlike Guaiacum in stimulating the capillary circulation, and promoting recovery in chronic forms of gouty inflammation; and as it does not disturb the bowels, it may often be administered when Guaiacum is not easily tolerated." (*Bentl. and Trim., 246.*)

The officinal preparations are an infusion of the root in boiling water; dose from one to two fluid ounces three or four times daily. A tincture of the root in proof spirit, dose from 1 to 2 fluid drachms. This is regarded as a good adjunct to stimulant and diaphoretic mixtures. Serpentaria is an ingredient in Tincture Cinchonæ Composition.

Infusion. 1407
Tincture. 1408

Chemical Composition.—"The principal constituents of serpentary root are, a volatile oil in the proportion of about 2 per cent. and a bitter principle. The volatile oil has the odour of the root, and the bitter principle (aristolochin), which was first made known by **Chevalier,** is described as an amorphous substance of a yellow colour, a bitter and slightly acrid taste, and is soluble in both water and alcohol. It requires further investigation. The medicinal properties of serpentaria are doubtless essentially, if not entirely, due to these two substances. But serpentary root also contains *tannic acid, resin, mucilage, sugar,* and some other unimportant ingredients." (*Bent. and Trim., 246.*)

1409

§ "It is met with in the bazars of Bombay, and is in general use by native practitioners under the name of *Kálá válá.* It appears to be used as a substitute for **Pavonia odorata** in Sanskrit prescriptions." (*Surgeon-Major W. Dymock, Bombay.*)

ARNEBIA, *Forsk.; Gen. Pl., II., 862.*

1410

Annual or perennial, hispid, spreading herbs, belonging to the Natural Order BORAGINEÆ. *Leaves* alternate. *Racemes* terminal, elongated, bracteate; *Flowers* subsessile, yellow. *Corolla-tube* elongated; lobes 5, distinct imbricate in bud. *Stamens* 5, dimorphic, in some flowers the stamens are below the mouth of the corolla, and the style protruding, in others protruding and style short. *Nuts* on a flat or nearly flat receptacle, scar basal, large, flat, or but little hollowed out, shortly produced up the inner surface, without a prominent margin.

There are in all some 12 species, of which 4 occur in India, but confined to the Panjáb, Kashmír, and Western Thibet.

1411

Arnebia tibetana, *Kurz; Fl. Br. Ind., IV., 176.*

Vern.—*Dimok*, BHOTI (**Aitchison**).

Habitat.—A native of North Kashmír and Western Thibet, altitude 7,000 to 12,000 feet; frequent.

MEDICINE. Root. 1412

Medicine.—§ "The scaly bark of the root-stock is employed as a dye and medicine for cough by the Bhotis of Ladak." (*Surgeon-Major J. E. T. Aitchison, Simla.*)

1413

Arnebia, sp.

Dr. Dymock informs me this root is imported into Bombay from Afghánistan, and used as a substitute for **Alkanet,** which see.

ARNICA, *Linn.; Gen. Pl., I. 440.*

1414

Arnica montana, *L.; DC. Prod., VI., 317;* COMPOSITÆ.

ARNICA.

Habitat.—Native of Western and Central Europe.

MEDICINE. 1415

Medicine.—Imported into India, being officinal in the *Pharmacopœia.* Used internally as a stimulant and externally as a sedative and resolvent. In British practice its use is limited to the application of the tincture to sprains, &c.

1416

Chemical Note.—§ "**Garrod's** experiments on the use of Tincture of Arnica for bruises indicate that the tincture has no more power in expediting the recovery of the skin to its normal condition than spirit of the same strength. The plant contains a bitter non-crystalline glucoside, *arnicin.*" (*Surgeon C. J. H. Warden, Prof. of Chemistry, Calcutta.*)

Arnotto, the seeds of **Bixa Orellana,** *Linn.*, which see.

1417

ARRACACIA, *Baner.; Gen. Pl., I., 884* (ARRACACHA, *DC. Prod., IV., 243*).

A genus of perennial herbs, containing some 12 species, belonging to the Natural Order UMBELLIFERÆ. They are natives of both Andean and South-West America, one species being cultivated in most warm temperate regions, forming an important article of food in Mexico.

Underground part thickened, tuberous, edible. *Leaves* pinnate or decompound; *segments* dentate or pinnatifid. *Umbels* compound; *involucral bracts* foliaceous, 1 or 0; *bracteoles* many, rarely foliaceous, entire. *Flowers* white, nearly allied to those of Conium. *Sepals* dentate, small. *Petals* subentire, point inflexed, broad or ovate. *Sylopods* conical, undulate at the margin. *Fruit* ovoid or ovoid-oblong, often pointed at the apex, compressed at right angles to the septum, and more or less constricted at the commissure. *Pericarp* transversely subterete or 5-gonal; *primary ridges* little or scarcely prominent, sometimes unequal. *Vitæ* many, often irregular or unequally confluent. *Carpophore* 2-partite. *Seeds* concave on the face, sulcate or involute.

The generic name is derived from the South American name *Arracacha.*

1418

Arracacia esculenta, *DC., Prod., IV., 244.*

THE ARRACACHA; PERUVIAN CARROT; APID, *Sp.*

Syn.—A. XANTHORRHIZA, *Bancroft.*

Habitat.—Supposed to be a native of the elevated regions of equatorial America, Pasto, and New Granada. It is now, however, met with in cultivation over most warm temperate parts of America, where the weather is free from extremes of cold or dry summer heat. Introduced into Jamaica; although many experiments have been made, it has hitherto failed in Europe. One or two plants are alive at the present day in India, but its introduction as a food-supply seems exceedingly doubtful.

Food.—**M. De Candolle**, in his *L'Origin. Cult. Pl.*, gives some interesting information regarding this plant. He says that the tuber compares well with potato, and the fecula is regarded as lighter and more pleasant to the taste. The lateral suckers which it throws out are used for propagation, and they are also more esteemed as an article of food than the central stem.

FOOD. Tubers. 1419

Fecula. 1420

In 1879 the Government of India, Revenue and Agriculture Department, obtained, through the Secretary of State, some plants of the *Arracacha*, but on reaching Calcutta only two were found alive. After a few days these were sent to Sikkim. On the way one died, and the other was retained at Mengpú (near Darjíling), but died in a few weeks.

Shortly after, the Revenue and Agriculture Department obtained seeds, which were sown in the Chajuri garden at Mussourie by **Mr. Duthie**, the Superintendent of the Saharanpore and Mussourie Botanical Gardens. Accounts of this trial have been given in the Annual Reports of these gardens, and been reproduced in various publications. Up to date, the introduction of *Arracacha* into India has failed. The following extracts from well-known authors may, however, prove useful to persons desirous of prosecuting experiments in India with this most useful plant:—

1421

"The root of this plant, for the sake of which it is cultivated, is a fleshy body, not unlike a parsnip in size and form, but more blunt, tender when boiled, and nutritious, with a flavour between the parsnip and a roasted chestnut. **Dr. Bancroft** compares it to a mixture of parsnip and potato. A fæcula, analogous to arrowroot, is obtained from it, by rasping in water, as starch is from the potato. It yields a large produce; according to **Boussingault**, as much as sixteen tons per acre, from land that will not bring more than nine or ten tons of potatoes. It requires deep rich soil, and is renewed annually by off-sets from the crown of the roots, which the Spaniards call its sons (Hijos). Otherwise its cultivation is similar to that of the potato.

"Several attempts have been made from time to time to introduce the arracacha into field culture in Europe; but it will not yield even to garden management, and all the experiments with it in Europe have terminated unsuccessfully. This seems to arise from the peculiar climate which it requires, and to which we have nothing analogous, unless in the south-west of Ireland. The mean temperature of the arracacha country is said by **Mr. Goudot**, who lived there for many years, to range from 64° to 82°, there being neither frost nor cold weather, nor dry summer heats, but an even damp climate. We are, therefore, unable to fulfil the first conditions demanded for this plant in field culture; and this circumstance, together with a great difficulty in preserving the roots through a winter, opposes an apparently insuperable bar to its introduction as a rival to our other eatable roots." (*Morton's Cyclopædia of Agriculture, 108.*)

1422

In a communication to Sir Joseph Hooker, **Mr. Henry Birchall** of Bogota says: "About 6,000 feet is nearly the upper level of any extensive cultivation of this plant, though it produces at points a good deal higher. It is rather difficult to obtain the seed, as from habit the peons invariably pull up the flowering plants, as they do not produce the edible root. I have several times missed getting the seed by the stupidity of the men who weeded the plantations.

FOOD.

"As regards cultivating from seed, my own experience is *nil;* but my neighbours assure me that by repeated replanting of the young plants at last the roots are developed.

"When this is attained the plant throws out a multitude of shoots from the crown. These being broken off are prepared by slicing the base neatly and then putting them in a hole dibbled about 5 or 6 inches deep, and require no further care than ordinary weeding, for which the rows and plants should be 3 feet apart.

"In our climate the root comes to perfection in eight to ten months, and the weight of a good specimen will be 8 to 10 lbs. No doubt if scientifically cultivated, and in properly loosened soil, much larger roots would be obtained. We do not even plough, but stick the seed in immediately after burning off the forest or the brushwood, as the case may be. It is cheaper to take new ground than to cultivate properly the old, as we have no command of skilled labour or good apparatus." (*Kew Reports, 1879, pp. 31, 32.*)

1423

"**Mr. Duthie** reports from Saharanpur, 30th May 1883: 'Of this valuable South American vegetable there are a few plants still left, and they are in a fairly healthy condition. They do not, however, appear to have formed to any extent the characteristic tubers which constitute the edible portion of the plants so highly valued in its native country. **M. DeCandolle,** in his recently published work on the origin of cultivated plants, observes that this vegetable bears comparison with potato, and yields a starch which is lighter and more agreeable. It has been tried in England and in several parts of Europe, but without success, the climate being evidently too damp for it. I intend to give it a trial at Arnigádh, where it will at any rate have a better chance of being looked after than it had at Chajuri.'

"**Mr. Morris,** the Director of Public Gardens and Plantations, Jamaica, has communicated to *The Planter's Gazette* for October 16, 1882, an important note on the little-known cultivation of this esculent in Jamaica. He states that it was introduced into the island in 1882 by **Dr. Bancroft**; it flourishes best in the Blue Mountain districts at elevations between 2,500 and 5,000 feet, with mean annual temperatures of 72° and 65° Fahr. respectively, and a mean annual rainfall of 100 inches.

"**Mr. Morris** adds: 'I believe the Arracacha is a most valuable food plant; and for my own part, I not only like it, but find that it becomes more palatable and desirable the longer it is used. If the natives of India take to it as an article of food, I can conceive nothing more likely to flourish in the hill districts, and to afford, with little labour, the means of sustaining life under adverse circumstances." (*Kew Reports, 1882, p. 17.*)

For further information, see *Botanical Magazine, t. 3092; Comptes Rendus, Nov. 24, 1845;* and *The Gardener's Chronicle, 1846, p. 235,* and *1848, p. 491. Grisebach's Flora of British West India Islands.*

1424

Arracacia moschata, *DC. Prod., IV., 244.*

THE SACCARACHA.

Habitat.—The species is said to frequent colder regions than the preceding, ascending in America to altitudes 8,400 feet. It smells like musk, hence the specific name; apparently this has not been introduced into Europe nor into India. The tubers are edible.

Arrowroot, see **Maranta arundinacea,** *Linn.;* SCITAMINEÆ.

ARSENIC.

Arsenic, White, or Arsenicum Album.

1425

Vern.—*Sanbul-khár, saféd-sanbul, sankhyá-sanbul, sankhyá,* HIND., DUK.; *Sankha visha, dárumuch, sámbala kshára,* SANS.; *Sanka, sámba lakshára, sammal-khár,* BENG.; *Somal, somal-khár,* GUJ., MAR.; *Vellai-páshánam,* TAM.; *Tella-páshánam, shenku-pashanam,* TEL.; *Vellap-páshánam,* MAL.; *Pháshána,* KAN.; *Shúk, turábul-hálik, sammulfár,* ARAB.; *Marge-mósh* or *marg-mósh,* PERS.; *Sudu-pásánam,* SINGH.

The word *sanbul* is used for white arsenic in the Deccan, but is also used for **Nardostachys Jatamansi** (Indian spikenard) in Arabic and Persian. (*Moodeen Sheriff.*)

§ "The Arabic name is *sammul-fár*, or the rat-poison, by which name it is sold in Muscat. The Indian names *sambul khár* and *sámbala kshára* are evidently corruptions of the Arabic name." (*Surgeon-Major C. T. Peters, M.B., South Afghánistan.*)

DYE & TAN. 1426

Dye and Tan.—"Arseniate of potash is used for preserving hides. Crude white arsenic is used as veterinary medicine in Burma." (*Prof. Romanis, Rangoon.*)

1427

Agricultural and Industrial Uses.—Five pounds of arsenic and one pound of soda are to be boiled in five gallons of water, until the arsenic is dissolved. To one measure of this solution, 160 measures of water are to be added, and the liquid thus formed is recommended by an American entomologist to be sprinkled upon plants infested by worms. The *Tropical Agriculturist, Vol. I., p. 602,* adds: ["*Query.*—Application of this solution for the destruction of grub. It might kill them without injuring the roots." (*Ed. T. A.*)] "Largely imported here for agricultural purposes." (*Dr. Dymock, Bombay.*)

MEDICINE. 1428

Medicine.—It is not necessary to discuss the imported and specially prepared European Arsenic compounds and drugs, since these may be found in the *Pharmacopœia;* the following notes should be understood to refer mainly to the indigenous drug. In the *Indian Pharmacopœia,* this substance is described as alterative, tonic, antiperiodic; in large doses powerfully poisonous. It has been used with much success in ague, neuralgia, and spasmodic affections, and in chronic skin diseases, including leprosy. In chronic rheumatism, cancer, uterine congestion, menorrhagia, snake-bite, and chronic catarrhal affections, it has proved an effectual remedy. (*Pharm. Ind.*)

It can be obtained pure in most bazars; is largely used at all dispensaries. Babu Rakaldas Ghose read a valuable paper before the Calcutta Medical Society, in which he stated that in obstinate cases of chronic intermittent fever which is seldom benefited by quinine, he found arsenic to do much good." (*Indian Medical Gazette, May 1881.*)

§ "Useful in ague in doses of $\frac{1}{30}$ grain." (*Assistant Surgeon Shib Chunder Bhuttacharji, Chanda, Central Provinces.*) "It is also considered as an aphrodisiac." (*Surgeon Anund Chunder Mukerji, Noakhally.*) "Much used in dispensary practice in malarious fevers and skin diseases." (*Surgeon G. Price, Shahabad.*) "Antiperiodic, tonic, relieves hemicrania, when applied locally, useful in rheumatism." (*Surgeon W. Barren, Bhuj, Cutch.*) "Invaluable in splenitis or ague in conjunction with quinine." (*Surgeon-Major Henry David Cook, Calicut, Malabar.*)

"Very good in bad cases of anæmia." (*Surgeon H. D. Masani, Karachi.*)

1429

Chinese Arsenic.—"*Sin-shih;* Arsenious acid, also called *Pih-sin* and *Hung-pe.* Of the specimens which I have received, some are apparently a natural mineral, constituting a translucent, crystalline mass, varying in colour from pure white to a yellowish brown or grey. Other specimens have the aspect of the ordinary massive white arsenic of European commerce.

"*Tsze-hwang*, yellow sulphuret of arsenic; native orpiment, *Pun-tsaou*. It occurs in the province of Yunan; probably also in Burma, as it has been shipped in considerable quantities from Moulmein. **Ainslie** states that it is exported from China to India.

"Orpiment is resorted to by the Chinese in cases of ague, but compounded in a manner so absurd as to render the dose extremely uncertain or even a nonentity.

"*Heung-hwang*, Native Red Sulphuret of Arsenic; Realgar: *Hiûm hoâm*, Cleyer, *Med. Simp.*, No. 176. It is found in the province of Yunan, in the south of China, and has been exported in small quantities to London from Canton. Realgar is also sometimes imported into England from Bombay.

"Small shallow cups, elegantly carved out of this mineral, and often highly polished, are used by the Chinese for administering certain medicines; by which means, when the inner surface of the cup is, as sometimes happens, in a somewhat disintegrated condition, it is evident that a minute dose of arsenic may be administered." (*Hanbury's Science Papers, pp. 220, 221.*)

1430

ARTABOTRYS, *R. Brown; Gen. Pl., I., 24.*

A genus of sarmentose or scandent shrubs, belonging to the Natural Order ANONACEÆ, comprising some 15 species, 10 of which are met with in India.

Leaves shining. *Flowers* solitary or fascicled, usually forming woody hooked recurved branches (peduncles). *Sepals* 3, valvate. *Petals* 6, 2 seriate; *base* concave, connivent; *limb* spreading, flat, subterete or clavate. *Stamens* oblong or cuneate; *connective* truncate or produced; *anther-cells* dorsal. *Torus* flat or convex. *Ovaries* few or many; *style* oblong or columnar; *ovules* 2, erect collateral. *Ripe carpels* berried.

1431

Artabotrys odoratissimus, *R. Br.; Fl. Br. Ind., I., 54.*

Syn.—UNONA ODORATISSIMA and HAMATA, *Roxb.; Fl. Ind., ii., 666.*

Vern.—*Madmánti, madan-mast*, DUK.; *Vilayati-chámpa*, BOMB.; *Manoranjitam*, TAM.; *Phala-sampenga, sakala-phala-sampenga, manoranjitam*, TEL.; *Madura-káméshvari, manuranjitam*, MAL.; *Manoranjatam*, KAN. (MADRAS).

Moodeen Sheriff says (that the Vythians assign a narcotive property to the flowers of this plant, *Madan-mast-ké-phúl*); hence the meaning of the Dukhni names—*intoxicated*. He adds, however, that the name is applied to the rhizomes of a Curcuma and also to a species of Aconitum. **Dr. Dymock**i nforms us that *Madan-mast* is the name by which the tubers of a species of Amrphophallus are sold in Bombay.

Habitat.—Southern parts of the Western Peninsula and in Ceylon; cultivated throughout India.

MEDICINE. 1432 PERFUMERY. 1433

I can find no record of the uses of this plant. The odour of the strongly-scented flowers is closely allied to that of the Ilang-Ilang (**Cananga odorata**).

1434

A. suaveolens, *Blume; Fl. Br. Ind., I., 55.*

Vern.—*Durie carban*, INDIAN ARCHIPELAGO.

Habitat.—A large, woody climber, met with in the forests of the Eastern Peninsula, from Sylhet to Malacca.

Medicine.—**Blume** informs us that the leaves of this plant are used to prepare an aromatic infusion, whose good effects have been extolled in the treatment of cholera. The seeds also afford a scented oil. These same properties are by some authors attributed to **Cananga odorata**, and it would appear that this plant, along with the two species of **Artabotrys** mentioned, are used for the same purposes. I have not been able to learn, however, that the remedial virtues for which they are famed in the Malaya are known to the people of India. It does not appear that the natives of India extract even the oil, although they frequently wear the flowers on their *pugris* or to deck their hair.

MEDICINE. Leaves. 1435 Infusion. 1436 Seeds. 1437 OIL. 1438 PERFUMERY. Flowers. 1439

Artanthe elongata, *Miq.*, see **Piper angustifolium,** *Ruiz.*; PIPERACEÆ.

ARTEMISIA, *Linn.*; *Gen. Pl.*, *II.*, *435*.

1440

A genus of herbs or shrubs belonging to the Natural Order COMPOSITÆ. It contains about 150 species, chiefly inhabitants of the north temperate regions.

Strongly scented. *Leaves* alternate, entire, serrate or 1-3-pinnatisect. *Heads* small, solitary or fascicled, racemose or panicled, never corymbose, heterogamous or homogamous, disciform; *ray flowers* female, 1-seriate fertile, very slender, 2-3 toothed; *disk flowers* hermaphrodite, fertile or sterile; *limb* 5-fid. *Inflorescence* ovoid, subglobose or hemispheric; *bracts* few-seriate, outer short, margins scarious; *receptacle* flat or raised, naked or hirsute. *Anther-bases* obtuse, entire. *Style arms* of hermaphrodite flowers with truncate usually penicillate tips, often connate in the sterile flowers. *Achenes* very minute, ellipsoid, oblong or subobovoid, faintly striate; *papus* O. (*Fl. Br. Ind.*, *III.*, *321.*)

The generic name is the classical name for wormwood, Lat. Artemisia, Gr. *ἀρτεμισία*.

Artemisia Absinthium, *Linn.*; *Fl. Br. Ind.*, *III.*, *328*.

1441

THE ABSINTHE; WORMWOOD.

Syn.—ABSINTHIUM VULGARE, *Gærtn.*; A. OFFICINALE, *Lam.*

Vern.—*Afsanthín*, PERS. and ARAB.; *Vilâyati-afsantín*, HIND., DUK.

Habitat.—An aromatic, silky, hoary herb, met with in Kashmír, altitude 5,000 to 7,000 feet; distributed to North Asia, Afghánistan, and westward to the Atlantic.

Botanic Diagnosis.—This belongs to the section Absinthium, or the species of Artemisia with heterogamous heads, ray flowers (female and disk-flowers hermaphrodite), and with the receptacle covered with long hairs.

A perennial species with hoary pubescence, *stem* erect, angular and ribbed. *Leaves* ovate or obovate, unequally 2-3-pinnatifidly cut into spreading linear-lanceolate obtuse segments, hoary on both surfaces, radical and lower cauline leaves narrowed into winged petioles. *Heads* $\frac{1}{8}$-$\frac{1}{4}$ inch diameter, dedicelled, hemispheric, in drooping secund racemes, terminating the branches; outer *involucre-bracts* oblong, hoary, narrowly scarious, inner orbicular broadly scarious; *receptacular hairs* long, straight.

History.—Several species of Absinthium (*ἀψίνθιον*, apsinthion) are referred to by **Dioscorides**, one of which appears to be this plant. The passage, "He hath made me drunk with wormwood" (*Lamentations of Jeremiah, Chap. III.*, *15*), doubtless refers to this plant, and it seems very probable that the extract, which is so largely consumed in France at the present day, was the preparation which produced the intoxication.

1442

Oil.—Wormwood yields by distillation a dark green or yellow oil, having a strong odour of the plant and an acrid taste. It is isomeric with camphor, and has the specific gravity 0·972. According to **Dr. Wright** it is chiefly composed of *absinthol* ($C_{10}H_{16}O$), and, when heated with phosphorus, pentasulphide, or zinc chloride, it splits into cymene ($C_{10}H_{14}$) and a resinous substance. The colouring substance is the Azulene of **Piesse.**

OIL. 1443

In large doses this oil is a violent narcotic poison; in a man, ½ oz. caused insensibility, convulsions, foaming at the mouth, and a tendency to vomit.

MEDICINE. 1444

Medicine.—According to **Braconnot**, 100 parts of this plant yield 1·5 of the volatile oil, 30 of bitter extractive matter, 2·5 of a very bitter resin, and 5 of a green resin. The plant yields also a large quantity of ash, chiefly carbonate of potash, known from remote times as salt of wormwood. The whole herb is an aromatic tonic, and formerly enjoyed a high reputation in debility of the digestive organs. It was also regarded as an anthelmintic, but, as **Christison** remarked, 'through the caprice of fashion, it is neglected at the present day. It is now chiefly used as a domestic and family medicine in Europe, but before the discovery of cinchona it was largely used in intermittents. It exercises a powerful influence over the nervous system, and its tendency to produce headache and other nervous disorders is well known by travellers in Kashmír and Ladák, who suffer severely when marching through the extensive tracts of country covered with this plant.

Infusion. 1445 Decoction. 1446 Poultice. 1447

It is chiefly used in the form of an infusion either in water or spirit: dose one to two ounces. A decoction is less narcotic than the infusion. The herb is sometimes prescribed in the form of a poultice or fomentation as an antiseptic and discutient.

ABSINTHE. 1448

Absinthe.—A liqueur, largely consumed in France; consists essentially of an alcoholic solution of oil of wormwood, containing a little angelica, anise, and marjoram. If carefully prepared it is of a bright green colour.

The effects of this liqueur are peculiar, and possess features not manifested by alcoholic excess. This has been described as absinthism. By the French Act of 1872, the trade in this liqueur has been put under severe restrictions. Its use has been prohibited in the French Army and Navy. Its effects appear to be exhilarating, but the habitual use, or excessive use, brings on gradual diminution of the intellectual faculties, ending in delirium and death. These effects resemble those produced by the oil. Absurd and extravagant statements appear, however, to have been made regarding the injurious effects of absinthe, but it is an established fact that its abuse is even more dangerous than the excessive consumption of alcoholic drinks. (*Bent. & Trim.; U. S. Dispens.; Smith's Dict.; Flück. & Hanb., Pharm.; Royle's Mat. Med., Ed. Harley, &c.*)

1449

Artemisia Abrotanum, a garden plant, sometimes seen in India; is the Southern-wood, or Old Man, of English writers. It is a native of the south of Europe, and is a favourite on account of its stimulating aromatic odour.

1450

A. Dracunculus, *Linn.; Fl. Br. Ind., III., 321.*

Habitat.—Western Thibet, altitude 14,000 to 16,000 feet; Lahoul, Afghánistan, West Asia, and South and Middle Russia.

FOOD. 1451

Food.—Introduced into Europe as a cultivated pot-herb nearly 300 years ago. It is the Tarragon of English gardeners; used to flavour dishes. Although a native of India, it does not appear to be put to any economic use.

A. indica, *Willd.;* Syn. **for A. vulgaris,** *Linn.,* which see.

1452

A. maritima, *Linn.; Fl. Br. Ind., III., 323.*

WORM-SEED, LEVANT WORM-SEED OR SANTONICA, *Eng.;* SEMEN CONTRA, SEMECINE, BARBOTINE, *Fr.;* WURM-SAMEN, ZITWERSAMEN, *Ger.*

Syn.—Artemisia, sp., in *Pharm. Ind.*

Santonica.

ARTEMISIA maritima.

Vern.—*Shih, sariqún, afsantín-ul-bahr* (αψίνθιον), ARAB., PERS.; *Kiramání owa*, BOMB.

Habitat.—Western Himálaya, from Kashmír to Kumaon, altitude 7,000 to 19,000 feet, Western Thibet; abundant in salt plains, altitude 9,000 to 14,000 feet. Commercially obtained from Russia.

Botanic Diagnosis.—This species belongs to the section Seriphidium, *i.e.*, with homogamous heads; flowers all fertile, receptacle naked; it is the only Indian representative of this section.

An exceedingly variable plant, with erect or sometimes drooping flower-heads. The whole plant hoary or tomentose, shrubby below; *stems* erect or ascending, much branched from the base. *Leaves* ovate, 2-pinnatisect; *segments* small, spreading, linear, obtuse, upper simple linear. *Heads* 3-8-flowered, ovoid or oblong, sub-erect in spicate fascicles; *involucre-bracts* linear-oblong, outer herbaceous, tomentose, inner scarious acute, glabrous.

1453

TRADE SUPPLIES.—In European commerce there are two forms of Worm-seed—*1st*, Aleppo, Alexandria, or Levant Worm-seed; and *2nd*, Barbary Worm-seed. The former comes from Persia, Asia Minor, or other parts of the East, and the Barbary from Palestine and Arabia.

Fluckiger and **Hanbury**, in their invaluable *Pharmacographia*, say: "The drug, which consists of the minute, unopened flower-heads, is collected in large quantities, as we are informed by **Bjorklund** (1867), on the vast plains or steppes of the Kirghis, in the northern part of Turkestan. It was formerly gathered about Sarepta, a German colony in the Government of Saratov; but from direct information we have (1872) received, it appears to be obtained there no longer.

"The emporium of worm-seed is the fair of Nishnei-Novgorod (July 15th to August 27th), whence the drug is conveyed to Moscow, St. Petersburg, and Western Europe." In 1864, 11,400 cwts. were imported into St. Petersburg. **Dr. Dymock** says it is brought from Afghánistan and Persia to Bombay in considerable quantities; value R2½ to R3 per maund of 37½ lbs.

MEDICINE. 1454

Medicine.—The flower-heads of this plant are largely used for their anthelmintic, deobstruent, and stomachic tonic properties. In the form of a poultice it is used to relieve the pain caused through the stings of insects and poisonous bites. Santonine is chiefly employed in the treatment of round and thread worms. It has the peculiar property of sometimes causing objects to appear yellow to patients under its action.

Worm-seed of the Bombay market has been examined by **Hanbury**, who considers that it does not materially differ from the Russian drug, but is slightly shaggy and mixed with tomentose stalks. "Arabic and Persian writers on materia medica generally describe Worm-seed under the name *Shíḥ*, giving as synonyms *Saríqún* and *Afsantín-ul-baḥr*." "In Bombay *shíb* is the recognised Arabic name of the drug amongst the hakims, who prescribe it, in doses of from 2 to 3 drachms, as an anthelmintic, and as a deobstruent and stomachic tonic. In the form of a poultice they useit to relieve the pain caused by the bites of scorpions and venomous reptiles."

§ "I have constantly observed santonine to be inert if administered when the worm is young. [This was in Rangoon.]" (*Honorary Surgeon P. Kinsley, Chicacole, Madras Presidency.*)

"Useful in gleet." (*Surgeon H. D. Masani, Karachi.*) "It renders the urine of a deep yellow colour." (*Brigade Surgeon G. A. Watson, Allahabad.*) "I have never seen the optical effect from santonine, though I frequently use the drug; in fact, I have made special enquiry on this point." (*G. B.*)

1455

Chemical Composition.—§"The small seed-like flower-heads contain an essential oil, which has an odour resembling cajuput oil and camphor. The anthelmintic properties of the acids are due to a glucoside santonin. According to **Hesse**, santonin is an hydriac of a crystallizable acid, santoninic acid, which, when heated to 12° C., is resolved into santonin and water. When heated with an alkali, santonin is not resolvable into santonin and water. (*Pharmacographia.*)" (*Surgeon C. J. H. Warden, Calcutta.*)

Santonin. 1456

Santoninum or **Santonin** is a crystalline substance prepared from worm-seed, about 1·2 to 1·4 per cent. being obtained. It occurs in flattened, shining, prismatic crystals, not altered by air but readily affected by light, turning yellow. It is odourless and tasteless, becoming in the mouth ultimately bitter. It is nearly insoluble in cold water, soluble in 250 parts of boiling water or 40 parts of alcohol at 15° C., or in 3 parts of boiling alcohol. When heated it melts at 170° C., forming, if rapidly cooled, an amorphous mass. The formula given for santonin is $C_{15}H_{18}O_3$; the principle of its isolation consists in that while not an acid it is capable of combining with bases; *i.e.*, santonin boiled with milk of lime. This forms santoninate of calcium, a soluble substance. On the addition of hydrochloric acid, santoninic acid $C_{15}H_{20}O_4$ is precipitated, and parting with H_2O santonine is thus produced.

Santoninic Acid. 1457

Fatal consequences have repeatedly followed the administration of large doses of this drug, the symptoms being giddiness, mental apathy, great paleness and coldness of the surface of the body, vomiting, profuse sweating, trembling, mydriasis, and finally, loss of consciousness, and convulsions. All objects appear yellow or even green, and the urine has been observed coloured after 16 minutes. The dose for an adult is from 2-4 grains; for a child two years old ¼ to ½ grain. (*U. S. Dispens.; Flück. and Hanb., Pharm., &c.*).

1458

Artemisia parviflora, *Roxb.; Fl. Br. Ind., III., 323.*

Vern.—*Kanyúrts*, PB.; *Burmar, basna tashang* LADAK.

Habitat.—Common in the higher regions of North-West Himálaya, in Lahoul, and Ladák.

FODDER. 1459

Fodder.—Browsed by goats and sheep.

1460

A. persica, *Boiss.; Fl. Br. Ind., III., 327.*

Vern.—*Shih, sariqún, afsantín-ul-bahr*, ARAB. and PERS.; *Pardesi da wano*, GUJ.; *Dawánd*, MAR.

Habitat.—Collected by **Bellew** in Afghánistan; it is also found in Western Thibet, altitude 9,000 to 14,000 feet.

Botanic Diagnosis.—Receptacle puberulous.

MEDICINE. 1461

Medicine.—**Bellew** states that the plant is used as a tonic, febrifuge and vermifuge.

1462

A. sacrorum, *Ledeb.; Fl. Br. Ind., III., 326.*

Vern.—*Tatwen, munyú, niurtsi, jau, chúmbar, zbiir, búrnak*, PB.; *Tatwen, burmack*, LADAK.

Habitat.—Western Thibet, Kanáwár, and the Thibetan region of Kumaon; altitude 9,000 to 17,000 feet.

MEDICINE. 1463

Medicine.—Said to be given medicinally to horses in affections of the head.

1464

A. scoparia, *Waldst. & Kit.; Fl. Br. Ind., III., 323.*

Vern.—*Jhau, lasaj, píla jau, biur, king khak durunga, lawange, doná, marúa*, PB.; *Chúri saroj, danti*, BAZAR NAME.

Habitat.—Found in the Upper Gangetic plain, and westward to Sind and the Panjáb, Western Himálaya; from Kashmír to Lahoul, altitude 5,000 to 7,000 feet; Western Thibet, altitude 7,000 to 12,000 feet.

MEDICINE. Infusion. 1465

Medicine.—The branches appear to be officinal in the Panjáb. The smoke is considered good for burns, and the infusion is given as a purgative.

FODDER. 1466

Fodder.—Browsed on by cattle and sheep.

Artemisia Sieversiana, *Willd.; Fl. Br. Ind., III., 323.*

1467

Vern.—*Afsantín, dauna,* PERS., ARAB., and BOMB.

Habitat.—Western Himálaya from Kashmír to Lahoul, altitude 8,000 to 10,000 feet; Western Thibet, China, and Russia.

MEDICINE. 1468

Medicine.—A plant very similar to **A. Absinthium,** *Linn.* It is said to be cultivated, at Bandora near Bombay, by Christians, from whom the fresh herb reaches the market. The Bombay imports of the drug are from Persia. Hakims prescribe this medicine in hypochondriasis, jaundice, dropsy, gout, scurvy; being used as a tonic, deobstruent, febrifuge, and anthelmintic, also as an emmenagogue. It is applied externally as a discutient and antiseptic. (*Dymock, Mat. Med., W. Ind., 361.*)

A. sternutatoria, *Roxb.*, see Centipeda orbicularis, *Lour.;* COMPOSITÆ.

A. vulgaris, *Linn.; Fl. Br. Ind., III., 325.*

1469

INDIAN WORMWOOD; FLEA-BANE.

Syn.—A. INDICA, *Willd.*

Vern.—*Nágdouná, májtari, mastaru, dona,* HIND.; *Nágdoná,* BENG.; *Tataur, púujan, banjírú, chambra, úbúsha, tarkhá,* PB.; *Búi mádarán, afsuntín,* PB. BAZAR NAMES; *Surband,* MAR.; *Titapat,* NEPAL; *Nágadamani, granthiparni,* SANS.

Dr. Moodeen Sheriff supplies the following note regarding the vernacular names for the forms of this species: "In Madras the native names are referred to two sections: (*a*) A. VULGARIS—*Marzanjósh,* ARAB.; *Marzangósh,* PERS.; *Douná,* HIND., DUK.; *Mar-i-kurondu,* TAM.; *Davanamu,* SANS., TEL., KAN. (*b*) A. INDICA—*Más-patrí,* DUK.; *Machi-pattarí,* TAM., TEL., MALA., and KAN.; *Garanthiparni,* SANS.; *Afsantíne-hindí,* ARAB. and PERS.; *Walkotondu,* SINGH.; *Marvá* is not a Hindustani synonym for *Douná,* as is supposed in some books; it is the name for **Origanum Majorana.**

Habitat.—Throughout the mountain tracts of India, altitude 5,000 to 12,000 feet; on the West Himálaya, Khásia Hills, Mánipur, and the mountains of North Burma. "A gregarious shrub, coming up on old cultivations between 3,000 and 6,000 feet in the Sikkim Hills, and often covering large tracts of land, until killed down by the tree growth which succeeds it." (*Gamble.*) Mount Abú, the Western Gháts from the Konkan southwards to Ceylon. Distributed to temperate Europe and Asia, Siam, Java, &c.

Botanic Diagnosis.—A tall, shrubby plant, with hoary pubescence or tomentose. *Stems* leafy, paniculately branched. *Leaves* large, ovate, lobed lacinate or 1-2 pinnatipartite, white tomentose beneath, rarely hoary or green on both surfaces; *lobes* acute, irregularly serrate or lobulate, lower petioled, upper sessile or with stipule-like petioles, uppermost linear-lanceolate, entire. *Heads,* $\frac{1}{8}$ to $\frac{1}{6}$ inch long, ovoid or subglobose, clustered or seriate, subsecund, in short or long suberect or horizontal panicled racemes; *involucre-bracts* woolly or glabrate, outer small, herbaceous, inner almost wholly scarious. *Corolla* glabrous. **Sir J. D. Hooker** adds to the above description that he is unable to separate, even as varieties, the following plants referred to by Indian authors: **A. indica, A. dubia, A.**

myriantha, A. paniculata, A. leptostachya, A. grata, A. Roxburghiana (of Calcutta Botanic Gardens).

One or two of the forms of this species, along with **A. Absinthium,** constitute the officinal worm-wood.

MEDICINE. Infusion. 1470

Medicine.—It has stomachic and tonic properties, and is used as a febrifuge. **Dr. Wight** states that the leaves and tops are used in nervous and spasmodic affections connected with debility; also an infusion of them is given as a fomentation in ulcers. (*Illust., II., 92.*) It may be used as an inferior substitute for cinchona in intermittent fevers; it is also employed in dyspepsia, and as an anthelmintic, and in liver diseases.

Amongst the vernacular names of this plant has been included the Mohammedan word *afsantín,* and it seems probable that this is one of the sources of the remedy known by that generic name.

Moxa. 1471

Tinder. 1472

The term *Moxa* is applied to a cube of inflammable tinder ignited in contact with the skin in order to produce cauterization. For this purpose, in many parts of the world, the tomentum from the leaves of wormwood is often used. The origin of the word *moxa* is obscure, but the practice of burning to produce revulsion, thereby relieving deep-seated inflammation, has been practised both by savage and civilised nations. The Mánipuris, and also the Angámi Nagás, rub between the palms of the hands the withered leaves of this plant, as also of a species of Anaphalis, until the tissue crumbles away, leaving only the tomentum in the hand. This they preserve as a tinder, which they ignite by rubbing rapidly a round stick by means of a string, made to revolve upon a small ball of the tinder placed in a hole in the ground (or simply between the hands), until the friction of the point ignites the tinder. I could not discover that they used this tinder as a *moxa,* although I made careful enquiries.

§ "It is used as a tonic, anthelmintic, antispasmodic, and expectorant, especially in diseases of children. The native women believe that its presence protects children from being eaten by witches." (*Brigade Surgeon J. H. Thornton, Monghyr.*) "Expressed juice is applied by native practitioners to the head of young children for the prevention of convulsions." (*Brigade Surgeon S. M. Shircore, Moorshedabad.*) "In Hindú medicine it is given in skin diseases and foul ulcers as an alterative." (*Civil Medical Officer U. C. Dutt, Serampore.*) "Tonic and antiperiodic." (*Surgeon W. Barren, Bhuj, Cutch.*)

MANURE. 1473

Manure.—Its ashes when burnt are considered to give a good manure for cultivation.

DOMESTIC. 1474

Domestic Uses.—Wormwood is used to prevent moths and other insects from infesting clothes and furniture. It is used as a symbol of bitter calamity, and is frequently mentioned in the Bible.

ARTHROCNEMUM, *Moq.; Gen. Pl., III., 65.*

1475

Arthrocnemum indicum, *Moq. DC. Prod., XIII., 2, 151;* CHENOPODIACEÆ.

Syn.—SALICORNIA INDICA, *Willd.; Icon., 737, non R. Br.* (*Pharm. Ind.*).

Vern.—*Jadu-pálang,* BENG.; *Machola, ghuri,* BOMB.; *Chil,* GUJ.; *Máchul,* MAR.; *Úmari,* TAM; *Koyyapippili,* TEL.

Habitat.—A gregarious weed, met with in the Sunderbunds and along the Coromandel Coast, also at Bombay.

MEDICINE. Barilla. 1476

Medicine.—**Roxburgh** urges that the preparation of "fossil alkali" or barilla from this plant should be encouraged on the Coromandel Coast, but he does not inform us whether it was actually prepared. Compare with remarks under **Caroxylon, Salicornia,** and **Barilla.**

ARTOCARPEÆ; *Gen. Pl., III., 346.*

1477

An important tribe of URTICACEÆ, referred by many botanists to the Sub-order MOREÆ, which see. It may be said to be represented by the Banian or Fig and the Jack-fruit.

ARTOCARPUS, *Forst.; Gen. Pl., III., 376.*

1478

A genus of evergreen trees with milky sap, belonging to the Tribe ARTOCARPEÆ, of the Natural Order URTICACEÆ. It comprises some 40 species, inhabitants of tropical Asia, Ceylon, the Malaya, China, and of the Pacific Islands.

Leaves alternate, large, coriaceous, penniveined, entire or lobed; *stipules* lateral, often very large. *Flowers* small, very numerous, monœcious, male and females forming distinct axillary and solitary, pedunculate, globose or cylindrical heads. *Male flowers* with a perianth of 2-3-4, lobed or partite; *segments* obtuse, concave, imbricate. *Stamen* 1, erect; *anther* slightly exserted; *ovary* rudimentary. *Female flowers*, perianth tubular, ovoid-oblong or linear; *bases* fleshy and consolidated into the infrutescence, apices free, with a minute opening, often 3-4 dentate. *Ovary* erect, included within the amalgamated perianths, free from each other, 1-locular, rarely 2-3 locular, with a solitary pendulous ovule; *style* terminal or lateral, simple or 2-3-fid; *stigma* various. *Nuts* enclosed within the persistent and consolidated perianth-tubes, forming the so-called flakes of the syncarpium. *Seeds* pendulous; *testa* membranous; *albumen* none; *embryo* erect or incurved; *cotyledons* equal, thick, fleshy or unequal; *radicle* short, superior.

The generic name is derived from ἄρτος, bread, and καρπός, fruit, in allusion to the species popularly known as the Bread-fruit tree.

Artocarpus Chaplasha, *Roxb., Fl. Ind., Ed. C.B.C., 634;* URTICACEÆ.

1479

Vern.—*Chaplash, chaplis*, BENG.; *Lut-ter*, NEPAL; *Chram*, GARO; *Sam*, ASS.; *Cham*, CACHAR; *Pani, toponi*, MAGH.; *Toungpeingnai, toungpeinné*, BURM.; *Kaita-dà*, AND.

Habitat.—A lofty, deciduous tree, met with in Eastern Bengal, Burma, and the Andaman Islands.

Botanic Diagnosis.—A tall, erect, straight, growing tree. *Leaves* in young tree pinnatifid, when old obovate, entire or remotely dentate; *stipules* long, axillary, caducous. Male and female *heads* on long peduncles. *Fruit* spherical.

Nearly allied to **A. Lakoocha**, with which compare.

CAOUTCHOUC 1480

Caoutchouc.—**Kurz** remarks that in Burma it yields a tenacious milky caoutchouc.

TIMBER. 1481

Structure of the Wood.—Yellow to brown, moderately hard, even-grained, rough, durable, seasons well. It seems to get harder and heavier as it gets older; two specimens from the Andaman Islands, cut in 1866 and stored since then in Calcutta, give respectively 46 and 52 lbs., and **Skinner** gives 63 lbs., but this is probably a mistake. (*Gamble.*)

It is much used for canoes; in Sikkim and Assam for planking, tea-boxes, and furniture. **Roxburgh** says that this wood is regarded as "superior to every other sort, particularly when employed under water."

A. hirsuta, *Lamk.; Brandis, For. Fl., 426.*

1482

Vern.—*Pát-phanas, rán-phanas*, MAR.; *Ayni, anjalli* or *angeli, aiyanepela*, TAM.; *Aini, ansjeni, ansjeli*, MAL.; *Hebalsu, heb, halasu, hesswa, hessain*, KAN

Habitat.—A lofty, evergreen tree of the forests of the Western Ghâts, ascending to 4,000 feet.

Botanic Diagnosis.—Young *shoots* hirsute; *leaves* alternate; *petiole* ovate-obtuse, entire, somewhat hairy underneath, particularly on the veins, 6-7 by 4-5 inches; *stipules* hirsute on the outside, lanceolate. *Male heads* cylindrical, pendulous, *female* globular and erect, both occurring in pairs in the axils of the leaves or leaf-scars.

CEMENT. 1483

The Juice as a Cement.—§"The concreted juice forms a waxy, tough, light-brown substance, which, when melted, is used as a cement to join broken earthenware and stoneware." (*Surgeon-Major W. Dymock, Bombay.*)

FOOD. Fruit. 1484

Food.—Produces a fruit, the size of a large orange, which contains a pulpy substance much relished by the natives.

TIMBER. 1485

Structure of the Wood.—Wood hard to very hard, yellowish brown, durable, seasons well. Weight about 35 lbs. per cubic foot.

Much used on the western coast for house and ship building, furniture, and other purposes. It "is a very large and handsome evergreen tree, whose massive trunk occasionally rises straight and clean-stemmed for 150 feet. It yields the Anjeli wood of commerce, and is equally valuable for ship and house building. A seasoned cubic foot weighs 40 lbs." "This is an excellent wood for making boxes, buildings and furniture generally, and, like its congeners, as ornamental as it is useful." (*Tropical Agriculturist, II., p. 883.*)

1486

Artocarpus incisa, *Linn.*

THE BREAD-FRUIT TREE of the South Sea Islands.

Vern.—*Rata-del* (or the foreign *del*), SINGH.

Habitat.—"Cultivated in South India, Ceylon, and Burma. Succeeds well in Bombay; a few good trees may be seen in the garden attached to the Albert and Victoria Museum. In Bengal the winter proves too severe for its growth." (*Roxburgh.*) Cultivated in the islands of the Asiatic Archipelago from Sumatra to the Marquesas Islands, when Europeans first began to visit them. It is probably a native of Java, Amboyna, and the neighbouring islands. (*DeCandolle, L'Origin. Cult. Pl., 238.*)

FRUIT. 1487

Recently efforts have been, and are continuing to be made, by the Agri.-Horticultural Society of Madras to establish and distribute the edible (seedless) varieties of this tree in Southern India. A large tree already flourishing in the garden of the late Sheriff of Madras has afforded the basis for propagation; and a stock of good sorts has been procured from Ceylon. The tree propagates by suckers or offshoots from the base of the stem rather than by cuttings. When the fruit is of good quality it rarely produces fertile seeds.

Botanic Diagnosis.—*Leaves* pinnatifid. *Male-heads* cylindric; *female* terminal, round. (*Roxburgh.*)

GUM. 1488

Gum.—Yields gum.

1489

A. integrifolia, *Linn.*

THE JACK-FRUIT TREE. (A name said by *DeCandolle, in L'Orig. Cult. Pl.*, to be derived from a common Indian name, *jaca* or *tsjaka.*)

Vern.—*Kánthál, katol, káthal, chakki, panasa, panas,* HIND.; *Kánthál, káthál,* BENG.; *Knáthál,* ASS.; *Kanthar,* SANTAL; *Poros,* KOL; *Panasa,* URIYA,; *Phanas,* MAR., BOMB.; *Pilá, pilápazham,* TAM.; *Panasa-pandu, pansa, véru-panasa,* TEL.; *Halsu, heb-helsu, halsina,* KAN.; *Teprong,* GARO; *Panasa,* SANS.; *Peingnai, pienné,* BURM.; *Cos,* SINGH.

Habitat.—A large tree, cultivated throughout India and Burma, except in the north. Supposed to be wild in the mountain forests of the Western Ghâts, ascending to 4,000 feet. (*Beddome; Wight.*) Its dome of dark foliage with the stem burdened with monster fruits (often 2½ feet in length) is perhaps one of the most characteristic features of the Indian village surroundings.

"It is both cultivated and found wild in the evergreen Sahyadri forests." (*Bombay Gazetteer, Vol. XV., 62.*) "It is also wild in the Eastern Ghâts." (*Central Provinces Gazetteer, p. 503.*) "Grows freely on the Eastern Ghâts, Rakhphalli taluka." (*Central Provinces Gazetteer, p. 503.*)

CULTIVATION.—"A pit is dug and filled with cowdung, and in this the jack seed is inserted in June or July. The cost of cultivation is *nil*, whilst the profits vary from 4 annas to R2 per tree, realized by the sale of the fruit." (*McCann's Dyes and Tans of Bengal.*) "The value of a jack tree in Surat is about R1-5 per annum." (*Bombay Gazetteer, II., 41.*) **DeCandolle** thinks that the cultivation of this tree is "probably not earlier than the Christian era. It was introduced into Jamaica by Admiral Rodney in 1782. It has also been introduced into Brazil, Mauritius, &c." (*L'Orig. Cult. Pl., 240.*)

1490

Botanic Diagnosis.—Glabrous or the young shoots with short stiff hairs, branchlets with annular raised lines, the scars of the caducous stipules. *Leaves* coriaceous, smooth, shining above, rough beneath, elliptic or obovate obtuse, mid-rib prominent beneath, with 7-8 lateral nerves on either side, 4-8 inches; *stipules* large, with a broad amplexicaul base, caducous. *Fruit* large, hanging on short stalks, oblong, fleshy, with a thick cylindrical receptacle; rind muricated. *Seeds* reniform, oily.

Gum.—The bark yields a very dark-looking gum, with a resinous fracture, soluble in water. (*Atkinson's Gums and Resins.*) The juice is used as a valuable bird-lime and as a cement.

GUM. 1491

A writer in the *Indian Agriculturist* describes certain experiments with the caoutchouc from this plant. He says it is elastic, leathery, water-resisting, and capable of removing pencil-marks; but, as remarked by the Editor, although the order **Artocarpeæ** of course yields caoutchouc, it is still a question, which experiment alone can decide, whether rubber of sufficient economic value could be obtained from this species. Each fruit yields about two ounces of milk, from which, according to the writer in the *Indian Agriculturist*, a drachm and a half of the caoutchouc-like substance can be obtained.

CAOUTCHOUC 1492 Fruit. 1493

Dye.—The wood, or its sawdust, yields on boiling a decoction used as a yellow dye, to colour the Burmese priest's robes, and to some extent it is in requisition as an ordinary yellow dye in Madras and other parts of India, and in Java. It is fixed with alum, and often intensified by a little turmeric. With indigo it gives a green, said to be used in Malda. *Kanthal* yellow is often used in dyeing silk. "In the Midnapur district a red dye for home use is produced by boiling the juice of green jack-fruits with *Ach* root and lime. With this red dye, cloth and the jute used for tying bombs are dyed." "Both fruit and wood are used as dyes in Bengal. According to **Mr. Liotard's** memorandum, it would seem that a dye is extracted in Oudh from the bark; and **Balfour** mentions that a yellow dye is obtained from the roots in Sumatra." (*McCann, Dyes and Tans of Bengal.*)

DYE. 1494 Wood. 1495 Yellow. 1496 Green. 1497 Red. 1498

Fibre.—The bark yields a fibre. § "A fibre extracted from the bark was sent to the Paris Exhibition from Sandoway; a cordage fibre is also obtained from the bark in Kumaon." (*T. N. Mukarji, Revenue and Agriculture Department, Calcutta.*)

FIBRE. Bark. 1499

The Jack-fruit Tree.

MEDICINE. Juice. 1500 Leaves. 1501 Root. 1502 Fruit unripe. 1503 Fruit ripe. 1504

Medicine.—THE JUICE of the plant is applied externally to glandular swellings and abcesses to promote suppuration. The young LEAVES are used in skin diseases, and the ROOT internally in diarrhœa.

§ "This is an important article of food, both when green as well as when ripe. The seeds contain a quantity of starchy matter, which may be separated by drying and pounding them. The unripe fruit is astringent, the ripe laxative, but rather difficult to digest, although very nutritious. The juice of the plant is used to promote absorption of glandular swellings." (*Surgeon D. Basu, Faridpur, Bengal.*)

FOOD. Fruit. 1505 Spirit. 1506 Seeds. 1507 Flour. 1508

Food.—The large fruit obtained from this tree would be more correctly described as a fruitescence, since, like the pine apple, it is an aggregation of the fruits produced by an assemblage of flowers. The individual fruits are often spoken of as flakes; they each consist of a seed surrounded by a pulpy mass of luscious tissue having a strong odour. The external rough skin of the fruitescence is rejected, and the yellow pulpy mass which surrounds the seeds eaten by the uatives of India, and by them regarded as one of the best Indian fruits. It is seldom eaten by Europeans. The average size of the fruit is from 12 to 18 inches long by 6 to 8 inches in diameter. Each contains from 50 to 80 or more flakes, of a soft juicy and sweet substance, which, if fermented and distilled, yields an alcoholic beverage, with a strong odour and peculiar flavour.

The seed, when roasted, is eaten as an article of food, and is said to resemble chestnuts. When ground to flour it very much resembles the Kashmír Singara-nut flour. A writer in the *Indian Agriculturist* says: "I believe it contains a very large percentage of starch, and as such could be utilized in a variety of forms." "If these seeds be taken to weigh a third of an ounce each, one fruit will give us 30 ounces of flour, and 20 fruits, the produce of one tree, 600 ounces=37 lbs. of flour."

"The fruit weighs up to 60 lbs., and is much used by the people. The roasted seeds are not unlike chestnuts, and are in bad seasons often the only food of the poorest hill people." (*Bombay Gazetteer, Vol. XV., Pt. I., 62.*)

§ "The fruit when unripe is cut into small pieces and cooked in curry with shrimps. The seeds of the ripe fruit, when roasted in hot ashes, are very palatable and nutritious, and in taste resemble somewhat the Spanish chestnuts." (*Mr. L. Liotard, Calcutta.*)

"The strong unpleasant odour of the ripe jack-fruit is probably due to the presence of butyarate of ethel." (*Surgeon C. J. H. Warden, Calcutta.*)

TIMBER. 1509

Structure of the Wood.—Heartwood yellow or rich yellowish-brown, darkening on exposure, compact, even-grained, moderately hard, seasons well, and takes a fine polish. Weight about 40 lbs. per cubic foot..

It is largely used for carpentry, boxes, and furniture, and is exported to Europe for cabinet-work, turning, and brush-backs.

"The trunk grows to a great girth. The wood is yellow when cut, but gradually darkens. It becomes beautifully mottled with time, and takes as fine a polish as mahogany. A seasoned cubic foot weighs 42 lbs. It is used for building and for furniture." (*Bombay Gazetteer, Vol. XV., Pt. I., 62.*) Mr. Gibson says this is a useful firewood tree, found, in Thána, growing in salt-marshes. The jack-trees in Bengal attain a great girth, but are not very lofty. Planks 20 or 24 inches across are often sawn out of jack-tree bolls. An average tree ceases to yield fruit when it has reached a circumference of 9 feet, and may at that time be sold at R20. (*Indian Agriculturist.*) According to the *Tropical Agriculturist*, large jack-trees will sell for as much as R100 each; they are used for canoes.

§ "Jack-wood is yellow, hard, takes an excellent polish, is beautifully marked, and is one of the handsomest furniture woods found in the country." (*Surgeon C. J. H. Warden, Prof. of Chemistry, Calcutta.*)

Sacred.—**Artocarpus integrifolia** is often seen on Buddhistic sculptures. In some instances it appears to have been mistaken for the custard apple. (See remarks under **Anonaceæ.**) 1510

Artocarpus Lakoocha, *Roxb.* 1511

Vern.—*Tiún, tiun dheu, daheo,* PB.; *Dahu, dhau, barhal, lakúch, dhává,* HIND.; *Lovi,* BOMB., DUK.; *Votamba,* MAR.; *Vonte,* KAN.; *Dháo,* KUMAON; *Dephal, dahu, dehua, lakúcha, mádár,* BENG.; *Dahu,* SANTAL, KOL.; *Dewa, chama, chamba,* ASS.; *Dawa,* CACHAR; *Barrár,* NEPAL; *Kamma régu, lakuchamu nakka-rénu,* TEL.; *Lakucha,* SANS.; *Myoukloukt, myauklót* or *mi-auk-tok,* BURM.; *Caunagona, etaheraliya,* SINGH.

Habitat.—A large tree met with in the outer hills of Kumaon, Sikkim, Eastern Bengal, Burma, and in the evergreen forests of the Western Ghâts and Ceylon.

Botanic Diagnosis.—Branchlets and under-side of the leaves with short, soft grey tomentum, not marked by the scars of the stipules as in **A. integrifolia.** *Leaves* coriaceous, oval or ovate-obtuse or shortly acuminate entire; *blade* 6-10 inches. *Flower-heads* globose, axillary, the male subsessile, the female shortly pedunculate. *Fruit* acid, of an irregular roundish shape, 3-4 inches in diameter, velvety yellow when ripe.

Caoutchouc.—A caoutchouc similar to that obtained from the preceding species is derived from this plant. CAOUTCHOUC 1512

Dye.—The root yields a yellow dye. Wood used in dyeing cloth yellow. (*Burm. Gaz., I., 138.*) DYE. 1513

Fibre.—A fibre is prepared from the bark; used for cordage. FIBRE. 1514

Food.—It flowers in March, and produces a fruit which is eaten by the natives. The male spadix is used in curry and also pickled. Mr. Mann says the bark is chewed in Assam as a substitute for betel-nuts. FOOD. 1515

The fruit is eaten in curries in Kanára (Bombay). (*Bombay Gazetteer, Vol. XV., Pt. I., 62.*)

Structure of the Wood.—Sapwood large, white, soft, perishable. Heartwood yellow, hard. It seasons well and takes a good polish. Weight 30 to 50 lbs. per cubic foot. TIMBER. 1516

Used for furniture and canoes.

A. nobilis, *Thw. Enum., Ceylon Pl., 262.* 1517

Vern.—*Del, aludel,* SINGH.

Habitat.—A large tree of Ceylon.

Food.—The seeds are roasted and eaten by the Singhalese. FOOD Fruit. 1518

Structure of the Wood.—Heartwood shining, moderately hard; pores filled with a white substance, giving the wood an elegant mottled appearance. TIMBER. 1519

Used for canoes and furniture.

ARUM, *Linn.; Gen. Pl., III., 967; Engl. in DC., Mono. Phaner., II., 580.* 1520

A genus of herbaceous plants, with tuberous corms, often edible, belonging to the Natural Order AROIDEÆ. The genus comprises some 20 species, inhabitants of Europe, the Mediterranean region and tropical Asia, extending from India to Afghánistan.

Leaves sagittate or hastate, base of the petiole sheathing. *Peduncles* most frequently solitary, short or long. *Spathe-tube* convolute; *blade* when opened out ovate or ovate-lanceolate; *spadix* sessile, shorter than the *spathe,* appendix

Aroids.

naked, frequently stalked and cylindrical, rarely clavate. *Inflorescence* monœcious, perianth none. *Female flowers* below forming a cylindrical mass, separated from the male by a tuft of hair like neuter flowers, which blend above into the male condition. *Stamens* 3-4; *anthers* sessile, opposite or sub-opposite, obovoid, dehiscing by a slit towards the apex, connective more or less prolonged; pollen vermiform. *Ovary* oblong-obtuse, 1-locular; *stigma* sessile; *ovules* 6 or many; orthotropous, erect; *funiculus* short; *placenta* parietal 2-3-seriate; *micropyle* superior. *Fruit* an obovoid many-seeded berry.

The generic name is derived directly from ἀρον, Gr., aros, aron, arum, Latin=the Cuckoo-pint. Some authors regard the Greek word, however, as derived from the Hebrew *Or*, fire or flame, probably referring to the burning or acrid character of the plants, or to the scarlet head of fruits.

Arum campanulatum, *Roxb.*; Syn. for **Amorphophallus campanulatus,** *Bl.*

A. Colocasia, *Willd.*; Syn. for **Colocasia antiquorum,** *Schott.*, which see.

A. cucullatum, *Lour.*; Syn. for **Alocasia cucullata,** *Schott.*, which see.

A. curvatum, *Roxb.*; Syn. for **Arisæma curvatum,** *Kunth*, which see.

A. cuspidatum, *Roxb.*; Syn. for **Arisæma cuspidatum,** *Engl.*, which see.

A. divaricatum, *Linn.*; Syn. for **Typhonium divaricatum,** *Dec.*, which see.

A. flagelliforme, *Lodd.*; Syn. for **Typhonium cuspidatum,** *Dec.*, which see.

A. fornicatum, *Roxb.*; Syn. for **Alocasia furnicata,** *Schott.*, which see.

A. gracile, *Roxb.*; Syn. for **Typhonium gracile,** *Schott.*, which see.

1521

A. Griffithii, *Schott.; N. E. Brown, Linnæan Soc. Jour., XVIII., 257.*

Habitat.—Afghánistan. (*Griffith; Aitchison, Kuram Mon Valley.*)

A. indicum, *Roxb.*; Syn. for **Alocasia indica,** *Schott.; Engler in DC. Phaner., II., 501.*

A. lyratum, *Roxb.*; Syn. for **Amorphophallus lyratus,** which see.

A. margaretiferum, *Roxb.; Dr. Dymock's Mat. Med., West. Ind., 664.* Syn. for **Plesmonium margaretiferum,** *Schott.; Engler, DC. Mono Phaner., p. 303*, which see.

A. montana, *Roxb.*; Syn. for **Alocasia montana,** *Schott.*, which see.

A. nymphæifolium, *Roxb.*; Syn. for **Colocasia Antiquorum,** *Schott.* VAR. **nymphæifolia,** which see.

A. odorum, *Roxb.*; Syn. for **Alocasia odora,** *C. Koch*, which see.

A. orixense, *Roxb.*; Syn. for **Typhonium trilobatum,** *Schott.*, which see.

A. rapiforme, *Roxb.*; Syn. for **Alocasia rapiformis,** *Schott.*, which see.

A. sessiliflorum, *Roxb.; N. E. Brown, Linnæan Soc. Jour., XVIII., 256; Wight, Ic., t. 800;* the *loth* of *Dr. Dymock's Materia Medica of Western India*, a Syn. for **Sauromatum sessiliflorum,** *Kunth*, which see.

A. speciosum, *Wall.; Stewart's Pb. Pl.*, Syn. for **Arisæma speciosum,** *Mart.; N. E. Brown, Linn. Soc. Jour., XVIII., 249*, which see.

Arum silvaticum, *Roxb.; Syn.* for **Synantheris silvatica,** *Schott.,* which see.

A. trilobatum, *Willd.,* in *Roxb.;* Syn. for **Typhonium divaricatum,** *DC.*

A. tortuosum, *Wall.; Stewart's Pb. Pl.,* Syn. for **Airsæma tortuosum,** *Schott.; Engler, DC. Mono. Phaner, II., 545,* which see.

A. viviparum, *Roxb.;* Syn. for **Remusatia vivipara,** *Schott.*

ARUNDINARIA, *Mich.; Gen. Pl., III., 1207.*

1522

(*Compare with* **Bamboo** *and also the definition given under* **Bambuseæ.**)

Arundinaria falcata, *Nees; Duthie's List of Grasses, 46;* GRAMINEÆ.

1523

HIMÁLAYAN BAMBOO.

Vern.—*Nirgal, nigál, ringál, nagre, narri, garri, gero, narkat, narqual,* HIND.; *Spiág, gurwa, spikso, pitso,* KANAWAR; *Kwei,* THIBET; *Prong,* N.-W. P.; *Titinigala,* NEPAL; *Prongnok,* LEPCHA.

Habitat.—Met with from the Ravi to Bhútán, above 4,500 feet in altitude in the western, but descending nearly to the plains in the eastern, Himálaya.

Fibre.—The leaves are used for roofing and baskets.

FIBRE. 1524

Structure of the Stems.—Six to 10 feet high, strong, annual; used for roofing and baskets.

Stem. 1525

A. Falconeri, *Benth. & Hook. f.; Brandis, For. Fl., 563.*

1526

Syn.—THAMNOCALAMUS FALCONERI, *H.f.*

Vern.—*Ringál.*

Habitat.—Deoban to Simla (*Brand.*); Kumaon, 7,000 to 8,500 feet, (S. and W.) Nepál (*Wall.*); (*Duthie's List of Grasses, 46.*)

A. Griffithiana, *Munro.*

1527

Habitat.—Met with in the Khásia Hills.

Botanic Diagnosis.—Stems 4 to 6 feet high; internodes woolly, sometimes prickly.

A. Hookeriana, *Munro.*

1528

Vern.—*Praong, prong,* LEPCHA; *Singhani,* NEPAL.

Habitat.—A bamboo, with stems 12 to 15 feet in height, common about Dumsong. Frequent in Sikkim at 4,000 to 7,000 feet in altitude (*Gamble.*)

Food.—The seeds are edible.

FOOD. 1529

A. intermedia, *Munro.*

1530

Habitat.—Sikkim, from 7,000 to 8,000 feet; stem from about 6 to 8 feet.

A. khasiana, *Munro.*

1531

Vern.—*Namlang,* KHASIA.

Habitat.—Met with in Khásia Hills; stem from 8 to 12 feet.

A. racemosa, *Munro.*

1532

Vern.—*Pummoon,* LEPCHA; *Pat-hioo, maling,* NEPAL; *Myooma,* BHUTIA.

Habitat.—A bamboo, 2 to 4 feet high, with blueish rough internodes, occurring in Sikkim and Nepal above 6,000 feet.

FIBRE. Mats. 1533 FODDER. 1534

Fibre and Timber.—It is extensively used for making mats and roofing, &c. An important fodder plant.

1535

Arundinaria spathiflora, *Trin.*

Syn.—THAMNOCALAMUS SPATHIFLORUS, *Munro; Brandis, For. Fl., 635.*
Vern.—*Ringál.*

Habitat.—Hattú, near Simla, 8,400 feet; Deoban range, 8,000 feet (*Brand.*); Garhwál, 8,500 feet; Kumaon (*Falc.*), Nepal. (*Duthie's List of Grasses, p. 46.*)

1536

A. Wightiana, *Nees.*

THE NILGHIRI AND WESTERN GHÂT HILL BAMBOO.

Vern.—*Chevari,* BOMB.; *Dalz. & Gibs., Bomb. Fl., 299; Brandis, 563.*

TIMBER. 1537

Timber.—A most useful Bamboo, yielding the walking-sticks of Mahableshwar: young stems are also eaten.

ARUNDO, *Linn.; Gen. Pl., III., 1179.*

1538

Arundo Donax, *Linn.;* GRAMINEÆ.

Syn.—DONAX ARUNDINACEUS, *Beauv.*; A. BENGALENSIS, *Retz.*
Vern.—*Sukna,* HOSHIARPUR.

Habitat.—Plains of the Panjáb and North-Western Provinces, also the lower Himálaya. (*Duthie's List of Grasses, 35.*)

A. Epigejos, *Linn.;* Syn. for **Calamagrostis Epigejos,** *Roth.,* which see.

A. Karka, *Roxb.;* Syn. for **Phragmites Roxburghii,** *Kunth,* which see.

1539

A. madagascariensis, *Kunth.*

Syn.—DONAX THOUARII, *Beauv.*

Habitat.—The Panjáb Himálaya, ascending to 8,000 feet in altitude.

1540

A. mauritanica, *Desf.*

Syn.—A. PLINIANA, *Tur.*; A. PLINII, *Viton*; PHRAGMITES GIGANTEA, *Gay.*
Habitat.—North-West Himálaya.

A. Phragmites, *Linn.;* Syn. for **Phraghmites communis,** *Trin.,* which see.

Asafœtida, see **Ferula Narthex,** *Boiss.;* UMBELLIFERÆ.

ASAGRÆA, *Lindl.;* (Reduced to **Schœnocaulon**) *Gen. Pl., III., 836.*

1541

Asagræa officinalis, *Lindl.;* LILIACEÆ.

CEVADILLA or SABADILLA.

Syn.—This plant has received various names,—VERATRUM OFFICINALIS, HELONIAS OFFICINALIS, and SCHŒNOCAULON OFFICINALE; but as it is not an Indian plant, it has been thought the more convenient course to leave it for the present under ASAGRÆA, its best known synonym.

Habitat.—A native of Mexico; imported into India.

MEDICINE. Veratria. 1542

Medicine.—Chiefly used in the preparation of the alkaloid Veratria, obtained from decoction of the seed: is useful externally to destroy pediculi. (See *U. S. Dispens., 15th Ed., 1252 and 1515.*)

Seeds. 1543

Chemical Composition.—§"Veratria, the active principle of sabadilla, is usually found in commerce as an amorphous, odourless, pale-grey powder, often containing a large percentage of resin. The taste is highly acrid and bitter, and when inhaled through the nostrils causes most painful and prolonged irritation. Veratria has been shown to be capable of existing in two isomeric modifications, the one soluble and the other insoluble in water. The alkaloid has been obtained in a crystalline form. Two other alkaloids have been isolated from the seeds, namely, *sabadilline* and *sabatrine*. Two volatile crystalline acids, *sabadillic* and *veratic* acids, are also contained in the seeds." (*Surgeon C. J. H. Warden, Calcutta.*)

ASARUM, *Linn.; Gen. Pl., III., 122.*

1544

A small genus belonging to the Natural Order ARISTOLOCHIACEÆ.

Leaves reniform. *Flowers* solitary, drooping bell-shaped 3-fid, appearing from between two opposite leaves. *Stamens* 12, inserted at the base of the style; *anthers* attached to the middle of the filaments. *Stigma* 6-lobed. *Capsule* 6-celled.

Asarum europœum, *Linn.*

1545

COMMON ASARABICA, ASARABACCA, or FOALFOOT, *Eng.*; CABARET or ASSARET, *Fr.*; HASELKRANT, *Ger.* (Balfour).

Vern.—*Taggar, tagar*, HIND.; *Upana*, SANS.; *Asárún*, ARAB.; *Mutriknujayvie* (?), TAM.; *Chepututaku* (?), TEL.; *Tagara*, BOMB., CUTCH.

Habitat.—Indigenous to temperate Europe and North Asia.

1546

History.—This plant has been used in medicine from very ancient times. **Ainslie**, in his *Materia Medica*, states: "The appellation *Asarún*, which has been given to this article by the Arabs and the Mahometan conquerors of India, **Moomina** informs us, was first bestowed on it by the Syrians, in whose country the plant at one time plentifully grew." All parts of the plant are acrid, but those used in medicine are the roots and the leaves, chiefly the latter.

1547

Description.—In the *U. S. Dispensatory* the root is described "as thick as a goose-quill, of a greyish colour, quadrangular, knotted and twisted, and sometimes furnished with radicles at each joint. It has a smell altogether analogous to that of pepper, an acrid taste, and affords a greyish powder." The leaves are "kidney-shaped, entire, somewhat hairy, of a shining deep green colour when fresh, nearly inodorous, with a taste slightly aromatic, bitter, acrid, and nauseous. Their powder is yellowish green."

1548

Chemistry.—It is further stated that in an analysis made by **Grager**, the root was found to contain "a liquid volatile oil, two concrete volatile substances, called respectively *asarum camphor* or *asarone* and *asarite*, a peculiar bitter principle called *asarin*, tannin, extractive, resin, starch, gluten, albumen, lignin, citric acid, and various salts." In the leaves he found asarin, tannin, extractive, chlorophyll, albumen, citric acid, and lignin.

MEDICINE.
Root.
1549
Leaves.
1550

Medicine.—Both the ROOT and the LEAVES were formerly much used in Europe as a powerful emetic and cathartic, the dose being 30 grains to a drachm, but as an emetic they have now been entirely superseded by ipecacuanha. They are now used only as an errhine. "One or two grains of the powdered root, snuffed up the nostrils, produce much irritation and a copious flow of mucus for several days until the desired effect is produced. The leaves are milder and generally preferred. They should be used in the quantity of three or four grains repeated every night until the desired effect is experienced. They have been strongly recommended in headache, chronic ophthalmia, and rheumatic and paralytic affections of the face, mouth, and throat, and are in great repute in Russia as a remedy for deranged state of health consequent on habits of intoxication." (*U. S. Dispens., 15th Ed., p. 1578.*) It is, however, not much used in India. **Ainslie** found that the "Tamil physicians occasionally prescribed the root as a powerful evacuant; they also employ the bruised and moistened leaves as an external application round the eyes in certain cases of ophthalmia." **Irvine**, in his *Mat. Med. of Patna*, says: "The dry plant, imported from Persia, is used as a stimulant. Dose ℈i." According to **O'Shaughnessy**, "the dried plant is sold in the Indian bazars under the name of *Asárún*. **Royle** states, however, that a hill plant, called *Tuggur*, is generally subsitituted for it." This is most probably **Valeriana Hardwickii**, for which *Asárún* and *Tagar* are Hindi names.

§ "Tonic and antiperiodic, applied locally to indolent ulcers." (*Surgeon W. Barren, Bhuj, Cutch.*)

1551

Asbestus or Asbestos.

Vern.—§ "*Shankha palita*, MAR. (literally wick made of shells)."

A fibrous mineral, now viewed as a variety of Hornblende, allied to augite, tremolite, and climolite, in which the proportion of alumina is less than usual. It contains a considerable quantity of magnesia, and is found in connection with serpentine. There are many varieties of asbestos, one in which the fibres are so long and flexible as to admit of being woven into cloth. This form is generally known as *Amianthus*.

MEDICINE. 1552 DOMESTIC. 1553

Medicine.—§ "Found in the Gokák Taluka in the Belgaum district in the Southern Maratha country, where it is used as an external application in ulcers, made into a paste, after rubbing it down with water, used especially in syphilitic ulcers." (*Surgeon-Major C. T. Peters, South Afghánistan.*) "Is obtainable in large quantities in the country to the south and west of the Kurum river, Afghánistan; used as medicine, and made into brooms and rough ropes, and padding for saddles." (*Surgeon-Major J. E. T. Aitchison, Simla.*)

Ropes. 1554

1555

ASCLEPIADEÆ.

"Herbs or shrubs, usually twining. *Leaves* opposite or obsolete, very rarely alternate, quite entire, exstipulate. *Inflorescence* various, usually an axillary umbelliform cyme; flowers regular, hermaphrodite, 5-merous. *Calyx* inferior, lobes or segments imbricate. *Corolla-lobes* or *segments* valvate or overlapping to the right, very rarely to the left; tube or throat often with a ring of hairs, scales, or processes (the outer or *corolline corona*). *Stamens* at the base of the corolla, filaments free in **Periploceæ**, with or without interposed glands; in other tribes, connate into a generally very short fleshy column, which usually bears a simple or compound ring or series of scales or processes (inner or *staminal corona*) that are attached to the filaments or to the back of the anthers, or to both; anthers crowning the column, connate or free, adnate by the connective to the stigma, 2-celled; tip often produced into an inflexed membrane; pollen forming one or two granular or waxy masses in each cell, the masses united in pairs or fours to a gland (*corpuscle*) which lies on the stigma. *Ovary* of two distinct superior carpels, enclosed within the staminal column; styles 2, short, uniting in the stigma, which is 5-angled, short, and included between the anthers, or is produced beyond them into a long or short simple or 2-fid column; ovules many, rarely few, 2-seriate in each carpel. *Fruit* of 2 follicles. *Seeds* compressed, usually flat ovoid, winged and surmounted with a dense long brush of hairs (*coma*) (absent in **Sarcolobus**); albumen copious, dense; embryo large, cotyledons flat, radicle short, inferior. *Distrib.*—Species about 1,000, chiefly tropical.

"The analysis of the plants of this order is most difficult, and in dried specimens never satisfactory, from the fleshiness and complexity of the coronal processes and anthers. I have spent many months over the Indian ones, and have kept pretty close to the generic limits adopted in the *Genera Plantarum*. I have, however, been obliged to abandon the tribe **Stapelicæ**, to suppress **Vincetoxicum**, and to propose several new genera.

SUB-ORDER I.—PERIPLOCEÆ.

Filaments usually free; anthers acuminate or with a terminal appendage; pollen-masses granular, in pairs in each cell.

TRIBE I. Periploceæ.

Characters of the Sub-order.

* Coronal scales or processes O.

Anthers with bearded appendages . . . 1. **Pentanura.**

** Coronal scales corolline, free, short, thick.

Corolla very small, rotate, lobes valvate . . 2. **Hemidesmus.**
Corolla small, rotate, lobes overlapping . . 3. **Cryptolepis.**
Corolla large, funnel-shaped, lobes overlapping . *3. **Cryptostegia.**

*** Coronal scales 5, free, close to or adnate to the filaments.

† *Coronal scales short, broad; filaments without interposed glands.*

A pubescent twining shrub; leaves opposite . 4. **Brachylepis.**
An erect tree; leaves alternate . . . . 5. **Utleria.**

†† *Coronal scales filiform or subulate.*

(α) *Filaments free, without interposed glands.*

Cymes stout, pubescent. Corolla-lobes short, broad . . . . 6. **Finlaysonia.**
Cymes slender, glabrous. Corolla-lobes slender, straight . . . . . . 7. **Atherostemon.**

(β) *Filaments free, with interposed teeth or glands.*

Cymes slender, glabrous. Corolla-lobes short, triangular . . . . 8. **Atherolepis.**
Cymes slender, glabrous. Corolla-lobes long, slender . . . . 9. **Atherandra.**
Cymes short, sessile. Corolla-lobes short, ovate 10. **Streptocaulon.**
Cymes loosely panicled. Corolla-lobes lanceolate . . . . 11. **Myriopteron.**

††† *Coronal scales short, broad; filaments connate, with interposed glands.*

Cymes peduncled. Corolla-lobes ovate, valvate 12. **Decalepis.**

**** Coronal scales connate into a lobed ring; filaments without interposed glands.

Corolla rotate, lobes overlapping . . . 13. **Periploca.**

SUB-ORDER II.—EUASCLEPIADEÆ.

Filaments connate; pollen-masses waxy.

TRIBE II. Secamoneæ.

Anthers with a membranous inflexed tip; pollen-masses in pairs in each cell (20 in all), sessile in fours (2 pairs) on the corpuscle.

Corolla rotate, lobes overlapping to the right . 14. **Secamone.**

Corolla rotate, lobes overlapping to the left . 15. **Toxocarpus.**
Corolla rotate, lobes valvate . . . . 16. **Genianthus.**

TRIBE III. **Cynancheæ.**

Anthers with a membranous inflexed tip; pollen-masses solitary in each cell (10 in all). Sessile or pedicelled in pairs on the corpuscle, pendulous.

* Corona single, corolline, 5-cleft . . . 17. **Glossonema.**
** Corona double, corolline and staminal . 18. **Oxystelma.**

*** Corona staminal, of 5 processes adnate to the anthers, or O.

† *Stem erect.*

Corolla valvate. Coronal processes laterally compressed . . . . 19. **Calotropis.**
Corolla valvate. Coronal processes spathulate . *19. **Asclepias.**
Corolla-lobes overlapping. Coronal processes short, fleshy . . . . . . 20. **Pentabothra.**

†† *Stem twining. Corolla-lobes overlapping.*

Corolla campanulate. Coronal processes ligulate 21. **Raphistemma.**
Corolla rotate. Coronal processes laterally compressed . . . . 22. **Pentatropis.**
Corolla funnel-shaped. Coronal processes laterally compressed . . . . . . 23. **Dæmia.**
Corolla campanulate. Coronal processes O . 24. **Adelostemma.**

**** Corona single, staminal, cupular or annular. Corolla rotate.

Corona of a 10-lobed ring, and 5 horny processes behind the anthers . . . . 25. **Holostemma.**
Corona annular. Leafy erect or twining herbs or shrubs . . . . 26. **Cynanchum.**
Corona annular. Leafless, straggling shrubs . 27. **Sarcostemma.**

TRIBE IV. **Marsdenieæ.**

Anthers with a membranous inflexed tip (absent in **Physostelma** and rarely in **Hoya**); pollen-masses solitary in each cell (10 in all), sessile, or pedicelled in pairs on the corpuscle, erect (rarely horizontal or pendulous in **Tylophora**).

* Corolla-lobes overlapping. Corona O, or corolline.

Stem twining. Corolla-lobes short. Corona O 28. **Sarcolobus.**
Stem pendulous. Corolla-lobes long. Stigma included . . . . 29. **Pentasacme.**
Stem twining. Corolla-lobes short. Corona on the corolla-tube . . . . . . 30. **Gymnema.**

** Corolla-lobes overlapping. Coronal processes on the staminal column, rarely O.

† *Corolla urceolate campanulate or salver-shaped.*

Corolla urceolate. Coronal processes minute or O. Stigma included . . . . . 31. **Gongronema.**
Corolla urceolate or salver-shaped. Coronal scales on the back of the anthers, simple . 32. **Marsdenia.**
Corolla rotate or salver-shaped. Coronal scales on the back of the anthers, notched . . 33. **Pergularia.**
Corolla salver-shaped, coriaceous. Coronal scales O (in Indian species) . . . . 34. **Stephanotis.**
Corolla-lobes long, doubled down inwards in bud 35. **Lygisma.**

†† *Corolla rotate.*

Cymes various. Column minute. Coronal processes fleshy . . . . 36. **Tylophora.**
Cymes umbelliform. Column large; coronal processes simple . . . . 37. **Treutlera.**
Cymes racemiform. Column minute, fleshy; coronal processes 2-fid . . . . . . 38. **Cosmostigma.**
Cymes umbelliform, pendulous. Coronal scales spreading, cuspidate . . . . . 39. **Dregea.**

*** Corolla valvate. Coronal processes adnate to the staminal column.

Corolla small, rotate. Column short, corona stellate. Follicles slender . . . . 40. **Heterostemma.**
Corolla large, rotate. Column short, corona stellate. Follicles stout . . . . 41. **Dittoceras.**
Corolla urceolate or disciform. Corona cupular, fleshy . . . . 42. **Oianthus.**
Corolla minute, urceolate. Coronal scales membranous, erect . . . . 43. **Dischidia.**
Corolla rotate. Corona very large, stellate . 44. **Hoya.**
Corolla cupular. Corona large, stellate . . 45. **Physostelma.**
Corolla-tube short, lobes long subulate. Ovary sunk in the calyx-tube . . . . 46. **Pycnorhachis.**

TRIBE V. **Ceropegieæ.**

Anthers incumbent on the stigma, without a membranous tip; pollen-masses one in each cell (10 in all) sessile in pairs on the corpuscle, erect or horizontal. Corolla-lobes valvate in all.

Corona double; corolline lining the corolla-tube and forming minute processes in the sinus of its lobes, staminal annular.

Calyx turbinate, 5-lobed. Corolla rotate . . 47. **Leptadenia.**
Calyx 5-partite. Corolla salver-shaped . . 48. **Orthanthera.**

** Corona staminal, simple or compound, annular 5-10 lobed, with 5 processes from its inner face which overlap the anthers.

† *Leafy herbs with terete stems and branches.*

Corolla rotate, lobes very narrow. Stem very slender, erect or twining . . . . . 49. **Brachystelma.**
Corolla-tube long. Stem stout or slender, erect or twining . . . . 50. **Ceropegia.**
Corolla rotate, stem erect and branches short, stout, fleshy . . . . 51. **Frerea.**

†† *Leafless herbs, with fleshy 4-angled stems and branches.*

Corolla rotate, lobes very narrow. Flowers lateral, sub-solitary . . . . . . 52. **Caralluma.**
Corolla rotate, lobes very broad. Flowers terminal, umbelled . . . . . . 53. **Boucerosia.**

Genus known by name only. **Odonanthera,** *Wight in Lindl. Veg. Kingd., 626." (Flora of British India, IV., pp. 1-4.)*

Distribution of the Indian Species.—There are a little over 1,000 species belonging to this order, known in the whole world, and they are mostly confined to tropical regions: 245 species are described in the *Flora of British India.* Of these a few are introduced and cultivated in the plains, and one or two are doubtful species. Excluding these,

1556

we have 236. Dividing India into four sections, the following indicates the distributions of the Indian species :—

I.—NORTH INDIA.—The Panjáb, Sind, and the North-West Provinces, and the corresponding frontier mountainous tracts, in all 18 species, or 7·62 per cent.

These may be still further classified :—

A	Confined to the plains	12
	Ascending to from 2,000 to 5,000 feet	3
	Ditto do. 5,000 to 10,000 feet	3
		18

Of this number, 7 are distributed into Afghánistan or Persia, and one not met with in Afghánistan, finds its way, however, into Spain.

II.—WEST INDIA.—Bombay and the greater part of Central India, and of the Central Provinces :—

B	In the plains	30
	Ascending to from 2,000 to 4,000 feet	5
		35

This is about 14·83 per cent.

III.—EAST INDIA.—Bengal, Assam, Burma, and Malacca :—

C	In the plains	77
	Ascending to from 2,000 to 5,000 feet	28
	Ditto do. 5000, to 10,000 feet	13
		118

This is about 50 per cent.

IV.—SOUTH INDIA.—Madras, Hyderabad, Mysore, and Ceylon :—

D	In the plains	23
	Ascending to from 2,000 to 5,000 feet	16
	Ditto do. 5,000 to 10,000 feet	3
		42

This is about 17·8 per cent.

To these must be added a few species which are not referred to either of the above, but which occur in two or more of the provinces,—in other words, occur throughout India :—

E	In the plains	6
	Ascending to from 2,000 to 5,000 feet	8
	Ditto do. 5,000 to 10,000 feet	4
	And ascending 9,000 to 12,000 feet	5
		23

This gives about 9·74 per cent.; but none of this last class find their way into the Afghán-Panjáb frontier, nor are distributed into Afghánistan.

We thus learn from the analysis of the Indian ASCLEPIADEÆ that they attain their maximum in the tropical plains of Bengal and Burma (50 per cent.), that Madras stands next in importance (17·8 per cent.), then Bombay (14·83 per cent.), and last of all the Panjáb (7·62 per cent.), the drier arid tracts of the Indian division of Asia which most resemble Afghánistan containing fewest species. Viewing the ASCLEPIADEÆ collectively we have the following results: tropical, 62·7 per cent.; extra-tropical (or warm temperate), 25·43 per cent.; and temperate, 11·9 per cent.

A. 1556

ASCLEPIAS, *Linn.; Gen. Pl., II., 754.*

1557

According to some authors the *Soma* plant of the Sanskrit authors is a species of ASCLEPIAS (see under **Sarcostemma** and also in *Max Müller's History of Sanskrit Literature*).

Asclepias acida, *Roxb.*; Syn. for **Sarcostemma brevistigma,** *Wight & Arn.*, which see.

A. asthmatica, *Willd.*; Syn. for **Tylophora asthmatica,** *Wight & Arn.*, which see.

A. curassavica, *Linn.*; ASCLEPIADEÆ.

1558

CURASSAVIAN SWALLOW-WORT or WEST INDIAN IPECACUANHA.

Vern.—*Káktundí*, PB.; *Kuraki, kákatundí*, BOMB.

Habitat.—Indigenous in the West Indies, but quite naturalised in India. Found as a weed in Bengal and various parts of India.

Medicine.—The root of this plant possesses emetic properties, and hence the West Indian colonists gave to it the name of *Bastard* or *Wild Ipecacuanha*. The expressed juice of the leaves acts successfully as an anthelmintic. It is also sudorific. The juice of the flowers is said to be a good styptic.

MEDICINE. Root. 1559 Juice. 1560 Leaves. 1561

In Jamaica it is called "blood flower" owing to its efficacy in dysentery. The root is regarded as purgative, and subsequently astringent. It is also a remedy in piles and gonorrhœa. (*Ainslie; Baden Powell, Panjáb Products, I., 361.*)

According to the *U. S. Dispensatory*, the root and expressed juice are emetic and also cathartic. The juice has been strongly recommended as anthelmintic, and according to **Dr. W. Hamilton,** it is useful in arresting hœmorrhages and in obstinate gonorrhœa. The medicine is, however, somewhat uncertain in its operation. (*U. S. Dispensatory, 15th Ed., p. 1579.*)

A. gigantea, *Roxb.*; Syn. for **Calotropis gigantea,** *R. Br.*, which see.

A. laurifolia, *Roxb.*; Syn. for **Genianthus laurifolia,** *Hook. f.*, which see.

A. pseudosarsa, *Roxb.*; Syn. for **Hemidesmus indicus,** *Br.*, which see.

A. rosea, *Roxb.*; Syn. for **Brystelma esculentum,** *Br.*, which see.

A. tenacissima, *Roxb.*; Syn. for **Marsdenia tenacissima,** *Wight & Arn.*

A. tinctoria, *Roxb.*; Syn. for **Marsdenia tinctoria,** *Br.*, which see.

A. tingens, *Buch.*; Syn. for **Gymnema tingens,** *Wight & Arn.*, which see.

A. tunicata, *Willd.*; Syn. for **Cynanchum pauciflorum,** *Br.*, which see.

A. volubilis, *Linn. f.*; (*Willd.* in *Roxburgh*); Syn. for **Dregea volubilis.**

Ash, The, see **Fraxinus floribunda,** *Wall.*; OLEACEÆ.

ASPARAGUS, *Linn.; Gen. Pl., III., 765.*

1562

Asparagus adscendens, *Roxb.*; LILIACEÆ.

Syn.—A. SATAWAR.

Vern.—*Sufèd-músli* or *sufèd-musli, satávar*, HIND.; *Khairuwa*, N.-W. P.; *Shaqáqule-hindi*, ARAB., PERS., and DUK.; *Sápheta musali, dholi musali*,

BOMB.; *Dholi musali, saphéd-musli, ujli-musli,* GUJ.; *Saféda musalí,* MAR.

This is the true *Saféd muslí* of commerce. **Dr. Moodeen Sheriff** first drew careful attention to the fact that various species of Asparagus, especially **A. sarmentosus,** were known as *saféd muslí,* but that only the roots of **A. adscendens** should be regarded as the true article. **Dr. Dymock** and other more recent authors have confirmed this opinion. (See **A. sarmentosus.**)

Habitat.—Found in Rohilkhand and other parts of India,

1563

Description.—In South India *saféd muslí* is the torn and dried roots of **A. sarmentosus.** These pieces are, however, quite unlike the shrivelled and decorticated tubers of the true plant. The tubers are about 2 to 2½ inches long by ¼ inch in diameter. They are of an ivory-white colour, often twisted, hairy, and brittle. When soaked in water they swell up and become spindle-shaped (*Dymock*).

MEDICINE. Safed musli. 1564

Medicine.—The tuberous roots of this species are used as a demulcent and tonic and as a substitute for *Salep;* and, indeed, by some authors they are regarded as superior to *Salep.*

"*Saféd musli* has an agreeable mucilaginous taste; it contains no starch. I have used it largely as an article of diet; it is far nicer than salep, and is generally relished by Europeans. To prepare it, take 200 grains of the powder, 200 grains of sugar, pour upon them slowly a large tea-cupful of boiling milk, stirring constantly all the time. Bombay is supplied with *saféd musli* from Rutlam in Gujarát." (*Dymock, Mat. Med., West. Ind.,* 685.)

§ "Tonic, demulcent; substitute for *salep misree.*" (*Surgeon W. Barren, Bhuj, Cutch.*)

"Said to be useful in diarrhœa, dysentery, and general debility." (*Surgeon Joseph Parker, M.D., Poona.*)

1565

Asparagus filicinus, *Ham.*

Vern.—*Alli palli, saunspaur, sensar pál, satsarra,* PB.

Habitat.—Occurs frequently in the Panjáb Himálaya from 3,000 to 8,500 feet.

MEDICINE. 1566

Medicine.—The root is considered tonic and astringent. In Kanáwár a sprig of this is put in the hands of small-pox patients as a curative measure. (*Stewart.*)

1567

A. officinalis, *Willd.*

THE ASPARAGUS.

Vern.—*Nág-doun, halyún,* HIND.; *Hillúa,* BENG.; *Halgún, mar-chobah, márgíyah,* PERS.; *Halgún, khasabul-hgyah, asbárá-ghús,* ARAB.

§ "*Haliyún,* PERS. The fruit is imported from Persia." (*Surgeon-Major W. Dymock, Bombay.*)

Habitat.—There are several wild Indian species used by the hill-people of Eastern India. Indian species have climbing or trailing stems, often spinose. The cultivated asparagus of Europe is a native of several parts of Great Britain near the sea. It is also very plentiful in the southern parts of Russia and Poland. It is frequent in Greece, and was formerly much esteemed as a vegetable by the Greeks and Romans. It appears to have been cultivated in the time of Cato the elder, 200 B.C., and **Pliny** mentions a form which grew in his time, of which three heads would weigh a pound. (*Treasury of Botany.*)

MEDICINE. Berries. 1568

Medicine.—According to **Dr. Honigberger,** the berries are used by the hakims in debility of the stomach, also in liver, spleen, and renal disorders. They consider them to be diuretic; tonic and aphrodisiac properties are also ascribed to them. **Dr. Irvine**, in his *Materia Medica of Patna (p. 39)*, says that the leaves, berries, and roots are used as demulcent and diuretic in native practice.

Roots. 1569

The *U. S. Dispensatory* says: "The ROOT, which is inodorous, and of a weak, sweetish taste, was formerly used as a diuretic aperient and purifier of the blood; and it is stated to be still employed to a considerable extent in France. It is given in the form of decoction, and made in the proportion of one or two ounces of the root to a quart of water." Considerable difference of opinion prevails, however, many authors considering that in the dried state at least the roots are wholly inert. "In the BERRIES, **H. Reinsch** has found a large proportion of glucose and a yellowish-red colouring matter, *Spargin*." "The sprouts themselves are not without effect, as the urine acquires a disagreeable odour very soon after they have been eaten. They have been accused of producing irritation, with a morbid flow of mucus, of the urinary passages."

Sprouts. 1570

Chemical Composition.—§"Asparagine was originally discovered in the juice of **A. officinalis**, but it has since been shown to exist ready formed in the juice of a large number of plants. It is especially abundant in the form of shoots developed in germination of leguminous seeds, and is found widely diffused in plants. (*Graham.*) Asparagine forms some marked crystals, which are nearly tasteless, but which possess diuretic properties, and impart a peculiar odour to the urine. (*Pharmacographia, p. 93.*) By the action of acids or alkalies asparagine is converted into aspartic acid, and also by the action of hydrating agents as albuminoid substances of animal and vegetable origin. It is contained in somewhat considerable quantity in best sugar molasses. (*Graham.*)" (*Surgeon C. J. H. Warden, Prof. of Chemistry, Calcutta.*)

1571

Food.—The blanched lower stems of this species are extensively eaten as a vegetable all over the world, and to a small extent in India, although they cannot be cultivated so successfully as in Europe. Indeed, the asparagus, chiefly seen at the Indian dinner table, is imported in tins from Europe and America. One of the chief recommendations to Indian-grown asparagus is that it is in season when all other vegetables are out of season. **Firminger** recommends that the seed be sown in August under shelter from the weather. At the close of the rains the seedlings should be about 10 inches high; they should then be planted in holes a foot wide and two or more deep, and well watered and kept constantly moist and gradually earthed up until the end of April or the beginning of May, when they will flower. The flowers should be carefully cut off, but the foliage should not be interfered with. When the rains set in they will require no further care, and in the following March may be forced for the table. This is done by removing the earth carefully and covering the roots with manure.

FOOD. 1572

Asparagus (punjabensis, in *Stewart's Panjáb Plants*).

1573

Vern.—*Sensar pál, chuti, kuchan, sanmali*, PB.

Habitat.—This plant is said to be common in parts of the plains of the Panjáb, east to the Sutlej, as well as in the Salt-Range, and on the Sutlej to 5,500 feet.

Medicine.—A sprig of this plant or of **A. filicinus**, according to **Stewart**, is put in the hands of small-pox patients as a curative measure. The leaves are officinal at Lahore.

MEDICINE. 1574

A. racemosus, *Willd.; Roxb., Fl. Ind., Ed. C.B.C., p. 291.*

1575

Vern.—*Satamúli*, SANS., BENG.; *Satáwar, shakákul*, HIND.; *Satáwar, bozandán, bozidán*, PB.; *Shaquáqul-e-misri*, DUK.; *Satávari*, GUJ.;

Satávari-múl, MAR.; *Sháqáqul*, PERS., ARAB.; *Tannir-muttan-kizhangu, skimai-shadavari*, TAM.; *Challa, challa-gaddu, pilli-píchara*, (fresh root), *sima-shatá-vari* (dry root), TEL.; *Shatávali*, MALA.; *Majjige-gadde*, KAN.; *Hatávari*, SINGH.

Habitat.—A climber, found all over India.

MEDICINE. 1576

Medicine.—The root of this plant is used medicinally as a refrigerant, demulcent, diuretic, aphrodisiac, antispasmodic, alterative, anti-diarrhœatic and antidysenteric. It is used chiefly as a demulcent in veterinary medicine. **Baden Powell** says that it prevents confluence of small-pox. It is used in impotence in the form of a preserve.

According to **Dr. Irvine** the root is used by native physicians as a stimulant and restorative. (*Mat., Med., Patna, p. 94.*)

§ "Not found of any use as a refrigerant." (*Surgeon C. J. W. Meadows, Burrisal.*)

1577

Asparagus sarmentosus, *Willd.*

Syn.—ASPARAGOPSIS SARMENTOSA, *Kunth; Dymock, Mat. Med., W. Ind.*, 685.

Vern.—*Shaqáqul*, or *shakákul*, ARAB., PERS., HIND.; *Satmulí*, BENG.; *Hatmuli*, ASS.; *Tilora*, SIND; *Shatávari*, BOMB., GUJ.; *Saféd-músli* (dry split roots), DUK.; *Satáva-ri-múl, zatar*, MAR.; *Kilávari, tan nír-vittán-kizhangu, tanni-muttán-kalangu* (fresh roots), TAM.; *Challa gaddalu, pilli-píchara*, TEL.; *Shatávari-kiz-hanna, shatávali*, MALA.; *Majjige-gadde*, KAN.; *Hatávari*, SINGH.; *Kanyo-mi*, BURM.

This is erroneously called *Safed musli* in some parts of India.

§ "This plant *Zatar*, MAR., does not produce *saféd musli*, but the roots fresh and candied are used medicinally under the name of *satawri*. The dry *saféd musli* of the bazar is quite a different article, probably from **Asparagus adscendens.**" (*Surgeon-Major W. Dymock, Bombay.*) "The native names of the fresh roots of **A. racemosue** and of **A. sarmeutosus** are in Madras almost the same." (*Honorary Surgeon Moodeen Sheriff, Khan Bahadur, Triplicane.*)

Habitat.—A climber, found in Upper India; common in the Konkans and the Deccan.

MEDICINE. Root. 1578

Medicine.—The root is simply mucilaginous, but is considered nourishing and aphrodisiac. Boiled with oil, it is applied to cutaneous diseases.

The roots of this plant are said to be used to adulterate or as substitutes for **Aconitum heterophyllum,** and are in that case sold under the name of *Atis* (**Aconitum heterophyllum**).

§ "The root juice boiled with *ghee* is given to children in debility and emaciation as a cooling and nourishing medicine." (*Native Surgeon Ruthnam Moodelliar, Chingleput, Madras.*)

"Direction, 15 grains per dose in gonorrhœa." (*Assistant Surgeon Nehal Sing, Saharanpore.*)

A. satawar (*Murray's Drugs of Sind*), see **A. adscendens,** *Roxb.*

1579

ASPHODELUS, *Linn.; Gen. Pl., III.*, 782.

Asphodelus fistulosus, *Linn.; Wight, Ic., t. 2062*; LILIACEÆ.

Vern.—*Piazi, bokat, bokát, binghar bij* (seed), PB.

Habitat.—Abundant as a field weed in most parts of the plains of the Panjáb, so much so near Jhelum as to be troublesome to the cultivator. (*Aitchison.*)

MEDICINE. Seed. 1580

Medicine.—The seed is officinal at Lahore. It is also said to be diuretic.

FOOD. 1581

Food.—It is eaten as a vegetable in times of scarcity.

Aspidium Filix-mas, *Swz.*, see **Nephrodium Filix-mas,** *Richard,* FILICES.

ASPLENIUM, *Linn.; Hook.'s Syn. Filic.,* 244.

1582

Asplenium (Anisogonium) esculentum, *Pr.*

Vern.—*Miwana-kola.* SINGH.

Habitat.—A common fern from the Himálaya to Ceylon, &c.

Food.—Dr. Trimen writes me that in Ceylon this is a well-known vegetable and curry-stuff, largely used by the natives.

FOOD. Vegetable. 1583

Asteracantha longifolia, *Nees;* see **Hygrophilla spinosa,** *T. And.;* ACANTHACEÆ.

ASTRAGALUS, *Linn.; Gen. Pl., I.,* 506.

1584

A genus of herbs or under-shrubs belonging to the Sub-order PAPILIONACEÆ, of the Natural Order LEGUMINOSÆ, and in the Tribe GALEGEÆ. There are in the world, as estimated by **Bunge,** 1,200 species. They belt the world in the north temperate zone, the head-quarters being in Western and Central Asia. India possesses 70 species.

The generic name is derived from Lat. Astragalus, and Gr. ἀστράγαλος, the ball of the ankle-joint, in allusion to the knotted and kneed nature of the procumbent stems of many species. They are in English generally known as the milk vetches.

Astragalus hamosus, *Linn.; Fl. Br. Ind., II.,* 122.

1585

Vern.—*Táj-e-bádshah, katilá, púrtúk, parang,* HIND.; *Giyáhe-qaisar,* PERS.; *Ákhil-ul-malik,* PB.; *Iklíl-ul-malik, asábeaul-malik,* ARAB.

Habitat.—An annual, growing in Beluchistan, Sind, and the Panjáb.

§ "*Akhlil-el-malik* is imported into Bombay from Persia." (*Surgeon-Major W. Dymock, Bombay.*)

Botanic Diagnosis.—Heads peduncled, dense; leaflets 13-25, oblong emarginate, pod long, cylindrical, glabrous, much recurved, nearly bilocular, and 16- to 18-seeded. (*Fl. Br. Ind., II., 122.*)

Dye.—**T. N. Mukarji,** in his *Amsterdam Exhibition Descriptive Catalogue,* says: "By dyers and calico-printers it is employed as an adjunct to dyeing substances, for producing a glaze on the coloured stuffs." This might be said of any member of the genus which yields gum tragacanth, but it would be interesting to have this record of actual use confirmed by specimens of the gum, and of the plant from which it was obtained. Gum tragacanth is imported into India.

DYE. 1586

Medicine.—It has emollient and demulcent properties, and is useful in the irritation of the mucous membranes. The pods are officinal, and are ground to be mixed with plasters.

MEDICINE. 1587

§ "Is laxative and used in nervous affections; made into a paste with vinegar, it is employed externally in headaches. It is said to be lactagogue, and to be used in catarrhal affections." (*Surgeon G. A. Emerson, Calcutta.*)

A. multiceps, *Wall.; Fl. Br. Ind., II.,* 134.

1588

Vern.—*Kandiára, kandei, kátar-kanda, pisar, sarmúl,* PB.; *Tinani, diddani,* AFG.

Habitat.—Found in the West Himálaya, altitude 10,000 to 12,000 feet; Simla, Kumaon, and Garhwál.

Botanic Diagnosis.—Main stems not produced; branchlets with densely crowded nodes, flowers 1-2 together in the leaf-axils, usually not peduncled; corolla twice as long as the calyx. (*Fl. Br. Ind.*)

MEDICINE. 1589

Medicine.—The seeds are given in colic, and also for leprosy.

FOOD. 1590

Food.—At times browsed by cattle. The calyces, which have a sweetish, pleasant taste, are said to be eaten in the Salt-range by the natives.

1591

Astragalus, ? sp.

Vern.—*Anjírá, láyí,* HIND.; *Anz rút,* ARAB.; *Kunjidah,* PERS.; *Gujar,* BOMB.

GUM. 1592

Gum.—A gum is exported from Persia into Bombay which **Dr. Dymock** regards as the true Sarcocolla of the ancients, and there would seem much to favour this idea. The gum is known as *anzarút,* ARAB., and *kunjídah,* PERS.; *anjíra,* HIND.; *gujar,* BOMB. **Meer Muhammad Husein,** in his *Makhzan-ul-Adwiya,* describes the plant which yields this gum as a small thorny shrub known as *shayakah,* a native of Persia and Turkistan.

MEDICINE. 1593

Medicine.—For some time Sarcocolla was supposed to be obtained from **Penæa (Sarcocolla) mucronata,** a native of the Cape of Good Hope. It is known to come from Persia, and it cannot therefore be obtained from a species of **Penæa** (the so-called Sarcocolla plants) which are found in the south of Africa. **Mr. Baden Powell** mentions **Peneæ** in his *Panjáb Products,* but, as pointed out by **Dr. Dymock,** the gum is entirely imported into India, coming from the Persian Gulf. The medicinal virtues of Sarcocolla have long been much admired by the natives of India, made either into an ointment and plaster, or into a medicated oil. It is one of the chief ingredients of the Parsi bone-setter's plaster (*lep*). The gum is described as aperient, and a resolvent of corrupt and phlegmatic humours, acting best when combined with myrabolans or Sagapenum. It is also supposed to be fattening, and is therefore eaten by the Egyptian women. This exceedingly useful gum, which is widely consumed in the East, does not seem to have attracted the attention of Europe to the extent which it deserves.

Dr. Irvine says that Sarcocolla is supposed to heal wounds rapidly. He gives the price at R1-4 a ℔.

§ "This gum, combined with bdellium, is commonly used as a local application for rheumatism and neuralgic pains." (*Surgeon J. Robb, Ahmedabad.*)

1594

Astragalus, sp.

The roots of a species of Astragalus are in Thibet made into a strong paper. (*Sir J. D. Hooker's Him. Journ., II., 162.*

1595

A. tribuloides, *Delille; Fl. Br. Ind., II., 122.*

Vern.—*Ogái,* PB.

Habitat.—Grows in the western and central parts of the Panjáb plains. Distributed through Afghánistan to Egypt.

Botanic Diagnosis.—Heads dense sessile; leaflets 13-15 oblong-lanceolate acute; pod short linear-oblong, densely pubescent, little recurved, 10-12 seeded, sub-bilocular.

MEDICINE. 1596

Medicine.—The seeds are used medicinally.

1597

A. virus, *Oliver.*

THE TRAGACANTH. (See **Bassora Gum.**)

Gum.—**Flückiger** and **Hanbury,** in the *Pharmacographia,* describe 10 species of Astragalus as yielding the tragacanth of commerce, none of which are met with in India. The above species is enumerated as one of the 10. (For further particulars see TRAGACANTH.)

Medicine.—The gum is officinal, being emollient and demulcent, useful in irritation of the mucous membranes, but especially of the pulmonary and genito-urinary organs. Imported into India and sold by druggists.

MEDICINE. Gum. 1598

§ "Tragacanth" or "*Katilla*" *katerá* is a valuable medicine in gonorrhœa. Its emollient and demulcent properties suggest its use in all cases of irritation of mucous membranes. During the hot season it is given to horses for its demulcent and refrigerant qualities." (*Surgeon R. L. Dutt, M.D., Pubna.*)

ATALANTIA, *Correa; Gen. Pl., I., 305.*

1599

Atalantia missionis, *Oliv.; Fl. Br. Ind., I., 513;* RUTACEÆ.

Syn.—LIMONIA MISSIONIS, *Wall.*

Vern.—*Pambúrú,* SINGH.

Habitat.—A small tree in the hotter parts of South India and Ceylon.

Structure of the Wood.—Wood yellowish white, sometimes variegated, moderately hard, close-grained. Annual rings marked by a white line and a belt of more numerous pores. Weight 48 lbs. per cubic foot.

TIMBER 1600

Used for furniture and cabinet-work.

A. monophylla, *Corr.; Fl. Br. Ind., I., 511.*

1601

THE WILD LIME.

Vern.—*Mákad-limbu, mákad-limbn,* MAR.; *Narguni,* URIYA; *Adavi-nimma,* TEL.; *Káttu-elumichcham-param, katyalu* (?), TAM.; *Kán-nimbe, adavi-nimbe,* KAN.; *Atavi-jambíra,* SANS.; *Mal-náraugá,* MAL.; *Jangli-nimbu,* DUK.; *Mátangnár,* S. KONKAN.

Habitat.—A large shrub or small tree of East Bengal, South India, and Ceylon.

Medicine.—Ainslie says that the berries of this thorny plant yield a warm oil, which is, in native medicine, considered as a valuable application in chronic rheumatism.

MEDICINE. Berries. 1602

Structure of the Wood.—Wood yellow, very hard and close-grained. Weight 65 lbs. per cubic foot. Numerous white concentric lines at varying distances.

TIMBER. 1603

The *Bombay Gazetteer* (*Vol. XV., Pt. I., 62*) says the wood of this plant is "close-grained and heavy, but is not generally used."

Recommended by Kurz as a substitute for box-wood.

Atis, The, see **Aconitum heterophyllum,** *Wall.;* RANUNCULACEÆ.

ATRIPLEX, *Linn.; Gen. Pl., III, 53.*

1604

"A genus of CHENOPODIACEÆ, with the foliage covered with a granular mealiness. The Oraches are chiefly distinguished by the two bracts or small leaves, enclosing the fruit, and enlarging after flowering; they are frequently dotted with large coloured warts, which give them a peculiar appearance. The genus possesses several species, which are very variable in form, according to soil and situation." (*Treasury of Botany.*)

The generic name is derived from ά, from, and τρέφω, to nourish. A genus of important fodder plants frequenting desert tracts near salt-marshes and on the sea-coast.

Atriplex halimoides, *Lindley;* Syn. for **A. Lindleyi,** *Moq.; DC. Prod., XIII., 2, 100.*

1605

"Over the greater part of the saline desert interior of Australia, reaching the south and west coasts. A dwarf bush, with its frequent companion **A. holocarpum**, among the very best for salt-bush pasture." (*Baron F. von Mueller, Select Extra-Tropical Plants, p. 39.*)

1606 **Atriplex hortensis,** *L.*, and **A. laciniata,** *L.; DC. Prod., XIII., 2, 91, 93.*

GARDEN ORACHE, THE MOUNTAIN SPINACH, *Eng.;* ARROCHE, *Fr.*

(FOR GARDEN SPINACH see **Spinacia oleracea.**)

Vern.—*Korake, suraka,* PB.

Habitat.—These two species inhabit the Western Himálaya in the temperate zone, also sub-montane tracts in the Panjáb, and in Afghánistan.

FOOD. 1607

Food.—The former is said to be a favourite vegetable on the Pesháwar valley. Is this the *malluach* or "mallows" of Job, xxx., 4? "Who cut up (*malluach*) mallows by the bushes and juniper roots for their meat." The garden orache is an erect-growing, hardy annual, with large hastate leaves, much cultivated in France for its large and succulent leaves, which are used as a spinach. It is, however, far inferior to the true spinach; there are several varieties, differing chiefly in the colour of the leaves and stems. The seeds are said to be unwholesome, exciting vomiting. (*Lindley's Vegetable Kingdom.*)

1608 **A. nummularia,** *Lindley; DC. Prod., XIII., P. II., 460.*

Habitat.—"From Queensland through the desert tracts to Victoria and South Australia."

FODDER. 1609

Fodder.—"One of the tallest and most fattening and wholesome of Australian pastoral salt-bushes, also highly recommendable for artificial rearing, as the spontaneously growing plants, by close occupation of the sheep and cattle runs, have largely disappeared, and as this useful bush even in many wide tracts of Australia does not exist. Sheep and cattle depastured on salt-bush country are said to remain free of fluke and get cured of this distoma-disease and of other allied ailments." (*Baron F. von Mueller, Select Extra-Tropical Plants, p. 40.*)

1610

The following account of the experiments connected with the efforts which are being made to introduce this most valuable plant into India will be found interesting:—

"The small plantation which was made last season continues to thrive. The plants are now from four to six feet high. They are remarkably healthy, and nearly all of them are in flower.

"The genus Atriplex differs from that of Chenopodium in having the flowers unisexual, and in some species of Atriplex the flowers are not only unisexual but diœcious, *i.e.*, some plants bear male flowers only and others only female ones. The salt-bush is described in the *Flora of Australia*, Vol. V. p. 170, as diœcious. A few of the plants in this garden, however, are distinctly monœcious, clusters of the broad fruiting bracts being rapidly developed beneath the terminal racemes of the withered male flowers. This is so far favourable for supplying a more bountiful supply of seed for distribution from our own plants. I have been daily expecting to receive a large supply from Australia.

"Up to date 468 plants have been distributed, and about 60 are left in stock.

"Inquiries have been made regarding the condition of plants despatched from this garden to different places in India. Those sent to the Cawnpore Farm all died about two months after they were planted. Of the 50 plants sent to **Mr. Ridley** at Lucknow, only two survived; these latter, he tells me, were planted out last November, and are now healthy plants about one foot high and with an equal spread. **Mr. W. Impey, C.S.**, writing from Cawnpore in March last, says: 'The **Atriplex nummularia** (plants) of last year are thriving very well. Some of the bushes are 3 to 4 feet high, and I have taken many cuttings from them.' Fifty plants were sent to Bara Banki. The President of the Local Committee informs me that they were planted in poor soil, where other trees or cultivation have hitherto failed. A few have died, and the remainder, though they have made considerable growth, are not thriving on the poorer soils as the plant was represented to be likely to do. **Mr. Dowie**, the Settlement Officer at Kurnaul, reports favourably on the plants sent to him on 31st December last.

"The salt-bush, being essentially a desert plant, should not be permanently transplanted until after the rainy season is over; this injunction applies more particularly to those parts of North-West India where the rains continue for any length of time. As soon as the plants have had sufficient time to establish themselves, no amount of rain is likely to injure them. If the seed is sown in pots during the hot weather, the seedling will be ready for transplanting in September or October." (*Report on Botanical Gardens, Saharunpur and Mussoorie, 1883-84, p. 8.*)

Atriplex spongiosum, *F. v. Mueller.*

FODDER. 1611

"Through a great part of Central Australia, extending to the west coast. Available, like the preceding and several other native species, for salt-bush culture. Unquestionably some of the shrubby extra-Australian species, particularly those of the Siberian and Californian steppes, could also be transferred advantageously to salt-bush country elsewhere, to increase its value, particularly for sheep pasture." (*Baron F. von Mueller, Select Extra-Tropical Plants, p. 40.*)

A. vesicarium, *Heward.*

FODDER. 1612

"In the interior of South-Eastern Australia, and also in Central Australia. Perhaps the most fattening and most relished of all the dwarf pastoral salt-bushes of Australia, holding out in the utmost extremes of drought, and not scorched even by sirocco-like blazes. Its vast abundance over extensive salt-bush plains of the Australian interior, to the exclusion of almost every other bush except **A. halimoides,** indicates the facility with which this species disseminates itself." (*Baron F. von Mueller, Select Extra-Tropical Plants, p. 40.*)

ATROPA, *Linn.; Gen. Pl., II., 900.*

1613

A genus of SOLANACEÆ, containing only one species, a native of the Western Himálaya, from Simla to Kashmír, altitude 6,000 to 12,000 feet, and distributed to Europe and North Persia.

A coarse, lurid, glabrous herb. *Leaves* entire, elliptic-lanceolate. *Pedicels* axillary, solitary, nodding. *Flowers* somewhat large, dirty purple or lurid yellow. *Calyx* large, deeply 5-lobed, scarcely larger in fruit. *Corolla* widely tubular campanulate; lobes 5, triangular, imbricate in bud; *stamens* attached near the base of the corolla, filaments linear; anthers oblong, dehiscing longitudinally. *Ovary* 2 celled; style linear, stigma obscurely 2-lobed. *Berry* globose. *Seeds* many, compressed; embryo peripheric. (*Fl. Br. Ind., IV., 241.*)

The generic name ἄτροπο, one of the three Fates—the goddess supposed to determine the life of man by spinning a thread: the name being given to this plant in allusion to its poisonous property. It is the Nightshade or Dwale of English writers.

Atropa Belladonna, *Linn.; Fl. Br. Ind., IV., 241.*

1614

DEADLY NIGHTSHADE.

Vern.—*Sag-angúr, angúr-shéfa,* HIND.; *Súchi,* PB.; *Girbúti,* BOMB.

ATROPA
Belladonna.

Deadly Nightshade.

"*Ustrung*, ARAB.; *Merdum seeah* (?), PERS.; *Yebruj*, BENG.; *Luckmuna, luckmunee*, HIND. These names are of very doubtful correctness, but are given on **Ainslie's** authority." (*O'Shaughnessy*.)

§ "*Girbúti*.—I never heard this name, nor have I seen the drug here." (*Dr. W. Dymock*.)

Habitat.—Simla to Kashmír, 6,000 to 12,000 feet; found wild in Kanáwár at 8,500 feet.

Dr. **Aitchison** writes me that the var. **lutescens**, *Jacq.*, "is a much more common plant from Kanáwár to Afghánistan," having the same properties as the type form of the species.

Botanic Diagnosis.—The leaves of the Indian plant are a little more acuminate in the Himálayan than in the European plant. This is probably what has given origin to **A. lutescens**, *Jacq.* **Matthiolus** calls this plant **Solanum majus**, and tells us that the Venetian ladies used water distilled from the plant as a cosmetic, hence the name *herba Belladonna*.

MEDICINE.
Leaves.
1615
Root.
1616

Medicine.—The officinal parts of this plant are its leaves and the dried root. They are powerfully sedative, anodyne, and antispasmodic. As an antispasmodic, it is a valuable medicine in the advanced stages of hooping-cough, spasmodic asthma, laryngismus stridulus, chorea, epilepsy, and spasmodic stricture of the urethra; as a sedative and anodyne, in various forms of neuralgia, rheumatism, tetanus, hydrophobia, delirium tremens, dysmenorrhœa, and other painful uterine affections, cancerous and other painful ulcerations; in cataract and other eye affections, in which it is desirable to dilate the pupil or to keep the edge of the iris free; it is invaluable in surgical practice. In rheumatic and scrofulous iritis it is a relieving agent.

The properties, preparations, and uses of this drug are too well known to require to be treated of here. The reader is referred to the following works: *Pharmacopœia of India, pp. 171-174; Flückiger and Hanbury's Pharmacographia, pp. 455 to 459; U. S. Dispensatory, 15th Ed., 281 to 285; Bentley and Trimen's Medicinal Plants, 193; Royle's Mat. Medica, Ed. Harley, pp. 488 to 496; &c., &c.*

It is a remarkable fact that while this most useful plant is exceedingly plentiful in many parts of the Western Himálaya, its medicinal virtues seem to have escaped the detection of the natives of India completely. Absolutely worthless drugs are carefully collected and exported to the plains of India, from the very localities in which Belladonna is abundant, and yet not a single leaf or root of this most valuable drug can be purchased, of Indian origin, in the native drug-shops of the plains. No mention is made of it by **Drs. Dymock, Moodeen Sheriff**, nor by **U. C. Dutt. O'Shaughnessy** refers to it briefly, and gives the paragraph which will be found under the vernacular names. He says it is "known in the bazars of Central Asia and the North of India." **Ainslie**, who gives the vernacular names republished by **O'Shaughnessy** and **Birdwood**, states clearly, however, that he has never seen the plant in India, and recommends its introduction.

It would therefore appear that the natives of India have been made familiar with the virtues of this plant in the form of an imported drug, while the Himálaya might supply the world with Belladonna.

1617

Chemical Composition.—§"The active principle of **A. Belladonna** is atropia, an alkaloid which is either identical with or very closely allied to daturine. It exists in all parts of the plant apparently in combination with malic acid. The quantity present in various parts of the herb has been determined by **Gunther** and others. The ripe seeds contain the largest percentage, while the root and stalk contain a very much smaller amount. Physiologically, atropine acts on the pupils, and on the system generally, in the same manner as daturine. A second alkaloid, Belladonin, has been discovered as existing in the plant, but according to **Blyth** it is probably a product of the decomposition of atropine. By the action of certain reagents atropine gives rise to various derivatives, tropine, atropic acid, and isatropic acid. According **to Biltz**, asparagine is contained in the Belladonna plant." (*Surgeon Warden, Professor of Chemistry, Calcutta.*)

MEDICINE. 1618

Special Opinions.—§ "Useful to diminish the secretion of milk; it checks excessive, and especially local, perspiration, as of the hands or feet, or of the head and face, in phthisis; it also checks salivation from mercury or other cause. It is antagonistic to Calabar bean, aconite, and to the poisonous principle of fungi (muscarin). (*Surgeon-Major E. G. Russell, Calcutta.*) "Useful in mercurial salivation." (*Surgeon H. D. Masani, Karachi.*) "I have found a drop of atropine occasionally dropped into the eye to give great relief in ocular neuralgia. I know of no better anodyne for external application in facial neuralgia than a combination of aconite and belladonna. I usually apply dry heat after their application to the painful part." (*Surgeon Joseph Parker, M.D., Poona.*)

Attar of Roses, see **Rosa.**

AUCUBA, *Thunb.; Gen. Pl., I., 950.*

1619

A genus containing 3 species of small trees or only 3 forms of 1, belonging to the Natural Order CORNACEÆ. A glabrous, branching shrub. *Leaves* opposite, petioled, ovate or lanceolate, obtusely serrate, leathery, shining, turning black on drying. *Flowers* small, diœcious. *Male calyx* small, 4-toothed. *Stamens* 4; *disk* quadrangular, fleshy. *Berry* ellipsoid, crowned by the calyx-teeth and style, smooth, shining, orange, yellow or scarlet.

The generic name is of Japanese origin. The presence of this plant is one of the most striking temperate Japanese features of the Eastern Himálaya and Mánipur as compared with the Western Himálaya.

Aucuba himalaica, *Hook. f.; Ill. Him. Pl., t. 12; Fl. Br. Ind., II., 747.*

1620

Vern.—*Phul amphi,* NEPAL; *Singna, tapathyer,* LEPCHA.

Habitat.—A small tree of the Sikkim Himálaya, Bhután, Nagá Hills and Manipur, between 5,000 and 9,000 feet.

Structure of the Wood.—Wood black when fresh cut, becoming lighter-coloured on exposure, hard and close-grained. Weight 55 lb. per cubic foot.

TIMBER. 1621

Auklandia Costus, *Falc.*, see **Saussurea Lappa,** *C.B.C.;* COMPOSITÆ.

AURUM.

1622

Aurum.

GOLD.

Vern.—(The metal) *Sóná,* HIND., DUK., BENG., GUJ., and MAR.; *Pau, thangan* TAM. and MAL.; *Bangáru,* TEL. and KAN.; *Zahab,* ARAB.; *Suvarnam,* SANS.; *Tar, tila* or *tilá,* PERS.; *Ran,* SINGH.; *Shue,* BURM.; (Gold leaf) *Sonehri-varaq, sóné-ká-varaq,* HIND.; *Soné-ká-tagat,* DUK.; *Sonár-pát, sonáli,* BENG.; *Tanga-réku, pou, réku,* TAM. and MALA.; *Bangáru-réku,* TEL. and KAN.; *Soné-cha-varuq, sona-nu-varaq,* GUJ.; *Suvarna-patram,* SANS.; *Ran-tahadu, ran-taqadu,* SINGH.; *Shue-saka,* BURM.

Medicine.—Is used in the form of leaf as a nervine tonic. Combined with silver leaf, arsenic, and other metals, in the form of confection called *májún,* or *maajún,* it is extensively employed by hakims.

MEDICIN 1623

The metal is first beaten into leaf free from any amalgam. It is then heated and rubbed with mercury some 13 or 14 times, when it is said to lose its metallic character, and becomes reduced to a reddish powder.

In this condition it is prescribed, and it is considered a valuable tonic and alterative, improving the memory and intellect. It is by the *kabirajes* prescribed in fever, consumption, insanity, impotence, and other nervous diseases. Dose one to two grains. (*U. C. Dutt.*)

"Gold leaves are used by Mussulman beggars and other hemp smokers. The hemp, *ganja*, is laid on the bowl of the pipe, the gold leaf is stretched across the mouth of the bowl, and on the leaf fire is placed; the foil is also sometimes plastered over sweetmeats." (*Bomb. Gaz., Vol. IV. 128.*)

§ "It is also used as a stimulant." (*Assistant Surgeon Bhugwan Das, Rawal Pindi.*) "Used to stop decayed teeth, and as plates to artificial teeth, false palates, &c." (*Brigade Surgeon G. A. Watson, Allahabad.*) "It is held in high repute for the treatment of consumption. The gold leaf is given with butter and sugar, or it is used in various combinations." *Surgeon-Major J. Robb,, Ahmedabad.*)

1624

DYE AUXILIARIES.

Auxiliaries used in dyeing, some of which cannot be viewed as mordants.

1625

1. **Lime.**—This is used in calico-printing with gums as a "resist paste." It is also used with sugar to promote the fermentation of indigo.

It is prepared from the following:—

(*a*) Limestone Rock, such as that obtained from the Khásia Hills.
(*b*) From *Kankar*, the calcareous tuberculated masses found in beds on the surface, or a little below the surface, of the soil from Behar northward to the Panjáb. In the North-West Provinces this is used for metalling the roads.
(*c*) By burning Land Shells collected in Bengal just after the rains.

1626

2. **Potash.**—This is chiefly obtained from the ashes of certain plants. The Common Millet is largely used for this purpose in the North-West Provinces. **Symplocos** and other bushy plants in the hills of Bengal; but in the plains of Bengal the ash of *Apáng* (**Achyranthes aspera,** *L.*) is largely used for this purpose—see **Alkaline Ashes.**

1627

3. **Reh,** an impure carbonate and sulphate of soda, found as a natural efflorescence on the soil, often rendering it uncultivable and burning up the vegetation. This is used, chiefly, like soap, to wash fabrics before they are dyed. (See **Reh** and also **Barilla.**)

1628

4. **Rassi.**—Carbonate of soda prepared from the preceding by precipitation of impurities.

1629

5. **Sajjí**, a mixture of carbonate of soda and potash or wood-ash. This is used chiefly in extracting the deeper red colours from safflower. (See **Barilla** and **Sajjí-mátí.**)

1630

6. **Saltpetre** is obtained like *Reh* as an efflorescence on the surface of the soil; it is chiefly used in wool-dyeing.

See also **Iron Sulphate, Ochre,** and **Proto-sulphate of Iron.**

1631

AVENA, *Linn.; Gen. Pl., III., 1160.*

A genus of grasses belonging to the Tribe Aveneæ of the Natural Order Gramineæ. There are said to be some 40 species in the whole world confined to the temperate regions. Annual or herbaceous plants. *Spikelets* 2-many-flowered, very rarely 1-flowered. *Glumes*, the inferior empty, equalling or overtopping the flowers. Lower *pale* large-awned, ending in 2 points, having lateral veins.

Meadow Oat Grass.

Awn dorsal-kneed and twisted. *Stamens* 3. *Ovary* hairy at the top; *styles* short, distinct, plumose. *Fruit* crested or rarely glabrous, furrowed, oblong or elongate-fusiform, enclosedwithin the glume and pale, the latter shortly adherent.

The generic name is the classical Latin name *Avena*, an Oat; *Avoine*, Fr.; *Avena*, Sp.; *Avea*, Port.; *Vena*, It. The genus was known to the Greeks under the name of *Bromos*, but there is no evidence that Oats were cultivated either by the Romans or by the Greeks. It seems probable that all the forms referred to this genus are but cultivated derivatives from a single prehistoric species, a native in all probability of east temperate Europe and of Tartary. They nowhere exist in a truly wild condition, but accompany cultivation, frequenting deserted fields, roadsides, or rubbish-heaps in a manner exhibited by no other cereal.

Avena fatua, *Linn.*

1632

THE WILD OAT.

Vern.—*Kuljud, ganer, gandal, jei*, HIND.; *Gozang, ganerjei, kásamm, yúpo, úpwa*, PB.

According to **Professor Buckman** in the *Treasury of Botany*, this is most probably the plant from which, by a continuous process of cultivation, the domesticated oat has been induced. He bases this opinion on a series of experiments from 1851 to 1860, in which he ultimately obtained a plant which could hardly be distinguished from the Tartarian or so-called potato oats. He also points out that shed oats gone wild on a field degenerate, the first indication of which being the production of hairs upon the grains, similar to those in **A. fatua**, a character which cereal oats never possess.

Habitat.—Inhabits the plains and hills of Northern India; common as a field weed in cereal crops throughout the plains; ascends the Himâlaya up to 9,500 and 11,500 feet.

Botanic Diagnosis.—Panicle erect, spikelets drooping, each of about 3 flowers; flower falling short of the glume, with fulvous hairs at the base; lower pale bifid at the end. Plant about 3 feet in height; root annual; upper glume 5-9-veined and awn much bent, the lower half twisted. The awn is also long, rigid, and sensitive to the changes of the atmosphere, as regards moisture. These peculiarities give the seed so much the appearance of a fly that it has been used in trout-fishing for this purpose. On coming in contact with the water, the long awns begin to twist about and deceive the fish by their apparent struggling. This property has also given origin to their being used as a hygrometer, the seeds jumping about when breathed upon, or when the atmosphere becomes moist.

MEDICINE. 1633

Medicine.—It is believed to produce poisonous and deleterious effects.

FODDER. 1634

Fodder.—**Stewart** and **Madden** say that in all the places where it grows it is pulled up or gathered for fodder, but is suspected of occasionally producing bad effects.

A. pratensis, *Linn.*

1635

MEADOW OAT GRASS; NARROW-LEAVED OAT GRASS.

Syn.—A. BROMOIDES, *Kunth.*

Habitat.—Reported to occur in Lahoul. In Europe this is described as a denizen of moors and poor clays; its specific name being thus inappropriate, as it is seldom met with on meadows.

Botanic Diagnosis.—Panicle erect, with simple or slightly divided branches; flowers erect, 3-6, exceeding the glumes, the upper of which are only 3-veined. Root fibrous; height nearly 2 feet.

FODDER. 1636

Fodder.—Baron von Mueller says it thrives well on dry, clayey soil, producing a sweet fodder; it is recommended for arid ground, particularly such as contains some lime, being thus as valuable as **Festuca ovina.**

1637

Avena pubescens, *L.*

DOWNY OAT GRASS.

Syn.—TRISETUM PUBESCENS, *R. & S.*

Habitat.—Royle found it at Simla. In Europe this is a common meadow grass in limestone pastures.

Botanic Diagnosis.—Panicles erect, nearly simple; flowers erect, 2 or 3, scarcely exceeding the glumes. A creeping plant with the lower leaves and sheaths hairy; height 1-2 feet.

FODDER. 1638

Fodder.—It is a sweet, nutritious, prolific, perennial grass, requiring dry but good soil containing lime. (*Mueller.*) The downy hairs which cover the surface of the leaves of this grass, when growing on poor soil, almost entirely disappear when it is cultivated on a richer soil. (*Loudon; Duthie.*)

1639

A. sativa, *Linn.*

OATS.

Vern.—*Jai, wilayati-jau, jávi,* HIND. and PB.

Under the subject of Oats, **DeCandolle** (*L'Orig. Cult. Pl., 299*) says there is no Sanskrit name for Oats, nor any in modern Indian languages. Again (*p. 300*), the European vernacular names prove an existence north-west of the Alps and on the borders of Europe, towards Tartary and the Caucasus. The most widely diffused name is the Latin Avena; Ancient Slav, *ovisu, ovesu, oosa;* Russian, *ovesu;* Lithuanian, *awiza;* Lettonian, *ausas;* Ostias, *abis.* The English word Oats comes, according to **A. Pictet,** from the Anglo-Saxon *ata* or *ate.* The Basque name *alba* or *oloa* argues a very ancient Iberian cultivation. The Keltic names are quite different: Irish *coirce, cuirce, corca;* Armorican, *kerch;* Tartar, *sulu;* Georgian, *kari;* Hungarian, *zab;* Croat, *zob;* Esthonian, *kaer,* are given by **Memnich** as generic names for oats.

1640

Habitat.—Of recent introduction into Indian agriculture; it was first grown in Northern India, under English auspices, round cantonments and stud depôts, for the supply of horses. The oat is cultivated in temperate regions throughout the globe even as far north as the arctic zone.

History.—The origin of this plant is unknown, but it is supposed by **Dr. Lindley** to have been originally a native of Northern Europe. This opinion is confirmed by **DeCandolle** in the passage quoted above. Plants gone wild from cultivation show an approach to the type of **A. strigosa,** *Schreb,* from which it is chiefly distinguished by the bristles at the end of the flowers. These might be presumed to have disappeared under cultivation. **A. strigosa** may, however, only prove a variety of **A. fatua,** in which case the origin of the domesticated oats would have to be traced to that species. See under **A. fatua.**

1641

Cultivation.—The cultivation of oats has not gained much extension. It is still confined to North India, where it is restricted chiefly to districts where horse-breeding is carried on, *viz.*, in the Meerut and Rohilkhand divisions of the North-West Provinces, and in the Hissar and Karnál districts of the Panjáb. In the Meerut Division the area annually under oats is 5,000 acres, and in Rohilkhand 3,000 acres. The total area under the crop in the 30 temporarily-settled districts of the North-West Provinces and Oudh, including of course the two divisions just mentioned, is returned at 9,781 acres.

AVENA sativa.

Oats.

Oats are grown as a rule on the better-class soils near village sites. The mode of cultivation differs in no way from that pursued with Barley; in fact they are often sown together. **Messrs. Duthie** and **Fuller** write: "With a copious supply of water it has been found that oats are an invaluable green fodder crop for the cold season, yielding as many as three cuttings, and then making sufficient growth to bear a thin crop of grain. A large area under oats is most successfully treated in this way each year at the Hissar Government cattle farm. When grown in this manner they class rather as a green fodder than as a grain crop." When grown for grain the outturn is (in Northern India) 18 maunds on irrigated and 10 maunds on unirrigated land per acre. There seems very little hope of the trade of India being much extended.

The produce probably comes from the northern parts of the Panjáb and the North-West Provinces.

TRADE RETURNS.

1642

The following are the Imports and Exports of Oats for the past five years ending 1883-84:—

Imports and Exports of Oats.

YEARS.	Imports.		Exports.	
	Cwt.	R	Cwt.	R
1879-80	48	247	84,095	3,47,574
1880-81	253	1,196	86,168	2,47,450
1881-82	738	5,130	75,967	1,91,791
1882-83	347	2,999	66,706	2,18,478
1883-84	537	4,174	87,725	2,91,909

The imports chiefly consist of oats brought by ships carrying horses.

The following table will show the quantity and value of the imports and exports of oats made in the year 1883-84:—

Detail of Imports and Exports in 1883-84.

Imports.	Cwt.	R	Exports.	Cwt.	R
Imported from—			Exported to—		
United Kingdom	285	2,453	Mauritius	87,195	2,89,853
			Natal	350	1,431
Australia	252	1,721	Other countries	180	625
TOTAL	537	4,174	TOTAL	87,725	2,91,909
Imported into—			Exported from—		
Bengal	51	375	Bengal	87,725	2,91,909
Bombay	354	2,451			
Sind	27	298			
Madras	105	1,050			
TOTAL	537	4,174	TOTAL	87,725	2,91,909

The above figures do not, of course, include the imports of oatmeal, which are included under Provisions and Oilman's stores.

"**Varieties of Oats generally Cultivated.**—The different kinds of oats are distinguished from each other by a variety of characteristics, such as colour, size, and form of the seeds, quality of the straw, the period of

AVENA sativa.

Oats.

ripening, liability to shed their seeds in high winds, and adaptation to particular soils and climates. There are three principal groups of oats, easily distinguishable by colour, *viz.*, white, black, and gray or dun. White oats are separable into two principal varieties, the late and early; and these again into several sub-varieties, characterized by certain peculiarities of growth.

"Early oats are best adapted for the higher class of soils, as the greater yield per acre more than compensates for the inferiority of the straw. Their earliness renders them very suitable for late districts; but the liability of some to shed their seeds in high winds, renders their cultivation in high-lying and exposed situations extremely hazardous.

"Late or common oats, as they are more generally termed in Scotland, are distinguished from the early variety by late ripening, thicker husk, and less meal; the latter being of better quality, lighter per bushel, not usually so prolific; the former, however, have a more vigorous constitution, and are better able to resist the effects of atmospheric changes, such as rains or droughts, and when ripe they are less liable to shed their seed in high winds; the straw is greatly superior as fodder; and lastly, they can be cultivated with greater success than the earlier varieties on inferior soils, and also those of a strong clayey nature.

"Black oats are of two kinds,—the one the Tartarian, having the ear only on one side of the straw; and the other the old or common black, with black seeds, but having a spreading ear, similar to the white varieties. Dun oats are to all appearance hybrids between the last mentioned variety and one or other of the white sorts, most probably the late or common white oat, as they have more of the characteristics of the last mentioned; such as hardiness, lateness, adaptation to clayey and cold-bottomed soils, and by the superior quality of the meal and straw." (*Morton's Cyclopædia of Agriculture, Vol. II., p. 482.*)

Of foreign oats Great Britain imported, in 1883, 15,248,467 cwt., and the annual imports show a steady increase.

The food value of oats is very great. The quantity of starch is nearest to that in barley. Oats are very rich in oil and fatty matter. The proportion of flesh-forming materials in good oats is larger than in wheat, barley, or Indian-corn. "Many people in Scotland live entirely on oatmeal, and their strong, muscular forms are undeniable proofs of the superior qualities of oats in supplying the materials from which the muscles are formed." *(Morton.)*

Four varieties of Scotch oats, free from husk, and dried at 212°, were analysed by **Professor Norton** and **Mr. Fromberg**, with the following results:—

	1.	2.	3.	4.
Starch	65·24	64·80	64·79	65·60
Sugar	4·51	1·58	2·09	0·80
Gum	2·10	2·41	2·12	2·28
Oil	5·44	6·97	6·41	7·38
Casein (avenine)	15·76	16·26	17·72	16·29
Albumen	0·46	1·29	1·76	2·17
Gluten	2·47	1·46	1·33	1·45
Epidermis	1·18	2·39	2·84	2·28
Alkaline salts and loss	2·84	1·84	0·94	1·75
	100·00	100·00	100·00	100·00

A. 1642

M. Boussingault, in his *Economic Rurale*, gives the following analysis of oats according to life-sustaining compounds :—

Nitrogenous ingredients	13·93
Non-nitrogenous ingredients	82·07
Inorganic ingredients	4·00
	100·00

And according to ultimate elements :—

Carbon	50·70
Hydrogen	6·40
Oxygen	36·70
Nitrogen	2·20
Ash	4·00
	100·00

For warm climates, oatmeal is not a proper article of human diet, as it heats the blood and produces eruptions on the skin; neither is it wholesome for persons engaged in sedentary employments.

The oat constitutes the most important article of food for horses in Great Britain.

Oat-straw has been supposed to be less nutritious than other kinds of straw; but it would appear this opinion is quite erroneous as regards the oat-straw grown in Britain. (*Morton's Cyclopædia, Article Oats, Vol. II., pp. 506-509; Atkinson, Him. Dist., 692; Duthie and Fuller's Field and Garden Crops, Part I., p. 14.*)

AVERRHOA, *Linn.; Gen. Pl., I., 277.*

1643

A genus containing 3 species of small trees, belonging to the Natural Order GERANIACEÆ. Two are cultivated in most tropical countries, and the third is indigenous to the New World, whence the cultivated species were most probably introduced into India by the Portuguese.

Averrhoa Bilimbi, *Linn.; Fl. Br. Ind., I., 439; Roxb., Fl. Ind., Ed. C.B.C., 387.*

1644

BILIMBI TREE.

Vern.—*Bilimbi,* (the fruit) *belambú,* HIND.; *Bilimbi, blimbi,* BENG.; *Belambú,* DUK.; *Blimbu,* GUJ.; *Bilambi,* MAR.; *Pulich-chakkáy, bilimbi-káy, koch-chit-tamarttai,* TAM.; *Pulusu-káya-lu, bili-bili-káyalu,* TEL.; *Vilunbikká, vilimbi, karichakka,* MAL.; *Kála-zoun-si, kala-zoun-ya-si,* BURM.

Habitat.—Cultivated in gardens on the plains of India. It flowers in the beginning of summer and ripens its fruit in about two or three months after. The fruit is cylindrical, about two inches long, and pulpy, and is very sour when green, but loses some of its acidity when ripe. It has become almost naturalized in India.

Food.—The fruit is generally used in pickle and in curry. The flowers are made into preserves.

FOOD. 1645

A. Carambola, *Linn.; Fl. Br. Ind., I., 439; Roxb., Fl. Ind., Ed. C.B.C., 387.*

1646

Vern.—*Karmal, khamrak, kamaranga,* HIND.; *Kámrángá, kamarak,* BENG.; *Kardai,* ASS.; *Tamarak, kamarakha,* GUJ.; *Kamarakha,* MAR.; *Khamaraka, karamara,* BOMB.; *Khamrak,* DUK.; *Tamarta, tamarttam-káy,* TAM.; *Karomonga, tamarta-káya,* TEL.; *Tamarat-túka,* MALA.;

Kamarak, KAN.; *Karmaranga*, SANS.; *Zoun-si, zoun-ya-si, saunggya*, BURM.

Habitat.—A small tree with sensitive leaflets, 15 to 20 feet in height; extensively cultivated in India for its apples, which when stewed are very palatable. It is found as far north as Lahore.

DYE. 1647

Dye.—§"The unripe apples are astringent, and are used as an acid in dyeing. The acid probably acts as a mordant." (*Deputy Surgeon-General G. Bidie, Madras.*)

MEDICINE 1648 Leaves. 1649 Root. 1650 Fruit. 1651

Medicine.—"The leaves, the root, and the fruit are used as cooling medicine." (*Amsterd. Cat.*) "The acid dried fruit is given in fevers." (*Irvine, Mat. Med., Patna, p. 55.*)

§"Kamranga" is met with in two forms in Bengal,—the sweetish acid and the extremely sour. The former is cooling and useful in feverishness. Both varieties have antiscorbutic properties. (*Surgeon R. L. Dutt, M.D. Pubna.*) "The fruit is an excellent antiscorbutic." *Surgeon-Major J. E. T. Aitchison, Simla.*) "It is highly cooling; if taken raw it brings on fever and chest complaints." (*Assistant Surgeon J. N. Dey, Jeypore.*) "The ripe fruit has a pleasant acid and sweetish taste, and is used for culinary purposes." (*Surgeon Shib Chunder Bhuttacharji, Central Provinces.*) "Fruit used for making pickles." (*Deputy Surgeon-General G. Bidie, C.I.E., Madras.*) "Fruit used in curries." (*Honorary Surgeon P. Kinsley, Chicacole, Ganjam, Madras.*)

FOOD 1652

Food.—It blossoms in the rainy season, and the fruit, which ripens in December and January, is about three inches long, and is eaten raw to a small extent by the natives. The flesh is soft, juicy, and refreshing. It is sometimes stewed in syrup with a little cinnamon, and is then very pleasant; it is also made into an agreeable jelly.

TIMBER. 1653

Structure of the Wood.—Light red, hard, close-grained. Weight about 40 lbs.

Mr. Home of the Forest Department says it is used in the Sunderbans for building purposes and for furniture.

1654

AVICENNIA, *Linn.; Gen. Pl., II., 1160.*

A genus of VERBENACEÆ, comprising in all some 3 or 4 species of bushes or small trees, frequenting the salt-marshes on the coast and in the tidal forests of rivers of India, Burma, the Andaman Islands, Africa, Australia, New Zealand, Tasmania, America, and the West Indies.

Branchlets fleshy. *Leaves* opposite, in the Indian plant coriaceous, elliptic-lanceolate. *Flowers* yellow, sessile, in rounded heads. *Calyx* of 5 sepals, supported by ovate-ciliate bracts. *Corolla-tube* short, limb of 4 nearly equal segments. *Capsule* compressed, ovate-mucronate, 2-valved, 1-seeded; radicle woolly; cotyledons thick, fleshy, folded. Like the true mangrove the seeds frequently germinate within the fruits while attached to the tree.

The genus is named in honour of **Avicenna**, an Arab physician, philosopher, mathematician, &c., who lived (at Bokhara?) between the years 980 and 1037.

1655

Avicennia officinalis, *Linn.*

THE WHITE MANGROVE.

Syn.—A. TOMENTOSA, *Jacq.*

Vern.—*Bina* (*bani in Gamble*), BENG.; *Mada, nalla-mada*, TEL.; *Tivar*, BOMB., MAR., SIND; *Oepata*, MAL.; *Lameb, thamé*, BURM.

Habitat.—A small tree or shrub of the salt marshes and the tidal forests of India and Burma, found also in Andaman Islands. **Roxburgh** says it is common near the mouths of rivers where the spring tides rise. It is found everywhere in the Sunderbans, often becoming a tree of considerable size; but on the Coromandel coast it is only a bush. **Kurz**

gives this species as frequent along the Burmese coast from Chitagong along with the next species.

Botanic Diagnosis.—"*Leaves* usually lanceolate and indistinctly white-tomentose beneath; *flowers* shortly spiked; *calyx-lobes* 1 line long; *style* very short." (*Kurz.*)

Dye.—The bark is used as a tanning agent. (*Birdwood, Bombay Prod.*) The ashes of the wood are used to wash cloth. (*Drury*). In Rio Janeiro the barks of various species of **Avicennia** are used in tanning leather.

DYE. Bark. 1656 Ashes. 1657 TAN. 1658

Food.—The kernels are bitter but edible.

FOOD. Kernels. 1659

Structure of the Wood.—Grey, with a darker heartwood, hard, heavy, consisting of numerous, narrow, well-marked, concentric layers. Weight 58 lbs. per cubic foot.

It is very brittle: used in India only for firewood. **Major Ford** says it is used for mills for husking paddy, rice-pounders, and oil-mills in the Andamans.

TIMBER. 1660

Avicennia tomentosa, *Roxb.; Kurz, For. Fl. Burm., II., 276; Roxb., Fl. Ind., Ed. C.B.C., 487.*

Vern.—*Bind*, HIND., BENG.; *Timmer, cheriá*, SIND; *Tivar*, MAR.; *Nalla-mada, mada-chettu*, TEL.; *Upputti*, MAL.

Habitat.—Common in India in low places near the mouths of rivers and in salt-marshes. In the lower parts of the Delta of the Ganges it grows to a tree of considerable magnitude. "Frequent in the tidal forests all along the coast of Burma from Chittagong down to Tenasserim." (*Kurz.*) It is abundant on Bombay and Malabar coasts.

Botanic Diagnosis.—"*Leaves* more or less obovate and usually indistinctly tawny-tomentose beneath; *flowers* in heads; *calyx-lobes* 2 lines long; *style* long and slender." (*Kurz.*) It seems doubtful how far this should be viewed as distinct from the preceding species, but if distinct the vernacular names and facts regarding the economic uses of the two plants would appear to have got completely mixed up together.

Medicine.—The roots possess aphrodisiac properties. The unripe seeds are used as poultice to hasten suppuration of boils and abscesses.

§ "Dhobies in the Madras Presidency use the ashes of the wood for washing and cleaning cotton cloths. It is also used in small-pox. The bark is used in Rio Janeiro for tanning." (*Surgeon H. W. Hill, Mánbhúm.*) "The bark is astringent." (*Surgeon-Major W. Dymock, Bombay.*) "Bark astringent, ashes used for washing and bleaching cloth; common in Madras." (*Deputy Surgeon-General G. Bidie, Madras.*)

MEDICINE. Root. 1661 Seeds. 1662 Ashes as Soap. 1663 TAN. Bark. 1664

Azadirachta, see **Melia.**

AZIMA, *Lam.; Gen. Pl., II., 681.*

1665

Azima tetracantha, *Lam.; Fl. Br. Ind., III., 620;* SALVADORACEÆ.

Syn.—MONETIA BARLERIOIDES, *L'Herit.; Roxb., Fl. Ind., Ed. C.B.C., 716;* FAGONIA MONTANA, *Miq.*

Vern.—*Kántagúr-kamai*, HIND.; *Trikanta-gati*, BENG.; *Sukká-pát*, DUK; *Sung-elley* or *sung-ilai, changan-chedi, muttu-chengan-chedi, nallo-chengan-chedi*, TAM.; *Tella-uppi, uppi-áku*, TEL.; *Kundali*, SANS.

Habitat.—A small, thorny shrub, growing plentifully in the Deccan and Ceylon. "On every part of the Coromandel coast it grows freely in all situations, and is in flower and fruit most part of the year." (*Roxb., Fl. Ind., Ed. C.B.C., 716.*) "Frequent in the dry forests and shrubberies of Ava and Prome." (*Kurz, II., 161.*)

FOOD. Berries. 1666 MEDICINE. Juice. 1667 Leaves. 1668 Bark. 1669 Decoction. 1670

Food.—The berries are white and are eaten.

Medicine.—The juice of the leaves is reported to relieve the cough of phthisis and asthma.

The bark is also used as an expectorant.

§ "A decoction of bark is given as an antiperiodic in ague with success. It is an astringent and tonic. The leaves are used for ulcers, and especially after small-pox, they are ground with turmeric and gingelly oil, and then applied on the surface of the body, removing the irritation of skin. (*Surgeon-Major Lionel Beech, Cocanada.*) "The root-bark is used in muscular rheumatism." (*Native Surgeon Ruthnam Moodelliar, Chingleput, Madras.*)

BACCAUREA, *Lour.; Gen. Pl., III., 283.*

1 **Baccaurea affinis,** *Müll.-Arg.;* EUPHORBIACEÆ.

2 **B. flaccida,** *Müll.-Arg.*

A small tree, chiefly met with in South India.

3 **B. parviflora,** *Müll.-Arg.*

Vern.—*Kanazo,* BURM. (*Kurz*).

4 **B. sapida,** *Müll.-Arg.; Gamble, Man. Tim., 354.*

Syn.—PIERARDIA SAPIDA, *Roxb., Fl. Ind., Ed. C.B.C., 323.*

Vern.—*Lutco,* HIND.; *Latká,* BENG.; *Kala bogoti,* NEPAL; *Sumbling,* LEPCHA; *Latecku,* ASS.; *Koli kuki,* KAN.; *Kanaizu,* MAGH.; *Kanazo,* BURM.; *Lutqua,* CHINESE.

Habitat.—A small or moderate-sized evergreen tree, met with in Eastern Bengal, Tippera, Burma, and the Andaman Islands.

DYE. Mordant. 5 Green dye. 6

Dye.—The leaves are used in Northern Bengal and Assam for dyeing. (*Gamble.*) The bark is used chiefly as a mordant in dyeing with madder and lac. "The Lepchas extract a *green* dye from the *leaves.*" (*Dr. Schlich.*) Regarding this statement, it is extremely doubtful whether any single plant yields a green dye; careful enquiry should be made to ascertain if the leaves are macerated along with the Lepcha indigo plant (**Marsdenia tinctoria**) to produce the green colour alluded to by **Dr. Schlich.** For information regarding other reputed simple green dyes, see **Hedyotis, Jatropha,** and **Vigna.**

FOOD. Fruit. 7

Food.—It produces a fruit of the size of a large gooseberry, yellow and smooth, with seeds embedded in a pulpy aril. It is acid and pleasant and esteemed by natives. In the Rangoon market it is generally plentiful. **Roxburgh** says that the Chinese gardeners in his employment knew the tree, and that in their country the fruit was called *lutqua.*

TIMBER. 8

Structure of the Wood.—Greyish brown, soft, liable to split badly. Weight about 42 lbs. per cubic foot.

9

BACCHARIS, *Linn.; Gen. Pl., II., 286.*

An American genus of herbs, shrubs, or trees, belonging to the COMPOSITÆ, and containing some 200 species. They are chiefly distinguished by the fact that the male flowers are on one plant and the female on another. The resinous species are largely used as firewood. An infusion of the seeds of one species is employed by the Brazilians as a sudorific and tonic, while another species yields a bitter principle, used in the treatment of fevers.

10 **Baccharis indica,** *Linn.,* Syn. for **Pluchea indica,** *Less.; Fl. Br. Ind., III., 272.*

Baccharis nitida, *Wall.; Syn.* for **Blumea chinensis,** *DC.; Fl. Br. Ind., III., 268.*

11

BALANITES, *Delile; Gen. Pl., I., 314.*

12

A small genus containing 2 or only 1 species, belonging to the Natural Order SIMARUBEÆ. Spiny shrubs or trees. *Leaves* coriaceous, 2-foliolate, entire. *Flowers* in axillary cymes, small, green. *Sepals* and *petals* 5; *disk* thick, conical. *Ovary* entire, 5-celled. *Seeds* solitary, pendulous from the apex of each cell.

Balanites Roxburghii, *Planch.; Fl. Br. Ind., I., 521;* SIMARUBEÆ.

13

Syn.—(?) Only a variety of B. ÆGYPTIACA met with in Africa; XIMENIA ÆGYPTIACA, *Roxb., Fl. Ind., Ed. C.B.C., 323.*

Vern.—*Hingan, hingu* or *hingen, ingua, hingol, hingota* or *hingot,* HIND.; *Hingon,* BENG.; *Gárrah,* GOND; *Hingana* or *hingan,* MAR.; *Egorea* or *igoreá, hinger,* GUJ.; *Hinganbet, hingan,* DUK.; *Hingánbet,* CUTCH; *Hingat,* ULWAR; *Hinganbet,* BOMB.; *Nanjundá,* TAM.; *Nanchuntá,* MALA.; *Gári, gára-pandu, gára-chettu* (plant), *ringri,* TEL.; *Ingudi-vrikshaka* (plant), *ingudi* or *ingudam,* SANS.

Habitat.—A small, thorny tree, growing in the drier parts of India, extending from Cawnpore to Sikkim, Behar, Gujarát, Khándesh, and the Deccan. It is found in Dehra Dún (*Royle*), and also in Burma.

Grows everywhere, often little more than a thorny bush in the Panch Mahâls, Gujarát. (*Bomb. Gaz., III., 200.*)

Properties and Uses—

Oil.—A fixed oil is expressed from the seed. The corresponding oil prepared from the African plant is known to the negroes as *Zachun.*

OIL. 14

§ "The oil is used as an application for the cure of o cline. It is referred to in the drama of *Shakuntalá.*" (*Sakhárám Arjun Rávat, L.M., Girgaum, Bombay.*)

Medicine.—The SEEDS, FRUIT, BARK, and LEAVES are used in native medicine. The seeds are given in coughs. The bark, unripe fruit, and leaves have anthelmintic properties attributed to them and are purgative. The bark is given to cattle as an anthelmintic, especially by the people of Western India. The unripe drupes have strong cathartic properties; they are also anthelmintic.

MEDICINE. Seed. 15 Fruit. 16 Bark. 17 Leaves. 18

§ "Seeds are expectorant, dose 2 to 30 grains. Fruit purgative, 1 to 20 grains." (*Surgeon W. Barren, Bhuj, Cutch.*)

"Seeds are useful in colic, dose half a seed." (*Joseph Parker, M.D., Poona.*)

Fodder.—The young twigs and the leaves are browsed by cattle.

FODDER. Leaves and Twigs. 19

The ripe fruit is oval, of a yellowish colour, composed of a sweet but disagreeable pulp surrounding the stone. In Western India, as also in Egypt, it is eaten as a fruit, and when fermented is said to yield an intoxicating liquor used by the negroes. **Baillon** gives the ripe fruit the name of Desert-date, and when unripe that of Egyptian myrobalan.

Structure of the Wood.—Yellowish white, moderately hard, no heartwood, no annual rings. Weight 48 lbs. per cubic foot.

TIMBER. 20

It is used for walking-sticks and for fuel, and by the Africans for house furniture.

Domestic Uses.—"The nut is employed in fireworks. A small hole is drilled in it, at which the kernel is extracted, and being filled with powder and fired, bursts with a very loud report, so exceedingly hard is the nut." (*Roxburgh.*) The nuts are made into crackers in the Panch Mahâls. The pulp of the fruit is used as a detergent to clean silk in Rájputana. (*Brandis.*) The bark yields a juice, used in the Panch Mahâls, Bombay, to poison fish. (*Bomb. Gaz., III., 200.*)

DOMESTIC. Crackers. 21 Soap Substitute. 22 Fish-poison. 23

24

BALANOPHORA, *Forst.; Gen. Pl., III., 235.*

A genus of leafless parasites which give their name to a small Natural Order, which may be briefly defined as parasitic herbs, fleshy, aphyllous, monœcious or diœcious, having scapes naked or scaly, and terminating in capitula of flowers, each having 2-8 generally 3-lobed perianth, with 3-many stamens inserted on the perianth, the ovary being inferior and 1-celled. In India we have only a few species belonging to this Natural Order, Balanophora itself containing the most important examples. The Indian species do not appear, however, to have been put to any economic use. In Dr. Dymock's *Mat. Med., West. India*, under *Gajpipal*, p. 719, occurs a brief notice of what in all probability is a species of Balanophora. It is sold in Bombay under the vernacular name of *gajpipal*, and is described as mucilaginous and astringent. Various other plants are also sold by Indian druggists as *gajpipal*. **Roxburgh** gives **Pothos officinalis** as the drug which bears that name, and in the Panjáb it appears that **Plantago amplexicaulis** is also known to the native druggists as *gajpipal*.

Gajpipal. 25

In various parts of the world the species of Balanophoreæ are known to possess astringent properties. The reddish juice of **Cupomorium coccineum** (the Fungus Melitensis) was formerly prescribed as an infallible styptic for hæmorrhage and diarrhœa. In Java, wax prepared from a species of Balanophora is made into candles.

§ "**Balanophora?** This is substituted in the shops of Bombay for **Scindapsus officinalis**. It is a parasite. Does not seem to possess any active properties." (*Sakhárám Arjun Rávat, L.M., Girgaum, Bombay.*)

BALATA.

BALATA GUM. 26

Balata gum. A caoutchouc-like substance, obtained in all probability from two or three species, chiefly **Achras Sapota**, *Linn.*, and several species of **Mimusops**, which see.

27

BALIOSPERMUM, *Blume; Gen. Pl., III., 324.*

A genus of EUPHORBIACEÆ, comprising some four species of Indian shrubs or bushes, belonging to the Tribe CROTONEÆ and the sub-tribe GELONIEÆ.

Leaves alternate, irregularly sinuato-dentate or sub-lobed, penniveined or at times tri-costate at the base. *Inflorescence* axillary racemes, flowers fascicled, rarely elongated and lax. *Flowers* monœcious, apetalous; *Sepals* 4-5, imbricate. *Stamens* 10-60; filaments thin, free; anthers terminal, loculi adnate. *Disk* entire. *Ovary* 3-4-locular; *style* short, fleshy, recurved, shortly 2-fid; *ovules* solitary in the loculi. *Capsule* 3-celled, separating into 3 cocci.

28

Baliospermum montanum, *Müll.-Arg.; Gamble, Man. Tim., 348;* EUPHORBIACEÆ.

Syn.—CROTON POLYANDRUM, *Roxb., Fl. Ind., Ed. C.B.C., 687*; C. ROXBURGHII, *Wall.*

Vern.—*Dánti, hakúm* or *hakún*, BENG., HIND.; *Dánti*, SANS.; *Habbussalátine-sahrái, habbussalátine-barri*, ARAB.; *Bédanjíre-khatái*, PERS.; *Konda-ámudam, naypawlum (Dr. Kinsley), adavi-ámudam*, TEL.; *Poguntig*, LEPCHA; *Jangli jamalgota*, N.-W. P.; *Dánti*, MAR.; *Jamálgotá, dántimul*, BOMB., GUJ., CUTCH. "The vernacular names of *B. montanum, Croton Tiglium, Jatropha glandulifera*, and *J. Curcas* are confounded with each other in most districts of India, particularly in the Madras Presidency." (*Moodeen Skeriff.*)

The root is sold as *dántimul* by native druggists.

Habitat.—One of the commonest shrubs of North and East Bengal. It extends to South India and Burma.

Properties and Uses—

MEDICINE. Seeds. 29

Medicine.—The SEEDS are used as a drastic purgative, but in overdoses are an acro-narcotic poison; they are sometimes used as a substitute for **Croton Tiglium,** and **Dr. Dymock** tells me they are often sold in the bazars under the vernacular names given for that plant (*Jamálgota*). They are also used externally as a stimulant and rubefacient. The OIL expressed from the seeds is a powerful hydragogue cathartic, and is useful for external application in rheumatism. **Madden** states that to the east of the Sutlej its leaves are in high repute for wounds. The sap is believed to corrode iron. The ROOT is considered cathartic and is used in dropsy, anasarca, and jaundice: it is generally administered with aromatics. (For the preparation of *Dánti-haritaki* confections, see *U. C. Dutt's Mat. Med., Hindús, p. 230.*)

OIL. 30

MEDICINE. Root. 31

§"A decoction of the LEAVES is said to be useful in asthma." (*Assistant Surgeon Bhagwan Dass, Rawal Pindi, Panjáb.*)

Leaves. 32

"Drastic, purgative, dose of powder 1 to 10 grains; of the oil 1 to 3 minims; used in dropsy." (*Surgeon W. Barren, Bhuj, Cutch.*)

"The root *Dántimul* is much used in Hindú medicine." (*Surgeon-Major W. Dymock, Bombay.*)

BALLOTA, *Linn.; Gen. Pl., II., 1212.*

33

Ballota limbata, *Benth.; DC. Prod., XII., 521;* LABIATÆ.

Vern.—*Búi, phútkanda, jandí, lana, kandiári, agzhan, awáni-búti,* PB.

Habitat.—A small prickly shrub, with yellow flowers, occurring on the Salt-range, Trans-Indus, and in the Jhelam basin, at times ascending to altitude 4,000 feet.

Medicine.—The JUICE of the leaves is applied to children's gums, and to ophthalmia in man and beast. (*Stewart.*)

MEDICINE. Juice. 34

Fodder.—Browsed by goats.

FODDER. 35

BALSAM.

Balsam, Canada, is obtained chiefly from **Pinus balsamea,** *L.;* see under **Abies.**

Balsam, Copaiba, is obtained from several species of **Copaifera,** natives of South America.

36

Balsam, Gurjan, see **Dipterocarpus.**

Balsam of Peru is obtained from **Myroxylon Pereiræ,** *Kotz.*, a native of Central America.

37

Balsam of Tolu is obtained from **Myroxylon Toluifera,** *H. B. K.*, a native of Central and South America.

38

BALSAMODENDRON, *Kunth.; Gen. Pl., I., 323.*

39

A genus of BURSERACEÆ, comprising some 10 species of balsamiferous, spiny, small trees or shrubs, inhabitants of North India and Arabia, and of tropical and South Africa.

Leaves alternate, 1-5-foliolate or imparipinnate; *leaflets* sessile, oblique, crenate-serrate. *Flowers* small, few fasciculate, polygamous, on racemose panicles. *Calyx* tubulo-urceolate, 4-toothed, persistent. *Petals* valvate. *Disk* erect, cupular. *Stamens* 6-8, usually 4 long and 4 short, inserted on the margin of the disk. *Ovary* sessile, surrounded by the disk and 4-celled; *style* short, *stigma* obtuse, 4-lobed; *ovules* 2 in each cell. *Drupe* indehiscent, ovoid, containing a 1-3-celled and -seeded stone or 1-3 distinct stones within the pericarp.

Baillon claims that by priority the correct name for this genus should be **Balsamea. Gleditsch Engler** (*Engler, Bot. Jahr., I., p. 42*) in his revision of the **Burseraceæ** concurs in this view. In a work on Indian

Economic Botany it seems desirable, however, to follow the *Flora of British India* in all matters of synonymy.

Balsamodendron Berryi, *Arnott.; Fl. Br. Ind., I., 129;* BURSERACEÆ.

GUM. 40

Habitat.—A tree of the forests on the east side of the Nilgiris.

Gum.—It is very fragrant and yields a gum-resin.

41

B. Kataf, *Kth.*

AFRICAN BDELLIUM.

Syn.—BALSAMEA ERYTHRÆA, *Engler;* AMYRIS KATAF, *Forsk.;* HEMPRICHIA ERYTHRÆA, *Ehrenburg.*

Vern.—*Bysabol,* BOMB.; *Muhishabole,* CUTCH; *Mhaisabol* (or *bésabol*), SANS.; *Habak-Hádí* (corrupted into *Habaghadi*), ARAB.

BDELLIUM. 42

Gum.—This gum-resin reaches Bombay from Berbera; the purer kinds very much resembling Myrrh, with which it has been confused by many authors. **Professor Oliver** refers it to **B. Playfairii**, thus making the gum-resin *hodthai* to be obtained from the same plant as *bysabol* (see **B. Playfairii**). *Bysabol* is darker and more reddish than the true Myrrh; it is but sparingly soluble in bisulphide of carbon, and the solution does not assume the violet shade characteristic of Myrrh on the addition of bromine. It has a much stronger or acrid taste, and a peculiar odour quite different from that of the true Myrrh. (*Kew Report, 1880, p. 50; Flück. and Hanb., Pharmacog., p. 146; Bentley and Trimen, Med. Plants, p. 60; Dymock's Materia Medica of Western India, p. 128.*)

§ "Emmenagogue. Invariably given after delivery, dose 1 to 6 grains; used locally for chronic ulcers." (*Surgeon W. Barren, Bhuj, Cutch.*)

43

B. Mukul, *Hook.; Fl. Br. Ind., I., 529.*

GUM GUGUL.

Vern.—*Guggul, gúgal, mukul, ranghan túrb,* BENG., HIND., DUK., GUJ., and SIND; (*Gogil,* HIND.;) *Gugal,* GUJ., CUTCH; *Guggala,* KAN., MAR.; *Maishákshi* (commonly *maisháchi*), *gukkal, gukkulu,* TAM.; *Mahi-sáksh;* or *máisákshi* (commonly *maisáchi*), *gugul,* TEL.; *Moql, moqle-arzaqi aflátan,* ARAB.; *Bóé-jahúdán,* PERS.; *Koushikaha, guggulu,* SANS.; *Gugula, jatayu* or *javáyu, ratadummula,* SINGH.

Habitat.—A small tree, found growing in the arid zones, Sind, Kattiawar, Rájputana, Khándesh.

GUM-RESIN. 44

Gum-resin.—Yields the gum-resin known as *Gugul* or as an *Indian Bdellium.* It occurs in vermicular or stalactitic pieces, is of a brown or dull green colour, and has a bitter, acid taste. It exudes from incisions on the bark made in the cold season. It swells when heated, diffusing a disagreeable odour.

Properties and Uses—

MEDICINE. 45

Medicine.—Indian Bdellium is used in native medicine as a demulcent, aperient, carminative, and alterative; especially useful in leprosy, rheumatism, and syphilitic disorders. It is also prescribed in nervous diseases. scrofulous affections, urinary disorders and skin diseases, and is employed in the preparation of an ointment used for bad ulcers.

It is known by the name of *gugul* or *mukul,* and is said to be moister and therefore not so brittle as myrrh, for which it is often used as a substitute, being much cheaper. The *Pharm. India* states that in general practice it is found useful in the form of an ointment in cleansing and stimulating indolent ulcers, and is a favourite in the treatment of Delhi sores, especially when combined with sulphur, catechu, and borax. *Gugul* has stimulating properties, and is sometimes given internally, especially in the treatment of horses.

SPECIAL OPINIONS.—§ "Applied as a hot paste to incipient abscesses, as an absorbent. Is used as an expectorant. Is aphrodisiac according to **Sk. Boali-Saina,** the 'king of Hakims.' Applied locally as a paste in hæmorrhoids." (*Surgeon G. A. Emerson, Calcutta.*) "Used externally and internally in muscular rheumatism, leprosy, piles, dysentery, gleet, scurvy, fistula, hysteria, anæmia, dyspepsia, and chronic diseases of the lungs. A preparation of *gugul* called *yogaraja gugul* is given internally for muscular rheumatism, and is said to be more effective in its action when administered with the decoction of *rasna;* given with the infusion of *adulsa* as an expectorant. There are 22 ingredients in the preparation called *yogaraja gugul.*" (*Surgeon W. Barren, Bhuj, Cutch.*) "Bdellium is sold in the native drug-shops in Madras, and is known as *Mahi-sakshi;* I have personally identified it." (*Deputy Surgeon-General G. Bidie, C.I.E., Madras.*) "The fumes of gugul are believed to be disinfectant." (*Surgeon R. L. Dutt, M.D., Pubna.*) "It is held in the highest repute in the treatment of rheumatism. It is given internally and applied locally. Internally it is given in the form of pill, with numerous other native drugs." (*Surgeon-Major J. Robb, Ahmedabad.*) "Repeatedly beaten with a hammer its efficacy is said to increase; it is extensively used here in sciatica and all rheumatic affections." (*Surgeon J. C. Penny, M.D., Amritsur.*) "In combination with other medicines (black pepper and colchicum) the gum is given in the form of confection in cases of rheumatism, hæmorrhoids, and flatulent dyspepsia. The dose is 1½ drachms." (*Lal Mahomed, Hospital Assistant, Hoshangabad, Central Provinces.*)

TIMBER. 46

Structure of the Wood.—Soft, white. Weight 20 lbs. per cubic foot.

Apparently not put to any purpose.

DOMESTIC. Cement. 47

Domestic Uses.—Mixed with mortar the gum forms an excellent cement; it is soluble in potash.

48

Balsamodendron Myrrha, *Nees.*

MYRRH.

Vern.—*Ból,* PERS., HIND., DUK.; *Gandha-rasaha, hirából, ból,* BENG.; *Hirábol, ból,* GUJ., CUTCH; *Bóla, gandha-rasaha, rasagandhaha,* SANS.; *Mur,* or *murr,* ARAB.; *Mor,* HEBREW; *Vellaip-pólam,* TAM.; *Bálimtra-pólam,* TEL.; *Bólá,* KAN.; *Bólam, gandarassa,* SINGH.

There are two important kinds of what may be called the true Myrrh; these are the African, or *Karam,* and the Arabian and Siam, or *Meetiya.*

Habitat.—A small tree of Arabia and of the African coast of the Red Sea. Often cultivated in Western India.

49

History.—There are several distinct substances which, in English, go by the name of Myrrh. There is the common British herbaceous plant belonging to the family of the Carrot (UMBELLIFERÆ) which, in all probability, derives its name, **Myrrhis odorata,** from the resemblance of the smell of its fresh green stems to that of the Eastern Myrrh gums. The Myrrh of the ancients is now pretty generally believed to have been the gum-resin known in India as *Heerabole* or *Myrrh,* a proportion at least of which is the produce of **Balsamodendron Myrrha,** *Nees.* Bdellium and *gum gugul* are sometimes known as "False Myrrh." Some authors think that the Myrrh of the ancients was also obtained from a species of the genus **Cistus,** the Rock Rose, a genus not represented in India. This idea is chiefly based upon the fact that the gum obtained from that plant is known at the present day by the name of "*Ladanum,*" a word supposed to be the same as the Hebrew "*Lát*" which has been translated as Myrrh. If this be correct, two distinct gum-resins have come to bear the same name in translations from the Hebrew writers.

BALSAMODENDRON **Myrrha.** — **Myrrh.**

THE GUM-RESIN. Myrrh. 50

COMMERCIAL FORMS.—Of the Myrrh of commerce there are two or three distinct varieties, and under each an assortment of stuffs of different commercial value. There is the east coast African Myrrh known in Bombay as *Karam*, and the South Arabian and Siam Myrrh, the *Meetiya*. Myrrh of good quality is also sent from Persia. The latter two forms are really only substitutes, however, for Myrrh. The *Karam* obtained from Africa may be said to be the true commercial Myrrh, but it is by no means proved that this is entirely obtained from **B. Myrrha,** *Nees.* The principal mart for Myrrh is in Bombay, the chief firms having their agents at Aden and Mukulla. These agents attend the great annual fair at Berbera, and exchange English and Indian goods for Myrrh and Bdellium. The bags of these, on arrival at Bombay, are said to contain, *1st*, a large proportion of roundish masses of fine Myrrh; *2nd*, a considerable proportion of small semi-transparent pieces of Myrrh; *3rd*, numerous pieces of dark-coloured Myrrh, mixed with refuse; *4th*, a small proportion of opaque gum-resin (*Guibourt's "Opaque Bdellium"*). The packages are assorted, the best qualities are re-shipped for Europe, as also the darker pieces, declared as second quality, while the refuse is exported to China. The best qualities of *Karam* Myrrh sell for R34 per maund of 37 lbs.; *Meetiya*, R16 to 25; and the so-called refuse, R8 per maund. (*Dymock's Materia Medica, W. Ind., pp. 124-131.*)

Myrrh is chiefly adulterated with inferior qualities, or with the gums and resins derived from other species of **Balsamodendron,** such as **B. Mukul,** *Hook.;* **B. pubescens,** *Stocks;* **B. Opobalsamum**; and sometimes also with **B. Roxburghii,** *Arn.*

Properties and Uses—

MEDICINE. 51

Medicinal Properties.—Myrrh is beneficial in dyspepsia, amenorrhœa, and chlorosis, and is a useful stimulant and astringent to all ulcerations or congestions of the mucous membrane. It is a useful application to old, foul and indolent ulcers, and a valued wash for the mouth and gums, and a gargle in ulcerated sore-throat. (*K. L. De, C.I.E., Rai Bahadur.*) It is a stimulant expectorant, much admired as a remedy for pulmonary affections, especially the asthma of the aged. (*Ind. Pharm.*) **Meer Mahommed Husain** says it is hot and dry, and that the best quality when broken shows white marks like those at the root of the finger-nails. Internally it is regarded as tonic and antispasmodic, emmenagogue, astringent and expectorant. Hakims use it for intestinal worms. It is "detergent, siccative, astringent, and aperient, a disperser of cold tumours, and one of the most important of medicines, as it preserves the humours from corruption." "Dissolved in women's or asses' milk, it is dropped into the eye in purulent ophthalmia." (*Dymock.*) It is said to cause abortion, and is useful in fever and epilepsy.

52

OFFICINAL PREPARATIONS.—It is an ingredient in Decoctum Aloes, Comp.; in Mistura Ferri, Comp.; in Pilula Aloes et Myrrhæ; in Pilula Asafœtidæ, Comp.; and in Pilula Rhei, Comp. It is also made into Tinctura Myrrhæ, P. B.

Dose in pill, powder, or emulsion, 10 to 30 grains; of tincture ½ to 1 fl. drachm.

53

SPECIAL OPINIONS.—§"*Múrú* (*Swahili-e-Africa*) is not found in Indian bazars, but is brought to Zanzibar from Mukulla and Arabia generally. In shape it is of a small, hard, black cone. Composition unknown; rubbed down on an earthen plate in water to the consistence of thin gruel and taken as a drink in flatulence and dyspepsia. Is very commonly given to children, and used internally and externally is said to allay the severe pain of orchitis (*Zanzibar*)." (*Surgeon-Major John Robb, Bombay.*) "The gum-resin, in combination with *gúr*, is given to increase the secretion of milk in women." (*Narain Misser, Hoshangabad, Central Pro-*

vinces.) "Myrrh, as obtained in the bazars, is very impure, but is used to check bronchial secretion. Owing to its impurity, not much reliance can be placed on it." (*A Surgeon.*) "In combination with dilute nitric acid, I have found Myrrh a very useful application to chronic and unhealthy ulcers. Its efficiency as an ingredient in expectorant mixtures is too well known to call for remark." (*Surgeon S. H. Browne, Hoshangabad, Central Provinces.*)

Balsamodendron Opobalsamum, *Kunth.; Brandis, For. Fl., 65.* 54

BALSAM OF BALM OF GILEAD.

Vern.—*Balásán, balasán-ká-tél,* HIND., DUK.; *Balasán, duhnul-balasán,* ARAB.; *Róghane-balasán,* PERS. The fruit: *Hábul bálasán* (correctly *habbul-balasán*), BOMB., ARAB.; *Tukhme-balasán,* PERS. The wood: *Úde-bálasán,* PERS., BOMB.

Habitat.—A small-branched tree found on both sides of the Red Sea, south of 22° north latitude. It is also recorded from several places on the Nubian coast and in Abyssinia. It is met with on the Asiatic side at Ghizandad in Arabia, at Aden, and Yemen. It is in all probability introduced into Palestine.

Properties and Uses—

GUM. 55

Gum.—The famous Balm of Gilead or Balsam of Mecca is imported into Bombay from Arabia. It is a greenish-yellow oleo-resin of the consistence of honey, used as a perfume and in medicine.

MEDICINE. 56

Medicine.—The wood (*Ood-i-Balasán*) and the fruit (*Tukm-i-Balesán*) are also imported, and are chiefly used as medicines by the Yunani Hakims of India. "The fruit is considered to be a powerful carminative and digestive; it is also praised as a stimulant, expectorant, and is usually administered in combination with tragacanth." (*Dymock.*)

§ "Mixed with oil of roses, balsam is used in earache. Made into a paste with lard, it is applied locally in scrofulous and cancerous sores." (*Surgeon G. A. Emerson, Calcutta.*)

B. Playfairii, *Hook. f.* 57

OPAQUE BDELLIUM.

Vern.—*Hotai,* or *hodthai,* SOMALI, DUK., ARAB.; *Meena-harma,* BOMB.

Habitat.—Met with in North-East Africa.

GUM-RESIN. 58

Gum-resin.—Yields an opaque, whitish gum-resin, which is used as a soap by the Arabs and Somalis to kill lice, and in Bombay in the cure of guinea-worm. (Compare with remarks under **B. Kataf.**)

§ "A recent chemical examination has shown that *Opaque Bdellium* and *Hotai* are far from being identical. *Dukh,* the Arabic name of the gum *Hotai.*" (*Surgeon-Major W. Dymock, Bombay.*)

B. pubescens, *Stocks.; Fl. Br. Ind., I., 529.* 59

Vern.—*Bayisa-gugul,* MAR.; *Bayi, bai,* BELUCH.

Habitat.—A small tree of Beluchistan and the hills separating that country from Sind as far south as Karachi.

GUM-RESIN. 60

Gum-resin.—It yields a small quantity of tasteless, inodorous, brittle gum, almost entirely soluble in water.

MEDICINE. 61

Medicine.—Dr. J. Newton reports that the gum obtained from this tree may be used in the form of ointment for cleansing and stimulating bad ulcers. It is a favourite application in Delhi sores, combined with sulphur, catechu, and borax; it is reported to stimulate healthy action.

B. Roxburghii, *Arn.; Fl. Br. Ind., I., 529.* 62

Vern.—*Gugala,* BENG.; *Gugal, mhaishabola,* BOMB.; *Gugar,* SIND; *Gugal,* GUJ.; *Gúkul,* TAM.

Habitat.—A small tree of East Bengal and Assam.

Properties and Uses—

GUM-RESIN. 63

Gum-resin.—It yields a gum-resin of a greenish colour, moist and easily broken, having a peculiar cedar-like odour; it is largely supplied to the Bombay market from Amráoti, and is much used by masons to mix with fine plaster. (*Dymock.*)

CEMENT. 64

OIL. 65

Oil.—**Baden Powell** mentions that the plant yields a medicinal oil. This is in all probability the gum-resin, which is quite liquid and not unlike an oil when fresh.

MEDICINE. 66

Medicine.—**Birdwood** mentions this plant in his list of drugs, but gives no information as to its medicinal properties.

BAMBOO.

The Bamboo. 67

Probably no plants are more valuable to the inhabitants of India than the graceful, gigantic grasses, popularly and collectively known as Bamboos. They constitute the Tribe BAMBUSEÆ of the Natural Order GRAMINEÆ,—the Grasses. The late **General Munro**, in a valuable paper upon Bambusæ published in the Transactions of the Linnæan Society of London, Vol. XXVI., 1870, gives an account of over 170 species. Previous to this, the literature of the subject consisted of scattered publications describing the bamboos of certain regions, the only complete paper having been written in 1839 by **Prof. Ruprecht**, in the Transactions of the Academy of St. Petersburg. Subsequent to **General Munro's** account of these exceedingly valuable plants, the late **Mr. Kurz**, of the Calcutta Botanic Gardens, in the Journal of the Asiatic Society of Bengal (Vol. XXXIX., 88, and XLII., 249), described a number of new or little-known species, and gave in the *Indian Forester*, Vol. I., pages 199, 335, under the title "Bamboo and its Uses," much interesting information of a practical nature. The difficulty in collecting complete sets of the leaves, flowers, and fruit of the same species of bamboo renders the study of the bamboos exceedingly troublesome. It must be admitted that there remains much to be done before we can be supposed to possess even an approximately complete knowledge of these most useful plants. **Bentham** and **Hooker**, in the *Genera Plantarum*, refer the members of the Bambusæ to 22 genera, of which 14 have representatives in India and the Malaya. Unfortunately, however, the information of an economic nature is uniformly published under the generic name Bamboo, or is, at most, associated with but one scientific name. It is, accordingly, almost impossible to refer the properties and uses of the various bamboos described by authors to their respective botanical species. On this account it has been deemed advisable to give, in this place, a popular account of the bamboo, instead of attempting to pursue the course usually adopted in this work. The following brief analysis of the more important genera of the Indian Bambuseæ may serve, however, to direct the reader's attention to their respective alphabetical headings, where fuller and more scientific details will be found :—

68

Tribe Bambuseæ.

Botanic Diagnosis.—Tall, bushy or arborescent grasses, with woody stems (the culms or halms of authors). *Leaves* flat; *sheath* large; *petiole* short or absent; *blade* articulated and generally caducous, leaving the sheath embracing the stem or branch. *Inflorescence* spicate, branched and spreading panicles of fertile spikelets rarely glomerulate; panicles and even spikelets sometimes bracteate. *Spikelets* pedicelled or subsessile, one to many-flowered. *Glumes* 2 or many, empty, often graduating above into the *palea*, the lower pairs of which may be empty or contain only abortive or incomplete flowers. *Palea* large, 2 outer and inner, standing opposite each other and protecting the floret, rarely absent, arranged distichously on the rachis; *outer* concave or involute, and generally keeled, those of the fertile being often quite different from the sterile flowers; *inner* flat or concave, 2-keeled and more delicate than the outer. (The palea afford perhaps the most important generic character.) The florets normally are composed of three whorls of organs arranged ternately: *1st, Lodicules* (or squamules)—3 hypogynous scales, 2 alternating with the palea, and 1 opposite or only 2, or absent—these may be said to correspond to the corolla; *2nd, stamens*—3, 6, or many; *3rd, ovary*—sessile or spuriously stalked, ovate or pearshaped, with a solitary *style* and *stigma*, 2-3-fid, rarely entire. The florets may be diclinous, hermaphrodite, or polygamous or even abortive, and it is by no means unusual to find the palea giving origin not to a floret but to a spikelet, this peculiarity producing branched and spreading inflorescences with pedunculate spikelets, so frequent a condition in the Bambuseæ.

Sub-tribe 1st, ARUNDINARIEÆ.

Stamens 3. *Palea*, 2-keeled. *Pericarp* thin, semi-adnate to the seed.

1. Arundinaria, *Mich.*

Classification of Bamboos. 69

Cæspitose shrubs with slender mostly annual stems, rarely arborescent. *Spikelets* many-flowered, mostly pedunculate, forming racemes or panicles, the branches occurring in the axils of small linear bracts which become large and amplexicaul in the species formerly referred to the genus Thamnocalamus; empty *glumes* 1-2, inferior.

Sub-tribe 2nd, EUBAMBUSEÆ.

Stamens 6. *Palea* 2-keeled. *Caryopsis* small, wheat-like. *Pericarp* thin, semi-adnate to the seed.

** Filaments free.*

2. Bambusa, *Schreb.*

Trees, rarely shrubs or more rarely scandent plants, growing in clumps; stems tall, woody. *Spikelets* 2 to many-flowered, generally sessile, in interrupted glomerulate panicles. Empty *glumes* 3-4, inferior. *Palea* boat-shaped, with ciliate keels or distinctly winged. Apex of the *ovary* hairy. *Style* deciduous, deeply 2-3-fid. *Embryo* conspicuous on the surface of the fruit. *Caryopsis* with a deep longitudinal furrow, often adherent to the palea.

*** Filaments united into a tube.*

3. Gigantochloa, *Kurz.*

Inflorescence and habit of Bambusa. *Spikelets* many-flowered. *Palea* boat-shaped, all 2-keeled.

4. Oxytenanthera, *Munro.*

Inflorescence and habit of Bambusa. *Spikelets* 1 to many-flowered, the terminal one only being fertile. *Palea* absent or indistinguishable from the glumes.

Sub-tribe 3rd, DENDROCALAMEÆ.

Stamens 6. *Palea* 2-keeled. *Caryopsis* often very large. *Pericarp* separating into an outer hard shell free from the seed.

5. Dendrocalamus, *Nees.*

Habit of Bambusa. *Spikelets* 2 to many-flowered; panicles distantly glomerulate, the flowers often only one, fertile. *Lodicules* none or very rarely represented by 1-2 rudimentary scales. *Inner palea* boat-shaped and 2-keeled. *Ovary* stipitate, hirsute on the apex; *style* long, filiform, entire or 2-3-fid at the apex, base persistent. *Caryopsis* terete, generally small; pericarp thick; position of the embryo generally not conspicuous.

Classification of Bamboos.
69

6. Melocalamus, *Benth.*

Habit of Bambusa. *Spikelets* few, 2-flowered, forming elongated and sparsely glomerulate panicles. *Ovary* glabrous. *Caryopsis* large (as large as a wood-apple); pericarp thick, fleshy. (**Kurz** by mistake placed the species of this genus under Pseudostachyum.)

7. Pseudotostachyum, *Munro.*

Sub-arborescent plants with the foliage of Bambusa. *Spikelets* few, one-flowered, in bracteate spikes, forming open panicles. Empty *glume* one, inferior; terminal glume also empty and globose. *Caryopsis* comparatively small, flattened globose; pericarp crustaceous.

8. Teinostachyum, *Munro.*

Arborescent, fruticose, or sub-scandent bamboos, with the foliage of Bambusa. *Spikelets* one-flowered, sub-spicate, forming branched panicles, bracteate. Empty *glumes* 1-2, inferior, and terminal glume also empty and acute. *Caryopsis* large, acuminately beaked; pericarp fleshy, and when mature separating into an outer cartilaginous coat and inner layer.

9. Cephalostachyum, *Munro.*

Bushy or arborescent bamboos. *Spikelets* one-flowered, in numerous terminal globose or glomerulate heads, protected by bracts. Empty *glumes* 1-2, inferior. *Caryopsis* oblong, beaked; pericarp thick.

Sub-tribe 4th, MELOCANNEÆ.

Stamens 6 to many. *Spikelets* one-flowered. *Palea* absent or the same as the glumes. *Pericarp* crustaceous or fleshy, free from the seed.

10. Dinochloa, *Buse.*

Lofty-climbing, woody bamboos. *Spikelets* small, round, forming glomerulate panicles. Empty *glumes* 3-4, inferior, obtuse, many-nerved. *Lodicules* none. *Stamens* 6, free. *Caryopsis* fleshy and berry-like, ovate-acuminate.

11. Melocanna, *Trin.*

Arborescent bamboos with the foliage of Bambusa. *Spikelets* bracteate, arranged in unilateral compressed spikes. Empty *glumes* many, inferior, mucronato-acuminate and not keeled, becoming convolute (resembling palea) above. *Lodicules* 2. *Stamens* 6, free or more or less connate. *Caryopsis* very large; pericarp thick and fleshy.

12. Ochlandra, *Thw.*

Arborescent bamboos. *Spikelets* large, densely capitate or sub-spicate. Empty *glumes* 3 to many, inferior. *Lodicules* very irregular. *Stamens* most frequently 6 or many, filaments variously connate. *Caryopsis* large, with a thick fleshy pericarp.

70

HABIT AND GROWTH OF THE BAMBOO.

Under each of the Genera briefly defined in the preceding pages, one or more species of gigantic or even arborescent grasses have been described by botanists, each of which may popularly be known as "Bamboo." Most authors, however, speak of **Bambusa arundinacea** as "the bamboo," an expression which is quite incorrect, since the spiny bamboo of South and West India is by no means either the most useful or most abundant species in India as a whole. It is quite customary also to read of **Bambusa vulgaris** as the "common bamboo," whereas in India, at least, this cultivated Eastern species is by no means a common plant. It would be more correct to speak of **Bambusa Tulda** as the "common bamboo," and, as far as Bengal is concerned, *it* certainly is the most abundant species, while **B. Balcooa** is nearly as plentiful, and is much more useful, having also a wider distribution. On these grounds it would, by the majority of the natives of India, be pronounced "the bamboo." The term "male bamboo" may be said to be applied to any solid bamboo used for spear or lance staves, walking-sticks, &c.; it is more particularly applicable to **Dendrocalamus strictus.** The stamens and pistil being on the plant, it is difficult to see why the term "male" should be given to a solid and "female" to a hollow bamboo, but they are expressions in frequent use in India.

Male Bamboo. 71

Peculiarities of habit. 72

As with all other grasses, the bamboo stem consists of a more or less hollow culm, with transverse solid joints called nodes. The thickness of the woody shell and the length of the internode varies exceedingly in the different species. One peculiarity is preserved by all bamboos, namely, the rapid growth of the young shoot. Running up to its full height before the branches are produced, the shoot at the same time attains its full thickness immediately on escaping from the ground. This is a most important provision, for, otherwise, the branched culm could never penetrate through the crowded mass of its associates. Having in about a month reached its full height, the shoot commences to produce its branches and branchlets, and thus weighted, it curves into the graceful plume which is the elegant and familiar feature of the plant. At the same time, the large sheathing leaf-scales of the young shoots give place to the mature and distichously-arranged leaves. These, owing to their horizontal position and the concavity of their upper surfaces, keep rustling and trembling with every passing breeze. As a rule the bamboo is gregarious, establishing itself so thoroughly over certain portions of wild forest-clad tracts that it exterminates all other forms of vegetation. Seen from a height nothing could be more lovely; but, to the traveller, who for days together may have to clear a path for himself, the interminable monotony, and the twilight shade and death-like stillness, broken only by the sighing of the grating culms, make the bamboo jungle dreary in the extreme. However, in mixed forests, an occasional bamboo clump has a most pleasing effect. It supplies the traveller, moreover, with some of his most essential materials of equipment. Indeed, in a bamboo tract tents may be dispensed with, for, through the expert handling of the bamboo, the camp follower, armed with a large knife, can, in less than an hour, erect a most comfortable hut and furnish it with beds, tables, and chairs, all constructed from the bamboo.

Popularly, bamboos may be divided into those which grow in separate clusters or clumps, and those which grow in a more continuous manner. The former are characteristic of the tropical, and the latter of the extra-tropical or temperate forests. The clumped forms give to the soil a curious undulated appearance, each elevated mass consisting of the old stems or rhizomes and the entangled and tufted roots with the earth gathered around them. In many localities ant-hills surround the bamboos to a height of several feet, ultimately proving fatal to the plant, which, however, for a time, appears to grow from the top of the elevated mounds. Each clump consists of from 30 to 100 culms, which attain a height of from 30 to 100 or even 150 feet. In the scattered forms the culms rarely rise singly from the rhizome, but form small clumps containing only a few culms, the clumps being so closely packed together as to form impenetrable jungles. These latter are smaller in thickness and height, and are generally solid; collectively they are known as hill bamboos. A few species are climbers, their festoons and pendulous boughs passing gracefully from tree to tree.

The home of the giant forms of bamboo may be said to be in the tropical and extra-tropical forests; on entering the temperate zones they dwindle down to mere under-shrubs, until ultimately they are scarcely distinguishable from other grasses.

73

THE STEM OR CULM.

For about two-thirds of its ower portion, the culm is unbranched, or possesses only very short and inconspicuous branches. As already stated, on escaping from the ground the shoot attains at once its full diameter, appearing like a great scaly cone, clad in large embracing sheaths. The shoots and leaf-scales afford the most important popular and practical characteristics for recognising the different species of bamboo. Solid-stemmed bamboos are, as a rule, much smaller than hollow ones, but bamboo culms may be said to range from the thickness of a goose's quill to more than a foot in diameter. Until the branches have been fully developed the culm is not mature; this generally occupies a variable but considerable period, the shoot attaining its full height in from 30 days to 2 or 3 months. The branches are produced from below upwards, and with their appearance the stem gradually matures. A good deal has been written as to the rate of growth of the shoot, but up to the present date exact and definite figures, even for the important species, cannot be obtained. It is probable that an average of 3 inches per day would not overstate the growth of the young shoots of the more important bamboos. This seems also, in the majority of species, to take place chiefly at night and to continue for a month pretty uniformly, being increased, if anything, during fine clear days, and retarded apparently in damp and cloudy weather. The period of sprouting is generally about the beginning of the rains. **Captain W. H. Sleemann** says: "In the rains of 1835, my bamboos at Jubbulpore had not thrown out their shoots at what I considered the usual time, and I asked my gardener the cause. He replied, 'We have had no *thunder* yet; as soon as the thunder comes, you will get shoots.' I asked him what possible connection there could be between the sound of thunder and the shooting of the bamboos. 'God only knows,' said he, 'but we know that till the thunder comes, the bamboos never shoot well.' The thunder came, and certainly the gardener's theory seemed to me to be confirmed by a very steady and abundant shooting of the bamboos." This same belief is very generally entertained by the natives of India, and, as remarked by **Mr. Kurz**, the observation may be a perfectly correct one, the phenomenon depending upon the greater amount of nitrogen compounds in the atmosphere during electric discharges. "Repeated cutting of too many bamboo-shoots considerably weakens the stock, while the cutting of full-grown halms does not more injure them than mowing does the grass. Indeed, it is believed that too much cutting of shoots results in early flowering of the stock itself, and such means in most cases death to the whole plant." (*Kurz, Indian Forester, I., 257.*) This statement is in keeping with a very general opinion (see *Bombay Gazetteer, XV., Pt. I., 63*) that the year before flowering, the large bamboos cease to send up shoots. Besides, it has an important bearing upon the question of the application of the bamboo for the manufacture of paper—young and not mature culms being necessary for that purpose. (*See page 378,—Bamboo as a Paper Material.*)

Rate of growth. 74

The number of shoots produced yearly from each clump varies according to the vigour of the individual and the peculiarities of the species. Kurz states that the larger species produce 12 to 20 and the smaller 30 to 50. "If we assume, say, only 10 per stock a year, we should get as many as 300 halms to the stock in 30 years, which is the mean age of most of the bamboo species, at which they begin to flower and die off; while 50 and fewer halms to a bamboo-stock is a very dense growth even in those primeval forests where the axe of man does not touch them."

PECULIARITIES OF BAMBOO STEMS.

It is by no means unusual to find the greatest variation in the colour of bamboo stems. Some are dark green, as in **Bambusa Tulda**; yellow and even striped yellow and green in **B. vulgaris**; bluish with rough internodes in **Arundinaria racemosa**; and pale glaucous with blue rings at the nodes in **A. Hookeriana.** Many species when young are covered with a tomentum of closely-adpressed hairs or whitish powder, which in many cases forms a useful character: it is best seen upon the young scales. Solidified buds are developed into formidable recurved spines in **Bambusa arundinacea** and **B. spinosa,** while many of the hill bamboos produce below the sheath on the lower half of the culm a whorl of rootlets which harden into spinescent bodies. These are popularly called the spiny bamboos. This root spinescent tendency was found to be developed to a formidable extent in the hill bamboo on the Burma-Mánipur frontier, especially on the Kassome hills beyond the Kaboo valley. Most bamboos show a tendency to flattening above the nodes, especially where buds are developed. This is apparently what has been taken advantage of in the production of what is known as the cultivated square bamboo of China. Interesting information regarding this curiosity will be found in the *Tropical Agriculturist* (*April 1884, p. 698, reproduced from N. C. Herald*), from which the following passage may be extracted: "Pre-eminence is assigned to the square variety of this most useful as well as ornamental plant, which has been a favourite in imperial gardens wherever its acclimatisation has been effected in the north. The EMPEROR KAO TSU once inquired of his attendants who were planting bamboos, concerning the various kinds. In reply he was informed respecting several remarkable species. Chékiang in particular furnished one that was an extraordinary curiosity, in that it was square, and for that quality and its perfect uprightness was much esteemed by officers and scholars. They also told him that it was used for many purposes of decoration and utility, including, among others, that of being made into ink-slabs."

Spiny Bamboos.
75

Spinescent Bamboos.
76

Square Bamboos.
77

DURABILITY OF THE BAMBOO.

78

This depends, in the first instance, upon the culms being cut when mature. Specific peculiarities render some culms more durable than others, as, for example, the thickness of the woody shell, and the amount of silicious matter deposited within the tissue. In this latter respect bamboos vary exceedingly. Long immersion in water greatly enhances the durability, rendering the stem less liable to the attacks of insects, owing to the sap, which the insects are fond of, being thus extracted.

FLOWERING OF THE BAMBOO.

79

A great deal has been written regarding this exceedingly curious and interesting peculiarity. The inflorescence exhibits many important variations, most of which have been accepted by botanists as generic characters. All the species commence to flower when in full leaf, but as the inflorescence expands the leaves as a rule fall off, until when in complete

Flowering of Bamboos.
79

flower the clump or certain portions of it are leafless. In some cases special flowering culms are produced, at other times every culm flowers, the flowering portion or the entire clump dying off after the seeds are mature. In a few instances the plant continues to flower as a perennial, while some bamboos are entirely annual, flowering and dying down to the ground every year. With all the larger species the flowering stage is reached after a prolonged period of vegetation, variously stated at from 25 to 35 years, and is almost regularly followed by the death of the whole stock. **Captain Sleeman**, in the *Trans. Agri.-Horti. Soc. of India, III., 139*, dated 1836, publishes a most interesting letter containing the main facts of all that is known even at the present day regarding this subject. It may be well to reproduce a portion of this most valuable letter :—

"All the large bamboos, whose clusters and avenues have formed the principal feature in the beauty of Dehra Doon, ever since the valley became known to us, or for the last quarter of a century, have run to seed and died this season; as well those transplanted from the original stock last season and those transplanted 20 years ago. This is the character of the bamboo: all the produce of the same seed will run to seed and die in the same season, without reference to the season in which they may have been transplanted from the original stock; and unless we have them from different stocks we shall always be liable to lose all that we have the same season, and to have our grounds deprived for eight or ten years together of what may have been their principal ornaments; for the bamboo does not in less time attain its full size and beauty. The shoots of the first season come up small, whether they be from the original stock or seed, or from transplants from the original stock. We may take from the original stock bamboos six inches in diameter with a sufficient portion of the roots, and transplant them; but the shoots of the first season, from this stock, will still be very small, those of the second season will be larger, those of the third larger again, and so on till in about eight or ten years they attain their full size. It is well known that bamboos do not increase in diameter after they come above ground; they shoot out as thick as they are to be, and increase only in length after they come up. What is the ordinary age of the bamboo I know not, but the people of the hill and jungly tracts of Central India calculate ages and events by the seeding of the hill bamboos; a man who has seen two *Kutungs*, or two seedings of the bamboo, is considered an old man—perhaps sixty years of age."

Bambusa arundinacea is the common species in the Dún, and all these flowered in 1880: assuming that **Captain Sleeman** alludes to that species, these dates would give 44 years as the period required to reach flowering.

Another correspondent, in the same volume, writes: "Field rice is selling at 16 seers local weight per rupee, and bamboo rice at 20 seers. This bamboo, I am told, does not bear seed every year, nor at any fixed periods. The sign of bearing showed itself last year, after, I am informed, a lapse of 20 years; and some very old people could not call to their recollection when, prior to the former event, bamboos had borne seed; perhaps encouraged by particular circumstances connected with elemental changes, they spontaneously fecundate; for it does not appear to me that the matured bamboo only bore seed, but young and old together, from the lowest stem to the highest points—a circumstance which does not admit of the belief that it follows the regular course by which nature governs the other orders of vegetation. This, of course, is a conclusion only from apparent causes, for I have not had time to investigate the fundamental cause or law by which it is influenced; and native traditions are rather incoherent and speculative to lead to

Flowering of Bamboos.
79

any satisfactory conclusion. Superstition, which seizes on everything, however trivial, as a material with which to manufacture a portent, assigned to the appearance of the seed a certainty of impending famine; for, say the Brahmans, '*When bamboos produce sustenance, we must look to heaven for food.*' But for the hundredth time, perhaps, is Brahminical prescience belied, for never was there a finer crop of rice on the fields than at present. It would not be surprising, however, if the common and intended meaning of the prophecy be hereafter denied, for bearing as it does a double meaning, like many of the responses of the Pythean oracle, the incorrect reading may be ascribed to ignorance."

"Each bamboo bears from 4 to 20 seers, which assigns to it in my opinion a character of extreme fruitfulness, considering the close and compact order in which the bamboos grow. Soon after bearing, the bamboo seems to have fulfilled its career and dies, but the roots again send forth offshoots to perpetuate the species on the same ground: nor is it in this manner only that it is propagated; for the seed germinates, as I have tried, and have not the least doubt that a plantation may be raised from it." (*J. B. Jones in Trans. Agri.-Horti. Soc. India, III., 143.*)

Beddome is of opinion that **Bambusa arundinacea** generally flowers at an age of about 32 years, he having ascertained that in Western India it flowered in 1804, 1836, and 1868; but **Dr. Brandis** adds that this species also flowered in Kanara in 1864. The most animated discussions have been published as to whether the bamboo flowers when it attains a definite age, or only at any period when mature, provided the circumstances of the season are favourable. In his *Himâlayan Journal* **Sir J. D. Hooker** seems to favour the latter theory, but there are many facts which go to support the former. Both may be true, and this is probably the wiser solution of the difficulty—that is to say, a bamboo may not flower before it has attained a certain age, but its flowering is not fixed so arbitrarily that it cannot be retarded or accelerated by climatic influences. It is an undoubted fact that the flowering of the bamboo is decidedly influenced by the causes which bring about famine, for the providential supply of food from this source has saved the lives of thousands of persons during several of the great famines of India. **Captain Sleeman** very wisely suggests that it would save the complete destruction of the bamboos of a district to introduce seedlings obtained from stock in other districts. It appears certain that it is immaterial whether cuttings are taken a few years or many years before the flowering; the parent or stock plant, as well as plants raised from cuttings, will all flower and die at the same time. Indeed, it has been shown that cuttings taken a year or so before the flowering, if unable to produce flowers, nevertheless die with the rest. This curious fact seems to indicate that the life of the plant is perfectly fixed and is renewed only by seed. The introduction of two or more batches of seedlings from remote districts into each forest would seem to be an expedient that might fairly well claim the attention of the forest authorities, but it would deprive the people, to a large extent, during times of scarcity, of the chance of obtaining a crop of bamboo grain. It would be intensely interesting to have the facts relating to bamboo flowering carefully recorded every year all over India. Do all the forests, for example, of a certain bamboo in India, exhibit a tendency to flower at once, or is there any relation between the periods of flowering of the same species in different parts of India? It seems likely that each of the recorded periods of flowering were in reality the times of flowering of different species of bamboo, and are thus relatively of little importance. The natives of some parts of India say the flowering can be averted by cutting down all the bamboo the year before the flowering is

Bamboo as a Paper Fibre.

expected. It is difficult to see how this can affect the plant. At most it can only retard the flowering for a year or so; and it is equally difficult to understand how it can be known when flowering is to take place.

80

Propagation of Bamboos.

This may be effected—

1st, **By Seed.**—This is the slowest but most effectual process. The ovary (surrounded by the lodicules and palea) drops from the plant and readily germinates, usually within the first week after reaching the ground. Some species germinate while the seed is still attached to the plant, the young seedling dropping from the parent when about 6 inches in size. Nothing is known as to the period of vitality of the bamboo seeds, but if carefully collected and matured in the usual way, they may be sent from one part of India to the other in good condition. This of course applies only to those which fall from the plant before germinating. Propagation by seed is the most certain plan, but the plant requires 10 to 12 years to attain a growth sufficient to admit of cropping.

2nd, **By Cuttings.**—This is the process most frequently adopted in India. The lower part, say 3 feet in length, of a growing half-mature stem is placed in the ground shortly after the commencement of the rains. This is most frequently cut off so as to leave, if possible, a portion of the rhizome attached. The cutting should be made a little below one of the nodes and buried so as to include this and the next node. Sometimes the cuttings are laid lengthwise along the ground on a specially-prepared soil, and the sproutings at each node with their rootlets are afterwards separated and transplanted to their final positions.

PROPERTIES AND USES OF THE BAMBOO.

Fibre as a Paper Material.

References.—*Routledge—On Bamboo considered as a Paper-making Material; Kew Reports, 1877, p. 35, and 1879, pp. 33, 34; Dr. King's Reports of the Botanic Gardens, Calcutta, for 1877, 1878, and 1879; Spons' Encyclop.; Official correspondence with Forest Department.*

Fibre. 81

"Of all the fibre-yielding plants known to botanical science, there is not one so well calculated to meet the pressing requirements of the paper-trade as 'Bamboo,' both as regards facility and economy of production, as well as the quality of the '*paper-stock*' which can be manufactured therefrom: grown under favourable conditions of climate and soil, there is no plant which will give so heavy a crop of available fibre to the acre, and no plant which requires so little care for its cultivation and continuous production." These are the opening sentences of **Mr. Routledge's** most useful and interesting little book on "Bamboo as a Paper-making Material," published in 1875. That young bamboo shoots can be used in this way is now an established fact, and great credit is due to **Mr. Routledge** for the energy and persistence with which he has advocated the claims of "Bamboo fibrous stock" to the paper manufacturers. While this is so, there are practical difficulties which seem likely to prevent bamboo from ever taking the place which the study of the prepared fibre apart from the plant would naturally suggest for it. The structural differences between monocotyledonous and dicotyledonous plants have repeatedly been pointed out, in connection with the subject of the fibres obtained respectively from certain members of these sub-kingdoms. In the former the bundles of vessels are isolated, and pursue a curiously-curved course through the stem growing at the upper extremity upwards and outwards, and at the lower downwards and outwards. This curvature gave rise

Bamboo as a Paper Fibre.
81

to the mistaken idea of internal growers or endogens. The vital difference between the two classes of stems is found chiefly, however, in the vessels, which in both cases take their origin at the growing point of the stem. In the monocotyledon they remain quite distinct from each other, being simply imbedded amongst loose cellular tissue. They have a pronounced tendency to elongate, but do not enlarge very much in thickness. In the dicotyledonous plants the fibro-vascular bundles, on the other hand, steadily increase in thickness, and as a consequence they coalesce together, forming zones of woody tissue. The young outer layer, including the bark, is thus the only portion of the dicotyledonous stem which can conveniently be reduced to fibre, whereas the great bulk of the monocotyledonous stem is amenable to the agencies now employed by the manufacturer for the production of fibre. It thus follows that not only have dicotyledonous stems to be subjected to an elaborate process for the purpose of separating wood from bark, and of liberating the fibres of which the bark is composed, but only a small percentage of the crop obtained from the field actually goes to fibre. These are powerful arguments, no doubt, in favour of bamboo and other monocotyledonous fibres for paper trade. The yield of flax per acre is about 5 cwt., hemp 7 cwt., jute 5 or 6 cwt., and cotton much less, while about 10 tons can, according to **Routledge,** be obtained from an acre of bamboo jungle. The paper manufacturer cannot afford to give the prices which can readily be commanded by any prepared fibre. Esparto, which may be directly placed in the pulping pans, might not incorrectly be taken as the type of fibre required by the paper trade. The increasing demands for this grass, which already exceed the supply, have forced upon the world the serious problem of finding a substitute, and for some time the greatest hopes were entertained that immense expanses of almost waste bamboo jungle in India would be rendered available, thus meeting the industrial wants and at the same time opening up a new source of revenue to the country.

82

Objections to Bamboo as a Paper Fibre.

It is very much to be regretted that the great expectations as to the future application of the bamboo to the requirements of the paper-trade were frustrated by practical obstacles which **Mr. Routledge** did not apparently foresee. These may be briefly stated:—

1st.—The young shoots only being serviceable for paper-making, three serious difficulties arise: (*a*) the bamboo shoots appear from June to July, and are in condition during August and September, but by the end of October they are too old; (*b*) the stock is found to suffer severely from the removal of the shoots; (*c*) each clump can yield only about three or four shoots a year. Experience has shown that this is about the full number which each clump can be supposed to yield. The shoots that are removed must not be cut close to the ground, otherwise the plant suffers still more severely.

2nd.—Experiments seem to have failed to induce the bamboo to produce a continuous supply of shoots throughout the year.

3rd.—It was found that a large percentage of old stems required to be left on the stools, otherwise the plant was in time killed, and that the same danger existed in the complete removal of the young shoots. This would necessitate a methodical working of the jungles, and thus considerably increase the charges of collection and transport. **Dr. King** has demonstrated (in *Reports of the Royal Botanic Gardens, Calcutta, 1877-80*) that if all the shoots be removed for three successive years the plant is killed. This danger may be averted for a time by systematic working of the clumps, but does not appear to be curable.

Bamboo as a Paper Fibre. 82

4th.—During the months in which the bamboo shoots appear, the climate of the most important bamboo forests is such that labour could not be obtained. In fact, bamboo forests occupy, as a rule, uninhabited districts, rendering the labour question, apart from the dangers to human life, one of the most serious difficulties.

5th.—The freight and transport charges incidental to all raw products which have to be conveyed for long distances are very considerable. In fact, owing to the scattered nature of the clumps which form bamboo jungles, human labour would be the only means of collecting the material to points from which it could be conveyed to the factory.

6th.—A most unexpected difficulty, which in itself almost renders the bamboo unsuitable for paper-making, exists also in the hard adpressed hairs which cover the scales and young stems. It has been found impossible to remove these, and they are not only dangerous to the men employed, but injure the paper seriously.

83

Experiments in Cultivating Bamboo as a Paper Fibre.

An experiment was, however, undertaken in Burma, and the terms of a concession discussed by Government some few years ago. **Mr. Routledge** found that paying R15 a thousand for the shoots landed at Rangoon, he could prepare fibre at a price that could be given by the paper manufacturer. A thousand green bamboo shoots weigh about 8 tons, and losing 75 per cent. in moisture they yield 2 tons of dry fibre. He thus paid about 30*s*. for the materials from which two tons of fibre were prepared. To this must be added the charges connected with the separation of the fibre and the shipping and freight to Europe.

A point of great importance, and one which must not be overlooked, is, that the bamboo shoots must be reduced to fibrous stock in India. Various proposals have been made to meet this difficulty. One, that floating machinery should be conveyed up and down the rivers to convenient places near the forests, and that the crushing, or the first stage of the process of manufacture, should be conducted in this manner on board flats. The ribbons of crushed bamboos are, however, very subject to destructive fermentation, being rapidly rendered useless. If not reduced at once to dry fibrous stock, the greatest possible care must be taken to see that the ribbons are conveyed in safety to the factory where they are to be reduced to fibre.

The idea of floating the young shoots down the rivers to some healthy situation, where a factory could be located, has also been discussed, and, if practicable from the manufacturer's point of view, this would only have to contend against the dangers and difficulties in the forests.

Another proposal is to have machinery which could be put down in the jungles, and moved from place to place as required. The obvious objection to this is the danger to human life from the malarious nature of the forests, and the expense and danger of conveying heavy machinery through wild tracts of country where there are no roads.

Mr. Routledge seems to be most in favour of the idea of selecting a healthy district where a large plantation could be opened out around the factory, and an elaborate system of irrigation might be adopted to force the growth of the bamboo. This seems to completely ignore questions of a vital nature; if the land were good and favourably situated with reference to facilities for exportation, it would never pay to put it under bamboo cultivation, and if not so favoured, it might be possible to make the bamboo grow, but quite another matter to make the factory a profitable undertaking.

Besides, there are many other fibres which in all probability would be

Bamboo as a Paper Fibre.
83

found much more profitable to cultivate in the way **Mr. Routledge** proposes than the bamboo. The Nepal paper plant, for example, would seem to be the fibre which should attract the most attention in the future. It has been long overlooked by Europeans, chiefly through the success of jute and other annual plants suited to the plains of India.

Mr. Routledge deserves every praise, however, for the manner in which he has urged the claims of bamboo as a paper material, and although the matter seems to have attracted less attention of late, it has not by any means been finally disposed of. Paper has been made both in India and China from the bamboo; indeed, in China, it is "the principal if not the only material for paper-making, and was there used as such when our forefathers were savages." (*Kurz.*)

In an official correspondence with the Government of India, **Mr. J. Sykes Gamble** writes regarding the Chinese method of preparing paper from the bamboo: "I would suggest that the Chinese method of making paper from pulp obtained from mature bamboos is more likely to pay than that proposed by **Mr. Routledge** of using only the young succulent shoots. The method of preparation of the fibre employed by the Chinese is given at page 255 of Vol. IV. of the *Indian Forester,* of which the following is an extract: The method of preparation from bamboo is as follows: The bamboo is stripped of its leaves and split into lengths of three or four feet, which are packed in bundles and placed in large water-tanks; each layer of bamboo is then covered with a layer of lime; water is poured on till the topmost layer is covered. After remaining in this condition three or four months, the bamboo becomes quite rotten, when it is pounded into pulp in a mortar, cleansed and mixed with clean water. This liquid is poured, in quantities sufficient for the size and thickness of the sheets required, upon square sieve-like moulds. These sheets (of which a skilful workman can make six in a minute) are allowed to dry, then taken from the mould and placed against a moderately-heated wall, and finally exposed to the sun to dry. The best quality is made from the shoots of the bamboo with alum added to the infusion; the second from the bamboo itself, though a higher grade of this quality is attained by the previous removal of the green portion."

It has been repeatedly proved by European manufacturers that the bamboo shoot can compete with other paper-yielding substances. Both in America and Europe, thousands of tons of bamboo fibre (supplied from the West Indies) have been made into excellent paper. "Efforts have been made in Brazil to utilise the fibre for textiles, in mixture with wool and silk." *(Spons' Encycl.)*

The obstacles to the establishment of bamboo as an industry are of a practical nature, but they are such as might be expected to give place to a pressing necessity for more fibre. So long as this is not the case, there does not appear to be much chance of the trade in bamboo fibre assuming the form of an established industry. For the convenience of those interested in the subject, the following extracts from **Mr. Routledge's** little book may be republished in this place: "An essential point in my system for treating 'Bamboo' to produce therefrom fibrous 'paper-stock,' consists in operating upon the stems of the plant when young, and preferably when fresh, as, and when, cut and collected.

"Brought, therefore, to a central factory in this condition, the stems are passed through heavy crushing rolls, in order to split and flatten them, and at the same time crush or smash the knots or nodes. The stems thus flattened are then passed through a second series of rolls, which are channelled or grooved, in order further to split or partially divide them longitudinally into strips or ribbons; these being cut transversely into convenient lengths, by a guillotine knife or shears, are delivered by a

Bamboo as a Paper Fibre.
83

carrier or automatic feeder direct to the boiling pans, or elsewhere, as desired.

"As the object of my process is to produce a fibrous or tow-like *Stock*, retaining as far as possible the normal or natural condition of the fibre, and not '*Half-stuff*' or '*Pulp*,' my system of treatment differs materially from the ordinary process of preparing fibres, more especially in the boiling and washing processes.

"Both of these processes I conduct in a battery, or series of vessels (16, 20, or more in number), such vessels being connected together by pipes or channels furnished with valves or cocks, so that communication between the individual vessels may be maintained, disconnected, and regulated as desired, in such manner that the vessels, being methodically charged in succession with the material to be operated upon, the heated leys (composed of caustic alkali) can be progressively conducted from vessel to vessel of the series, passing over and through the material placed therein.

"The leys are thus used again and again (each successive change or charge of ley carrying forward the extractive matters it has dissolved from the fibre with which it has been in contact) until exhausted or neutralised (when they are discharged), fresh leys being methodically and successively supplied, until by degrees the extractive matters combined with the fibre or fibrous material have been rendered sufficiently soluble, when hot water for washing or rinsing is in the same continuous manner run successsively from vessel to vessel, over and through the material contained therein, until the extractive matters rendered soluble by the previous alkaline baths have been carried forward and discharged, leaving the residuary fibre sufficiently cleansed.

"By this system of boiling in continuity, until all the effective alkali in the leys is exhausted or neutralised, I realise an economy of from 30 per cent. to 40 per cent. of soda over the ordinary process of boiling, and by the subsequent washing or rinsing in the same continuous manner, without removing the material from the vessels, the normal structure of the fibre is in a great measure retained, waste is minimised, and thus, while being thoroughly cleansed and freed from extraneous matter, the strength and staple of the fibre are preserved; a considerable saving of fuel results from the heated liquors being used again and again, less steam being required, as also less water, while at the same time economy of both labour and power is effected over the ordinary system.

"Assuming the boiling and succeeding washing processes to be concluded, and the material ('Bamboo') in one of the vessels of the series in its regular succession to be found sufficiently treated and cleansed, a final cooling water is run on and through the fibre, which is then drained and the contents of the vessel (disconnected for the time being from the series) emptied into a waggon running on a railway, by which it is conducted to a press or otherwise to abstract all the remaining moisture possible.

"The dry, or semi-dry, fibre is then submitted to the action of a willow, or devil, by means of which it is opened or teazed out, and converted readily into a tow-like condition, when it is dried by a current of heated air induced by a fan-blast, and finally baled up for storage or transport, in a similar manner to cotton or jute.

"In this condition of '*paper-stock*' it may be kept an indefinite length of time without injury, and when received by the paper manufacturer, requires merely soaking down and bleaching to fit it for making into paper, either by itself, or used as a blend with other materials, as desired." (*Mr. Routledge's "Bamboo considered as a Paper-making Material," pp. 12-14.*)

Much has been written for and against the idea of practically utilising our vast bamboo forests as a supply for paper, but the present position may not incorrectly be described as a controversy of opinions which have not been put to a practical test. Experiments of sufficient magnitude have not as yet been undertaken to ascertain whether or not **Mr. Routledge's** proposals can be practically carried out in India. If cheap machinery could be invented suitable for the preparation of the fibrous stock, one would be inclined to hope that the smaller proprietors of bamboo jungles might take to preparing the fibre. Indeed, past experience would seem to jusify the opinion that the native and not the European is the proper person to look to for the cultivation and preparation of fibres.

MEDICINE.

MEDICINE Tabashir. 84

In the interior of the hollow stems of some bamboos, chiefly **Bambusa arundinaceæ,** a silicious and crystalline substance is found, known in the bazars of India as *Tabáshír* or

BAMBOO MANNA.

Vern.—*Bans-lóchan, bans-kapúr,* HIND.; *Báns-kápúr,* BENG.; *Thstoriyá,* ASS.; *Vansa-lochana, venú-lavanam,* SANS.; *Tabáshír,* ARAB. and PERS.; *Banasa-lóchana, banasa-mítha,* MAR.; *Váns-kapúr, vás-nu-mítha,* GUJ.; *Munga-luppu,* TAM.; *Veduruppu,* TEL.; *Moleuppa,* MALA.; *Bidaruppu, tavakshírá,* KAN.; *Uná lunu, una-kapura,* SINGH.; *Vá-chhá, vathegá-kiyo, vathegasá, vasan,* BURM.

There are two varieties of *tabáshír* known in the bazars, *viz., kabúdi,* blue; and *safed,* white: the former is only pale blue.

The following interesting historical account of this substance is extracted from **Sir George Birdwood's** *Bombay Products:* "*Tabásheer* is an article of the greatest antiquarian interest, as **Salmasius, Sprengel,** and **Fee** are of opinion that it is referred to, and not sugar, by the ancients,—**Dioscorides** and **Pliny,** for example, where they mention *σάκχαρον* and *Saccharum.* **Salmasius** states that the Saccharum of the ancients, as described by them, had none of the properties of sugar, and was used in ways sugar never could be; and in another place that the *σάκχαρον* of the Greeks was *tabasheer* 'beyond all controversy.' Against this dictum the line in Lucan has been cited—'*Quique bibunt tenera dulces ab arundine succos,*' as if the bamboo could be a '*tenera arundo.*' But **Salmasius** quotes this very line, and yet goes on to show by arguments one finds it difficult to refute, although common sense would reject the conclusion, that cane-sugar was unknown to the ancients. One would think **Pliny's** description left little room for doubt; yet **Salmasius,** by means of a comma, alters its whole meaning. The passage is as follows: '*Saccharon et Arabia fert, sed laudatius India; est autem mel in arundinibus collectum, gummium modo candidum, dentibus fragile, amplissimum nucis avellanæ magnitudine, ad medicinæ tantum usum.*' But says **Salmasius,** '*ita haæc distinguenda, collectum gummium modo, non ut est vulgo gummium modo candidum. Haæc omnia prorsum quadrant in tabascir, vel saccharum mambu,*'—'It is white, brittle to the teeth, is collected in reeds, is sweet' (!) 'and useful in medicine.' **Dioscorides** says: 'What is called *σάκχαρον* is a kind of concrete honey, found in reeds in India and Arabia Felix, in consistence like salt, and brittle between the teeth like salt. Takendissolved in water it is borne by the stomach,' &c. It is difficult to deny that sugar is not here meant, and very hard to allow that *tabasheer* is. **Pliny,** copying from **Dioscorides,** as is plain, perhaps confused *tabasheer* with sugar in his description, and thus has involved the passage in obscurity. The President of the Bombay Branch of the Royal Asiatic Society has suggested to the compiler a reading of **Pliny** as ingenious as that of **Salmasius,** and probably more just, inasmuch as it supports the common-sense view in the 'Sugar Controversy.' Placing a full

MEDICINE.
Tabashir.
84

stop where the first semicolon occurs, the **Honorable Mr. Frere** reads the passage as follows: '*Saccharon et Arabia fert sed laudatius India. Est autem mel in arundinibus collectum,' &c.* As if **Pliny**, on mentioning, at once dismissed so familiar an article as 'Saccharon,' and then went on to describe in detail so rare a substance as *tabasheer* must have been. **Fee, Sprengel,** and **Humboldt** simply follow **Salmasius, Humboldt** very diffidently. A passage from his *Prolegomena de distributione Geographica Plantarum* (quoted in his 'Cosmos'), states an opinion, all, on reading the whole controversy on sugar, will probably acquiesce in, and is on other accounts worth introducing here: '*Confudisse videntur veteres saccharum verum cum Tebaschiro Bambusae, tum quia utraque in arundinibus inveniunter, tum etiam quia vox sanscradana sacharkara, quae hodie (ut Pers. Schakar et Hind. Schukur) pro saccharo nostro adhibetur, observante Boppio, ex auctoritate Amarasinhæ, proprie nil dulce (madu) significat, sed quicquid lapidosum et arenaceum est, ac vel calculum vesicæ. Verisimile igitur vocem sacharkara duntaxat tebaschirum (succur nombu) indicasse, posterius in saccharum nostrum humilioris arundinis (ikschu, kandeschu, kanda) ex similitudine aspectus translatum esse. Vox Bambusae ex mambu derivatur; ex kanda nostratium voces candis zuckerkand. In tebaschiro agnosciter Persarum schir, h. e. lac, Sanscr. Kschiram.*' The Sanscrit name for *tabascher* is *tvakkschirá*, bark-milk. **Herodotus** (Book XIV., ch. 194), writing of the Gyzantians, observes that in their country a 'vast deal of honey is made by bees; very much more, however, by the skill of men.' In a note **Rawlinson** states: 'Bees still abound in the country, and honey is an important article of commerce. A substitute for honey is likewise prepared from the juice of the palm.' **Sprengel** states that the sugar-cane is first mentioned by **Abulfaidil**, 13th century, and sugar by **Moses Chorenensis**, A.D. 462; and notwithstanding that it must, the writer would apprehend, be mentioned in Hindú books of a far earlier date, it is not a little remarkable that a Hindí name of sugar is *Cheene*.'" (*Birdwood's Economic Products of the Presidency of Bombay, pp. 95-96.*)

Tabáshír is largely used by Hindús and Mohammedans, and is considered "cooling, tonic, aphrodisiac, and pectoral. It is an ingredient in many compound medicines which are given in different lung diseases. Sanskrit writers describe it as sweet, *i.e.*, not bitter" (*Dymock, Mat. Med., W., Ind. 697*). "The most complete account of its varieties, history, formation, and properties has been published by **Sir David Brewster** (*Philosoph. Trans., 1819,* and *Edin. Journ. of Science, Vol. VIII., p. 286*), and in the same paper are embodied some earned remarks by **Prof. H. H. Wilson** on its nomenclature and the uses to which it is applied by the natives, drawn from Sanscrit works." "It is highly prized in native practice as a stimulant and aphrodisiac; but from its composition we are warranted in believing that as a medicinal agent it is inert." (*Pharm. of India.*)

The deposit called *Bansolochan* (or *tabáshír*) is supposed to be efficacious in paralytic complaints, flatulency, and poisoning cases. It is highly prized in native practice as a stimulant and aphrodisiac. It is supposed to be cool and to remove thirst, and therefore useful in fever, jaundice, and pulmonary affections. The Baigas of the Central Provinces are said to be very clever at detecting bamboos in which this *bansolochan or tabáshír* is likely to be found. The substance holds a high reputation as a febrifuge, and accordingly fetches, in these Provinces, a good price. (*Gazetteer.*)

It seems probable that *tabáshír* is an after-product from the natural sap of the bamboo, which gives to the young shoots their peculiar flavour. This has not been clearly established, however, nor indeed has the process of the excretion been looked carefully into by modern botanists, and accordingly rather conflicting statements occur in the writings of authors upon this subject. My friend **Mr. Peppe,** of the Opium Department, informs me that he knows a native merchant who made a large fortune upon *tabáshír.* He examined very carefully the bamboos that were found to contain the salt, and came to the conclusion that its formation was due to the action of an insect that perforated the bamboo. He tried to imitate this action, with the result that he found that by making a small perforation above a point in half-mature bamboos the salt formed freely. This he practised systematically and made a considerable sum of money before he finally glutted the market with *tabáshír.*

Bamboo Sugar. 85

The following extract from **La Maout** and **Decaisne's** *System of Botany* seems to convey an entirely mistaken idea: "The young shoots of these two trees (**Bambusa arundinacea** and **B. verticillata**) contain a sugary pith, which the Indians seek eagerly; when they have acquired more solidity a liquid flows spontaneously from their nodes, and is converted by the action of the sun into drops of true sugar. The internodes of the stems often contain silicious concretions, of an opaline nature, named *tabasheer* (a substance presenting remarkable optical properties)." What could possibly be alluded to as the "true sugar" it is difficult to see; certainly no Indian bamboo has ever been observed to do so, and the only excretion is *tabáshír*, which, as stated, is deposited in the interior of the nodes. **Kurz** in his paper on "Bamboo and its uses," however, speaks of the fluid within the bamboo stems: "The water which often accumulates in the bamboo-joints, especially of very hollow kinds, is used against bowel complaints, with what success I cannot say, but all I can add is, that this water in the bamboo halms, like that found in the pitchers of the *Nepenthes*, has often quenched my thirst during my tours in the Java hills." This fluid sap within young bamboo stems may be the source of the "true sugar" alluded to in the above passage, but in India at least neither has the fluid sap been seen in such abundance as described by Kurz, nor has the spontaneous excretion of sugar on the outside of the stems ever been recorded by Indian travellers.

Bamboo Sap. 86

§ "There are two varieties sold in the bazars, a white and a bluish white. Mixed with honey it is used locally in aphthæ. Some hakims say that if this drug be used for any length of time it is apt to induce impotence." (*Surgeon G. A. Emerson, Calcutta.*) "The white variety has been calcined." (*Surgeon-Major W. Dymock, Bombay.*)

"Cooling medicine, generally given in fever to assuage thirst; also expectorant, 5 to 20 grains. The cooling *churna* is prepared by mixing together 8 parts of *vansa lochana*, 16 of *pimplee*, 4 cardamoms, one of cinnamon, with a sufficient quantity of honey or ghee. Dose of the powder one to two scruples." (*Surgeon W. Barren, Bhuj, Cutch, Bombay.*) "Used as a medicinal ingredient in cases of diarrhœa, dysentery, &c." (*Mr. H. Z. Darrah, Assam.*)

Chemical Composition of Tabáshír.

The most complete analysis yet published is that of **Prof. T. Thomson** of Glasgow (*Records of Gen. Science, Feb. 1836*), "who found its constituents to be—in 100 parts, Silica, 90·50; Potash, 1·10; Peroxide of Iron, 0·90; Aluminia, 0·40; moisture, 4·87; loss, 2·23."

In addition to *tabáshír*, other parts of the bamboo are sometimes used medicinally. **Kurz** says: "The stiff, fragile, very fugaceous hairs or rather bristles on the sheaths of the shoots are used for poisoning. They are put in the meal, or more usually in the coffee, to be partaken, and are said to cause death, not suddenly, but the action is very slow and the victim succumbs only after many months." (*Kurz in Indian Forester, I., 239.*)

Poison. 87

In addition to many other important uses, "the bamboo" is supposed to act as an emmenagogue, a decoction of the leaves and shoots being

Decoction. 88

used both in India and China to assist the lochial discharge after child-birth.

Splints. 89

In places where ordinary surgical appliances are not available, the leaf-sheaths or carefully-cleaned sections of stems may be used as splints. It is by no means an unusual thing to find a bamboo joint used as an artificial limb, the stump of the leg being simply inserted at the open end of the bamboo.

FOOD.

FOOD. The shoots. 90

It has already been stated that, in times of scarcity, bamboo grain has saved the lives of thousands of human beings. This was the case in the Orissa famine of 1812. A similar event took place in Canara in 1864, when it was estimated that 50,000 persons came from Dharwar and Belgaum districts to collect the seed. In 1866 bamboo grain sold in Maldah at 13 seers to the rupee, rice being 10. Many other instances are on record of the providential flowering of the bamboo having saved the lives of starving people, but while this is so, it is impossible to follow **Mr. Kurz** and others who have advocated the extended cultivation of bamboo as a means of averting famine. "Here we have," says **Kurz,** "at once a key in dealing with the mitigation of famine in India." For this purpose he recommends the extended cultivation of **Bambusa arundinacea, B. Tulda, and B. vulgaris,** suggesting the encouragement, in these proposed famine relief plantations, of tuberous wild plants, such as **Dioscorea** (yams), **Tacca, Amorphophalus** *(Ol),* **Colocasia** *(Kutchu),* &c. He is of opinion that the extended cultivation of bamboo over wild tracts of country would increase the humidity and thus prevent the tanks and streams from being dried up. In this way a larger amount of fish might be produced. There cannot be a doubt but that cultivation in any form would greatly improve the condition of barren tracts of country, and thus lessen the tendency to famine; but there would seem to be no special claim in favour of the bamboo, the more so since the crop of grain, which occurs only after thirty years, brings with it a plague of rats which injures the country for years after. Moreover, when grain is plentiful, the bamboo seed is not much eaten. "It is a very unsafe aliment, being apt to produce diarrhœa and dysentery." (*Dr. Bidie.*)

The young shoots constitute a most important article of food all over India, nearly every bamboo being eaten in this stage; but the larger species are those most generally used. Freed from the sheaths and hairs, they are cut up into small pieces and eaten in curries. They are also pickled or made into preserves. The very young shoots of the smaller species, if boiled in water with a little salt, resemble an inferior quality of asparagus. "They are eaten in Assam with great relish." (*Mr. H. Z. Darrah, Assam.*)

TIMBER.

TIMBER. 91

Bamboos form "the most important portion of the minor forest produce of all forest divisions, and one that increases in value every year." (*Atkinson.*) It would occupy a volume even to enumerate by name all the uses to which the mature bamboo stems are put. Suffice it to say that to the inhabitants of the regions where the bamboo luxuriates, it affords all the materials required for the erection and furnishing of the ordinary dwelling-house. Certain species are more serviceable for posts, and others are more adapted for matting and basket-work, but with one or two species every requirement may be met. For the construction of the mats of which the walls of huts are made, **B. Tulda** is the species most frequently used, and a strip from the outer green layer of this stem forms at once a most convenient and useful rope to tie the parts of the house which require to be made fast. **B. Balcooa,** on the other hand, having a thicker and more durable shell, is generally used for posts, boat-oars, and masts and all other purposes requiring greater strength and durability. **Dendrocalamus strictus** and one or two allied species of male bamboos are those resorted to for walking-sticks and spear-shafts, good solid bamboos being for these purposes in considerable demand.

TIMBER.

DOMESTIC USES.

DOMESTIC Fine Mats. 92 Common Mats. 93 Bamboo String. 94

Every person in India is familiar with the simple yet clever way in which the bamboo is cut up and split into bands of every size or thickness so as to allow of its being used in the manufacture of mats of any degree of quality, from the exceedingly fine mats made at Midnapur in Bengal, to the ordinary coarse mat extensively used in house-building. Thin strips of bamboo tied with strings are made into elegant door *tatties* (or curtains). Hollow bamboos are beaten here and there and cut at the nodes, lengthwise, and thereafter opened out and flattened into slabs which may be used for the seats of chairs, tops of tables, beds, or other articles of furniture. In fact, everything necessary for the erection and furnishing of a comfortable house can be obtained from the bamboo. The large Karen houses, each of which constitutes a village by itself, and which is large enough to contain as many as 200 to 300 persons, are entirely constructed of bamboo. Fishermen frequently build bamboo houses over the rivers. The greater part of the people in Eastern India and the Malay live entirely in bamboo houses. Bamboo bridges are frequent all over India. A complicated mass of uprights, in all directions, supports a pathway consisting of bamboo mats covered with a sprinkling of earth, and resting upon a few horizontal bamboos attached to the uprights. To the new comer, bridges of this nature seem most insecure, but if in good condition they may be ridden over with perfect safety.

Domestic and Agricultural Vessels. 95 Spoons and Knives. 96 Ladles. 97 Agricultural Implements. 98 Industrial Appliances. 99

The larger hollow species are well suited for aqueducts, water-pails, pots, cups, and other vessels. Pieces of thick bamboos from three to six feet in length, with the partitions perforated, so as to form long pails, are carried by hill watermen, suspended over the back by a bamboo string passing across the forehead, instead of the water-skin used by the *bhisti* of the plains. From these long tubes the water escapes with a gurgling noise, but it may be carried for days without either getting warm or being in any way spoiled. A single joint of a green bamboo is frequently used as a cooking-pot, the rice and water being placed inside and the mouth covered up; this primitive pot is placed on the fire until the rice is cooked. Spoons and knives are meanwhile being cut by the cook from the nearest bamboo clump. A simple ladle is made by cutting down to a handle the upper portion of a joint, leaving about 2 or 3 inches of the bottom as a large spoon or ladle. By the aid of this ladle the food when cooked is divided, and by means of the knife made of the hard outer portion of the stem, fish or other animal food may be cut up. In fact, every domestic appliance may be made of bamboo, including the pail used in milking the cows and the churn with which the butter is prepared. In the Nagá country a section of a bamboo is used for stamping out circular rice-biscuits required in certain religious observances. All sorts of agricultural implements are also made of the bamboo, and the appliances for spinning cotton and wool, and also for reeling silk, are often constructed entirely of the same material.

Fishing Appliances. 100 Traps. 101 Hooks. 102

The fisherman makes his oars, masts, fishing appliances, baskets, and even his hooks, of bamboo. One of the most curious hooks perhaps in the world is the one in common use in Bengal and Assam. This consists of a short piece of well-seasoned bamboo, say 3 inches in length and $\frac{1}{8}$ of an inch in thickness. The string is attached to the middle of the twig, which is then bent into the shape of the letter U. The bait

Harpoons. 103

generally used is the common green grasshopper, the head of which is plucked off and rejected. The two points of the bent bamboo twig are now inserted into the open end of the body and the baited hook dropped into the water. The upper end of the string is attached to a small piece of bamboo, about a foot in length, and left floating in the water with the baited hook suspended. The fisherman, from his dug-out, drops these lines in likely positions and rows himself about from one to the other. When the fish cuts the bait the hook jumps open in its mouth, the extremities getting amongst its gills. Large fish are often caught in this way, the pain and inconvenience of the hook apparently preventing the fish from offering the resistance which would at once set it at liberty. The common and characteristic harpoon of Bengal consists of a piece of **Dendrocalamus strictus** about 6 feet long, cut into 8 or 10 long pieces about as thick as the little finger. These are smoothed and rounded up to within a foot of the top, where the bamboo is firmly bound with string or wire to prevent its splitting further. The point of each of these portions is armed with a metal-pointed cap. The fisherman, rattling this instrument against the side of the dug-out, alarms the large fish from their hiding-places amongst the weeds; and no sooner is a fish visible than with great adroitness the harpoon is thrown, and the prongs spreading out as it enters the water, so large a space is covered as to leave the fish but a poor chance of escape if once the fisherman has been allowed to come sufficiently near.

All sorts of ingenious contrivances are made for catching fish, and in the majority of cases they are constructed from the bamboo. Perhaps none are more beautiful than the small and delicate basket traps which are placed here and there in walls of aquatic plants built artificially across portions of tanks. Excellent fishing-rods are also made of the solid bamboo: these are in universal use all over India. Jointed rods of European manufacture have not as yet found their way to India, except in the hands of a few European anglers.

Fishing-rods. 104

Spear-shafts. 105

Walking-sticks. 106

Bows and Arrows. 107

Poles. 108

Articles of Warfare. Spears. 109

Although the bamboo is not suited for the construction of boats or canoes, it is by no means unusual to find a raft, composed of one or two large bamboos lashed together, used by the fishermen on lakes. Timber is also largely floated down the rivers upon bamboo rafts.

Bamboo is extensively used for making spear-shafts, bows and arrows, poles for carrying loads, &c.. The spiny bamboos were formerly planted in ditches around forts as a protection. The Nagás and other hill tribes use the hardened outer woody portion as knives and spears. The jungles and forests around villages are often covered for miles with these formidable weapons. Short sharp bamboo knives called *pangís* are buried amongst the leaves along the foot-path in such a position as to go right through the foot of the unfortunate traveller. Often three of these are arranged, two sloping forward and the other facing the traveller on his approaching the village. The foot is by accident placed between these, and being cut by the one in front, is rapidly withdrawn, only to have the other two violently driven in from behind. Sometimes thousands of these *pangís*, both visible and invisible, cover the entire surface of the ground,—so much so that the village is unapproachable to any person but the inhabitants, who are familiar with every turning that has to be taken to escape this formidable bamboo defence. Pits are also dug, the bottom of which is full of these knives, pointing in every possible direction. The mouth of the pit is cleverly covered with leaves, and the animal or man who places his foot upon this trap falls to a fearful and certain death. Crude scabbards are also made of the bamboo, and handles for swords, knives, and axes.

Knives. 110

Pangis. 111

Musical Instruments: Fifes. 112

All sorts of curious musical instruments are made of the bamboo—from the fife to the crude violin with its two or three strings. The strings are prepared from the green outer layer of the stem carefully cut, and when tightened give out a dull musical tone. In Mánipur and the Nagá country the hill tribes prepare an exceedingly curious jew's-harp from the bamboo. This consists of a thin piece of bamboo not unlike the common musical pitchfork in size and shape, only that it has three instead of two arms, and is not more than ⅛ of an inch in thickness. This is placed in the mouth just as with the jew's-harp, and a monotonous music is produced. (*Mr. McCabe, Deputy Commissioner, Nagá Hills.*) Perhaps the most amusing musical contrivance is the bamboo Æolian harp made in the Malay Peninsula. The bamboos in a village clump or far away in the jungles are perforated here and there in such a way as to keep whistling in all tones at once as the wind blows through the culms. The sound produced in this way has been described as at times soft and liquid like the notes of a flute, and again deep and full like that of the organ. "A kind of very curious whistle is used by the Chinese for driving away evil spirits, &c. Several holes are pierced in a piece of bamboo, two of the natural knots being left, one of which offers an opening out in a slope; to each extremity are fastened two long strips of paper from 15 to 18 feet in length and 6 to 8 inches wide. A string is attached to a groove made in the bamboo, and when there is a little wind, this curious kite is sent aloft, remaining in the air as long as the wind is strong enough to keep it up. In this position a monotonous whistling is produced, resembling at times the noise of a jet of steam, or the sighing of the wind in trees."

Violins. 113

Violin strings. 114

Jew's-harp. 115

"The *auklong* of the Malays is a very agreeable instrument. It consists of a number of hollow bamboo-joints, of various but selected length and thickness, which are cut out below and hang down from a bamboo frame. These give various swinging tones and strength, according to their size, on being beaten with a bamboo staff. On the occasion of festivities, such as a marriage, circumcision, &c., Malays greatly use the green halms of bamboo (especially the larger sorts), and have them put in specially-prepared fires. The air enclosed in the joints gets heated, and the joints burst with a heavy report, which varies in strength from that of a pistol to that of a small gun, according to the sort of bamboo used, smaller halms being usually added which keep up a continuous rattling and crackling noise." (*The Indian Forester, Vol. I., pp. 234-35.*)

BAMBUSA, *Schreb.; Gen. Pl., III., 1210.*

116

A genus of large bamboos growing in clumps or clusters, comprising some 24 species.

Erect arborescent, rarely scandent plants. *Leaves* shortly petiole, but with a large articulate sheath. *Flowers* in a few species occurring on leaf-bearing stems, generally on leafless short stems, a few of which appear and die every year, or the entire clump reaches maturity at once and after flowering dies. It often happens that the bamboos of an entire district flower and die at once, followed by a dense mass of seedlings; this is especially true of B. arundinacea. *Spikelets* 2-many-flowered, generally sessile, interrupted. Inferior *glumes* 3-4, empty. Palea 2-keeled, distinctly ciliate or winged. *Stamens* 6, free. *Style* deciduous, deeply 2-3-fid. *Caryopsis* small, wheat-like, with a membranous pericarp closely adnate to the seed.

The generic name Bambusa is a Latinised form of the Mahratta name Bambu.

Bambusa affinis, *Munro, 93;* Gramineæ.

117

Vern.—*Theeshe, thaikwa,* Burm.

Habitat.—Found in Martaban, and said by **Munro** to be scandent; by **Kurz** to be a small tufted species, attaining a height of from 15 to 20 feet.

The Spiny Bamboo of Western India.

118

Bambusa arundinacea, *Retz.*

THE SPINY BAMBOO OF CENTRAL, SOUTH, AND WEST INDIA.

Syn.—B. ORIENTALIS, *Nees;* ARUNDO BAMBOS, *Linn.;* BAMBOS ARUNDINACEA, *Pers.*

Vern.—*Báns, kattang, magar báns, nal báns,* HIND.; *Báns, behúr báns,* BENG.; *Bnáh,* ASS.; *Katanga,* KOL.; *Mat,* SANTALI; *Wah-kanteh,* GARO; *Bariala,* CHITTAGONG; *Magar, nál,* PB.; *Wans,* GUJ.; *Kalak, padai,* KONKAN (THANA); *Vas,* PANCH MAHALS; *Mandgay,* BOMB.; *Bháns, chánsá* (if small), *bambú* (if large), DUK.; *Kati wadúr,* GOND.; *Vansa, kíchaka,* SANS.; *Qasab,* ARAB.; *Nai,* PERS.; *Mangal,* TAM.; *Mulkas, kanka* (Upper Godaveri District), *bónga, veduru, bonga-veduru, pente-veduru* (Madras), TEL.; *Bidungulu,* KAN.; *Wa-nah,* MAGH.; *Kyakatwa,* BURM.; *Kattú-úna, úna,* SINGH.

§ In towns all kinds of Bamboos are called *kallak* in Marathi. (*Dr. Dymock.*)

Habitat.—A common bamboo in Central and South India and Burma. Cultivated in many places in North-West India and in Bengal.

Botanic Diagnosis.—Stems tall, green, spinescent, growing in clumps of 30 to 100 each, attaining a height of from 30 to 50 feet; walls of the culm thick, cavity small; lower branches spreading; spines strong, sharp, curved either in pairs at the base of a branch or in threes, the middle one the largest. *Leaves* small, thin, lanceolate, 4-8 inches long and $\frac{1}{3}$ to $\frac{2}{3}$ broad, generally glabrous, but sometimes with scattered hairs underneath; nerves 5 to 6 on either side of the midrib; *spikelets* mostly sessile, in dense half-whorled clusters. Flowering *glume* thickened and mucronate at the apex, glabrous, not ciliate at the edges. *Ovary* glabrous, style deeply 2- or 3-fid. Flowering appears to take place after long intervals, probably at the age of 30 years (*Brandis*). On this subject **Mr. Duthie** writes: "The simultaneous flowering and subsequent dying of almost every individual plant of this species in certain districts and at certain stated times has been an interesting subject for observation. There seems to be no particular age at which the flowering takes place; the event is probably to a great extent influenced by the nature of the season." (For further information regarding the flowering of this and other species, see under **Bamboo.**)

By the older writers on Indian economic products the properties of the Bamboos as a whole seem to have been referred to this species, and it is accordingly difficult to separate some of the popular botanical and vernacular synonyms. The following remarks under the heading of Medicine may be understood to belong to more than one species, and might not incorrectly be referred to the peculiar species met with in each district in India:—

Properties and Uses—

MEDICINE. 119 Surgical Appliances. 120

Medicine.—"In addition to the many important uses to which the Bamboo is applied in tropical life, it forms by no means an insignificant article of the Indian Materia Medica. Its supposed virtues are set forth at length in the *Taleef Shereef* (Art. *Bans,* p. 28, No. 114). A belief in the emmenagogue properties of the leaves is common alike in India and China; but neither in this nor in any other character does it appear worthy of attention as a medicine. In positions where ordinary surgical appliances are not at hand, it is well to bear in mind that, with very little manipulation, splints of any required length or size can be obtained with little delay from the stems of the bamboo. For this purpose the older drier stems are to be preferred, the younger yielding somewhat on pressure." (*Pharmacopœia of India, pp. 256-7.*) It is by no means unusual to find a joint of the bamboo being used in India as an artificial limb, the stump being simply inserted into the open end of the bamboo.

§"The tender leaves of this plant are used with black pepper and common salt to check diarrhœa in cattle." (*Brigade Surgeon J. H. Thornton, B.A., M.B., Monghyr.*) "The most efficacious application for dislodgement of worms in ulcers is a poultice made by pounding the young shoots of the bamboo. The juice is first poured on the vermin, and the ligneous mass is applied and secured by a bandage." (*Honorary Surgeon P. Kinsley, Chicacole, Ganjam District, Madras Presidency.*) "The leaf-bud is used in the shape of a decoction to encourage the free discharge of the menses or lochia when this is scanty." (*Native Surgeon Ruthnam Moodelliar, Chingleput, Madras Presidency.*) "Is used in leprosy, fevers, and hœmoptysis." (*Surgeon-Major D. R. Thompson, M.D., Madras.*)

MEDICINE. Leaves. 121 Poultice of shoots. 122 Leaf-bud. 123

A silicious concretion known most frequently under the Persian name *Tabáshír* is obtained chiefly from the interior of the stems of this species. It is much used by the natives as a drug. (See under **Bamboo.**)

Food.—The seed resembles unhusked rice, and is eaten by the poorer classes like that cereal. As it appears at the very season when drought occurs, and other crops have generally failed, it is of some advantage to the poor. The young shoots are cut when tender, and eaten like asparagus. This remark applies to nearly every species of bamboo, the young shoots being known as *bans-ka-kulli*.

FOOD. Grain. 124 Shoots. 125

Fodder.—The leaves and twigs form an important fodder, this species being largely consumed by elephants.

FODDER. 126

Timber.—This bamboo is of good quality and strong, and is used for all purposes.

TIMBER. 127

Domestic and Sacred Uses.—This and other species of bamboo are frequently represented upon Buddhistic sculptures. The stems are used very extensively for domestic purposes. (See under **Bamboo.**)

DOMESTIC. 128

Bambusa baccifera, *Roxb.; Fl. Ind., Ed. C.B.C., 305;* **Syn.** for **Melocanna bambusoides,** which see.

B. Balcooa, *Roxb.; Fl. Ind., Ed. C.B.C., 305.*

129

Sometimes called THE FEMALE BAMBOO.

Vern.—*Balku,* BENG.; *Betwa,* CACHAR; *Bhaluká,* ASS. and CACHAR; *Bling,* LEPCHA.

References.—*Munro, Linn. Soc. Trans., XXXVI., 100; Brandis, For. Fl., 567; Voigt's Hort. Suburb. Calc., 718; Gamble, Man. Timb., 428.*

Habitat.—A native of the plains of the eastern side of India, extending from Bengal into Assam and Cachar. This is the large and characteristic bamboo of the villages of Bengal. **Dr. Brandis** thinks that the bamboo below Simla, ascending to 5,500 feet, may belong to this species.

Botanic Diagnosis.—Differs chiefly from **B. Tulda** in its larger leaves, not pubescent, and possessed of distinct transverse veins. Scales ovate or obovate, with distinct longitudinal nerves. The spikelets are also only $\frac{1}{3}$ to $\frac{1}{2}$ inch long, and the joints of the rachis short and glabrous.

Timber.—A bamboo, with stems often 50 to 70 feet in height, stouter and taller than in **B. Tulda.** This is the best Bengal species for building, scaffolding, and other works which require both size and strength. Long immersion in water tends to make it firmer and proof against the attacks of Bostrichi. (*Roxb.*)

TIMBER. 130

Bambusa Brandisii, *Munro, 109.*

Syn.—DENDROCALAMUS BRANDISII, *Kurz, II., 560.*

Vern.—*Ora,* BENG.; *Turgu-wah,* MAGH.; *Keyllowa, wabo,* BURM.

Habitat.—A gigantic species, met with in Chittagong and Burma, up to 4,000 feet.

Botanic Diagnosis.—Young shoots with adpressed tawny hairs; auricles waved, decurrent, fringed inside; ligule narrow. Angle of the inner palea minutely ciliate. (*Kurz.*)

TIMBER. 131

Timber.—Stems often become 126 feet high and 30 inches in circumference.

Bambusa Falconeri, *Munro, 95; Brandis, For. Fl., 568.*

Vern.—*Chye, kag.*

TIMBER. 132

Habitat.—Found in the North-West Himálaya.

Botanic Diagnosis.—The final identification of the large bamboo at the foot of the North-West Himálaya is a subject still very obscure. (See note under **B. Balcooa.**) The present species was described from a flowering specimen collected by the late **Dr. Falconer** in Dehra Dún. Spikelets lanceolate, wholly glabrous, nearly 1 inch long and about 10-flowered. Flowering glumes mucronate, with numerous broad, prominent nerves, somewhat like those of **B. Tulda.** (*Brandis.*)

TIMBER. 133

B. khasiana, *Munro; Munro, 97.*

Vern.—*Tumar*, KHASIA.

Habitat.—Met with in the Khásia Hills.

B. nana, *Roxb.; Gen. Pl., III., 1211.*

Vern.—*Pilawpinanwa*, BURM.

Habitat.—**Kurz** says of this species that it is rarely cultivated in and around Rangoon.

TIMBER. 134

Botanic Diagnosis.—Young shoots with the sheaths not or only obscurely auricled at the mouth. A small bamboo with small leaves, whitish beneath.

B. nutans, *Wall.; Munro, 92; Brandis, For. Fl., 567.*

Vern.—*Nal-báns*, BENG.; *Mahlbans*, NEPAL; *Mahlu*, LEPCHA; *Jiushing*, BHUTIA; *Bidúli, mukial*, ASS.; *Pichle*, SYLHET.

Habitat.—A most beautiful species, largely planted near villages in Nepal, Sikkim, Khásia Hills, Assam, Sylhet, and Bhután, ascending to from 5,000 to 7,000 feet.

Botanic Diagnosis.—Closely allied to **B. Tulda**, the leaves being of medium size and with soft pubescence beneath. Spicules long, with elongated, articulated, and clavate joints to the rachis.

TIMBER. 135

Timber.—It is a small species with almost solid stems. (*Munro.*) "The culm is of large diameter, with a broad hollow part, but the wood is hard." (*Gamble, Trees and Shrubs of Darjiling.*)

136

B. orientalis, *Nees; Beddome, Flor. Sylv., t. ccxxxi.*

Habitat.—A bamboo met with in South India.

137

B. pallida, *Munro.*

Vern.—*Burwal, bakhal*, CACHAR; *Usken*, KHASIA.

Habitat.—A bamboo with stems about 50 feet long; met with in Eastern Bengal and Assam.

138

B. polymorpha, *Munro; Kurz, ii, 553.*

Vern.—*Kya-thoungwa* or *kyathaungwa*, BURM.

Habitat.—Common in the upper mixed forests of the Pegu Yoma and Martaban.

Botanic Diagnosis.—An unarmed bamboo with large strongly-fringed auricles, with the sheaths of the young shoots green and yellow, adpressed bristles white. Anthers purple. Stigma white. Inner palea with the angles quite smooth. (*Kurz.*)

Bambusa spinosa, *Roxb.; Fl. Ind., Ed. C.B.C., 305.*

139

THE SPINY BAMBOO OF EASTERN INDIA.

Syn.—**Dr. Brandis** and also **Dr. Kurz** regard this merely as a form of **B. arundinacea**, peculiar to the eastern side of India; they can find no characters to separate these spiny bamboos. **Dr. Roxburgh** treats them as quite distinct species. In this opinion he is supported by **General Munro**, who distinguishes the two plants, giving the characters which will be found below for **B. spinosa**.

Vern.—*Bur, behar báns*, HIND. (*Duthie*); *Behor*, BENG.; *Koto*, ASS.; *Kinkoit*, CACHAR; *Yakatwa*, BURM.

Habitat.—A native of Bengal, Assam, and Burma; also of the north-eastern division of the Madras Presidency. Cultivated in the North-West Provinces and other parts of India.

Botanic Diagnosis.—"A paler coloured and more striated panicle, smaller and more coriaceous spiculæ, with fewer flowers, generally smaller leaves which are often hairy on the under-side, and with the petiole sometimes remarkably swollen at the base." (*Munro in Trans. Linn. Soc., XXVI., 105.*)

TIMBER. 140

Timber.—"This beautiful, middling-sized, very elegant species, I have only found in the vicinity of Calcutta, where now and then some of the oldest are found to blossom, about the beginning of the rains, in June." "Like the other species, this is employed for various useful purposes; and as it grows to a pretty large size and with a smaller cavity than any of the others, it is strong and well adapted for a variety of uses." (*Roxb., Fl. Ind., Ed. C.B.C., 306.*)

B. teres, *Ham.*

141

A native of Bengal and Assam.

B. Thouarsii, *Kth.*; Syn. for **B. vulgaris**, *Wendl.*, which see.

B. Tulda, *Roxb.; Fl. Ind., Ed. C.B.C., 304.*

142

THE COMMON BAMBOO OF BENGAL.

Vern.—*Peka*, HIND.; *Tulda, jowa, mitenga, matela, dyowa báns*, BENG.; *Mák*, SANTAL; *Pepesiman*, KOL.; *Makor*, MAL. (S. P.); *Wahghi*, GARO; *Madaewah*, MAGH.; *Theiwa, thoukwa*, or *thaikwa*, BURM.

References.—*Brandis, For. Fl., 566; Munro, Trans. Linn. Jour., XXVI., 91; Kurz, For. Fl., Brit. Burm., II., 552; Gamble, Man. Timb., 427; Voigt, Hort. Suburb. Calc., 718.*

Habitat.—This is the common Bamboo of Bengal, where it grows in great abundance everywhere, flowering in May. "Not uncommon in the deciduous forests of Pegu, generally occupying lower and moister stretches of ground in company with *tinwa* (**Cephalostachyum pergracile,** *Munro*), the dry hills surrounding being covered with **Dendrocalamus strictus.**" (*Brandis.*)

Botanic Diagnosis.—*Leaves* middle-sized, pale and soft, pubescent beneath, transverse veins none. *Spikelets* terete, 1-2 inches long, joints of rachis elongated, thickened into a hairy disc, under the flowering glume; scales cuneate, thickened at the base, but without prominent nerves.

FIBRE. Mats, &c. 143

Fibre.—Largely used for mats, baskets, fans, and window-blinds. This is in fact one of the most useful plants in Bengal.

FOOD. Shoots. 144

Food.--The young shoots are pickled when only about 2 feet high; they are tender. (*Roxb.*)

TIMBER. 145

Structure of the Wood.—"The wood is strong, and the halms are used for roofing, scaffolding, mats, and other purposes." (*Gamble.*) Found more durable if soaked in water previous to being used. This is regarded in Bengal as one of the best quality of bamboos. Both **Roxburgh** and **Voigt** mention several varieties. The following extract will be found to give the more important forms: "*Jowa báns* (*piábansh ?*) of the Bengalis, is only a large variety of this species, and used chiefly for scaffolding and building the larger and better sorts of houses of the natives. It differs from **Tulda** proper in the greater length and thickness of the joints. *Basini báns* of the Bengalis is another variety of **Tulda**. It has a larger cavity, and is used chiefly to make baskets. *Behoor báns* is of a small size, very solid and strong, much bent to one side, and armed with numerous strong thorns, which renders it very fit for hedges. A staff of this species must be placed in the hand of every young Brahmin when invested with the sacerdotal cord, otherwise they say the ceremony cannot be performed." (*Roxb., Fl. Ind., Ed. C.B.C., 305.*)

Jowa bans. 146

Basini bans. 147

Behur bans. 148

149

Bambusa vulgaris, *Wendl.*

THE YELLOW AND GREEN STRIPED BAMBOO.

Syn.—B. THOUARSII, *Kunth.*; B. VULGARIS, *Schrad.*; B. ARUNDINACEA, *Aiton.*

Vern.—*Kallak, vansa kalaka*, BOMB.; *Basini bans*, BENG.; *Una*, SINGH.

References.—*Brandis, For. Fl., 568; Thwaites, En. Ceylon Pl., 375; Dalz. and Gibs., Bomb. Fl., 299; Munro, 106; Beddome, Fl. Sylv., CCXXXII.; Gamble, Man. Timb., 428; Lisboa, Bomb. Pl., 137.*

Habitat.—Cultivated throughout India; said to be a native of the southern and central parts of Ceylon. It is also supposed to be a native of Sylhet and Chittagong, is naturalised in the West Indies, and cultivated in South America.

Botanic Diagnosis.—Leaves with distinct transverse veins. Spikelets laterally compressed, flowers distichous. Empty glumes 2; flowering-glumes ovate-lanceolate, narrowed at the base; longitudinal veins prominent near the apex, indistinct below, mucronate, and ciliate at the apex, fimbriate; keels of palea conspicuous near the top of the flowering glume. Anthers penicillate at the apex, with short hairs; style slender, filiform, 2-3-fid at the end.

TIMBER. 150

Timber.—Stems 20 to 50 feet, yellow, or striped yellow and green. Joints 4 inches in diameter and more, with thin walls.

"Much used by Cinghalese for temporary buildings and other purposes. The flowers, which are very rarely produced, very much resemble those of the next species (**B. arundinacea**); but their outer paleæ are somewhat longer, and terminate in subulate points." (*Thw., En. Ceylon Pl., 375.*)

Banana, or **Plantain,** see **Musa.**

Barberry, see **Berberis vulgaris,** *Linn.*; BERBERIDEÆ.

BARILLA.

151

Barilla (a crude carbonate of soda) is in India obtained from two sources: (*a*) as an efflorescence on the soil, and (*b*) from the ashes of certain salt-worts or plants containing sodium.

The former is scarcely known in Europe and should receive the name *sajji-máti*, while the carbonate of soda obtained from plants is, strictly speaking, the barilla of commerce or the *khár-sajji* of Indian bazars.

of Sodium Carbonate. **BARILLA.**

CARBONATE OF SODA OBTAINED AS AN EFFLORESCENCE.

SAJJI-MATI. 152

(*a*) **Sajjí** and **Sajjí-mátí.**—The efflorescence known as *reh* is exceedingly abundant in India, occurring over many large tracts of country, often rendering the soil quite sterile. This has within recent years received the most careful attention both of the Agricultural Departments and of the Geological Survey. Under "*Reh*" this subject will be dealt with at greater detail, and it is necessary here to refer to it only in so far as it is connected with the subject of *sajjí* or barilla. The decomposition of rocks through the action of the atmospheric oxygen and carbonic acid gives origin ultimately to soluble sulphates, carbonates, and chlorides. These are carried away by the rivers. In the fresh-water alluvial plains of India such salts have accumulated during the lapse of centuries, to an extent sufficient to give origin, by chemical changes in the soil, to the so-called *reh* efflorescence, the heat of the sun drawing the ultimate salts to the surface. In an interesting report on this subject, published by **Dr. W. Center**, Panjáb Chemical Examiner, the process of capillary attraction, or the drawing of the salts to the surface in the form of an efflorescence, is carefully gone into. "These salts, however, are not deposited, as they exist in solution, as new laws come into play. The chief of these is that during evaporation the least soluble salt that can be formed is first deposited; but this is modified by two other laws,—the tendency of certain compounds to form double salts, and the tendency of substances with the same crystalline form to crystallize out together. The efflorescences thus produced consist of three groups: 1st, the neutral, which contain no carbonate of soda (these consist chiefly of sodium chloride and sulphate, and frequently magnesium sulphate); 2nd, the alkaline, which contain carbonate of soda, and alkaline chlorides and sulphates, but no lime or magnesian salts; 3rd, the nitrous efflorescences. These generally contain no alkaline carbonate, but consist chiefly of nitrate of lime and alkaline chlorides. Others contain alkaline nitrate, chloride and sulphate. They are developed where the soil has become loaded with organic nitrogenous matter."

"*Reh* is thus not a special salt or mixture of salts, but a very variable compound. It is really the most easily soluble salt in the earth-water, remaining in solution after the deposition of carbonate of lime, &c., on evaporation. The ingredients and their relative proportions are found to vary in different places, exactly as the well-waters at different spots differ in saline contents, and in the same area there is a close relation between the two."

Mr. Medlicott, Superintendent of the Geological Survey (*Records of the Survey, Vol. XIII., 273*), has also contributed greatly to our knowledge of *usar* (sterile) and *kalar* (saline) lands, and has thrown much light on the explanation of *reh* efflorescences. He has shown that the relative proportion of common salt to sodium sulphate varies from 4 to 24 per cent. In one district one salt predominates, in another a second is more abundant. Common soda (carbonate of soda—*sajjí*) is sometimes present, and such earths are locally used by the washermen in place of soap. This is, however, a much less frequent *reh* efflorescence than sodium sulphate; but it seems likely that the so-called *sajjí-mátí* of our Indian bazars is a specially selected *reh*, containing as its principal ingredient carbonate of soda. *Sajjí*, strictly speaking, means pure carbonate of soda.

BENGAL 153

Occurrence of Sajjí-mátí.—In Bengal *sajjí-mátí* is said to be found in Behar north of the Ganges at Patna. **O'Shaughnessy** mentions that it exists in great abundance in the neighbourhood of Monghyr. In a recent correspondence conducted by the Government of India, Revenue and Agricultural Department, and at the instance of **Sir J. D. Hooker**, the

SAJJI-MATI. MADRAS. 154

Commissioner of Salt Revenue, Madras, reports: "As soils containing a large percentage (from 30 to 50 per cent.) of carbonate of soda, abound all over the country and are habitually collected by the people for use as *dhobies'* earth, and in dyeing, and for the manufacture of soap and of glass bangles, I see no advantage in undertaking the manufacture of barilla from alkaline plants." This efflorescence referred to is common in Mysore and Travancore, and a small internal trade exists in conveying it to the town of Madras. It is purified by a simple process of lixiviation.

N.-W. P. 155

This same practice prevails in some parts of the North-West Provinces where *sajjí-mátí* occurs; it is said to be prepared at Ghazipur. In the correspondence referred to it is stated that the Lucknow Paper-mills have, at the suggestion of the Agricultural Department, commenced to manufacture their own caustic soda from the *reh* earth. By this improvement they have effected an annual saving of R11,000; they have at the same time abandoned the importation of European caustic soda.

PANJAB. 156

In the Panjáb *sajjí-mátí* is said (*Official Correspondence*) to be prepared at Multan, Gugaria, Jhang, and Shahpur (*sajjí-khár?*). In Bombay *sajjí-mátí* or carbonate of soda efflorescences is met with. The Collector of Ahmedabad reports that barilla is not manufactured from plants in his district, but that the efflorescence known as *oos* or *khár* is used in washing clothes and also in the manufacture of soap and glass: it is worth about an anna per basket.

BOMBAY. 157

In the Northern Division of Bombay this *oos* efflorescence is said to cover a large area of land; it is used extensively for soap and glass making. Both the Paper-mills and the Soap-manufacturing Company of Bombay import, however, their caustic soda from England, having found the locally-prepared article more expensive.

Chemistry of Sajjí-mátí.—Speaking of sodium efflorescences, **Mr. Medlicott** (*Official Correspondence*) says: "The direct derivation of soda from the *reh* salts would no doubt be an easy process (for a chemical manufacture); sometimes soda forms a principal constituent of the *reh*, and it is then freely utilised by the *dhobies;* but the commonest constituent of the *reh* is sodic sulphate, which is the first result in the process of 'soda' manufacture, by the treatment of common salt with sulphuric acid. Thus the most expensive part of the process would be saved, and the further conversion of sodic sulphate into sodic carbonate ('soda') by the action of lime and fuel is simple." **Mr. Pedler**, Professor of Chemistry, Calcutta, says that the samples of *sajjí-mátí* which he has examined have consisted mainly of carbonate of soda. He adds: "As a commercial article, I believe *sajjí-mátí* is usually more valuable than the barilla obtained from Spain, for while barilla rarely contains more than 25 or 30 per cent. of carbonate of soda, *sajjí-mátí* is known sometimes to contain as much as 50 per cent. of that substance." (*Official Correspondence, December 15th, 1884, No. 3816.*)

BARILLA.

Carbonate of Soda obtained from the Ash of certain plants—Barilla.

SAJJI-KHAR. 158

(*b*) **Khár-sajjí**, or **Sajjí-khár**, or **Barilla.**—This is carbonate of soda obtained from the ashes (*khár*) of certain salt-worts. This must not be confused with pearl-ash or the form of potassium carbonate obtained from most other plants. In the correspondence to which reference has already been made this mistake occurs, lists of pearl-ash plants having been enumerated as those from which barilla is obtained. The manufacture of barilla first assumed commercial importance in Spain, and was an article of considerable value until **Le Blanc** discovered his method of preparing soda from common sea-salt. Since then it has considerably declined. Before this important discovery the demand for barilla caused attention to be directed to India as a country to which the trade might possibly be

Barilla or Sajjí-khár.

BARILLA.

SAJJI-KHAR. MADRAS. 159

extended. **Roxburgh** at the beginning of the century recommended the cultivation of one or two plants on the coast of Madras, but there is no evidence of this having been acted upon.

PANJAB. 160

Mr. Baden Powell (in his *Panjáb Products, Vol. I., 86*) has given a most instructive account of barilla manufacture as practised in the Panjáb. The process by which this substance is prepared is carried on during the month of October and the three following months. The plant after being cut down is allowed to dry. The next step is to dig a pit of a hemispherical shape, about 6 feet in circumference and 3 feet deep. One or more vessels with holes perforated are inverted and placed in the bottom of the pit, the holes being kept closed when the operation begins. The dry plants are gradually burned, and during the process a liquid substance is found to run down into the inverted vessels. After this has taken place, the residue is stirred up by means of a flat piece of wood and kept covered over for three or four days till it cools. Care must be taken not to allow water to get to the molten liquid, otherwise the whole mass would blow up. In the inverted vessels will be found a pure form of *khár-sajjí*, and in the bottom of the pit an impure form containing a mixture of ashes. This process differs only very slightly from that followed in Spain. In the latter country the plants are burned on iron bars placed across the mouth of the pit, and vessels to separate the substance into pure and impure barilla are not placed in the bottom.

In the correspondence to which repeated reference has already been made, the Panjáb Government has supplied some interesting information regarding the present condition of the manufacture of barilla. The industry exists only to a limited extent in Montgomery and Jhang, and not at all in Jhelum, Rawal Pindi, Gujarát, and Mozafergarh. "In Shahpur and Multan, however, the manufacture of *sajjí* is considerable. The Deputy Commissioner of Shahpur reports that the outturn is from eight to ten thousand maunds a year, and the revenue derived by Government by the lease of *sajjí*-producing lands amounts at present to over R9,500 per annum. The price, too, from various causes has risen from R1-2 to about R1-10 per maund since 1865.

"The income derived in the Multan district is also increasing, and though not so high as it was ten years ago, is higher this year than in any year since 1880." The Deputy Commissioner of Multan says that in his district the plants are cut in the months of January and February, and not in October and November as stated in **Baden Powell's** *Panjáb Products*. He adds: "I can find no evidence that the introduction of soda salts manufactured by purely chemical processes has injuriously affected the trade in barilla." He adds that the land on which barilla-yielding plants grow was leased for 1883-84, and realised " R7,907, which is higher than that realised in any of the past ten years, except 1875-76, 1877-78, 1878-79, and 1879-80."

The Settlement Report of Shahpur district contains an interesting account of *sajjí* manufacture. The Deputy Commissioner says in reference to Colonel Davis' report: "The account of *sajjí* manufacture given by Colonel Davis in 1865 seems to contain all the information required, and this industry is now in about precisely the same condition as it was then. As far as I have been able to ascertain, the introduction of soda salts manufactured by purely chemical processes has not affected it at all injuriously On the contrary, the price of *sajjí* has lately risen to R1-8 and R1-12 per maund, but this is said to be chiefly due to the fact that owing to recent droughts the growth of the plants has been less flourishing than formerly. The sums realised from farming the monopoly of manufacturing this alkali amount still to upwards of R8,000. The income under the head *sajjí* last year was a little over R9,500. The quantity of *sajjí*

BARILLA. **Sodium Carbonate.**

SAJJI-KHAR. PANJAB.

manufactured in this district is said to be about 10,000 maunds, but the plant itself is also highly esteemed as a fodder for camels, and the farmers of *sajjí* do not allow camel-owners to take the plant for fodder gratis."

The following extracts from the Settlement Reports of Jhang and Montgomery might also be here given: "**Caroxylon Griffithii** is the *khár*. There is a considerable disagreement as to what plant or plants *sajjí* is made from. In the Jhang district *sajjí* is made from *khár* only. I have made repeated enquiries and have always received the same answer, that *sajjí* is made from *khár*, but that sometimes, as sugar is sanded, and as a variety of jams are partly made from turnips and decayed figs, so is the bulk of the *sajjí* increased by burning *lana* with the *khár*. I have been constantly in camp at the time the *khár* is cut, but I have never seen a single bundle of cut *lana*, and such adulteration is very uncommon. All four plants are excellent grazing for camels."

In Montgomery "a good deal of misapprehension seems to exist about the *lana* plant. There are three kinds of *lana*: '*Khangan khár*' (**Caroxylon Griffithii**); '*Góra lana*;' and '*Methar lana*' (Salsolas). There is also a plant called '*Phesak lani*' (**Suæda nudiflora**). *Sajjí* (barilla), an impure carbonate of soda, is made from the first two. No *sajjí* is made from the others. The best *sajjí*, called '*Lóta sajjí*,' is made from '*Khangan khár*;' an inferior quality, known as '*Bhútni sajjí*,' from '*Góra lana*.' All four plants can be seen in the Montogomery civil station."

SIND. 161

In the same correspondence the Commissioner of Sind reports that there are no soda salts manufactured by purely chemical processes in Sind, but that there is a substance called *khár* manufactured from a plant called "*lani*," which grows wild all over the province and springs up spontaneously after a copious fall of rain. The *khár* or salt obtained from this plant is commonly used in Sind for dyeing, washing, and soap-making purposes, and in the manufacture of common glass. The Commissioner gives the following account of the process adopted in manufacturing this salt from the "*lani*" plant, which, it will be observed, is very similar to that pursued in Spain: "The '*lani*' plant is cut and gathered together in heaps. A circular pit varying from one and a half to two or three feet in depth and diameter, according to the convenience of the individual manufacturer and the quantity to be manufactured, is then dug in a clean level piece of ground. A fire is kindled near the pit and the freshly-cut plant thrown on it. The action of the fire causes the juice of the plant to exude and run into the pit. Fresh quantities of the plant are thrown on the fire from time to time, until the pit is almost filled with the liquid exudation. The mass is then stirred with a pole for from two to three hours, after which the pit is covered over, and on the third day, when the liquid has cooled down and solidified, it is dug out and broken into pieces for use."

Mr. Erskine adds that the manufacture flourishes most near Kutchee in Khelat, about 5,500 maunds of *khár* being annually imported into Jacobabad; that the quantity manufactured in Shikárpur, and in Thar and Párkar, is roughly estimated at 5,550 maunds and 3,000 maunds respectively every year; that the demand for the article has not been affected by the manufacture of soda salts by chemical processes, and that its price varies between R1 and annas 8 a maund." In another part of the same correspondence: "The Political Resident at Aden reports that **Salsola (Suæda) nudiflora**, vulgarly called 'Aden Balsam,' grows freely in the plain in the neighbourhood of Aden, and that before the purchase of Shekh Othman large quantities of the bush were wastefully burnt to produce salt, but that the shrub is now preserved within British limits. He observes that the bush seems to possess great vitality and fecundity; that it

ADEN. 162

is called by the Arabs '*asl*,' and the barilla made therefrom is named '*hotmi*;' that the Indians style it indifferently *khár*, *khár-sají*, and *sají-khár*; that the method of manufacture is primitive and resembles that described in the correspondence accompanying the letter from the Government of India, except that iron rods are not placed over the holes wherein the plant is consumed, and that advantage will be taken of the Spanish method in working the industry, which it is proposed to do shortly under Government supervision." Major Hunter adds: "Soda salts manufactured by purely chemical processes are only imported into Aden to the extent of ten or twelve hundredweights per annum, and do not affect the local manufacture in any way. In Aden barilla is produced in circular cakes having a diameter of about eighteen inches and a maximum thickness of eight inches. The value may be roughly quoted at from five to eight annas per 28 lbs. It is anticipated that a certain amount of profit will be gained by the Municipality, to whom the bushes belong, either by the manufacture of barilla under supervision, or by the sale of the right to produce it."

BARILLA-YIELDING PLANTS. 163

The following are the Indian plants reported to yield Barilla :—

1. **Anthrocnemum indicum**, *Moq.*; COROMANDEL COAST.
2. **Caroxylon fœtidum**, *Moq.*; SIND AND PANJÁB.
3. „ **Griffithii**, *Moq.*; Regarded as one of the best plants in the PANJÁB.
4. **Salicornia brachiata**, *Roxb.*; SUNDERBUNS AND COROMANDEL.
5. **Salsola brachiata**, *Pall.*; AFGHÁNISTAN.
6. „ **Kali**, *Willd.*; SIND AND PANJÁB.
7. **Suæda fruticosa**, *Forsk.*; SIND AND PANJÁB.
8. „ **indica**, *Moq.*; SUNDERBUNS AND COROMANDEL.
9. „ **nudiflora**, *Moq.*; ADEN; PONDICHERRY.

For further particulars regarding the above plants, consult their positions in this work.

It seems highly desirable that the distinction into *sajjí-mátí* and *sajjí-khár* urged in the above remarks should be clearly observed in all future enquiries into this subject. Care should also be taken not to confuse with these Pearl-ash, the *khár* or *kshára*, so extensively prepared all over India (see **Alkaline Earths**). The former are crude salts of sodium, the latter of potassium. Information regarding iodine-yielding plants will be found under **Kelp**.

BARLERIA, *Linn.*; *Gen. Pl., II., 1091.*

164

A genus of under-shrubs or herbs (generally spiny), belonging to the Natural Order ACANTHACEÆ, comprising some 60 species, chiefly natives of the old world—26 occurring in India.

Leaves opposite, entire. *Flowers* showy, purple, blue, yellow or white; sessile, solitary or in dense or sub-lax spikes. *Sepals* 4, in opposite pairs, outer pair very much the larger, anterior often emarginate bifid or deeply 2-lobed. *Corolla-tube* elongated, sometimes very long, funnel-shaped upwards; lobes 5, sub-equal, ovate or elliptic, imbricate in bud. *Stamens* 5, 2, having oblong 2-celled anthers and 2 small, rudimentary or rarely with a few grains of pollen, an abortive 5th sometimes present. *Disc* large. *Ovary* 4-ovulate; *style* long, shortly bifid or sub-entire. *Capsule* ovoid or oblong, 2-or 4-seeded below the middle.

The generic name is in honour of a Dominican traveller, the **Rev. J. Barrelier, M.D.**

Barleria cristata, *Linn.*; *Fl. Br. Ind., IV., 488.*

165

Syn.—B. DICHOTOMA, *Roxb.*; *Fl. Ind., Ed. C.B.C., 471*; B. CRISTATA, *Willd. in Roxb., Fl. Ind.*; *Wight, Icon., t. 453.*

BARLERIA prionitis.

The Barleria.

Vern.—*Jhánti* and *sada-jati*, BENG.; *Jhinli*, ASS.; *Tadrelú* (Bazar name, *bánsá siyáh*), PB.; *Gorp-jiba*, *kála-bánsa*, N.-W. P.; *Koileka*, URIYA; *Jhinti*, SANS.

Habitat.—A small elegant shrub, often met with in gardens, and found wild on the sub-tropical Himálaya, Sikkim, Khásia Hills, the mountains of Burma, of Central India, and of Madras, at an altitude of 4,000 feet. Distributed to the Malaya and China.

MEDICINE. 166 Seeds. 167 Roots and Leaves. 168

Properties and Uses—

Medicine.—The seeds are supposed to be an antidote for snake-bite, and the roots and leaves are used to reduce swellings, and an infusion is given in coughs. (*Madden; Stewart; Atkinson; &c.*)

169

Barleria longiflora, *Linn. f.; Fl. Br. Ind., IV., 485.*

Syn.—B. LONGIFLORA, *Willd., in Roxb., Fl. Ind., Ed. C.B.C., 471; of Nees in Wall., Pl. As. Rar., III., 93; and in DC. Prod., XI., 235.* B. LONGIFOLIA, *T. Anders.* Should not be confused with BARLERIA LONGIFOLIA, *Linn., Amœn. Acad., IV., 320,* or with RUELLIA LONGIFOLIA, *Roxb., Fl. Ind., Ed. C.B.C., 475,* which is HYGROPHILA SPINOSA, *T. Anders.* This mistake has been made by **Murray** in *Plants and Drugs of Sind,* and by other authors.

Habitat.—A small unarmed shrub, met with in the South Deccan Peninsula.

170

B. montana, *Nees; Fl. Br. Ind., IV., 487.*

Vern.—*Kolistá, ikharí*, BOMB.

Habitat.—A herbaceous species, met with in the Deccan, frequent in the Western Ghâts; extends from Jubbulpore to Travancore.

171

B. prionitis, *Linn.; Fl. Br. Ind. IV., 482.*

Syn.—B. PRIONITIS, *Willd., in Roxb., Fl. Ind., Ed. C.B.C., 470; Wight's Icon., t. 452; Dalz. & Gibs., Bomb. Fl., 189.*

Vern.—*Katsareyá*, HIND.; *Kántájáti*, BENG.; *Dasakarantod*, URIYA; *Kal sunda*, or *kula sunda*, *korhánti*, *vajra danti*, BOMB.; *Vajra-danti*, MARWAR; *Kántá shelio*, GUJ.; *Piwala koranta* or *koreta*, MAR.; *Lál-phúl-ké-kólsé-ká-pattá*, DUK.; *Vajra danl*, CUTCH; *Shemmuli*, *varamulli* (?), TAM.; *Mulu-goranta*, TEL.; *Karuntaka*, *vajradanti*, SANS.; *Kattúkurandú*, SINGH.

Habitat.—A small, spiny bush, with plentiful buff-coloured flowers; met with in tropical India, abundant in Bombay, Madras, Assam, Sylhet, and Ceylon. Sometimes planted as a hedge.

Properties and Uses—

GUM. 172

Gum.—Referred to by **Mr. Baden Powell** (*Panj. Prod., I., 412*) as one of the beautiful dark red-brown or black gums, apparently contributed by Madras to the Panjáb Exhibition of 1864.

§ "The gum alluded to above by **Mr. Baden Powell** is most probably a preparation from the juice. When fresh it is yellow, but afterwards turns black." (*Surgeon-Major W. Dymock, Bombay.*) "This plant is not known to yield any gum." (*Assistant Surgeon Sakhárám Arjun Rávat, Bombay.*)

MEDICINE. Juice. 173

Medicine.—Scarcely any new information has come to light since **Dr. Ainslie** wrote—"The juice of this leaf, which is slightly bitter, and rather pleasant to the taste, is a favourite medicine of the Hindús of Lower India, in the catarrhal affections of children which are accompanied with fever and much phlegm. It is generally administered in a little honey or sugar and water, in the quantity of two table-spoonfuls twice daily." **Dr. Dymock** adds—"The natives apply the juice of the leaves to their feet in the rainy season to harden them, and thus prevent the laceration and cracking of the sole which would otherwise occur."

Thwaites, in his *Enumeration of Ceylon Plants*, says this is employed as medicine by the Singhalese.

Leaves. 174

§ "The leaves are exhibited in syphilitic affections as an alterative. (*Surgeon-Major W. D. Stewart, Cuttack.*) "Used in fevers and catarrhal affections." (*Surgeon H. W. Hill, Mánbhúm.*) "Useful in coughs and infantile diarrhœa." (*Surgeon-Major D. R. Thompson, M.D., C.I.E., Madras.*) "Used as a diaphoretic and expectorant by natives." (*Deputy Surgeon-General George Bidie, C.I.E., Madras.*)

Barleria strigosa, *Willd.; Fl. Br. Ind., IV., 489.*

175

Syn.—B. CÆRULEA, *Roxb.; Fl. Ind., Ed. C.B.C., 471.*

Vern.—*Dasí*, BENG.; *Raila baha*, SANTAL; *Wáhiti*, BOMB.

Habitat.—Much cultivated in India, but wild in the lower hills of Bengal up to an altitude of 4,000 feet; in Orissa and Chutia Nagpur, extending to the Western Ghâts (*var.* **terminalis**), also in **Sikkim** and **Assam.**

Medicine.—The **Rev. A. Campbell**, of Pachumbá, Chutia Nagpur, sends me a specimen of a **Barleria** which appears to be this species along with the following note: "This plant is called *Raila baha* by the Santals, and a preparation from the root is by them given in severe spasmodic coughs."

MEDICINE. Root. 176

Barley, see **Hordeum vulgare.**

BAROSMA, *Willd.; Gen. Pl., I., 290.*

Barosma betulina, *Bart. et Wendl.;* RUTACEÆ.

177

THE BUCHU.

Habitat.—A native of South Africa; the dried leaves are imported into India and sold by all chemists. In addition to the above species the drug is obtained also from **B. crenulata**, *Hook.;* **B. serratifolia**, *Willd.*

Medicine.—It is aromatic, stimulant, and tonic, chiefly used in disorders of the genito-urinary organs.

MEDICINE. 178

BARRINGTONIA, *Forst.; Gen. Pl., I., 720.*

179

A genus of trees belonging to the Natural Order MYRTACEÆ; it contains some 20 species, inhabitants of tropical Asia, Africa, Australia, and Polynesia, frequent near the sea.

Leaves alternate, often crowded near the extremities of the branches. *Flowers* in terminal or lateral racemes, or occasionally interrupted spikes. *Calyx-tube* scarcely produced above the ovary; lobes 2-4-valvate or 3-5-imbricate. *Petals* 4, rarely 5, much imbricate, somewhat adnate to the base of the staminal tube. *Stamens* many, in several series, connate into a tube at the base. *Ovary* inferior, 2-4-celled, crowded with an annular disc; *style* long, simple, stigma small; *ovules* 2-8 in each cell, pendulous. *Fruit* fibrous or somewhat berried.

The generic name is in honour of the English Antiquary, the **Hon. Daines Barrington, F.R.S.**; it is the typical genus of the **Barringtoniceæ**, sometimes called the ANCHOY PEAR FAMILY.

Barringtonia acutangula, *Gærtn.; Fl. Br. Ind., II., 508;* MYRTACEÆ.

180

Sometimes called INDIAN OAK.

Vern.—*Ijál, samundar-phal, panniári, ingar*, HIND., DUK.; *Ijar*, MONGHYR; *Hijál, samundar*, BENG.; *Nichula, hijjala*, SANS.; *Hinjol*, SANTAL; *Saprung*, KÓL.; *Kinjolo, hinjara*, URIYA; *Hendol*, ASS.; *Kanapa* or *kanapa-chettu, batta, kurpá, kadamic*, TEL.; *Hole kauva*, KAN.; *Ingar, ijal, samundar-phal, tuwar* or *twara, kanapachethi* (the fruit is known as *samudra phala* or *samand ar-phal*), BOMB.; *Piwar*, *tiwar, newar, niwar, datte-phal,* MAR.; *Samundra phula,* CUTCH; *Nivar,* KONKAN; *Ella-midella,* SINGH.; *Kyaitha, Kyéni kyéni-kyibeng,* BURM.

Habitat.—A moderate-sized, evergreen tree, met with in the Sub-Himálayan tract from the Jumna eastward; in Oudh, Bengal, Central and South India, and Burma. Distributed from Ceylon and Singapore to the Malaya and North-West Australia. One of the most plentiful trees in Bengal, especially near the coast or beyond the tidal range. Common in the swamp forests of Pegu and Tenasserim, frequent in Kanára, Bombay, along the banks of streams and in moist places.

Properties and Uses—

TAN. Bark. 181

Tan.—The bark is used for tanning in Burma.

MEDICINE. Leaves. 182 Fruit. 183 Root. 184 Seeds. 185

Medicine.—The LEAVES and the FRUIT are used in native medicine. The ROOT is bitter and supposed to be similar to Cinchona in its properties. It is also held to be cooling and aperient. The SEEDS are warm and dry, used as an aromatic in colic and in parturition; also in ophthalmia. (*Baden Powell.*) "*Samundar-phal* is faintly aromatic and very bitter, and is considered by the natives to be warm, stimulating, and emetic; in Bombay it is often prescribed alone or in combination with other medicines as an external application in colds. A few grains are often given as an emetic to children suffering from catarrh, and seldom fail to induce vomiting." (*Dymoch, Mat. Med., W. Ind., 266.*) "The fruit rubbed in water is administered as an emetic." (*Lisboa, Useful Pl., Bomb.*) The kernels powdered and prepared with sago and butter are said to be used in diarrhœa.

MEDICINE. Juice. 186

Chemical Composition.—The seeds, according to **Dr. Dymock**, are about the size of a nutmeg. They are easily softened by immersion in water. "The bulk of the seed consists of starch."

187

Special Opinions.—§ "A few grains of the seed, with the juice of fresh ginger, are given to children as an expectorant and emetic. It appears to be a very efficient remedy." (*Surgeon-Major W. Dymock, Bombay.*) "The juice of the leaves is given in diarrhœa. The powdered seeds are used as snuff in headache." (*Civil Medical Officer U. C. Dutt, Serampore.*) "The powdered fruit is an ingredient along with *mál kangoni* (**Celastrus**) in a cosmetic; it is rubbed on the skin in cases of fever attended with nervous symptoms. Mixed with dry ginger it is also rubbed on the skin to check profuse sweating." (*Assistant Surgeon Sakhárám Arjun Rávat, L.M., Girgaum, Bombay.*)

Snuff. 188 A Cosmetic. 189

FISH-POISON Bark. 190

Poison.—The bark is used to stupefy fish in most parts of India. (*Bomb. Gaz., XV., Pt. I., 63.*)

TIMBER. 191

Structure of the Wood.—White, shining, warps in seasoning, moderately hard, even-grained, said to be durable. The radial section is beautifully mottled with the medullary rays, which appear as irregular plates. Weight 46 lbs. per cubic foot. (*Gamble.*) "The wood is reddish, and, though tough and strong, is not in general use. A seasoned cubic foot weighs 56 lbs." (*Bomb. Gaz., XV., Pt. I., 63.*)

It is used for boat-building, well-work, carts, rice-pounders, and cabinet-making. **Beddome** says the wood turns black when buried in mud.

192

Barringtonia pterocarpa, *Kurz; Fl. Br. Ind., II., 509.*

Vern.—*Kyétha,* BURM.

Habitat.—A small evergreen tree of Pegu and Tenasserim.

193

B. racemosa, *Blume; Fl. Br. Ind., II., 507.*

Syn.—B. RACEMOSA, *Roxb.; Fl. Ind., Ed. C.B.C., 446; Wight, Ic., t. 152.*

Vern.—*Samudra, cuddapah,* TAM., MAL.; *Samudra phal,* BENG.; *Ijjul,* HIND.; *Nivar,* KONKAN; *Kywégyi, Kyai-beng,* BURM.; *Deya-midella,* SINGH.

Habitat.—A moderate-sized, evergreen tree, with spikes of pink flowers, common on the Eastern and Western Coasts, from the Konkan to the Sunderbuns, Burma, Andaman Islands, Ceylon, and Malacca.

Properties and Uses—

Medicine.—The ROOT of the plant resembles Cinchona in medicinal virtues. It has deobstruent and cooling properties. The FRUIT is efficacious in coughs, asthma, and diarrhœa. The SEEDS are used in colic and ophthalmia.

MEDICINE. Root. 194 Fruit. 195 Seed. 196

The pulverised fruit is used as snuff, and combined with other remedies is applied externally in diseases of the skin. (*Treasury of Botany.*)

§ "The powder of fruit is used in skin diseases." (*Deputy Surgeon-General G. Bidie, C.I E., Madras.*)

Structure of the Wood.—Wood white, very soft, porous. Weight 27 lbs. per cubic foot. Skinner gives 53 lbs. and says it is used for house and cart building, and that it has been tried for railway sleepers.

TIMBER. 197

Barringtonia speciosa, *Forst.; Fl. Br. Ind., II., 507; Wight, Icon., t. 547.*

198

Vern.—*Kyi, Kyaigyee*, BURM.; *Dod-dá*, ANDAMANS.

Habitat.—A small, glabrous tree, with entire leaves; a native of the Andaman Islands, Singapore, and Ceylon; occurs also on the Southern Deccan Peninsula, but not wild.

Properties and Uses—

Oil.—In the Moluccas a lamp-oil is said to be expressed from the seeds of this plant. (*Treasury of Botany.*)

OIL. 199

Medicine.—Several brief notices have appeared regarding the properties of this plant. (*See Indian Forester, X., 75; and the Report of the Chemical Examiner, British Burma, August 1883.*) The active principle of the bark appears to be a volatile oil combined with a resin. The drug is simply narcotic; it stupefies fish without killing them.

MEDICINE. Bark. 200

Domestic Uses.—When dry the fruits are sometimes used as fishing-floats. (*Smith's Econ. Dict.*)

DOMESTIC. Fruits used as fishing-floats. 201

BASELLA, *Linn.; Gen. Pl., III., 76.*

202

A genus of CHENOPODIACEÆ containing only one species, which, however, in cultivation assumes 2 or 3 distinct forms bearing specific names. A herbaceous, succulent, glabrous climber, freely branched. *Leaves* ovate, oblong or cordate, alternate, subsessile or petioled, acute or obtuse, entire. *Flowers* in short axillary spikes, or simple elongated spikes, or branched, white or red. *Flowers* hermaphrodite, sessile; *perianth* fleshy, compressed to about the middle, cut into 5 erect obtuse teeth. *Stamens* 5, inserted in the mouth of the tube. *Ovary* globose; styles 3, connate at the base; *seed* solitary; *albumen* very small, *embryo* coiled up.

The generic name is said to be the Malabar name of the plant; it is sometimes spoken of as the Malabar Nightshade. For convenience of reference to the economic facts, the names **B. alba** and **B. rubra** have been retained.

Basella alba, *L.; Wight, Icon., t. 896;* CHENOPODIACEÆ.

203

INDIAN SPINACH.

Vern.—*Pói* (cultivated), *bon-pói* (wild), *myal-ki-bhāji, sufēd-bachlá*, HIND.; *Sufēd-bachlá-ki-bhāji*, DUK.; *Pói*, CUTCH, SIND; *Vasla-kire*, TAM.; *Alu-bachehali, karu-bach-chali, polam-bachchali, pedda-bach-chali* (a variety), TEL.; *Basella-kira*, MAL.

Habitat.—Cultivated in almost every part of India, especially in lower Bengal and Assam.

Properties and Uses—

DYE. 204

Dye.—It yields a very rich purple dye, which is, however, difficult to fix (*Drury*). This is said to be obtained chiefly from the form which received the name of **B. cordifolia.**

MEDICINE. Leaves. 205

Medicine.—Murray mentions this plant amongst his drugs, but says nothing about its medicinal properties.

§ "The leaves are made into a pulp used to hasten suppuration." (*Surgeon C. J. W .Meadows, Burrisal.*) "Cooling properties." (*Surgeon W. Barren, Bhuj, Cutch, Bombay.*)

FOOD. Leaves and Stems. 206

Food.—The succulent leaves and stems are used as a pot-herb (made into curry) by natives of all classes. Indeed, this forms a most important article of food; scarcely a village exists, in Bengal at least, where a hedge-row covered with this favourite pot-herb may not be seen.

§ "It is a very wholesome vegetable and makes a good spinage. It is much better than the ordinary (*ság*) greens of the country." (*Surgeon K. D. Ghose, M.D., Khulna.*) "Both this and the next form have similar properties, and are much used as vegetables." (*Surgeon-Major W. Dymock, Bombay.*) "Contains a good deal of mucilage and is used as substitute for spinach." (*Surgeon-Major P. N. Mukerji, Cuttack, Orissa.*)

207

Basella rubra, *Linn.*

Vern.—*Pói, lál-bachlú*, HIND.; *Rukto-púi, púisák*, BENG.; *Lál-bachlé-ki-bháji*, DUK.; *Shivappu-vasla-kire*, TAM.; *Alla-batsalla, pedda-mattu-neatku-batsala, erra-allu-bach-chali*, TEL.; *Putiká*, SANS.; *Chovvauna-basella-kira*, MAL.

Habitat.—Met with in Bengal, and indeed throughout India, under cultivation.

Properties and Uses—

MEDICINE. Juice of the Leaves. 208

Medicine.—The juice of the leaves is used in native practice in catarrhal affections of children.

§ "Leaves made into pulp to hasten suppuration." (*Surgeon C. J. W. Meadows, Burrisal.*) "Demulcent and diuretic, useful in gonorrhœa balanites." (*Assistant Surgeon J. N. Dey, Jeypur.*)

FOOD. 209

Food.—As with the preceding form, this is cultivated as a pot-herb.

210

Basket-work & Wicker-work, List of the more important plants used for—

In this class the fibres have not been specially prepared, or spun and woven, but either entire or after having been treated in a required manner they are worked into baskets or mats by hand.

Alnus nitida (baskets).
Andropogon muricatus (mats).
Arundinaria falcata (baskets).
A. racemosa (mats).
Bambusa arundinacea (baskets).
B. Tulda (mats).
Borassus flabelliformis (mats, baskets).
Calamus Rotang (baskets).
Caryota urens (baskets).
Cocos nucifera (mats).
Corypha umbraculifera (mats.)
Cyperus Pongarie (mats).
C. tegetum (floor-mats).
Dendrocalamus strictus (baskets).
Hibiscus tiliaceus (mats).
Indigofera atropurpurea (baskets).
I. heterantha (baskets).
Juncus effusus (mats).
Macrochloa (Stipa) tenacissima (mats, baskets).
Maranta dichotoma (*Shital-pati* mats).
Melocanna bambusoides (mats).
Moringa pterygosperma (mats).
Nannorhops Ritchieana (matting, baskets, leaves used for).
Pandanus odoratissimus (matting).
Parrotia Jacquemontiana (baskets).
Phœnix farinifera (mats).
P. sylvestris (mats, baskets).
Phragmites Roxburghii (*durma* mats).
Pseudostachyum polymorphum (baskets, mats).
Rhus Cotinus (basket-making)

Saccharum Sara (mats, leaves used for).
S. Munja (mats).
Saccharum spontaneum (mats, grass used for).
Salix babylonica (baskets).
Salix daphnoides (baskets).
S. tetrasperma (baskets).
S. Wallichiana (baskets).
Tamarix dioica (baskets).
Typha angustifolia (mats).
T. elephantina (mats).

BASSIA, *Linn.; Gen. Pl., II., 658.*

II

A genus of trees belonging to the Natural Order SAPOTACEÆ, comprising some 30 species, inhabitants of India and the Malay.

Leaves petioled, coriaceous, silky or tomentose beneath when young; *stipules* caducous. *Pedicels* axillary, fascicled among the sub-terminal tufts of leaves or in the axils of fallen leaves. *Calyx-segments* 4, 2-seriate, 2 outer valvate, enclosing the inner (except in **B. butyracea**). *Corolla-tube* campanulate; lobes 6-12. *Stamens* at least twice as many as the petals; *anthers* lanceolate-acute, connective, often mucronate or excurrent. *Ovary* villous, 4-12-celled. *Berry* globose. *Seeds* ellipsoid, hilum long, sometimes large; albumen none; radicle very small.

The genus is named in honour of **Fernando Bassi**, a former Curator of the Botanic Gardens at Bologna.

Bassia butyracea, *Roxb.; Fl. Br. Ind., III., 546;* SAPOTACEÆ.

212

THE INDIAN BUTTER TREE.

Vern.—*Chiúra* or *chyúra* or *chára, chaiúra, bhulel,* KUMAON; *Cheulí,* OUDH; *Phalwara* or *phulvárá,* HIND.; *Chúri,* NEPAL; *Yel, yel pote,* LEPCHA.

The butter from the fruits is called *chiúra-ke-pina* (Almora), and *phalel, phulel* or *phalwára, phulwa* (in the plains).

Habitat.—A deciduous tree of the Sub-Himálayan tract, from Kumaon to Bhután, between 1,000 and 5,000 feet.

In **Mr. Atkinson's** manuscript this tree is stated to be very abundant at Pithoragurh, where the bees feed on its fragrant flowers; hence the honey is highly esteemed. It is also common in the valley of the Kálí.

Botanic Diagnosis.—Leaves obovate or obovate-oblong; calyx-lobes 5, much imbricate but not 2-seriate; corolla-tube not fleshy, lobes spreading; stamens 30-40, filaments glabrous, as long as the anthers.

Properties and Uses—

OIL. 213

Oil.—The seeds on expression yield a concrete oil known as *phulwa.* This is extracted by beating the seeds to a consistence of cream, and placing the mass thus obtained in a cloth bag, upon which a weight is laid until all the oil or fat is expressed. This becomes of the consistence of hog's lard, is inodorous, and of a delicate white colour; it contains 34 parts of fluid oil and 6 parts of vegetable matter. (*Mr. E. Solly.*) It dissolves readily in warm alcohol, leaving the vegetable impurities undissolved. At 95° it retains its consistency, but melts completely at 120°. (*Roxburgh in Asiatic Researches, VIII., 447.*) This vegetable butter, being cheaper than *ghí,* is sometimes used as an adulterant. It is burned in lamps, and as it burns with a bright light without smoke or smell, it may be utilised in the manufacture of candles. It makes excellent soap. Its oil has many properties which should commend it to the attention of the candle and soap makers, and it is surprising that it has not taken a better position during the past half century. (*Don, Prod. Nepal, 146; Royle's Illustr., p. 15; Trail in Proceed. Corres. Commerce and Agriculture, Royal Asiat. Soc., p. 115; also a complete and interesting account by the Editor, Jour. Agri.-Hort. Soc. of India, Vol. I., 19.*

BASSIA latifolia.

The Mahuá,

MEDICINE. Butter. 214

Medicine.—The butter is highly valued on account of its efficacy in rheumatism, especially in contraction of the limbs and other painful affections. It seems deserving of further attention. As a hair-oil, perfumed with *attar* of roses and other sweet-scented oils, it is largely used as a valuable preservative to the hair. Made into perfumed ointments it is also extensively used by the wealthier classes. It is an excellent emollient for chapped hands. (*Pharm. of India; Roxburgh; Voigt; Baden Powell, Pb. Pr., I., 423; Atkinson, Him. Dist., 715; Year-Book of Pharm., 1878, 258.*)

FOOD. Pulp of Fruit. 215 Cake after expression of Oil. 216 Syrup. 217 Sugar. 218

Food.—The pulp of the fruit is eaten, and also the cake left after the expression of the oil. The flowers are not eaten, but from them a sweet syrup is prepared which is boiled down into sugar. The sugar thus prepared resembles in appearance that prepared from the date-palm. "The grain is very small, and, as *gúr*, would fetch considerably less than the finer specimens of cane-sugar. It was, however, equal, if not superior, to ordinary date-sugar, of which such abundant supplies reach Calcutta." (*Jour., Agri.-Hort. Soc. of India, I., 22.*) As already stated, the oil is both eaten and used as an adulterant for *ghí*.

TIMBER. 219

Structure of the wood.—Wood light brown, hard; annual rings marked by a dark line. Weight 52 lbs. per cubic foot.

220

Bassia latifolia, *Roxb.; Fl. Br. Ind., III., 544.*

THE BUTTER OR MAHUÁ TREE.

Vern.—*Mahwá, mahuá, mahulá, maul, janglí-mohá, janglí-mohvá, mowá*, HIND., OUDH; *Mahwá, banmahuva, mahúla, maul*, BENG.; *Moha*, URIYA; *Mandukum*, KÓL.; *Mohul*, BHUMIJ and MAL. (S. P.); *Matkom*, SANTAL; *Mahurá*, BHIL (SURAT); *Mahu*, BAIGAS; *Irúp, irrip, irhu*, GOND; *Mohu*, KURKU; *Mhowa*, C. P.; *Mová, mahua, mohá*, BOMB.; *Janglí móhá, moha*, DUK.; *Mahuda* or *mahura*, GUJ.; *Mowda, ránácha-móhácha-jháda, ránácha-ippécha-jháda, moho, mora, maha*, MAR.; *Illupi, elupa, kat illipi, káthi-iluppai, káttuiluppai, káttu-írrupai*, TAM.; *Ippi, ippa, yeppa, adavi-ippe-chettu*, TEL.; *Hogne, hippe, kádu-ippe-gida*, KAN.; *Poonam, káttirippa bonam*, MALA.; *Madhuka, atavi madhuka vriksha*, SANS.; *Darakhte-gulchakáne-sahrái*, PERS.; *Kansan*, BURM.; *Quindah* (the oil).

Habitat.—A large, deciduous tree, indigenous in the forests of the Central Provinces; it may in fact be said to extend from Kangra, Kumaon, and Oudh, through the Central Provinces and Chutia Nagpur to the Western Ghâts, and distributed in the south-east to Ava. It is plentiful in many parts of the Bombay Presidency, especially in Gujarát. It forms gregarious forests, generally associated with the *Sál;* abundant where met with, it may be described as forming scattered and isolated forests over the region indicated. It gradually disappears towards Calcutta, and is only sparingly met with in the Madras Presidency, its place being taken by **B. longifolia.** **Dr. Stewart** does not regard the plant as indigenous to the Panjáb.

Botanic Diagnosis.—Leaves elliptic or oblong-elliptic, shortly acuminate; calyx-lobes 4, the 2 outer sub-valvate including the others, rusty tomentose; corolla-tube fleshy, lobes erect; anthers 20-30, 3-seriate, subsessile. It attains a height of 40 to 60 feet.

The *Mahuá* thrives on dry, stony ground. It is protected by the natives, but is not artificially planted. It sheds its leaves from February to April. The cream-coloured flowers, clustering near the ends of the branches, appear in March and April, and are soon followed by the new leaf-buds. The fruits are green when unripe, and reddish yellow or orange when ripe, fleshy, one to two inches in length, with one to four seeds, which ripen about three months after the flowers have fallen. The tree is valued for its flowers, its fruit, its seed, and its timber; and is of considerable economic importance to a large proportion of the poorer classes of the natives of India. (*Liotard.*)

Properties and Uses—

Gum.—It yields a white milky gum from incisions and from cracks in the bark. The discharge of gum is facilitated by a process of ringing the trees, practised in Chutia Nagpur during the fruiting season. The gum does not seem to be of any economic value.

GUM. 221

Dye.—The BARK is often used as an adjunct in dyeing where dark colours or black are desired; along with the leaves it is also sometimes employed as a tan.

DYE. 222
TAN. 223

THE OIL.

Oil.—A greenish-yellow oil eaten by the Gonds and other Central Indian tribes is extracted from the kernel of the fruit; it is used to adulterate *ghí*. This is sometimes called *Dolí* oil, especially in Western India, the same name being applied to the seeds. It is called *Madhuka Sára* in Sanskrit, and is recommended as a medicine. It is often sold in the form of cakes, which keep fresh for a few months in cold climates, but in the plains of India they soon become rancid, separating into a clear oil and a brown fatty substance. The cakes are sold as *Illipi Butter.*

OIL. Cakes. 224
Liquid. 225
Fatty Matter. 226
Clear Oil. 227

To extract the oil the kernels are taken out from the smooth, chestnut-coloured pericarp, by being bruised, rubbed, and subjected to a moderate pressure. They are then ground and the oil obtained by cold expression. In the Central Provinces, the kernels are pounded and boiled, and then wrapped up in two or three folds of cloth and the oil thereafter expressed. In the western tracts of Bengal and in the Central Provinces, besides being used for lighting, this oil forms a very inexpensive substitute for *ghí.* It is a useful oil for soap, and is largely used by the poorer classes as a lamp oil.

CANDLE AND SOAP TRADE.—The following interesting passage is extracted from *Drury's Useful Plants of India:* "In 1848 a quantity of *Mahwah* oil was forwarded to the Secretary of the East India and China Association, with the view of ascertaining its market value and applicability for the manufacture of candles and soap. The managing director of Price's Patent-Candle Company stated in reply: 'I beg to inform you that the *Mowah* oil, of which you furnished us samples, is worth in this country, for the manufacture of candles, £8 per ton less than Petersburg tallow. We have tried a great many experiments upon it, and found it to be of the same value as cocoanut oil, and its being harder makes up for the colour being inferior. Large quantities could be used in this country at about £35 per ton."

CANDLE & SOAP. 228

In Gujarát, "Soap is manufactured by Musalmans. This is made by mixing alkali soda and lime in water, and allowing them to soak for some hours. The water is then drawn off and a quantity of *mahuda* oil, *doliu*, is added, and the whole boiled in large brick caldrons. When ready the mixture is run off into shallow brick troughs and left to cool. It is then gathered into a large heap, pounded with heavy wooden mallets, and cut into round cakes. According to the amount of *mahuda* oil it contains, soap varies in price from 1½*d.* to 3*d.* (1-2 annas) the cake." (*Bombay Gazetteer, III., 76.*) In Ahmedabad, soap is made from the oil of this tree, called *doliu* oil. The oil is largely burned by the Bhils and other hill tribes. In the Deccan the oil is used for making country soap.

Gujarat Soap. 229
Ahmedabad Soap. 230
Other Mahua Soaps. 231

The Gazetteer of the Central Provinces remarks that for the purpose of preparing the " oil the exports of the seed might be largely increased." "The seed of the *mahua* (which succeeds the flower from which the spirit is made) is extensively used for the manufacture of oil for burning; and the failure of the *mahua* crop is usually followed by a high price of oil

BASSIA latifolia.

The Mahuá.

throughout the year in which the failure occurs." (*Oudh Gazetteer, III., 71.*)

MEDICINAL PROPERTIES.

MEDICINE. Flowers. 232

Medicine.—The FLOWERS are used in coughs in the form of a decoction. The medicinal properties attributed to this plant are stimulant; demulcent, and emollient, heating, astringent, tonic, and nutritive. The SEEDS yield, on expression, a thick concrete OIL, which is recommended to be applied to the head in cephalalgia. The oil is much valued by hill tribes in the treatment of skin diseases. The residuum or CAKE, left after the expression of the oil, is employed as an emetic and also as a detergent.

Seeds. 233 Oil. 234 Cake. 235 Spirit. 236

According to the *Pharmacopœia of India,* "the SPIRIT distilled from the flowers has a strong smoky odour, somewhat resembling Irish whisky, and rather a pungent fœtid flavour, which, however, disappears with age. The freshly-distilled spirit proves very deleterious, exciting gastric irritation and other unpleasant effects." **Dr. U. C. Dutt** says this spirit is described by **Susruta** as heating, astringent, tonic, and appetising. The *Pharmacopœia* adds that **Dr. Dutt** reports "having used the weaker (diluted ?) spirit extensively; and in his opinion it is less injurious to the digestive system than rum, more resembling beer in its effects on the constitution and nutrition of the body. This view is coincident with that of **Dr. W. Wright.** It is evidently a powerful diffusible stimulant, and when matured by age may be used as such, when brandy and other agents of the same class are not available." (*Pharm. Ind., 131.*) The LEAVES are boiled in water, and given as a cure for several diseases; they make a good embrocation. "The MILK of the green fruit, and of the tender bark, is given as a medicine." (*Voigt.*)

Leaves. 237 Milky sap. 238 Bark. 239

Dr. Irvine (*Mat. Med., Patna*) says that the BARK is used in decoction as an astringent and tonic. "The bark is sometimes used as a remedy for rheumatic affections." (*Mysore Catalogue, Calcutta Exhibition.*) **Voigt** says it is rubbed on the body as a cure for itch.

Fish-poison. 240 Rat and Insect poison. 241

The residue cake, after the extraction of the oil, is said to be used to poison fish. This seems doubtful, but the statement is made by several writers. The smoke produced in burning the cake is reputed to kill insects and rats.

242

Special Opinions.—§ "Used as a detergent in the southern districts of Madras." (*Hon. Surgeon E. A. Morris, Negapatam.*) "The expressed oil is applied to the arms of children to allay the extreme itchiness sometimes caused by the presence of intestinal worms." (*A Surgeon, Alighar.*) "The flowers mixed with milk are used in impotence due to general debility; they are given in doses of about one ounce with eight ounces of fresh milk, and are often an efficient remedy. The dried flowers are used as a fomentation in cases of orchitis for their sedative effect." (*Hospital Assistant Lal Mahomed, Hoshangabad, Central Provinces.*) "The flowers of the *mowa* appear to impart their peculiar odour to the secretions of the body when eaten. This is notably the case in cattle, the milk being flavoured when they are allowed to feed on *mowa.*" (*Surgeon S. H. Browne, M.D., Hoshangabad, Central Provinces.*) "The flowers are sometimes boiled and eaten by the lower classes. I know a case of dangerous vomiting with brain symptoms caused by eating an excess quantity of flowers. The spirit, if carefully prepared and re-distilled, is not deleterious." (*Surgeon Shib Chunder Bhuttacharji, Chanda, Central Provinces.*)

FOOD.

FOOD. Flowers. 243

Food.—The FRUIT is sometimes eaten, but the principal edible structure is the succulently-developed FLOWERS (*i.e.,* corollas); these are eaten raw or cooked, or in the form of sweetmeats. SUGAR may also be prepared from the flowers. In many parts of the country they are baked into CAKES. The SEEDS also may be eaten, but it is chiefly valuable for the oil which it yields on expression, the cake being utilised as an article of food both for men and animals. The flowers afford both food and drink to a large number of persons during a great part of the year, *viz.*, from March to September. After having been steeped in water, and allowed to ferment, a SPIRIT is distilled from them which is largely consumed by the inhabitants of the mountainous tracts of the central table-land of India. In a note by **Mr. Liotard**, published by the Revenue and Agricultural Department, an interesting abstract of information regarding *mahuá* will be found, from which the following passage may be republished here: "When the buds appear, the people clear the jungle from below the trees; and when the flowers fall, women and children, and sometimes men, may be seen busily occupied in the early mornings gathering large quantities. It is reckoned that each tree during the season gives from 6 to 8 maunds of flowers, the quantity varying according to the size of the tree and the nature of the season. This is used in two ways: (1) as an article of food, and (2) as a material for the manufacture of a spirituous liquor.

Fruits. 244
Sugar. 245
Cakes. 246
Seeds. 247
Spirit. 248

"*As an article of food* it possesses, when fresh, a peculiar luscious taste, with an odour somewhat suggestive of mice. When dried the flavour has some resemblance to that of inferior kinds of figs, and forms an important addition to the food-supply of the poorer classes of parts of the country in which the tree grows in abundance. Under the Mahratta rule it is said to have been a common practice to cut down the *mahuá* trees in the Bhil country so as to afflict the lawless hill tribes and reduce them to straits. This shows how much the people depend on the produce of these trees for food. The flowers are used either freshly gathered or after being sun-dried. They are eaten cooked or uncooked, often with parched grain or with the seeds of the *sál* tree, or with leaves of other plants. Jackals, bears, wild pigs, and deer are very fond of *mahuá*.

"*For the manufacture of spirits*, the flowers when dried are sold by the hill people, at various rates, either to the village distillers or to the *baniahs* by whom they are exported. The dried flowers are immersed in water for four days; they are then fermented, and thereafter distilled. The liquor produced from a single distillation is extremely weak, ranging from 60° to 90° under proof. But a second distillation is sometimes resorted to, especially where still-head duty is levied irrespective of strength, and in this case a spirit averaging 25° below proof is obtained. The distillation is practised in the Panjáb to a small extent; in Rájputana every village apparently has its spirit-shop for the sale of the distilled liquor; in the North-West Provinces and Oudh the liquor is made in the eastern and southern districts, and is of common use among certain classes; in the western districts of Bengal it is abundantly distilled; so also in the Central Provinces, and in parts of the Bombay Presidency, especially in the northern and southern divisions."

PANJAB. 249

ABSTRACT OF PROVINCIAL REPORTS REGARDING THE VALUE OF MAHUÁ.

An indefinite series of extracts from published works might be given to illustrate how exceedingly important the *mahuá* tree is to the hill tribes of India. In the *Bombay Gazetteer (XII., 26)* will be found the following passage: "Its chief value lies in the pulpy bell-shaped flower, which, when dried, is eaten by the natives, and is distilled into the common spirit of the country. Almost every animal, wild or domestic, eats the fresh flowers. It is an important article of trade, and during the hot months is the chief means of subsistence to the Bhils and other hill tribes. The wood is hard and lasting, but the tree is too valuable to be cut

BOMBAY. 250

The Mahuá.

BOMBAY.

for timber. The seed, when allowed to form, is enclosed in a thick walnut-like pod. It yields an excellent oil good for food and burning, and also for skin diseases. The leaves and bark make useful embrocations. Altogether, the *moha* is one of the most valuable of Khándesh trees, but as it grows in the wildest forests, most of the produce is lost or supports wild animals only. In the open country a few good *moha* trees are a small fortune."

"The *mahudá*, with its strongly-veined leaves and its heavy sickly-smelling flowers, is in every respect a noble tree and of great value to the district. For months in the year its flowers and fruits are meat and drink to many of the poorer classes, and its timber is of excellent quality." (*Bombay Gazetteer, III., 198.*) "In Gujarát and Rájputana, every village has its spirit-shop for the sale of the distilled liquor from the flowers. In the Island of Caranja, opposite to Bombay, the Government duty on the spirits distilled (chiefly from this flower) amounts to at least £60,000 per annum; I rather think that £80,000 is more generally the sum. The Parsís are the great distillers and sellers of it in all the country between Surat and Bombay, and they usually push their distilleries and shops into the heart of the forest, which lines the eastern border and hills of those countries. The spirit produced from **Bassia** is, when carefully distilled, much like good Irish whisky, having a strong, smoky, and rather fœtid flavour; this latter disappears with age." (*Dr. Gibson, in Hooker's Journ. Bot., 1853, p. 90.*)

CENTRAL PROVINCES. 251

In the Central Provinces the poor people draw half their sustenance from the fleshy flowers at certain seasons of the year. "The spirit most used in the Central Provinces is the *dárú* distilled from *mhowá*." (*C. P. Gaz., Intro., p. cxlv.*) **Mr. J. G. Nicholls**, Commissioner of Excise, on enquiries instituted with regard to the *mahuá* of the Central Provinces, obtained valuable information. Summarising the reports which he received, it would be within the mark to say that in the Central Provinces alone over 1,400,000 persons use the *mahuá* as a regular article of food. The following extract from **Mr. Nicholls'** paper on the value of *mahuá* (or *mohuá*) (taken from the *Indian Forester, V., 475*) will be found highly instructive: "From my enquiries, I am led to believe that one maund for the annual consumption of each individual is a moderate and quite safe estimate. But one maund of *mohwá* sets free more than an equal weight of grain, probably one and a half maunds."

A saving through Mahua. 252

Mr. Fernandez, Assistant Conservator of Forests, has given this matter his special attention. He calculates that a frugal family will save 30 per cent. of grain on the average annual consumption of cereals by the partial and seasonable substitution of *mohwá* flower and oil.

"I will first estimate the saving to be only 1¼ maunds of grain for each person, 5 maunds being the annual average consumption of each individual of the population. This represents 17,50,000 maunds of grain saved; or we may calculate in another way: that this supplementary source of food-supply sets free so much of our arable land as would be required to produce an extra 17,50,000 maunds of cereals, to be more profitably employed in the cultivation of cotton, linseed, and the more valuable agricultural products, resulting in a still larger balance of trade in favour of these Provinces.

"But so as to be sure of avoiding an over-estimate, suppose the saving to be only 12,50,000 maunds a year; this sets free so many maunds of the highest priced grain mostly for export out of the provinces. Calculating at 20 seers to the rupee, this would represent one quarter of a million of pounds sterling as the value of the *mohwá* crop to these Provinces in ordinary years, exclusive of whatis used for distillation. The latter want

The Mahuá.

BASSIA latifolia.

CENTRAL PROVINCES.

will always, in ordinary years, be first provided for, because good prices would be forthcoming.

"The *mohwá* used for distillation yields a revenue of close on ten lakhs of rupees a year, raised, in a way, by self-imposed taxation on classes who could not otherwise (save by the salt tax) be made to pay their quota towards the expenses of the State.

"At the lowest valuation, in ordinary years, the *mohwá* produce is worth to the country at large not less than 35 lakhs of rupees.

Annual value R35,00,000.

"In times of scarcity its economic as well as its monetary value rises with the intensity of the distress: it often becomes of vital importance. It must be remembered that failure of agricultural crops in these Provinces is more frequently the result of excessive rainfall than of drought. The *mohwá* crop would be unaffected by an excessive rainfall in the period of the monsoon. It suffers from drought to some extent, more so from frost. Where, but for this supplementary source of food-supply, we should experience famine, with it we should only have distress. Without it, I think we should always have chronic scarcity in the wildest parts of the Provinces.

"We are now called on to part with a portion of our crop for export to Bombay. It is understood that, besides the demand in that direction for the purpose of distillation and for consumption as food, it is also required for use in the manufactures in connection with ship-painting and caulking. I have no reliable information on this point. I mention what I have been told for what it may be worth. But at any time demands may arise at the bidding of the chemist and the manufacturer in quarters where the purchasing power would be so great as to draw off much of our crop. It is not unlikely that the distillers of the North-Western Provinces will soon begin to indent for *mohwá* on our northern districts.

"Taking the lowest valuation of the crop, *i.e.*, 35 lakhs of rupees, and capitalising this at 15 years' purchase, we get the present value of the bearing trees, as flower and oil-producing sources, represented by $5\frac{1}{2}$ millions sterling.

"But to replace the present existing trees in full bearing, would require much more than fifteen years. Considering this and the cost of artificially-stimulated reproduction, together with the incidental disturbances of the normal conditions of life, double this amount would scarcely compensate the Provinces for their sudden destruction.

"This goodly endowment from the hands of bountiful Nature, this inheritance may not with impunity be wrecked or impaired. It should be held as a great trust, to be left, at the least, intact, by the present generation, for the support and enjoyment of generations yet unborn."

OUDH. 253

In Oudh it is principally found in the western half of Pertabgarh. The flower withers in April, and drops from the tree during the night. There are calculated to be 434,570 trees in the district. Assuming each tree to yield 20 seers, this, at the average price, would give a value of R1,44,856. As a rule the *mahuá* crop is good only once in every three years. (*Oudh Gaz., III., 71-72.*)

R1,44,856.

BENGAL. Palamau. 254

The following extract from the "Statistical Account of Bengal" gives the substance of **Mr. Forbes**' Settlement Report of Palamau with regard to *mahuá*: "The most important of all the indigenous jungle products is the flower of the *mahuá* tree, as the abundance or deficiency of this crop affects the market price of all other foods throughout the year. The total number of *mahuá* trees in Palamau, from which fruit was regularly gathered, was estimated by **Mr. Forbes** in 1869 at 113,885, of which 18,492 belonged to Government farms and were specially dealt with at the time of the settlement. All were of indigenous growth, and it appears not to be the practice to rear trees artificially."

BASSIA latifolia.

The Mahuá.

BENGAL.

"*Mahuá* blossoms are rarely eaten fresh, but are dried on a smooth floor of cow-dung and mud, until they shrivel to a quarter of their original size, and take a light-brown colour, so as to resemble raisins. They are usually prepared by boiling. This takes all the flavour out of the flower, and it is therefore eaten with the seeds of the *sál* tree called *sarráyi*, or some acid leaves or herbs, to give it a relish. Those who can afford to do so eat *mahuá* fried in *ghí* or butter. The yield of a *mahuá* tree varies very much in different seasons. A large tree will bear in a good season from 4 maunds 2 seers to 4 maunds 29 seers of ripe blossom; but the average yield is about 2 maunds 28 seers, which when dry does not weigh more than 1 maund 14 seers. Of late years the price of *mahuá* blossom has risen. It used to sell at 3 maunds for the rupee, but at the present market value, about a rupee and a half is paid for 2 maunds. During the distress of 1869, the price rose to 14 and 11 seers per rupee. The fruit of the *mahuá* tree begins to form immediately after the fall of the blossoms, and ripens in June. The weight of the yield of fruit generally equals that of the crop of blossom. Natives never gather the fruit, or even shake the tree to make it fall, the belief being that if this were done the tree would not bear in the following year. When ripe, the *mahuá* fruit is about as big as a peach, and is made up of three separate envelopes, with a white nut or kernel inside. The two outer skins are either eaten raw or cooked as vegetable, and the inner coating is dried and ground up into a kind of meal. Of the kernel itself an oil is made; four seers of kernels making one seer of oil, which is largely used both for cooking and for adulterating *ghí*. Before, however, it can be used for the latter purpose, it must be clarified with butter-milk, to prevent its offensive smell from being detected in the *ghí*. The oil sells at 9 seers for the rupee. The amount annually made is small, and it can rarely be purchased two months after the manufacturing season is over." (*Dr. W. W. Hunter's Statistical Account of Bengal, Vol. XVI., pp. 243-44.*)

Hazaribagh. 255

In **Mr. V. Ball**'s account of *mahuá* in the *Journal of the Asiatic Society, No. II. of 1867*, incorporated in *Hunter's Statistical Account of Bengal*, will be found the following interesting account of the *mahuá* tree in the Hazaribagh District: "The duty of collecting the fallen blossoms is chiefly performed by women and children; at dawn they may be seen leaving their villages with empty baskets and a supply of water for the day's use. Before the crop has commenced to fall, they take the precaution to burn away the grass and leaves at the foot of the tree, so that none of the blossoms may be hidden when they fall. The gleaners generally remain under the trees all day, alternately sleeping and collecting the crop; the male members of the family, visiting the trees once or twice during the day, bear off the produce in *banghis*. It often happens that the people who collect come from a considerable distance, in which case they erect with the branches of *sál* a temporary encampment of huts, in which they live until the crop is all gathered in. In front of each of these huts a piece of ground is made quite smooth and hard, for the purpose of spreading out the flower to dry. When perfectly dry, the blossoms have a reddish-brown colour, having lost three fourths of their original dimensions, and about half their original weight. It is the custom with some of the natives, before spreading them out to dry, to pull off the little ring of foliaceous lobes which crowns the fleshy corolla. It is very difficult to collect trustworthy statistics regarding the amount of yield of the *mahuá* trees. I have been told, and it has been repeated to me several times, that a first-class tree will yield two maunds a day, and that this will continue for fifteen days. This estimate, I believe, is more than double the real facts. The rent of the trees varies much according to their abundance in the district, the quality

BENGAL.

of the previous rice crop, and various other circumstances affecting the demand and supply. In parts of Házáribágh I have known ten small trees to be let for a rupee, while a single fine larger one would sometimes bring the same amount. In Mánbhúm, I have been pointed out trees for which a sum of from two to three rupees was charged, but I have also heard of trees being hired in the same district for four annas." "Two maunds of *mahuá* are stated by some to furnish a month's food to a family consisting of a father, mother, and three children. It is, however, seldom eaten alone, being much more frequently mixed with the seeds of *sál* or with some of the leaves of the plants which are collectively called *ság*. The cooking is performed as follows: The *sál* seeds, having been previously well dried in the sun, are roasted and then boiled alone; the *mahuá* flowers are then also boiled, and the water is thrown away. So far having been cooked separately, they are then mixed and re-heated; sometimes a small quantity of rice is added. It is the custom to cook but once a day, and each member of the family helps himself whenever he feels hungry." (*Dr. W. W. Hunter's Statistical Account of Bengal, Vol. XVI., pp. 48-49.*)

Monghyr. 256

Mr. Lockwood (formerly Magistrate and Collector of Monghyr) published in the Linnæan Society's Journal (*Vol. XVII., 89*) a most instructive account of *Mahuá*, the facts of which may be said to be applicable to every district in which the plant occurs. The following extract may therefore be republished here: "During the season of scarcity which prevailed at Behar in the year 1873-74, the *mahwa* crop, which was unusually abundant, kept thousands of poor people from starving; and all famine officers will recall its peculiar odour as they passed through the villages where it had been collected. The residue of the *mahwa* which is not eaten is taken to the distilleries, and there, with the aid of rude pot-stills, is converted into a strong-smelling spirit, which bears considerable resemblance to whisky. The Government holds a monopoly of spirit manufacture; and when I first went to Monghyr in 1873, the custom was to charge a duty of eight shillings for every cwt. of the raw material as it entered the distillery, on the supposition that so much *mahwa* would only yield three gallons of proof spirit. Subsequently, in consequence of experiments made by the officers under me, this duty was somewhat raised; but in England I find that over six gallons of proof spirit can be produced from a cwt. of *mahwa*. The Government of India should be made aware of this fact; and it would probably be advantageous to introduce patent stills in the place of the rude machines now in use.

Proof Spirit (6 gallons from 1 cwt. of Mahua). 257

"The amount of *mahwa*, which nominally paid Government duty yearly in Monghyr, was 1,750 tons; but with patent stills under Government control, the *mahwa* would probably yield a much larger revenue to the State. An Italian gentleman, who was living at Monghyr when I was there, took out a patent for removing, by a very simple process, the essential oil, or whatever it is, which gives the *mahwa* spirit its peculiar smell; and for some time I thought he would make a rapid fortune; orders poured in on him from Calcutta, and the demand promised to be immense. But just as the inventor had taken up a whole side of the Government distillery and got all his preparations complete, the rum-distillers in Calcutta petitioned the Board of Revenue, and a prohibitive duty was imposed, which completely putan end to the manufacture of scentless *mahwa* spirit. A sample was sent to the Chemical Examiner at Calcutta; and he reported that the spirit was pure and wholesome, and came very near good foreign brandy.

Scentless spirit from Mahua. 258

"But not only are the *mahwa* flowers good for distilling spirit; they are still more useful for feeding cattle. My father, the Rector of Kingham, has been feeding his pigs on the *mahwa* which I brought home, and

For feeding Cattle. 259

BASSIA latifolia.

The Mahuá.

BENGAL.

Mahwa Pork is beginning to be celebrated in his neighbourhood. Indeed, so favourably has it been received, that I have been requested to procure considerable quantities, both for distilling spirit and for feeding cattle. The Bassia family is the only family I know which yields a flower in sufficient quantities for feeding cattle and distilling spirit on a large scale. Potatoes, maize, and barley, which are principally used, are costly in production and uncertain in their yield, but the *mahwa* crop never fails. The oldest inhabitant in Monghyr had never heard of a season when the *mahwa* crop was not abundant; for, whether the fruit subsequently forms or not, the corolla is certain to be there, and certain to fall in great profusion. The extraordinary keeping qualities of *mahwa* form also a further recommendation to its introduction into England. Before leaving India, I had a ton shovelled into sacks and put on board a vessel in Calcutta. They were gathered in April 1876, and, after being kept for nearly two years, are as good as when first dried. No weevil, apparently, attacks these flowers as they attack grain.

Keeping qualities of the Mahua. 260

"India would benefit greatly if *mahwa* flowers met with a demand in England. The vast forests of *mahwa* trees, which now yield little profit to their owners, would soon become a source of wealth; and the collection of the corollas would give work to thousands of poor people who at present inhabit the rocky country where the *mahwa* grows.

"To sum up the merits of the *mahwa* flowers for distilling purposes and feeding cattle, they are: 1, cheapness; 2, unlimited supply; 3, certain yield; 4, nourishing qualities; 5, good keeping qualities."

TRADE IN MAHUA.

External.

261

An effort has lately been made to establish a European trade in *mahuá*, mainly as a source of spirits of wine and for the purpose of feeding cattle. The following quotations furnished by the Department of Finance and Commerce give all that can be published, however, of a definite nature regarding this new trade:—

Export of Mahuá or Mowra flowers from British India.

	Quantity.	Value.	Country to which exported.	Quantity.	Value.
1883-84.	Cwt.	R		Cwt.	R
Bengal . .	42,215	99,785	France	269,215	6,70,399
Bombay .	227,114	5,70,879	Other Countries .	114	265
Total .	269,329	6,70,664	Total .	269,329	6,70,664
1884-85. (8 months.)					
Bengal . .	195	449	France . . .	30,193	60,663
Bombay . .	30,177	60,468	Other Countries .	179	254
Total .	30,372	60,917	Total .	30,372	60,917

With the view to regulating the trade in *mahuá* spirit, the Bombay Government have passed certain legislative measures which have had the effect of making a State monopoly of the purchase of the flowers. Meanwhile the export trade has received an unforeseen check in the attitude taken by the French Government in prohibiting the importation into France of *mahuá* flowers. It was found that *mahuá* spirit was being used as an

MAHUA TRADE.

adulterant for brandy,—a new trade which would have materially influenced French interests. This difficulty is much to be regretted, since it may have the effect of retarding the development of the more legitimate trade in this valuable product. Without in any way disturbing the present relations of the *mahuá* crop as a source of food to the hill tribes, a large and profitable export trade might easily enough be established in *mahuá* flowers and seeds.

INTERNAL TRADE.

262

Mahuá is one of the most important articles of export from Kaira. Definite information regarding the internal trade as a whole cannot be obtained, but in the preceding pages enough has been given to show that the *mahuá* is an exceedingly valuable plant, and a large local trade is done in both the flowers and the seeds.

§ "The flowers are very largely exported to France for the manufacture of inferior brandy." (*Surgeon K. D. Ghose, Bankura.*) "The spirit is much drunk in Jeypur (Vizagapatam district)." (*Surgeon-Major J. Byers Thomas, Waltair, Vizagapatam.*) "Fruits and flowers are eaten by the poor in the Konkan, and in Gujarát a spirit is distilled." (*Surgeon W. Barren, Bhuj, Cutch.*)

TIMBER. 263

Structure of the Wood.—Sapwood large; heartwood reddish brown, from hard to very hard. Annual rings indistinct. A cubic foot of seasoned wood weighs 61 to 68 lbs.

It is not much used, owing to the flowers being too valuable to allow of the tree being cut for timber; it has been tried for railway sleepers in the Central Provinces, and **Beddome** says it is used for the naves of wheels, for door and window frames and panels, for furniture and country vessels. **Mr. C. F. Manson** describes the *mahuá* as the most generally useful tree of the Santal Pergunnahs. **Cleghorn** says it is "a strong wood, but never felled by the natives."

DOMESTIC. Leaves as Plates. 264

Domestic Uses.—In many parts of India the broad leaves are used as plates.

265

Bassia longifolia, *Willd.; Fl. Br. Ind., III., 544.*

THE MOWA OR MAHUÁ TREE OF SOUTH INDIA.

Vern.—*Mohá, mohuá,* HIND.; *Mohuvá,* BENG.; *Móhá,* DUK.; *Madhúka,* SANS.; *Darakhte-gulchakán* PERS.; *Mahwa, mohi,* BOMB.; *Mahuda,* CUTCH; *Móhácha-jháda, ippicha-jháda,* MAR.; *Mahudá, mová-nu-jháda,* GUJ.; *Illupi, elupa, iluppai, iruppai,* TAM.; *Ippi, yeppa, ippe-chettu, pinna-ippa* or simply *ippa, ippa-pú (flower),* TEL.; *Hippe, ippigidá,* KAN.; *Ellupi, irippa,* MAL.; *Mí,* SINGH.; *Kan zannu, kánsó,* BURM.

Habitat.—A large evergreen tree of South India and Ceylon. Common in Kanára. **Mr. Baden Powell** (quoting **Mr. Barnes'** settlement report) must have mistaken this plant for the preceding, as he describes it (*Pb. Prod., I., 422*) as common in Kangra district, Panjáb. It is entirely a South Indian plant, being common in Mysore, Malabar, the Anamallays, and the Circars.

Botanic Diagnosis.—Leaves lanceolate, narrowed at both ends, glabrous, distinctly nerved; anthers 16, 2-seriate, subsessile, tips 3-toothed; young fruit globose, densely hirsute.

This should be carefully distinguished from the next species, the character of the fruits readily separating them.

Properties and Uses—

GUM. 266

Gum.—Yields an inferior gum known as *Ellopa.* **Ainslie** informs us that this is used in Madras as a remedy in rheumatic affections. **Roxburgh** remarks that there is frequently to be found a drop of whitish, soft, tasteless resin on the apices of the flowers before they open.

OIL. 267

Oil.—An oil is expressed from the seeds. It is yellow and semi-solid; used for burning, for soap, and to adulterate *ghi*. It is said to be well adapted for the soap trade; it retains its solid form under 95° F. It is seldom sold in the bazar, but is used for private consumption. It is suitable for the manufacture of candles. In Kanára, candles and soap are made from the oil of this species. (*Bomb. Gaz., XV., 63.*) The crushed seeds of this and of the preceding species, after separation of the oil, are baked into cakes and sold as a detergent; these cakes are largely used for washing the hair.

Cake a Detergent. 268

Medicine.—The PLANT has astringent and emollient properties assigned to it. The LEAVES, the BARK, and the JUICE of the bark and of the young FRUIT are used medicinally. As with **B. latifolia**, this species yields two important products—a fixed concrete OIL and a SPIRIT, the former obtained by expression of the SEEDS, the latter by distillation of the FLOWERS. The oil is said to be good for skin diseases; owing to the rapidity with which it becomes rancid it is not of much pharmaceutical value in the plains of India. The flowers are said to act as a mild laxative. (*Mysore Catalogue.*)

MEDICINE. Leaves. 269 Flowers. 270 Seed. 271 Bark. 272 Juice. 273 Spirit. 274

Food.—The economic uses of this tree in the South are similar to those of **B. latifolia** in the central table-land of India.

FOOD. Flowers. 275 Fruits. 276 Seeds. 277 OIL. 278

Structure of the Wood.—Heartwood red, moderately hard, close-grained. Weight 61 lbs. per cubic foot.

TIMBER. 279

Beddome says it is very flexible and durable; is valued for ships' keels, for trenails, and for planking below the water line; and that it is used for carts, furniture, and bridges. **Cleghorn** says the wood is good "for trenails; it is comparatively free from the attacks of the *Teredo navalis;* it is procurable among the logs brought down the Godavari. It is valued for all purposes, in situations where it is not exposed to air, as planking of ships below the water line, frames on which well-walls are built, &c."

280

Bassia malabarica, *Bedd.; Fl. Br. Ind., III., 544.*

Habitat.—A middle-sized tree, native of South Kanára, Malabar, and the Anamallays; up to 4,000 feet, abundant. (*Beddome.*)

Botanic Diagnosis.—Branchlets glabrous, leaves lanceolate or oblong-obtuse, or scarcely acute, glabrescent, distinctly nerved; stamens 16, in 2 series, subsessile, connective, excurrent, lanceolate linear; young fruit oblong-lanceolate, glabrous.

A very nearly allied species to the Ceylon **B. neriifolia**, *Moon.*

Bassia Mottleyana, *DeVriese.; Fl. Br. Ind., III., 546.*

Gutta-percha. 281

A tree met with in Malacca and Borneo, known as *Kotian*, and said to yield a copious milky juice, which hardens into a kind of Gutta-percha, which see.

B. Parkii, *Don.*

Shea Butter. 282

A tropical West African tree; yields a fat known in commerce as the Galam or Shea Butter. This substance was first described by Mungo Park.

BASSORA.

283

Bassora Gum.—A group of high-coloured gums resembling tragacanth, but very inferior, the colour being most objectionable. These are collectively known in commerce as Bassora gum, because the gum of this class which first attracted attention is supposed to have been exported from Bassora; they are also sometimes called Hog-tragacanth or Hog-gums. In India they are collectively known under the generic name *katíra.*

A gum exported from Calcutta to America, and which in America received the name of "Gum-Hogg," has recently attracted considerable attention, as it has been found very useful in marbling paper and the edges of books. This seems to be the gum of **Cochlospermum Gossypium.** An analysis of it has appeared in several publications. It is not soluble in water, but instead swells into a soft transparent mass. The filtrate after removal of this mass gives a faint precipitate with solutions of sub-acetate of lead, but no reaction with oxalate of ammonia. It is neutral, and has neither taste nor smell. Alcohol and ether have no solvent action on the insoluble transparent mass, but this is soon dissolved on being boiled with dilute sulphuric acid, the resulting solution showing no reaction with tincture of iodine, nor with Trommer's test for sugar. When boiled with a weak solution of an alkali or alkaline carbonate, it is found to be speedily converted into a uniform thick mucilage of a pinkish colour. On this alkaline solution being neutralised with an acid, it is found to remain soluble, while it has lost its objectionable colour.

These are the main characteristics of the so-called Bassora gums which may be distinguished from the Tragacanth series by the following characters: Tragacanth is only slightly soluble in water, but, owing to its wonderful affinity for water, it will absorb as much as fifty times its weight, swelling into a thick mucilage. The filtrate obtained on separating this mucilage will be found to yield "an abundant precipitate with acetate of lead, and to mix clearly with a concentrated solution of ferric chloride or of borax, in these respects differing from solutions of gum arabic. On the other hand, it agrees with the latter, in that it is thrown down as a transparent jelly by alcohol, and rendered turbid by oxalate of ammonium." "Tragacanth is readily soluble in alkaline liquids even in ammonia water, and at the same time it assumes a yellow colour." (*Flück. and Hanb.'s Pharmacog.*) The mucilaginous mass is tinged blue on addition of a little of the test solution of iodine. This constitutes a convenient reaction to distinguish Tragacanthin from Bussorin, the latter remaining unaffected by the iodine. Chemically, *Tragacanthin* and *Bussorin* are in all probability identical, being represented by the formula $C_{12} H_{20} O_{10}$. The blue reaction appears to be due to the presence of a small proportion of starch, held mechanically by the Tragacanthin. This blue reaction with iodine on the mucilage, and the immediate turbidity of the filtrate with oxalate of ammonium, are the most characteristic tests for tragacanth. **Giraud** views the presence of a pectic principle as the most characteristic chemical feature of tragacanth.

The India Bassora gums or Hog-gums are as follow :—

(*a*) *Pale-coloured.*

284

1. **Cochlospermum Gossypium.**—This is the *kúmbi* or *gabdi* of Hindustan. The tree is very abundant in the forests of the North-West Himálaya, extending across the central table-land of India to the west coast and to Prome in Burma. This seems to be the Gum-Hogg referred to as exported from Calcutta to America, in which a future trade seems possible to spring into existence.

285

2. **Sterculia urens.**—This is the *gúhú* or *gúlar* and *kúrí* of Hind., and *odla* of Assam. A common tree in the sub-Himálayan tracts from the Ganges eastward. This is certainly one of the best of Indian Bassora gums, and various reports have from time to time been obtained regarding it. Samples were sent from Chanda in 1873 to London, one broker reporting that it was worth 20*s*. a cwt. Samples were also sent from Haiderabad and valued at 30*s*. to 45*s*. a cwt.

3. Other species of Sterculia yield gums identical with the above.

286

4. **Saccopetalum tomentosum,**—said to yield a clear gum tragacanth.

(*b*) *Dark-coloured.*

These are decidedly inferior to those mentioned in group (*a*).

Bassora Gum.

287

1. **Moringa pterygosperma.**—This is the Horse-radish tree of Bengal, or *sajna.* It yields an abundance of dark-coloured Bassora gum which rapidly decays into a black powder. This is one of the gums often called *mochá-rás,* and also false tragacanth or gum-hog of European writers.

288

2. **Bombax malabaricum.**—This is the *sémul* or red-silk cotton-tree. It yields *mochá-rás,* a gum which is often declared as gum-hog. It is superior to the last mentioned, but of little or no commercial value.

289

3. **Ailanthus excelsa.**—This is the *maharukh* of Hind. and Mar. This much resembles moringa gum, but occurs in deep, dark red, large, rounded tears, instead of masses.

290

4. **Stereospermum suaveolens.**—The *páral* or *párlú.* A tree of the Sub-Himálayan tracts. It yields a dark-coloured massive gum of the Bassora series, of which very little is known, as it rarely occurs in the bazars.

291

To this list may be added the gum of **Odina Wodier,** for although this appears to be a soluble and not a bassora gum, it is perhaps one of the gums most abunduntly pressed on the market, either as an adulterant for gum arabic, or as a substitute for bassora and tragacanth. Perhaps no tree yields, in Bengal at least, a larger supply of gum than this; and although it is not so plentiful as **Moringa pterygosperma,** the gum is more frequently met with in bazars than almost any other gums. But for the fact that it is mixed with whitewash for walls, it may be said to be worthless as far as has yet been discovered, and it is mentioned in this place mainly as a caution, since there seems little doubt but that it is one of the principal substances used to adulterate more valuable gums.

292

European Gums generally known as Gum-hog.—The gum which chiefly goes by this name is obtained from **Symphonia globulifera,** a member of the Gamboge family and a native of the West Indies and of America; this is said to have received its name from the hogs being in Jamaica observed to rub themselves on the gum as it issues from the tree. By many writers the gum from the almond tree is also spoken of as gum-hog, and there are reports of a considerable trade between Persia and Bombay in a gum presumed to be obtained from that tree. Much confusion prevails, however, regarding the gums derived from the genus Prunus. One feels disposed to assume that the character of the gums of all the species of that genus would be more or less alike, and yet the greatest confliction exists in the literature of this subject. The Persian gum-hog, **Dr. Dymock** says, is a cheap substitute for more soluble gums. From this remark one would be almost justified in assuming that it was a soluble and not an insoluble gum, and had therefore been incorrectly called gum-hog. **Stewart** and **Baden Powell** say that apricot gum is soluble, and while **Dr. Cook** does not allude to cherry and almond gum, he places all the other gums obtained from the genus Prunus among his true gums, and not in the tragacanth series. Cherry gum has, however, been carefully examined, and pronounced to be much more nearly related to tragacanth and bassora than to the true gums. **John** detected in it a principle very similar to *bassorin,* but **Berzelius** places this as a mucilage nearer the mucilages from flax seed and quince seed than the gelatinous mass obtained by saturating tragacanth or bassora gum in water. **Giraud** has carefully examined this mucilage, and states that it differs from tragacanthin in the absence of a pectic principle; and he further adds that it contains as much as 20 per cent. of cellulose. In most countries there are three or four trees which yield gums of the Bassora series. Those met with in India have already been discussed, and in European commerce the gum of **Symphonia globulifera** and the gum of the almond tree are those generally met with.

This comparative account of the bassora or gum-hog substances with the tragacanth gums may be concluded by recapitulating briefly (what will be found gone into at greater detail under **Acacia Senegal**) the accepted theory of the chemical nature of the true gums and of the bassora and tragacanth gums. The former are now viewed as compounds formed by an organic acid by union with an alkali obtained on the organic acid percolating through the cell-wall. The latter, on the other hand, are much more nearly related to cellulose. In fact, according to **Von Mohl**, confirmed by **Wigaud**, they are metamorphoses of cellular tissue; hence their chemical relation to cellulose.

Gum Tragacanth is imported into India and may be had in every bazar. It is obtained from one or two species of Astragalus, a genus of Leguminosæ, of which we have many representatives on the temperate Himálaya, none of which appear to yield tragacanth. For further information see **Tragacanth.**

Bassorine, see **Orchis mascula,** *L.*; ORCHIDEÆ.

BATATAS, *Chois.*; *Gen. Pl., II.*, 872.

293

A group of CONVOLVULACEÆ now reduced to IPOMŒA.

Batatas edulis, *Chois.*, see **Ipomœa Batatas,** *Lamk.*; and
B. paniculata, *Chois.*, see **Ipomœa digitata,** *Linn.*; CONVOLVULACEÆ.

BAUHINIA, *Linn.*; *Gen. Pl., I.*, 572.

294

A genus of arborescent or scandent plants, belonging to the LEGUMINOSÆ, in the Sub-Order CÆSALPINIEÆ, and comprising some 130 species, diffused throughout the tropics.

Unarmed plants with simple, usually deeply-cleft leaves, rarely entire or bijugate. *Flowers* showy, in copious simple or panicled often corymbose racemes. *Calyx-tube* with the disk produced to the top, sometimes long and cylindrical, sometimes short and turbinate; limb entire and spathaceous or cleft into 2-5 teeth. *Petals* 5, sub-equal, usually with a distinct claw. *Stamens* 10 or reduced to 5 or 3; if fewer than 10, with sterile filaments absent or present; filaments free, filiform; anthers versatile, dehiscing longitudinally. *Ovary* stalked, many-ovuled; style long or short, stigma small or large and peltate, subterminal or oblique. *Pod* linear or rarely oblong, flat, continuous within, dehiscent or indehiscent. *Seeds* albuminous.

The generic name was given in honour of the botanists **John** and **Oaspar Bauhin,** the brothers being commemorated by the two-lobed nature of the leaves.

Bauhinia acuminata, *Linn.*; *Fl. Br. Ind., II.*, 276.

295

Vern.—*Kánchan*, BENG.; *Kachnár, kachnál*, HIND., DUK.; *Kánchana, kánsana* (variety), TEL.; *Mahahlega byu*, BURM.

Habitat.—An erect shrub, with elegant white flowers, met with in the North-West Provinces, Bengal, Burma, South India, and Ceylon.

Botanic Diagnosis.—Flowers in close axillary racemes, petals as long as the calyx-limb, which is cleft into 5 subulate teeth at the tip, pod with a rib on each side of the upper structure.

This species, and also **B. tomentosa,** belongs to the section PAULETIA. Erect shrubs or small trees, with large showy flowers and connate leaflets. *Stamens* 10, all fertile. *Calyx* with a very short tube and spathaceous limb. *Pod* narrow, dehiscent.

OIL. 296

Oil.—Mentioned as an oil-yielding plant in *Spons' Encyclopædia.*

B. anguina, *Roxb.*; *Fl. Br. Ind., II.*, 284.

297

THE SNAKE CLIMBER.

Vern.—*Nag-pút*, SYLHET; *Naiwilli*, NEPAL; *Suhatúngrungrik*, L.

Habitat.—A climber of North and East Bengal, Sikkim, Chittagong, Martaban, Burma, and South India. It also occurs on the Western Peninsula and is distributed to the Malay.

Botanic Diagnosis.—Fertile stamens 3. Calyx-tube scarcely any. Flowers minute, in copiously-panicled racemes. A climbing shrub with copious circinnate tendrils, having the stem bent upon itself in a remarkable manner alternately concave.

Properties and Uses—

FIBRE. 298

Fibre.—Its bark is used in rope-making.

TIMBER. 299

Structure of the Wood.—Soft and porous. The stems are bent generally in alternate folds and with a straight thick margin.

DOMESTIC. Serpent charms. 300

Domestic Uses.—"The most regularly serpentine pieces of the stems and large branches are carried about by our numerous mendicants to keep off serpents." (*Roxb., Fl. Ind., Ed., C.B.C., 347.*)

301

Bauhinia macrostachya, *Wall.; Fl. Br. Ind., II., 281.*

Syn.—B. SCANDENS, *Roxb., Fl. Ind., Ed. C.B.C., 346; Wight, Ic., t. 264, non Linn.*

Vern.—*Gunda-gilla,* BENG.

Habitat.—An extensive climber, found in the forests of Sylhet and Assam.

Botanic Diagnosis.—A cirrhose plant. Leaves 9-nerved, pubescence thin grey; pedicels moderately long; calyx-tube turbinate, very oblique; sepals deltoid; petals much exserted.

This belongs to the section PHANERA or erect or scandent species with usually 3 or sometimes 4-5 stamens. Calyx-tube mostly produced; limb usually 5-cleft, sometimes spathaceous. The following species belong amongst others to this section: **B. ornata,** *Kurz;* **B. purpurea,** *Linn.;* **B. retusa,** *Ham.;* **B. Vahlii,** *W. & A.;* and **B. variegata,** *Linn.*

Properties and Uses—

FIBRE. 302

Fibre.—The BARK yields a strong fibre. "The line made from the fibre sent by **Major Jenkins** sustained, for forty-five minutes, 168 lbs., having stretched six inches only in three feet, and therefore is about the same strength with our best *sunn* hemp. But, whether from the mode of preparation or the nature of the material, it is so harsh and stubborn, and the fibres stick so close together, that the heckles tear it to pieces and injure its strength." (*Royle's Fibrous Plants of India, p. 297.*)

Slow-match. 303

At the Panjáb Exhibition "there was a sample from the foot of the Kangra Hills, where it is described as used to make ropes for bedding, and the bark, which burns or smoulders slowly, is used for a slow-match." (*Baden Powell, Panjáb Products, I., 510.*)

304

B. malabarica, *Roxb.; Fl. Br. Ind., II., 277; Brandis, 159.*

Vern.—*Amlí, amlósa,* HIND.; *Karmai,* BENG.; *Gourubati,* URIYA; *Laba,* KOL.; *Amli taki,* NEPAL; *Kattra,* ASS.; *Cheppura, basavana páda* KAN.; *Korala,* MAR.; *Kundapula, dhondel, kangali,* GOND; *Ambotha, chapa,* KURKU; *Pulla dondur, puli shinta, pulhari,* TEL.; *Cheppuru,* KAN.; *Apta,* BERAR; *Bwaygyin, bwéchin,* BURM. (*Amlí* is also the HIND. and DUK. name for **Tamarindus indica.**)

Habitat.—A moderate-sized, bushy, deciduous tree, met with in the Sub-Himálayan tract (Kumaon, 1,000 feet in altitude, and ascending to 4,000 feet in Behar) from the Ganges to Assam, in Bengal, Burma, and South India.

Botanic Diagnosis.—Leaves 7-to 9-nerved, slightly cordate, deeply bifid; flowers in short, mostly simple corymbs; bracts minute, lower pedicels $1\frac{1}{2}$-2 times the calyx; calyx-limb 5-cleft, style produced.

This, as also **B. racemosa**, belongs to the section PILEOSTIGMA. Erect shrubs or climbers with small flowers and connate leaflets. Fertile stamens 10, calyx with a short tube and spathaceous or 5-cleft limb. Pod narrow, indehiscent.

Properties and Uses—

Food.—The leaves are very acrid, but are eaten by people in Burma. (*Brandis.*)

FOOD. 305

"The young shoots which appear just before the rains are used as a vegetable in the Konkan; when cooked they are slightly bitter but very palatable." (*Dymock.*)

§ "*Korala*, MAR.—The tender leaves are eaten as a vegetable. They are not believed to be acrid in the tender state. The name *Amli* is in Bombay applied to **Tamarindus indica** and not to **Bauhinia malabarica.**" (*Assistant Surgeon Sakhárám Arjún Rávat, Bombay.*)

Structure of the Wood.—Light-reddish brown, with irregular masses of black or purplish wood near the centre; moderately hard. Numerous narrow, wavy, white, concentric bands of softer tissue, alternate with bands of harder and red-coloured wood of equal width, in which the numerous fine, uniform, and equi-distant medullary rays are distinctly visible. Weight about 48 lbs. per cubic foot.

TIMBER. 306

It is rarely used.

Bauhinia ornata, *Kurz; Fl. Br. Ind., II., 281.*

307

Vern.—*Myaukhelga*, BURM.

Habitat.—Pegu.

Botanic Diagnosis.—An elegant species, clothed with deciduous, bright, ferruginous, silky pubescence. Leaves 9-11-nerved, pedicels long; flowers small; calyx-tube short, turbinate; sepals 5, rather exceeding the tube; petals slightly exserted.

B. purpurea, *Linn.; Fl. Br. Ind., II., 284; Brandis, For. Fl., 160.*

308

Vern.—*Koiral, karár, karalli, gray*, PB.; *Koliár, kaliár, kaniár, kandan, khairwal, kwillar, koilari, sona*, HIND.; *Khwairalo*, NEPAL; *Kachik*, LEPCHA; *Deva kanchan, rakta kanchan, koiral*, BENG.; *Buruju*, KOL.; *Koinar*, LOHARDUGGA; *Singyara*, SANTAL; *Kundrow*, MAL. (S. P.); *Kodwari*, GOND; *Koliari*, KURKU; *Rakta chandan, atmatti, ragta kánchan, deva kanchana*, MAR.; *Peuya áre, mandareh*, TAM.; *Kánchan(?), pedda-áre, bódanta-chettu*, TEL.; *Sarúl, surúl, kanchivála*, KAN.; *Kánchan*, SANS.; *Mahalay kani, mahahlegani*, BURM.

Habitat.—A moderate-sized, deciduous tree of the Sub-Himálayan tract, from the Indus eastward, Central and South India, and Burma.

Botanic Diagnosis.—Leaves 9-11-nerved, pubescent grey, pedicels short; sepals not fully distinct, exceeding the turbinate tube; petals oblanceolate, glabrous, exserted.

Properties and Uses—

Gum.—Yields a gum called *Sem-ki-gónd*.

GUM. 309

Dye.—The bark is used for dyeing and tanning.

DYE & TAN. 310

Fibre.—A fibre may be prepared from the bark.

FIBRE. 311

Medicine.—The BARK of this plant is astringent, the ROOT carminative, and the FLOWERS laxative.

MEDICINE. Bark. 312 Root. 313 Flowers. 314

§ "A decoction of the astringent bark is recommended as a useful wash in ulcers." (*Civil Medical Officer U. C. Dutt, Serampore.*) "Bark acts as an astringent in diarrhœa, the flowers are laxative, the roots tonic." (*Surgeon W. Barren, Bhuj, Cutch.*)

Food.—Dr. Stewart says that the FLOWERS are used as a pot-herb in curries, and that they are also made into pickles; the leaves are given to cattle as fodder.

FOOD. Flowers. 315 FODDER. Leaves. 316

BAUHINIA racemosa.

Bauhinia.

TIMBER. 317

Structure of the Wood.—Pinkish white, turning dark brown on exposure, moderately hard. Weight 40 to 50 lbs. per cubic foot.

Used for agricultural implements and in construction.

318

Bauhinia racemosa, *Lam.; Fl.Br. Ind., II., 276; Gamble, Man. Timb., 139.*

Syn.—B. PARVIFLORA, *Vahl.; Roxb., Fl. Ind., II, 323.*

Vern.—*Kachnál, gúriál, thaur, ashta, makkúna, maula, dhorára, marvil, ghila,* HIND.; *Banraj, banraji,* BENG.; *Kaimu,* KOL.; *Gatonli,* ORAON; *Katmanli,* KHARWAR and LOHARDUGGA; *Beriju,* SANTAL; *Ambru,* MAL. (S. P.); *Ambhota,* URIYA; *Ashta, Makkúna,* OUDH; *Mahauli,* BANDA; *Maula, Dhorára,* C. P.; *Kosúndra, taur,* PB.; *Dhondri, dhundera, astra, bosha,* GOND; *Jhinja,* AJMERE; *Amba bhósa,* BHIL; *Bossai,* KURKU; *Are-ká-jhár,* DUK.; *Ati* (?), *archi* (?), *areka, áre-maram,* TAM.; *Ari, áre, adda,* TEL.; *Apta, apatá, kanrája, seyára,* MAR.; *Aptá,* THANA; *Asindro, asíndri, asotri,* PANCH MAHALS; *Aupta,* KAN.; *Svetakanchan,* SANS.; *Hpalan, palan,* BURM.

Habitat.—A small, crooked, deciduous tree, met with in the Sub-Himálayan tract from the Ravi eastwards, ascending to 5,000 feet; in Oudh, Bengal, Burma, and Central and South India. Distributed to China, the Malay isles, and Timor.

Botanic Diagnosis.—Leaves small, deeply cleft, 7-9-nerved; flowers in lax, simple racemes; calyx-limb entire, stigma sessile.

Properties and Uses—

GUM. 319

Gum.—It yields a gum of which little is at present known.

FIBRE. 320

Fibre.—A strong fibre is made from the inner bark; used for cordage, but is not durable in water. It yields a good bast.

Specimens were shown at the late Calcutta International Exhibition from various parts of India, notably from Salem District, Madras. May this not be the undetermined bast fibre described by **Royle** under the name of *Asta patu,* sent from Bírbhúm to the Exhibition of 1851?

§ "Used here for making country ropes; leaves used for making cigarettes." (*Surgeon-Major W. Dymock, Bombay.*)

MEDICINE. Gum. 321

Medicine.—The GUM of this plant is used medicinally in South India. (*T. L. Stewart.*)

Leaves. 322

§ "A decoction of the leaves is used to relieve headache in malarious fevers." (*Surgeon-Major W. Dymock, Bombay.*)

FOOD. Seeds. 323

Food.—The seeds are eaten by the people in some parts of the country. In parts of Northern India the leaves are eaten by buffaloes.

Leaves. 324

§ "The leaves are pickled by the Burmese." (*J. C. Hardinge, Esq., Rangoon.*)

FODDER. Leaves. 325

TIMBER. 326

Structure of the Wood.—Light brown, hard, with irregularly-shaped masses of darker-coloured and harder wood near the centre. Weight 40 to 56 lbs. per cubic foot. "The wood is strong and close-grained." (*Bomb. Gaz., XVII., 63.*)

Good, but not used, owing to the plant never growing big enough. It is sometimes burned as firewood.

DOMESTIC. Cigarette Covers. Leaves. 327

Domestic Uses.—The leaves are made into cigarette covers in the Panch Mahâls, Gujarát. These are in Thána called *bidis.* In Thána alone the right to pluck the leaves used as *bidis* fetches an annual rent of R1,500. The leaves of **Diospyros melanoxylon** are also used for the same purpose. The trade in *bidis* in Khándesh is small.

Slow-match. 328

Matchlockmen make their matches of the bark of this tree; it burns long and slowly, without the help of saltpetre or any other combustible. To prepare the bark it is boiled, dried, and beaten. (*Roxb., Fl. Ind., Ed. C.B.C., p. 345.*)

Sacred plant. 329

"A sacred plant of the Hindús; worshipped on the Dasera festival." (*Bomb. Gaz., X., 401.*)

Bauhinia retusa, *Ham.; Fl. Br. Ind., II., 279; Gamble, 161.*

330

Syn.—B. EMARGINATA, *Wall.*; PHANERA RETUSA, *Benth.*

Vern.—*Kurál*, PB.; *Kandla, kanalla, kuayral, gwayral, kanlao, semla*, HIND.; *Laba*, KOL.; *Twar*, ORAON; *Katman*, KHARWAR; *Thaur*, GOND; *Kaimu*, LOHARDUGGA; *Tewar*, PALAMAU; *Nirpa*, TEL.

Habitat.—A moderate-sized, deciduous tree of the North-West Himálaya, from the Beas eastward, ascending to 4,500 feet; Simla, Garhwál, Kumaon, and Central India.

Botanic Diagnosis.—Leaves rigidly coriaceous, rather broader than long, 4-6 inches long, 9-nerved, glabrous beneath, usually deeply cordate; calyx-tube turbinate, very short.

Properties and Uses—

GUM. 331

Gum.—It yields a clear gum called *Semla gónd*, almost exactly resembling gum arabic. It is eaten by the poorer classes, and is used to waterproof terraced roofs. **Roxburgh** says: "From wounds made in the bark a brownish mild gum is produced." It is used as a medicine either alone or in combination with other medicines. The annual export from Dehra Dún is about 2,500 maunds.

MEDICINE. 332

§ "Is used as an external application to sores. It is considered as an emmenagogue and diuretic by some native practitioners." (*Surgeon G. A. Emerson, Calcutta.*)

TIMBER. 333

Structure of the Wood.—Reddish white, with irregularly-shaped darker masses near the centre, hard. Weight 58 lbs. per cubic foot.

Not used.

B. scandens, *Roxb.*; syn. for **B. macrostachya,** *Wall.*

B. tomentosa, *Linn.; Fl. Br. Ind., II., 275; Roxb., Fl. Ind., Ed. C.B.C., 345.*

334

Vern.—*Kachnár*, HIND.; *Asundro*, GUJ.; *Chámal*, KONKAN; *Piválákún, chan, aptú*, MAR.; *Kánchini*, TAM., TEL.; *Usamaduga*, MADRAS; *Maha-hlœ-ga-wa*, BURM.; *Kaha-pettang*, SINGH. The vernacular names *kachnár, kachnál*, and *kánchan* or *kánchini* are applied to more than one species of Bauhinia. (*Moodeen Sheriff.*)

Habitat.—North-West Provinces and throughout India to Ceylon and Penang. Distributed to China and tropical Africa.

Botanic Diagnosis.—An erect shrub with downy branches. Flowers usually in axillary pairs, petals much longer than the entire calyx-limb, pod stalked, not ribbed, near the upper suture.

Properties and Uses—

FIBRE. 335

Fibre.—From the bark a fibre is prepared.

OIL. 336

Oil.—**Balfour** simply mentions this plant among his oils, without describing it.

MEDICINE. Flower. 337 Fruit. 338 Seed. 339 Bark. 340

Medicine.—As a medicine, the plant is antidysenteric, anthelmintic, and useful in liver complaints. **Ainslie** says that the dried buds and young flowers are prescribed in dysenteric affections. According to **Rheede** the decoction of the root-bark is useful in inflammation of the liver. (*Dymock, Mat. Med., W. Ind., 224.*)

§ "Applied locally in aphthæ. The fruit is diuretic; an infusion of the bark is used as an astringent gargle. The seeds made into a paste with vinegar are said to be efficacious as a local application to wounds inflicted by poisonous animals." (*Surgeon G. A. Emerson, Calcutta.*) "Hakeems administer the dried leaves and young flowers in dysenteric affections. A decoction of the bark of the root is used in cases of liver, also as a vermifuge." (*Surgeon H. W. Hill, Mánbhúm.*)

TIMBER. 341

Structure of the Wood.—Tough, close-grained, with a black heartwood; when full-grown it is very soft.

342

Bauhinia Vahlii, *W. & A.; Fl. Br. Ind., II., 279.*

Syn.—B. RACEMOSA, *Vahl.; Roxb., Fl. Ind., Ed. C.B.C., 346.*

Vern.—*Malghán, maljan, malú, maurain, jallaur* or *jallur,* HIND.; *Chehur,* BENG.; *Sihár, mahalan, maúl,* C. P.; *Borla,* NEPAL; *Sungung rik,* LEPCHA; *Jom,* SANTAL; *Lama, rung,* KOL.; *Shioli,* URIYA; *Maulan,* KHARWAR; *Taur,* PB.; *Chambolli,* DUK.; *Chambúra, chambúli, chárbor,* MAR.; *Paur, bela,* GOND; *Adda,* TEL.

Habitat.—This is one of the most extensive, as it is the most abundant and most useful, of Indian climbing Bauhinias. It is found all along the lower Himálaya (ascending in Kumaon and on Paresnath in Behar to 2,500 feet) from the Chenab eastward, North and Central India, and Tenasserim.

Botanic Diagnosis.—Tendrils strong, woody; pubescence dense grey or ferruginous; leaf-lobes obtuse, pedicels long; calyx-tube cylindrical; petals much exserted, densely pilose.

Properties and Uses—

GUM. 343

Gum.—Yields a copious gum which seems to be of little use.

FIBRE. Bark. 344

Ropes. 345

Fibre.—The uses of this climber are, perhaps, more numerous than those of any other forest plant; the strong cordage prepared from its BARK is an important article with the hill tribes. Specimens of this fibre were exhibited at the London Exhibition of 1851 under the name of *Patwa* or *mawal.* A large collection of strong red ropes made from it were also displayed at the late Calcutta International Exhibition. In the Kew Report for 1879 it is stated that a sample of this fibre was submitted by **Sir J. D. Hooker** to **Mr. Routledge** of the Ford Paper Works, Sunderland, who reports as follows: "Excellent strong fibre; hemp character and tough. Green yield 60 per cent., bleached 54·7 per cent." **Capt. Huddleston** in his Report on Hemp in Garhwál, 1840, gives the following facts: "The '*malloo*' is a large creeper, forty or fifty yards in length, and of considerable thickness, from the bark of which a very strong rope is made. The natives chiefly use it for tying up their cattle, and sewing their straw mats with the fresh bark; it also makes capital matches for guns, and muzzles for oxen and calves." It is "cut generally in July and August, though it may be cut all seasons, and the outer bark being stripped off is thrown away, the inner coating being used for ropes, as wanted, by being previously soaked in water and twisted when wet. A large creeper will produce a maund of fibre, called '*seloo.*' The bark before being used is boiled and beaten with mallets, which renders it soft and pliable for being made into ropes and string for *charpoys.* Though this fibre makes very strong ropes, it is not over-durable, and rots if kept constantly in water; it will last about 18 months, but requires occasional soaking, and I am informed that when coated with tar it does not last much longer. The fibre is not collected for sale, but only for the natives' own use as they may require it; but any quantity, I imagine, might be obtained, and at cheap rates."

Royle, in his *Himálayan Botany* and also in his *Fibrous Plants of India,* gives an account of this fibre, quoting the above extract. No additional information has since appeared, and the fibre is still unknown to the European industries.

MEDICINE. Seeds. 346

Leaves. 347

Medicine.—The SEEDS are said to possess tonic and aphrodisiac properties. The LEAVES are regarded as demulcent and mucilaginous remedies.

FOOD. Seeds. 348

Pods. 349

Food.—The SEEDS are eaten raw; when ripe, they taste like cashewnuts. (*Roxb., Fl. Ind., Ed. C.B.C., p. 346.*) They are also eaten fried. The young PODS are cooked and eaten by the hill tribes. "The seeds taken from the huge pods of **B. racemosa** are eaten in the hills. The pods look like pieces of thick undressed leather, about a foot long and an inch or two broad; they are placed over the ashes of a fire till they roast and split open; the flat soft seeds are taken out and eaten: the flavour is pleasant, but the seed is not wholesome." (*Baden Powell's Panjáb Products, Vol. I., p. 265.*)

TIMBER. 350

Structure of the Wood.—Porous, in broad, irregularly-broken concentric layers, alternating with red, juicy, bark-like tissue; the pith is cross-shaped. The foliage is very dense and the stems do great damage to the trees they climb over; it is very prevalent in *sál* forests, and in many provinces is being systematically exterminated.

DOMESTIC. Leaf-bellows. 351 Plates. 352 Cups. 353 Umbrellas. 354 Rain-hats. 355

Domestic Uses.—In the *Kew Report for 1881* it is stated that the leaves of this plant and not those of **Cochlospermum Gossypium** are those used in the construction of the crude leaf-bellows in Sikkim. They are sewn together and used as plates, cups, rough table-cloths, umbrellas, and rain-hats and caps. "The leaves, which are heart-shaped, and above a span in breadth and the same in length, are made into 'chattas' (umbrellas), are sewed together with twigs into baskets for holding pepper, turmeric, and ginger, and are brought to Sreenuggur in great quantities for sale, being used by the poor instead of dishes to eat off, and the *buneeahs* wrap up their goods with them. A load of the leaves fetches about 2 annas." (*Capt. Huddleston's Report on Hemp in Garhwál, 1840.*) In Chutia Nagpur the Santals cut off the dry loops of tendrils (those which have failed to catch any object), and make with these finger-rings worn as a charm against dropsy. (*Rev. A. Campbell.*)

Bauhinia variegata, *Linn.; Fl. Br. Ind., II., 284.*

356

Vern.—*Kachnár, koliar, kurál, padrián, khwairaal, guriál, gwiar, bariál, kaniár, kándan, khairwál,* HIND.; *Rakta kánchan,* BENG.; *Kurmang,* MECHI; *Síngya,* KOL.; *Kundol,* BHUMIJ; *Jingya,* SANTAL; *Taki,* NEPAL; *Rha,* LEPCHA; *Kachnár,* C. P.; *Kanchan, ragtákanchan,* MAR.; *Kanchan,* KONKAN; *Kovidara,* BOMB.; *Segapu-munthari,* TAM.; *Kanchivala-do,* KAN.; *Borara,* URIYA; *Bwaycheng, bwéchin,* BURM.

Habitat.—A moderate-sized, deciduous tree, found in the Sub-Himálayan tract, from the Indus eastward, and throughout the forests of India and Burma. Common everywhere, ascending to 4,000 feet in altitude, preferring the low hills of India, but largely cultivated as an ornamental tree throughout the plains. Often completely covered with large purple and white flowers which appear in the beginning of the hot season.

Botanic Diagnosis.—Leaves 9-11-nerved, pubescence grey, pedicels short; calyx-limb entire, spathaceous, equalling the cylindrical tube; petals glabrous-obovate, clawed, much exserted.

Closely allied to **B. purpurea.**

Properties and Uses—

GUM. 357

Gum.—This tree, like most other members of the genus, yields the gum known as *Sem* or *Semla gónd.* It is a brown-coloured gum. *Sem-ki-gónd* is, in fact, a sort of generic name for the gum obtained from the species of Bauhinia. It swells in water like cherry-tree gum, a very small proportion only being soluble.

DYE & TAN. Bark. 358

Dye.—The bark is used in dyeing and tanning. (*Bomb. Gaz., XV., Part I., 64.*)

§ "The bark is used by dyers in Madras." (*Deputy Surgeon-General G. Bidie, C.I.E., Madras.*)

OIL. 359

Oil.—The seeds are said to yield an oil.

MEDICINE. Flowers. 360 Roots. 361 Cataplasm. 362

Medicine.—The ROOT in decoction is given in dyspepsia and flatulency; the FLOWERS with sugar as a gentle laxative; and the bark, flowers, or root, triturated in rice-water, as a CATAPLASM to promote suppuration.

Bark. 363 Buds. 364 FOOD. Seeds. 365 Buds. 366 TIMBER. 367

The BARK is described as alterative, tonic, and astringent, useful in scrofula, skin diseases, and ulcers. It is also used to remove intestinal worms and to prevent the decomposition of the blood and humours; on this account it is useful in leprosy and scrofula. "The DRIED BUDS are used as a remedy for piles and dysentery. They are considered by natives cool and astringent, and are useful in diarrhœa and worms." (*Baden Powell's Pb. Prod., I., 344.*)

Food.—It flowers in February-March; the seeds ripen two months later. The buds are eaten as vegetables when prepared with animal food (*Drury.*)

Structure of the Wood.—Grey, moderately hard, with irregular masses of darker and harder wood in the centre. Hard and serviceable, but seldom used owing to the small size. Weight 40 to 50 lbs. Used for agricultural implements.

SACRED. 368

Sacred and Domestic Uses.—Often seen on Buddhistic sculptures.

BDELLIUM.

Bdellium, a myrrh-like resin, of which there are three kinds:—

369

1st.—Indian, the produce of **Balsamodendron Mukul,** *Hook.*, met with in Sind, Rájputana, Khándesh, Berar, Beluchistan, and Arabia. This substance is also obtained from **B. Roxburghii,** and, in Beluchistan, from **B. pubescens.** Mukul or Gugul (Indian Bdellium) from Coromandel is the produce of **Boswellia glabra,** and that from the Western Himálaya is the produce of **Boswellia serrata.**

370

2nd.—African Bdellium. This is now believed to be the produce of **Hemprichia erythræa,** *Ehrenb.* (a synonym for **Balsamodendron Kataf,** *Kunth*). This substance to a certain extent resembles Myrrh, but is of a darker colour. It is twice the price of the Indian Bdellium. Both this and the preceding are given to buffaloes to increase their milk.

371

3rd.—The Opaque Bdellium. This is the produce of **Balsamodendron Playfairii,** *Hook.*, which see.

372

BEADS.

Beads.—Articles of personal ornament (and chiefly natural objects used for this purpose) may be enumerated under the above heading. They are Beads, Rosaries, Garlands, Necklaces, Earrings, &c., and may be classed:—

1st. Those which belong to the Mineral Kingdom, such as glass and stone beads used by the mass of the people (*i.e.,* not including those which would be pronounced as jewels), alabaster, metal ornaments, &c.

2nd. Animal Kingdom.—Coral, pearls of the cheaper kind, ivory, shells, fish and other bones, feathers, skins, &c.

3rd. Vegetable Kingdom.—Flowers, fruits, seeds, specially-prepared pieces of wood or other natural botanical structures.

A complete list of the objects used for the above purposes would be highly interesting and instructive. But such a list may be viewed as having an ethnological rather than an economic interest, and would, therefore, be somewhat out of place in the present publication. The subject is, however, replete with interest, and as a considerable trade is done in certain articles which must be enumerated here, it has been thought desirable to give the leading facts which can be collected together in a limited space. It is hoped that at least one object may be served by the publication of even an incomplete list of this nature,—namely, the creation of an interest in a subject which the advances of civilization are certain to obscure more and more every day. The first attempts made by savage races at clothing and adornment were most probably decoration by means of natural objects. A careful study of the shells, bones, seeds, fruits, and flowers, used for this purpose by aboriginal tribes at the present day, would throw a flood of light upon many obscure anthropological subjects destined to be obliterated with the advances of foreign trade in glass beads and cheap European ornaments.

I.—BEADS AND OTHER ORNAMENTS WHICH BELONG TO THE MINERAL KINGDOM.

373

Glass Beads and False Pearls.

An enquiry was instituted into this subject on the suggestion of the Government of India in the Department of Revenue and Agriculture. **Mr. W. J. Wilson** published a report in February 1883 of considerable interest. He divided glass beads into two primary sections: (*a*) those imported into India and China, and (*b*) those manufactured in India. Of the former he established seven sections.

(*a*) Foreign Beads.

European beads. 374

1st. Pound Beads.—These are made chiefly in Venice. The glass is drawn into tubes, cut into small pieces, and by means of sand the edges are rounded off and polished. There are said to be 20 standard sizes of pound beads, of all colours. Black is the favourite colour, but in Rájputana light blue is in great demand. Red, blue, amber, pink, and white are also used: the smaller sizes are in the greatest demand.

They are used for a variety of purposes. The larger ones are made into necklaces, wristlets, and rosaries, while the smaller are employed in the decoration of shoes, hookah stems, toys, lac-bangles, carpets, &c.

2nd. Seed Beads.—These are smaller than the preceding.

3rd. Broken Beads.—These are like pound beads, only longer and the ends not rounded off.

4th. Pigeon-egg Beads.—These are about one inch in length and five eighths of an inch in diameter. They are chiefly used to decorate horses and cattle.

5th. Cut-glass Beads.—These are met with chiefly in Central India and Sind.

6th. Spotted Beads.—Are in demand in the Central Provinces.

7th. Round Beads.—Are not much used.

The following are the imports of beads for the past five years, and also an analysis of those for the year 1883-84, showing the countries from which the imports are obtained and the provinces to which imported:—

Imports of Beads (Glass) and False Pearls.

Years.	Quantity in Cwt.	Value in Rupees.
1879-80	13,751	8,79,895
1880-81	15,483	11,68,060
1881-82	16,724	11,84,148
1882-83	19,897	12,79,023
1883-84	23,243	16,18,728

BEADS.

Cheap Manufactured Articles used as

Analysis of the Imports for 1883-84.

Presidency or Province into which imported.	Quantity in Cwt.	Value in Rupees.	Countries whence imported.	Quantity in Cwt.	Value in Rupees.
Bengal	10,029	5,05,568	United Kingdom	637	4,70,970
Bombay	12,972	10,08,231	Austria	1,752	2,46,993
Sind	19	2,200	Belgium	109	3,717
Madras	87	15,562	France	350	77,397
British Burma	136	87,167	Italy	17,959	7,23,386
			Egypt	18	6,285
			Ceylon	566	23,787
			China—Hongkong	1,792	61,576
			Straits Settlements	40	3,101
			Other countries	20	1,516
TOTAL	23,243	16,18,728	TOTAL	23,243	16,18,728

The above analysis of the imports shows that Italy is the country which meets the major portion of the Indian demand for glass beads. The imports for the past five years indicate a steady increase, those for 1883-84 being very nearly twice as much as for 1879-80.

China beads. 375

The beads which come from China to India are chiefly round, and vary from a quarter to half an inch in diameter. They are ruby or green coloured, and are commonly met with in the Central Provinces and in Rájputana.

(*b*) MANUFACTURED IN INDIA.

Indian beads. 376

Indian glass beads are said to be manufactured in Kaira and Surat in Bombay, Jaipur and Bundi in Rájputana, Saugor in the Central Provinces, Jaunpur in the North-West Provinces, and Delhi and Multan in the Panjáb. They are described as of seven kinds:—

(1) Imitations of imported beads.
(2) The Saugor beads—round, flat, about a quarter inch in diameter and one eighth to three sixteenths of an inch in length.
(3) Pigeon-egg beads made in Kaira.
(4) Large flat beads made in Kaira.
(5) Spherical beads made in Surat.
(6) Small ring beads made at Delhi and Multan.
(7) Flat beads made at Lucknow.

No information can be given as to the extent of this Indian industry, but it seems probable that old glass is largely used in the manufacture of beads. When made from indigenous materials the beads are always very coarse and badly coloured. The *Kánch* or country glass made from *reh* soil is either green or black. **Dr. Owen**, in his Catalogue of the Jeypore articles shown at the Calcutta International Exhibition, says: "Glass beads, as imitations of emeralds, rubies, sapphires, and turquoises, are very well made, which are then cut in facets by lapidaries. These latter were once largely exported from Jeypore and engaged several families, but have fallen into the background for some years, as the competition with European-made beads was found too strong."

Lac beads. 377

A certain amount of lac beads are regularly made, and in lower Bengal a very large trade exists in glass and lac bangles (see GLASS); the former are generally green or black, and the latter are often ornamented by European beads, attached while the lac is still soft.

(*c*) STONES, ALABASTER, &c.

Stone beads. 378

A very large trade is done in the cheaper kind of stones. These are collected on the mountains of India and Burma, and are also brought across the frontier from the northern Himálaya. A considerable trade is done at Simla in beads, necklaces, &c. The principal stones are turquoise, rubies, onyxes, cornelians, emeralds, jadestone (false), serpentines, agates, jaspers, marbles, &c., &c.

(*d*) METAL BEADS.

Metal beads. 379

Small beads made of various metals are commonly met with, the more elegant being gold beads used along with precious stones or coral.

II.—OBJECTS OF ORNAMENT BELONGING TO THE ANIMAL KINGDOM.

The most important objects belonging to this section are of course pearls and coral. As these will be found discussed in their alphabetical positions, it is not necessary to do more than mention them by name in this place. Shells are largely used for this purpose, none perhaps more extensively than the common

COWRIE.

Cowries. 380

These are imported into Bombay chiefly from the Laccadive and Maldive Islands, and from Zanzibar. From time immemorial they have been used as coins by the Hindús; the currency being—

4 Cowries = 1 Ganda.
20 Gundas = 1 Pan.
16 Pans = 1 Káhan.

The present rate is 24 gandas or 96 cowries to a pice and 4 pice to an anna and 16 annas to the rupee: hence $96 \times 64 = 6{,}144$ cowries to the rupee.

They are extensively used as articles of adornment for cattle and horses, and amongst the hill tribes are also used for personal ornament. In the Nagá Hills they are cut lengthwise, the back being removed and rejected. Cut in this way they are sewn over garments, chiefly in rows, upon a piece of black cloth worn by the men, and forming a sort of kilt. Formerly the number of rows of white cowries denoted the deeds of daring committed by the wearer. He was permitted to wear one row for his first murder, another for the second, and another for the third. After that he might wear as many rows as he chose, but most preferred the triple line. By modern usage, however, all full-grown males wear a black kilt with three rows of cowries. (For further information see COWRIE.)

THE CONCH OR CHANK-SHELLS.

Conch. 381

These are fished up from deep water by divers in the Gulf of Manaar, on the coast opposite Juffnapatam in Ceylon, and also from Travancore, Tuticorin, &c., &c.

A curious trade exists in Dacca in cutting these into rings, armlets, &c. In the Nagá Hills they are cut up into beads. One half the shells are suspended from the back of the neck, the point being directed downwards, and the remainder are cut up into long pieces or beads forming the front part of the chain. The conch is of course extensively used for the horns blown at temples. (For further information see CONCH.)

Bones. 382

A number of small bivalve shells are used as ornaments by the Andaman Islanders, as also bones of various animals, including human bones. The reader is referred to an interesting paper regarding these by Professor Allen Thomson, F.R.S., in the *Journ. Anthrop. Inst., XI.,* 295.

FEATHERS, SKINS, &c.

Feathers. 383

Feathers of various birds are used as personal ornaments, and also pieces of skins, furs, &c. The available information is, however, too imperfect to admit of these being gone into in detail. The large black and white feathers of the horn-bill are much prized by the Angámi Nagás, and the tail feathers of the wild cock by the Gáros. The blue and green feathers of a woodpecker are used as ear ornaments by the Angámis. The beak of the horn-bill is attached to the helmet of the Mishmi chief or head-man, and bands of bear's-skin are used in the construction of the helmets worn by many of the Assam tribes. Goats' and human hair, black and dyed red by madder, are extensively used by Assam tribes for decorative purposes. The boar tusk, curving downwards and terminated above by a tuft of red hair, is the most fashionable earring amongst certain Nagás, just as a bunch of cotton-wool 2 or 3 inches in diameter inserted into a greatly dilated ear-perforation, is admired by others, especially the tribes in North Mánipur.

Hair. 384

Teeth and Tusks. 385

Cotton-wool. 386

III.—BEADS AND OTHER ORNAMENTS WHICH BELONG TO THE VEGETABLE KINGDOM.

Natural Vegetable beads. 387

Certain parts of the following plants ar used as beads, rosaries garlands, &c. Fuller information will be found in their respective alphabetical positions in this work, but the abstract given below may be found useful to persons desirous of studying such objects collectively and from an ethnological point of view.

1. **Abrus precatorius**, The Crab's-eyes or *rati* seeds.—The fact of this red shining seed with its black eye-spot being used for rosaries, suggested the specific name precatorius. They are strung together along with sheels and black seeds in necklaces, and are also largely used in the decoration of boxes, baskets, &c.

2. **Adenanthera pavonina**, The Red-wood or *rakta-kanchan*.—The brilliant scarlet seeds of this tree are larger than the preceding, flattened and devoid of the black eye spot, otherwise they are very much alike. They are strung and worn by the women in many parts of India.

3. **Adhatoda Vasica**, The *Baxas* of Bengal.—The wood of this plant is made into small beads resembling those made from **Ægle Marmelos, Cajanus iudicus**, and **Flacourtia Ramontchi**.

4. **Ægicerus majus.**—The pretty white flowers of this shrub are made into garlands on the western coast.

5. **Ægle Marmelos**, The *Bel*.—Beads are made from the rind as well as from the wood. Strung with the fibre of **Agave americana**, they are worn by the Sudras to denote that they are not Mohammedans.

6. **Æschynomene aspera**, The *Sola*.—Prepared pieces of pith are sometimes worn by the aboriginal tribes as ear-ornaments. Garlands of beads of the pith or *sola*, coloured and tinselled, are used to decorate idols and worn bv brides and bridegrooms.

7. **Allium sativum**, The Garlic.—A necklet of the cloves or young bulbs is worn by children as a charm against whooping-cough.

8. **Aquillaria Agallocha**, Eagle-wood.—Beads made of this odoriferous wood are occasionally seen.

9. **Areca Catechu**, Betel-nut palm.—Polished beads are made from the betel-nut: they are rarely worn entire, but are turned into fancy shapes.

10. **Bamboo.**—A ring of specially-prepared bamboo is placed in the ear-perforation by the Tankúl Nagás of Mánipur.

11. **Borassus flabelliformis.**—The leaves are cut up into neat bracelets and worn by Santal girls.

Personal Ornaments. BEADS.

Vegetable-beads.

12. **Bauhinia Vahlii.**—The dried tendrils are worn as finger-rings by the Santals, as a charm in dropsy.

13. **Butea frondosa,** The *Palas.*—The beautiful bright orange flowers of this tree are sometimes, and were formerly extensively, worn in the ears by the Hindú women.

14. **Cajanus indicus,** The *Urhar.*—The wood is made into small beads (see **Adhatoda Vasica**).

15. **Cæsalpinia Bonducella.**—"Necklaces of the seeds strung upon red silk are worn by pregnant women as a charm to prevent abortion." (*Dymock.*)

16. **Calotropis gigantea,** The *Akanda* or *Madar.*—The purple-coloured corona of the flowers of this plant are, in Bengal, separated from the rest of the flower and strung into garlands. A garland of the flowers is also used in the worship of *Máruti,* the monkey-god.

17. **Canna indica,** The Indian Shot, or *Kiwára,* or *Lál-sarbo-jayá.*—The black seeds of this plant are sometimes strung as beads along with the red crab's-eye seeds.

18. **Carissa diffusa.**—Flowers strung into garlands and worn in the hair by women on the western coast.

19. **Caryota urens.**—The dark-coloured oval seeds of this palm are used as buttons, and by the Mohammedans are sometimes strung as beads.

20. **Coix lachryma,** Job's-tears.—There are two principal forms of this grain, one almost round and either white or black. This form is sometimes, though less frequently, used for ornamental purposes than the next, but it constitutes an important article of food amongst the hill tribes on the eastern frontier of India. The second form is tubular, about ½ an inch long. This is extensively used for decorative purposes, the dresses worn by the Karen women being often completely covered by pretty designs of this grain. It is also used by the Nagá and other Assam tribes in the construction of earrings and other simple and elegant articles of personal adornment.

21. **Corypha umbraculifera,** The *Bazarbatú*-nuts imported into Bombay; also exported from N. Kanara by Arabs from the Persian Gulf who trade along the western coast. Price R20 to R25 a *candy* of 616 lbs. These are worn as beads by the Hindú devotees.

22. **Cotton-wool,** in large bundles often 2-3 inches in diameter.—Cotton-wool is worn in ear-perforations by the northern Mánipur Nagás, and also certain classes of the Nagás proper. Similar tufts are also used in decorating the hair. As a modern degeneration it is by no means an unusual thing to find two or three empty cartridge cases placed in the ear instead of the cotton decorations—the brass ends being turned forward.

23. **Dalbergia Sissoo.**—The green seeds are worn by Santal girls as pendants from the ear.

24. **Daphne papyracea.**—Garlands of the flowers are used in religious ceremonies in the Panjáb Himálaya, at Chumba, &c.

25. **Diospyros, sp.**—**Gamble** says that the Burmese use the wood for earrings.

26. **Elæocarpus Ganitrus** or *Rudraksha.*—The five-grooved and elegantly-tubercled nuts are worn as a necklace by the followers of *Siva* in order to obtain *Sivalocke* (the heaven wherein the god *Siva* resides), and in order to gain his graces. They are also supposed to preserve the health. Considerable importance is attached to the number of facets on the nuts. (*Pb. Notes and Queries, March 1885, p. 63.*) Imitations of these nuts are made in Eagle-wood.

27. **Elæocarpus lanceolatus,** The *Utrasum* Beads.—These are said to be imported from Java.

Vegetable-beads.

28. **Elæocarpus tuberculatus.**—As with the two preceding, the nuts of this tree are used as beads.

29. **Entada scandens.**—The large seeds of this climber are worn as charms and made into small ornaments, snuff-boxes, &c. They are also largely used by Indian washermen to crimp linen, hence are often called the *Dhobis'*-nut.

30. **Euonymus grandiflorus,** The *Siki* Nut.—These are strung as necklaces.

31. **Euonymus fimbriatus.**—The red seeds are strung into ornaments for the head.

32. **Flacourtia Ramontchi,** The *Bunj* or *boinch* (see **Adhatoda Vasica**).

33. **Ficus glomerata.**—The fruits are strung and put round a pregnant woman's neck on a particular day in the eighth month. (*Lisboa, Useful Plants, Bombay.*)

34. **Gyrocarpus Jacquini,** *Zaitun.*—The seeds are made into rosaries and necklaces.

35. **Hibiscus rosa-sinensis,** The Shoe-flower.—The flowers are strung into garlands, and, combined with the yellow Indian Marigold (*genda*), are used in Bengal, being specially in demand as an offering to the goddess *Káli.*

36. **Ipomœa bilobata,** The *Dopati-latá.*—In Bombay it is said garlands of this creeper are hung around the huts occupied by women on the sixth day after confinement to protect the new-born babe. (*Lisboa, Useful Plants, Bombay.*)

37. **Jasminum grandiflorum,** The *Jatí,* or Spanish Jasmine.—The flowers are generally used to make durbar and wedding garlands. (*Voigt.*)

38. **Jasminum Sambac,** *chamba.*—The fragrant flowers much used as a hair ornament by women in the Bombay Presidency.

39. **Linum usitatissimum,** The common *Flax.*—§ "Some necklaces said to be composed of sections of the stems of this plant were sent to me from Calcutta (Bazar) along with others which I sent to Kew.) (*Mr. J. F. Duthie.*)

40. **Mangifera indica,** The mango tree.—The leaves are strung into garlands which hang about Hindú temples. No marriage or burial ceremony of the Hindús in Western India is complete without these garlands.

41. **Melia Azedarach,** The *Ním* or Bead Tree.—The stone from this succulent fruit is used all over India as a bead. These beads are perforated and strung into necklaces and rosaries. During the prevalence of epidemics of small-pox, &c., they are suspended as a charm over doors and verandahs to keep off infection.

42. **Mimusops Elengi,** The *Bakul.*—The flowers are strung into garlands. The tree is sacred to *Siva.*

43. **Nerium odorum,** The Sweet-scented Oleander (*kanér, karabis, &c.*). —There are two varieties of this plant, one red and the other white flowered. The flowers are used in garlands.

44. **Nelumbium speciosum,** The Sacred Lotus or *Padma.*—Designs of this flower are frequent in Hindú and Buddhistic sculptures, an inverted lotus forming the dome of all Buddhist and Jain temples. It is sacred to *Lakshmi.* The dry nuts are strung as beads and the flowers in garlands.

45. **Nyctanthes Arbor-tristis,** The *Singhar* or *harsinghar.*—The natives collect the flowers and string them as necklaces or wear them in the hair. (*Drury.*)

46. **Ocimum sanctum,** The *Tulsi* or Sacred Basil.—The root or woody stem is cut and made into beads worn by the Vaishna-vas, the rosary consisting of 108 beads. The plant is sacred to *Vishnú.*

Vegetable-beads.

47. **Ocimum Basilicnm,** The Sweet-scented Basil.—The wood is used like the preceding.

48. **Oroxylum indicum** (*Calosanthes indica*), The *Sona* or *ullu.*—The large flat-winged seeds of this plant are strung as ornaments to temples.

49. **Pandanus odoratissimus.**—The sweet-smelling spathes which enclose the male flowers and also the male flowers of this tree are perhaps the commonest hair ornament of the western coast Hindú women.

50. **Putranjiva Roxburghii,** The *Joti.*—The black nuts of this plant are made into necklaces and rosaries and are worn by Brahmins and put round the necks of children to ward off disease caused by evil spirits; hence the name *putra-jiv*=life of a child.

51. **Reeds.**—Pieces of reeds are worn in the ears by some of the Assam tribes. They are also used to enlarge the ear-perforations, being by bent like the letter N or W.

52. **Samadera indica.**—The seeds are strung together and tied round children's necks as a preventive to asthma and affections of the chest. (*Drury.*)

53. **Symplocos spicata,** The *Búri* of Sylhet.—**Roxburgh** says the seeds of this plant are very hard, about the size of a pea, and resemble a minute pitcher; when perforated they are strung like beads and by the natives are put round the necks of their children to prevent evil.

54. **Tabernæmontana coronaria,** The *Tagar* of Bengal, or *Chandni* of Upper India.—The flowers are strung as garlands, and they are also presented as offerings to the gods.

55. **Tamarix articulata,** The *Farás.*—The wood is made into small ornaments.

56. **Tagetes erecta,** The Indian Yellow Marigold or *Genda.*—Garlands of this flower are largely given by the presiding Brahmans to worshippers. It is also extensively used in the decoration of houses; along with the red leaf-like bracts of **Euphorbia pulcherrima,** this constitutes the Christmas decorations of Calcutta. The evergreens used on such occasions consist of **Polyalthia longifolia,** the *debdaru,* also mango leaves, plantain stems, and bamboo twigs.

57. **Vanda Roxburghii.**—The leaves are split and worn by Santal girls as anklets; hence the Santal name *dáré banki.* (*Rev. A. Campbell.*)

58. **Vateria indica,** The Indian Copal Tree.—The resin is made into beads, which very much resemble the true amber. (*Roxb.*)

Bear's Grease. Used medicinally as an emollient in rheumatism. 388

BEAUMONTIA, *Wall.; Gen. Pl., I., 721.* 389

A genus of evergreen climbing trees or shrubs belonging to the Natural Order APOCYNACEÆ, and containing only 4 species, inhabitants of India and the Malaya.

Leaves opposite, nerves distant, arched. *Flowers* very large, white, in terminal cymes; bracts leafy. *Calyx* 5-partite, glandular or not within. *Corolla-tube* very short, throat large, bell- or funnel-shaped, naked; lobes broad, overlapping to right. *Stamens* at the top of the tube, included in the throat; filaments thickened at the top; anthers horny, sagittate, conniving over and adhering to the stigma; cells spurred at the base. *Disc* deeply 5-lobed. *Ovary* 2-celled, cells many-ovuled; style filiform, top clavate, stigma fusiform. *Fruit* long, thick, woody, at length dividing into 2 horizontally spreading follicles. *Seeds* compressed, ovoid or oblong, top contracted, crowned with a pencil of hairs; cotyledon thick or thin, radicle short, superior.

Beaumontia grandiflora, *Wall.; Fl. Br. Ind., III., 660;* APOCYNACEÆ. 390

Syn.—ECHITES GRANDIFLORA, *Roxb., Fl. Ind., Ed. C.B.C., 246.*

Vern.—*Barbari,* NEPAL.

Habitat.—An extensive climber of East and North Bengal, with large showy lemon-white flowers. It is found from Nepal eastward to Sikkim, Sylhet, and Chittagong; ascending to 4,000 feet.

FIBRE. 391

Fibre.—A fibre is prepared from the young twigs.

BEES.

392

Bees of India.—" Bees of the genus **Apis** (the hive or honey bee) abound all over India Burma, and Ceylon, and they are found on the higher regions along the northern boundary of Bhután and the frontier of Thibet. They are but imperfectly known to European entomologists. A few important varieties have been discovered by the writer, while of others only the worker is known. The habits of the known kinds have not been systematically studied under cultivation. There is also much confusion in the nomenclature, but the enquiries of **Dr. A. Gerstacker** have done much to clear this up. **Gerstacker** considers the thirteen species described by **Fabricius, Latreille, Klug, Guerin,** and **Smith,** mostly mere colour varieties, comprising only three species whch form two distinct groups—the type of one group being **A. dorsata,** and of the other **A. mellifica.** The larger Indian varieties of the second group, which the writer is presently examining, were unknown to **Gerstacker.**

Group I, Forms of. 393 A. dorsata. 394 A. zonata. 395 A. bicolor. 396

" *Group 1.* **Apis dorsata.**—The insects of this group are **A. dorsata,** *Fab.,* (**A. nigripenis,** *Lat.,*) **A. zonata,** *Guér.* (**A. zonata,** *Smith*), and **A. bicolor,** *Klug.*

" **Description.**—The bees of this group differ from **A. mellifica** in being larger, in building 4½ cells to the inch, in the shape of the abdomen, in having 13 rows of bristles forming the pollen basket, in the relative positions of the eyes and ocelli, and in a very slightly different arrangement of nervures of the anterior wings. It would seem that this bee does not build larger cells for drones than for workers, and that the drone is similar in shape and size to the worker, differing principally in the head, which resembles the head of the drone of **A. mellifica.** It builds one large comb 3 to 5 feet long, 2 feet or more deep; the brood comb is 1¼ inch thick, and the store comb much thicker. Although both **A. dorsata** and **A. florea** are normally single-comb bees; under exceptionally favourable circumstances they build a second comb and their single combs are built much larger than otherwise usual,—*e.g.,* **A. dorsata,** building in rock cavities; and a comb of **A. florea** built in a dwelling-house was found to be about 5 feet in area in addition to being in some places double, the comb of this bee being usually single and perhaps less than one foot in area. Probably in all these very large nests there are several queens, and they are not comparable to single stocks of **A. mellifica.** The arrangement of the stores and brood is the same as in other species. **A. dorsata** as found in India, is exceedingly constant in size and colour; it is found in forests, but frequently builds in towns. It is reputed to be very vicious, but unless disturbed it does not attack, and could be handled by some of the measures usually employed by bee-keepers.

" **Habitat.**—**A. dorsata** is found all over India, but not at great heights above sea-level; it is said to be found at 2,000 feet or more in Bhutan, but may justly be termed a tropical insect indigenous to the plains.

Combs. 397 Wax. 398

" **Economic information.**—The large size of the comb and bee has excited hopes of this insect proving, under cultivation, of great economic value, and European bee-keepers have endeavoured to obtain stocks of it. **Mr. Benton,** a dealer in foreign bees, went to Ceylon for the purpose, but he was unfortunate in his efforts, for the queens died. He states he does not consider them so vicious as reputed when once hived, but he gave up the attempt to cultivate the species. Several years previous the writer undertook to obtain stocks, if likely to prove useful in Europe, but did not hive any, as it was considered better to first investigate the economic value of other Indian species. The reasons against any attempt to cultivate **A. dorsata** in a hive are—(1) The bee builds naturally in the open. (2) It builds normally only one comb, so that the honey cannot be removed without removing the brood also. (3) Although it builds a very large comb, this comb is not so great in cubic capacity, normally, as the combs built by a stock of **A. mellifica,** which is readily cultivated and well understood already. (4) It is only found in a tropical climate, and in this respect differs from **A. mellifica** and **A. indica,** the most productive varieties of which are apparently indigenous to localities having more or less severe winters. **A. dorsata** probably might be cultivated in a semi-wild state in the forests, and the produce largely increased by this means. The present practice of indiscriminately robbing every stock found of all its comb stores and brood might be replaced by a more rational mode of procedure; for, although not hived, many of the processes applied in the economic management of **A. mellifica** might be applied to the semi-wild **A. dorsata.** The bees might be fed to stimulate breeding or prevent starvation. Excessive swarming might be interfered with. Certain stocks might be selected to breed from, as in the old system of bee-keeping. It might be found practicable to remove only portions of the comb, and the bees might be induced to build on or in artificial structures more accessible than the branches of trees.

Honey. 399

"Large quantities of both wax and honey are taken in the forests from **A. dorsata**; this wax appears to be bought up by dealers, and some is exported. The honey is sold and mostly consumed locally, but is commonly of very inferior quality, being contaminated by pollen, the juices of larvæ, &c. It is also commonly thin and liable to fermentation. The use of a simple extractor, care being taken to ripen when necessary, and to grade it instead of mixing good and bad together,—these and other simple improvements would greatly increase the value of the honey. It appears highly probable that most of the honey produced by bees building in the open air is thin and requires ripening by evaporation to remove its liability to fermentation. Of 60 to 70 specimens sent to the Calcutta Exhibition, very few were free from fermetation.

Group II, Forms of. 400

"*Group 2.* **Apis indica, Apis florea.**—The bees of this group agree with **A. mellifica** in having nine rows of bristles to the pollen baskets, and in the division of the anterior wings, relative position of eyes, in building drone comb, in the drones being widely different from the workers in shape, &c. In fact, as described by entomologists, they differ from **A. mellifica** mainly in size and colour. This group includes **A. indica,** *Fab.* (**A. socialis,** *Lat.*), **A. socialis** and **dorsata,** *Lepelletier.* (**A. delesserti,** *Guerin*); **A. peronii,** *Lat.*; **A. perrottetii,** *Quér.*; **A. nigrocineta,** *Smith*; and **A. florea,** *Fab.* (**A. indica,** *Lat.*, **A. lobata,** *Smith*). **Dr. Gerstacker** regards the last as a distinct species, and the others as being colour varieties of another species, the **A. indica** of **Fabricius.**

401

402

403

A. FLOREA 404

"**Description.—A. florea** is very constant in colour, size, and shape all over India.

"It is the smallest known specimen of the genus **Apis.** Its worker cells are 9 to the inch, and its drone cells about 6; the drone is relatively to the worker much larger than in **A. mellifica,** and has a thumb-like projection on the metatarsi of the posterior legs. This drone also differs in some other structural respects from that of **A. mellifica.**

"Like **A. dorsata,** this species builds in the open, a single comb, and is only found in the plains.

Comb. 405

"**Habitat.**—Its comb is usually built attached to a branch and commonly in bushes, but sometimes under the cornices of houses and inside buildings. Its comb is often only as large as a man's hand; at other times it may, as already stated above under **A. dorsata,** be greatly extended and in part duplicated.

Wax. 406
Honey. 407

"**Economic information.**—The honey is small in quantity, and that of the small combs built in the open air is commonly very thin, but that found in large sheltered combs is similar to the honey produced by **A. mellifica.** The honey and wax of this species is not of commercial importance; they are often collected, but seldom offered for sale.

A. INDICA. 408

"**Description.**—**A. indica** is described as much smaller than **A. mellifica.** but it is very imperfectly known; the writer has found that some varieties are larger than many of the European forms, and that **A. indica** hitherto known to entomologists includes only some varieties, and these the smallest and least valuable. **A. indica** differs very widely in size and colour with locality, those from the most elevated northern regions being much darker and larger than from the plains.

Combs. 409
Wax. 410
Honey. 411

"The smaller forms of **A. indica** build 6 cells to an inch, producing but little surplus honey, and swarming early and frequently, so that in the plains stocks are light and of little economic value. The Bhután variety is much larger, building $5\frac{1}{2}$ cells to the inch, and forming heavier stocks. The varieties found in the Hazára District, Panjáb, and north of Simla, on the Thibet frontier, are as large or even larger than **A. ligustica,** and appear from the reports received to be at least as productive as **A. mellifica.** The varieties of **A. indica** found in the plains generally and at Landour, Chumba, Mussoorie, in Burma, Ceylon, Assam, the Khásia Hills, Bengal proper, Orissa, Kurnool, and other parts of the Madras Presidency, Central India, the Murree Hills, &c., are small. The varieties met with in the plains are lighter coloured than those of the higher regions; the latter have darker bodies than the former, and also dark wings. All build worker cells 6 to the inch. This species is cultivated or rather encouraged in most parts of India for the sake of its honey.

412

It is the variety of the small bee which is cultivated at the hill stations, several Europeans having been very successful. The varieties found in the plains are in some cases more prone to sting than those of the higher regions. The Bhután variety builds more comb than the smaller varieties; it is exceedingly easy to handle, but is not so courageous as **A. mellifica** and **A. indica** of the plains; the sentinels at the hive-door run in as soon as alarmed, instead of coming out and defending the hive. All the above varieties so far as known are inferior to **A. ligustica** under cultivation, as they permit the presence of insect vermin in their hives, and are therefore very liable to the ravages of moth. They appear much more prone to swarm than **A. mellifica.** The large variety of the Hazára District, Panjáb, and the cultivated variety of Bashahr, are probably as productive as **A. mellifica.** An attempt is being made to obtain stocks for observation as to productiveness, temper, and resistance to moth and other vermin, so as to bring this economically valuable variety under cultivation. The productiveness of **A. indica** appears least in the plains, being there very little, and greatest in the higher regions; the greatest yield reported from a cultivated stock is 30 lbs. of honey.

413

414

"**Economic information.**—Large quantities of honey are obtained from **A. indica** in the higher regions; the honey differs in appearance with the season's pasturage, that obtained in the autumn being usually light coloured. Much of the honey is inferior in quality from its liability to fermentation, the mode of extracting it, and the fact that it is not graded. It varies in price from 2 annas a seer at some hill stations, where it is plentiful, to 8 annas a lb. In the bazars of the towns comb honey of the best kind produced in very small quantities by the European methods of cultivation fetches R1 a lb. No doubt wax from **A. indica** is sold, but probably the greater portion of the wax taken in the forests and that exported is from **A. dorsata.** The production of wax and honey in India, although it attains considerable value in the aggregate, admits of enormous expansion by the introduction of improved modes of cultivation; and as there is great demand for good honey, bee culture would be exceedingly profitable.

415

"**Successful Rearing of A. mellifica.**—The culture of the small varieties of **A. indica** will, no doubt, be replaced by that of the best varieties of **A. mellifica,** or the large varieties of **A. indica.** The prevalence of moth during the rainy season, and the absence of the long winter rest of Europe, will render it preferable to cultivate a species which, like **A. ligustica,** is specially able to protect itself against moth. The introduction of European species, although previous attempts have failed, has at last been accomplished, and the production of honey and wax will be developed by cultivation of European or sufficiently productive Indian species,—the principal points requiring special attention being more frequent superseding of queens, where there is little or no winter rest, and care to stimulate breeding at such times so as to profit by the early pasturage. **A. ligustica** has been successfully introduced into India, a queen imported into Calcutta in November 1882 died at the end of March 1885; and she was laying abundantly almost the whole of this time." (*J. G. Douglas, Esq., Telegraph Department, Calcutta.*)

416

BEESHA, *Kunth.; Gen. Pl., III., 1215.*

A genus of Bamboos reduced by the *Genera Plantarum* to OCHLANDRA, *Thw.*, which see, and also under BAMBUSÆ. The following are the species formerly referred to this genus.

417

Beesha Rheedii, *Kunth.; Munro, 144; Beddome, ccxxxiv.;* GRAMINEÆ.

Vern.—*Bish-báns,* BENG.; *Pagu-tulla, vay, vaysha,* CHITTAGONG; *Bísha,* MAL.

Habitat.—A bamboo met with in Malabar and Cochin; stems 16 feet high.

418

B. stridula, *Munro.*

Vern.—*Batta,* SINGH.

Habitat.—Met with in Bombay and Ceylon; stems 6 to 18 feet high.

419

B. travancorica, *Beddome.*

Vern.—*Irúl,* TRAVANCORE.

Habitat.—Met with in the Hills of Tinnevelly and Travancore, 3,000 to 5,500 feet; stems 6 to 8 feet high; A densely-gregarious species.

Beet and Beet-root, see Beta vulgaris, *Moq.;* CHENOPODIACEÆ.

420

BEGONIACEÆ.

A natural order of herbaceous plants, referred to two genera,—Begonia having 398 species, and HILLEBRANDIA 1 species.

In India there are over 64 species of the former genus. They inhabit all moist tropical countries except Australia. The affinities of the Natural Order are very obscure; they are most nearly related to Cucurbitaceæ and Datisceæ. The discovery of the genus Hillebrandia (in the Sandwich Islands) has suggested a close affinity to Saxifrageæ. They are highly ornamental plants and great favourites of the modern foliage cultivator, but they are of no economic value. The following extrac from the *Flora of British India* gives the diagnostic characters of the order :—

"Succulent herbs or undershrubs; stem often reduced to a rhizome or tuber. *Leaves* alternate (sometimes falsely whorled in **B. verticillata**) more or less unequal-sided, entire, toothed or lobed; stipules 2, free, frequently deciduous. *Peduncles* axillary, divided into dichotomous cymes, the branches and bracts at their divisions generally opposite. *Flowers* white, rose or yellow, showy, sometimes small monœcious. *Male : perianth* (of the only Indian genus) of 2 outer valvate opposite sepaloid segments, and 2-0 inner smaller segments; stamens indefinite often very many, free or monadelphous anthers narrowly obovoid. *Female : perianth* (of the only Indian genus) of 5-2 segments. *Ovary* inferior (in Hillebrandia, half-superior 2-3-4 celled; placent. s vertical, axile (at the time of æstivation), divided or simple; styles 2-4, combined at the base of the stigmas, branched or tortuous; ovules very many. *Fruit* capsular, more rarely succulent, often winged, variously dehiscing or irregularly breaking up. *Seeds* very many, minute globose or narrow cylindric, testa reticulated; albumen very scanty or o. (*Flora of British India, Vol. II., 635.*)

421

BEGONIA, *Linn.; Gen. Pl., I., 841.*

Begonia Rex, *Putzeys.*, and other species; *Fl. Br. Ind., II., 635;* BEGONIACEÆ.

FOOD. 422

Food.—Many species of this herbaceous genus having succulent stems are used as pot-herbs, and when fresh have a pleasant acid taste.

Speaking of his companion while ascending the Kaklang Pass, Sikkim, **Sir J. D. Hooker** says: "The great yellow-flowered **Begonia** was abundant, and he cut its juicy stalks to make sauce (as we do apple-sauce) for some pork which he expected to get at Bhomsong; the taste is acid and very pleasant." (*Hooker's Himálayan Journal, Vol. I., pp. 292-93.*) The natives of Chittagong, where the plant is plentiful, use the leaves as a pot-herb. (*Roxb., Fl. Ind., Ed. C.B.C., p. 676.*) It is used by some of the tea-planters of Assam as a substitute for Rhubarb.

MEDICINE. Juice. 423

Medicine.—Several species, such as **B. silhetensis,** *C. B. Clarke;* **B. picta,** *Sm.;* **B. rubro-venia,** *Hook.;* **B. laciniata,** *Roxb.;* **B. Rex,** *Putzeys.* The juice is poisonous to leeches, and may therefore be used to kill them when found in the nostrils of animals. See **Anagallis arvensis,** *Linn.*, and **Leeches.**

Hair-wash. 424

§ "When clarified with soda bicarb., the juice makes an excellent application for the hair." (*Mr. G. F. Poynde, Roorkee.*)

425

BEILSCHMIEDIA, *Nees; Gen. Pl., III., 152.*

A genus of trees belonging to the LAURINEÆ, comprising some 20 species, inhabitants of tropical Africa, Asia, Australia, New Zealand, and America.

Leaves sub-opposite or alternate. *Flowers* bisexual in short axillary racemes. *Perianth* deeply 6-cleft, deciduous. Outer circle of 6 perfect stamens, opposite to the perianth segments and generally alternating with small glands; anthers introrse, the inner circle of 3 perfect stamens, with lateral, semi-extrorse anthers alternating with 3 short staminodia; anthers 2-celled, valves opening upwards. *Ovary* incompletely 2-celled, with 3 ovules; style filiform, stigma discoid. Fruit a dry oblong-seeded berry, base incompletely 2-celled. (*Brandis, 378.*)

426

Beilschmiedia Roxburghiana, *Nees;* LAURACEÆ.

Syn.—LAURUS BILOCULARIS, *Roxb.; Fl. Ind., Ed. C.B.C., 341.*

Vern.—*Konháiah,* OUDH; *Tarsing,* NEPAL; *Kanyu,* LEPCHA; *Tapchi,* GARO; *Serai-guti,* ASS.; *Shatoobeng,* BURM.

TIMBER. 427

Habitat.—An evergreen tree found in Eastern Himálaya up to 8,000 feet in Eastern Bengal, Burma, and the Andaman Islands.

Structure of the Wood.—White, moderately hard, even-grained; heartwood with red and green streaks. Annual rings marked by sharp lines. Weight about 37 lbs. per cubic foot.

TIMBER. 428

It is used in Assam for boats; in Darjeeling for building, tea-boxes, and other purposes.

Beleric myrobalan, see **Terminalia belerica,** *Roxb.;* COMBRETACEÆ.

Belladona, see **Atropa Belladona,** *Linn.;* SOLANACEÆ.

BENINCASA, *Savi.; Gen. Pl., I., 824.*

429

A genus containing only one species—an extensive climber, belonging to the CUCURBITACEÆ, most probably a native of tropical Asia, Africa, and America, but cultivated in all tropical countries.

Softly hairy; tendrils 2-fid. Leaves cordate, reniform-orbicular, more or less 5-lobed; petiole without glands. *Flowers* large, yellow, monœcious, all solitary, without bracts. *Male:* calyx-tube campanulate, lobes 5, leaf-like serrate; petals 5, nearly separate, obovate; stamens 3, inserted near the mouth of the tube; anthers exserted, free, one 1-celled, two 2-celled, cells sigmoid. *Female:* calyx and corolla as in the male; ovary oblong, densely hairy; style thick, with 3 flexuose stigmas; ovules numerous, horizontal; placentas 3. Fruit large, fleshy, oblong, pubescent, indehiscent; seeds many, oblong, compressed-margined.

The genus is named after an Italian nobleman, **Count Benincasa.**

Benincasa cerifera, *Savi.; Fl. Br. Ind., II., 616.*

430

THE WHITE GOURD MELON.

Syn.—CUCURBITA PEPO, *Roxb.*, includes this plant as well as C. PEPO, *DC.*

Vern.—*Péthá, chal-kumra, gol kaddú,* PB.; *Kumrá, chál-kumrá,* BENG.; *Gól-kaddú, kudímah, kóndhá, kumrhá, kumrá, pétha, phúthiá,* HIND.; *Kumhrá, bhunja,* KUMAON; *Kohalá,* MAR.; *Kúshmánd, kohula,* CUTCH; *Bhúru kolu, koholú,* GUJ.; *Kohala, koholen, gólkadú,* BOMB.; *Gol-kuddú,* SIND.; *Kaliyána-púshinik-káy,* TAM.; *Búrda-gúmúdú, búdide gummadi, pendli-gummadi-káya,* TEL.; *Kumpalanná, kumpalam,* MAL.; *Búde-kumbala-káyi,* KAN.; *Kúshmánda, kúsh-pándaha,* SANS.; *Majdabh,* ARAB., PERS.; *Kyauk-pa-yon,* BURM.

References.—*Roxb., Fl. Ind., Ed. C.B.C. (in part), 700; Voigt, Hort. Sub. Calc., 57; Duthie and Fuller's Field and Garden Crops, N.-W. P., p. XLV; Dymock's Mat. Med., W. Ind., 287; U. C. Dutt, Mat. Med., Hind., 167; Official Correspondence, Home Dept., 1880, p. 313; De Candolle, L'Origin. Cult. Pl., p. 213; Baden Powell's Pb. Prod., 265; Atkinson's Him. Dist., Vol. X., 700, &c.*

Habitat.—Cultivated in India; according to **DeCandolle** it is a native of Japan and Java.

Botanic Diagnosis.—This plant is so like the Pumpkin that the earlier botanists took it for one. To distinguish it, however, from **Cucurbita Pepo,** *DC.*, the following characters may be given: Softly hairy. *Male:* flowers large, solitary; *petals* 5, nearly free; *stamens* 3, inserted near the mouth of the tube; *anthers* free, exserted. *Fruit* 1 to 1½ feet, cylindric, without ribs, hairy when young, and bright green, ultimately becoming smooth and covered with a bluish-white waxy bloom; flesh white.

CULTIVATION.—**Duthie** and **Fuller** say that this plant is restricted as a rule to little highly-manured patches in the vicinity of village sites. In Bengal it is frequently seen creeping over huts,—in fact, the oval fruits with the white mealiness constitute a striking feature of the Bengal village.

431

Properties and Uses—

Oil.—The fruit of this plant excretes upon its surface a waxy substance which resembles the bloom found on plums and cucumbers. This is said to be produced in sufficient quantity to be collected and made into candles.

OIL. Waxy substance. 432

The White Gourd Melon.

Oil from seed. 433

The seeds also yield a mild, bland, pale-coloured oil. As this plant has been very much confused by botanists with **Cucurbita Pepo,** *DC.*, it is probable that some of the native names given above are incorrectly applied to this species. It would be very important to have specimens of the plants, from which oils have been prepared, supplied along with these oils so as to admit of final determination. The greatest possible ambiguity exists in the literature of this subject.

MEDICINE. Fruit, Juice. 434

Medicine.—The fruit possesses alterative and styptic properties, and is popularly known as a valuable anti-mercurial. It is also said to be cooling. It is considered tonic, nutritive, and diuretic, and a specific for hæmoptysis and other hæmorrhages from internal organs. For this purpose the FRESH JUICE from the fruit is administered, while a slice of the fruit is at the same time applied to the temples. According to the Sanskrit authors, it is useful in insanity, epilepsy, and other nervous diseases; the fresh juice is given either with sugar or as an adjunct to other medicines for these diseases.

"It would appear that the older Sanskrit writers were not acquainted with its peculiar action on the circulatory system by which it rapidly puts a check to hæmorrhage from the lungs. The Rája Nirghantu, the oldest work on therapeutics, gives a long account of its virtues, but does not allude to its uses in phthisis or hæmoptysis. Neither does Susruta mention it in his chapters on the treatment of hæmorrhage and phthisis, though the plant is alluded to by him elsewhere. The more recent compilations, such as Chakradatta, Sangraha, Sarangadhara, &c., give numerous preparations of the article and detail its uses." "In preparing this medicine" in the form of a confection "old ripe gourds are selected. Those not at least a year old are not approved. They are longitudinally divided into two halves and the pulp scraped out in thin flakes by an iron comb or scratcher. The watery juice that oozes out abundantly during this process is preserved, the seeds being rejected. The pulp is boiled in the above-mentioned juice, till soft. It is then tied up tightly in a cloth and the fluid portion allowed to strain through it. The softened and drained pulp is dried in the sun and the watery portion preserved for future use. Fifty *tolas* of the prepared pulp are fried in sixteen *tolas* of clarified butter, and again boiled in the juice of the fruit, till reduced to the consistence of honey. To this are added fifty *tolas* of refined sugar, and the whole is heated over a gentle fire, till the mass assumes such a consistence as to adhere to the ladle." The pot is then removed from the fire, and a number of flavouring demulcents added, such as pepper, ginger, cumin, cardamoms, cinnamon, &c., the mixture being stirred until cold. Dose from one to two *tolas*, according to the age and strength of patient. (*U. C. Dutt.*) "The SEEDS possess anthelmintic properties, and are useful in cases of tænia. The expressed OIL of the seeds, in doses of half an ounce, repeated once or twice at an interval of two hours, and followed by an aperient, is said to be equally efficacious." May be used as a substitute for male fern. (*Official Correspondence from Bombay Committee regarding the revision of the Pharmacopœia of India.*)

Seeds. 435

OIL. 436

437

Special Opinions.—§ "The fresh juice is often used as a vehicle to administer pearl shell for the cure of phthisis in the first stage." (*Assistant Surgeon Sakhárám Arjún Rávat, Girgaum, Bombay.*) "It is considered a specific in pulmonary consumption. A native preparation made from the ripe fruit called *Kushandakhanda* is considered very efficacious in phthisis pulmonalis, and I have seen people benefited by it." (*Surgeon K. D. Ghose, Bankura.*) "This is so universally believed to be useful in pulmonary consumption that some trials should be made in order to discover whether it has any effect on the bacillus of phthisis discovered by **Dr. Koch.** I have seen it produce a decided effect in arresting pulmonary tuberculosis."

(*Surgeon K. D. Ghose, M.D., Khulna.*) "Preserve is given in piles and in dyspepsia as an antibilious food." (*Surgeon-Major W. Moir, Meerut.*) "This forms one of the chief ingredients of the vapour bath used in syphilitic eruptions." (*Assistant Surgeon Anund Chunder Mukerji, Noakhally.*) "The expressed juice of the mature fruit possesses purgative and alterative properties. It is used in cases where the system has been affected by mercury." (*Brigade Surgeon J. H. Thornton, B.A., M.B., Monghyr.*) "The preserve of the white melon is an easily digestible and highly nutritious food in wasting diseases, as consumption." (*Surgeon-Major R. L. Dutt, Pubna.*) "Much used in diabetes with successful results, the juice of the cortical portoin (4 oz.), combined with 100 grains of each of powdered saffron and bran of red rice, given morning and evening, with strict diet." (*Surgeon E. W. Savinge, Rajamundry, Godavari District.*) "The most common way in which the juice is used is in the shape of a confection with sugar, &c., as a cooling and fattening medicine." (*Native Surgeon Ruthnam T. Moodelliar, Chingleput, Madras Presidency.*) "Useful in pills given with *surun*. Antidote for mercurial poisoning administered in the form of *pak*." (*Surgeon W. Barren, Bhuj, Cutch.*)

FOOD. **Vegetable.** **438**

Food.—The white gourd melon is used in the following ways: (*a*) as a vegetable, (*b*) as a curry, and (*c*) as a sweetmeat called *heshim*. "This species is used principally in making a sweetmeat, which consists of pieces of this gourd coated with sugar; it is said to have cooling properties." (*Baden Powell's Panjáb Products, p. 265.*)

Sweetmeat. **439**

Ben oil, the oil obtained from the seeds of **Moringa aptera,** *Gærtn.*, which see.

Benzoin or **Benjamin,** see **Styrax Benzoin,** *Dryand.*; STYRACEÆ.

BERBERIDEÆ.

439*a*

A natural order of herbs, bushes, or climbers, comprising about 100 species, referred to 19 genera,—inhabitants of the temperate regions. In India there are only 17 species referred to 6 genera. The following descriptive account and analysis of the order, extracted from the *Flora of British India*, may be found useful:—

"Usually shrubby, sometimes climbing, glabrous plants. *Leaves* simple or compound, with articulate segments; buds scaly; *stipules* very rare (**Berberis**). *Flowers* often globose, regular, solitary or in simple or compound racemes, usually yellow or white. *Sepals* and *petals* free, hypogynous, very caducous, 2-many-seriate, in 3 rarely 4-6-nary whorls, imbricate, or the sepals rarely valvate. *Stamens* 4-6 (rarely 8) opposite the petals, free or connate; anthers adnate, erect, dehiscing by lateral or dorsal slits, or by 2 revolute or ascending lids or valves. *Carpels* 1-3, rarely more, oblong; style short or 0, stigma dilated or conic or oblong; ovules usually indefinite on the ventral suture or covering the walls of the ovary, anatropous, rarely orthotropous. *Ripe carpels* dry or fleshy, dehiscent or not. *Seeds* with a crustaceous fleshy or bony testa; albumen copious, dense; embryo minute or long, straight or curved, radicle next the hilum."

Tribe I.—Lardizabaleæ. *Stem* usually climbing. *Flowers* unisexual or polygamous. *Carpels* 3. *Seeds* usually large, testa bony.

An erect shrub, leaves pinnate . . **1. Decaisnea.**
Climbing shrubs, leaves digitate.
Stamens monadelphous . . . **2. Parvatia.**
Stamens free . . . . . . **3. Holbœllia.**

Tribe II.—Berbereæ. Stem O or erect. *Flowers* hermaphrodite. *Carpel* 1. *Seeds* usually small.

Ovules erect basal. Shrubs.
Fruit berried . . . . . . **4. Berberis.**
Ovules superposed along the veniral suture.
Leaves decompound. Ovules few . . **5. Epimedium.**
Leaves simple, palmate. Ovules many . **6. Podophyllum.**

440

BERBERIS, *Linn.; Gen. Pl., I., 43.*

A genus of shrubs containing some 50 species, the characteristic members of BERBERIDEÆ.

Wood yellow. *Leaves* pinnate or simple and then fascicled in the axils of 3-5 partite spines. *Flowers* yellow, hermaphrodite, fascicled, racemed or solitary. *Sepals* 6, with 2-3 appressed bracts, imbricate in 2 series. *Petals* 6, imbricate in 2 series, usually with two basal glands inside. *Stamens* 6, free; anther-cells opening by recurved valves. *Ovary* simple; stigma peltate sessile or on a short style; ovules few, basal, erect. *Berry* few-seeded.

441

Berberis angulosa, *Wall.; Fl. Br. Ind., I., III.*

Vern.—*Chutra*, NEPAL.

Habitat.—A large, erect shrub of the inner ranges of East Kumaun, Nepal and Sikkim, above 11,000 feet.

TIMBER. 442

Structure of the Wood.—Dark grey or yellowish brown, hard. Weight about 50 lbs. per cubic foot.

443

B. aristata, *DC.; Fl. Br. Ind., I., 110.*

THE BARBERRY.

Vern.—*Chitra, chotra, dár-hald, rasvat, kashmal*, HIND.; *Súmlú, simlu, kasmal, chitra*, PB.; *Chitra*, NEPAL; *Tsema*, BHUTIA; *Chitra, zarishk*, PERS.

Moodeen Sheriff gives the following vernacular names arranged under three heads:—

(*a*) **Berries.**—*Zarishk*, HIND., PERS.; *Zarish*, DUK.; *Anbar-báris, ambarbáris*, ARAB.

(*b*) **Extract.**—*Rasvat*, HIND.; *Fil-zahrah, pil-zahrah*, PERS.: *Huzizehindi, fil-zahraj*, ARAB.

(*c*) **Wood** or **Root.**—*Dár-hald, dár-chób*, PERS., HIND.; *Dár-hald*, ARAB.

444

Berberis aristata, B. asiatica, B. Lycium, and **B. vulgaris** are with difficulty distinguished from each other, and in consequence they have been mistaken for each other all over India. The same vernacular names are probably applied to each of these plants and the same properties attributed to all. Considerable ambiguity therefore exists in the published statements regarding these Barberries.

Habitat.—B. aristata is an inhabitant of the temperate Himálaya between 6,000 and 10,000 feet in altitude, extending from Bhután to Kanáwár, the Nilgiri Hills, Ceylon, &c.

Botanic Diagnosis.—An erect, much-branched bush; leaves evergreen or nearly so, obovate or oblong entire, or with few distant spinous teeth; flowers in compound, often corymbose, racemes; berries tapering into a short style; stigma small, subglobose.

There are two varieties in addition to the type from B. **aristata**: *1st*, **floribunda**; *2nd*, **micrantha.**

Properties and Uses—

DYE & TAN. Root & Stem. 445

Dye.—A yellow dye, obtained from the root and stem, is used in tanning and colouring leather. The wood is generally known as *dára-*

halada ; the extract as *rasota, rusot, vasavanti,* or *ruswul* (see also under **B. Lycium)** ; the fruit as *ambarabárisa* (see *Dymock's Mat. Med., Western India*). **Professor Solly,** in *Agri.-Horticultural Society of India, IV., pages 272-279,* writes that the colour exists chiefly in the bark and in the young wood immediately below the bark, and that in old wood the proportion is small, but much superior in quality. In India it appears the root only is used; this doubtless contains colouring matter, but, according to the Professor, not of so good a quality. Barberry is perhaps one of the best tanning dyes in India. The supply is quite inexhaustible; some five or six species occur everywhere in great abundance along the entire Himálaya ; they are temperate bushes, growing on exposed hill-sides between 6,000 and 10,000 feet in altitude, and often constitute thickets of many miles in length. They are equally plentiful on the Nilgiris and in Ceylon.

Oil.—The seed yields an oil.

OIL. 446

Medicine.—The FRUIT or berry is given as a cooling laxative to children. The STEMS are said to be diaphoretic and laxative in rheumatism. The DRIED EXTRACT of the root is extensively used as a purgative for children, and especially as an application in ophthalmia. It is also an excellent application for sun-blindness. The ROOT-BARK abounds in the characteristic bitter principle ; it acts as a tonic and antiperiodic. It is a valuable medicine in intermittent and remittent fevers, and in general debility consequent on fevers. It is also used internally in native practice as a stomachic and in diarrhœa, &c. The berries are useful as an antiscorbutic.

MEDICINE. Fruit. 447 Stems. 448 Extract. 449

Special Opinions.—§ "Instead of the root-bark of **B. aristata,** in my practice I have used the root itself and found it to be quite equal, if not superior, to the former. Its advantages are that it is about fifty times cheaper and more abundant. The root is one of the few really good medicines in India, and deserves the special attention of the profession. As an antiperiodic and antipyretic it is at least quite equal to quinine and Warburg's tincture, respectively; and as a diaphoretic, decidedly superior to James's powder. It is of the greatest service in relieving pyrexia and in converting the continued and remittent fevers into the intermittent, and also in preventing the return of the paroxysms of the latter. In addition to its cheapness, its advantages over Warburg's tincture and quinine are, that however repeatedly it may be used it neither produces a great depression of the system nor has any bad effects on the stomach, bowels, brain, or the organs of hearing. Unlike the alkaloids of cinchona, it can be employed beneficially during an attack of fever. A very good preparation of the root is the decoction, twelve ounces of which is equal to one bottle of Warburg's tincture. If administered during a paroxysm, in two doses (℥ vi each) at the interval of two or three hours, it relieves the fever by producing a copious perspiration; six drachms of the tincture of the root is equal to one bottle of Warburg's tincture. If used in two doses with water during a paroxysm, this produces precisely the same effect as the decoction. There is very little difference between the actions of the tincture and decoction of the root, but the former is preferable to the latter for two reasons,—*viz.,* the smallness of its dose, and the fact that the tincture can be prepared in a large quantity and kept ready for use. To ensure the full antiperiodic effect, the drug should not only be employed during the paroxysm, but also in the same dose every fourth or fifth hour in the intermission ; the cure is completed by the continued use of the drug in smaller doses for four or five days more after the fever ceases to return. Used in the manner explained above, the tincture and decoction have proved successful in many cases of malarious and jungle fevers, in a few of which quinine and also arsenic had previously failed. The watery extract and simple powder of the root are very inferior preparations, and generally

450

very indifferent in their actions. The great and continuous heat which is required to prepare the extract seems to a large extent to destroy its efficacy. The wood of **B. aristata**, particularly that of the stem, is also possessed of the same medicinal properties as the root, but much inferior to the latter. The species of Berberis owe their actions to an active principle called *Berberine*.

"*Preparations from the root.*—Decoction, tincture, and watery extract. *Decoction:* Take of the root, in shavings or coarse powder, six ounces, water two pints and a half; boil on a slow fire till the liquid is reduced to one pint. *Tincture:* Take of the root, in shavings or coarse powder, six ounces, proof spirit one pint; macerate for seven days with occasional agitation, strain and add more proof spirit to make one pint. *Extract:* Take the shavings or coarse powder of the root in any quantity, boil with water till the liquor thickens, strain and evaporate on a sand-bath to the consistence of an extract. *Doses* of the decoction, from two to six fluid ounces; of the tincture, from two to six fluid drachms; and of the extract, from one to two drachms." (*Honorary Surgeon Moodeen Sheriff, Khan Bahadur, Madras.*)

"The extract (*Roswat*) mixed with opium and lime-juice is a most useful external application in painful eye affections." (*Surgeon J. Anderson, M.B., Bijnor.*) "I invariably use this drug in the treatment of indolent ulcers, and have never had occasion to change it for any other local application." (*Surgeon Joseph Parker, M.D., Poona.*) "The tincture of the root-bark, officinal in the Indian Pharmacopæia, is found useful in enlargement of the liver or of the spleen in 30-drop doses 3 times daily." (*Assistant Surgeon Nilruttan Banerji, Etawah.*) "A good febrifuge and antiperiodic, not required during intermission." (*Surgeon W. Forsyth, Dinagepore.*) "It is known here as *Dáru Húldar;* the extract as *Rasvati.*" (*Surgeon-Major J. Robb, Ahmedabad.*)

FOOD. 451

Food.—The oblong fruits are dried in the sun like raisins; are purplish or pinkish and wrinkled; they are eaten and are regarded as palatable.

TIMBER. 452

Structure of the Wood.—Yellow, hard. Weight 52 lbs. per cubic foot. Used for fuel.

453

Berberis asiatica, *Roxb.; Fl. Br. Ind., I., 110.*

Vern.—*Kilmora*, KUMAON; *Máte-kissi, chitra*, NEPAL.

Habitat.—Dry valleys of the Himálaya, altitude 3,000 to 7,500 feet; from Bhután to Garhwál, Behar (on Parasnath hill), altitude 3,500 feet.

Botanic Diagnosis.—Bark pale, spines 5-fid, small, leaves orbicular or broad obovate, sub-entire or coarsely spinous lacunose, white beneath; racemes short corymbose, berries with a distinct style; stigma capitate.

Properties and Uses—

MEDICINE. 454

Medicine.—The medicinal properties of this species are similar to those of the preceding.

FOOD. 455

Food.—The fruit is used in the same way as that of **B. aristata,** *DC.*, and **B. Lycium,** *Royle.*

456

B. coriacea, *Brandis; Gamble, Man. Timb., 14.*

Vern.—*Kashmal*, SIMLA.

Habitat.—A large, erect, thorny shrub of the North-West Himálaya, above 8,000 feet; often forming alone or with other shrubs large extents of scrubby jungle,—*e.g.*, in the valley south of Nágkanda near Simla.

TIMBER. 457

Structure of the Wood.—Yellow, moderately hard. Weight about 54 lbs. per cubic foot.

Berberis Lycium, *Royle; Fl. Br. Ind., I., 110.*

458

Vern.—*Kashmal, chitra,* HIND.; *Kushmul,* N.-W. P.; *Kasmal,* SIMLA; *Darhalad* (the wood), BOMB.; *Kasmal-rasout,* CUTCH; *Zirishk* (the fruit), PERS., and *ambarbáris,* ARAB.; *Raswanti* or *rasout* (the extract).

The Sanskrit name *Darvi* is, in South India, given to **Coscinium fenestratum,** *Colebrooke,* but in Northern India it is applied to a species of **Berberis.** The name *rasout* is generally given to the extract from the wood or root of this and of **B. asiatica** and **B. aristata.** **Dr. Royle,** in a paper to the Linnæan Society of London, proved that this *Rasout* was the *Lycium* of the ancients. Lycium (λύκιον) is mentioned by **Dioscorides, Pliny, Celsus, Galen,** and **Scribonius Largus,** and by many of the later Greek writers as well as by the Arabian physicians. It was held in high esteem as a drug, and was used in the treatment of chronic ophthalmia.

Habitat.—An inhabitant of the Western Himálaya in dry hot places, altitude 3,000 to 9,000 feet, from Garhwál to Hazára.

Botanic Diagnosis.—Bark white; leaves sub-sessile, sub-persistent, lanceolate or narrow obovate-oblong, usually quite entire, pale, not lacunose glaucose beneath; raceme elongate, berries ovoid, style conspicuous, stigma capitate.

Properties and Uses—

Oil.—The seed yields an oil.

OIL. 459

Medicine.—The medicinal extract from the root, known under the name of *Rasout,* is highly esteemed as a febrifuge and as a local application in eye diseases. In chronic ophthalmia it has been used with success, when combined with opium and alum. **Dr. O'Shaughnessy** expresses his opinion on the medicinal uses of this drug in the following terms: "*Rasout* is best given as a febrifuge in half-drachm doses, diffused through water, and repeated thrice daily or even more frequently. It occasions a feeling of agreeable warmth at the epigastrium, increases appetite, promotes digestion, and acts as a very gentle but certain aperient. The skin is invariably moist during its operation."

MEDICINE. Extract. 460

Some difference of opinion prevails as to whether *rasout* should be regarded as a special preparation from the root of this species only, or from **B. asiatica, B. aristata,** as well as **B. Lycium.** The extract has been used by a few European practitioners and found useful in the treatment of chronic ophthalmia. It was employed for this purpose by **Mr. Walker,** of Edinburgh, who found it very efficient. The preparation used by him consisted of equal quantities of Lycium and burnt alum, with half the quantity of opium. It was applied, mixed with lemon-juice, to the consistence of cream, over the eyelids and eyebrows. (*U. S. Dispens., 15th Ed.*) It has also been frequently used and favourably reported on by European doctors in India. A tincture of the root-bark is often recommended in the treatment of fevers.

Special Opinions.—§ "In hæmorrhoids *Rasout* is a very popular remedy in doses of from 10 to 30 grains." (*Assistant Surgeon Mokund Lall, Agra.*) "The watery extract is a bitter tonic and febrifuge in doses of half a drachm. In combination with equal parts of alum and opium, it is used as a *lép* to the eyelids in ophthalmia, often acting like a charm, subduing swelling and allaying irritability. Previously to its application the eyelids should be fomented with *tukmi-páni* or *ním-páni.*" (*Surgeon C. M. Russell, Sarun, Bengal.*) "Similar in action to the sulphate of Berberiæ; useful in eye diseases." (*Surgeon W. Barren, Bhuj, Cutch, Bombay.*)

461

"Is taken internally in 5 to 15 grain doses with butter in bleeding piles. Its solution, 1 drachm to 4 oz. of water, is used as a wash for piles. Its ointment, made with camphor and butter, is applied to pimples and boils, being supposed to suppress them." (*Surgeon J. C. Penny, M.D., Amritsur.*) "Is an excellent tonic and febrifuge, especially in the low fevers of aged people; the tincture in ½-drachm doses." (*Surgeon D. Picachy, Purneah.*) "The Nilgiri barberry has been used in the treatment of ague with good results." (*Surgeon-General W. R. Cornish, Madras.*)

TIMBER. 462

Structure of the Wood.—Yellow, moderately hard. Weight 52 lbs. per cubic foot.

463

Berberis nepalensis, *Spreng.; Fl. Br. Ind., I., 109.*

Syn.—B. PINNATA, *Roxb., Fl. Ind., Ed. C.B.C.*; MAHONIA NEPALENSIS, *DC.*

Vern.—*Amúdanda, chiror,* PB.; *Chatri, milkisse, jamnemunda,* NEPAL.

Habitat.—A shrub or small tree with large pinnate leaves, common on the outer Himálaya, from the Ravi eastward to the Khásía and Nagá Hills, Tenasserim and the Nilgiris, at altitudes above 5,000 feet.

Botanic Diagnosis.—Leaves pinnate; leaflets opposite, oblong ovate or lanceolate, spinous-toothed, palmately 3-5-nerved; racemes dense-flowered.

Properties and Uses—

DYE. 464

Dye.—Used, to a small extent, by the Bhutias and Nagás as a yellow dye.

TIMBER. 465

Structure of the Wood.—Bright yellow, hard. Weight 49 lbs. per cubic foot.

Has a handsome colour and might be useful for inlaying.

B. vulgaris, *Linn.; Fl. Br. Ind., I., 109.*

THE TRUE BARBERRY, *Eng.*; EPINE-VINETTE, VINETTIER, ECORCE DE RACINE DE BERBERIDES, *Fr.*; FAUERACH, GEMEINER, SAUERDORN, BERBERRITZE, BERBERITZEN (SAURCH), WURZELRINDE, *Germ.*; BERBERO, *It., Sp.*

Vern.—*Zirishk, kashmal, chachar* or *chochar,* PB.; *Bedana, cutch,* PERS.; *Ambar-baris,* ARAB.

Habitat.—A deciduous thorny shrub on the Himálaya from Nepal westward, in shady forests, above 8,000 feet; Afghánistan and Beluchistan to Europe.

Properties and Uses—

DYE. 466

Dye.—A yellow dye is extracted from the roots; along with alkaline ley is used in Poland for colouring leather.

MEDICINE 467

Medicine.—The Barberry is regarded as officinal in the Panjáb, being given as diuretic, and for the relief of heat, thirst, and nausea. It is astringent, refrigerant, and antibilious. In small doses it is tonic, in larger cathartic. It was formerly given in jaundice, probably on the principle of signature, the yellow colour suggesting its supposed efficacy.

§ "Cooling laxative medicine. In the form of decoction it is useful in scarlet fever and brain affections." (*Surgeon W. Barren, Bhuj, Cutch.*) "Dried like raisins or currants, the berries greatly resemble the latter." (*Surgeon-Major J. E. T. Aitchison, Simla.*) "Diuretic, demulcent in dysentery." (*Assistant Surgeon Nehal Sing, Saharunpore.*)

468

Chemical Composition.—"**Dr. Græger** found in the ripe fruit 15·58 per cent. of integuments and seeds, 17·20 of soluble solid constituents, and 67·22 of water. The constituents of the juice in 100 parts of fresh berries were 5·92 parts of malic acid, 4·67 of sugar, 6·61 of gum, 67·16 of water and 0·06 salts of potassia and lime. (*A. J. P., Jan. 1, 1873, p. 14*). The *root* and *inner bark* have been used for dyeing yellow. The bark of the root is greyish on the outside, yellow within, very bitter, and stains the saliva when chewed. **Brandis** found in 100 parts of the root 6·63 of bitter, yellow extractive (impure berberine), 1·55 of brown colouring matter, 0·35 of gum, 0·20 of starch, 0·10 of cerin, 0·07 of stearin, 0·03 of chlorophyll, 0·55 of a sub-resin, 55·40 of lignin, and 35·00 of water."

"To a second alkaloid found in barberry bark, the names of *vinetine*, *oxyacanthine*, and *berbine* have been applied. To procure it the mother-liquor of berberine is precipitated by carbonate of sodium, the precipitate treated with dilute hydrochloric acid, and the liquid filtered and precipitated by ammonia. The impure alkaloid thus obtained may be purified by washing with water, drying, exhausting with ether, evaporating, dissolving the residue in dilute hydrochloric acid, and finally precipitating by ammonia. Vinetine is a white amorphous powder, crystallizable from its alcoholic and ethereal solutions, purely bitter, fusible unchanged at 139·50° C. (283° F.), insoluble or but slightly soluble in water, sparingly dissolved by cold but freely by hot alcohol and ether, and freely soluble in alcohol. It forms soluble salts with the acids, and its chloride is white." (*U. S. Dispens., 15th Ed., 1586.*)

FOOD. Fruits 469

Food.—The dried fruits, under the name of *zirish-tursh zarishke-trush* (sour currants), are imported from Cabul, Herat, and Kandadar into the Panjáb. They form a pleasant acid preserve; the unripe ones are pickled as a substitute for capers.

TIMBER. 470

Structure of the Wood.—Lemon-yellow, moderately hard and even-grained. Weight 55 lbs. per cubic foot.

A good firewood.

471

BERCHEMIA, *Neck.; Gen. Pl., I., 377.*

Berchemia floribunda, *Wall.; Fl. Br. Ind., I., 637;* RHAMNEÆ.

Vern.—*Kala lag*, KUMAON; *Chiaduk*, NEPAL; *Rungyeong rik*, LEPCHA.

Habitat.—A large, erect or climbing shrub or small tree, found in the Himálaya from the Jhelum to Bhután, and on the Khásia Hills.

TIMBER. 472

Structure of the Wood.—Yellow, turning grey on exposure, porous.

Bergamot, see **Mentha citrata;** LABIATÆ.

Bergamotte, or **Lime,** see **Citrus Limetta;** and

Bergera Kœnigii, *Linn.*, see **Murraya Kœnigii,** *Spreng.;* RUTACEÆ.

473

BERRYA, *Roxb.; Gen. Pl., I., 232.*

A genus of TILIACEÆ, containing only one species, a large tree. *Leaves* alternate, ovate, acuminate, glabrous, base cordate, 5-7-nerved. *Panicles* large, many-flowered, terminal and axillary. *Calyx* campanulate, irregularly 3-5-lobed. *Petals* 5, spathulate. *Stamens* many, inserted on a short torus; anthers didymous, lobes divergent, opening lengthwise. *Staminodes* 0. *Ovary* 3-4-lobed, cells 4-ovuled; style consolidated, stigma lobed; ovules horizontal. *Fruit* loculicidally 3-4-valved, each valve 2-winged. *Seeds* pilose, albumen fleshy; cotyledons flat, leafy, radicle superior next the hilum.

The generic name is in honour of the late **Dr. Andrew Berry**, a Madras botanist.

474

Berrya Ammonilla, *Roxb.; Fl. Br. Ind., I., 383.*

THE TRINCOMALI WOOD.

Vern.—*Sarala-dévadaru*, TEL.; *Hpet-woon, petwun*, BURM.; *Halmil-lila* or *halmilla*, SINGH.

Habitat.—A large tree found in South India, Burma, and Ceylon.

Properties and Uses—

FIBRE. 475

Fibre.—In the *Amsterdam Catalogue* a fibre from this tree is mentioned as having been sent from Burma.

TIMBER. 476

Structure of the Wood.—Heartwood dark red, very hard, close-grained, but apt to split; it has, even when old, a smooth, rather damp feel. The wood is very durable. **Mr. Gamble** reports that a specimen, which had been 50 years in Calcutta, was found to be perfectly sound and good on being cut into. Weight 48 to 65 lbs. per cubic foot.

It is used for carts, agricultural implements, and spear-handles, and in Madras for masula boats, and is much esteemed for toughness and flexibility. In Ceylon "the wood of this fine tree is very valuable for building and other purposes." (*Thwaites, Enum., Ceyl. Pl., 32.*)

477

Berrya Ammonilla, *Roxb.*, var. mollis.

Vern.—*Hpekwoon*, BURM.

TIMBER. 478

"Is found on elevated ground; the wood, which is red, is much prized for axles, the poles of carts and of ploughs and spear-handles; it is also sawn up for building purposes." (*Br. Burm. Gaz., I., 127.*)

Berthelotia lanceolata, see **Pluchea lanceolata,** *Oliv.*; COMPOSITÆ.

479

BETA, *Linn.; Gen. Pl., III., 52.*

A genus of herbaceous plants belonging to the CHENOPODIACEÆ, comprising some 12 or 13 species.

Glabrous herbs with fleshy radicle leaves. *Flowers* small-ternate, or glomerulate, rarely solitary, glomerules axillary or on simple or paniculate terminal spikes. *Flowers* hermaphrodite. *Perianth* 5-partite, persistent, and adherent to the base of the ovary. *Stamens* 5, perigynous, filaments subulate; anthers oblong. *Ovary* semi-inferior and surrounded by the staminal and perianth fleshy ring; stigma 2-3, rarely more, short subulate, connate at the base, papilose on the inner surface. *Seed* horizontal, attached laterally; testa membranous.

The generic name is the classical Roman name for the cultivated species. The ancient Greeks, who used the leaves and roots, called the plant *Teutlion*, also *Sevkles* or *Sfekelie*, a word which very much resembles the Arab *Selg*, *Silq*. The latter word has apparently been adopted by the Portuguese, who call it *Selga*. The Celtic word *Bett* = red may be the source from which the word Beta was derived. (**DeCandolle.**)

480

Beta maritima, *L.*

THE BEET-ROOT.

Syn.—B. VULGARIS, *Moq.*

Vern.—*Pálak*, HIND.; *Palak, bit pálang* or *pálang ság*, BENG.; *Pálanki*, SANS. (*according to U.C. Dutt*).

Habitat.—Two or three distinct forms are very extensively cultivated over the greater part of India as a cold-season crop. The principal are the *Red Beet* (**B. vulgaris**) and *White Beet* (**B. Cicla**). These are chiefly grown by Europeans, the root being extensively used as a vegetable. The so-called Indian *Beet* (**B. bengalensis,** *Roxb.*) is an erect-branched species, cultivated by the natives on account of the leaves, which are eaten as a vegetable in stews, curries, &c.

FOOD. Sugar. 481

Food.—The manufacture of sugar from the beet-root has, within recent years, become one of the most important industries of Europe. The white root is chiefly used for this purpose. In 1830 the extraction of beet sugar commenced in Germany and France, but it has now spread all over the Continent and to Canada, the United States, and New

Beet Sugar.

BETA maritima.

Zealand. In fact, it is cultivated in most countries where the mean temperature is about 62° to 65° F. A moist hot atmosphere is unfavourable, hence of course India is precluded from ever becoming a beet-sugar-producing country. The plant grows freely enough in the cold season, but as a garden crop only. **Mr. Duthie,** in his annual report of the Saharanpur botanic gardens for 1884, says: "By constant selection of the darkest-coloured roots for seed-stock, it has improved so that the roots are now hardly distinguishable from those raised from the best imported seed." Few plants are more easily modified than beet by careful cultivation, but while, as **Mr. Duthie** says, it is possible to produce an acclimatised stock which will yield seed as good as that imported from Europe, the plant is not likely, however, to be cultivated in India as a field crop, either to feed cattle or as a source of sugar. The interest in beet-sugar, as far as India is concerned, consists entirely in the fact that it affects materially our cane-sugar industry, and must necessarily continue to do so. France, Austria, and Germany, in order to foster and develope the beet-sugar trade, instituted a protective system of giving bounties to home refiners, and at the same time heavy importation duties were levied upon all foreign sugars. This system naturally led to a vast extension of beet cultivation and of refining operations. Over-production soon caused ruinous reduction in prices of sugar, cane-sugar falling in the exact ratio with beet. This naturally resulted in the bankruptcy of numbers of beet-growers, and of some of the largest refiners, a financial crisis having occurred in Vienna in consequence of these failures. The area under beet may now undergo some contraction, and probably will do so, prices improving in consequence; but unless this actually takes place, a prolonged low price like what now prevails must prove disastrous to the cane-sugar industries of the East and West Indies. Already the beet-sugar trade has materially affected the cane-sugar of India, and the extension of cane-sugar cultivation in Fiji, Queensland, and other places is not calculated to lessen the danger.

482

In **Mr. Giffen's** report to the Board of Trade (London, 1884) will be found much interesting and valuable information which cannot be too carefully studied by our cane-sugar producers: "The total sugar crop of the world at the present time may be put, in round figures, at 6,000,000 tons. The known increase in 30 years has been very nearly half that amount." **Mr. Giffen** further writes: "As bearing on recent controversies, it may also be of interest to point out that, since the date of giving my evidence, British cane-sugar appears to have increased quite as much in proportion as beet-root sugar. In 1877-79, the production of British cane-sugar was 403,000 tons per annum, and its proportion to the total 12 per cent.; in the following three years the production was 419,000 tons per annum, and its proportion to the total was still 12 per cent. Possibly later figures may show a different result, but if there has been any change it must have been quite recent. For about 15 years it will be seen the proportion of British cane-sugar in the total production has been the same as it is now, *viz.*, 12 per cent. The remarkable growth of beet-root sugar, in recent years, would thus seem to have been mainly in competition with foreign cane-sugar. Though the production of that sugar in amount has steadily increased, its proportion to the whole has fallen from 60 per cent. 20 years ago, to 40 per cent. at the present time; British cane-sugar, on the contrary, has not only increased in amount, but has increased so rapidly, for 15 years at least, as to maintain its former proportion to the total production." "The production of beet sugar being about 2,000,000 tons, the proportion of beet to cane in the sugar production of the world is thus about one third." (*Report, Board of Trade, London, 1884.*)

483

In his Review of the Sea-Borne Foreign Trade of British India for

The White Birch.

1884, **Mr. J. E. O'Conor** has given an interesting *résumé* of the present position of the Indian sugar trade. In March 1882 the import duty of 5 per cent. on sugar was "taken off, with other import duties," "and the remission was vehemently opposed by the representative in the Legislative Council of the mercantile community of Calcutta, on the ground that it would assuredly bring about the extinction of the sugar industry in Bengal. The prediction, so far, has been singularly falsified, and if the trade should collapse now, after having had, for two full years since the abolition of the duty, a far more flourishing existence than it had ever previously known, its decay must be attributed to other and wholly different causes than the removal of a protective duty."

Alcohol. 484

In addition to sugar, alcohol is also prepared from beet-root. This is effected in three different ways: (*a*) by rasping the roots and submitting them to pressure, thereafter fermenting the expressed juice; (*b*) by maceration with water and heat; (*c*) by direct distillation of the roots. For full details of the preparation of beet sugar and beet spirit the reader is referred to *Spons' Encyclopædia, p. 1831. Tropical Agriculture by Simmonds* will also be found to contain (*p. 213*) an interesting account of beet sugar. An effort was made to introduce the cultivation of beet, for the purpose of sugar manufacture, into Kashmír, but the scheme came to nothing (*Stewart*).

MEDICINE. 485

Medicine.—The seeds have cooling and diaphoretic properties. **Bellew** says that the fresh leaves are applied to burns and bruises.

Betel leaf, see **Piper Betle,** *Linn.*; PIPERACEÆ.

Betel nut, see **Areca Catechu,** *Linn.*; PALMÆ.

486

BETULA, *Linn.; Gen. Pl., III., 404.*

A genus of small trees belonging to tribe BETULEÆ, of the Natural Order CUPULIFERÆ, comprising some 25 species, inhabitants of the cold temperate regions of Europe, Asia, and America.

Deciduous trees with serrate leaves, having resinous dots beneath. *Flowers* all in catkins; scales of the barren catkins ternate, the middle one bearing the stamens. *Perianth* absent; scales of the fertile catkin 3-lobed, 3-flowered, membranous, deciduous. *Female flowers*, 3 in the axil of each bract; bracts deciduous on fruiting, and generally membranous. *Ovary* 2-celled, each with 1 ovule; styles 2, filiform. *Fruit* naked, indehiscent, 1-celled, 1-seeded, membranous, winged.

The generic name is derived from *Betu*, CELTIC, or *Beithe*, GAEL.; the *Birch*, ENG.; and *Betula* or *Betulla*, LATIN; *Bhurja*, SANS.; and *Bhuj* and *Burich*, hill names in the Panjáb Himálaya.

487

Betula alba, *L.*

THE EUROPEAN OF WHITE BIRCH.

Habitat.—The white birch is common throughout Europe, Siberia, Asia Minor, and North America. It approaches nearer to the pole than any other tree, and frequents alpine regions where plants are scarce.

Properties and Uses—

This is perhaps one of the most useful trees of Northern Europe. A few of its properties may be here enumerated by way of comparison with our common Indian species. The wood is too soft to be employed as a building material, but it is valued by cartwrights, upholsterers, and turners for its tenacity. It is much used for firewood, and its charcoal is in high demand. The bark is impermeable to water, and very durable if submerged or kept below the soil, hence it is put to a variety of purposes, such as for utensils, shoes, cords, boxes, snuff-boxes, and for preserving roofs from moisture. A variety (or distinct species), **B. papyracea,** is made

Bark. 488

into the light and portable birch canoes of Canada; these are formed of slabs of birch bark bound together by the root-fibres of the white fir.

Birch bark contains an astringent principle used in tanning leather, a resinous balsamic oil which, by distillation, becomes empyreumatic and is used in the preparation of russia leather. The cellular part of the bark is rich in edible starch, and thus forms a valuable source of food to the Samoiedes and the Kamtschatkans. This sap is sugary before the sprouting of the leaves, and is considered an excellent antiscorbutic in North America, and both vinegar and beer are prepared from it. (*La Maout and Decaine's System of Botany.*)

Starch. 489
Sugar. 490

"Birch-tar is made to a small extent in Russia, where it is called Dagget, from the wood of **Betula alba,** *L.* It contains an abundance of pyro-catechin, and is esteemed on account of its peculiar odour, well known in the russia leather. A purified oil of birch-tar is sold by the Leipzig distillers." (*Flück. and Hanb., Pharmacog., 623.*) "The extraction of birch-bark oil is an industry of some importance in North Europe and Siberia, and is conducted in the following manner: An iron pot is filled up with bark, and covered with a close-fitting lid, through which is inserted an iron pipe. On this is inverted a smaller pot, and the rims are carefully fitted together and well luted with clay. The two are then turned upside down, so that the pot with the bark in it is uppermost. The apparatus is half sunken in the ground, well banked with a mixture of sand and clay, and a wood fire is kindled around it. When this distillation has continued long enough, the luting is removed, and the pots are separated, when the lower one is found to contain a thin oil floating on pyroligneous acid, or, when the bark has been impure, on pitch. The yield of pure birch-bark oil is about one third by weight of the white bark used." (*Spons' Encyclop.*) This property is apparently unknown to the natives of India.

Tar. 491

Medicine.—In Europe and America, birch oil has been found useful as a local application in chronic eczema. "The young shoots and leaves secrete a resinous substance, having acid properties, which, combined with soda, is said to produce the effects of a tonic laxative. The inner bark, which is bitterish and astringent, has been employed in intermittent fevers." "The leaves, which have a peculiar, aromatic, agreeable odour, and a bitter taste, have been employed, in the form of infusion, in gout, rheumatism, dropsy, and cutaneous diseases." (*U. S. Dispens., 15th Ed., 1587.*)

OIL. 492
RESIN. 493
Bark. 494
Leaves. 495

Betula acuminata, *Wall.; Brandis, For. Fl., 458; Gamble, Man. Timb., 372.*

496

Vern.—*Púya udish, hambar máya, makshéri, sheori, shag,* PB.; *Bhújpattra, háur, shául,* HIND.; *Haoul,* KUMAON; *Shakshin,* THIBET; *Saver, sauer, payong, útis,* NEPAL; *Hlosungli,* LEPCHA; *Dingleen,* KHASIA.

Habitat.—A large tree, met with in the Himálaya, from 6,000 to 8,000 feet, in the Khásia Hills, the mountains of Mánipur, and the Nagá Hills to Martaban.

Properties and Uses—

Fibre.—The bark when mature peels off in larger slabs than in any of the other species, and is therefore not so serviceable for the purposes to which the others are put.

FIBRE. 497

Food.—On the mountain tracts of North-East Mánipur, bordering on the Nagá Hills, the Lahúpas cut off the bark in large slabs just before the leaves appear. The inner layer of these slabs is carefully separated from the liber and sun-dried. This is either eaten like biscuits, or it is reduced to flour and cooked as an article of food. The tree is much prized by these naked savages, and in early spring yields a considerable

FOOD. Bark. 498
Prepared Biscuits. 499

portion of their diet. This remarkable fact does not appear to have been observed by any traveller, previous to my exploration in 1880 of the hill tracts of Mánipur, and apparently the nutritious properties of the bark have not been discovered by other Indian hill tribes. (See remarks under **B. alba.**)

TIMBER. 500

Structure of the Wood.—White, moderately hard, close-grained. Weight 41 lbs. per cubic foot.

It is very little used, but **Wallich** says it is hard and esteemed in Nepal for all purposes where strength and durability are required. "The wood is close-grained and takes a fine satin polish. It is particularly good for door panels, and the examples in the Government House at Naini Tál show that it is a valuable acquisition for ornamental work." (*Atkinson's Him. Dist.* (*X., N.-W. P. Gaz.*), *818.*)

501

Betula Bhojpattra, *Wall.; Brandis, For. Fl., 457; Gamble, Man. Timb., 372.*

THE INDIAN BIRCH TREE; INDIAN PAPER BIRCH.

Syn.—B. JACQUEMONTII, *Spach.*

Vern.—*Bhújpattra* or *bhujpatar*, HIND.; *Búrj, burzal, bhúj, phurs*, PB.; *Shák* or *shag, pád, phatak, takpa*, LADAK, LAHOUL, PITI, KANAWAR; *Takpa*, BHUTIA; *Phuspat*, NEPAL, TUZ., BHOTE; *Bhúrjapatra, bhojpatra*, BOMB.; *Bhuja patra*, CUTCH; *Bhojapatra*, GUJ.; *Bhurja putra*, SANS.; *Bhujapatri chettu*, TEL.

Habitat.—A moderate-sized, deciduous tree, found in the higher ranges of the Himálaya, forming the upper edge of arborescent vegetation, and ascending to 14,000 feet.

Properties and Uses—

FIBRE. Paper substitute. 502

Young twigs. 503

Fibre.—The bark is used as a substitute for paper by some of the hill tribes, and supposed by them to be more durable than paper. It is brought down to the plains and largely used in the manufacture of hookah tubes. The young branches are plaited into twig bridges. "The bark is well known as the material upon which the ancient Sanskrit manuscripts of Northern India are written. **Dr. Buhler**, in his account of a tour in Kashmír in search of Sanskrit manuscripts, says: 'The Bhurja MSS. are written on specially prepared thin sheets of the inner bark of the Himálayan birch, and invariably in Sârada characters. The lines run always parallel to the narrow side of the leaf, and the MSS. present therefore the appearance of European books, not Indian MSS., which owe their form to an imitation of the Talapattras. The Himálaya seems to contain an inexhaustible supply of birch-bark, which in Kashmír and other hill countries is used both instead of paper by the shopkeepers in the bazars, and for lining the roofs of houses, in order to make them water-tight. It is also exported to India, where, in many places, it is likewise used for wrapping up parcels, and plays an important part in the manufacture of the flexible pipe-stems used by hookah-smokers. To give an idea of the quantities which are brought into Srinagar, I may mention that on one single day I counted fourteen large barges with birch-bark on the river, and that I have never moved about without seeing some boats laden with it. None of the boats carried, I should say, less than three or four tons weight."

Books. 504

Pipe-stems. 505

"The use of the birch-bark for literary purposes is attested by the earliest classical Sanskrit writers. **Kalidasa** mentions it in his dramas and epics; **Susruta, Varahamihira** (circa 500-550 A.D.), know it likewise. **Akbar** introduced the manufacture of paper, and thus created an industry for which Kashmír is now famous in India. From that time the use of birch-bark for the purpose of writing was discontinued, and the method of preparing it has been lost. The preparation of the ink which was used for Bhurja MSS. is known. It was made by converting almonds into charcoal and boiling the coal thus obtained with *gomútra* (*Urina bovis*); this ink is not affected by damp or water. (Journal, (Bombay Branch), Royal Asiatic Society, Vol. XII., No. XXXIV.A.)" (*Dr. Dymock, Mat. Med., W. Ind., 602.*)

"The bark peels off in large sheets, and is used for umbrellas, for writing upon, and for the flexible tubes of hookahs. Every consignment of the ornamental papier-maché ware of Kashmír reaches the Panjáb packed in wrappers of birch-bark. The houses of Kashmír are often wrapped with it." (*Baden Powell, Panjáb Products, I., p. 569.*) "The bark is used for *chatta* or rude umbrellas, and for covering tubes of *hookahs*, or native smoking-pipes, and being of a sacred character it is burnt on the funereal pile. Hindú pilgrims visiting the shrine of Amrnath in Kashmír divest themselves of their ordinary clothes before entering the shrine, covering their bodies with the *bhojpattra*. It is now brought to the plains for lining the tubes of *hookahs*, and the leaves or bark are used to cover the baskets of Ganges water sold by itinerant pilgrims." (*Balfour's Cyclop.*)

Umbrellas. 506

Clothing. 507

Medicine.—The bark of the black birch is valuable for its aromatic and antiseptic properties. (*Murray.*)

MEDICINE. Bark. 508

Special Opinions.—§ "The decoction of the bark is used as a wash in otorrhœa and poisoned wounds." (*U. C. Dutt, Civil Medical Officer, Serampore.*) "The infusion of the bark is used as a carminative; it is prescribed also in hysteria." (*Surgeon W. Barren, Bhuj, Cutch.*) "Much used to write medicinal charms upon." (*Surgeon-Major W. Dymock, Bombay.*)

Charms. 509

Fodder.—The leaves are lopped for cattle fodder.

FODDER. Leaves. 510

Structure of the Wood.—White with a pinkish tinge, tough, even-grained, moderately hard. Weight about 44 lbs. per cubic foot.

TIMBER 511

It is extensively used in the inner arid Himálaya for building; it is elastic, seasons well, and does not warp.

"Wood good: used for cups, common turnery, and for fuel by travellers in the higher ranges." (*Baden Powell, Panjáb Products, p. 969.*)

Cups. 512

Fuel. 513

Betula cylindrostachys, *Wall.*

514

Vern.—*Shaoul*, KUMAON; *Sauer*, NEPAL; *Sungli*, LEPCHA.

Habitat.—A tall, deciduous tree, met with in Kumaon, Nepal, Darjiling Hills, from the Terai, up to 6,000 feet.

Structure of the Wood.—Red, hard, heavy, strong, and seasons well. Weight 52 lbs. per cubic foot.

TIMBER. 515

Seldom used except for firewood and charcoal, for which purposes it is very good. Experiments made by **Mr. Whitty** with several kinds of wood fuel for the Darjiling-Himálayan Railway showed that this was the best for locomotive purposes.

Railway fuel. 516

Bhang, see **Cannabis sativa**, *Linn.*; URTICACEÆ.

Bile of certain animals, see **Fel.**

BIOTA, *Endl.; Gen. Pl., III., 427.*

517

This genus has, by the *Genera Plantarum*, been reduced to THUYA, *Linn.*, which see.

Biota orientalis, *Endl.*; CONIFERÆ.

518

THE ABOR VITÆ.

Syn.—THUYA ORIENTALIS, *Linn.*

The Arnatto Dye.

Birch, Indian, see **Betula Bhojpattra,** *Wall.*

Birch oil, see **Betula alba,** *L.*

519

BISCHOFIA.

A genus of EUPHORBIACEÆ, containing only one species, a large, glabrous tree with trifoliate leaves and caducous stipules. *Flowers* diœcious or monœcious, in axillary panicles. *Calyx* of 5-valvate segments, those of the male flowers concave, enclosing the stamens at first, afterwards reflexed; those of the female flowers lanceolate. *Petals* none. *Stamens* 5, opposite the segments, and inserted round a raised circular central body (the rudimentary ovary), *filaments* very short. *Ovary* 3-celled, 2 ovules in each cell; *styles* linear, entire. *Fruit* a globose drupe, enclosing 3 indehiscent, 1-2-seeded cocci. (*Brandis, For. Fl., 445.*)

The genus is named after **Dr. A. Bischof.**

520

Bischofia javanica, *Bl.; Brandis, For. Fl., 446.*

Syn.—ANDRACHNE TRIFOLIATA, *Roxb., Fl. Ind., Ed. C.B.C., 1703.*

Vern.—*Kein, korsa, irum,* HIND.; *Kainjal,* NEPAL; *Sinong,* LEPCHA; *Taisoh, urúm,* MECHI; *Uriam,* ASS.; *Bolsuru,* GARO; *Joki,* CACHAR; *Boke,* BOMB.; *Thondi,* TAM.; *Govarnellu,* HASSAN; *Modagerri vembu,* TINNEVELLY; *Yagine* (?), BURM.

Habitat.—A deciduous tree met with in Kumaon, Garhwál, Oudh, Gorakhpur, Bengal, South India, and Burma.

TIMBER. 521

Structure of the Wood.—Red, rough, moderately hard, with a small darker-coloured heartwood. Weight $47\frac{1}{2}$ lbs. per cubic foot.

In Assam it is esteemed one of the best timbers and used for bridges and other works of construction. **Beddome** says it is used by planters in the Nilgiris for building, and is sometimes called *Red Cedar.*

Bitch or **Bish,** see **Aconitum ferox** and **A. Napellus,** *Wall.;* RANUNCULACEÆ.

Bitter-sweet, see **Solanum ducamara,** *Linn.;* SOLANACEÆ.

522

BIXA, *Linn.; Gen. Pl., III., 125.*

A genus of BIXINEÆ, containing one or at most only two species of large spreading bushes.

Leaves simple; stipules minute. *Flowers* in terminal panicles, 2-sexual. *Sepals* 5, imbricate, deciduous. *Petals* 5, contorted in bud. *Anthers* opening by 2 terminal pores. *Ovary* 1-celled; style slender, curved, stigma notched; ovules many, on 2 parietal placentas. *Capsule* loculicidally 2-valved, placentas on the valves. *Seeds* many, funicle thick, testa pulpy; albumen fleshy; embryo large, cotyledons flat.

The generic name is supposed to be derived from the vernacular name given to the plant by the Indians of the Isthmus of Darien.

523

Bixa Orellana, *Linn.; Fl. Br. Ind., I., 190.*

THE ARNATTO or ARNOTTO DYE; ROCOU (derived from *Urucu,* the Brazilian name), *Fr.*

Vern.—*Latkan, latkhan, watkana,* HIND., BENG.; *Koug kuombi,* SANTAL; *Jarat, jolandhar,* ASS.; *Gúlbas,* URIYA; *Powasi,* CHITTAGONG; *Reipom,* MANIPUR; *Shál-ké-pandú-ká-jhár,* DUK.; *Kisri, kesari, kesuri, sendri* or *shendri,* MAR., BOMB.; *Jáphara-chettu, jafra-vittulu-chettu, kurungu-múnji-vittulu-chettu,* TEL.; *Japhra-maram, jafra-virai-maram, kurungu-munjil-varai-maram,* TAM.; *Kuppa-mankala, rangamali, rangamali-hannu* (the fruit), KAN.; *Thidin, thi-deng,* BURM.

Habitat.—A graceful shrub, with handsome white or pinkish flowers and echinate red capsules; originally a native of America, now largely cultivated in India for the red or orange dye obtained from the pulp which surrounds the seed. Found in Pegu and Tenasserim. "Cultivated and escaped." (*Br. Burm. Gaz., I., 136.*) Extensively cultivated by the better class of ryots in Raipur, Central Provinces.

Botanic Diagnosis.—Two forms of this plant are equally plentiful in India, the one with white flowers and greenish capsules, and the other with pink flowers and red capsules. These cannot be regarded botanically as varieties, but they are recognisable, and, curiously enough, the natives of India regard the former as indigenous, while they readily admit that the latter is an introduction. **Roxburgh** even seems to have regarded the white-flowered form as indigenous, but modern botanists do not support this view. **Dr. Buchanan Hamilton,** a contemporary of **Roxburgh's,** published, in 1833, the following interesting account, from which it would appear **Dr. Buchanan** regarded the Arnatto as a recent introduction: "The **Bixa,** an American plant, is now rapidly spreading over Bengal, the inhabitants having found it a useful yellow dye, which they employ to give their cloths a temporary colour in the *Dolyatra* or festival of Krishna. With this also they colour the water, which, on the same occasion, they throw at each other with squirts. For these purposes it is well qualified, as the colour easily washes out, and the infusion has a pleasant smell. (Compare with facts given under **Abir.**) By them it is called *Lotkan,* and they say that before it grew commonly in the country, the dry fruit was brought from Patna. Probably some other fruit was then brought, and its use has been superseded by that of the **Bixa,** to which the natives have given the old name, as there can be no doubt of its being an American plant, and its fruit could scarcely have been brought here from the West Indies. In many parts it is called European Turmeric." (*Buchanan's Statistics of Dinajpur, p. 155.*)

There seems no doubt whatever that both forms of the plant were originally introduced from America, the white-flowered form having in all probability been longer in India. While plentiful everywhere around gardens and villages, it has nowhere gone wild, and is thus scarcely naturalised in India. It was used as a source of war paint in the West India Islands and Brazil before the discovery of America.

Properties and Uses—

DYE. 524

Dye.—The pulp gives a beautiful flesh colour, largely used in dyeing silks. It is altered by certain combinations into orange, deep orange, or red, the brighter orange and red colours being obtained in combination with red powder of **Mallotus philippinensis.** The dye is exported to Europe mainly from the West Indies, and is used chiefly to colour cheese and other edible articles, such as chocolate, &c. **Mr. Lisboa** says that milkmen sometimes use it to colour buffalo's milk so as to pass it off as cow's milk. (*Useful Plants of Bombay.*)

Butter and Cheese dye 525

Preparation of the Dye.—It may be extracted from the seeds direct, or the pulpy matter may by boiling be separated from the seeds and made into cakes like those of lac or indigo. In this form it is generally sold in Europe. "The mode in which it is obtained is by pouring hot water over the pulp and seeds, and leaving them to macerate and then separating them by pounding with a wooden pestle. The seeds are removed by straining the mass through a sieve, and the pulp being allowed to settle, the water is gently poured off, and the pulp put into shallow vessels, in which it is gradually dried in the shade. After acquiring a proper consistence it is made into cylindrical rolls or balls, and placed in an airy place to dry, after which it is sent to market. It used to be most common in this form of small rolls, each 2 or 3 ozs. in weight, hard, dry, and compact; brownish without and red within. The other process

Rolls. 526

BIXA Orellana.

The Arnatto Dye.

of manufacture is that pursued in Cayenne. The pulp and seeds together are bruised in wooden vessels, and hot water poured over them; they are then left to soak for several days, and afterwards passed through a close sieve to separate the seeds. The matter is then left to ferment for about a week, when the water is gently poured off, and the solid part left to dry in the shade. When it has acquired the consistence of solid paste, it is formed into cakes of 3 or 4 lbs. weight, which are wrapped in the leaves of the banana and known in commerce as flag Arnatto. This variety is of a bright yellow colour, rather soft to the touch, and of considerable solidity."

Cakes. 527

Paint. 528

"**Labat** informs us that the Indians prepare an Arnatto greatly superior to that which is brought to us, of a bright shining red colour, almost equal to carmine. For this purpose, instead of steeping and fermenting the seeds in water, they rub them with the hands, previously dipped in oil, till the pulp comes off and is reduced to a clear paste, which is scraped off from the hands with a knife, and laid on a clean leaf in the shade to dry. Mixed with lemon-juice and gum, it makes the crimson paint with which Indians adorn their bodies." (*Tropical Agriculture by Simmonds, pp. 388-89.*)

529

EUROPEAN PROCESSES.—Regarding the extraction of the dye, **Ure** writes: "**Leblond** proposed simply to wash the seeds of the **Bixa** till they are entirely deprived of their colour, which lies wholly on their surface; to precipitate the colour by means of vinegar or lemon-juice, and to boil it up in the ordinary manner, or to drain it in bags as is practised with indigo. The experiments which **Vanquelin** made on the seeds of the **Bixa**, imported by **Leblond**, confirmed the efficacy of the process which he proposed; and the dyers ascertained that the Arnatto obtained in this manner was worth at least four times more than that of commerce; that, moreover, it was more easily employed; that it required less solvent; that it gives less trouble in the copper, and furnishes a purer colour.

"Arnatto dissolves better and more readily in alcohol than in water when it is introduced into the yellow varnishes for communicating an orange tint."

530

CHEMICAL REACTIONS.—"The decoction of Arnatto in water has a strong peculiar odour and a disagreeable taste. Its colour is yellowish red, and it remains a little turbid. An alkaline solution renders its orange-yellow clearer and more agreeable, while a small quantity of a whitish substance is separated from it, which remains suspended in the liquid. If Arnatto be boiled in water along with an alkali, it dissolves much better than when alone, and the liquid has an orange hue." "The acids form with this liquor an orange-coloured precipitate, soluble in alkalis, which communicate to it a deep orange colour. The supernatant liquor retains only a pale-yellow hue." (*Ure's Dictionary.*)

531

SPECIAL OPINION.—§ "The pulpy part of the seed forms the arnatto of commerce. It is imported into England from Mexico, Brazil, &c., in two forms—in masses of 5 to 20 lbs., and as a homogeneous paste in casks of 4 to 5 cwt. The paste has the consistence of butter, and the odour of urine, which, it is stated, is added to keep it moist and improve the colour. At Cayenne, where the dye is largely manufactured, the ripe fruit is crushed and allowed to remain in water for several weeks. The mixture is then strained through coarse cloth; on the liquor standing, the colouring matter gradually settles. This is then collected and evaporated until it is of a pasty consistence. Improvements have been introduced in the manufacture, and the seeds, instead of being crushed, are washed with water and fermentation stopped by some re-agent. The colouring matter yielded by this method is in a fine state of division, and is known as *Bixin*. It is made into tablets. The colouring matter of arnatto is stated to consist of two colouring principles,—*orellin*, which is yellow, soluble in water, and which gives a yellow colour to cloth when mordanted; and *bixin*, which when pure forms a cinnabar-red powder, insoluble in water, but easily dissolved by alkaline solutions. Arnatto is employed only to a limited extent in dye-works, but it is often used to colour varnishes, cheese, butter, &c." (*Surgeon C. J. H. Warden, Prof. of Chemistry, Calcutta.*)

532 MORDANTS AND AUXILIARIES.—The mordant used with arnatto is most frequently crude pearl-ash; the alkali facilitates its solution, but the quantity of alkali used must be regulated according to the depth of colour required. The colour is, however, fleeting: it is chiefly used for silk, and seldom or never for woollen fabrics. After dyeing the silk with arnatto the colour may be deepened or reddened by means of vinegar, alum, or lemon-juice. The Mánipurís are said to use the fruit of **Garcinia pedunculata** for this purpose, a fruit which it is reported has at the same time the power of fixing the colour. This statement requires confirmation, since the dye is generally regarded as fleeting. The leaves of **Symplocos grandiflora** are used in Assam as a mordant with this dye (*G. Mann, Esq.*). The yellow tendency of the colour, produced through alkalis may be reduced on the addition of acids, the more natural red being produced, and restored again by further treatment with alkalis. Arnatto is entirely insoluble in acids, the colouring matter being precipitated; hence the necessity of using an alkali as the solvent as a first stage in the process of dyeing. **Dr. McCann** says: "The bark of this plant is used in Kuch Behar as a mordant in dyeing with Morinda."

Bark. 533

Fibre.—Bark yields a good cordage. (*Dymock.*) This is said to be used in the West Indies.

FIBRE. 534

Medicine.—Astringent and slightly purgative, also a good remedy for dysentery and kidney diseases. The pulp (a well-known colouring matter) surrounding the seeds is astringent. (*Roxburgh.*) The seeds are cordial, astringent, and febrifuge. (*Lindl.*)

MEDICINE. Seed pulp. 535 Seeds. 536

Structure of the Wood.—Wood pinkish-white, soft, even-grained.

TIMBER. 537

The friction of two pieces of this wood is said to readily produce fire; for this purpose it is used by the West Indians.

Tinder. 538

Bloodwood, Indian, or Jarul, see **Lagerstrœmia Flos-Reginæ,** *Retz.*; LYTHRACEÆ.

BLUMEA, *DC.*; *Gen. Pl., II., 289.*

359

A genus of annual or perennial, woolly or pubescent herbs, belonging to the COMPOSITÆ. This may be regarded as the Groundsels of India; they are only separable from LAGGERA by the tailed anther-cells.

Leaves alternate, usually toothed or lobed. *Heads* corymbose panicled or fascicled, rarely racemed, heterogamous, disciform, purple-rose or yellow, outer flowers many-seriate; female fertile, filiform, 2-3 toothed; disk flowers hermaphrodite, few fertile, tubular and slender, limb 5-toothed. Involucre ovoid or campanulate; bracts many-seriate, narrow, acute, soft or herbaceous, outer smaller; receptacle flat, naked. *Anther bases* sagittate, tails small, slender. *Style arms* of hermaphrodite flowers flattened or almost filiform, rarely connate with the adjoining anthers. *Achenes* small, subterete or angled, ribbed or not; pappus 1-seriate, slender, often caducous.

The genus contains about 60 species, natives of tropical and subtropical Asia, Africa, and Australia. It is named in honour of the distinguished Dutch botanist **Dr. Blume**, who, in 1828, published a *Flora of Java*. **Dr. Dymock** says that in Bombay the vernacular name *Bhamburda* is applied to all Blumeas.

Blumea aurita, *DC.*, see **Laggera aurita,** *Schultz-Bip.*; COMPOSITÆ.

540

Blumea balsamifera, *DC.; Fl. Br. Ind., III., 270.*

Syn.—CONYZA BALSAMIFERA, *Linn.*

Vern.—*Kakaróndá,* HIND.; *Kalahád,* GUJ.; *Bhamaruda,* MAR.; *Pon ma thein,* BURM.

Habitat.—A sub-bushy plant, met with on the tropical Himálaya, from Nepal to Sikkim, altitude 1,000 to 4,000 feet, extending to Assam, Khásia Hills, Chittagong, Burma, and the Straits.

Botanic Diagnosis.—A tomentose or villous woolly plant, stem tall, corymbosely branched above, leaves 4-8 inches, coriaceous elliptic or oblanceolate, usually silky above, serrate, sometimes pinnatifid, narrowed into a usually auricled short petiole; heads $\frac{1}{4}$-$\frac{1}{3}$ inch, sessile or peduncled in rounded clusters on the stout branches of a large spreading or pyramidal panicle; involucre bracts tomentose, receptacle glabrous; achenes 10-ribbed, silky; pappus red.

This belongs to the fifth section of the genus which is characterised by having numerous heads, large or small, forming narrow or broad terminal branched corymbs or panicles. Shrubs or small trees with large leaves.

MEDICINE. Plant. 541

Medicine.—The whole plant smells strongly of camphor, which may, indeed, be prepared from it. A warm infusion acts as a pleasant sudorific, and it is a useful expectorant in decoction.

542

B. densiflora, *DC.; Fl. Br. Ind., III., 269.*

THE NGAI CAMPHOR.

Syn.—B. GRANDIS, *DC.*

Vern.—*Pung-ma-theing, phúm-masin,* BURM.

Habitat.—Found in the tropical Himálaya, from Sikkim to Assam, Mishmi, the Nagá Hills, and the Khásia Mountains; also met with in the Tenasserim province of Burma.

Botanic Diagnosis.—Stem stout, panicle and leaves beneath densely tomentose or clothed with thick white felted wool, leaves 8-18 inches, broadly elliptic or elliptic-lanceolate, narrowed into a long-winged, sometimes appendaged petiole, puberulous above, serrate-toothed or pinnatifid, heads $\frac{1}{4}$ inch diam. Sessile, in rounded clusters on a large branched panicle; involucre-bracts narrow, rather rigid, receptacle narrow glabrous; corolla-lobes of hermaphrodite flowers hairy; achenes 10-ribbed, pubescent; pappus red.

This belongs to the same section as the preceding species.

Properties and Uses—

CAMPHOR. 543

Camphor.—A few years ago **Mr. E. O'Riley** prepared Camphor from this plant, which was pronounced identical with that imported from China. For further particulars see **Camphor.**

544

B. eriantha, *DC.; Fl. Br. Ind., III., 266.*

Botanic Diagnosis.—Pubescent or tomentose or clothed with scattered long hairs, rarely silky-villous, stems 1 foot, slender, dichotomously branched from the base; leaves 1-3 in., acutely and irregularly toothed, lower petioled obovate obtuse, upper sessile, obovate or oblong acute, heads small, $\frac{1}{4}$-$\frac{1}{3}$ in., mostly on the long slender peduncles of dichotomous cymes, rarely fascicled; peduncles and involucre clothed with long silky hairs, receptacle glabrous; achenes very minute, angles obtuse, sparingly silky.

This belongs to the fourth section of the genus or Blumeas with few heads, rarely many, $\frac{1}{4}$-$\frac{1}{2}$ in., usually peduncled and forming loose axillary and terminal corymbs often clustered. Pappus white.

Properties and Uses—

MEDICINE. 545

Medicine.—A specimen of what appears to be an extreme form of this species (or a new and undescribed species) was forwarded to me for identification by **Dr. Dymock** of Bombay. It is an erect plant with curious tufts of woolly hairs at the bottom of the stem. It appears that in Bombay the plant is used as a flea or insect powder (see **B. lacera**).

546

Blumea lacera, *DC.*; *Fl. Br. Ind., III.*, *263.*

Vern.—*Kakróndá, kukkurbandá, jangli-múli,* HIND.; *Kukursungá, bura-súksúng,* BENG.; *Nimúrdi,* BOMB.; *Jangli-kásní, jungli-mulli, divári-mulli,* DUK.; *Nárak-karandai, kattu-mullángi,* TAM.; *Káru-pógáku, advi-mulangi,* TEL.; *Kukuradru,* SANS.; *Kamáfitús,* ARAB.; *Mai-yagán,* BURM. *Kakróndá* and other vernacular names are applied to more than one allied species of **Blumea** and **Laggera** without much regard to the colour of their flowers. (*Moodeen Sheriff.*)

Habitat.—A common weed throughout the plains of India, from the North-West (ascending to 2,000 feet in the Himálaya) to Travancore, Singapore, and Ceylon.

Botanic Diagnosis.—This species is placed by the *Flora of British India* in the second section or species with many villous heads, $\frac{1}{4}$-$\frac{1}{2}$ in. in diameter, the heads being more or less clustered and forming dense oblong spikes or contracted panicles at the top of the stem, only exceptionally arranged in loose open corymbs. It smells strongly of turpentine, a character which, when taken along with the glabrous receptacle and yellow flowers, readily separates it from its nearest allies.

Stem erect, simple or branched, very leafy; leaves petioled, obovate-toothed or serrate, rarely lobulate; heads $\frac{1}{8}$ in., in short axillary cymes and collected into terminal spiciform panicles, rarely corymbose; involucre bracts narrow, acuminate, hairy; receptacle glabrous; corolla yellow; lobes of the hermaphrodite flowers nearly glabrous; achenes sub-4-gonous, not ribbed, glabrate.

The above diagnostic characters have been reproduced from the *Flora of British India*, in the hope that they may enable economic botanists to remove the ambiguity which still rests on the species of BLUMEA used for medicinal purposes. The yellow flowers of this species should at once separate it from **Laggera aurita**, with which it has been confused. **Dymock** says of the Bombay drug: "I am inclined to identify *kakronda* with **B. lacera.**" This opinion is supported by the fact that the author of the *Makhzan* describes the flowers of *kakróndá* as yellow. **Moodeen Sheriff** refers *kakróndá* to **Blumea (Laggera) aurita,** a plant with pink flowers. Through the kindness of **Dr. Dymock,** however, I had the pleasure of examining a specimen of the plant which he viewed as **B. lacera,** and regarding which he contributed a note to the *Pharmaceutical Journal, June 7th, 1884*. Along with my friend **Mr. C. B. Clarke** I have carefully examined this Bombay plant, and it appears to be a new species of Blumea not yet described, or an extreme form of **B. eriantha,** *DC.* It is certainly not **B. lacera,** *DC.*, and accordingly the vernacular names and economic information, as far as Bombay is concerned, should be removed from this position. It does not follow, however, that the *kakróndá* of Madras or of other parts of India is the same as in Bombay, and accordingly it has been deemed advisable to retain, for the present, the economic information in this position.

MEDICINE. 547

Medicine.—*Kakróndá* is used as a febrifuge, and also to stop bleeding, being regarded as deobstruent and stimulant. Mixed with black pepper it is given in cholera. An astringent eye-wash is made from the leaves.

"**B. lacera** is a perennial plant, with obovate, deeply serrated leaves and yellow groundsel-like flowers, the whole plant being thickly clothed with long silky hairs. The natives of the Konkan, near Bombay, call it *Nimúrdi*, and make use of it to drive away fleas and other insects. 150 lbs. of the fresh herb in flower was submitted to distillation in the usual manner with water, and yielded about 2 ounces of a light-yellow essential oil, having a specific gravity of 0·9144 at 80° F., and an extraordinary rotating power, 100 mm. turning the ray 66° to the left. **Mr. D. S. Kemp**, who made the observation, checked it by examining a 10 per cent. solution in alcohol, which gave 6·6.

"This Blumea is of interest as the possible source of an insect powder. I am forwarding a supply of the plant and a specimen of the oil to **Mr. Holmes** for experiment and also for identification, as the genus is a difficult one." (*Dymock in Pharm. Jour., June 7th, 1884.*)

Special Opinions.—§ "The expressed juice of the leaves is a useful anthelmintic, especially in cases of thread-worm, either taken internally or used locally." (*Surgeon J. Anderson, M.B., Bijnor.*) "Used by many Hospital Assistants and highly thought of by them as a febrifuge and astringent." "Is an invaluable remedy in Tinea tarsi. The juice of the fresh leaves is used as *Kajole* after removing the scales from the roots of the eyelashes. (*Asst. Surgeon Bolly Chand Sen, Campbell Medical School, Sealdah, Calcutta.*) "The fresh root held in the mouth is said to relieve dryness." (*U. C. Dutt, Civil Medical Officer, Serampore.*)

548

BOAT- AND SHIP-BUILDING—Woods used for.

A further list of woods of this nature, see CANOES.

Acacia arabica.
Albizzia Lebbek (in South India for boats).
Alseodaphne, sp.
Amoora Rohituka and **spectabilis.**
Anacardium occidentale.
Anogeissus latifolia (ships).
Artocarpus hirsuta (ships).
Barringtonia acutangula.
Bassia latifolia (country vessels)
B. longifolia (ships' keels).
Beilschmieda Roxburghiana (boats).
Berrya Ammonilla (used in Madras for masula boats).
Calophyllum inophyllum (masts, spars).
C. polyanthum (masts, spars, boats).
C. spectabile (masts, spars).
Capparis aphylla (knees of boats).
Carapa moluccensis (native boats).
Cassia siamea.
Celtis australis (oars).
Ceriops Candolleana (knees of boats).
Cinnamomum glanduliferum (boat-building),
Cordia Myxa (boat-building).
Dalbergia Sissoo (boats).
Dillenia indica.
D. pentagyna (ships).
Dolichandrone stipulata (oars and paddles).
Drimycarpus racemosus (boats).
Dysoxylum Hamiltoni (boats).
Eriolæna Candollei (paddles).
Eucalyptus Globulus (ships).
Fagræa fragrans (boats, anchors).
Fraxinus floribunda (oars).
Gmelina arborea (boats).
Grewia oppositifolia (oar shafts).
G. tiliæfolia (masts, oars).
Heritiera littoralis (boats).
Hibiscus tiliaceus (light boats).
Hopea, sp. (boat hulls).
Kydia calycina (oars).
Lagerstrœmia Flos-Reginæ (ship-building, boats).
L. microcarpa (ships).
Melanorrhœa usitata (anchor stocks).
Melia Azadirachta (ships).
Miliusa velutina (oars).
Morus cuspidata (boat oars).
Nectandra Rodiæi (ships).
Pentace burmanica (boats).
Pinus longifolia (bottoms of boats).
P. Merkusii (mast pieces).
Podocarpus bracteata (oars, masts).
P. latifolia.
Polyaithia cerasoides (boats).
Populus euphratica (boats).
Pterocarpus Marsupium (boats).
Salvadora oleoides (knee timbers of boats).
Sandoricum indicum (boats).
Shorea robusta.
S. stellata (boats).
Swietenia Mahagoni (ships).
Swintonia Schwenckii (boats).
Tectona grandis (ships).
Terminalia tomentosa (boats).
Thespesia populnea (boats).
Vateria indica (masts of native vessels).
Xylia dolabriformis (boats).

BŒHMERIEÆ.

549

An important tribe of fibre-yielding plants belonging to the Natural Order URTICACEÆ, in the sub-order URTICEÆ. To enable the reader to understand the position of the BŒHMERIEÆ, in the following pages will be found a brief account of the properties and uses of the URTICEÆ as a whole and an analysis of the genera, followed by a more detailed account of the BŒHMERIEÆ and of the genus BŒHMERA itself.

Affinities of the Urticeæ.

550

These may be popularly defined as the Nettle family. They have of course their closest affinities to the other sub-orders of the URTICACEÆ, but, as pointed out by **Weddell**, they may also be viewed as having many affinities to the TILIACEÆ, just as EUPHORBIACEÆ may be regarded as approaching the structural peculiarities of MALVACEÆ. To the general observer the coarsely serrate, hairy and opposite leaves of URTICEÆ suggest a strong external resemblance to many LABIATÆ. The affinity to TILIACEÆ is, however, more than in mere external appearance, since URTICACEÆ and TILIACEÆ may be viewed as affording the great majority of our bast or liber fibres such as Flax, Rhea, Jute, &c. This indicates a structural agreement which is fully illustrated by many other characters, such as the form, veination, and corrugation of the leaves, the stipules, definite inflorescence, valvate æstivation, 2-lobed anthers, and smooth pollen. Of the URTICACEÆ the sub-order URTICEÆ bear out this resemblance in the most marked degree.

Habitat of the Urticeæ.

551

The URTICEÆ are chiefly tropical as far as the distribution of genera is concerned. Europe is poorest in species, and so in India are the temperate altitudes of the Himálaya. But what is lost in the number of genera is compensated for by the great prevalence of individuals. URTICEÆ and PARIETARIA follow closely the haunts of man in the temperate regions, and in these situations cover relatively as much space as do the more numerous forms which inhabit the tropical and extra-tropical regions.

Economic uses of the Urticeæ.

552

But for the valuable fibres obtained from URTICEÆ the properties of the whole sub-order might be described as unimportant. The stinging hairs have been used as counter-irritants. There is little to justify the belief in the virtues of the calcareous salts contained in many nettles or of the nitrate of potash in the pellitories. The young twigs of certain species of URTICA, POUZOLZIA, DEBREGEASIA, and ELATOSTEMA are eaten as pot-herbs to a small extent, as are also the tubers of **Pouzolsia tuberosa.** From an industrial point of view, however, the liber fibres are exceedingly valuable. The various species of BŒHMERIA, but more particularly **B. nivea,** yield the rhea fibre of commerce—a fibre which we are accustomed to hear requires only the aid of some contrivance by which it can be conveniently and cheaply separated from the bark and deprived of its gummy substance, to become one of the most valuable textile fibres in the world. Next in importance may be mentioned the fibres from the species of VILLEBRUNEA, the *Bon rhea* of Assam; after these MAOUTIA, the *Poya* fibre; the fibres derived from the species of GIRARDINIA or Nilgiri Nettles, and those from URTICA, the true Nettles. All these and probably many others are deserving of an extended investigation, for there seems every likelihood that sooner or later one or more of them will meet the demand for new textile fibres. For cordage there are several deserving of the most careful examination such as the fibres of DEBREGEASIA, ELATOSTEMA,

FORSKOHLEA, GIRARDINIA, LAPORTEA, PILEA, POUZOLZIA, and SARCOCHLAMYS.

The subject of rhea fibre has been before the public since **Roxburgh** first drew attention to it in 1811. Many important experiments have been performed, but at the present moment the enquiry might not inaccurately be described as paralysed through the hitherto insurmountable obstacle which the tenacity of the fibre has offered, together with the difficulty of freeing it from the gummy substance. It would seem, however, that a serious mistake has been made in not having, coincidently with these mechanical experiments, instituted a thorough enquiry into the subject of the allied species which afford rhea or nettle fibres of varying economic value. *Spons' Encyclopædia* justly remarks : "Much remains yet to be done in identifying the varions BŒHMERIAS, which cover a very wide range, and in deciding which species or varieties will yield the most and best fibre adapted to Western wants" (*p. 932*). There are some 18 forms of BŒHMERIA itself found in a wild state in India, and including all the allied genera, there are no less than 45 species of plants, most of which doubtless yield fibres—45 plants so closely allied that they are all popularly viewed as wild forms of rhea. There are also 10 nettle plants which yield fibres, and in addition 31 fibre-yielding plants related to the two great groups which may popularly be said to be represented by the Rhea and the Nettle. Thus, belonging to the sub-order URTICEÆ there are in India 70 or 80 fibre-yielding plants all of which (with perhaps five or six exceptions) are utterly unknown to the European textile industries. This being so it would seem that had the experiments with rhea fibre been associated with a strict enquiry into the relative value of the fibres afforded by the entire sub-order URTICEÆ, a rhea fibre plant might have been discovered which would have rendered expensive and complex machinery unnecessary. (See *Bon-rhea,* first para. under **B. nivea.**) Anticipating this to be the line of enquiry likely to be taken up in the future, the brief botanical analysis of the genera of URTICEÆ, which will be found in subsequent pages, may be found useful. From the conflicting information in the writings of authors on Indian Economic science, due to the species of URTICEÆ not having been scientifically determined, it is very much to be regretted, however, that the description of the fibres under each of these genera is so exceedingly imperfect.

553

Structural variations of the Urticeæ.

The classification of the members of the URTICEÆ depends upon variations in the following characters:—

The **Stem** may be herbaceous, erect or creeping, or it may be subfruticose or woody and even arborescent.

The **Leaves** may be opposite or alternate, symmetrical or unsymmetrical at the base, equal or unequal, one being either very much smaller or even abortive, producing, in originally opposite leaves, an apparently alternate condition. They may be penninerved, 3-costate (*i.e.*, 3-nerved from the base), 3-plicostate (*i.e.*, two lateral nerves springing from the midrib so as to make the leaf appear 3-costate), or they may be 5-costate. The margin may be entire or variously toothed or incised. The surface may be glabrous or hairy, and the hairs may be stinging or not; it may be smooth, bulate (*i.e.*, corrugated or crumpled), or rough from the presence of variously-shaped ciystoliths. The leaves are stipulate (except in some PARIETARIEÆ, where the stipules are rudimentary or abortive); the stipules may be free, one on either side of the petiole or interpetiolar (united into one between the petioles of opposite leaves) or intrapetiolar (*i.e.*, axillary), free or united into an entire or more or less bifid ligulate body; the stipules may also be caducous or persistent.

The **Inflorescence** may be described as definite axillary cymes, solitary or grouped in the axils, simple or ramified, composed of simple or compound racemes or spikes or compressed into capitula, symmetrical or unilateral. The axis or receptacle may accordingly be elongated, filiform or flattened, or concave, resembling the receptacle of the fig.

The **Flowers** are declinous [*i.e.*, unisexual, and therefore either monœcious (on one plant) or diœcious (male flowers on one individual and female on another)], or they may be polygamous (*i.e.*, declinouse and hermaphrodite flowers on the same individual plant). The flowers may be sessile and grouped together, forming glomeruli or pedunculate, the peduncule or pedicel having often one or two joints, especially in the male flowers. The inflorescence may be naked or bracteate, the bracts being either small or large and foliaceous, free or connate, often forming an involucre in the capitulate forms. The sepals are generally regular in the male flowers, 5- 4- 3- or even 2- or 1-merous, free or more or less connate, valvate or imbricate. In the female flowers they are less regular and generally fewer in number and more frequently connate, even when they are free in the male flowers.

The **Stamens** are generally of the same number as the sepals of the male flowers, opposite and often uncoiling elastically: in the female flowers they are occasionally represented by hypogynous staminodes or abortive stamens. When the perianth is adnate to the ovary a sort of perigynous condition is produced, but the ovary is not, strictly speaking, inferior, since the union of the perianth tube to it is only partial and easily separable.

The **Pistil** is rudimentary in the male flowers, but the shape and form or hairiness of this rudiment is often of importance. In the females it varies considerably: it is flat, smooth, glabrous, or granulated; it is free from the perianth or united to it, the persistent perianth tube often becoming succulent and causing the achene to appear like a drupe. The style may arise from the apex or not; it may be short or long, elongated or filiform or capitate, and papillose or hairy.

ANALYSIS OF THE INDIAN GENERA OF URTICEÆ.

I. STINGING NETTLES. 554

Tribe I. Urereæ.—Herbs, under-shrubs, rarely trees, *with stinging hairs*, and opposite, decussate, or alternate spiral leaves. Flowers in cymes; male perianth 4-5-merous, rarely 2-3-merous; ovary rudimentary; female perianth 2-5-lobed or partite, *free from the ovary.*

* *Achene erect; leaves opposite, stipules lateral, free or united and interpetiolar.*

1. **Urtica.**—*Male perianth* 4-merous, exterior segments small. *Stigma* papillose-capitate.

** *Achene oblique; leaves alternate; stipules intrapetiolar (axillary), very frequently united.*

2. **Fleurya.**—Annual herbs. *Flowers* glomerulate, forming racemose spikes or panicles. *Female perianth* of 4 segments, equal or unequal, one large, hooded, and furnished with a stinging hair. *Stigma* ovate or linear.

3. **Laportea.**—Under-shrubs or trees or perennial herbs. *Flowers* glomerulate panicled, rarely racemes. *Female perianth* 4-lobed, equal or unequal, persisting, almost unchanged around the fruit, reflexed. *Stigma* filiform.

4. **Girardinia.**—Erect herbs or almost under-shrubs. *Flowers* in glomerulate spikes or sub-paniculate spikes; when fruiting coveredwith

long stinging hairs. *Female perianth* ovoid-tubular, bifid, the upper lip 2-3 dentate, the lower almost abortive. *Stigma* subulate.

II. STINGLESS NETTLES. 555

Tribe II. Procrideæ.—Herbs, rarely woody below, *unarmed (e.g., hairs not stinging)*; leaves opposite or by abortion *alternate and then often distichous.* Flowers forming capitate cymes or arranged upon a discoid receptacle; male perianth 4-5-merous, rarely 2-3-merous; ovary rudimentary; female perianth 3-5-partite, *free from the ovary.* Staminodes sometimes present.

* *Leaves opposite, one unequal or imperfect.*

5. **Pilea.**—Under-shrubs or herbs, erect or prostrate. *Flowers* forming cymose-capitula or lax racemes. *Female perianth* 3-partite, one large, hooded, glandular or scaly, staminodes often at the base of the sepals; achene included with the succulent calyx or exserted. *Stigma* short, penicillate.

** *Leaves alternate distichous, unsymmetrical, a large leaf usually alternating with a small bract-like or abortive one.*

6. **Elatostema.**—Under-shrubs or perennial or annual herbs. *Flowers* collected on a regular or irregular discoid receptacle. *Female perianth* small, abortive or absent. *Stigma* sessile—a brush of caducous hairs.

7. **Procris.**—Sub-succulent and often epiphytic under-shrubs, erect, usually glabrous. *Flowers,* males forming glomerules, arranged in lax cymes, rarely capitula; females capitulate, collected upon a globose fleshy receptacle. *Female perianth* small or 3-4-partite, becoming fleshy and enclosing the ovary.

III. STINGLESS NETTLES. 556

Tribe III. Bœhmerieæ.—Under-shrubs or trees, rarely herbs, unarmed, with opposite or alternate leaves. Flowers collected into glomeruli or scattered ex-involucrate or with small scarious bracts, forming axillary solitary or ramified spikes or cymes; male perianth 4-5-merous, rarely 2-3-merous; ovary rudimentary; female perianth most frequently tubular, mouth contracted, 2-4-toothed, including and sometimes adherent to the ovary or very short or even absent.

* *Fruiting perianth membranaceous or moist, achene included, free.*

8. **Bœhmeria.**—Under-shrubs; leaves opposite or alternate. *Flowers* monœcious or diœcious, male and female in separate inflorescences constituting sessile glomeruli in the axils, or forming secund spikes, or arranged in branched panicles. *Stigma* filiform, persistent.

9. **Chamabaina.**—A diffuse herb with opposite leaves. *Flowers* forming axillary glomeruli. *Stigma* ovate, persistent.

10. **Pouzolzia.**—Herbs sometimes woody at the base; leaves alternate, rarely opposite. *Flowers* occasionally diœcious, arranged in axillary glomeruli or spikes, in the monœcious forms male and female flowers are often on the same inflorescence; male perianth 4-5-merous, rarely 3-merous. *Stigma* filiform, deciduous.

11. **Distemon.**—Herbs with alternate leaves. *Flowers* glomerulate, arranged in simple spikes; male perianth 2-merous, rarely 3-merous. *Stigma* linear, deciduous.

** *Fruiting perianth very often fleshy, free or adnate to the achene.*

12. **Sarcochlamys.**—A shrub with alternate leaves, rough above, white tomentose below. *Flowers* glomerulate, forming axillary solitary or paired spikes, males lax, females dense flowered; male perianth 5-partite, abortive. Ovary lanate. Fruiting perianth oblique, accrete, gibbous, mouth contracted, lateral and dentate. *Stigma* short, penicillate.

*** *Fruiting perianth adnate to the ovary.*

13. **Villebrunea.**—Under-shrubs or trees, with alternate leaves. *Flowers* diœcious, forming capitulate glomeruli, sessile in the axils of the leaves or fascicled, lax, dichotomous cymes; fruiting perianth thin, fleshy. *Stigma* sub-pellate, sessile, penicillate, ciliate.

14. **Debregeasia.**—Shrubs with alternate leaves. *Flowers* axillary, sessile glomeruli, or numerous cymes. *Female perianth* minutely toothed at the contracted mouth, in fruit becoming succulent. *Stigma* penicillate.

**** *Female perianth minute or absent.*

15. **Maoutia.**—An under-shrub with alternate leaves. *Flowers* glomerulate, small-panicled. *Female perianth* minute or absent. *Stigma* penicillate.

16. **Phenax.**—Delicate under-shrubs. *Flowers* glomerulate, sessile in the axils, bracts prominent, ferruginous. *Female perianth* absent. *Stigma* elongated.

IV. STINGLESS NETTLES. 557

Tribe IV. Parietarieæ.—Herbs or under-shrubs, rarely shrubs, unarmed. Leaves alternate, entire. Flowers diclinous or polygamous, 1-3, rarely numerous, included within an involucre of free or connate bracts; bracts sometimes only 2. Femal perianth tubular, free.

17. **Parietaria.**—Herbs often diffuse; stipules small or absent. *Cyme* axillary, 3-8-flowered, androgynous or polygamous; *bracts* free, herbaceous, involucrate. *Style* elongated, crowned with a papillose stigma.

V. STINGLESS NETTLES. 558

Tribe V. Forskohleæ.—Herbs with non-stinging hairs; leaves alternate or opposite. Flowers diclinous, grouped in the axils of the leaves, generally involucrate. Male flowers irregular; female perianth free from the achene, which it completely encloses, or absent.

18. **Forskohlea.**—Under-shrubs or tough herbs, covered with hooked hairs; leaves alternate, more rarely opposite, crenate or dentate; stipules lateral, free. *Flowers* contained within a campanulate or tubular involucre; perianth of both males and females tubular below, obtusely 3-dentate, densely lanate within.

19. **Droguetia.**—Differs from **Forskohlea** in the flowers being generally solitary or arranged in terminal spikes. *Male perianth* campanulate, shortly toothed.

Note.—For further information regarding the above genera, consult their respective alphabetical positions.

559

BŒHMERIA, *Jacq.; Gen. Pl., III., 387.*

A genus of URTICACEÆ, comprising about 45 species of small trees or shrubs, inhabiting the sub-tropical and tropical regions of Asia and America. There are some 18 species met with in India, the most prevalent of which may be said to extend from Nepal through Sikkim to Assam, the Khásia Hills, Cachar, Burma, and Ceylon. Only three species can be said to be more generally distributed, reaching the outer North-West Himálaya as far west as Garhwal and extending to the plains and lower hills of Western India, while none occur in plains and hills of the Panjáb proper.

Leaves opposite or alternate, sprinkled with inconspicuous punctiform cystoliths equal or unequal, dentate (very rarely 2-lobed), 3-nerved petiolate; *stipules* axillary, free or less frequently connate, deciduous. *Flowers* minute, uniesxual, aggregated into elongated axillary solitary heads or clusters (glomeruli), scariously bracteate or forming axillary spikes or branching racemes or cymose panicles. *Flowers* monœcious or diœcious; *male perianth* 4-partite or -lobed (very rarely 3- or 5-parted); lobes leafy, ovate, sub-acuminate or mucronate, valvate in bud. *Stamens* 4, opposite the perianth lobes and inserted below the clavate or sub-globose rudiment of the gynæcium, glabrous or shortly lanate at the base. *Female perianth* gamophyllous, tubular or saccate, compressed or ventricose, 2-4-dentate at the contracted mouth. *Ovary* sessile or stalked, enclosed by the perianth tube and sometimes even adnate to it, tapering into the elongated filiform persistent *style; stigma* papillose on one side of the style; *ovule* orthotropous, solitary, sub-eerect or ascending. *Achene* enclosed in the mucrescent perianth and often cohering to it; pericarp crustaceous and thin or nut-like; *albumen* more or less copious; *cotyledons* of the fleshy embryo elliptical, usually a little longer than the conical radicle.

560

Affinities of Bœhmeria.—This genus belongs to the Tribe URTICEÆ and to the Sub-tribe BŒHMERIEÆ; the Bœhmerieæ may be referred to four series of allied genera. All the species of Bœhmeria receive popularly the name of the Rhea or Grass-cloth fibre plants, and, indeed, the bushy or herbaceous members of two or three other allied genera equally fall within that designation, since they all yield delicate, white, silvery and exceedingly strong fibres. It seems likely, however, that the true rhea fibre is the produce alone of **B. nivea.**

561

B. squamigera has been formed into the genus CHAMABAINIA, on account of the stigma being capitate instead of linear—a distinction which seems scarcely worthy of such importance; the leaves are also opposite. In POUZOLSIA the style is filiform, articulated and caducous, and the fruit enclosed by the winged or costate calyx. In **Distemon** the leaves are alternate, the male flowers 2-merous, rarely 3-merous, and the stigma linear deciduous. The members of these three genera, with **Bœhmeria** itself, may be regarded as the series of true Rhea-fibre plants—the **Eubœhmeria** of botanists, characterised by having the tubular female perianth free or adherent to the ovary, dry or membranous in fruit, and with 2 or 4 apical teeth.

562

In the second series, **Sarcochlamydeæ,** the female calyx is free, with a lobed or dentate mouth, fleshy and succulent around the fruit. In India we have only one species belonging to this series, *viz.*, **Sarcochlamys pulcherrima,** *Gaud.;* this is met with in Assam, the Khásia Hills, Sylhet, Chittagong, and Sumatra.

563

In the third series, **Villebrunieæ,** the following genera are represented in India: **Villebrunea,** 3 species; **Debregeasia,** 3 species. These are recognised by having the female calyx adnate to the ovary, with a short dentate or sub-entire limb.

564

The fourth series, **Maoutieæ,** is characterised by the calyx being rudimentary or absent. There is only one Indian species of any importance belonging to this series, *viz.*, **Maoutia Puya,** *Wedd.*, the *Poi* fibre of Assam.

Note.—The above remarks regarding the various genera of Indian rhea-fibre plants have been given in this place in the hope that they may prove useful to persons desirous of discovering the correct botanical sources of the fibres which in the different provinces of India go by the name of rhea. For fuller details consult the genera mentioned in their respective alphabetical positions in this work, and compare with brief botanic diagnosis of these genera already given in the preceding pages.

565

Bœhmeria caudata, *Poir. (non Swartz.).*

Syn.—A form of **B. platyphylla**?, *Don*, and not a distinct species, probably *var.* **macrostachya.**

Habitat.—A large shrub frequent in Chittagong and Ava (*Kurz*).

Botanic Diagnosis.—*Leaves* opposite, sharply crenate-serrate; stipules lanceolate, acuminate. Female perianth elliptical, obovate or roundish.

FIBRE. 566

B. comosa, *Wedd.; DC. Prod., XVI., I., 205.*

Syn.—B. DIFFUSA, *Wedd.; Kurz, For. Fl. Burm., II., 423;* U. COMOSA, *Ham.*

Habitat.—A leaf-shedding small shrub, about 2-4 feet high, frequent in the mixed open forests all over Burma, ascending to 3,000 feet in altitude, and extending west to the Khásia Hills, Sikkim, and Nepal.

Botanic Diagnosis.—*Leaves* 2-6 inches long, ovate lanceolate, long acuminate, crenate-serrate; stipules linear lanceolate, deciduous. *Glomerules* axillary spicate. *Female perianth* compressed lanceolate to obovate, 2-4-toothed; *stigmas* twice as long as the tube.

FIBRE. 567

Bœhmeria cuspidata, *Bl. (non Wedd.).*

FIBRE. 568

Habitat.—Nepal.

A species apparently of no importance and very little known.

B. Didymogyne, *Wedd.; DC. Prod., XVI., I., 204; Kurz, II., 423.*

FIBRE. 569

Syn.—DIDYMOGYNE BŒHMERIOIDES, *Wedd.*

Habitat.—A herbaceous, glabrous bush, said to be found in Moulmein.

Botanic Diagnosis.—*Leaves* alternate, 2-4 inches long, crenate-serrate from the middle. *Female perianth* becoming oblong narrowed upwards, enclosing 2 carpels, each with a distinct style (*according to Weddell*).

B. Helferii, *Bl.; D.C. Prod., XVI., I., 204; Kurz, II., 423.*

FIBRE. 570

Habitat.—A bush with branches having adpressed pubescence, met with in Tenasserim. (*Kurz, Burm. Fl., II., 423.*)

B. lobata.

FIBRE. 571

The *ullah* sold for hemp at Almora, and is common in Garhwál and Kumaon (*Baden Powell*). I have been unable to recognise the plant referred to by **Mr. Baden Powell**, but it must be of considerable importance, since it is mentioned by several writers upon Indian Economic Science. The name **B. lobata** does not occur in botanical works.

B. macrophylla, *Don; Brandis, For. Fl., 403.*

572

Syn.—URTICA PENDULIFLORA, *Wall.*

Vern.—*Saochála, golka,* KUMAON; *Kamli,* NEPAL.

Habitat.—This broad-leaved shrub is met with from Kumaon eastward through Nepal and Sikkim to the Khásia Hills, altitude 4,000 feet. Flowers in August to September.

Botanic Diagnosis.—*Branches* 4-angled, with short adpressed hairs. *Leaves* opposite, long lanceolate, pustulate-rugate above, the pustules terminated by a gland, softly pubescent beneath, obtusely serrate 3-costate, the lateral nerves extending through little more than the lower half, the remainder penninerved from the midrib; stipules lanceolate, hairy on the midrib; petioles strigose, 1 in. long. *Flowers* monœcious in long drooping axillary spikes, the clustered flowers in the axils of lanceolate bracts.

Fibre.—Its bark yields a beautiful fibre, much prized for fishing-nets.

FIBRE. 573

B. malabarica, *Wedd.; DC. Prod., XVI., I., 203; Kurz, For. Fl., Burm., II., 422.*

574

Syn.—URTICA MALABARICA, *Wall.*

Vern.—*Takbret,* LEPCHA; *Maha-deya-dúl,* SINGH.

Habitat.—A shrub 4 feet in height, or sometimes a small tree 20 feet high, met with in the Carnatic, the Konkan, Sylhet, the Khásia mountains and lower Himálaya, extending to the tropical forests of Arracan. Plentiful in the moister tropical and extra-tropical forests of India and Burma; very common in Ceylon.

FIBRE. 575

Fibre.—The liber yields a strong fibre. **Kurz** says, "The liber of this and of most Bœhmerias yields a strong fibre." **Thwaites** says that the Singhalese make fishing-lines from the fibre.

576

Bœhmeria nivea, *Hook. & Arn.; Wight, Ic., t. 688; Hooker's Jour. Bot., III. (1851), 315, t. 8.*

Syn.—URTICA NIVEA, *Linn. var.* CANDICANS, *Wedd. in DC. Prod., XVI., I., 207;* B. TENACISSIMA, *Gaudich;* B. CANDICANS, *Hassk.;* URTICA CANDICANS, *Burm.;* U. TENACISSIMA, *Roxb., Fl. Ind., Ed. C.B.C., 656.*

References.—*Brandis, For. Fl., 402; Spons' Encyclopædia, 921; Drury, U. P., 81; Atkinson's Him. Dist., 797; Baden Powell's Pb. Prod., I., 503; Hem Chunder Kerr's Report on Jute, pp. 4-5; Lindley and Moore's Treasury of Botany; Report on Rhea Fibre by Dr. Forbes Watson, 1875; reprinted with a lecture, 1884; Rhea, by W. H. Cogswell, Agri.-Hort. Soc. of India, Vol. VII., Part II., 1884; The Fibrous Plants of India, Dr. Royle, 1855; Cyclopædia of India, Dr. Balfour; The Ramie, Theo. Moerman; The Indian Forester, Feb. 1884; "The Tropical Agriculturist," Feb. 1884; Records of Govt. of India, Rev. and Agri. Dept.*

Comm. Names.—RHEA, CHINA-GRASS, *Eng.;* RAMIE, ORTIE BLANCHE SANS DARDS DE CHINE, *Fr.;* RAMEH, RAMIE, *Java, Malay.*

Vern.—*Schou* or *schu* or *tchou* (the plant), *schou-ma* (fibre of the schou), CHINESE; *Tsjo, siri, so, mao, karao, akaso,* JAP.; *Klooi, caloee, ghoni,* SIAM AND SUM.; *Kankhúra,* BENG.; *Rhea,* ASS.; *Poah,* NEPÁL; ? *Goun,* BURM.

For *Bon-rhea,* Ass., see **Villebrunea appendiculata,** *Wedd.; DC. Prod., XVI., I, 235*[25]. **Kurz** regards the *Bon-rhea* as the China-grass cloth, which would thus be quite distinct from the Rhea fibre proper. If this be correct, we have in India been trying to produce from the wrong plant a fibre to compete with the Chinese grass-cloth. This might account for the fact that the samples of Indian rhea fibre exported to Europe have uniformly been pronounced inferior to the China fibre. It seems highly desirable that the grass-cloth of China should be carefully looked into with the object of confirming the opinion which generally prevails that it is obtained from the same species as the Rhea fibre of India. (*Compare with pages 461 and 469.*)

Habitat.—A shrub indigenous in India, and probably also in China, Japan, and the Indian Archipelago.

Botanic Diagnosis.—*Branches* terete, herbaceous, and with the petiole tomentose from long, soft, spreading hairs. *Leaves* alternate, broad ovate 3-6 in. long, acuminate, dentate, with large triangular slightly curved teeth, base truncate and tapering suddenly into the petiole, which is half the length of the blade or longer; upper surface of the leaf rough, pubescent, the under white, densely matted with closely adpressed hairs. *Flowers* green, monœcious in axillary panicles; panicles in pairs, shorter than the petiole, bearing numerous sessile flower-heads along their entire length. *Female panicles* lax-branched, with rounded glomeruli (covered with the long styles), occurring in pairs in the axils of the upper, male in the axils of the lower leaves. *Style* much exserted, hairy. *Ovary* enclosed completely by the tubular, hairy, 4-toothed female perianth.

Many unfortunate mistakes occur in the literature of this species, some of which have greatly tended to retard the development of the rhea fibre industry. The plant has been confused with many other widely different species. **Baillon**, for example, in his *Natural History of Plants,* Vol. III., p. 503, gives an illustration of a plant which, apparently by mistake, is said to be **B. nivea**; the leaves are opposite instead of alternate, and the inflorescence is not that of this species. *The American Agriculturist,* January 1884, reproduces an old plate of **Mauotia Puya** as an illustration of **Bœhmeria nivea,** &c.

RHEA FIBRE.

Cultivation and Preparation.

577

Where Cultivated.—Assam, Eastern and Northern Bengal, also in Saharunpore and Calcutta Botanic Gardens; introduced by Agri-Horticultural Societies into Madras and Rangoon for experimental purposes.

It has also been cultivated in Natal, Mauritius, Algeria, in the Island of Corsica, South France, the Channel Islands, and even in Great Britain.

578

Soil.—The rhea plant is exceedingly hardy, and thrives in almost any description of soil. But preference should be given to a rich, light, sandy loam, well worked and sufficiently shady. The subsoil should be good, as the roots penetrate 12 to 14 inches deep in search of nutrition.

579

Climate.—For profitable working, a situation should be chosen which would promote the quickest growth of the stems, and yield the greatest number of cuttings with the best quality of fibre. A situation fulfilling these conditions would most probably be found in a tropical climate with a moist atmosphere and fairly good rainfall. It would succeed in almost any part of the tropical plains of India.

580

Preparation of the Soil.—The land, if not naturally rich, should be manured; it should also be ploughed to a considerable depth, and tilled lightly so as to remove the weeds. Furrows or small trenches 3 feet apart should then be made, and the land kept ready to receive rhea roots or cuttings by the end of the rainy season. An analysis of rhea shows that the most favourable manure should contain nitrate of soda, sea-salt, and lime. Valuable information as to the cultivation of the grass-cloth plant in China, and the extraction of the fibre, appears at pages 359-362 of **Dr. Forbes Royle's** *Fibrous Plants of India*, 1855, having been translated from a Chinese treatise into French by **M. Stanislas Julien** and retranslated into English by **Dr. Royle.**

581

Planting and Care of the Crop.—Rhea is easily propagated. It grows readily from root or stem cuttings and from seed. Supposing the mode of propagation by root-cuttings to be adopted, the young lateral shoots with their roots should be cut off and planted in furrows before the end of the rainy season, to a depth of 3 inches; a little watering may be necessary should the weather be dry. It will be found that plants will grow rapidly to a height of 4 or 5 feet; that the roots will become stronger every year, the plant being perennial. The first crop may be ready in two months from the date of planting out, especially in favourable situations. There are many advantages in a rhea crop: it is perennial, and does not therefore require to be renewed every year. It resists variations in temperature owing to the roots penetrating into the subsoil. Year by year the roots spread, becoming stronger and more productive. The crop is never destroyed by caterpillars or other insects, owing to the quantity of tannin which the bark contains; and lastly, three or four cuttings may be taken off the same ground every year. But it has a serious disadvantage in that it is one of the most exhausting crops known, requiring the land to be left fallow before anything else can be put on the same field after the removal of the crop.

582

Cutting the Rhea.—Some experience is necessary to decide the right time for cutting. As a general rule, care should be taken to effect the cutting before the plant becomes covered with a hard or woody bark, the formation of which is indicated by the green skin turning brown, the discoloration commencing at the bottom of the stem. A practical way of finding whether the plant is ready for cutting is to pass the hand down the stems from top to bottom. If the leaves break off crisply, a crop of cuttings may be taken off the plants. **Dr. Forbes Watson** says that the

BŒHMERIA nivea.

Rhea Fibre.

CULTIVATION.

plants are ready for cutting when 3½ to 4 feet in height. "If the length is not more than 2 feet, the fibre is very fine, but the chances are you get more waste, and not such a good percentage of fibre. In the long stems the fibre is not so fine as in the medium ones." Care should be taken, however, not to remove more than can be treated for extraction of the fibre within the 24 hours. "Experience," says **Mr. Theo. Moerman**, "has enabled us to establish the fact that the fibre of the second cutting is superior to the first, and that in every instance it is preferable to cut the stalks before the plant flowers and before it is completely mature in order to obtain a finer and softer fibre."

583

Outturn and Cost of Production.—About four or five cuttings can be had from the same ground a year. The best crops are those cut in June to August; the February crop yields the strongest fibre. **Major Hannay** reported that in Assam "The average crop of one Assam *poorah* (1⅓ acres) well manured, and with a full crop of stems or reeds, was from 10 to 12 maunds." (*Calcutta Review, 1854.*) But he omitted to explain whether this was the weight of stem or of fibre, or whether it was the yield of one or more cuttings. Another writer in the *Review* added, however, to this statement, the notice of an experiment made in the vicinity of Calcutta in 1854, and said: "A plot of ground containing 550 square yards gave on an average cutting 301¾ lbs. of sticks, from which was obtained 11 lbs. of fibre. Now, 550 yards is almost one ninth of an acre; but not to overstate the returns, this may be estimated at one eighth. Hence 11×8=88 lbs. per acre, which, again, multiplied by 4, the number of cuttings, would give yearly per acre 352 lbs. of fibre."

584

Dr. Forbes Watson says: "I am aware that there are some notable statements which have been founded upon experiments made in Algiers. Estimates have been made, showing that you could get forty tons per acre, but I think these require to be verified before we can accept them. Anyway, I do not see that we can conclude at the present—I hope I shall be mistaken—that each crop will yield more than 250 lbs. per acre. You may, however, obtain three crops, or even four, in the year, which would bring it to 1,000 lbs. per acre." **Theo. Moerman**, in his little book on "The Ramie," says that the annual yield of fibre per acre is five or six times greater than the quantity which the cotton plant produces in the best seasons and in the most favourable climates. **Mr. J. Bruckner**, of New Orleans, estimated from personal experience that each cutting of the Ramie, after the plant has reached the height of 3 or 4 feet, produces from 600 to 800 lbs. of retted disintegrated fibre per acre. Supposing the crop in question to give three cuttings in the year, the total outturn per acre would be from 1,800 to 2,400 lbs. of fibre. **M. Edouard Nicolle** of Jersey, however, affirms (says **Mr. Moerman**) that in his Ramie plantations he obtains annually at Jersey three crops which yield a total of "11,250 lbs. of raw fibre (or bark separated from the centre wood of the stalks), which gives him from 5,000 to 7,875 lbs. of fine fibre ready to be combed out and used in filatures." There must apparently be some mistake, for further on **Mr. Moerman** makes **M. Nicolle** say that he obtained a total annual return of 5,625 lbs. of the fibre containing bark, which "are equal to a minimum of 3,375 lbs. of well-retted and thoroughly-cleaned fibre ready for use at filatures."

585

In China, according to *Spons' Encycl.* (p. 922), "The stems are gathered for industrial purposes in the first year when about 1 foot high. In the tenth month of every year, before cutting the offsets, the ground is covered with a thick layer of horse or cow dung; in the second month the manure is raked off, to allow the new shoots to come up freely. In the second year the stems are again cut. At the end of three years, the roots are very strong, and send up many shoots. Cropping then takes place three times a year, the stems being cut when the suckers from the root-stock are about ½ inch high. The first harvest is got in at about the beginning of the fifth month; the second in the middle of the sixth or beginning of the seventh month; and the third, in the middle of the eighth or beginning of the ninth month. The stems of the second crop grow fastest and yield the best fibre. After the crop the stocks are covered with manure and immediately watered. A well-cared-for plantation lasts for 80 to 100 years. The principal points to be investigated in order to determine the best methods of growing the plants on a commercial scale are as follows: (1) Influences of irrigation and manuring, especially the effect of returning to the soil the waste portions of the plant. (2) The variation of the amount and quality of the fibre according to the season. (3) The comparative quality of the fibre of short stems (3 feet) and that of full-grown stems (5-8 feet). (4) The effect of the density of growth upon the thickness and the straightness and branchiness of the stems, and upon the yield per acre, especially in connection with the prospect of a greater number of crops annually und the condition of limited height. (5) The best and cheapest methods of gathering, stripping, and sorting the stems."

586

Separation of Fibre.—The modes by which this is accomplished by manual labour and by machinery will be found under another heading (see page 642), but it may not be out of place to say something here as to the condition of the stems most favourable for the extraction of the fibre. They require to be acted upon while green, and at most within a few hours after they are cut. **Major-General Hyde**, who presided at a meeting of the Society of Arts (*London, 12th December 1883*), at which **Dr. Forbes Watson** delivered a lecture on Rhea with special reference to Messrs. Death and Ellwood's patent "Universal Fibre extractor," in summing up the discussions which followed the lecture, while referring to certain experiments performed by **Mr. Greig**, said: "The fibre was placed in a shed, and remained there until Monday morning, and on Monday morning the mass, as high as that table, was like a large mass of isinglass glued up together with the fibre in it; nothing could be done with it, and it had to be thrown away. That showed the absolute necessity of attacking the stem the instant it was cut, with a running stream of water to carry away the gum whilst it was in its natural state. It was then easily attacked, but let it wait or dry in any way, then the difficulty commenced and increased. The colour of the fibre was also darkened in proportion with the delay in removing the juice."

THE GUM OF RHEA FIBRE.

GUM. 587

When the experiments with **Mr. Greig's** machine were concluded, all the rollers, &c., were found to be thickly covered with a very hard varnish,—so hard that it could only be taken off by a chipping chisel. It had the appearance of lac. The analysis of this dry juice has been published as follows: "The juice contains 62 per cent., by weight, of oxalate of lime; and, besides this, some alumina, oxide of iron, and other mineral matters which dissolve in hydrochloric acid; the residue, insoluble in dilute hydrochloric acid, consists of colouring and resinous matter, and forms 2·5 per cent. by weight of the dry juice." (*Foot-note to Dr. Forbes Watson's Lecture before the Society of Arts, p. 13.*)

VALUE OF THE PREPARED FIBRE.

Prepared Fibre. 588

China-grass fetches about £49 to £50 a ton in London; Indian rhea fibre a slightly lower figure. According to **Dr. Forbes Watson**, Messrs. Death and Ellwood's *"Universal fibre extractor"* could turn out the fibre at "from £7 to £9 a ton, calculated at 100 lbs. of fibre for the working day per machine." "Such being the case, the result will be that China-grass may be introduced at a much cheaper price than hitherto. What that price will be I cannot say, but I think it will be possible to sell at £30 to £35 a ton, possibly less." (*For Death and Ellwood's Machine see page 481.*) **Mr. Collyer**, in the discussion which followed **Dr. Forbes Watson's** lecture at the Society of Arts, said that "for rhea at £30 a ton there was no limit, practically, to the quantity which could be sold; at £40 it would go slowly; at £50, with the present price of wool, it was barred." A manufacturer remarked—"If you bring it down to £35 you will sell a lot; if you bring it to £30 nobody knows the quantity we can use." **Mr. Haworth**, at the same meeting, said that a larger quantity of rhea would one day be sold than of jute at the present day.

HISTORY OF THE RHEA INDUSTRY.

HISTORY. 589

In the *Ramayana* mention is made of the nettle-cloth, and it is praised for its beauty and fineness. There is therefore some *primâ facie* evidence that *a* nettle fibre has been known for several centuries in India. So early as in the reign of Queen Elizabeth in England, **Lobel** the botanist relates that in Calcutta, in the East Indies, the people manufactured from the fibres of a species of nettle a very fine and delicate tissue. Later, these fine cloths were imported into Europe, but principally from Java to the Netherlands, where this cloth was in great demand under the name of *neteldoek*, a name indicating the origin of these cloths. The word *netel* means nettle and *doek* tissue. From that time attempts were made, and with success, to imitate with flax fibre the beautiful and fine tissue of the ramie, of which, after all, it is but a weak counterfeit. (*Theo. Moerman.*)

Dr. Roxburgh, without apparently being aware of the existence of rhea in Assam and parts of Bengal, and of the fact that it was being cultivated and used by the natives there, procured from Sumatra in 1803 four plants of the *Caloee*, and planted them in the Botanical Gardens, Calcutta. He gave the plant the name **Urtica tenacissima.** These imported plants grew and multiplied so rapidly that shortly after he had several thousands. About this time the discovery was made by **Dr. Buchanan Hamilton** that the *konkura* of Rungpore and Dinagepore was identical with the plants **Dr. Roxburgh** was cultivating. In 1810, **Dr. Buchanan** sent to England three bales of fibre from the plants grown by **Dr. Roxburgh.** The experiments made with this fibre showed that a cord spun from it sustained a weight of 252 lbs. against 84 lbs. required by Her Majesty's Dockyard to be borne by Russian hemp of the same size. In 1814 more bales of the fibre were sent by **Dr. Buchanan** to the Court of Directors (England). In 1816 the Court sent out several of the machines then recently patented by Messrs. Hill and Bundy to be used in the preparation of rhea. From this date, however, the interest in rhea fibre seems to have fallen off until 1840, when the discovery by **Colonel Jenkins** of the same plant growing wild in Assam again caused attention to be directed to it. A few specimens from Assam were sent to the Agri.-Horticultural Society of Calcutta, and from cuttings thus obtained plants were grown in the Society's Garden. From this date the Society received contributions from several writers, from time to time, giving new facts regarding the growth and preparation of the fibre in Northern India. **Dr. McGowan** furnished information and samples from China, and **Dr. Falconer** and afterwards **Sir William Hooker** identified Rhea as the same plant from which the Chinese grass-cloth is prepared. (*Compare with remarks at pages 464 and 479.*)

HISTORY.

In 1851 several specimens of rhea in various stages of preparation were forwarded to the London Exhibition; they attracted considerable attention and were awarded no less than three prize medals. The following year a consignment of the fibre from Assam was forwarded by the Government of India to the Court of Directors; it was experimented with by **Dr. Forbes Royle**, the result being that its average strength, as compared with Russian hemp, was declared to be in the ratio of 280 to 160.

EFFORTS TO EXTEND RHEA CULTIVATION.

590

In 1854 the Court of Directors asked the Government of India to furnish 10 tons of the raw fibre, but owing to its limited cultivation only one third the quantity could be supplied. **Sir Fredrick Halliday**, then Lieutenant-Governor of Bengal, directed the purchase, during the ensuing three years, of a quantity of fibre up to 10 tons a year, in order to encourage the cultivation. These purchases were transmitted to London and sold. The fibre had by this time become known in England and France, and as it was thought that its further development might be safely left to private enterprise, the experimental consignments were discontinued.

The demand continued satisfactory, though on a rather small scale, but was supplied chiefly by China and only to a very slight extent by India.

In 1872, however, the fibre seems to have been making rapid progress; China supplied through London between 200 to 300 tons, valued at about £80 a ton. In that year a sudden change occurred: the demand fell off and the price came down to from £30 to £40 a ton for the China, and from £19 to £30 a ton for the Indian fibre. Rhea waste began to command a readier sale than the combed fibre, for it was found by the manufacturers that in the waste state it was procurable at a smaller cost and therefore more profitable, since in the end (owing to the want of proper extraction) both waste and combed fibre had to be treated with the same care, trouble, and expense. (*Journ. Soc. Arts.*)

In 1880 the Rajah of Dinagepore intended undertaking the cultivation of rhea in his estate. He tried to purchase a supply of roots from the cultivators; but as soon as the news of the project spread over the districts, exorbitant prices were demanded. The Rajah then procured 25 maunds of roots from Saharanpur, 11 maunds from the Calcutta Botanic Garden, and 11 maunds from the Bally Paper Mills. With the two latter supplies, 10 bighas were planted in May and June, and it was the intention of the Rajah to plant a hundred acres of land with rhea. The results of these experiments have not as yet been made public.

In 1881 Messrs. Burrows, Thomson, and Mylne, estate-holders in the Shahabad District (Bengal), intended, among other things, to induce their tenants to grow rhea and to prepare the fibre as a domestic industry. They wrote: "We see no reason why its preparation by hand should not become as successful in India as it is in China. Certain kinds of available and cheap labour are as plentiful in the former as in the latter country. The ***parda-nasheen*** women and girls, in vast numbers, of poor high-caste families, confined as they are by custom to their houses, cannot assist the male members of the family in any out-door work, or contribute to the general earnings, except to a very small extent by cotton-spinning, and the demand for this homespun yarn decreases as mills are adopted to produce it cheaper and better." In 1882 it was ascertained that these gentlemen were growing the rhea plant and were trying several methods of preparing the fibre; that they had sent to England some of the fibre, treated in a way likely to suit the manufacturers of England and France, and were awaiting the result before encouraging an extended cultivation of the plant. The final result has not as yet been made public.

Rhea Fibre.

More recently rhea has in Europe acquired a position of considerably augmented importance as an industrial product. Large plantations have now been organised in Italy. Portugal has already planted a million roots, and Spain has taken important steps in the matter. France seems to have given the lead in the movement, and during 1882 several million root-plants were imported. The plantations in Algiers and Egypt have also been materially increased. (*Journal of Society of Arts.*)

PROPERTIES AND USES OF RHEA FIBRE.

USES OF RHEA. 591

Rhea has been recognised as pre-eminent amongst fibres for strength, fineness, and lustre. Experiments made by **Dr. Forbes Royle** as to strength showed that its average power, as compared with Russian hemp, was in the ratio of 280 to 160. Its fineness has been demonstrated by **Dr. Forbes Watson,** who showed that "the mean diameter of the ultimate fibres of flax is about $\frac{1}{2080}$ of an inch, of jute $\frac{1}{1900}$, of hemp $\frac{1}{2100}$, of rhea from Assam about $\frac{1}{2160}$, and of Chinese rhea $\frac{1}{2260}$, of an inch. The length of the fibre varies from 2·36 inch to 7·87 inch, and even 9·84 inch; the mean diameter is about 0·002 of an inch (*Spons' Encycl.*). Regarding silkiness, jute is the only fibre known commercially which can compete with rhea, but jute is far inferior to it in strength and durability. Rhea has, besides, a high resisting power when submitted to the influence of moisture and variations of atmospheric condition. This power may, to some extent, be tested by the action of high-pressure steam on fibres. Experiments were carried out under the direction of **Dr. Forbes Watson** with this object: the fibres of rhea and of other plants were exposed for two hours to steam of about two atmospheres and then boiled in water for three hours, and the loss in weight ascertained. They were then again exposed to the action of steam at the same pressure for four hours and the loss in weight again ascertained. "The percentage loss of a specimen of Chinese rhea amounted only to 0·89, and of Assam rhea to 1·51, while flax lost 3·50 per cent., Italian hemp 6·18, Russian hemp 8·44, and jute even 21·39 per cent." **Dr. Forbes Watson** says: "A very characteristic, and in some respects unfavourable, quality of the Rhea is the comparative stiffness and brittleness of the fibre, and most of the difficulties which in spinning and manufacturing it have to be overcome are due to this circumstance. It is this stiffness which prevents rhea, although so strong in its usual condition, from sustaining as easily as other fibres the effect of a sharp bend or kink. Thus, if a knot be tied with a small bundle of fibres, the rhea will break very readily, much more so than flax, for instance, although all fibres will break more readily under such conditions. Another consequence of this stiffness is that the fibre does not twist easily, and the yarn spun from rhea is often very rough, notwithstanding the smoothness and silkiness of the individual filaments. This roughness is due to the projecting ends of the ultimate fibres, turned outside by the twist which the yarn receives in spinning. On the other hand, the stiffness or hairiness has also certain advantages, as, in consequence of this, rhea readily combines with wool. Thus rhea, in virtue of its quality, has a wide range of affinity with other fibres, though it is not perfectly similar to any of them. This explains why its experimental applications cover such a wide field. It has been actually tried as a substitute for cotton, hemp, flax, wool, and silk." In his more recent lecture before the Society of Arts upon this fibre, **Dr. Forbes Watson** says: "Now, what is rhea good for? It is difficult to say what it is *not* good for. It is the strongest fibre in nature."

MIXED FABRICS. 592

RHEA WITH COTTON.—The first trials in the use of rhea with cotton were made in 1862 in England and France. Rhea fibre from China was cut into length of two inches and treated with alkalies and oil, giving a

USES OF RHEA.

material suitable for admixture with cotton. This cottonised rhea was the subject of various experiments; it was mixed with cotton, was spun, and the yarn woven into different fabrics and dyed and printed without any difficulty. The fabrics so made gained in strength and acquired a certain amount of gloss. But considering the matter in a commercial point of view, it may be said that rhea will never pay as a mixture with cotton: it will always remain too valuable a material to be used as an admixture with or even as a substitute for cotton, the cost of extraction being prohibitive of such use. This was true some few years ago, but it seems probable that new machinery will lower the price of rhea until admixture with cotton will be possible. **Mr. W. Haworth**, speaking of the use of rhea with cotton, says: "Rhea would make the warps of the finest cotton goods, and the wefts could be made of Sea Island or other fine cotton. It could be used for the finest materials up to the coarsest."

593

RHEA WITH FLAX.—The probability of its being used with flax occurred to the early experimenters, but experience soon showed that it was necessary to overcome technical difficulties before rhea could be spun successfully on flax machinery. These were subsequently overcome, and **Moerman** in his pamphlet on rhea mentions the fact that he examined specimens spun in some French and Belgian mills by flax machinery on cold-water frames, and that they were smooth and glossy, the gloss being secured by passing the fabric between cylinders. **Dr. Forbes Watson** in his report (1875) wrote: "If, as seems probable, rhea could be worked up on the same machinery as flax, the development of the rhea trade would be immensely facilitated, inasmuch as there would then be an immediate and practically unlimited field for its consumption." In his lecture (1883) he says: "Many years ago, one of the largest flax-spinners in the kingdom spent a considerable sum—£20,000, I believe—in trying to use China grass in the place of flax; but the experiment was given up, owing to the hairy character of the yarns produced. It is, however, quite possible to prepare rhea in a way which would enable it to be spun on flax machines; and we find table-cloths and beautiful fabrics of this material equal to anything that could be produced from flax."

594

RHEA WITH WOOL.—In combination with wool, rhea seems to have a chance of success, and its application in this manner attracted most attention and for a time achieved the greatest share of success, since it was less costly than wool and bore a striking similarity to it. "The prepared rhea, or China grass, cut up into suitable lengths, has, in fact," says **Dr. Forbes Watson**, "been found capable of being spun on worsted machinery, and then used like mohair or other long-stapled wools, for the manufacture of certain kinds of fabrics which depend for their effect on the gloss of the material. These fabrics were made, as a rule, with cotton warps, rhea yarn of comparatively little twist being used as weft. The use was mainly for ladies' dresses, and at first it seemed as if the success was complete. But after a certain time the inferiority of the new fabrics for ladies' dresses became manifest. Although everything that could be desired as regarded appearance and finish, there was the fatal objection that in wear they became easily creased, as the vegetable rhea fibre is wanting in the great elasticity possessed by wool. In view of such an inferiority, the prices then ruling for rhea made its use for this purpose no longer remunerative. The new trade collapsed as rapidly as it had sprung up, and since 1872 the matter is again one of experiment. The creasing, however, is to be got over by mixture with wool or by the use of very thick cotton warps, and fabrics of a new kind have been manufactured on a small scale, and have found a ready sale."

Rhea Fibre.

Dr. Forbes Watson's report is full of useful information, and the following passage may also be quoted: "There is sufficient evidence that at prices of the raw material permanently lowered"—by more efficient and less costly modes of extraction and preparation of the fibre, as well as by extended cultivation of the plant—"there would be a larger field for the use of rhea as a substitute for long-stapled wool. Even if its use for ladies' dresses were not again resumed, there are hangings, carriage linings, carpets, and other manufactures for which the suitability of rhea has been established, and for which its application continues to engage the attention of some of our most eminent manufacturers. There are several circumstances favouring the use of rhea in this line rather than in competition with flax. The material competed with is higher priced than flax, the better classes of wool varying from £130 to £280 per ton, whilst those which in their raw state are lower priced contain such a proportion of dirt that the price for the really available fibre is here also, in reality, not much lower. There is also the circumstance that the rhea combing waste, or noils, has been found very suitable for mixture in bulk with rough kinds of wool, and capable of being used for blankets, as also, possibly, for giving strength to shoddy, and for a variety of other rough purposes."

595

RHEA WITH SILK.—As a mixture with silk, rhea has a formidable rival in jute; and although the subject of using rhea as a substitute for or admixture with silk has been repeatedly taken up in England and in Lyons, and by the application of rhea it has been found possible to imitate, to a certain extent, the effects of silk in certain mixed fabrics, the special use of rhea for this object has never acquired any real footing. **Dr. Forbes Watson** says, however, that "rhea is prepared in various ways, so as to leave the gloss upon it, giving it all the appearance of silk, and it is certainly far superior even for mixing with silk than jute."

596

RHEA AND HEMP.—In Assam and Bengal where the rhea grows, the use to which it is commonly put is the same as that for which hemp in Europe is used,—*i.e.*, it is employed for nets, fishing-lines, and other purposes for which strength, lightness, and power of resisting water are essential. Viewed as a material for such use, rhea figures prominently in its chances of success. Hemp is, it is true, lower priced than rhea; but it suffers a greater loss in weight in the process of heckling than rhea, while the latter is superior in strength and in resistance to water; and lighter cordage of it would do the same work as heavier ones of hemp. "For many purposes," says **Dr. Forbes Watson**, "such as ships' rigging, the increase in lightness is in itself an important consideration, apart from the saving of the material." "On all these grounds rhea may be substituted with advantage for hemp, even if it be at a considerably higher price than hemp. The same may be said of its cognate use for canvas and sail-cloth instead of flax. In that case also the superior strength of rhea results in the double advantage of a saving in material and of greater lightness, and would enable it to compete successfully with flax, even if this latter were considerably cheaper per ton."

Ropes and Cords.
597

RHEA AS A ROPE AND CORD FIBRE.—The great strength of the fibre, its lightness and power of endurance under water, are qualities which place it in the first rank of fibres suitable for ropes and cables.

Nets
598

LOCAL APPLICATIONS OF THE FIBRE.—In Upper Assam the *doombs* or fishermen cultivate the rhea plant, and extract the fibre by manual labour, employing it in the construction of their fishing-nets.

In the Rungpore and Dinagepore districts a limited amount of rhea is regularly cultivated in some localities, especially along the banks of the

Attri and Teesta rivers, where fishermen reside. The cultivators find a ready and remunerative sale for the fibre; but they have seldom over a few square yards under plant; and although it is cultivated all over these districts, the cultivation is only practised on a small scale. In Bhagulpore, people of the *dhanook* caste are said to prepare rhea fibre and to sell it to the silk and tusser weavers in the district, the inference to be drawn being that the weavers mix rhea with silk.

Used in India as a silk substitute. 599

As a Paper Material, Rhea is, of course, not likely to be of much use, owing to its value and high price. But some of the waste can undoubtedly be used for this purpose, chiefly as an admixture to impart strength and cohesion to very inferior materials.

Waste. 600

When dried, the leaves are very fibrous and may be used as a paper material. **Theo. Moerman**, in his little book on "The Ramie," mentions that about 6,750 lbs. of dried leaves may be obtained from an acre.

Minor uses of Rhea Fibre.—"Amongst the minor uses of rhea may be mentioned the fact that it is sometimes used for packings of steam-engines. As a curiosity, it deserves to be mentioned that rhea fibre is now in use for polishing ivory, such as billiard balls, &c." Experiments have recently been made to turn rhea fibre into a material closely resembling leather to be used as a substitute for leather bands.

Steam-packings. 601
Polishing Ivory, 602
Leather substitute. 603

METHODS of TREATING and SEPARATING the FIBRE.

Manual Labour.

604

The real difficulty in the way of an extended utilisation of the rhea fibre is the decortication of the stems, or, in other words, the extraction of the fibres, at a reasonable rate, and in a condition fit for commerce. **Dr. Forbes Watson**, in a lecture before the Society of Arts, explained the constituents of the rhea stem. He said: "You will observe, on breaking this sample of green rhea, I succeed in getting off a certain quantity of green fibre, tearing it down in this manner. I wish to refer, in the first place, to the composition of the component parts of this bark. The outside portion consists of a film to which a very distinguished chemist has applied the term cutose. Below that there is a bark which contains the green colouring matter of the plant, that is called vasculose, and next to that comes the fibre itself. That fibre, and the bark attached to it, is united to the stem by another principle which is called pectose." The difficulty consists in getting the bark and the other matters separated from the fibre. To accomplish this various contrivances and machinery have been specially introduced and patented.

In China, Borneo, and Sumatra the following system is adopted: "The stalks are cut and collected in bundles, and are then thrown into still pools and kept there for several days until the process of retting, so as to cause the bark to separate easily from the wooden parts, is sufficiently advanced. At this stage the bundles of stalks are removed from the water, and all the cortical bark or raw fibre is immediately collected. To do this the bark on the stalks is split in the centre; two fingers are inserted between the wood and the bark and slipped along the whole length of the stalk between the wood and the bark, which brings out the fibre in two strips or ribbons. These strips are spread out on fields to complete and finish off the process of retting by exposure to the dew; but those who are more skilful collect these strips of bark into bundles, and again for a second time throw them into water to effect a cleansing by a fresh and more complete process of retting. By this second steeping another fermentation and decomposition is brought about of the sap or pith incrusted in the fibres. This process completes more thoroughly the retting which is not effected by the simple exposure to the dew.

BŒHMERIA nivea.

Rhea Fibre.

MACHINERY.

After this second retting, it only remains to work up and comb out the fibre, and thus prepare it to be spun to any quality of fineness." (*Theo. Moerman.*)

In Java the natives do not, apparently, resort to retting in pools. The same author explains the mode adopted, thus: "After dividing the stalks into halves lengthwise, they remove the bark, from which they then separate the epidermis and the adhesive portions by scraping it with a knife until the fibre begins to appear. This is white, with a slight shade of green. They content themselves with washing this fibre several times in water, and then dry it; but this manipulation, as will be easily understood, is not sufficient to entirely get rid of the glutinous matter which adheres to the fibres." In Borneo and Sumatra the following mode is in practice: The stalks are collected in bundles and are exposed for four or five days to the action of water. This destroys the thin bark and much of the gummy matter, and partially separates the fibre, which is then taken out, dried, and exposed to the dew for several days.

In Upper Assam the following method is practised: "The operator holds the stalk in both hands nearly in the middle, and pressing the forefinger and thumb of both hands firmly, gives it a peculiar twist, and breaks through the inner pith. Then passing the fingers of his right and left hand rapidly, alternately, towards each end, the bark with fibre is completely separated from the stalk, in two strands. The strands of bark and fibre are now made up into bundles of convenient size, tied at the smaller end with a shred of fibre, and put into clean water for a few hours, which I think causes the tannin or colouring matter to wash out. The cleaning process is as follows: the bundles by means of the tie at the smaller end are put on a hook fastened in a post at a convenient height for the operator, who takes each strand separately by the larger end in his left hand, passes the thumb of his right hand quickly along the inner side, by which operation the outer bark is completely separated from the fibre; and the riband of fibre is then thoroughly cleaned by two or three scrapings with a small knife. This completes the operation, with some loss, however, say one fifth; and if quickly dried in the sun it might at once be made up for exportation; but the appearance of the fibre is much improved by exposure (immediately after cleaning) on the grass to a night's heavy dew, in September or October, or a shower of rain during the rainy season. After drying the colour improves, and there is no risk from mildew on the voyage homewards." (*Major Hannay, in the Fibrous Plants of India, by J. Forbes Royle, 1855, p. 363.*)

Machinery used in separation of the Fibre.

RHEA MACHINERY. 605

The rhea fibre (or China grass) having been made known in Europe at the beginning of the present century, the attention of experts seems immediately to have been turned to the question of improvements in the extraction and preparation of the fibre. The first patent was taken out by a **Mr. James Lee** "for separating the fibre by mechanical means and without the aid of water-retting." No tangible results, however, seem to have been obtained so far as can be ascertained from the employment of this machinery. Meanwhile attention continued to be devoted to the question, and among other inventions may be cited the chemical process of Messrs. L. W. Wright & Co., for which these gentlemen obtained a patent in 1849. Their process "consists, essentially, in a very ingenious arrangement for boiling the stems in an alkaline solution, after they have previously been steeped for 24 hours in water of a temperature of 90°. The fibre is then thoroughly washed with pure water, and finally subjected to the action of a current of high-pressure steam till nearly dry." At the London International Exhibition of 1851

MACHINERY.

these gentlemen exhibited samples of rhea prepared by their process and received a silver medal. To other exhibitors were also awarded prizes; but still the question of the preparation of the fibre remained unsolved.

In 1869 the Government of India turned its attention to the utilisation of the rhea fibre, and issued a Resolution in which it expressed the conviction that the value of the rhea fibre was undoubted, that all the conditions necessary for its cultivation on a large scale were present in India, and that the only obstacle to the development of an extensive trade in this product was want of suitable machinery for the separation of the fibre from the stems and bark of the plant in its green or freshly-cut state. To encourage the invention of such machinery an announcement was made by the Government of India in June 1870 that a public competition would be held and a prize of £5,000 would be given for the best machinery. No fewer than 32 competitors entered their names, but at the last moment only one of them, **Mr. Greig,** of Edinburgh, appeared in India. The trial took place in August 1872 at Saharunpur, where a plantation of rhea had been established for the purpose. It was found that the cost of preparing the clean fibre by this machinery amounted to more than £15 a ton, and at the same time the fibre was pronounced defective in quality and was valued at £28 a ton only in England, and declared suitable for cordage only. Under these circumstances the full amount of the prize was not awarded, but in consideration of the fact that the machine was a *bonâ fide* and meritorious attempt to meet the requirements of the case, a donation of £1,500 was given to the inventor.

The following year (1873) fresh trials were arranged to take place in England under the superintendence of **Dr. Forbes Watson,** with a supply of rhea stems from the south of France. A notification was issued by the India Office, and 200 applicants responded. The trial did not, however, prove a success, as the supply of plants was less and of poorer quality than had been expected. In the latter end of the same year, a fresh offer of plants was made by **Dr. Forbes Watson** to those who wished to continue their experiments, and upwards of 100 asked for fresh supplies. These were procured from the district of Vaucluse (France) and made over to the applicants. The results have not been made public.

Meanwhile the demand for rhea fibre in Europe seemed to continue. Having reconsidered the matter, the Government of India, in a Resolution dated August 1877, renewed the offer of rewards. The terms now offered were that a reward of R50,000 would be given to the inventor of the best machine or process which would separate the bark and fibre from the stem, and the fibre from the bark of the **Bœhmeria nivea,** and a further reward of R10,000 to the inventor of the next best machine or process, provided it was adjudged to possess merit, and to be capable of adaptation to practical uses. The machine or process required was to be "capable of producing, by animal, water, or steam power, a ton of dressed fibre of a quality which shall average in value not less than £45 per ton in the English market, at a total cost, including all processes of preparation and all needful allowance for wear and tear, and not more than £15 per ton laid down at any port of shipment in India, and £30 in England after payment of all the charges usual in trade before goods reach the hands of the manufacturer." The machinery was to be simple, strong, durable and inexpensive, and suited for erection in plantations where rhea was grown. The competition was to take place at Saharanpur, the Government agreeing to provide accommodation for the competing machines, as well as affording the motive power required. The Government was also to pay for the transport of all machines from the sea-coast to Saharanpur up to a limit of one ton for each machine,

Rhea Fibre.

MACHINERY.

and to allow a free second class ticket by rail to that station to any person in charge of a machine.

The trials were fixed to commence on the 15th September 1879, and a Committee of Judges was appointed to conduct them. Twenty-four applications for permission to compete were received; but only ten competitors ultimately arrived at Saharanpur, and, of these, three withdrew from the competition. The trials were held in September and October 1879.

The fibre turned out by each of the competing machines was carefully packed and despatched to the Secretary of State, with a view to its being tested and reported on by experts in the trade in England. The reports received from the Secretary of State (August 1880) stated that the samples were far inferior to the fibre imported into England from China, the value of which at that period was £50 a ton. As no competitor had produced a fibre of a value even approaching the amount fixed in the Resolution of August 1877, the Committee did not recommend the grant of either of the prizes to any of the competitors. They were, however, of opinion that some of the machines possessed sufficient merit to warrant the grant of a reward to the owners, and the gentlemen mentioned by them as deserving of remuneration were Messrs. Nagoua, Vander Ploeg, and Cameron. The fibre turned out by Mr. Vander Ploeg was valued less highly than that produced by Messrs. Nagoua, and Cameron; but the Committee attributed this to the fact that he aimed at producing the fibre in a finished state fit for the spinner (a condition in which it was understood that the English dealer did not require it), and not to the inability of his machines to yield as good fibre as those of Messrs. Nagoua, and Cameron. The Committee remarked also that there was little novelty in Mr. Cameron's process, and that it was only an improvement on a method by which fibre was actually extracted from various plants by the natives of India. The same method was also applied in many of the Indian jails for the extraction of aloe fibre. The process was simple enough, and might be employed by the natives without special instruction, and any kind of stem, green or dry, short or long, could be treated by it; but it would be difficult of application in a rhea plantation, where the stems of many acres of land would have to be worked off quickly. Having regard to these circumstances, the Committee recommended that a grant of R5,000 each be made to Messrs. Nagoua, and Vander Ploeg, and another of R1,000 to Mr. Cameron.

The Government of India reviewed the above facts in a Resolution, dated March 1881, and decided, in concurrence with the Committee, that, as none of the fibre produced came up to the conditions prescribed, the prizes offered in 1877 could not be awarded. At the same time the Government of India agreed in the Committee's opinion that some recognition of their efforts was due to the three gentlemen whose machines yielded the best results or appeared to possess superior merit, and sanctioned the grant to them of the sums recommended by the Committee. The Government of India further stated that, "From the low valuation put by the English firms on the samples of fibre produced at the late competition, it does not seem probable that Indian rhea fibre will be able, for the present at least, to compete successfully with the Chinese product; while the experience which has been so far gained also points to the conclusion that in most parts of India the cultivation of rhea cannot be undertaken with profit. Rhea is naturally an equatorial plant, and it requires a moist air, a rich soil and plenty of water, while extremes of temperature are unfavourable to it. Such conditions may be found in parts of Burma, in Upper Assam, and in some districts of Eastern and Northern Bengal;

MACHINERY.

and, if rhea can be grown in such places with only so much care as is required in an ordinary well-farmed field for a rather superior crop, it is possible that it may succeed commercially. Until, however, private enterprise has shown that the cultivation of the plant can be undertaken with profit in these or other parts of the country, and that real need has arisen for an improved method of preparing the fibre in order to stimulate its production, the Government of India thinks it inadvisable to renew the offer, which it has now made for the second time without result, of rewards for suitable machines. But in order to aid persons who are anxious to try the cultivation of the plant in localities which are *primâ facie* suitable, the Government will be willing to place roots at their disposal. A plot of about two or three acres will, therefore, continue to be kept under rhea in the Botanical Gardens at Howrah for the supply of roots to intending growers."

A sample of China grass, valued at £50 a ton in the English market, was deposited in the Economic Museum at Calcutta, and, in accordance with the recommendation of the Committee, specimens of the fibre produced by the several competitors at the trials at Saharanpur, with the valuations of the experts noted on them, were also deposited in the Museum for inspection by the public. It seems remarkable that so many fruitless attempts should have been made in India and scarcely any effort put forth to ascertain why it was that the China grass-cloth was uniformly superior to the Indian article. This would have settled the question as to whether rhea is in reality the same thing as China grass-cloth. If rhea and grass-cloth were found to be actually produced from the same plant, this enquiry would naturally have brought to light a more accurate account of the Chinese mode of separation of the fibre than we as yet possess. It is remarkable, however, that Chinese grass-cloth should be much finer than rhea; that on being boiled it should lose only ·89, while rhea under the same treatment parts with 1·51 of its weight. These and other facts, in addition to the pronounced superior quality and therefore higher price paid for China grass-cloth as compared with rhea, would seem to confirm the suspicion that these two fibres may after all be obtained from different plants.

This remark is made purely as a suggestion, but it seems highly desirable that we should not only thoroughly examine all the plants met with in India which afford rhea-like fibres, as well as re-examine the plant from which the Chinese grass-cloth is obtained, before much more money is spent on experiments with new machinery.

The withdrawal of the stimulus afforded by the Government prizes did not, however, damp inventive ardour; and among other new machines may be noted those of Messieurs Fairer and Frémy, and of Mr. H. C. Smith, commonly known as Messrs. Death and Ellwood's Universal Fibre Extractor.

Messieurs Fairer and Frémy's invention consists in subjecting the plant to the action of steam for a period varying from 10 to 25 minutes, according to the length of time the plant has been cut. After steaming, the fibre and its adjuncts are easily stripped from the wood. It then appears in strips or ribands, containing the objectionable gum and outer bark. To remove these, the strips are subjected to a chemical process in baths. This dissolves out the cutose, vaculose and pectose, and releases the fibre in its clean, silky, white condition, ready for the spinner.

Messrs. Death and Ellwood give the following specification of their machine: "The universal fibre-cleaning machine, invented by Mr. H. C. Smith, manufactured and improved by Messrs. Death and Ellwood of Leicester, and brought to public notice by the General Fibre Company of London, is a very simple, compact, and well-designed machine. It

BŒHMERIA nivea.

Rhea Fibre.

MACHINERY.

consists of a cast-iron drum, perfectly balanced, on which eight gun-metal beaters are bolted. The drum revolves in front of a table or feed-plate fixed below the centre of the drum so as to give a scraping action when the beaters pass it. The feed-plate is adjustable to and from the beaters by set screws, so that a fine or thick fibre can be cleaned. Immediately below the feed-table is a jet pipe which throws a strong, thin, flat sheet of water against the whole width of the drum. These are the essential parts of the machine, and they are mounted on a cast-iron frame, which carries them as well as a trough to receive and let out water, refuse and waste, and to prevent the water being thrown about. Two men feed the machine; each taking from three to five leaves or stems at a time, places the thick ends upon the feed-table and pushes them against the revolving drum provided with beaters. These smash the woody parts of stems, disengage the pulpy matters of leaves, loosen all refuse matter, and by their action draw the crushed stems or leaves under the drum: here the sheet of water presses the stems or leaves against the beaters, a beating and scraping action continues, and the sheet of water, acting as a cleanser as well as an elastic cushion or backing to the fibre while it is struck by the beaters, ensures a thorough cleaning. The stems or leaves are allowed to pass half way into the machine, and when withdrawn all extractive matter has gone and clean fibre is obtained. This is held in the hands of the operators, who then pass and withdraw the thin ends in the same way. The result is clean pure fibre, which is then hung up to dry, and when dry is ready to be baled at once. The cost of a single machine is £55, that of a double one complete is £100. A semi-portable engine to work two of the machines is supplied by the General Fibre Company of London for £82-10. On comparatively small plantations, instead of the steam-engine, bullock gear can be used which, for a single machine, is supplied at £30 by the Company." (*Extracted from Hanlon and Liotard's report to Government of Bengal.*) **Dr. Forbes Watson** gives, in his lecture on rhea, delivered before the Society of Arts (*London, 1883*), an interesting history of the circumstances which suggested to Mr. Smith's mind the idea of the Universal Fibre Extractor: "What first suggested it to his mind was noticing the great aloes, the stems of which grow up to 30 or 40 feet. Mr. Smith observed during the monsoon in Mauritius that where the inner leaves were dashed against these great stems, they were broken up, the result being that the filth got washed away and the fibres were left hanging. This suggested to his mind theidea of a machine in which a rush of water would play the same part."

Some few months ago a series of experiments were performed, under the joint direction of the Government of India and Government of Bengal, with fibre-extracting machines suitable for all fibres. Some nine exhibitors came forward, but the committee awarded the prize of R2,000 to Messrs. Death and Ellwood. The committee consisted of Messrs. J. W. Hanlon and L. Liotard, assisted by the Agri.-Horticultural Society of India. In concluding their report upon "The Universal Fibre-cleaning machine," the members of the committee, in recommending the prize to be awarded to Messrs. Death and Ellwood, say: "We are satisfied that as an extractor of fibres Messrs. Death and Ellwood's machine is a distinct advance in mechanism of this class, that it extracts fibres in their natural colour, and in good merchantable condition, that it operates on all plants with the same facility, and that it is suited to the requirements of this country, and is likely to prove of great service to its fibre industry." **Dr. Forbes Watson**, speaking of this machine in his lecture before the Society of Arts, 1883, says: "It is provided with what are called beaters,—that is to say, a certain number of projecting ribs,—and it revolves in front of a feeding-table at a great rate, being worked at 600 revolutions a minute.

"This operation goes on in front of the feeding-table as it is called, and this constitutes the whole machine, as regards the mechanical portion, with the exception of the water. Below, and at an angle of about 45°, a strong flattened jet of water passes, and I will tell you what the effect of that is. The cylinder, remember, is rapidly revolving; you feed in at the side here, the beaters catch and break up the stalks into very small pieces, and the jet of water, coming from below, meets the fibre, and keeps it up against the beaters, so that it is really beaten in a stream of water. The result of this is, you not only get the fibre cleared of a large portion of its gum, but you have next to no waste, and what little there is is excellent for many purposes—it can be made use of, as most other waste products can. This explains the secret of the success of the invention, and how it solves the problem of a machine for cleaning rhea."

Food.—When green the leaves are very much liked by cattle, and are nutritious. When salted they will curdle milk like rennet. (*Lindley's Vegetable Kingdom.*)

FOOD. Leaves. 606

I am indebted to **Mr. L. Liotard,** of the Revenue and Agricultural Department, for much assistance in collecting many of the extracts compiled in the preceding pages regarding rhea fibre.

Bœhmeria platyphylla, *Don; Brandis, For. Fl., 403.*

Syn.—B. MACROSTACHYA, *Wedd.*; URTICA MACROSTACHYA, *Wall.*

Vern.—*Gargela* (KUMAON), HIND.; *Kamli,* NEPAL.

Habitat.—A large shrub or small tree, met with in the outer Himálaya up to 7,000 feet, in the Khásia Hills, East Bengal, South India, and Ceylon.

Botanic Diagnosis.—*Branches* 4-sided. Leaves opposite broad ovate; petiole ¼ the length of the leaf or longer. Styles hairy exserted (rarely shorter than the female perianth tube). One of the commonest and most variable species in the genus.

Structure of the Wood.—Moderately hard, reddish brown, with occasional concentric bands of darker and lighter colour.

FIBRE. 607 Timber. 608

The following may be enumerated as the principal Indian varieties:—

α, Hamiltoniana, *as in DC. Prod., XVI., I., 213.*

Syn.—B. HAMILTONIANA, *Wedd., in Ann. Sec., Nat.; Kurz, Fl. Burm. II., p. 424*; URTICA HAMILTONIANA, *Wall.*

Vern.—*Tuksur,* LEPCHA; *Sapsha,* BURM. (*Kurz*).

Habitat.—An evergreen, small tree, often 20 feet in height, met with in the lower tropical Himálaya from Sikkim and Bhután eastward to Burma. "Plentiful in the tropical forests, especially along choungs of the eastern slopes of the Pegu Yomah and Martaban east of Tounghoo." (*Kurz.*)

Botanic Diagnosis.—*Leaves* opposite, long acuminate, minutely toothed 4-6 in. long, 3-nerved with a gland at the basilar nerve axil. *Styles* shorter than the perianth. **Kurz** regards this as distinct, but the only character in favour of this view is apparently the short style.

Fibre.—"Strong cordage can be obtained from the liber." (*Kurz.*)

FIBRE. 609

β, macrostachya, *Wedd. in DC. Prod., XVI., I., 211.*

Syn.—SPLITGERBERA MACROSTACHYA, *Wight, Ic., t. 1977*; BŒHMERIA MAURITIANA, *Wedd.*; B. WIGHTIANA, *Wedd.*; URTICA CAUDATA, *Poir.*

Habitat.—A large bush met with on the Nilgiri Hills and Ceylon, with long petiolate leaves and female spikes generally undivided.

FIBRE. 610

γ, **rotundifolia,** *Wedd. in DC. Prod., XVI., I., 212.*

Syn.—B. ROTUNDIFOLIA, *Ham. et Don, Prod. Nepal, p. 60.*

Vern.—

FIBRE. 611 **Habitat.**—A small bush met with in the Konkan, in Nepal, the Khásia Hills, and Ceylon, ascending to 1,500 feet in altitude, with rotundate abruptly acuminate leaves.

δ, **scabrella,** *Wedd. in DC. Prod., XVI., I., 211.*

Syn.—URTICA SCABRELLA, *Roxb., Fl. Ind., Ed. C.B.C., 685; Wight, Ic., t. 691;* U. CAUDATA, *Nic.;* B. OURANTHA, *Miq.;* B. SCABRELLA, *Gaudich.*

FIBRE. 612 **Habitat.**—A shrubby, spreading form, met with in Nepal, Assam, Khásia Hills, Chittagong, the Nilgiri hills, Ceylon, and Java, with small cordate, serrate, rough leaves and flower-spikes erect, as long as the leaf or shorter; male ones crowded, short, and in the lower axils; female ones above and generally solitary.

Apparently not put to any economic use, although all the species of this genus are known to yield good fibres. Flowers at the end of the rains, and the seeds ripen during the cold season. (*Roxb.*)

FIBRE. 613 ε, **Zeylanica,** *Wedd.*

Common in the Central Provinces and Ceylon up to an elevation of 6,000 feet.

614 **Bœhmeria polystachya,** *Wedd.*

Syn.—URTICA POLYSTACHYA, *Wall.*

Habitat.—A Nepal species apparently not put to any economic purpose. It is also met with in East Kumaun.

615 **B. rugulosa,** *Wedd.; Brandis, For. Fl., 403.*

Syn.—URTICA RUGULOSA, and U. VENOSA, *Wall;* B. NERVOSA, *Madden.*

Vern.—*Geti, gainti,* HIND.; *Dar,* NEPAL; *Sedeng,* LEPCHA.

Habitat.—A small tree with greyish-brown branches, met with in Garhwál, Kumaon, Nepal, Sikkim, and Bhután.

Botanic Diagnosis.—Branches terete when young, as also the petioles and under sides of the leaves, hoary. Leaves alternate, elliptic-lanceolate, 3-5 in. long, obtusely dentate, with 3 longitudinal nerves from the base to the apex, each penniveined, the lateral branching veins on the inside anastomosing with each other, those on the outside with an intramarginal vein; petiole many times shorter than the leaf. Flowers diœcious in round sessile clusters, each cluster in the axil of a cordate membranous bract. Brandis says the leaves very much resemble those of **Sarcochlamys pulcherrima,** but that it is readily distinguished by the long simple flower-spikes.

TIMBER. 616 **Structure of the Wood.**—Red, moderately hard, even-grained, durable, seasons well. A nice wood, easy to cut and work. Weight 41 lbs. per cubic foot.

It is used in Kumaon and Nepal for making bowls; in Sikkim for milk-pails, churns, and other dairy utensils. The Lepchas make cups, bowls, and tobacco-boxes of it.

B. salicifolia, *Don;* syn. for **Debregeasia bicolor,** which see.

617 **B. travancorica,** *Bedd.*

Habitat.—A small tree of the Wynaad, South Kanara Ghâts, and the Travancore hills up to 4,500 feet.

BŒRHAAVIA, *Linn.; Gen. Pl., III., 5.*

618

A genus of spreading, herbaceous plants, belonging to the NYCTAGINEÆ, comprising some 30 species, widely dispersed throughout the warm regions of the globe.

Annual or perennial plants, woody below, glabrously glandular or pubescent, branches few, spreading. *Leaves* opposite subsessile or petiolate, equal or unequal, entire or sinuate, fleshy. *Flowers* few and small, in umbellate or capitulate panicles, sessile or pedunculate, flowers articulated to the peduncle, bracts often deciduous; the young fruit frequently covered with glandular hairs. *Perianth-tube* short or long; base ovoid, contracted above the fruit; limb infundibuliform, margin 5-lobed; lobes distinct plicate, deciduous. *Stamens* 1-5 exserted filaments thin, unequal, free above, connate below, anthers didymous. *Ovary* stipitate, oblique; style erect, attenuated into the peltate stigma. Fruit (when young) obovoid 5-costate or 5-angled, glandular; ripe fruit oblong; 1-celled, 1-seeded; embryo usually conduplicate.

Bœrhaavia diffusa, *Linn.; Wight, Ic., t. 874;* NYCTAGINEÆ.

619

THE SPREADING HOG-WEED.

Syn.—B. PROCUMBENS, *Roxb.; Fl. Ind., Ed. C.B.C., 49;* B. ERECTA, *Gartn.: Roxb., Fl. Ind.; DC. Prod., XIII., I., 452.*

Vern.—*Sánt*, HIND.; *Gádha púrna, punarnabá. seveta punarnabá*, BENG.; *Punarnavá, visha kharpara, sothaghni* (? *sindika*), SANS.; *Jan-tóps*, SINGH.; *Punarnavá, khâpará, ghetuli*, BOMB.; *Vakha khaparo, dholi sáturdi, moto satodo*, GUJ.; *Punarnuwa (satodiputchee)*, CUTCH; *Vasu*, MAR.; *Thikrí-ká-jhâr*, DUK.; *Nakbel*, SIND; *Mukaratte-kire, mukúk-rattai*, TAM.; *Atika mamidi*, TEL.

Habitat.—A troublesome weed found all over India.

Botanic Diagnosis.—There are two well-marked varieties of this plant, one with white and the other with red flowers. In Bengali, the former is called *shwet-púrna* and the latter *gudha púrna.* This is, perhaps, one of the most abundant and troublesome of weeds, changing its appearance completely according as it is found growing on the top of a ruined wall or on an exposed situation in poor soil, or under shade and in good soil. All the forms are doubtless referable to one species. Some are short, erect, branched; others tall, straggling, or even climbers.

Properties and Uses—

FOOD. 620

Food.—The Rev. A. Campbell, Santal Missionary, Gobindpur, has furnished me with a most interesting series of specimens of this plant. The small bushy form found in the wild state is used, it appears, by the Santals as a medicine, but the plants which spring up in their vegetable gardens are cultivated as pot-herbs. They do not sow or propagate the plant; it exists in a state of semi-cultivation only, but at the same time it greatly improves, becoming a climber and producing large succulent leaves. I have received specimens from **Mr. Campbell** quite six feet in length, the whole plant so completely altered that, but for the flower and fruit, it is recognised with difficulty. In this half-cultivated condition it occurs in every Santal village, and constitutes a considerable article of food. The cultivation of this plant as a pot-herb is a fact which does not seem to have attracted much attention; it is alluded to in a few words by **Balfour**, and is included in his list of green vegetables used in the Madras Presidency. We have here what may be viewed as the first approach to the cultivation of a herbaceous wild plant as an article of food. From its succulent nature it seems highly probable that, under careful management, considerable improvement might be effected. With the present outcry for new fodder plants before us, it is worth suggesting that there would seem to be some hope of finding in this hardy indigenous plant a useful addition to our list of fodder plants. Indeed, the cultivation of an indigenous plant, such as this, seems much more hopeful than fruitless attempts to

Pot-herbs. 621

introduce delicate exotics, which at most are capable of being cultivated only in special or peculiar, and therefore limited tracts of country, or during certain seasons of the year. **Bœrhaavia** is a perennial which, in its wild state, luxuriates on the poorest waste lands. With the slightest effort, a field of it might be raised which would continue to yield green fodder throughout the year, and it therefore seems worth ascertaining whether cattle would thrive on such a crop. Lanan in his "*Hortus Jamaicensis*" says that in Jamaica the leaves are given to hogs, hence the English name Hog-weed. It is given to cattle in Bengal as a medicinal food and is supposed to increase the quantity of milk. (*Babu T. N. Mukerji.*) (See *Ainslie, Balfour, Drury, &c.*)

Fodder. 622

MEDICINE. Root. 623 Plant. 624 Leaves. 625

Medicine.—THE ROOT, used in infusion or given in powder, acts as a laxative, diuretic, anthelmintic, and cooling medicine. It has been found to be a very good expectorant, and has been prescribed in several cases of asthma with marked success. Taken in large doses it acts as an emetic. (*Ainslie, O'Shaughnessy, 512; Pharm. of Ind.; &c.*) The root is said to be a strong emetic. (*Bomb. Gaz., IV., 14.*) "In Goa the herb is esteemed as a diuretic in gonorrhœa. In Bombay THE PLANT is much used as an external application to dropsical swellings." A poultice of THE LEAVES is reported to be useful in abscesses. (*Dymock, Mat. Med., W. Ind., 540.*) "One of its Sanskrit synonyms—*sothagni*—means cure for dropsy. A decoction of *punarnavá* root is recommended to be given with the addition of powdered *chiretá* and ginger in anasarca." "AN OIL, prepared with a decoction of the root and a number of the usual aromatics in the form of a paste, is rubbed on the body in general anasarca, complicated with jaundice. It is called *Punarnavá taila.*" (*U. C. Dutt, Mat. Med., Hindús, 222.*) "The Peruvians give an infusion of the **Bœrhaavia scandens** in cases of gonorrhœa." (*Ainslie, II., 205.*)

Oil. 626

627

Special Opinions.—§ "The root of this is much used here in the cure of bronchitic asthma; smoking is not allowed." (*Surgeon-Major P. N. Mukerji, Cuttack, Orissa.*) "Expectorant, antispasmodic, and tonic; dose of the infusion 1 to 2 oz." (*Surgeon W. Barren, Bhuj, Cutch.*) "An infusion of the dry herb with nitrate of potash has been found by me to be very efficacious in dropsical affections. In slight cases a dish of the fresh herb boiled and salted, and eaten with bread (*chappaties*), without any other treatment, seems to do good." (*Asst. Surgeon Nobin Chunder Dutt, Durbhunga.*) "The white variety is preferred to the red. The root is a good medicine for dropsy and asthma." (*Surgeon-Major Bankabehary Gupta, Pooree.*) "The root bruised in water is a common application to the feet in cases of general debility." (*Sakhárám Arjun Rávat, L.M., Girgaum, Bombay.*) "Assistant Surgeon Moti Lal Mookerji extols this plant as a diuretic, especially in dropsy." (*Surgeon-Major A. Sanders, Chittagong.*)

628

BOMBAX, *Linn.; Gen. Pl., I., 210.*

A genus of trees belonging to the MALVACEÆ, and comprising some 10 species,—2 natives of tropical Asia, 1 of tropical Africa, and the remainder of tropical America.

Leaves digitate, deciduous. *Peduncles* axillary or subterminal, solitary or clustered, 1-flowered. *Flowers* appearing before the leaves. *Bracts* none. *Calyx* leathery, cup-shaped, truncate or 5-7 lobed. *Petals* obovate. *Stamens* pentadelphous, bundles opposite the petals, and divided above into numerous filaments; anthers reniform, 1-celled. *Ovary* 5-celled; style clavate, stigmas 5; ovulse many in each cell; *capsule* loculicidally 5-valved, valves leathery, woolly within. *Seeds* woolly, testa thin, albumen scanty; cotyledons contor-triplicate.

629

Bombax insigne, *Wall.; Fl. Br. Ind., I., 349.*

Vern.—*Semul-tula*, BENG.; *Saitu*, MAGH.

Habitat.—A large tree, trunk without prickles met with in Chittagong, Burma, and the Andaman Islands.

Botanic Diagnosis.—Trunk without prickles; leaflets 7-9, obovate cuspidate-acuminate, glaucous beneath; filaments slender, ⅔ the length of the petals.

Properties and Uses—

GUM. 630

Gum.—It yields a brown gum.

TIMBER. 631

Structure of the Wood.—Similar to that of **B. malabaricum**, but pores smaller and more scanty. The wood is also more durable than that of **B. malabaricum.** The specimen from the Andamans had been 12 years in Calcutta in the rough, and was only slightly discoloured on being cut up. (*Gamble.*)

632

Bombax malabaricum, *DC.; Fl.Br. Ind.; I., 349; Wight, Ic., t. 29.*

SILK COTTON-TREE.

Syn.—B. HEPTAPHYLLA, *Cav.; Roxb., Fl. Ind., Ed. C.B.C., 574;* SALAMALIA MALABARICA, *Schott.;* GOSSAMPINUS RUBRA, *Ham.*

Vern.—*Semul* or *sémal, shembal, semur, pagun, somr, ragat-sénbal, ragat-sémal, kánti-sénbal,* HIND.; *Rokto-simul, simul,* BENG.; *Simbal,* HAZARA; *Shirlan,* SUTLEJ; *Dél,* KOL.; *Edel,* SANTAL; *Simur,* MAL. (S. P.); *Bouro,* SIMURI, URIYA; *Boichú, panchu,* GARO; *Sunglú,* LEPCHA; *Semar* or *semur,* C. P.; *Saur, saer, somr, semul, shembal,* BOMB.; *Sávara, simlo, samar, kante savar, kanteri samar, shevari, tamari, savari,* MAR.; *Rato-shemalo, shemolo, shimlo, shimul, shimar,* GUJ.; *Kántón-ká-khatyán, kántón-ká-sémul, lál-khatyán,* DUK.; *Mundla-búraga-chettú,* TEL.; *Pulá, mul-ilava-maram, mulilavu,* TAM.; *Pula-maram, mul-lilava, mullila-púla,* MAL.; *Mullu-búragamará, burla,* KAN.; *Wallaiki,* GOND; *Katseori,* BHIL; *Lapaing,* MAGH.; *Sálmali, mochá,* SANS.; *Kattu-imbúl,* SINGH.; *Letpan, didu, lepán-bin,* BURM.

Habitat.—A very large, deciduous tree, with branches in whorls, spreading horizontally, and the stem with large thorny buttresses. Met with throughout the hotter forests of India and Burma. It is abundant on the eastern side of India, ascending the mountains to 4,000 feet in altitude. Distributed to Java and Sumatra. It is the largest and most characteristic tree of eastern Rájputana. (*Rájputana Gaz., p. 25.*)

Botanic Diagnosis.—Trunk and branches covered with large corky prickles; leaflets 5-7, quite entire, cuspidate, base tapering; filaments ligulate, half the length of the petals; capsule oblong-obtuse.

Properties and Uses—

GUM. Mocharas. 633

Gum.—*Mócharas* (*i.e.,* the juice—*ras* of the *mócha*), *mochras, mocherus, mucherus,* and various other forms of the word, are names given to a brown, astringent, gum-like substance, frequently seen in Indian bazars. It occurs in the form of light or dark brown tears, which are often hollow, much resembling galls. It is sometimes called *supárí-ká-phul* (*i.e.,* flowers of the *supárí,* or betel-nut palm). It is difficult to account for this latter name; the word *phul* is certainly used very frequently with a wide meaning, so much so that it would be quite easy to understand its being applied to the large gall-like tears of this gum. It is much more difficult to account for the supposition that they were the *phul* (flowers) of the *supárí,* unless we imagine that, as with Catechu at the present day, this astringent gum was formerly eaten in *pán* along with the betel-nut. It seems quite satisfactorily proved that **Dr. Birdwood** was mistaken when he stated that he believed the *mócharas* was "a kind of gall produced on the **Areca Catechu**" (*Bombay Prod., 10*). "**Dr. Birdwood** affirms that he has himself gathered precisely identical excrescences from **Areca Catechu.**" (*Dr. Cooke's Report on Gums, 1874, p. 40.*) It is a remarkable fact that no one has as yet confirmed **Dr. Birdwood's** observation regarding these gall-like excrescences occurring upon **Areca Catechu,**

but while this is so, his statement regarding them has passed into the literature of the subject. (See *Baden Powell, Pb. Prod., I., 319, 397; Cooke's Report on Gums and Resins, 40; Atkinson's Gums and Gum-resins, 27, &c.*) **Dr. Dymock**, however, in his recent work on the *Materia Medica of Western India*, attributes *mócharas* to **Bombax**, but makes no mention of **Birdwood's** astringent gall-like excrescences, while discussing the properties of the Betel-nut palm. **Dr. Birdwood** affirms that "all his attempts to obtain gum of any kind from **Bombax** completely failed in Bombay, and he has no hesitation in saying that the red cotton-tree affords no gum whatever." (*Cooke's Report.*) **Dr. Stewart** in his *Punjáb Products* (*p. 24*) says: "The gum which exudes ftom the bark is given often with **Ægie** for dysentery and diarrhœa." Several other more recent observers have, however, collected and described the gum obtained from **Bombax malabaricum**, and there would thus seem no ground for doubting that **Bombax** is the chief source of the *mócharas* of our bazars. Indeed, it is highly probable that the other gums sold under that name are only substitutes or adulterants. In an interesting letter addressed to the *Indian Forester* (*Vol. VIII., 153*) **Mr. Baden Powell** gives a detailed account of a tree of **Bombax malabaricum** which in his private garden at Lahore yielded a quantity of *mócharas.* It appears that if the tree is artificially wounded this substance will not be produced. The formation of the gum is due to some functional disease, and commences below the bark like a large swelling. After removing the mass of dark-coloured and decayed *mócharas*, **Mr. Powell** watched closely the formation of new gum. He says: "To my surprise, it issued in various-shaped masses, or worm-like pieces, as if one squeezed oil-paint out of a tube; this gradually curled up or coagulated into a mass, as chance would have it. It consisted of a rather firm, slightly translucent, dirty-whitish-yellow jelly. To the taste it was almost insipid, but with a slight roughness, indicating astringency. It proved wholly insoluble in cold water, and nearly so in boiling water, though I think it went into a pulp under such treatment. It did not appear either soluble in pure spirits of wine, but imparted a red colour to the liquid."

"This jelly, when dried by the air and heat of the sun, acquired a dark-brown colour; the surface dried first, and the inner part gradually shrunk afterwards, accounting for the blister-like irregular pieces." **Mr. Atkinson**, in his *Himálayan Districts*, says: "The gum of this tree is known as *mócharas*."

634

Dr. Moodeen Sheriff, in his admirable "Supplement to the Pharmacopœia of India," attributes a portion at least of the *mócharas* to **Bombax malabaricum**, but makes no mention of the so-called *supárí-ka-phul*. He states that there are two varieties of the *mócharas*. "Both occur in very irregular, nodular, smooth, and shell-like pieces, opaque and dark brown in colour, the difference being, one is very hard and broken with difficulty, and the other is brittle and easily broken, and less astringent in taste. The latter is the inferior of the two, and is the produce of **Bombax malabaricum.** No gum is produced from this tree on making incisions (however deep), but occasionally a very small quantity of it is exuded spontaneously. It is of a yellowish red or flesh colour at the beginning for some days, and then becomes deep brown. After some months it gradually and occasionally acquires the form I have described." This account was published in 1869, and it entirely concurs with **Mr. Baden Powell's** personal observations published in the *Indian Forester* in 1882, from which an extract has already been given. The bulk of the evidence which has since come to light goes a long way to show that both forms of *mócharas* described by **Moodeen Sheriff** are in all probability derived from **Bombax.**

The following extract from **Dr. Dymock's** *Materia Medica of Western*

India (published 1883) will be found to convey the main facts known regarding *mócharas*:—

"DESCRIPTION.—When first exuded it is a whitish fungous mass which gradually turns red, and finally dries into brittle mahogany-coloured tears. The larger tears are hollow in the centre, the cavity being produced during the gradual drying of the jelly-like mass which first exudes. Dry *mócharas* when soaked in water swells up, and resumes very much the appearance of the fresh exudation. The taste is purely astringent like tannin. 635

"MICROSCOPIC STRUCTURE.—*Mócharas* is not a simple juice, but the product of a diseased action, which consists in a proliferation of the parenchyme cells of the bark. Upon making a section of the diseased part, a number of small cavities are seen which contain a semi-transparent jelly-like substance, consisting of oblong cells with botryoidal nuclei. At the margin of the cavity the columns of healthy cells are seen breaking up, and the cells separating to join the jelly-like mass; this gradually increases in size and finds its way to the surface to be extruded as *mócharas*. Upon its first appearance it is of an opaque, yellowish-white colour, firm externally, but semi-fluid internally, and there is no central cavity. The cause of the diseased condition of the bark which produces *mócharas* has not been determined. 636

"COMMERCE.—*Mócharas* is collected by Bheels and wandering tribes in Western India. It is sold by all the druggists. Value R4 per Surat maund of 37½ lbs. The gum of Moringa (*shégva*) is frequently mixed with *mócharas*; though similar in colour, it may readily be distinguished by its weight and solidity." 637

Fibre.—THE INNER BARK of the tree yields a good fibre suitable for cordage. The seeds yield the so-called red SILK-COTTON or *simal* cotton, a fibre too short and too soft to be spun, but largely used for stuffing pillows, &c. It has also been talked of as a paper fibre. The smoothness of the cotton prevents cohesion or felting, and hence in the textile industries this fibre could only be used to mix with others, imparting a silky gloss to the fabric. A writer in *The Tropical Agriculturist*, speaking of the *imbul* (which is either this tree or the white silk-cotton tree), says: "I believe this product (tree cotton) will become far more important than Ceará rubber. Civilization is rapidly opening its mind to the fact that *pulan* (the Singhalese for cotton-wool) makes a sufficiently soft bed, at a comparatively small cost, while in it the manufacturers of gun-cotton have found a cheaper and equally efficient raw material, as a succedaneum, to that used formerly."

FIBRE. Bark. 638
Cotton. 639
Gun-cotton. 640

The Kapok Fibre.—The demand for new fibres has recently directed attention to the subject of silk-cottons, and it seems that the produce of two, if not three, very different trees have, in Government reports and public newspapers, been confused with each other. Much of what has or can be written regarding one of these fibres is probably applicable to the others, but it seems desirable that they should be carefully distinguished. In a correspondence regarding silk substitutes from Messrs. Manning, Collyer, & Co., London, forwarded to the Government of India by Her Majesty's Secretary of State, silk-cotton or *kapok* was incidentally discussed. Samples of silk-cotton were accordingly forwarded for examination, with the result that Messrs. Manning, Collyer, & Co. reported that the *semul* cotton supplied them was better known as *kapok*, and that there was but a small demand in England for the article. A considerable trade, they added, exists "in Holland, where, however, the longer-stapled qualities from Java are much preferred." The sample supplied "being on the seed will lose very considerably in cleaning, and the present estimated value is 2*d*. to 2¼*d*. per lb.; possibly with regular 641 supplies rather more might be obtained, say 3*d.* a lb." **Mr. Collyer,** speaking at a meeting of the Society of Arts (London, 1883), said that the Dutch were far in advance of us in using silk-cotton. At the Amsterdam Exhibition and in Holland, **Mr. Collyer** remarks that silk-cotton fetches 8*d.* a lb., whereas in England only 2*d.* can be got. The quality of the fibre was, however, better. In a further article in the *Journal of the Society of Arts,* **Mr. Collyer** goes into the subject of silk-cottons, and mentions several species, but appears to have overlooked entirely the *simul* tree of India—**Bombax malabaricum.**

Through the kindness of **Professor W. T. Thiselton Dyer** of Kew, I have seen a circular on the *kapok* fibre issued by Messrs. J. C. Kütgen & Co., Rotterdam (dated November 1883). From this interesting paper it appears that the chief use of the fibre is in upholstery. The *kapok* is, however, quite distinct from the *simal,* and since the former fetches a much higher price than the latter, it seems desirable that the two should be carefully distinguished in all experimental or commercial consignments. The *kapok* is obtained from **Eriodendron anfractuosum,** the white silk-cotton, while the *semal* is the fibre from the seeds of **Bombax malabaricum,** the red silk-cotton. Since both trees occur abundantly in India, to participate in the new and apparently considerable trade in *kapok,* all that seems necessary is for India to direct its attention to the correct plant.

The following brief notice of the Indian silk-cottons may help to remove ambiguity, but the reader is referred for fuller details to the accounts given under each in their respective alphabetical positions in this work. The plants are enumerated in the order of probable merit.

642

1st—**Eriodendron anfractuosum,** *DC.*

THE KAPOK, OR WHITE SILK-COTTON.

This is particularly plentiful in the Konkan, but it grows in most parts of India, and its cultivation could be extended. As a road-side tree, while affording shade, it might be made to yield a distinct revenue to the country.

643

2nd—**Bombax malabaricum,** *DC.*

THE SIMAL, OR RED SILK-COTTON.

This is the commonest of the silk-cotton trees, occurring throughout the peninsula, but more particularly in the eastern side, and ascending the hills to 4,000 feet in altitude.

644

3rd—**Cochlospermum Gossypium,** *DC.*

THE KAMBI, OR GALGAL.

A common tree of the lower hills of India from Garhwál, Bundelkhand, Behar, Orissa, and westwards to the Deccan. It has large yellow flowers, and is not uncommon in cultivation throughout the country, especially in South India. It does not appear that the samples of this form of silk-cotton have been consigned to Europe and declared as such, so that its peculiar merits have not been definitely determined.

645

4th—**Calotropis gigantea**—the *Madar* and other ASCLEPIADACEÆ and APOCYNACEÆ—yield silky hairs—the coma of the seeds. These are generally classed as silk-cottons, but with the exception of *madar* none of these fibres have as yet been experimented with. The natives of India regard the *madar* silk-cotton as much cooler than *simal,* and affirm that it has a soothing effect.

OIL. 646

Oil.—Cooke, in his *Oils and Oil-seeds,* makes mention of this plant as yielding an oil, but gives no other information about it.

FOOD. 647

Food.—The flower-buds are eaten as a pot-herb.

The Assistant Commissioner of Balaghat, in **Mr. Liotard's** note on

Mahuá, says that this constitutes a regular article of food. "Of the minor forest produce, about 5,000 maunds of *simal* are used as food." Monkeys eat the young flower-buds. (*Ulwar Gaz., 32.*)

Fodder.—The leaves and twigs are lopped for fodder.

FODDER. 648

Medicine.—THE GUM or dried juice, *mócha-ras*, which the tree yields, is used as an aphrodisiac. This gum contains a large proportion of tannic and gallic acids, and may be successfully employed in cases requiring astringents. It has also tonic and alterative properties; it is regarded as a styptic, and is used in diarrhœa, dysentery, and menorrhagia. In Rewa Kantha, Gujarát, this gum is known as *kamarkas*; it is "ground to powder and drunk in milk as a tonic." (*Bomb. Gaz., VI., 14.*) The gum of the *semul* tree, *mócha-ras*, is given to children as a laxative, and the dried flowers are used as demulcent (*Irvine*). The dry flowers, with poppy seeds, goat's milk, and sugar, are boiled and inspissated, and of this conserve two drachms are given three times a day in hæmorrhoids. (*Medical Topography of Dacca, by Dr. J. Taylor, 56.*)

MEDICINE. 649

GUM. 650

Flowers. 651

Root. 652

A decoction of the root gives a gummy substance, used in the Deccan as a tonic medicine. May not part of the *mócharas* sold by our druggists be this resinous extract? The roots have stimulant and tonic properties attributed to them. They have come to bear the name of *musla*, but this must not be confused with *safèd-musli*. The *Pharmacopœia of India*, while not exactly making this mistake, publishes a note regarding *safèd-musli* under **Bombax**, and then proceeds to say that the roots sold under that name appear not to belong to **Bombax**, but to be the roots of some monocotyeldonous plant. Both *musla-simal* and *safèd-musli* exist, however, and have separate properties attributed to them. The *Ulwar Gazetteer* says (page 32): "The roots of this plant are called *musla*, and they are much used in medicine." "*Musli-sembal* is a light woody fibrous root of a brownish colour, with a thin epidermis, easily detached, and a very fibrous thick tuber. It acts as a stimulant and tonic, and some consider it in large doses emetic. It is said to contain 10 per cent. of resin." (*Baden Powell, Panjáb Products, I., 333.*) The young roots dried in the shade and powdered form the chief ingredient in the *musla-semul*, a medicine highly thought of as an aphrodisiac; it is also given in impotence.

Extract from Root. 653

THE BARK and THE ROOT are also emetic. THE LEAVES are made into a paste and used as an external application.

Bark. 654

Special Opinions.—§ "Its gum (*mócharas*) is useful in diarrhœa of children, dose 20-30 grains, with equal parts of sugar." (*Surgeon J. Anderson, M.B., Bijnor.*) "The tap-root of the young plant is used for gonorrhœa and dysentery." (*Surgeon-Major P. N. Mukerji, Cuttack, Orissa.*) "The leaves singed and beaten or rubbed with water to a pulp make a useful application for glandular swellings." (*Mr. W. Forsyth, Civil Medical Officer, Dinajpore.*)

Leaves. 655

Structure of the Wood.—White when fresh cut, turning dark on exposure; very soft, perishable; no heartwood; no annual rings. The wood of old trees is often of a dull-red colour. It is not durable, except under water, when it lasts tolerably well.

TIMBER. 656

It is used for planking, packing-cases, and tea-boxes, toys, scabbards, fishing-floats, coffins, and the lining of wells. In the Konkan, Bengal, and Burma, the trunk is often hollowed out to make canoes and water-troughs.

Dr. Buchanan says it is the timber commonly employed by the natives of Behar "for making doors and window-shutters; for it lasts well in such situations, and is very strong to resist the attacks of robbers." (*Statistics of Dinajpore.*) It is used as firewood in the Konkan. (*Bomb. Gaz., X., 40.*)

DOMESTIC. Tinder. 657

Domestic Uses.—The cotton is made into tinder. The tree is often mentioned in the Vedas. It is the Yamadruma, or tree of Yama, the Indian god of Death.

Bonduc, see **Cæsalpinia Bonducella,** *Roxb.;* LEGUMINOSÆ.

658

BORAGINEÆ.

Herbs, shrubs or trees, often hispid or scabrous. *Leaves* alternate, very rarely opposite, exstipulate, mostly entire. *Flowers* usually in dichotomous scorpioid cymes, rarely solitary and axillary. *Calyx* inferior, 5-, rarely 6-8-toothed or -lobed, usually persistent in fruit. *Corolla* gamopetalous, often with scales in the throat, rarely 4-6-lobed, imbricate (rarely twisted) in the bud. *Stamens* as many as the corolla-lobes, alternate with them, upon the corolla-tube. *Ovary* superior; cells 2, 2-ovuled, or 4-1-ovuled; style terminal or from between the ovary-lobes, long or short, stigma capitate or 2-lobed, rarely the style twice bifid; ovules sub-erect from the inner basal angle of the cell. *Fruit* drupaceous or dividing into 2-4 nutlets. *Seeds* erect or oblique, testa membranous, albumen fleshy, copious, sparing or 0; embryo straight or curved, radicle superior. Species 1,200 throughout the world.

Tribe I. Cordieæ. Trees or shrubs. *Style* terminal on the entire ovary, twice bipartite. *Drupe* 4-1-seeded; albumen 0; cotyledons plicate longitudinally.

Calyx-teeth very short, irregular . . . . . 1. **Cordia.**

Tribe II. Ehretieæ. *Style* terminal on the entire ovary, simple, bipartite, or styles 2. *Drupe* with 2 2-celled or 4 1-celled pyrenes, or of 4-1 nuts.

Trees or shrubs. Style 2-fid . . . . . 2. **Ehretia.**
Prostrate herb. Styles 2 . . . . . 3. **Coldenia.**
Virgate shrub. Style 1, stigma capitate. . 4. **Rhabdia.**

Tribe III. Heliotropieæ. *Style* terminal on the entire ovary, depressed—conic at the apex, or with a horizontal ring below the stigmas. *Fruit* as of Ehretieæ.

Shrubs, often scandent. Style short, shortly 2-lobed . . . . . . 5. **Tournefortia.**
Herbs, style dilated at the apex or above the base 6. **Heliotropium.**

Tribe IV. Borageæ. Herbs. *Style* simple or bifid, rising from between the ovary lobes (except in *Trichodesma*). *Nutlets* 4, rarely 3-1 by suppression (2 in *Rochelia*); albumen 0.

Sub-tribe I. Cynoglosseæ. *Nutlets* attached to a convex or conical carpophore, scar continued to the apex of the nutlets, which are often depressed, produced, or saccate at the base.

* *Fruiting calyx enlarged, enclosing the nutlets.*

Anthers conically convenient, lanceolate, subexsert 7. **Trichodesma.**

** *Nutlets depressed, their bases produced downwards.*

Stamens included. Nutlets obovoid, scar punctiform 8. **Actinocarya.**
Stamens included. Margins of nutlets reflexed over their backs . . . . 9. **Omphalodes.**
Stamens included. Nutlets obovoid, glochidiate . 10. **Cynoglossum.**
Stamens exserted; anthers large, linear oblong . 11. **Lindefolia.**
Stamens exserted; anthers small, shortly oblong . 12. **Solenanthus.**

*** *Nutlets connate, forming a pyramidal fruit, margined, hardly produced downwards.*

Racemes ebracteate. Margin of the nutlets glochidiate, often reflexed . . . . . 13. **Paracaryum.**

Racemes bracteate. Margin of the nutlets glochidiate, scarcely reflexed . . . . . 14. **Echinospermum.**

Sub-tribe II. Eritrichieæ. *Nutlets* attached to a convex or conical carpophore, scar in the middle or lower half of the nutlets, which are not depressed at the base, but are produced at the apex above the scar, free round the base of the style.

* *Scar in the basal half of the nutlet.*

Racemes ebracteate. Nutlets 4 . . . . 15. **Eritrichium.**
Racemes bracteate. Nutlets 2, 1-seeded . . . 16. **Rochelia.**

** *Scar in the middle of the inner face of the nutlets.*

† *Scar small, without a prominent, thickened, incurved margin.*

Flowers axillary, subsessile. Fruiting calyx enlarged . . . . 17. **Asperugo.**
Almost stemless. Fruiting calyx not enlarged . 18. **Microula.**

†† *Scar depressed, with a thickened incurved margin.*

Flowers axillary, pedicelled . . . 19. **Bothriospermum.**
Flowers axillary, subsessile . . . 20. **Gastrocotyle.**

Sub-tribe III. Anchuseæ. *Nutlets* on a flat or nearly flat receptacle; scar basal, prominent, hollowed out, with a prominent thickened margin.

* *Corolla-throat closed by 5 scales.*

Corolla-tube straight . . . . . 21. **Anchusa.**
Corolla-tube curved . . . . . 22. **Lycopsis.**

** *Corolla-throat naked or hairy within, but without scales.*

Racemes dense. Calyx large . . . . 23. **Nonnea.**

Sub-tribe IV. Lithospermeæ. *Nutlets* on a flat or nearly flat receptacle; scar basal, but little hollowed out, without a prominent margin.

* *Racemes ebracteate, corolla-lobes distinct.*

Corolla-tube cylindric. Anthers included . . 24. **Mertensia.**
Corolla-tube cylindric. Anthers exserted . . 25. **Moltkia.**
Corolla-tube short. Nutlets tetrahedral . . 26. **Trigonotis.**
Corolla-tube short. Nutlets ovoid-oblong . . 27. **Myosotis.**

** *Racemes bracteate, corolla-lobes distinct.*

Corolla-throat naked or with small scales . . 28. **Lithospermum.**
Corolla-throat densely filled with hairs . . 29. **Sericostoma.**
Hispid spreading herbs. Corolla yellow, tube elongate . . . . 30. **Arnebia.**
Sub-erect herbs. Corolla purple, tube elongate . 31. **Macrotomia.**

*** *Corolla-lobes reduced to minute teeth.*

Anthers lanceolate, connivent in a cone . . 32. **Onosma.**

The preceding extract from the *Flora of British India* will doubtless be found useful to the student of Economic Botany. It will at least serve to direct his attention to the names of Boraginaceous genera, to identify which it will, however, be necessary to consult the *Flora*, since an analysis can at most isolate the tribes or more marked genera, and is of use only when the reader possesses a perfect and typical specimen. As formed at the present day, the Boragineæ may briefly be said to embrace two very different groups of plants, the one tropical or warm-temperate trees, shrubs, or herbs, and the other temperate or extra-tropical herbs. This is not absolutely correct, but it is so far so as to make a statement of the distribution of the Indian Borageworts somewhat misleading. The herbaceous or what may be called the true or more typical Boragineæ (the members of the tribe Borageæ) are almost entirely temperate. They abound in the southern part of Europe, the Levant, and the temperate regions of Asia. They are less frequent in northern latitudes, and almost disappear from the tropics. This fact is so well known—the Forget-me-nots being viewed as a most typical feature of the temperate regions—that an analysis of the Indian species of Boragineæ will assign to the plains a very misleading proportion; but it must be borne in mind that these belong chiefly to the tribes Cordieæ, Ehretieæ, and Heliotropieæ, which might be viewed as constituting a separate order, and indeed the Cordieæ have been treated as such by many authors.

In Boragineæ there are in all 1,200 species, of which India possesses 139. Of the latter, 51 or 36·7 per cent. are confined to the plains; 25 or 20·1 per cent. ascend to 5,000 feet in altitude; 30 or 21·6 per cent. to 10,000 feet; and 30 or 21·6 per cent. are met with above that altitude. Thus 51 are tropical and 98 temperate. This is an approximately correct statement only, since some of those included in the second group ascend from the plains to the hills and thus overlap the division into tropical and temperate.

Their distribution over India shows a corresponding temperate character, the majority occurring in the North-West Provinces and the Panjáb, the plains of which are much colder than the plains of the eastern side of India, and have accordingly a much larger number of species, especially of cold-season annuals. The eastern side of India has 24 species or 17·2 per cent.; South India 14 or 10·0 per cent.; Western India 11 or 7·9 per cent.; Sind 6 or 4·3 per cent; and the Panjáb and North-West Provinces 59 species or 42·4 per cent. Distributed over two or more of these divisions,—that is, occurring throughout India,—there are 25 species or 17·9 per cent. Nearly all the species thrown into this last group are distributed to North India (*i.e.*, Panjáb and North-West Provinces, &c.), so that it is quite clear that the region of Boragineæ, as far as India is concerned, must be viewed as the hills, plains, and mountains of the northern section of the empire.

659

The affinities of the Boragineæ are with Labiatæ and Verbenaceæ, but they form so well-marked an assemblage that it is not necessary to enter into this subject. The Cordiaceæ differ from the more typical Boragineæ in being arborescent, and in having a twice-forked terminal style, baccate fruit, and plaited cotyledons.

660

Properties and Uses.—Few are of any very great importance. Many species contain a mucilage to which is often combined a bitter astringent principle. The Comfrey root (**Symphytum officinale**) was formerly used in hæmoptysis. The Boragos are regarded as diuretics. Cynoglossum yields a poisonous narcotic root. The sweetly-scented Heliotrope (**Heliotropium peruvianum**) belongs to this natural order, and so of course do the favourite Forget-me-nots (Myosotis). The fruits of some of the Ehretieæ are edible, and the roots of Anchusa, Onosma, Alkanna, and Arnebia afford the dye Alkanet. One of the most curious and interesting members of this order is the drug sold in India under the name of *Gaozabán*, which recent investigation seems to have proved to be a species of Echium; it is imported from Persia. Some doubt also prevails as to the drug *Rattanjot*. While various substances are sold under that name, the true article appears to be the root of some Boraginaceous plant, very probably a species of Onosma.

661

BORAGO, *Linn.; Gen. Pl., II., 854.*

Borago indica, *Linn.* (as in *Roxb., Fl. Ind., Ed. C.B.C., 854.*

Syn. for **Trichodesma indicum,** *Br.*, which see.

Borago zeylanica, *Linn.*, see **Trichodesma zeylanicum,** *Br.; Fl. Br. Ind., IV., 154.*

BORASSUS, *Linn.; Gen. Pl., III., 939.*

662

An erect, graceful palm, with a terminal crown of fan-shaped leaves, belonging to the tribe BORASSEÆ of the Natural Order PALMÆ. It is a native of Africa, but at the present day exists in a state of cultivation throughout India. It is almost unnecessary to give an enumeration of the generic characters of this well-known palm. There is only one species, and, as far as the plains are concerned, it is the only palm with fan-shaped leaves. The flowers are diœcious, occurring in panicled spikes; male thick, cylindrical; flowers fasciculate in the axils of broad, whorled, imbricate, connate bracts; ovary 3-celled.

The generic name is βόρασος, the Greek name for a palm fruit.

Borassus flabelliformis, *Linn.;* PALMÆ.

663

THE PALMYRA PALM; BRAB TREE.

Vern.—*Tál, tála, tár, tári,* HIND.; *Tál,* BENG.; *Tale,* SANTAL; *Tád, tád-nu-jháda,* GUJ.; *Dral* or *tád,* SURAT; *Tár-ká-jhár, tár,* DUK.; *Táticha-jháda, táda, talat-mád,* MAR.; *Potu táti* (the male tree), *penti táti* (the female), *táti-chettu,* TEL.; *Panai-maram* or *panna-maram, panam, pannie, pampai* (?) TAM.; *Paná,* MAL.; *Táll, tále, pané-mara,* KAN.; *Tála,* SANS.; *Darakhte-tári,* PERS.; *Tal-gass, Tal,* SINGH.; *Htan, tan,* BURM.

Habitat.—A tall palm with cylindrical stem, cultivated throughout tropical India, and beyond the tropics in Bengal and the southern part of the North-West Provinces. The young stems are covered with dry leaves, or rather with the lower part of the petioles, while the old stems are marked with the hard, black, long and narrow scars of the fallen petioles. In Upper India it is chiefly seen on embankments around tanks, but in Bengal it luxuriates in the mixed cocoanut and date-palm jungles. **Brandis** says it extends up both sides of the Persian Gulf, attaining about the same latitude as in North-West India. It is also cultivated in Prome, in Ceylon, and in the Indian Archipelago. "It thrives in this district, although it never grows spontaneously; and is finely adapted for covering the naked sides of tanks which are now almost entirely useless." (*Buchanan's Statistics of Dinajpore, p. 150.*)

664

Synopsis of the Economic Uses.—Every part of this plant is made use of in some way or other. A Tamil poem enumerates some 800 uses to which the various parts are put. *The Tropical Agriculturist* (June 1884) publishes a list of the more important of these uses, enumerated by **Mr. Robert O. D. Asbury** of Jaffna, Ceylon, arranged in seven groups as follows: Group I., Wooden utensils; Group II., Food materials; Group III., Leaves; Group IV., Fibre; Group V., School things and toys; Group VI., Toddy-drawers' utensils; Group VII., Miscellaneous.

THE GUM.

GUM. 665

Gum.—A gum, obtained from this palm, is said to have been sent from Madras to the Panjáb Exhibition; it is black and has a shining fracture.

FIBRE.

FIBRE. 666

Fibre.—The fibre extracted from the leaf-stalks is used for rope and twine-making, and may also be used for paper. This fibre is strong and wiry, and is about 2 feet long. In Ceylon it is extracted and the ropes and string, largely used for cattle yokes and other agricultural purposes, are made of it. In Madras it is also made into rope and twine. In Bengal the trees are too scattered to admit of an extended trade in this fibre. The long cord-like and dark-coloured fibro-vascular bundles are carefully extracted, however, while preparing dug-outs, &c. By the

Wood-fibre. 667

fishermen these are made into invisible fish-traps. For this purpose they are platted into a long tapering tube, the meshes of which are 2 inches in size. This tube is placed in a dividing wall of weeds run across the tank; it thus forms what appears a natural and apparently easy passage from one expanse of water to another. At the end of the passage is placed, however, a noose made also of the *tár* fibre. In darting through the passage this trap is so arranged that the fish must run its head into the noose and is thus firmly secured. The fishermen put the *tár* fibre through some process of preparation, but it is not spun or twisted in any way, a single thread or fibro-vascular bundle being used.

Coir. 668 Baskets. 669 Braid. 670 Mats. 671 Thatch. 672

Coir or fibre from the pericarp is doubtless prepared in many parts of India, but no definite information can be obtained. The leaves are made into fans or worked into boxes and baskets and into many minor objects; amongst these may be mentioned a braid platted of thin strips of the leaves, and used for ornamental purposes. They are also extensively used for thatching huts.

§ "Ropes for country craft are made from the leaf and leaf-stalks, mats are also made from them, and the hoods which labourers wear in the monsoon, called in Mahratta *Khori*. The leaves are also used for thatching huts." (*Surgeon-Major W. Dymock, Bombay.*)

MEDICINE.

MEDICINE. Juice. 673 Ash. 674 Bud. 675 Root. 676 Toddy Poultice. 677

Medicine.—The JUICE of this plant is used as a stimulant and antiphlegmatic. When freshly drawn, it is exceedingly sweet, and, if taken regularly for several mornings in succession, acts as a laxative. It is also useful in inflammatory affections and dropsy. The fermented juice called *tári* or toddy is intoxicating. The ASH of the dry spadix is an *antacid* in heartburn; **U. C. Dutt** says that it is regarded by the Hindús as useful in spleen. The terminal BUD of the *tál* tree is regarded as nutritive, diuretic, and tonic. The ROOT is regarded as cooling and restorative. (*U. C. Dutt.*) "A useful stimulating application, called TODDY POULTICE, is prepared by adding fresh-drawn toddy to rice-flour till it has the consistence of a soft poultice, and this being subjected to a gentle fire, fermentation takes place. This, spread on a cloth and applied to the affected part, acts as a valuable stimulant application to gangrenous ulcerations, carbuncles, and indolent ulcers." (*Drury's Useful Plants.*)

Petioles. 678 Fruit-pulp. 679

"The juice of the FRESH PETIOLES is given as a stimulant antiphlegmatic, and is used by native physicians as an adjunct to stimulating drugs in the low stages of intermittent and remittent fevers." "The PULP of the ripe fruit is applied externally in skin diseases." (*Babu T. N. Mukerji, in his Amsterdam Catalogue.*)

Cotton. 680

The light-brown, COTTON-LIKE SUBSTANCE from the outside of the base of the fronds, is employed by the Singhalese doctors as a styptic to arrest hœmorrhage from superficial wounds.

681

Special Opinions.—§ "Vinegar, toddy, and a spirituous liquor are made from this tree. The juice, slightly fermented, is used in diabetes. The ash of the spadix is given internally in bilious affections." (*Surgeon G. A. Emerson, Calcutta.*) "The expressed juice of the leaf-stalk and young root is used in cases of gastric catarrh and to check hiccup. The fresh juice is diuretic and used in gonorrhœa. The fermented juice is uncertain in its action and sometimes acts as a drastic purgative." (*Brigade Surgeon J. H. Thornton, B.A., M.B., Monghyr.*) "Fresh juice is cooling and is considered as a luxury in the hot season." (*Assistant Surgeon Shib Chunder Bhuttacharji, Chanda, Central Provinces.*) "An extract of the green leaves is used internally in secondary syphilis." (*Surgeon-Major J. J. L. Ratton, M.D., Salem.*) "The fresh juice obtained by cutting the spadix is a good diuretic and is useful in cases of dropsy The fermented

MEDICINE.

juice (toddy) is used as yeast in baking bread; it is very intoxicating." (*Surgeon J. Anderson, M.B., Bijnor.*) "Water contained in the cavities of the pulp, when unripe, is used as a remedy for nausea and vomiting. The water is sweetish in taste." (*Assistant Surgeon Anund Chunder Mukerji, Noakhally.*) "The palm seeds when immature contain a milky fluid which is sweetish and cooling. It is often given to prevent hiccup and sickness." (*Surgeon-Major R. L. Dutt, M.D., Pubna.*) "The fruit is cooling and useful in relieving thirst in fever." (*Surgeon-Major A. S. G. Jayakar, Muskat, Arabia.*) "Mixed with aromatics the unfermented saccharine juice taken in the mornings is a good tonic in emaciation of the body where the patient can digest it." (*Native Surgeon T. Ruthnam Moodelliar, Chingleput, Madras.*) "The ash of dry spadix is largely used with other drugs by *kaberajes* as an antiperiodic; it is feebly so." (*Assistant Surgeon Devendro Nath Roy, Sealdah, Calcutta.*)

FOOD.

The Juice, Toddy, and Sugar.

FOOD. 682

THE JUICE. 683

By far the most important product of this plant is the juice—*Ras*—obtained on tapping the flower-stalk. This, before sunrise, is sweet and agreeable to the taste, and while fresh is either consumed as a beverage or boiled down to sugar. In the Madras Presidency the quantity of jaggery sugar made from the juice of this palm is very considerable. After sunrise the juice rapidly ferments, however, and is then converted into toddy—*tári*—an intoxicating drink. **Dr. Ainslie** (*Mat. Ind., I., p. 451*) describes four kinds of toddy which were prepared in his time, but makes no mention of the date-palm toddy. He gives preference to cocoanut-palm juice, after that to palmyra, then the toddy from **Caryota urens**, and last of all that from the *Ním* tree. In most parts of India toddy is extracted from some palm or other, but in Bengal one might almost say the date palm was exclusively used for this purpose. The Palmyra, on the other hand, is the toddy palm of South India, of the Konkan, of Burma, and of Ceylon.

Definite information cannot, however, be obtained regarding the amount of Palmyra toddy, or of the sugar actually prepared in India, since, in the returns given for this substance, separate records are not kept of the trees from which the palm toddy and sugar are obtained. In another part of this work, under the heading "Toddy," further details will be given, but the following abstract may be found useful:—

Juice or Ras. 684

(1) The fresh juice is called *ras.* If not consumed before sunrise it turns milky, and rapidly ferments.

Molasses or Jaggery. 685

(2) The fresh juice if boiled down yields molasses or *jaggery* from which sugar may be refined. The juice collected for this purpose has a small piece of lime placed in it to prevent fermentation while suspended from the tree.

Refined Palmyra Sugar. 686

(3) The fermentation is accelerated by placing in the liquid what are known as fermentation-seed,—that is, rice saturated with old or fermented *ras.* The fermented liquid is called *toddy* or *tári.*

Toddy or Tari. 687

(4) If distilled, palm-wine or *arak* is the result.

Distilled Spirit or Arrack. 688

(5) By destructive distillation a good quality of vinegar is produced from the juice.

Vinegar. 689

The various methods of extracting the juice, and of preparing either of the five substances above briefly enumerated, will be best shown by republishing from standard authors an account of the industry as practised in Madras, Bombay, Ceylon, Burma, and Bengal:—

"The mode of procuring the vinous sap is as follows: The spadix or young flowering branch is cut off near the top, and an earthen *chatty*

THE JUICE.

MADRAS. 690

or pitcher then tied on to the stump; into this the juice runs. Every morning it is emptied and replaced, the stump being again cut, the vessel placed as before, and so on, until the whole has been gradually exhausted and cut away. It is known in Tamil as the *Pannungkhulloo.* It is from this liquor that sugar is extracted, and by the same process as that described for procuring the toddy, except that the inside of the earthen vessel or receiver is powdered with chunam, which prevents any fermentation; the juice is then boiled down, and dried by exposure. Some few trees that from unknown causes do not flower in spring, put out their flowers in the cold season, and give a scanty supply; but in spring many are rendered artificially barren by breaking off the flowering-bud as it begins to form. These also flower in the winter season, and are called *Basanti.* They do not give above 2½ maunds of juice, but this is of as much value as the 6 maunds which a tree gives in spring. Either the male or female will answer for the spring or winter crop, but the female alone will yield juice in the rainy season. When this is wanted, the fruit is allowed to form, and afterwards the point of the spadix or stem which supports the clusters is cut and allowed to bleed." (*Drury, Useful Plants, p. 83.*)

BOMBAY. 691

In the Konkan, Thána District, "the fan palm is the chief liquor-bearing tree. It grows wild all over the district, and is found by tens of thousands in the coast sub-divisions. The trees are of different sexes, the male being called *talai,* and the female *tád.* The juice of both is equally good. The trees are also known as *shilotri, dongri,* and *thalzáni,* according as they have been planted by the owner or grow on uplands or on lowlands. Fan palms artificially reared grow rather more quickly than wild ones. The ground is not ploughed, but a hole, about a foot deep, is made, and the seed buried in it in *Jeshth* (May-June). No watering is necessary, and the only tending the plant requires is the heaping of earth round the base of the stem to quicken the growth. In about twelve years it is ready for tapping, and will yield liquor for about fifty years, or, as the saying is, to the grandson of the man who planted it. In the case of the male palm, *talai,* the juice is drawn from the *lendis,* which are finger-like growths, from twelve to fifteen inches long, given out in clusters at the top of the tree. Some of the fingers in the cluster are single, others spring in threes from a common base. Each finger is beaten with a piece of stick called a *tapurni,* three times in three lines along its whole length, and all the fingers of the cluster are tied together. In three or four days, the points of the fingers are cut by the *áut,* a sharply-curved knife with a keen flat and broad blade. The points are cut daily for about a fortnight, when the juice begins to come. Under the tips of the fingers earthen pots are placed into which the juice is allowed to drop, and to keep off the crows a sheath of straw is bound round the *lendis* so as to close the mouth of the jar. The female tree gives out spikes from twelve to fifteen inches long, with the fruit seated all round the sides of the spike, as in a head of Indian-corn. The spikes are known as *sapat koti, gangra,* and *pendi,* according as the juice issues when the berries, *tádgolás,* are still minute, fairly grown, or very large. In trees which yield juice while the berries are still very small, *sapat koti,* the spike, is beaten, and on the third day its point is cut, and the sides rubbed with the hand so as to brush off the incipient fruit. In ten or twelve days the juice begins to drop. In trees which yield juice when the spike is fairly grown, *gangra,* the spike, must be beaten on the interstices between the berries with a long stone, called a *dagdi gunda,* or, if the interstices are very fine, with an iron pin called *lokhandi gunda.* On the third day the tip is cut, and in about fifteen days the juice begins to flow. In trees which

THE JUICE.

BOMBAY.

yield juice when the fruit is large, *pendi*, the parts of the spike visible between the berries, are beaten in the same way, and a month afterwards the end of the spike is cut daily for about a fortnight, when the juice generally begins to come. As the *gangra* and *pendi* are cut, the fruit on the sides has to be gradually removed. A fan palm tree will yield from six to sixteen pints (three to eight *shers*) of juice every twenty-four hours. Almost the whole is given off during the night. When the juice has begun to flow, the fingers of the male tree and the spike of the female tree must have their points cut morning and evening. The distillation of palm juice is simple. The juice is put into an earthen jar, *madka*, and allowed to stand for five days. It is then placed over a fire, and the spirit, rising as vapour, passes through a pipe into another jar into which it is precipitated in a liquid form by the action of cold water. One hundred *shers* of juice yield about twenty-five *shers* of spirit." (*Bomb. Gaz., Vol. XIII., Pt. I., pp. 22-23.*)

In Kolába District, "with few exceptions these palms are self-sown and no care is taken of them, except that a few thorns are sometimes set round seedlings to keep cattle away. The tree is full-grown at twenty-five or thirty years. It is tapped for about thirty years more, and is said to live about forty years after it has grown too old to be tapped. Both the male and female trees are tapped. The spathe, *pogi*, of the male tree is called *lendi*. Vigorous trees throw out from three to five spathes a year, some in November, *sargacha hangam*, and the rest in February, *bhár kála*. Trees that are not in full vigour throw out spathes in November only. The spathe is gently bruised with a piece of wood, the bruised parts bound together, a slice is cut off the point of the spathe by the drawer's sharp and broad-bladed knife, *áut*, and a pot is tied over the end to catch the juice. The tree is then tapped twice a day, a little slice being cut off the end of the spathe at each tapping. Under this process each spathe lasts, according to its length, from a month to a month and a half. The tapping season continues from October to May. The drawer is paid at the rate of 1*s.* (8 annas) a month for each tree. Each tree yields about $3\frac{3}{8}$ pints ($1\frac{1}{2}$ *shers*) a day, which at $\frac{5}{16}d.$ the pint (6 pies the *sher*) is worth $1\frac{1}{8}d.$ (9 pies), or 2*s.* $9\frac{3}{4}d.$ (R1-6-6) a month. Taking five months as the average time during which tapping lasts, the approximate gross profits are 14*s.* (R7). Deducting from this 6*s.* (R3) paid to Government and 5*s.* (R2-8) to the *Bhandári*, the net profit on each tree is about 3*s.* (R1-8). This was the state of affairs before 1879-80, when the tree-tax was raised to 12*s.* (R6); since this change the tapping of palmyra trees has ceased, except in Alibág. Palmyra juice can be distilled, but this is never done, as the supply of cocoa-palm liquor is in excess of the demand." (*Bomb. Gaz., Vol. XI., p. 29.*)

Little more than 20 years ago, the Bombay Government, becoming alarmed at the amount of spirituous liquor which was consumed, gave orders that, in Surat, large numbers of this noble tree and of the date palm should be destroyed. But in 1868 the total number of toddy-yielding trees was estimated at 1,243,711, of which 47,810 were palmyra palms. (*Bomb. Gaz., II., 39.*)

The following extracts from *Simmonds' Tropical Agriculture*—originally written by **Mr. W. Fergusson**—will be found to convey the more important facts regarding the extraction of toddy as practised in Ceylon:—

CEYLON. 692

"At the season when the inflorescence begins to appear, when the spathes have had time to burst, the 'toddy-drawer' is at work in the palmyra groves. His practised eye soon fixes on those trees fit for the 'scalping knife,' and if they have not dropped the foot-stalk

THE JUICE.

CEYLON.

of the leaves, the first operation, if the trees are valuable, is to wrench them off." "An expert climber can draw toddy from about forty trees in a few hours. In Jaffna a distinction is made between toddy and sweet toddy; the former, called by the Tamils 'culloo,' is the fermented, the latter the unfermented, juice."

"The juice of the palmyra is richer in saccharine matter than that of most other palms, in consequence, perhaps, of the tree more generally growing in dry sandy soil, and in a dry climate. The great fault of the jaggery made at Jaffna seems to arise from the too free application of lime, a small quantity of which is absolutely necessary to prevent fermentation."

"According to **Forbes**, three quarts of toddy will make 1 lb. of jaggery. **Malcolm** remarks that jaggery resembles maple sugar, and that in the neighbourhood of Ava 1 lb. sells for the third of a penny. In Jaffna 3 lbs. are sold for *2d.* The usual process of making jaggery, as pursued at Jaffna, is exceedingly simple. The sweet toddy is boiled until it becomes a thick syrup, a small quantity of scraped cocoanut kernel is thrown in, that it may be ascertained by the feel if the syrup has reached the proper consistency, and then it is poured into small baskets of palmyra leaf, where it cools and hardens into jaggery. In these small plaited palmyra baskets it is kept for home consumption, sent coastwise, chiefly to Colombo, or exported beyond seas to be refined. To make *vellum* or crystallized jaggery, which is extensively used as a medicine, the process is nearly the same as for the common sugar, only the syrup is not boiled for so long a period."

"Toddy serves extensively as yeast, and throughout Ceylon no other is employed by the bakers; large quantities of it are also converted into vinegar, used for pickling gherkins, limes, the undeveloped leaves of the cocoanut and palmyra trees, and other substances; but by far the greatest quantity is boiled down for jaggery or sugar. About 1,000 tons are said to be manufactured of it in Ceylon." (*Tropical Agriculture, Simmonds, 265.*)

BURMA. 693

Dr. Brandis says: "The most valuable produce of the tree is the sweet sap which runs from the peduncles cut before flowering, and collected in bamboo tubes or in earthen pots tied to the cut peduncle. Nearly all the sugar made in Burma, and a large proportion of the sugar made in South India and the Konkan, is the produce of this palm."

BENGAL. 694

The practice of extracting juice from the *tál* palm is almost unknown in Bengal, or at all events it is rarely if ever done, the date palm taking its place. Sugar is accordingly not made from this palm in Bengal, but "sugar-candy manufactured from it is imported into Calcutta from Ceylon, Madras, and the Archipelago. This is chiefly used in medicine as a remedy for cough and pulmonary affections." (*Babu T. N. Mukerji.*)

The Fruit and Seed.

695

The tree flowers in March, and the young fruits ripen in April and May and the mature fruits in July and August. These are about 5-7 inches in diameter, green when young, but becoming brownish black, shaded with yellow, as they mature. They form large clusters in the axils of the upper leaves. Normally each fruit contains three nuts or by abortion only one or two. The pericarp consists of three distinct layers, *viz.*, the epicarp or outer skin of the fruit, the mesocarp or fibrous and succulent layer within the epicarp, and last of all the stony endocarp or shell of each nut. Within the shell occurs al arge solitary seed, which consists of a thin seed-coat, in contact with the shell on the one side, and with a layer of albuminous matter on the other. When young, the interior of the albumen is filled with a jelly-like fluid. As it matures, this albumen becomes hard and firmly attached to the shell, the liquid being deposited with the growth of the embryo. Ultimately the embryo fills completely the central space, taking the place of the liquid now deposited as albumen.

It has been thought desirable to give the above detailed description of the structure of this fruit in order to remove the ambiguity which exists regarding the economic uses of its various parts.

Seeds. 696 Talsans. 697 Preserves. 698

THE UNRIPE FRUIT.—About April to May a certain number of the fruits are removed from the trees. The epicarp and mesocarp are removed and rejected, the shell is split open and the seed obtained. This constitutes the edible structure sold in Bengal under the name of *talsans.* The soft albumenous layer and the jelly-like fluid contained within it are eaten fresh, being regarded as cool and refreshing. They are sometimes cut into small pieces and flavoured with sugar and rose-water; in this condition they are viewed as a delicacy. In India it is very rarely the case that either the fresh seed or the above preparation from it is eaten by Europeans.

Fruit. 699 Pulp. 700

THE RIPE FRUIT.—In July and August, when the fruits are ripe, they are removed from the tree. The mesocarp or succulent and fibrous layer, after being passed through a preparatory process, is eaten as an article of food. "The yellow pulp surrounding the seeds of the ripe fruit is sweet, heavy, and indigestible. It is extracted by rubbing the seeds over a wooden scratcher, and with the addition of a little lime it settles into a jelly, which is a ready mode of taking the pulp. It is also made into cakes with flour and other ingredients." (*U. C. Dutt, Hindú Mat. Ind., 249.*) By seed in the above passage should be understood nut; the fibrous tissue which ramifies through the succulent mesocarp is attached to the endocarp or shell of the nut. The succulent pulp scraped away from this tissue has a peculiar odour, and is sweetish; it is either eaten raw, or is mashed and strained with a little flour and sugar, completely mixed up to form a mass, and is then made into small flat cakes and fried in *ghí* or mustard oil; the cakes are known as *pátáli*, or *pithá.*

Patali Cakes. 701

"In order to make the first kind of cake (*pátáli*), the scraped pulp is mixed with lime, and cocoanut spread evenly on a plate in which it is allowed to stand for an hour, after which it is found in a solid state, owing to the effect of the lime on the pulp. In order to make *pithá*, the pulp is mixed with rice or wheaten flour and then friedi n oil. In Bengal *tál* pulp is not preserved, does not form an important article of food, and there is no trade in it. In short, the *tál* occupies a very unimportant place among the Bengal fruits." (*Babu T. N. Mukerji, Revenue and Agricultural Department.*)

Punatu. 702

In Ceylon this pulp is known as *Punatú.* "The pulp of the fruit is preserved for use in the following manner: The ripe fruits are put into baskets containing water, and are then squeezed by the hand till the pulp forms a jelly. Layers of this jelly are spread on palmyra-leaf mats to dry on stages. Layer after layer is deposited to the number of about fifteen. These are left in the sun about a fortnight or three weeks, only covered at night, and protected from the dew and rain. The best sort is called *Pimatos,* and the tough withery kind made from the remaining fruits gathered at the end of the season, which is much in favour, *Tot Punatú. Punatú* is sold by the *mat* at 3*s.* to 6*s.* each, and is the chief food of the islanders of Ceylon, and of the poorer classes of the peninsula, for several months of the year." (*Tropical Agriculture, Simmonds, 267.*)

VEGETABLE. Young Seedling. 703

GERMINATED SEED.—After scraping off the succulent tissue of the mesocarp, the nuts are found to be perfectly solid, and so hard that it is almost impossible to break them. If thrown aside in a heap or buried in

Embryo. 704 Flour. 705

the earth for two or three months, however, they germinate. The very young seedling or tip of the root and young stem are eaten as a vegetable or pickled. The most valuable part is, however, the seed or rather embryo within the nut. This is removed by splitting open the nut and removing the large embryo, which is often as much as an inch and a half in length. This is either eaten dry or after being roasted or cooked in various ways, or it is reduced to a flour not unlike tapioca. This forms an important article of food in most parts of India where the palm is grown to any very considerable extent. "The developed embryo is sweet in taste, and is considered nourishing. It is sometimes preserved in sugar, but in Bengal it is not an article of commerce. It is called *tal-ati* (*phapal*) in Bengali. The tap-root and young plants are not edible, but are used in medicine as a stimulant." (*Babu T. N. Mukerji.*)

In Ceylon they are known as *Kelingoe.* "The nuts are collected and buried in heaps in the ground. When dug up after the space of three months, the young shoots are called *kelingoes;* they supply the inhabitants with a nourishing aliment. In size, colour, and shape they resemble a parsnip, and look like a cold potato. In its fresh state it will keep good for a couple of months, and when well dried in the sun, for a whole year. In this state they are called *odials.* When reduced to flour or meal, the favourite *cool* or gruel is made of it." (*Tropical Agriculture, Simmonds, 267.*)

Root. 706 Cabbage. 707

According to **Balfour, Drury,** &c., the root is stated to be used as an article of food and to afford a kind of tapioca. I have not been able to have this statement confirmed, and suspect that the germinating seedling is what is meant. The young leaf-bud or cabbage, as with most other palms, may be eaten, but the tree would be killed were this practice followed; it is thus scarcely correct to enumerate this amongst the properties of the plant.

TIMBER.

TIMBER. 708

Structure of the Wood.—The outer shell of hard wood consists of an almost solid mass of thick fibro-vascular bundles, more scattered in the male than in the female trees. The centre is soft, but only rarely hollow. **Brandis** says: "Forked and branching stems are occasionally found."

Shell. 709 Inner part. 710

The outer hard woody shell is the part used as timber. The trees after being felled are cut lengthwise into two, the soft fibrous part removed, and the hard outer portion adapted to the purpose for which it is intended. From the structure of the fibres it splits easily, but is stated to support a greater cross strain than any other known wood. Iron nails, however, rapidly decay it, so that, except for posts, it is not generally serviceable for house-building. The hollowed-out halves are used as water-pipes, gutters, or open water-channels. They are made into dug-out canoes. The swollen, rounded, and lower end forms the front of the canoe, and the tapering end has either a piece of the original wood, 6 inches in length, or a lump of mud, placed in it to close the mouth. The rounded end from which the mass of rootlets spring requires no protection, for it is nearly as hard as the outer shell of the stem proper. The timber is used for posts, rafters, and a number of minor purposes; it is, in fact, the timber most used, of all the Palm family, for house-building and other domestic purposes. A small export trade is done in the wood for making walking-sticks, umbrella-handles, rulers, and other small and ornamental purposes. In India it is often made into shuttles.

Small articles. 711 Sticks. 712 Shuttles. 713

A rule exists in many parts of India (an unwritten law), that for every palmyra palm that is felled another must be planted. This is a very fortunate arrangement, for it would be difficult to find a tree regarding the uses of which so much might be written. **Mr. Vincent,** in his report of the forests of Ceylon, says of it: "The Tamils throughout the Jaffna peninsula derive no small portion of their food from the palmyra products, whilst a large number may be said to live on the tree entirely. In spring they make jaggery (a kind of sugar); during the rest of the year they live on the money so earned, and on *Punatto* and *Kelingoes*."

DOMESTIC AND SACRED USES.

It has already been stated that in a Tamil poem 800 articles are described as prepared from the palmyra palm; a large number of these are minor domestic appliances which need not be enumerated in this work; a few of the more important may, however, be mentioned:—

DOMESTIC. Fans. 714 Punkhas. 715 Baskets. 716 Umbrellas. 717 Braid. 718 Books. 719 Paper. 720

Domestic.—*The Leaves.*—These are made into fans and large *punkhas*, variously lacquered or painted, and into baskets of many forms and designs, both for domestic and ornamental purposes. In Madras neat work-baskets are made of palmyra leaf. A single leaf is often held over the head as a kind of umbrella. Strips of the leaf, carefully cut, smoothed, and slowly dried in the sun and rubbed with oil, were formerly used in place of paper for writing letters and books on, and to this day are so used in Orissa and South India. For this purpose a steel pen or style is employed. During the operation of writing, the leaf is held in the left hand and the letters are scratched upon the surface. In order that the characters may be better seen, ink made of lamp-black or some other colouring substance, and gum, is rubbed over the surface. "On such slips all the letters and edicts of the Dutch Government used to be written, and sent round open and unsealed. When a single slip was not sufficient, several were bound together by means of a hole made at one end, and a thread on which they were strung. If a book had to be made for the use of the Wihares or any other purpose, they sought for broad and handsome slips of *talapat* leaves, upon which they engraved the characters very elegantly and accurately, with the addition of various figures delineated upon them by way of ornament. All the slips had then two holes made in them, and were strung upon an elegantly twisted silken cord, and covered with two thin wooden boards. By means of the cord the leaves are held even together, and by being drawn out when required for use, they are separated from each other at pleasure. In the finer binding of these kind of books the boards are lacquered, the edges of the leaves cut smooth and gilded, and the title is written on the upper board; the two cords are fastened by a knot or jewel, secured at a little distance from the boards, so as to prevent the book from falling to pieces, but sufficiently distant to admit of the upper leaves being turned back while the lower ones are read. The more elegant books are in general wrapped up in silk cloth, and bound round by a riband, in which the Burmese have the art to weave the title of the book. The palmyra books are never much beyond 2 feet in length and 2 inches in breadth, as the parchment-like ribs between the little ribs will not admit of their increase in size." (*Mr. W. Ferguson's account, reprinted in Tropical Agriculture by Simmonds.*)

School-books. 721

In the road-side schools of Bengal and most parts of India long strips of palmyra palm-leaf constitute the note-books and exercise-books used by the boys. They are carried to and from school generally wrapped up in the little piece of matting upon which the pupil sits. Instead of a style, however, they use a reed pen, covering the strips of palm leaf all over with large black characters. When the lesson or exercise is finished, these strips are taken to the nearest tank and washed clean again.

It is almost impossible to enumerate all the purposes to which the palmyra is put; suffice it to say that a very large number of the articles of domestic use are, in the rural districts of India, constructed from some part

Caps. 722 Hats. 723 Jugs. 724 Plates. 725 Water-pails. 726 Baskets. 727 Toys. 728 Mats. 729 Cement. 730

of this most useful palm. Caps and rain-hats, cups and rice-jugs, plates, water-pails, water-baskets, cooly-baskets, baskets for storing grain, oil-press baskets, betel-nut baskets, clothes-baskets, sieves, books, toys, and other miscellaneous articles, mats, punkhas, screens, fences, and thatching, are all frequently made of this substance, to a greater extent of course in some districts than in others, depending upon the prevalence of the palm.

The juice.—"Amongst a variety of purposes to which it is put, is that of being mixed with the white of eggs, and with lime from burnt coral or shells. The result is a tenacious mortar, capable of receiving so beautiful a polish that it can with difficulty be distinguished from the finest white marble." (*Tropical Agriculture, Simmonds, 266.*)

SACRED USES.—The palmyra palm is one of the trees looked upon by the Hindús with veneration. It is accordingly planted for the public good, the following being the trees the planting of which secures the kingdom of heaven: 1, Pipal; 2, Champaka; 3, Nagakesara; 7, Tál; and 12, Narikela. According to some authors it is the *Kalpa* or *Ilpa*, the *Kalpadrum, Kalpakataru* or *Kalpavriksha* of the Vedic writings, being regarded as the symbol of vegetation, of universal life, and of immortality. It is also by some authors viewed as the sacred tree of Buddha. It is frequently seen on Buddhistic sculptures.

The existence of the names which are now applied to this palm in the Sanskrit writings is no absolute proof that the modern usage or adaptation of these names is correct. Botanical evidence is entirely opposed to the *tál* palm being a native of India, the tree having been introduced from Africa. Moreover, this palm does not grow so far to the north as the Panjáb even in the present day, so that the names which are found in the Vedic writings of a date prior to the invasion of Northern India in all probability refer to a totally different plant. This is the more probable when it is recollected that **Nannorhops Ritchieana,** *Wendl.*, a palm which in its fan-shaped leaves and in other respects resembles the palmyra, is a native of the Trans-Indus mountains of Western Sind, and of the Salt-Range, passing to Beluchistan, Afghánistan, and Persia. Although rarely more than 20 to 25 feet in height, the uses to which this palm is put, and the fact of its luxuriating over low arid mountain tracts where few trees of any description are found, might naturally have combined to assign to it a high place in the esteem of the Vedic poets. But without hazarding any very definite opinion as to what may have been the *Kalpa* tree of the ancients, there seems very little doubt as to the Palmyra palm having been introduced into India long after the Vedic invasion.

BORAX.

731

This is the Borate of Sodium, or rather Biborate of Sodium, $Na_2B_4O_7.10H_2O$. BORAX or BORATE DE SOUDE, *Fr.*; BORAX, BORSAURES NATRON, *Germ.*; BORACE, *It.*; BORAX, *Sp.*

Vern.—*Sohágá, tinkál,* HIND.; *Sohágá* or *suhágá,* BENG.; *Sohágah,* DUK.; *Kuddia-khár, tankan-khár,* GUJ.; *Vengáram, puskara,* SINGH.; *Lakhiya, letkhya,* BURM.; *Venkáram* or *vengáram,* TAM.; *Velligáram, elegáram,* TEL.; *Ponkáram, vellakáram,* MAL.; *Biligárá,* KAN.; *Tan-kana,* SANS.; *Búrakies-sághah* or *buruq-es-sághah, bóraq, milhus-sághah,* ARAB.; *Tinkár tankár,* PERS.; *Sohága, tinkár* or *tinkal, tsalé* (one variety being *chú tsalé* or water of borax, and the other *tsalé mentog* or flower borax; *Baden Powell, Pb. Prod., I, 94*), PB.; *Vavut,* KASHMIR; *Sal, shal, chú-sal,* THIBETAN. (**Dr. Aitchison** adds that it is called *sál* when collected from the soil, and *chú-sal* when from water.)

The word Borax is of Arabic origin, and Tincal (which by **Balfour** and most other authors is given as an old English name for Borax), is a Europeanised corruption of the Thibetan name *Tschuchal* (*chú-sal*), or of the Persian *Tankár* and the Sanskrit *Tankana;* indeed, the word *Tinkál* is of common use on the Panjáb frontier. *Tanna-khar,* TURKI; *Pang-sha* and *Yueh-ship,* CHINESE. It would seem probable that the article was first consigned to Europe from South India, and with it the name *Tinkál.*

§ "بورق: *Búrak* properly means that which is put in dough to make it inflated and shining—*Papri-lon* or *Papri khan—i.e.,* Carbonate of Soda and Potash. *Búrak-es-sághah* is Borax, because it polishes silver. There are also other kinds of *Búrak. Tankár* is a Persian word, and means Borax; it is probably derived from the same source as the Sanskrit *Tankana.* The Persians also call Borax *Bureh.*" (*Surgeon-Major W. Dymock, Bombay.*)

HISTORY OF BORAX.

The word "Borax," as stated above, has come commercially to mean Biborate of Soda, but various other borates are met with in trade, and Boracic acid itself has come into use as a source of borax or as a substitute for it; hence the word "Borax" should be employed with some caution.

NATIVE BORAX. 732

(*a*) Borax proper is a native borate of sodium found along with common salt on the shores of certain lakes in the Panjáb, frontier of Thibet, and in Thibet itself. It is probably also met with in Persia and on the China-Thibetan frontier. Outside this limit it is found in California, in Peru, and in Ceylon.

ARTIFICIAL BORAX. 733

(*b*) One of the most important sources is the artificially-prepared borax from the Lagoons of Monte Cerboli in Tuscany. From the volcanic fissures of that region hot aqueous vapour is emitted. This is collected in artificial basins called lagoons. In course of time the water condensing in these basins is found to be charged with boracic acid. This is removed by crystallization, and, by the action of carbonate of soda, is in solution converted into Borax. The discovery of this process of making artificial borax is due to **Cartier** and **Payen,** and it is regularly practised in France.

In England the Italian boracic acid is neutralised by mixing the dry acid with soda ash and exposing the mixture to the heat of a reverberatory furnace. By this latter process ammonia is liberated and collected as a by-product.

(*c*) The borates of lime or double borates of lime and soda. These occur in immense reniform blocks, and are generally associated with gypsum and common salt. They are almost completely soluble in acids.

(*d*) Borate of magnesia is also a convenient source of boracic acid, containing about 70 per cent. when pure. This is found generally in nodules associated with gypsum and potash salts.

The supply of boracic acid being a monopoly in the hands of Count Lardarel, some years ago (1855) an effort was made to open up the Indian trade in native borax. An address was submitted to Lord Dalhousie, in which it was pointed out that the imports into England of Italian boracic acid were at that time 1,100 tons, and only 300 to 600 tons of Indian borax. An enquiry was accordingly instituted, which resulted in some interesting facts regarding the Ladak borax having been brought to light, but down to the present date no appreciable development of the trade seems to have taken place. In *Cunningham's Ladak* (pp. 239-40) occurs an account of the Pugá borax and sulphur mines. **Captain W. C. Hay** visited the Pugá valley, and the following passages may be reprinted from his report: "It is a small valley, which may roughly be calculated at two miles in length, and three quarters of a mile in breadth (*i.e.,* the portion from whence the *sohágá* or *tincál* is collected); it extends east and west, and has a fine stream running through it into the River Indus; but the portion producing the borate of soda is, if not

PANJAB BORAX. Puga. 734

BORAX. **Natural and Artificial,**

PUNJAB BORAX.

watered by, still under the influence of, thermal springs, varying in fou places, where I took the temperature, from 130, 140, 150, to 167 degrees,—the temperature of the streams into which these empty being in July 56 degrees.

"I ascertained, as nearly as I could, that the entire produce of the valley might be roughly calculated at 20,000 *kutcha* maunds (a *kutcha* maund is equal to about 32 lbs.), the greater portion of which found its way to Rámpúr in Bishahr; some to Kúlú *viá* Mandi to the lower hills, and a small quantity *viá* Chamba to Núrpúr. Nearly all that going *viá* Rámpúr is taken into the lower hills in the neighbourhood of Sabáthu, Bhají, &c., where wood is procurable, and where, during winter, it is refined by the carriers who go there to graze their flocks. It thus becomes borax, in which state it nearly all finds its way to Jagádri* in the plains, and thence, I presume, goes down the River Jumna or Ganges. It is probable that little, if any, finds its way to England.

"Pugá is not, however, the only place where the *sohágá* is produced; there is another locality near Rodok yielding it, from which the route to the plains is *viá* the Nite Pass. This borax is said to be of a very superior quality, nearly pure, and requiring little or no cleaning; but it is produced from a portion of Thibet in Changthán, subject to China. Doubtless, other localities exist, if the jealousy of the factors could be overcome, and enable us to explore. Nearly all the Trans-Himálaya lakes seem to contain salts of various descriptions, well worthy of chemical analysis; to this I shall advert in a future paragraph.

"The transport of this tincal is almost entirely effected on goats and sheep, being the animals at present best adapted to the mountainous pathways. The trade being to a certain extent precarious, the profits the merchants demand to protect themselves from loss would, at a first view appear large; when, however, the severity of the climate which they have to encounter, and the losses from snow falling over precipices, &c., are taken into consideration, it is not so exorbitant.

"The price of three sheep-loads at Pugá I have stated to be one rupee; the average journey of a laden sheep being about a *kós* per diem, it takes nearly one month to reach Kúlú from Pugá, where the same sells for eight rupees, and if cleaned as borax, it sells at Sultánpúr (Kúlú) at five rupees the *kutcha* or *kachchá* maund; and if taken to the lower hills at Kudli, Sisova, and Teki, at six rupees the *kutcha* maund. After it is purchased by the Jagádri merchants, I cannot say what expenses attend it, but the difficulties are over, and the prices here quoted clearly show the immense risk that is run on the first month's journey, compared to the second from Sultánpúr to the lower hills, which occupies upwards of a fortnight and sometimes a month, as the sheep get out of condition, and are soon tired after the long journey.

"At present the people depend entirely upon falls of snow, as rain never falls in those regions, and they suppose that snow is necessary to produce the *sohágá*, which probably might be equally well produced by flooding. The time, I am informed, required for its reproduction, is only ten or twelve days; but the sun in July and August is so very powerful that probably a succession of evaporations might be caused; this would form ground for a chemical report."

Kashmír and Rampur. 735

Davies, in his *Report on the Trade and Resources* of the countries on the north-western boundary of British India, says that "Borax goes to Kashmír, but in larger quantities to Rámpúr, and from thence to Kurrachee."

* At Jagádri the process of refining is extensively carried on.

from the Pugá Valley. BORAX.

Simla. 736

Lord Hay, then Deputy Commissioner of Simla, in a report upon the Panjáb Borax, says: "The people who are engaged in the *sohágá* trade are chiefly Kanáwaris and Khampos (a class of wandering traders) of Lahoul, Thibet, and Spiti. In the summer months they resort to the Pugá mines and other places, to which the *sohágá* found in Tartary is brought, and return in the autumn before the passes are closed to the lower hills, where they remain during the winter pasturing their flocks, refining their *sohágá,* effecting sales of it to the Simla merchants, and making purchases of miscellaneous goods to take back with them in the ensuing summer.

"The refining process is exceedingly simple, and consists of dissolving the crude borax in two parts of hot or ten parts of cold water, and then allowing it to crystallize." "Formerly it was the custom to cover over the crude borax with *ghí* to prevent efflorescence; this practice has been, I believe, discontinued of late years." *

"To Rámpúr and Sultánpúr, about 2,500 maunds, or 90 tons, are annually brought. Last year it sold at Simla for nine rupees a maund or £25 a ton, and at Jagádri it is now selling for twelve rupees or £37 a ton.

Jagadri. 737

"The trade of borax with Kúlú is almost entirely confined to the merchants of Jagádri."

In a letter from **Mr. Edgeworth** to the Secretary to the Chief Commissioner, Panjáb (*Feb. 1854*), occurs the following interesting information: "From Jagádri to Furruckabad it is taken on hackeries, 25 maunds on each, for hire of which R50 are paid, and from thence by water; the price of boat-hire varying considerably. These statistics, however, would be a guide to any European merchant wishing to engage in the trade."

"To give an idea of the increase in the borax trade with India during the last few years, it is only necessary to mention that while in the year 1846-47, when the price was R9 a maund, only 1,731 maunds were exported from Calcutta, during the last six months of 1854 the large amount of 10,896 maunds, at R22 per maund, have been shipped for Europe."

[The above extracts from **Captain Hay's** report, from **Lord Hay's** report, and from **Mr. Edgeworth's** letter, are reprinted from *Mr. Baden Powell's Panjáb Products* (*Vol. I., pp. 90 to 95*).]

N.-W. P. BORAX. 738

Mr. Atkinson, in his *Economic Minerals of the North-Western Provinces,* gives some interesting information regarding borax, from which the following note regarding the purification of the substance will be found useful: "The borax is pounded and placed in shallow tubes, and then covered with water to the extent of a few inches; to this is added a solution of about two pounds of lime dissolved in two parts of water for every ten maunds (820 pounds) of borax, and the whole mass is well stirred every six hours. Next day it is drained on sieves or cloth, and after this is again dissolved in $2\frac{1}{2}$ times its weight of boiling water, and about sixteen pounds of lime added for the above quantity. It is then filtered, evaporation takes place, and subsequently it is crystallized in funnel-shaped vessels, usually of *kansa,* an alloy of copper and zinc, or lead. The loss in weight is about 20 per cent." (*p. 34*).

Barmdeo. 739

In a report published in 1877 by the Secretary to the Government of the North-Western Provinces and Oudh will be found some interesting information regarding the Thibetan trade in Borax. The Secretary goes into the subject of the amount of uncleaned borax brought into Barmdeo,

* Dr. Dymock of Bombay informs me that this practice is not discontinued, and that the Hakims prefer it: it reaches Bombay by way of Kurrachi.

BORAX.

Panjáb Trade

with the view of ascertaining whether it was possible to educate the Bhotias to select and clean their borax before carrying it across the frontier. By a process of sifting, the borax is referred to two classes,—*chauki*, or large crystals of borax; and *reg*, or borax dust. The former is so pure that it requires no further cleansing, but the latter has to be boiled once or twice in order to separate the dirt. The result shows that of 100 maunds of the article as imported into the North-West Provinces, 60 maunds of *chauki* are separate, and the 40 maunds of *reg* become reduced by first boiling to 10 maunds of *kunj* and 30 of *kandi;* the latter by further boiling yields 5 maunds more of the purified borax or *kunj* and 25 maunds of dirt, so that of 100 maunds of the article as imported, 25 maunds are rejected as dirt. It seemed absurd that these 25 maunds should be carried across the frontier when the borax might easily enough be purified in Thibet. The sifting at least could be effected, but the boiling might, from difficulty in fuel, be impossible. It was found, however, that there were other difficulties connected with the system of monopoly; the Bhotias who carry it across the Himálayas do not bring it from the borax-fields.

Chauki. 740
Reg. 741
Kunj. 742
Khand. 743
Refuse. 744

The region of Indian borax may be said to commence in the west, at the valley of Pugá in Ladak, passing east to the lakes of Rudokh. Along this tract of country, and extending considerably to the east, a chain of salt lakes occurs, most of which in all probability afford borax. To the south of Lhasa, at the Yamdok Cho, borax is known to have been collected from time immemorial. Holes are dug in the arid soil of many parts of the deserts of Tartary, wherein tincal collects, and is periodically gathered.

Trade in Borax.

745

Borax is chiefly imported into India through the North-West and the Panjáb Himálaya. The following statement shows the principal imports:—

From Thibet into the North-West Provinces.

Years.	Weight in Maunds.	Value in Rupees.	
1882-83	21,527	1,72,216	
1883-84	33,856	3,37,938	

From Ladak and Thibet into the Panjáb.

Years	Weight in Maunds.	Value in Rupees.	
1882-83	{ 9,179	{ 63,896	Ladak.
	{ 2,588	{ 15,528	Thibet.
Total	11,767	79,424	
1883-84	{ 9,088	{ 73,081	Ladak.
	{ 3,33	{ 20,028	Thibet.
Total	12,426	93,109	

brought into N.-W. Provinces.

BORAX.

TRADE STATISTICS.

Thus it would appear that the total imports into India across the frontier in 1883-84 amounted to 46,282 maunds (=33,058 cwt.), valued at R4,31,047. The foreign exports of that year were 16,216 cwt., valued at R3,58,518, showing that whereas only 16,216 were exported, there were consumed in the country 16,842 cwt. During the year there were also imported from foreign countries 38 cwt., valued at R1,493 These figures give some idea of the importance of borax as an article of internal trade, and they show at the same time its enhanced value from its entrance into India until it is exported to foreign countries.

§ "Three kinds of Borax are met with in the Bombay market,—*viz.*, European, Cawnpore (Thibetan), and Kurrachi (Teliya Tankankar). The European can often be purchased at the same price as the impure Thibetan; it is imported in casks. The Thibetan occurs in circular cakes, thin at the edges, as if crystallized in a basin. The Teliya Tankankar is in thin flaky crystals with a greasy surface." (*Surgeon-Major W. Dymock, Bombay.*)

For the past few years the exports have been steadily decreasing—a natural consequence of the discovery of extensive beds of borax in America, and of the greatly extended trade in the artificially-prepared article. Indeed, India cannot hope to compete in the foreign trade in borax, but the internal consumption, which is very considerable, will always make the trans-frontier imports of importance. It is noteworthy that while the external trade has been falling off year by year, the internal trade seems to have correspondingly increased.

The following analysis of the exports to foreign countries for the past year shows the province from which imported and the country to which exported :—

Analysis of the Trade in Borax for 1883-84.

Presidency from which exported.	Weight in Cwt.	Value in Rupees.	Country to which exported.	Weight in Cwt.	Value in Rupees.
Bengal . .	16,095	3,54,699	United Kingdom .	14,134	2,92,585
Bombay . .	121	3,819	Arabia . .	38	1,130
			China—Hongkong .	1,713	56,424
			Straits Settlements .	258	6,147
			Turkey in Asia .	36	1,057
			Other countries .	37	1,175
TOTAL .	16,216	3,58,518	TOTAL .	16,216	3,58,518

746

PROPERTIES AND TESTS FOR BORAX.—A salt occurring in colourless, transparent, shining, monoclinic prisms, odourless, slightly efflorescent, having a cooling, sweet taste, with an alkaline reaction. It is a detergent to the mouth, clearing the throat. It has the composition $Na_2B_4O_7$ with 10 molecules of water of crystallization. This is what is called prismatic borax, but if crystallized at 79° it forms octohedra, having only 5 molecules of water. When a solution of borax is evaporated at 100° C., the salt is left as a transparent, amorphous, brittle mass, containing only four molecules of water. When heated a dry powder of borax begins to lose its water, then melts; on further heating it swells up, forming a porous mass; and at a red-heat fuses, forming a colourless glass, from which water of crystallization has been completely expelled. This forms the borax beads now so extensively used in chemical analysis. If touched with a metallic salt, the borax bead will, in the reducing flame of the blow-pipe, become coloured *red* with sub-oxide of copper, *green* with ferrous oxide, &c.; and in the oxidizing flame, *red* with ferric oxide, *violet* with manganese salts, *blue* with cobalt oxide, &c. These beautiful reactions form convenient tests for the metallic salts.

Borax itself may be readily detected by a number of delicate tests. An aqueous solution, on the addition of sulphuric acid, should deposit shining crystalline scales which will be found to impart a brilliant green colour to the flame of a spirit-lamp. This convenient test will establish the presence of the merest trace of borax. The chief adulterants are phosphate of sodium and alum. The former may be readily detected by the fact that it will quickly effloresce in the heat of a drying-room, and the latter by the brilliant cobalt blue produced before the blow-pipe when a piece of alum is touched with a solution of cobalt chloride.

Borax has the property of rendering cream of tartar soluble, but for this purpose boracic acid may be substituted for borax. One of the most curious properties of borax has recently attracted much attention—namely, its power of destroying fermentation. **Schnetzler** (*Pharm. Journ., 3rd Series, V., 846, abstracted into the Year-Book of Pharm., 1875*) demonstrated the action of this substance upon the protoplasm of vegetable cells, and proved beyond doubt that the yeast plant is rapidly destroyed, fermentation being therefore impossible in the presence of borax. He further showed that, in consequence of this fact, both animal and vegetable matter might be preserved for years without undergoing putrefaction. This fact justifies the theory of the value of borax as an antiseptic lotion.

The Uses of Borax.

DYE. 747

Dye.—It is used as a mordant in dyeing, especially in calico-printing along with turmeric.

MEDICINE. 748

Medicine.—"Borax was known to the ancient Hindús from a very remote period and is mentioned by **Susruta.**" It is viewed by the native doctors as a tonic, and is regarded as useful in loss of appetite, painful dyspepsia, cough, asthma, &c. (*U. C. Dutt.*) As an antiseptic lotion and as a stimulating wash for hot eruptions on the body and scaly skin diseases, borax may be said to be an established remedy. It exerts a peculiar detergent action on the mucous membrane; it is accordingly regarded as a useful drug in aphthous and other ulcerations of the mouth, and in pruritus, not only of the external body, but also of the urethra and vagina. Internally, it is very little used by European physicians, but has been prescribed in dropsical affections and epilepsy. It is supposed to possess a powerful influence over the uterus, promoting menstruation and facilitating parturition; it has also been used in dysmenorrhœa, and as an astringent in uterine hœmorrhage it has been used with alleged benefit.

Special Opinions.—§ "Used as a detergent in various affections of the skin, also as an ingredient of spleen powders." (*Brigade Surgeon S. M. Shircore, Moorshedabad.*) "It is used as a germicide in thrush and ringworm, but not so efficacious as boracic acid." (*Brigade Surgeon G. A. Watson, Allahabad.*) "Bazar borax used in hospital practice, in the form of ointment, in psoriasis and eczema and as a lotion in pruritus and herpes circinatus. Dissolved in acetic acid it forms the common solution used for ringworm. It is found very useful in allaying the irritation of prurigo and erythema." (*Assistant Surgeon Jaswant Rai, Mooltan.*) "Efficacious application and gargle for aphthous sores." (*Assistant Surgeon Shib Chunder Bhuttacharji, Chanda, Central Provinces.*) "Useful as a local application in aphthæ, sore nipples, ulcers, ringworm, &c." (*Brigade Surgeon J. H. Thornton, B.A., M.B., Monghyr.*) "Borax is considered by the Hakims a powerful promoter of digestion, if taken in 10 to 15 grain doses, about an hour after meals with a little water." (*Assistant Surgeon Mokund Lall, Agra.*) "A handful or so to the bath relieves lichen tropicus." (*Surgeon-Major G. Y. Hunter, Karachi.*) "Useful with honey in thrush." (*Surgeon-Major C. R. G. Parker, Pallaveram, Madras.*) "I constantly use this for sore mouth in conjunction with glycerine as a gargle." (*Surgeon-Major H. D. Cook, Calicut, Malabar.*) "Burnt borax is used in dyspepsia attended with acidity—dose 10 grains." (*Surgeon-Major E. C. Bensley, Rajshahye.*) "It is also found to be of use in parasitic skin diseases, especially the ringworm or eczema caused by the parasite called *chambal.*" (*Assistant Surgeon Bhugwan Dass, Rawal Pindi, Panjáb.*) "A very useful hæmostatic in menorrhagia, a good lotion in thrush and ringworm, and pruriginous eruptions." (*Brigade Surgeon W. R. Rice, M.D., Jubbulpore.*) "Mixed with other substances, as bark, charcoal, &c., as a dentrifice, acts in whitening the teeth." (*Honorary Surgeon P. Kinsley, Chicacole, Ganjam, Madras.*)

749

Industrial Uses.—The most important use of this substance is unquestionably in "the glazing of all descriptions of pottery and china-ware, as well as for enamelling clock and watch faces, iron plates, &c." (*Spons' Encycl.*) It is also largely used in the process of soldering oxidizable metals, its action being to clean the surfaces by fusing away the oxides into a borax bead. It is extensively used by Indian goldsmiths and in the manufacture of artificial gems. It is also, along with shell-lac, made into a useful varnish. The dentist finds it valuable in making plates for artificial teeth. Plumbago and other pots are found to last much longer if painted with borax. For household purposes its uses are practically unlimited, it being in some respects superior to soda. As a substitute for or addition to soap, it cleanses fabrics without injuring the colours.

Borax is sold in every Indian bazar, and appears to be used for a variety of purposes which have not been carefully investigated and made public.

750

BOSWELLIA, *Roxb.; Gen. Pl., I., 322.*

A genus of balsamiferous trees, belonging to the Natural Order BURSERACEÆ; there are in all about six species, natives of India and tropical Africa,—one species, with two distinct varieties, being plentiful at the foot of the Western Himálaya, Central India, Rájputana, the Deccan, the Circars, and the Konkan.

Bark frequently papyraceous. *Leaves* alternate, exstipulate, imparipinnate, deciduous, with opposite sessile usually serrate leaflets. *Flowers* small, white, hermaphrodite, in axillary racemes or panieles. *Calyx* small, 5-toothed, persistent. *Petals* 5, distinct, narrowed at the base, imbricate. *Disk* annular-crenate. *Stamens* 10, 5 long, 5 short, inserted at the base of the disk. *Ovary* sessile, 3-celled; style short, stigma 3-lobed; ovules 2 in each cell, pendulous. *Drupe* trigonous, containing 3 1-seeded pyrenes which finally separate. *Seeds* compressed, pendulous.

The genus is named in honour of **Dr. John Boswell** of Edinburgh.

Boswellia (Species not satisfactorily determined).

751

THE TRUE FRANKINCENSE OF OLIBANUM of European commerce.

Vern.—*Kundur, lubán, thus*, ARAB., PERS., HIND.; *Kunduru*, SANS.; *Visesh, esesh*, BOMB.; *Parangi-shámbiráni, kunurakkam-pishin*, TAM.; *Parangi-sámbráni*, TEL.

GUM. 752

Olibanum. 753

Gum.—It is probable that several species yield Olibanum, of which **B. Carterii** is perhaps one of the most important. They are trees inhabiting the Somali coast of Africa to Cape Guardafui, and also the south coast of Arabia.

The Arabs, as early as the tenth century, carried Olibanum to India, and the Indian names for it have, through the lapse of time, become almost hopelessly mixed up with those given to the Indian species of this genus, and also with those given to the Balsamodendrons. It is impossible, therefore, to fix definitely the names of the balsamiferous plants. Mohammedan writers distinguish several kinds of the imported or African and Arabian Olibanum :—

Kundur Zakar. 754

1st.—Kundur Zakar, or male Frankincense. This is esteemed the best quality, and consists of deep yellow tears. It should burn readily and not emit much smoke.

Kundur Unsa. 755

2nd.—Kundur Unsá, or female Frankincense.

Kundur Madharaj. 756

3rd.—Kundur Madharaj. This consists of artificially-prepared tears made by shaking the moist exudation in a basket.

Kashfa. 757

4th.—Kishár or *qishár Kundur*, or *Kashfa.* This consists of the bark of the tree coated with the exudation. This is the *Dhúp* of the Bombay market, and, under that name, forms a distinct article of commerce.

Dukak Kundur. 758

5th.—Dukák or *daqáq Kundar*, or dust of Olibanum. This meets the demand of the Indian and Chinese markets, the finer qualities of Olibanum being exported from Bombay, after assortment, to Europe. (*Surgeon-Major Dymock.*)

The *Pharmacographia* gives an enumeration of the plants supposed to yield Olibanum or gum-resins which have been or may be mistaken for that substance, of which the following may be given as an abstract :—

759

I.—**B. Carterii**, *Birdw.*—This includes three forms—

(*a*) *Meddu* or *Mohr madow*, yielding the *Lubán bedowi* or *Lubán sheheri* of **Playfair. Hildebrandt** describes this as a tree indigenous to the limestone range of Ahl or Serrutin, the northern part of the Somali country. This is the plant represented by **Bentley** and **Trimen** in their *Medicinal Plants*, figure 58.

(*b*) A form sent by **Playfair**, along with the preceding, having almost entire leaflets, velvety below, glabrous above.

Maghrayt d'sheehaz of the Maharas.

760

II.—**B. Bhau-Dajiana**, *Birdw.*—Very nearly allied to, if indeed specifically distinct from, **B. Carterii.**

761

III.—A species which yields *Lubán bedowi.* It is a native of Bunder Murayah, Somali country; never found on the hills close to the sea, but further inland and on the highest ground.

762

IV.—**B. neglecta**, *S. Le M. Moore.* The vernacular name of this tree is given as *Murlo* or *Mohr add.*

763

V.—**B. Frereana**, *Birdw.*—This is a well-marked species known to the natives as *Yegaar.* It yields the fragrant resin sold as *Lubán Meyeti* or *Lubán-matí.* This the authors of the *Pharmacographia* regard as most probably the substance originally known as Elemi. *Lubán-matí* differs from the samples of true Olibanum in not containing gum; it may be described as composed of resin and an essential oil. (*Flückiger in Pharm. Journal, 3rd Series, VIII., 805.*) **Dr. Dymock** says this is sold in Bombay as *Pándhri Esesh.* It is the plant which yields the stalactitic Olibanum, a substance which differs chiefly from the other forms in the absence of soluble gum.

764

VI.—**B. papyrifera**, *Endl. Richard* (?). This is the *makur* of Sennaar and the mountainous regions, on the Abyssinian rivers Takazze and Mareb, ascending to 4,000 feet above the level of the sea. It appears not to grow in the outer parts of North-Eastern Africa. While this yields a resin, there is not the slightest "reason for attributing any commercial Olibanum" to it. It is probably more nearly allied to *Lubán-matí* than to Olibanum.

VII.—**B. serrata,** *Roxb.* (see page 515).

HISTORY OF OLIBANUM.

HISTORY. 765

This substance, being an essential ingredient in incense, has been known from extreme antiquity. Many centuries before Christ the drug was one of the most important articles of trade which the Phœnicians and Egyptians carried on with Arabia. Frankincense is mentioned by **Herodotus** (*B.C. 484*) and by **Theophrastus** (*B.C. 394-287*) as an article produced in the country of the Sabæans (the south shores of Arabia ?), where it was found in 1844-46 by **Carter**. The Arabs seem to have procured it from the ancient Sabæans, and it is an interesting fact that they took it into China amongst other articles as early as the tenth century, and that this trade has existed down to the present date. The Arabs also brought it to India, where it was known to the Sanskrit writers under the name *Kundura*, a word derived from the Arabic and Persian name *Kundur*. **Diodorus** (*B.C. 50*) refers to frankincense as one of the products of the rich country owned by the Arabs. **Strabo** (*B.C. 54 to A.D. 24*) mentions frankincense and balsam as met with in the country of the Sabæans, and **Pliny** (*A.D. 23-79*) says there is no country which bringeth forth frankincense but Arabia. **Arrian** (*A.D. 90*) describes Makulla as the coast of the country of frankincense. **Ptolemy, Dioscorides, Marco Polo, Garcia de Orta, Celsius, Linnæus,** and many others mention this gum-resin." (*Birdwood, Bomb Prod.; Flück. & Hanb., Pharmacographia; Dymock, Mat. Mad.; and U. C. Dutt, Mat. Med. of the Hindús; &c.*)

MEDICINE.

MEDICINE. **Olibanum.** 766

Description of Olibanum.—Olibanum, as met with in European commerce, may be described as a dry gum-resin, consisting of tears often an inch in length, and of an ovate or oblong, clavate or stalactitic form, and mixed with impurities. The pieces are light yellow to brown, or pale green, or colourless. The odour is balsamic and resinous, especially while being burned. In taste it is bitter and terebinthinous, dissolving in the mouth. By heat it softens without actually fusing, decomposing at high temperatures.

"Olibanum is considered by the Mohammedans to be hot and dry, and to have dessicative, astringent, and detergent properties. It is used internally and externally in much the same way as we use the products of the Pines and Firs. Recently olibanum has been made officinal in the *Pharmacopœia of India*, where it is recommended in chronic pulmonary affections, such as bronchorrhœa and chronic laryngitis, employed both internally and in the form of fumigation. In the same work an ointment has been introduced which is said to be a good stimulant application to carbuncles, ulcerations, boils, &c. I have found that a good imitation of commercial Burgundy Pitch may be made by incorporating melted olibanum with water in a steam bath; a sufficiently good quality for this purpose can be purchased for R12 per cwt." (*Surgeon-Major W. Dymock, Bombay.*)

767

Chemical Composition.—The following extract from the *Pharmacographia* will be found to contain all that is known of the chemistry of this substance: "Cold water quickly changes olibanum into a soft whitish pulp, which when rubbed down into a mortar forms an emulsion. Immersed in spirit of wine, a tear of olibanum is not altered much in form, but it becomes of an almost pure opaque white. In the first case the water dissolves the gum, while in the second the alcohol removes the resin. We find that pure olibanum treated with spirit of wine leaves 27 to 35 of gum,* which forms a thick mucilage with three parts of water. Dissolved in 5 parts of water it yields a neutral solution, which is precipitated by perchloride of iron as well as by silicate of sodium, but not by neutral acetate of lead. It is consequently a gum of the same class as gum arabic, if not identical with it. Its solution contains the same amount of lime as gum arabic affords.

"The resin of olibanum has been examined by **Hlasiwetz (1867)**, according to whom it is a uniform substance having the composition $C^{20}H^{30}O^{3}$. We find that it is not soluble in alkalis, nor have we succeeded in converting it into a crystalline body by the action of dilute alcohol. It is not uniformly distributed throughout the tears; if they are broken after having been acted upon by dilute alcohol, it now and then happens that a clear stratification is perceptible, showing a concentric arrangement.

"Olibanum contains an essential oil, of which **Braconnot** (1808) obtained 5 per cent., **Stenhouse** (1840) 4 per cent., and **Kurbatow** (1871-1874) 7 per cent. According to **Stenhouse** it has a sp. gr. of 0·866, a boiling point of 179·4° C., and an odour resembling that of turpentine, but more agreeable. **Kurbatow** separated this oil into two portions, the one of which has the formula $C^{10}H^{16}$, boils at 158° C., and combines with HCl. to form crystals; the other contains oxygen. The bitter principle of olibanum forms an amorphous brown mass.

"The resin of olibanum submitted to destructive distillation affords no umbelliferone. Heated with strong nitric acid it developes no peculiar colour, but at length camphoretic acid (see **Camphor**) is formed, which may be also obtained from many resins and essential oils if submitted to the same oxidizing agent." (*Pharmacographia, Fluck. and Hanb., pp. 138, 139.*)

768

TRADE IN OLIBANUM.

"Bombay is the centre of the Olibanum trade. The houses which deal in gum have agents in Arabia and Africa who buy it up and forward it here in a mixed condition. It passes through the Custom House as *Esesh*, and is next sorted into four or five different qualities. The first, consisting of all the large clean tears, is destined for the European market. The intermediate qualities, and the last, which is only the dust and refuse, supply the Indian and China requirements. The *Kishar Kundur* or *Kashfa* of the Arabs forms a distinct article of commerce under the Indian name of *Dhúp*. The method of collecting Olibanum in Africa has been described by **Cruttenden** (*Trans., Bomb. Geograph. Soc., VII., 1846, 121*). **Carter** in the same publication has described the collection of the drug in Southern Arabia. In both localities a simple incision in the tumid bark is made and the product collected as soon as it becomes sufficiently hard. The collection is carried on from March to September in Africa, and from May to December in Arabia.

African Olibanum. 769
Indian Olibanum. 770

"Olibanum is shipped from Makulla, Aden, and other neighbouring ports to Bombay; as already mentioned, it is there sorted for the different markets. The trade is in the hands of Khojas and Bunnias. The price varies from R4 per cwt. for the dust to R20 per cwt. for the finest tears. Bombay exports from 25,000 to 30,000 cwt. annually. Nearly four fifths of this quantity go to Europe, and the rest to China." (*Dymock, Mat. Med., W. Ind., 122.*)

It would appear that a certain amount of the Olibanum met with in commerce is exported direct from Egypt to Europe. This is the so-

* I obtained 32·14 per cent. from the finest tears of the kind called *Fasous bedowi*, with which I was presented by Captain Hunter of Aden.—F. A. F.

called African Olibanum, the term "Indian Olibanum" being in the trade applied to the same article which reaches Europe *viâ* India, coming originally from the Red Sea ports to India, and thereafter being re-exported to Europe. It seems a mistake to suppose that any of the Indian article is obtained from the indigenous Indian plant.

Boswellia serrata, *Roxb., ex Colebr., in Asiat. Res., ix., 379, t. 5; Fl. Br. Ind., I., 528;* BURSERACEÆ.

771

Sometimes called THE INDIAN OLIBANUM TREE.

Syn.—B. THURIFERA, *Roxb. ex Flem.;* B. GLABRA, *Roxb.;* B. THURIFERA, *Colebr.* (as in *Gamble's Manual of Timbers*); LIBANUS THURIFERA, *Colebr.*

Vern.—(The gum-resin) *Salhe, salei,* or *sálai, sálgá, sél-gónd, kundur, salpe, lubán,* HIND.; *Lubán, salai, kundro,* BENG.; *Saleya,* LOHARDUGGA; *Salga,* SANTAL; *Anduku, anduga, gúggar, dúmsal,* KUMAON; *Salla, bor-salei. ganga,* GOND.; *Silái* or *salái* (at NAGPUR), C. P.; *Sálar,* ULWAR; *Salai, salga, guggula, sálaya-dhup, salaphali,* BOMB.; *Salaphali,* MAR.; *Kundur,* DUK.; *Dhúp, mukul salai, gugali,* GUJ.; *Saliya gugul,* CUTCH; *Kungli, gúgúlu, kúndrikam morada, kundurukkam-pishin, parangi-shámbi-ráni,* TAM.; *Parangi-sámbráni, anduga-pisunu, anduku, ándu,* TEL.; *Vella-kundirukkam,* MAL.; *Chittu,* KAN.; *Salasi-niryása sallaki, kunduru guggulu,* SANS.; *Bastaj, kundur, lubán,* ARAB.; *Kundur,* PERS.; *Thabi-ben,* BURM.; *Kundrikam,* SINGH.

It is probable that the name *Gugul* should have been restricted to this plant, but modern use has extended it to include **Balsamodendron Mukul.** There are two varieties, both of which yield the gum-resin incorrectly called Indian Olibanum :—

Var. 1st—serrata proper.

Habitat.—A moderate-sized gregarious tree of the intermediate, northern, and southern dry zones, Sub-Himálayan tract from the Sutlej to Nepal, the drier forests of Central India from Berar to Rájputana, and southward to the Deccan, the Circars, and the Konkan. Frequent on the eastern slopes of the Pegu Yomah and Martaban, Burma. (*Kurz.*)

Botanic Diagnosis.—This is **B. thurifera,** *Roxb.,* and is characterised by the leaflets being sessile, pubescent, coarsely crenate-serrate; racemes axillary, shorter than the leaves.

Properties and Uses—

Gum-resin.—The gum-resin, *Sálai gugul,* occurs as a transparent golden yellow, semi-fluid substance, which slowly hardens with lime. **Moodeen Sheriff** says that when it is found in the soft massive form it is known as *Gandah ferozah;* in tears (? true olibanum) it is known as *kundur.* It is pungent, having a slightly aromatic taste and balsamic resinous odour. It becomes opaque when immersed in alcohol or in water, the proportion of resin to gum being much smaller than in frankincense. The opaque, soft, whitish mass produced by water when rubbed in a mortar forms an emulsion. Indian Olibanum is consumed almost entirely in Central and Northern India, and is never exported.

GUM-RESIN. Gandah ferozah. 772 Kundur. 773

In the Upper Godaveri it yields plentifully the resin Olibanum (*C. P. Gazetteer, 503*). A sweet-scented gum, "burnt in religious ceremonies and sometimes used to strengthen lime" in Rewa Kantha, Gujarát. (*Bombay Gazetteer, VI., 13.*) A very common tree on all trappean hills, conspicuous by its white and scaly bark. No such substance as frankincense is extracted from it in Khándesh. The gummy wood is, however, used for torches.

Sir J. D. Hooker, in his *Himálayan Journals* (*Vol. I., 29*), says that while travelling on the mountain tracts of Behar he came across a small forest of this tree near Belcuppé. The gum was flowing abundantly from

BOSWELLIA serrata.

Salái Gugul, or Indian Olibanum.

the trunk, very fragrant and transparent. **Dr. Irvine**, in his *Topography of Ajmere* (*p. 135*), says that the tree is very plentiful in the Ajmir hills. The *gundabirosa* is the prepared gum-resin, and is similar in appearance and qualities to Venice turpentine. It is brought from Mewar, Harowtee, and the Shekhawattee hills, and is considered stimulating. An oil is distilled from it said to cure gonorrhœa. The gum-resin is also made into ointment. It is much used in painting, especially by the *lakheries* men who paint with coloured lac (?).

Care must be taken not to confuse this gum-resin with the Olibanum or Frankincense of commerce, or with **Mukul** (see **Boswellia**, sp., **and Balsamodendron Mukul**). The Sanskrit name *Kunduru*, derived from the Arabic word *Kundur*, is most probably wrongly applied to the gum-resin of this species. It should be restricted to frankincense, a substance which reaches India from Arabia and Africa. The true Sanskrit name for this plant is most probably *Sallaki*, from which is derived the Hindi word *Salái*. It would also appear that this is the *Guggulu* of Sanskrit writers, which is described as moist, viscid, fragrant, and of golden colour when freshly exuded. Gum-gugul of the present day is Indian Bdellium (**Balsamodendron Mukul**). (*Surgeon-Major Dymock, Mat. Med., W. Ind., 123.*)

MEDICINE. Gum-resin. 774

Medicine.—Very little of a definite nature is known of the medicinal virtues of this gum. It is probable that all that has been written on the subject should be considered as applying exclusively to imported Olibanum. **Dr. Dymock** says that the *Guggulu* of the Sanskrits was regarded as a demulcent, aperient, alterative, and a purifier of the blood. The gum at the present day is used in rheumatism, nervous diseases, scrofulous affections, urinary disorders, and skin diseases, and is generally combined with aromatics. It is regarded as a diaphoretic and astringent, and is used in the preparation of ointment for sores. It is also prescribed with clarified butter in syphilitic diseases; with cocoanut oil for sores, and as a stimulant in pulmonary disease. Mixed with gum acacia it is used as a corrective for foul breath; taken for any length of time in ʒi doses it is said to reduce obesity.

Special Opinions.—§ "The gum-resin is used to promote the absorption of bubo, and is applied locally. The oil in 10 or 20 minim doses is useful in gonorrhœa, taken in demulcent drinks." (*Surgeon C. M. Russell, Sarun, Bengal.*) "Refrigerant, diuretic, emmenagogue, and clolic; doses 5 to 40 grains, used in aphthæ, placenta prema, amenorrhœa, dysmenorrhœa, sore nipple, gonorrhœa, ringworm." (*Choonna Lall, Hospital Assistant, Jubbulpore.*) "Astringent, applied in the form of an ointment to chronic ulcers, diseased bones, buboes, &c." (*Surgeon W. Barren, Bhuj, Cutch.*)

FOOD. Flowers and Nuts. 775

Food.—"The flowers and seed-nut are eaten by the Bhils." (*Bombay Gaz., XII., 27.*)

TIMBER. 776

Structure of the Wood.—Wood rough, white when fresh cut, darkening on exposure, moderately hard. It is not durable, but it has been reported that five sleepers made of it and soaked for some time in a tank filled with the leaves of *Bahera* (**Terminalia belerica**), put down in June 1876 on the Holkar and Neemuch State Railway, are still (1881) perfectly sound and good. (*Indore Forest Report, 1876-77*, quoted in *Indian Agriculturist* of May 1878.) The timber is recommended for tea-boxes. (*Indian Forester, IX., 377.*)

It is used for fuel and for making charcoal, which in Nimar is employed for iron-smelting. This "is a common, and though not very large, a very beautiful tree (in Panch Maháls). Its narrow-pointed leaflets and drooping branches give it something the look of the English garden acacia. Its grey flakey bark is noticeable. It yields a cheap resin, and besides for fuel, its wood is used in making platters." (*Bombay Gazetteer, III., 199.*)

Var. 2nd—glabra, sp., *Roxb.* 777

Vern.—*Kundur,* HIND., PERS., ARAB.; *Salhi,* PB.; *Farangí-aúd, kundur,* DUK.; *Parangí-shámbiráni, kundurukam-pishin,* TAM.; *Guggilapu-chettu* (tree), *gugil, anduga-pisunu, parangi-sámbráni,* TEL.; *Mannakungiliyam, valanku-chámbráni,* MALA.; *Bringilobán,* BURM.

Habitat.—A moderate-sized tree of North-West India. Leaflets nearly or quite glabrous, and generally entire or nearly so; racemes terminal, sub-panicled.

It seems probable that this form yields the solid rounded pieces or tears described by some authors as of Indian origin, owing to its drying more rapidly than the gum-resin from **B. serrata.** **Royle** describes picking tears off the trees, and states that these burn rapidly with a bright light, diffusing a pleasant odour.

For further particulars regarding the Boswellias and Frankincense, the reader is referred to *Dr. Dymock's Materia Medica of Western India* (from which much of the above information has been obtained); to *Dr. Birdwood's Monograph of the Genus Boswellia in the Linnæan Society's Transactions, XXVII.*; to "Olibanum" in *Flückiger and Hanbury's Pharmacographia (p 133, Ed. 1879)*; *Bentley and Trimen's Medicinal Plants, 58*; *Pharmacopœia of India, 52*; *Royle's Illustrations of the Botany of the Himálaya, p. 177*; *Ainslie, Vol. I., p. 136*; *Moodeen Sheriff's Supplement to the Indian Pharmacopœia*; *Spons' Encyclopædia.*

BOTRYCHIUM, *Sw.; Syn. Fil., 447.* 778

A genus of ferns belonging to the sub-order OPHIOGLOSSACEÆ, distinguished by having the fructification in a compound or raceform panicle, forming a separate branch of the frond.

Botrychium virginianum, *Swartz.; Syn. Fil, 448; Clarke's Ferns, N. Ind., in Trans. Lin. Soc. 1880, p. 588.* 779

Habitat.—The Himálaya from Kumaon to Bhután, the Khásia Hills; very common at altitudes from 5,000 to 8,000 feet.

FOOD. 780

Food.—Sir J. D. Hooker, in his *Himálayan Journals,* says (*Vol. I., 293*): "This large succulent fern grows plentifully at the Raklang Pass in Sikkim; it is boiled and eaten both here and in New Zealand."

BOUCEROSIA, *W. & A.; Gen. Pl., II., 782.* 781

Fleshy, leafless herbs, with thick 4-angled stems, angles toothed. *Flowers* terminal, rather large, solitary or umbelled, more or less purple. *Sepals* narrow. *Corolla* campanulate or rotate, lobes 5, short, broad, valvate. *Corona* annular, adnate to the column, 5-lobed, lobes 2-fid, subulate, erect or spreading, with a linear fleshy process on the inner face at the sinus inflexed over the anther. *Column* minute, short; *anther-tips* inappendiculate; pollen-masses one in each cell, sessile, erect, suborbicular, compressed. *Stigma* low, conical, 5-angled, tip truncate, depressed. *Follicles* slender, straight, terete, smooth. *Seeds* flat, winged, comose.

Boucerosia Aucheriana, *Dcne.; Fl. Br. Ind., IV., 78;* ASCLEPIADEÆ. 782

Vern.—*Charúngli, chungi, pawanne, pamanke,* PB.

Habitat.—Found in the western part of the outer Himálaya, in the Salt-Range and Trans-Indus, ascending to 3,000 feet.

MEDICINE. Stems. 783

Medicine.—The juicy stems are considered stomachic, carminative, and tonic. **Bellew** states that they are also used as vermifuge, and **Masson** mentions that, dried and powdered, they are taken as stimulants.

Boucerosia edulis, *Edge.*, see **Caralluma edulis,** *Benth.*

784

BOUEA, *Meissn.; Gen. Pl., I., 420.*

A genus of trees belonging to the Natural Order ANACARDIACEÆ, containing some five species, natives of tropical Asia and the Malay Archipelago.

Leaves opposite, petioled, coriaceous, glabrous, quite entire. *Flowers* small, in axillary and terminal panicles, polygamous. *Sepals* 3-5, deciduous, valvate. *Petals* 3-5, imbricate. *Disk* very small. *Stamens* 3-5, inserted within the disk, all fertile. *Ovary* sessile, style short, terminal stigma obscurely unequally 3-lobed; ovule ascending from the wall of the cavity. *Drupe* fleshy; stone thin, fibrous, 1-celled, 1-seeded. *Seed* sub-erect; cotyledons fleshy; radicle very short, inferior.

785

Bouea burmanica, *Griff.; Fl. Br. Ind., II., 21.*

Syn.—B. OPPOSITIFOLIA, *Meissn.; Kurz, i., 306;* MANGIFERA OPPOSITIFOLIA, *Roxb.* (*Fl. Ind., Ed. C.B.C,. 215*).

Vern.—*Meriam, mayan,* or *mai-een,* BURM.

Habitat.—A moderate-sized, evergreen tree, met with in Burma and the Andaman Islands.

FOOD. 786

Food.—The tree has an edible fruit, for which it is often cultivated.

TIMBER. 787

Structure of the Wood.—Grey, hard, with a dark reddish-brown heartwood. Weight 55 lbs. per cubic foot.

It is not specially used, but it is said by **Roxburgh** to be very durable.

788

BOWS AND ARROWS, Timbers used for.

Acacia Catechu.
Areca Catechu.
Cephalostachyum capitatum.
Dendrocalamus strictus and other Bamboos.
Dolichandrone stipulata.
Garcinia speciosa.
Grewia oppositifolia.
Grewia vestita.
Lagerstrœmia tomentosa.
Parrotia Jacquemontiana,
Reeds, various species used for arrows.
Shorea siamensis.
Taxus baccata.

Boxwood, see **Buxus.**

789

BOXWOOD, Timbers used as substitutes for.

Atalantia monophylla.
Cratæva religiosa.
Celastrus spinosus.
Chloroxylon Swietenia.
Dodonæa viscosa.
Gardenia gummifera.
G. latifolia.
Hemicyclia sepiaria.
Homonoya symphylliæfolia.
Ixora parviflora.
Memecylon edule.
Murraya exotica.
Olea ferruginea.
Psidium Guyava.
Punica Granatum.
Santalum album.
Sonneratia acida.
Viburnum erubescens.

Brachyramphus sonchifolius, *DC.*, see **Lactuca remotiflora,** *DC.;* COMPOSITÆ.

790

BRAGANTIA, *Lour.; Gen. Pl., III., 122.*

A genus of small bushes belonging to the Natural Order ARISTOLOCHIACEÆ, and containing 3 species, natives of India and the Malaya.

Leaves alternate, shortly petiolate, 3-5 nerved, oblong-lanceolate or obovate. *Flowers* forming a lax cymose-corymb or short racemose cyme. *Perianth* single, bell-shaped, adnate to the ovary and articulated to the top; teeth 3, equal, deciduous. *Stamens* 6-12, arranged in one series on a disk around the base of the style and slightly adherent to it; *filaments* free or slightly united below, often united by threes. *Ovary* inferior, 4-locular, style short, bearing 3-many stigmatic arms; ovules many, arranged on 2 series and pendulous. *Fruit* somewhat like a siliqua, slender, about 4 inches long, terete. *Seeds* oblong, 3-angled.

The wood of the stem is very peculiar, differing in a remarkable degree from that of the ordinary exogens.

Bragantia Wallichii, *R. Br.; Wight, Ic., t. 520; DC. Prodr., XV., pt. I., 430.* 791

Vern.—*Alpam*, MAL.

Habitat.—A small bush, with decumbent branches, met with in the southern half of the Bombay Presidency, near the coast, and in Madras and Ceylon.

Medicine.—"The whole plant, mixed with oil and reduced to an ointment, is said to be very efficacious in the treatment of psora or inveterate ulcers. Like other plants belonging to the same natural order, it is supposed to have virtues in the cure of snake-bites. he juice of the leaves, mixed with the *Vussumbú* (**Acorus Calamus**) root, the root itself rubbed up with lime-juice, and made into a poultice and externally applied, are the chief modes of administering it among the natives." (*Drury's Useful Plants, 86.*) This is regarded as one of the most powerful antidotes to poison known on the west coast. A Malabar proverb says: "As soon as the *Alpam* root enters the body, poison leaves it." **B. tomentosa,** *Blume*, a native of Japan, according to **Horsfield,** is bitter and is regarded as an emmenagogue." (Compare with *Pharm. of India, p. 99; Drury's Useful Plants; Fra. Bartolome's Voyage to the East Indies, p. 416.*)

MEDICINE. Plant. 792 Leaves. 793 Root. 794

Bran.—A coarse product of wheat, separated from the latter in the milling process. See **Triticum sativum.** 795

BRASSAIOPIS, *Dcne. & Planch.; Gen. Pl., I., 945.* 796

Large shrubs or trees, glabrous and tomentose, armed or not. *Leaves* digitate, or palmate or angled; stipules connate within the petiole, not prominent. *Umbels* in large compound panicles, young parts at least stellately tomentose; bracts not large, often persistent; pedicels rising from a dense cluster of persistent bracteoles, not jointed under the flower; flowers often polygamous. *Calyx* 5-toothed. *Petals* 5, valvate. *Stamens* 5. *Ovary* 2-celled; styles 2, united, long or short. *Fruit* broadly globose or turbinate, 2- or by abortion 1-seeded. *Seed* not compressed; albumen ruminated.

Brassaiopis mitis, *C. B. Clarke; Fl. Br. Ind., II., 736;* ARALIACEÆ.

Vern.—*Moqchini*, NEPAL; *Suntong*, LEPCHA.

Habitat.—A small tree of Sikkim Himálaya, above 5,000 feet; common at Darjíling.

Structure of the Wood.—Soft, white, spongy. Weight 24 lbs. per cubic foot.

TIMBER. 797

B. speciosa, *Dcne. & Planch.; Fl. Br. Ind., II., 737.*

Syn.—B. FLORIBUNDA, *Seem.*

Habitat.—A small tree, met with from Nepál to Assam and Chittagong.

Structure of the Wood.—White, soft, resembling the preceding.

TIMBER. 98

799

BRASSICA, *Linn.; Gen. Pl., I., 84.*

A remarkable group of plants, belonging to the Natural Order CRUCIFERÆ. There are supposed to be some 80 species, natives of the temperate regions of the Old World; but under cultivation the forms have been increased almost indefinitely.

Glabrous or hispid herbs; root-stock often woody. *Leaves* large, pinnatifid or lyrate, rarely entire. *Flowers* yellow, in long racemes. *Sepals* erect or spreading, lateral, usually saccate at the base. *Pods* elongate, terete or angular, often with an indehiscent 1-seeded beak; valves convex, 1-3-nerved, lateral nerves flexuous; style beaked or ensiform; stigma truncate or 2-lobed. *Seeds* 1-seriate, globose or sub-compressed; cotyledons incumbent, concave or conduplicate, the radicle within the longitudinal fold.

The generic name is derived from the Latin *Brassica* and the Celtic *Bresic*, or *Braisscagh*. Nearly all the species of this interesting and most valuable genus exist entirely in a state of cultivation. They are anti-scorbutic. It may be stated that no plant with a 4-merous condition of the corolla, and with 4 long and 2 short stamens, is known to be poisonous. These are the eye-marks of the CRUCIFERÆ, a family which yields the majority of the vegetables used by the inhabitants of temperate countries. Of the Cruciferous genera, **Brassica** is the most important. To it belong the mustard, the cabbage, the cauliflower, the broccoli, the borecole, and the turnip, with their innumerable varieties. The following are the important Indian wild or cultivated species, with their principal culinary forms.

The various species of **Brassica** met with in India may conveniently be referred to two important sections:

1st.—Those which may be regarded as of Asiatic origin,—*i.e.*, are either indigenous to Asia or were introduced at an early date.

2nd.—Modern introductions, or the species which may be viewed as most probably European forms of the genus.

The former includes the varieties and races of **Brassica campestris, B. juncea,** and **B. chinensis**; the latter, **B. alba, B. nigra, B. oleracea,** and **B. rapa.** The European members are easily recognised, and there can be no possible confusion regarding them; but unfortunately it is quite otherwise with the Asiatic forms. **Messrs. Duthie** and **Fuller** (*Field and Garden Crops*) have, however, contributed a most valuable paper on this subject, which has gone a long way to clearing up many of the doubtful points regarding these plants. I take the liberty to reproduce here their most useful analysis of the botanical characters of the Asiatic species:—

* Foliage usually glaucous and smooth, rarely hispid; leaves amplexicaul, auricled; seeds yellow or brown.

† *Corymbs few-flowered; sepals erect; pods very thick, not torulose, 2-3-4-valved; seeds large, yellow or brown.*

Pods erect, 2-valved . . . **B. glauca,** *Roxb.*
Pods pendulous, 3-4-valved . . . **B. trilocularis.**
Pods erect, 4-valved . . . **B. quadrivalvis.**

†† *Corymbs many-flowered; sepals spreading; pods stoutish, somewhat torulose; seeds brown or reddish brown, rather large, minutely rugose.*

Pods not torulose, slender, with a long tapering beak; seeds dark brown . **B. dichotoma,** *Roxb.*

Pods somewhat torulose. short, with sharp beak; seeds reddish brown . **B. glauca,** *Royle, var.* **Toria.**

** Foliage usually bright green and more or less hispid; leaves stalked or the upper ones sessile, not amplexicaul; pods thin, torulose; seeds small, dark brown or reddish brown, distinctly reticulated. **B. juncea** and **B. chinensis.**

From an agricultural point of view the Asiatic forms may be referred to three important sections:

(*a*) *Sarsón,* may be called the Indian rape and colza series.
(*b*) *Tória,* is exported either as mustard or as rape.
(*c*) *Rái,* commonly known as Indian mustard.

How far it is possible to separate the *sarsóns* into groups corresponding to the rape and the colza of Europe may be doubted, but the line which separates these from *tória* and from *rái* is well marked and should be carefully observed. The various forms of *sarsón* are seldom grown alone, but constitute an element in mixed crops, being generally sown, in Upper India at least, along with nearly every crop of wheat and barley. On the other hand, *tória* is nearly always grown by itself.

If not destroyed by blight, *sarsón* or *tória* (*lahi*) is a most valuable crop,—much more so in fact than either wheat or barley. The danger of blight is, however, very considerable, and few small cultivators care to risk an entire failure in the hope of a good crop, which, if good, is of course very remunerative; hence the rapes and also mustard are generally grown as mixed crops. In India, as a whole, mustard (*rái*) is not so extensively cultivated as rape. It is a mixed crop, the weight of oil obtained is only one fourth instead of one third, and it is also less esteemed as an article of food. It is a little difficult to determine what is meant by mustard in the returns of Indian export trade, but that the amount of true mustard must be very small seems beyond doubt; indeed, it is highly probable that neither **Brassica alba** nor **Brassica nigra**—the true mustards of European commerce—are ever exported from India. It seems quite likely that the better classes of each of the above Asiatic crops are exported as Mustard and the inferior as Rape.

Brassica alba, *H.f. & T.T.; Fl. Br. Ind., I., 157.*

800

THE WHITE MUSTARD.

Vern.—*Sufëd-rái, sufëd-ráyán,* HIND., DUK.; *Dhóp-rai,* BENG.; *Pandhora-mohare,* MAR.; *Ujlo-rái,* GUJ.; *Vellai-kadugu,* TAM.; *Tella-avalu,* TEL.; *Vella-katuka,* MALA.; *Bili-sásave,* KAN.; *Siddhártha, shvetasarshapa,* SANS.; *Khardale-abyaz,* ARAB.; *Sipandáne-supid,* PERS.; *Suddu-abbe,* SINGH.; *Aphiyu-munniyé-zi,* BURM.

Habitat.—This is supposed to be a native of the more southern portions of Europe and of Western Asia.

Botanic Diagnosis.—This is the plant which yields the so-called "White Mustard." It is by no means a common plant, but may be recognised by its large yellow flowers, and spreading, hispid, few-seeded pods, with a long empty and flat beak. (Compare with **B. Nigra.**)

Properties and Uses—

FOOD. Seeds. 801

The seeds, large and white.

Flour. 802

The flour, rarely used alone.

Oil. 803

The oil, little known.

The plant is also eaten as salad, the seeds being sown thickly, and the young seedling plants cut when about 2 inches high; the leaves and young sprouts are also eaten as a green vegetable. The cake is much used in Europe to feed sheep. It is regarded as fattening for sheep; black oil-cake is not considered so good for this purpose.

Leaves and sprouts. 804 Oil-cake. 805

"When triturated with water, the seeds form a yellowish emulsion of very pungent taste, but it is inodorous, and does not, under any circumstances, yield a volatile oil. The powdered seeds made into a paste with cold water act as a highly stimulating cataplasm. The entire seeds yield to cold water an abundance of mucilage." *(Flück. and Hanb., Pharmacog.)*

MEDICINE. 806

Medicine.—White mustard alone must be regarded as useless, but it is invariably mixed with black in the preparation of mustard flour.

Chemistry.—The vesiccating or stimulating properties of white mustard are due to sulphocyanate of acrinyl. It does not pre-exist in the seed and cannot be obtained by distillation. (For further information see **B. nigra.**)

807

Brassica campestris, *Linn; Fl. Br. Ind., I., 156.*

To this species belong the turnip, the rape, coleseed, colza, and other forms known in Europe. The Indian examples may be said to be represented by *sarsón* and *tória.* It is necessary to make some attempt at referring the Indian forms to their sub-species and varieties, but until the subject has been more carefully investigated, very little of a definite and trustworthy nature can be given. The greatest possible difficulty exists in separating the economic and trade facts which have appeared in Indian works and reports on the subject; mistakes are almost unavoidable. But before attempting to establish the varieties of the Indian members of this species, it may be found convenient to give in this place a brief indication of the arrangement adopted in Europe, so that comparison may be possible with the Indian plants.

European Forms of Brassica campestris.

SUB-SPECIES I—**campestris** proper.

Colza. 808

COLZA, WILD NAVEW OR NAVETTE, COLESEED, SWEDISH TURNIP, *Eng.;* CHOU DES CHAMPS, NAVETTE, *Fr.*

Botanic Diagnosis.—*Root* tuberous. *Leaves* glaucous, radical, hispid, upper glabrous. *Racemes* close, the open flowers rising above the buds, and caducous before the corymb lengthens into the raceme.

This is the wild coleseed of the fields of England and of many parts of Europe. It is sometimes cultivated in France for its seed—the *colsa, colsa,* or *colsat,* the *chou oleifére* of the French. It is unfortunate that in England, and, indeed, in many parts of the Continent, the name coleseed or colza has been applied to rape as a synonymous term. They are perfectly distinct; the seed-produce of colza is much greater in quantity though inferior in value to rape. Colza is much grown in France and Belgium, but by the British farmer it is supposed to exhaust the soil. The Swedish turnip is a cultivated form of this plant, bearing the same relation to the normal form which the *khol-rabi* does to the cabbage.

It seems very probable that the hairy plants which **Roxburgh** called **Sinapis dichotoma** and **S. trilocularis,** as also **Brassica quadrilocularis,** *H.f. & T.*, should be viewed as forms of this sub-species and not of **Napus,** as stated by **Messrs. Duthie** and **Fuller.** The nature of the seed and of the oil yielded by these forms corresponds with that of the French and Belgium plant, thus affording additional corroboration to the above suggestion, based on the agreement of the botanical characters which exists between them. As cultivated crops, however, they are quite distinct from rape. It would therefore be of considerable practical advantage to have the returns of these forms declared separately from those which more properly should be called Indian rape. This would be of importance even although it must be admitted that dried specimens of the former plant can hardly be distinguished from the latter.

RAPE. 809

SUB-SPECIES 2—**Napus.**

810

RAPE, NAVEW OR COLESEED, *Eng.;* CHOU NAVET, *Fr.;* DER RÜBEN KOHL, *Germ.*

Botanic Diagnosis.—*Root* fusiform. *Leaves* all glabrous and glaucous. *Raceme* elongated at the time when the flowers expand. (*Note.*—The character of the raceme of this and of the preceding sub-species would appear to have been reversed in some recent works on this genus.)

Although **Napus** is largely seen as a weed of cultivation in corn-fields, especially in Scotland, it may be viewed as an escape from cultivation. As a crop the seeds yield the rape oil of commerce.

It seems probable that to this sub-species should be referred **Sinapis glabra,** *Roxb.,* and **S. glauca,** *Royle.*

SUB-SPECIES 3—**Rapa.**

THE TURNIP. 811

THE TURNIP, *Eng.;* CHOU A FEUILLES RUDES, *Fr.;* DER RÜBEN KOHL, *Germ.*

Vern.—*Shalgham,* HIND., BENG., PERS.; *Shaljam,* ARAB.

Botanic Diagnosis.—*Root* tuberous. *Radical* leaves green, not glaucous, hispid; stem leaves glaucous and glabrous. *Flowers* falling off before the corymb lengthens into a raceme.

Habitat.—Grown as a garden crop all over India.

Properties and Uses—

The young leaves used as food.

The root largely used as food.

The seeds are chiefly used for propagation, but an oil is also prepared from them. The common cultivated Turnip may almost be said to be acclimatised in India, and to have gained great favour with the natives as a vegetable. The Brahmans and Baniyas have a prejudice against it, however, from a suspicion of a relation to beef or animal matter.

Indian Forms.

Speaking of the forms of **B. campestris** met with in the North-West Provinces, **Mr. Atkinson,** in the manuscript which he has most obligingly placed at my disposal, says: "The varieties known as **S. glcauca** and **S. dichotoma** are well marked and are known by different names in almost every district, and appear to be entitled to specific notice." These forms, he goes on to say, are "extensively cultivated in almost every district in the N.-W. Provinces on account of the oil obtained from the seed. They are, however, sparsely grown in the Benares District, because there they are peculiarly subject to the attack of *lábó,* a small black fly."

Var. 1—dichotoma, sp., *Roxb.*

812

It seems probable, as already indicated, that, owing to the hairiness of the leaves, this should be viewed as a form of the sub-species **B. campestris proper;** thus corresponding to the Swedish turnip and colza series.

KÁLÍ SARSÓN (The Indian commercial name).

Syn.—SINAPIS DICHOTOMA, *Roxb.;* S. BRASSICATA, *Roxb.*

Vern.—*Káli sarsón, kálé-rái, sursi, jariya, lahota, laita, jadiya,* HIND.; *Sursha, shurshi* or *sursi, sanshi, káli-sarsón, sáda-rái,* BENG.; *Sarsu,* RAJ.; *Kálé-ráyan,* DUK.; *Surah,* CUTCH; *Sarsawa, kála-rái,* GUJ.; *Sherasa, kála-mohare,* MAR.; *Karuppu-kadugu,* TAM.; *Nalla-áválu,* TEL.; *Karuppa-katuka,* MALA.; *Sarsive, kappu-sasoe,* KAN.; *Sarshapa, kála-sarshapa,* SANS.; *Khardale-asvad,* ARAB.; *Sipandáne-siyah,* PERS.; *Kalu-abbé,* SINGH.; *Amé-mnnniyén-zi,* BURM.

Botanic Diagnosis.—Upper leaves lyrate or entire, amplexicaul, lower auricled, deeply pinnatifid; the ground ones being more or less hairy;

used for culinary purposes. Pods sub-cylindrical, 2-3 inches long, with a long tapering beak; seeds small, dark or light brown, smooth or minutely rugose.

I am inclined to think a serious mistake has been made by European authors in regarding this variety as identical with **S. glauca,** *Roxb.* The latter plant yields a decidedly superior oil, and both seeds and plant are readily distinguished by the most ordinary native, and their properties narrated with precision.

FOOD. Leaves. 813 Seed. 814 Oil-cake. 815 OIL. 816

Oil.—Colza oil is used by the natives of India chiefly to anoint the body and for illuminating purposes.

§ "On this side of India the oil (*Sarasin,* GUJ.) is used for pickles and culinary purposes: the oil-cake is given to cattle." (*Assistant Surgeon Sakhárám Arjun Rávat, Girgaum, Bombay.*) It seems probable that the above remark should be transferred to the next variety. (*G. Watt.*)

817

Var. 2—glauca, sp., *Roxb.*

This seems to belong to the sub-species **Napus.**

RAPE-SEED, RARA-SARSON (the Indian commercial name). **Roxburgh** calls this "White Mustard."

Syn.—SINAPIS GLAUCA, *Roxb.*

Vern.—*Sarsón, sarsón-zard, bára-lai, shetashirsa, banga-sarsón, pílá sarsón, rára-rada, rára-sarsón, píli-rái,* HIND.; *Shwet-rái,* BENG.; *Rájiká, tuverika,* SANS.; *Sarashire, raira,* GUJ.; *Pílé-ráyán,* DUK.

§ "Raira, *i.e.,* like mustard." (*Dr. W. Dymock, Bombay.*)

Botanic Diagnosis.—The leaves are amplexicaul, the lower deeply pinnatifid, the ground ones being quite glabrous; used for culinary purposes. Pods very thick, laterally compressed, $\frac{1}{3}$ to $\frac{1}{2}$ inch in length, with a broad flattened beak; seeds round, smooth, light yellow or white, but occasionally deeper coloured.

FOOD. Leaves. 818 Seed. 819 Oil-cake. 820 OIL. 821

Oil.—It is highly probable that a great part of the so-called mustard oil of India should be transferred to this position under the name of *sarson* or Indian rape oil. How far the oil from the next variety, *toria,* may also belong to this class cannot at present be determined, but it has been deemed advisable to give here an abstract of all the available information regarding Indian rape oil. Speaking of this plant, **Roxburgh** says: "The entire seed is used for various economical purposes; an oil is also expressed from it, which is much used in the diet of the Hindús." "Rape oil is expressed after the ordinary fashion of the oil-presser or *teli,* who returns to the cultivator one third of the weight of the seed in oil." (*Duthie and Fuller.*) "The oil is expressed in the same manner as *til* by means of a large wooden mortar and pestle worked by cattle. In Bulandshahar the seed sells at 12 seers for the rupee; one maund yields 13 seers of oil and 27 seers of cake at a cost of 12 annas. The oil sells at $3\frac{1}{4}$ seers per rupee and the cake at 35. The oil is used in the preparation of condiments, such as pickles, preserves, curries, and for other culinary purposes. It is also used for burning, though, from its strong odour, it is not a favourite with Europeans. It is, however, capable of purification by agitation in a leaden vessel or with sulphuric acid and water." (*Mr. Atkinson's MSS.*) The oil which is most esteemed as an article of food amongst the Hindús is that obtained from **B. juncea,** which see.

Special Opinions.—§ "Combined with camphor, it forms an efficacious embrocation in muscular rheumatism, stiff neck, &c. Pure oil, commonly used by natives of Bengal to anoint the body before bathing, strengthens the skin and keeps it cool and healthy. Sometimes used by suicides to dissolve opium, thus hastening the effect of the narcotic. The seeds pounded and mixed with hot water form an efficient counter-irritant poultice." (*Assistant Surgeon Shib Chunder Bhuttacharji, Central Provinces.*) "The oil is also used as an external application in dengue fever with great benefit." (*Surgeon-Major S. A. G. Jayakar, Muskat, Arabia.*) "Is much used for rubbing on the chest in bronchitis, especially of children." (*Surgeon-Major P. N. Mukerji, Cuttack, Orissa.*) "Similar action to mustard, but less effective." (*Surgeon W. Barren, Bhuj, Cutch.*)

Var. 3—Toria, *Duthie and Fuller.*

822

It seems probable that this should be viewed as a variety of the subspecies **Napus**; it has some resemblance to the summer rape of Germany, the *navette d'ete* of France.

Syn.—SINAPIS GLAUCA, *Royle.*

Vern.—*Tori, tóriya, khetiya, lahi* or *lai,* also *dain,* and *dáin-lai,* HIND.; *Tuverika,* SANS.

Botanic Diagnosis.—The whole plant quite smooth and glaucous, 2-3 feet in height. Lower leaves lyrate or pinnatifid, upper amplexicaul, lanceolate, entire. The leaves are used for culinary purposes. Flowers bright yellow, sepals spreading. Pods rather slender, 1¼ to 1¾ long, transversely compressed, and more or less torulose; beak about ¼ inch long. Seeds small, roundish or semi-compressed, reddish brown, finely rugose.

FOOD. Leaves. 823 Seeds. 824 Oil-cake. 825 OIL. 826

No definite information can be obtained regarding the oil expressed from the seeds of this form, except that the oil is regularly expressed, and that this plant is the staple mustard crop of the hills.

As a rule, it is grown alone and is produced in the greatest abundance in the districts which border on the Himálayan Terai. In the North-West Provinces and Oudh, it occupies annually about 35,000 acres, in the 30 temporarily-settled districts, yielding 4 to 6 maunds of seed per acre.

RAPE-SEED.

827

From the preceding remarks the reader may have gathered that the Indian forms of **Brassica campestris** may, with at least a certain degree of accuracy, be referred to three primary sections:—

Section I.—Colza, which corresponds to **Roxburgh's Sinapis dichotoma,** and the abnormal forms of that plant which have come to be known as **B. trilocularis** and **B. quadrilocularis.**

Section II.—Rape or **Sinapis glauca,** *Roxb.*

Section III.—*Toria,* or another form, most probably of rape, which has received the name of **S. glauca,** *Royle.*

There is every reason to suppose that II. and III are commercially known as rape, although perhaps the last may occasionally be classed as mustard. These three forms individually represent agricultural products of the greatest importance to India. They would seem sufficiently distinct to have justified their retention at least as varieties, very much corresponding to the original species. The natives display a highly-developed power of observation in this direction; they have long become perfectly familiar with these plants, and can, as a rule, name them with unerring certainty. In our trade returns, however, rape is always spoken of as if there was only one plant concerned. It has hence become necessary to discuss the subject of rape collectively, although it must be remarked that it is very desirable that an effort be made to distinguish the three forms, and, if possible, to publish separate returns for each.

Food.—Whether or not it may be found correct to botanically sub-divide the Indian forms of **B. campestris** into two or three primary sections, resembling rape and colza, and to identify these sections with the corresponding European forms, it cannot be doubted but that such a classification would serve a commercial purpose. It would separate the oil which in Indian commerce is called rape oil, from that which might with

FOOD. Seed. 828 Oil. 829 Oil-cake. 830

advantage, in order to remove confusion, receive the name of colza, as well as both these from mustard oil and the other oils obtained from the remaining members of the genus. It will be enough, however, to suggest this separation; subsequent research may reveal further corrections and subdivisions, for there are many points which it is difficult to settle definitely in the present state of information. Perhaps the best botanical character that can be cited in support of the proposed separation is the glabrous nature of the ground leaves of the forms above referred to as *Navet* (rape), and the more or less hairy ground leaves of **S. dichotoma,** corresponding with those of *Navette* (colza). The seeds in the former are smooth and light, in the latter smooth or rough, but dark coloured. Rape oil (**S. glauca**) is regarded as better in quality than (colza oil) the oil from **S. dichotoma,** the latter being used chiefly to anoint the body, while the former is largely used in cookery, and is exported to Europe for illuminating purposes and to meet a demand in the India-rubber manufacture. As already stated, in the trade returns of the exportation of rape oil and seed from India, apparently both the above are included as different qualities of rape, if not also the oil expressed from **B. juncea** and **Eruca sativa.**

In his *Panjáb Products,* **Mr. Baden Powell** has mistaken these plants; he identifies *sarson* or rape with **Sinapis juncea,** mustard with **S. campestris,** of which he apparently views **S. alba** and **S. nigra** as varieties. Regarding **Mr. Atkinson** as correct, I have in substance followed the admirable division given by him in his *Himálayan Districts,* page 770, also in the private MSS. from which one or two passages have already been extracted.

In European commerce, rape and colza are names which unfortunately have come to be used almost synonymously. The separation here recommended of the probably corresponding Indian forms has been deemed advisable, chiefly with a view to more clearly identifying the Indian oils allied to mustard. The oils obtained from these are even more distinct than the oils from the European plants, and their respective properties are well understood and appreciated by the natives of India. Some such separation seems highly desirable. **Simmonds,** in his *Tropical Agriculture* (1877), remarks of Indian so-called rape-seed, that "the prices in the London market in the beginning of 1877 were, for Calcutta brown, 59*s.* 6*d.* to 60*s.* per quarter, and for Ferozepore 59*s.*" Under mustard he seems to include **S. chinensis, S. dichotoma, S. pekinensis, S. ramosa, S. glauca,** and **S. juncea** as the mustard-yielding species of Asia. The majority of these plants are those which yield the so-called rape-seed as exported from India; **Brassica (Sinapis) juncea** alone falling within those pronounced to be mustard. In fact it is probable, as already stated, that the bulk of the seed exported from India as mustard is obtained from **B. juncea,** and not from **B. alba** and **B. nigra,** the true mustards.

"In India, rape-seed is very commonly sown mixed with mustard-seed, and almost as an auxiliary with grain crops. It prefers loams, and does not thrive on clay soils. The sowing takes place in October, and the harvest in the following February, the plants being cut somewhat prematurely, otherwise the pods would burst and much of the seed be lost. The latter is ripened by exposure to the sun for 3 or 4 days on the threshing-floor, and is then easily dislodged." "The Indian seed known as 'Guzerat Rape,' largely crushed at Dantzic, is found to yield $3\frac{1}{2}$ per cent. more oil than European seed, and leaves a cake richer in fatty matter and albuminoids; it is shipped from Bombay and brings the highest price of any." (*Spons' Encycl.*)

Gujarat Rape. 831

A good deal has been written regarding the superiority of this so-called Guzerat rape. It seems to be a superior quality of var. **Toria** or of var.

glauca. In the *Bombay Gazetteer* occurs the following notice of this form : "Rape-seed, *sarsón,* Brassica Napus, holds the first place among oil-seeds and the third place among crops in general (in Kadi sub-division). Land intended for it is left fallow for four months and ploughed twenty times before the seed is sown. The crop does not require any watering. The seed is sown through drills in November at the rate of from 2 to 3 *sérs* to the *bigha* and reaped in March, and the average yield varies from 400 to 800 lbs. When the crop is grown in *bajarváda* land, the yield is small and rarely exceeds 200 lbs. The rape-seed grown in this division is of a better description than any in Gujarát, and has a larger·grain. The produce forms one of the chief articles of export." (*Bomb Gaz., VII., 97.*) The Kew Report of 1877 says : "Guzerát rape-seed has been crushed at Dantzig, and is found to yield 3·5 per cent. more oil than rape; the cake also yields 10 per cent. fatty matter and 34 per cent. albuminoids, both being in excess of the amounts yielded by ordinary rape."

In an official correspondence with the Home Department regarding a proposed future edition of the *Pharmacopœia of India*, the Bombay Committee describe a Gujarát plant under the name **B. juncea**; from the botanical characters given, one is compelled to believe this must be **B. campestris,** *var.* **Toria**. The leaves are said to be glabrous and attenuated at the base, the upper ones lanceolate and entire. The seeds are oblong, light brown, and minutely reticulated. These are certainly not the characters of the hairy, petiolate, non-attenuated leaves of **B. juncea,** nor the characters of the seed of that plant. If this presumption proves correct, it is a remarkable fact that the medicinal virtues of the mustard should be attributed to a form of rape. It is probable, however, that the botanical description is not that of the plant for which the medicinal virtues have been given. A chemical analysis of the Gujarát seed would put at rest all possible misunderstandings. (See Chemical note regarding mustard and rape.)

TRADE RETURNS.

832

INTERNAL TRADE.

In the returns of Internal Trade, the quotations are sometimes given as "Rape," mustard not being mentioned; sometimes as "Mustard," and apparently no rape; and again as "Rape and Mustard" jointly. It is thus impossible to separate these so as to show the relation of each to the foreign exports which are published separately for Rape and for Mustard. Of rape and mustard, the North-West Provinces and Oudh exported 1,545,327 cwt., valued at R50,66,068 during the year 1883-84; Bombay imported 1,120,345 cwt., valued at R66,85,583; and Calcutta imported 2,773,621 cwt., the bulk of which was borne by the East Indian Railway.

EXTERNAL TRADE.

The following table shows the exports of rape-seed to other countries by sea during the six years ending 1883-84 :—

YEARS.	Quantity in Cwt.	Value in Rupees.
1878-79	2,165,475	1,36,67,869
1879-80	1,380,572	85,37,717
1880-81	1,255,580	67,10,338
1881-82	1,935,621	1,03,19,272
1882-83	2,821,420	1,57,05,233
1883-84	3,945,727	2,44,14,331

BRASSICA juncea.

Indian Mustard.

The following analysis of the exports of rape-seed for the year 1883-84 shows the Presidencies or Provinces whence exported, and the countries to which consigned :—

Presidency or Province from which exported.	Quantity in Cwt.	Value in Rupees.	Country to which exported.	Quantity in Cwt.	Value in Rupees.
Bengal . .	1,692,023	89,34,889	United Kingdom .	1,970,395	1,07,95,473
Bombay . .	1,302,623	97,12,685	Austria . .	2,400	18,300
Sind . .	887,970	57,29,168	Belgium . .	671,879	45,35,799
Madras . .	63,111	3,37,589	Denmark . .	4,910	32,730
			France . .	776,204	55,02,402
			Germany . .	304,754	21,73,768
			Italy . . .	34,821	28,51,680
			Egypt . . .	178,792	10,92,683
			Aden . . .	1,400	10,675
			Other countries .	172	821
TOTAL .	3,945,727	2,44,14,331	TOTAL .	3,945,727	2,44,14,331

833

Brassica juncea, *H. f. & T.; Fl. Br. Ind., I., 157.*

THE RÁI OR INDIAN MUSTARD.

Syn.—SINAPIS RAMOSA, *Roxb.;* S. CUNEIFOLIA, *Roxb.;* S. RUGOSA, *Roxb.;* S. NURCEA, *Linn.;* S. JUNCEA, *Linn.*

Vern.—*Rái, sarsón, sarsón-lahi, gohna-sarsón, bari-rái, barlái, bádsháhi-rái, sháhzáda-rái, khas-rai,* HIND.; *Rái sarishá,* BENG.; *Asúr,* KASHMIR; *Rái,* GUJ., KUTCH; *Rái, sarsón, rájiká,* BOMB.; *Mohari, ráyán,* MAR.; *Rájiká,* SANS.; *Abba,* SINGH.

Habitat.—Cultivated abundantly in India; it extends westward to Egypt, and eastward to China. This is in fact the plant which in India bears the name of mustard, and takes the place of **B. nigra** in all warm countries. In the North-West Provinces and Oudh it is generally sown on borders of fields of wheat, barley, or peas, sometimes broadcast, 3 lbs. per acre of seed being required, and yielding an outturn of 3 to 4 maunds to the acre. It is largely grown in the south of Russia and in the steppes north-east of the Caspian, flourishing on saline soils. At Sarepta, Saratoo, it has been largely cultivated for a century, and it is also grown in Central Africa.

Botanic Diagnosis.—"A tall, erect annual, 3-5 feet in height, with bright green foliage, rarely glaucous, more or less hispid towards the base; stems much-branched, smooth, terete, often tinged purplish red, especially at the joints. Leaves not amplexicaul, the lower ones stalked, lyrate or pinnatifid, margin variously serrate-dentate, often very hispid, especially when young; petioles channelled, upper leaves sub-sessile, linear-lanceolate, smooth, dentate, or the uppermost quite entire. Racemes terminal; flowers stalked; pedicels elongated in fruit, divaricate, calyx with linear boat-shaped spreading sepals. Petals small, bright yellow. Pods slender, 1-2 inch long, sub-compressed torulose; beak about $\frac{1}{8}$ the length of the pod; valves with a prominent midrib. Seeds small, sub-globose, dark or reddish brown, with a rough reticulated testa." *(Duthie and Fuller.)*

§ "The measurement of the seeds might be given, as our brown mustard is very much larger than the Guzarat mustard, and has ovate-lanceolate or runcinate toothed leaves." *(Surgeon-Major W. Dymock, Bombay.)* The seeds of this species are much smaller than those of any of the preceding forms. *(G. Watt.)*

834

Cultivation of Indian Mustard.—In Bengal this is a much more important crop than rape. In the Mánbhúm district (see *Hunter's Statistical Account of Bengal*) it is sown in dry land, in October, and cut in February. It is sometimes sown alone or as a mixed crop with peas, *musarí*, barley, &c., grown on high lands. In Cuttack it is described as sown in October and reaped in January, and as luxuriating on soils where salt is deposited. In Julpaigori "Mustard is extensively grown as an oil-seed, and, next to rice, is the most important crop of the district. It is sown broadcast on highlands in November and December and is reaped in March and April. The young leaves of the plant are used as a vegetable." In the Administration Report of Bengal for 1882-83 it is stated that "in Bengal proper, mustard seed is of greater importance than linseed. Of all descriptions of oil, mustard oil is the most largely consumed and most relished by the people. Poor lands and lands recently reclaimed from jungle are generally sown with it, the yield being considerable in comparison with the small amount of labour required for cultivating and preparing the land." "This species is not cultivated in the North-West Provinces to the same extent as **B. campestris**, *sarsón*, though it is the staple crop of Kumaon. The seeds are exported for their oil." (*Mr. Atkinson's MSS.*) Detailed information regarding the relative amount of this crop, as compared with rape, cannot be obtained from the provincial reports, but it may be stated that it is a much more important crop in Bengal than in the other provinces.

Properties and Uses—

FOOD. Leaves. 835 Seeds. 836 FODDER. 837

Food.—The leaves are used as a vegetable. In Kumaon the plant is cultivated chiefly for its leaves, which are eaten. (*Atkinson.*) When the supply of fodder happens to run short in January or February, the mustard crop is frequently cut green and given to cattle.

The seeds are small, round, dark, distinctly reticulated. About 15 to 20 occur in each cell of the pod; in these respects **B. juncea** seems recognisable from the other members of the genus, most of which have large light-coloured or yellow seeds, generally smooth, with rarely more than half the number of seeds in the pod.

Ground into flour, they are used largely as an adulterant with the true mustard. The seeds, whole or broken, are often used to flavour curries. By pressure they yield more than 20 per cent. of a fixed oil which is used in Russia in place of olive oil.

OIL. 838

Indian Mustard Oil.—**Roxburgh** apparently regarded this oil as inferior to rape oil. This does not appear to be the case. It is of a much purer kind than that from **B. campestris**; it has not the peculiar rancid smell characteristic of rape and colza; it is clearer in colour and is used almost entirely as an article of food, being the oil most generally employed in the plains of India for that purpose. This seems to be the oil called mustard oil so largely prepared in our jails by convict labour. The seeds are reported to yield from 20 to 25 per cent. of oil.

FLOUR. 839

Mustard Flour.—As has been already stated, this plant may be called the Indian Mustard. In point of structure it is perhaps more nearly allied to the true Mustard than to any other member of the genus. Its properties seem also very similar, and, in fact, the seeds are largely used to adulterate, or as a substitute for, mustard in the preparation of "Mustard flour."

MEDICINE. Seeds. 840

Medicine.—"The seeds commonly met with in the bazars of India, which, from their colour, may be denominated *Brown Mustard Seed*, possess properties similar to those of the Black and White Mustard Seed, for which they may be employed as an efficient substitute, especially in the preparation of mustard poultices." (*Pharm. of Ind.*)

Under the name of **B. juncea** the Bombay Pharmacopœia Committee give the following: "Externally used in internal congestions, in spasmodic, neuralgic, and rheumatic affections, and in morbid states of the cerebro-spinal system, as an emetic in ebrietas and other cases where it is desirable simply to empty the stomach without inducing a depressing influence in the system. In native practice, for external use, it is often combined with moringa bark or garlic, which greatly increases its activity. Taken internally in moderate quantities it acts as a digestive." (Compare with the note regarding above under Rape, page .) The *United States Dispensatory* says: "The mustard flour which the seeds yield is of a very fine yellow, and affords on distillation the oil of black mustard." (*15th Ed., p. 1305.*) "The seeds closely resemble those of **B. nigra,** and afford when distilled the same essential oil." (*Flück. and Hanb., Pharmacog., 68.*)

841

Brassica nigra, *Koch; Fl. Br. Ind., I., 156.*

THE BLACK or TRUE MUSTARD, *Eng.;* MOUTARDE NOIRE, *Fr.;* MUSTERT, SEUFSAMEN, *Ger.;* SENAPA, *It.;* MOSTARDA, *Por.*

Syn.—SINAPIS ERYSIMOIDES, *Roxb.;* SINAPIS NIGRA, *Linn.*

Vern.—*Rái, kálí rái, tírá, tárá míra, lahi, banárasi rái, jag rái, asl-rái, ghorrái, makra-rái, &c.,* HIND.; *Rái sarishá,* BENG.; *Rái, káli rái,* GUJ.; *Rái, sarsan,* BOMB.; *Kadagho,* TAM.; *Avalo,* TEL.; *Bile sasive, karisasive, sasive,* KAN.; *Rájiká* (?), *sarshap,* SANS.; *Sárshaf* (the name by which it is known in Indian hospitals), PERS.; *Khirdal* or *khardál,* ARAB.; *Ganaba,* SINGH.; *Kiditsai,* CHINESE.

§ "Madras vernacular names are the same as those given in page 523 under var. 1—**dichotoma,** sp., *Roxb.*" (*Moodeen Sheriff.*)

Habitat.—Cultivated in various parts of India and Thibet, chiefly on the hills. It is found wild over the whole of Europe, excepting in the extreme north.

Botanic Diagnosis.—This may be distinguished from **B. alba** by its stem-clasping or adpressed and nearly glabrous short pods.

History.—Mustard was well known to the ancients. It is mentioned by **Theophratus, Dioscorides, Pliny;** and it has been cultivated as an article of food in Europe since the thirteenth century. Its essential oil was first noticed in 1660.

FOOD. Leaves. 842 Seeds. 843

Food.—The leaves are all petioled, the lower lyrate and the upper entire. They are used for culinary purposes.

The seeds are about $\frac{1}{25}$ to $\frac{1}{50}$ of an inch oblong, and dark-coloured, with a reticulated surface.

OIL. Bland. 844 Essential. 845

True Mustard Oil.—A bland oil, expressed from the seed, is used for various economic purposes. About 23 per cent. is usually expressed. The oil is inodorous, non-drying, and solidifies at 0° F. It consists essentially of glycerides, of stearic, oleic, erucic, and brassic acids, the last being homologous with oleic acid. An essential oil is obtained through the action of water. (See *Chemical Composition.*)

MEDICINE. 846

Medicine.—The seeds of this plant are used in medicine as poultice, being a useful and simple rubefacient and vesicant. Mustard poultices prove highly serviceable in cases of febrile and inflammatory diseases, internal congestions, spasmodic, neuralgic, and rheumatic affections. Mustard flour in water is highly recommended as a speedy and safe emetic. The bland oil is largely prescribed by native doctors.

The seeds or flour act as a digestive condiment if taken moderately. If swallowed whole they operate as a laxative, and for this purpose are sometimes prescribed in dyspepsia and other complaints attended with torpid bowels.

847

Chemical Composition.—"Both black and white mustard seeds contain a fixed non-drying oil which is obtained by expression, the amount varying

True Mustard.

BRASSICA nigra.

Chemical Composition. 847

from 25 to 35 per cent.; this forms the mustard oil of commerce. Mustard oil contains erucic, stearic, oleic, and sinapoleic acids. White mustard-seed oil, in addition to these acids, contains tenic acid. (*Flückiger.*) On distillation black mustard yields a volatile oil, the essential oil of mustard. This volatile oil does not exist ready formed in the seeds, but is a product of the action of the myrosin on myronate of potash, now called sinigrin. While the seeds are dry these bodies do not come in contact, but directly water is added, the myrosin decomposes the sinigrin into sulphocyanid of allyl, essential oil of mustard, sugar, and a potash salt of sulphuric acid. It is to the essential oil of mustard that the pungent smell and taste of mustard is due. Applied to the skin it causes almost instant vesication. Myrosin is an albuminous principle, and at a temperature of 140° Fahr. coagulates, and then ceases to have the power of decomposing sinigrin. When, therefore, the pungency of mustard is required, boiling water should never be employed in the preparation. White mustard seeds do not yield this volatile oil on distillation with water. The seeds contain, in addition to myrosin, a crystalline principle of sulphosinapisin—sinalbin. In the presence of water and myrosin this body splits up into sulphocyanate of acrinyl, sulphate of sinapine, and glucose. The vesicating properties of white mustard are due to the first mentioned of these bodies." (*Surgeon C. J. H. Warden, Prof. of Chemistry, Calcutta.*)

Special Opinions.—§ "The pure fresh oil is a stimulant and mild counter-irritant when applied externally. As such it is very useful in mild attacks of sore throat, internal congestion, and chronic muscular rheumatism. The oil is also used as an article of diet and is rubbed on the skin before bathing." (*Surgeon D. Basu, Faridpore.*) "Mustard oil with camphor may be rubbed in rheumatism with advantage. Mustard poultices should be removed when the skin is reddened, otherwise troublesome vesication is caused, intractable ulcers resulting." "The small black variety called *benarasy rai* is as good a rubefacient as English mustard." (*Bolly Chund Sen, Teacher of Medicine.*) "In common oil-mills in jails a maund of good seed yields 13 seers of oil. The oil rubbed on the feet and the bridge of the nose cuts short a head-cold in one night. I have never seen it fail. In slight bronchitic affections of children it makes a very useful mild counter-irritant application to the chest. It is also a very useful application in ordinary sore throat." (*Surgeon K. D. Ghose, Khoolna.*) "The oil rubbed over the chest in children has a great effect in relieving bronchial irritation. In influenza the oil rubbed on the feet after a foot-bath gives immediate relief. A little rubbed on the nose stops the running within a few hours." (*Surgeon K. D. Ghose, Bankura.*) "Mustard oil is very useful as a liniment to the chest in cases of bronchitis." (*Hony. Surgeon P. Kinsley, Ganjam, Madras.*)

"Mustard oil—

(1) Is used by natives to anoint the body before bathing. It prevents excessive perspiration and prickly heat, also protects the skin from the direct rays of the sun.

(2) Is used to anoint infants; after oiling they are exposed to the sun. This process is said to render the skin tolerant of the excessive heat.

(3) As a substitute for lard or *ghee*, it is extensively used in cooking.

(4) Internally, a few drops taken after meals promote digestion and act as a mild cholagogue and diuretic.

(5) The oil is very efficacious as a stimulating liniment in cough, catarrh, &c." (*Surgeon L. Dutt, Pubna.*)

848

MUSTARD.

The majority of the plants to which Europeans in India give the name of mustard should be transferred bodily to rape and its associates, to which they are certainly much more nearly allied. The true mustard is very scarce in India, and seems to have been introduced. **Ainslie** fixes its introduction within the present century, and the first time **Roxburgh** saw the plant was when raised from seed sent him from the Wynaad in South India. It is nowhere extensively cultivated, but is met with chiefly on the hills, and it is more than probable that it existed on the Himálaya from remote times, although unknown to the fathers of Indian botanical science. It is quite likely, however, that the ancient Sanskrit writers had not seen the true black and white mustard, and that the word *rájiká* may have originally denoted a form of **Brassica juncea**, and the word *siddhártha* a form of **B. campestris.** Nowadays these names are chiefly applied to the true black and white mustard, **B. nigra** and **B. alba,** respectively. **Brassica juncea** is the principal source of Indian mustard.

Mustard Flour.
849

The seeds of the black and white mustard are ground into what is known as mustard flour. The French mustard flour is much darker in colour than the English, because the seeds are not first husked. It is much more acrid and pungent, for the husk contains the principal store of pungency. Mustard flour is never prepared in India, or, at all events, never used as a condiment, except in making pickles from green mangoes and other sub-acid fruits. The seeds are ground and used as a poultice, and the expressed oil is also used medicinally. In Japan and China, mustard is regarded as a medicine of great importance. The ancient Hindús do not appear to have known the essential oil of mustard. This oil, as already stated, does not exist in the seeds, but is chemically produced by the action of water, as, for example, when a seed or a little of the flour is put into the mouth. Chemically, mustard seed consists of a bland fixed oil (obtained by pressure) and a peculiar inodorous substance called myroncic acid, together with a third substance which has been called myrosyne. By the action of water upon these substances the essential oil is produced, which is known chemically as pyrosyne.

White mustard is much inferior commercially, but is generally mixed with black mustard. It is said to be cultivated at Ferozpur, but is scarcely known in India. The white oil-cake is a valued food for sheep.

In the preparation of mustard flour, the relative quantities of black and white mustard used are commonly two parts of black to three of white, but the proportions vary. In Russia, **B. juncea** is ground into mustard flour, and so may most of the other Indian species; but they yield an inferior article to the true mustard flour of commerce, and, as already indicated, their true position is with the rape and colza of Europe. It is much to be regretted that the true mustard **B. nigra** and **B. alba,** the rape **B. Napus** (or in India **B. glauca**), the colza **B. campestris** *proper* (or in India **B. dichotoma**), and **B. juncea,** if not also **Eruca sativa,** have become hopelessly confused in our trade reports under the common name of rape and mustard. A considerable injury has thereby been done, and a check given to the development of foreign trade in these seeds. It will require time and careful observation to remove this fully, and to identify and distinguish the commercial products.

The quantity of pure mustard produced in India cannot at present be very great. From the confusion referred to above, it is impossible to arrive at any very definite information, since we cannot determine how far the term "Mustard" may be confined to the products of **Brassica alba** and **nigra.** The true mustard is cultivated chiefly on the hills, and

The Cabbage.

BRASSICA oleracea.

is used in medicine or for culinary purposes. In the official Catalogue of the Paris Exhibition of 1867, it is stated that 3,000 tons of flour, equal to 2,000,000 francs worth, were annually produced in France.

TRADE RETURNS.

850

The Annual Statement of the Trade and Navigation of British India with Foreign Countries gives the following figures as the exports from India for the past five years under the head of "Mustard":—

Exportation of Mustard.

YEARS.	Quantity in Cwt.	Value in Rupees.
1879-80	2,369	15,181
1880-81	17,448	1,03,240
1881-82	24,346	1,44,508
1882-83	23,145	1,37,750
1883-84	10,111	64,513

The following analysis of the exports of mustard for the year 1883-84 is interesting, as showing the relative quantities produced in the various provinces and the more important foreign countries to which exported:—

Presidency from which exported.	Quantity in Cwt.	Value in Rupees.	Country to which exported.	Quantity in Cwt.	Value in Rupees.
Bengal	471	2,860	Belgium	357	2,176
Bombay	9,260	59,061	France	7,903	49,980
Madras	380	2,592	Mauritius	399	2,552
			United States	675	4,725
			Ceylon	200	1,391
			Straits Settlements	237	1,477
			Other countries	340	2,212
TOTAL	10,111	64,513	TOTAL	10,111	64,513

Brassica oleracea, *Linn.; DC. Prod., I., 213.*

851

THE CABBAGE AND ITS ASSOCIATES.

Vern.—*Kópi*, BENG.; *Kóbi, karam-kallá*, HIND.; *Kóbiá*, GUJ.; *Karam-ki-bhájí, karam-ká-ság*, DUK.; *Gós-kíre*, TAM.; *Gós-kúra*, TEL.

Habitat.—A much-valued cold season vegetable, introduced by Europeans into India. A large, coarse form, extensively cultivated by the natives, has become perfectly acclimatised, and the early cabbages met with in the market are the young heads of this plant.

FOOD. 852

Medicine.—The seeds are diuretic, laxative, stomachic, and anthelmintic. The leaves form a good application in gout and rheumatism.

MEDICINE. 853

To this species the Cabbage and all its associates belong; these are supposed to have been produced by cultivating the European Wild Colewort or Wild Cabbage.

The following are the principal forms:—

B. (oleracea) sylvestris—The Wild Colewort.
B. (oleracea) acephala—The Green Kale or Borecole.
B. (oleracea) bulleata—The Savoy Cabbage.
B. (oleracea) gemmifera—The Brussels Sprout.
B. (oleracea) capitata—The Red and White Cabbage.
B. (oleracea) caulo-rapa—The Turnip-stemmed Cabbage or *Kol Rabi.*
B. (oleracea) botrytis—The Cauliflower and Brocoli.

Brassica quadrivalvis, *H. f. & T. T.,* see **B. trilocularis,** *H. f. & T. T.*

854

B. Tournefortii, *Gouan.; Fl. Br. Ind., I., 156.*

Habitat.—Is said to be cultivated between Ajmír and Delhi, but is unknown commercially.

The flowers are pale yellow, and the seeds large and compressed.

855

B. trilocularis, *H. f. & T. T.; Fl. Br. Ind., I., 156.*

Will probably prove a cultivated form of **B. campestris** as already indicated, being very nearly allied to the hairy form known as var. **dichotoma.** The seeds are large and white. An interesting series of specimens prepared by **Mr. Duthie,** Superintendent of the Botanic Gardens, Saharanpur, has been placed in the Calcutta Botanic Gardens herbarium. These seem to prove that the number of the valves in the fruit is of little or no importance, but depends more upon treatment than upon specific peculiarities.

856

BRAYERA, *Kunth.; Gen. Pl., I., 622.*

Brayera anthelmintica, *Kunth.;* ROSACEÆ. *Dc. Prod., II., 588.*

Vern.—*Cusso* or *Konsso* (?); *Kabsún, kafsún,* ARAB. and HIND.

Habitat.—Native of Abyssinia; imported into India, and sold by druggists.

MEDICINE. 857

Medicine.—The dried flowers and tops are anthelmintic.

§ "A bazar commercial article in Bombay; it comes direct from Africa." (*Surgeon-Major W. Dymock, Bombay.*)

Bread-fruit tree, see **Artocarpus incisa,** *Linn.;* URTICACEÆ.

858

BREYNIA, *Forst.; Gen. Pl., III., 276.*

Breynia rhamnoides, *Müll.-Arg.;* EUPHORBIACEÆ. *Dc. Prod., XV., pl. 2, 440.*

Syn.—PHYLLANTHUS SEPIARIA, *Roxb.;* MELANTHUS RHAMNOIDES, *Wight, Ic., t. 1898.*

Vern.—*Tíkkar,* OUDH.

TIMBER. 859

Habitat.—A large shrub or small tree; common in the Oudh forests, and in Banda, Bengal, and South India.

860

BRICK-CLAYS.

Brick-clays.

"TERRE À BRIQUES, *Fr.;* ZIEGELERDE, *Ger.;* ARGILLA DA FARMATTONI, *Ital.*

"In the neighbourhood of most of the large rivers in India, clays are to be found more or less suitable for brick-making; but little selection of good deposits has as yet been exercised, except in the larger cities, or in connection with railway works. As a rule, Indian-made bricks do not bear any reputation for strength or durability, though in many cases their inferiority is due to the system of manufacture rather than to the material available.

"The largest brick factory in India is the Government one at Akra near Calcutta. Within the last fifteen years the demand for first-rate bricks in Madras has increased enormously, and this has been met by very excellent productions from the numerous seams of clay in the alluvial deposits of that part of the Coromandel. On the west coast, and particularly at Cannanore, the clays have been largely utilised in the brick and tile productions of the Basel Mission. (*See 'Manual of the Geology of India,' Pt. III., p. 569.*)" (*Contributed by Superintendent, Geological Survey of India.*)

BRIDGES. 861

BRIDGES—Timber used in the construction of.

[*Note.*—Nearly every timber might be used for this purpose, but the followiug are those specially mentioned by authors.]

Afzelia bijuga.
Albizzia procera.
Alnus nitida (hooked sticks for rope bridges).
Bassia longifolia.
Bischoffia javanica.
Calamus montanus.
Calophyllum tomentosum.
Cedrela serrata.
Cedrus Deodara.
Eucalyptus Globulus.
Fagrœa fragrans.
Garcinia speciosa.
Hardwickia binata.
Mesua ferrea.
Mimusops littoralis.
Quercus annulata (used for the same purposes as **Q. lamellosa**).
Q. lamellosa.
Parrotia Jacquemontiana.
Pinus Gerardiana.
Salyx daphnoides.
Schima Wallichii.
Shorea robusta.
Tectona grandis.
Xylia dolabriformis.

862

BRIEDELIA, *Willd.; Gen. Pl., III., 267.*

A genus of trees, shrubs, or climbers, belonging to the Natural Order EUPHORBIACEÆ. There are 25 species, natives of tropical Asia, Africa, the Malaya, and Australia.

Leaves alternate, entire, short-petioled, generally distichous, with prominent parallel and lateral nerves. *Flowers* monœcious, subsessile, in axillary clusters; bracts small, scale-like, in male flowers numerous, subsessile, in the female few or solitary and often petiolate. *Calyx-tube* turbinate, segments 5-valvate in bud. *Petals* 5, small, scale-like, stalked or spathulate blade often dentate. In *male flowers, stamens* 5, inserted on a central column placed on a flat sinnate disc. *Female flowers*, ovary 2-celled, the base enclosed in the calyx-tube, and surrounded by an inner membranous, cup-shaped or tubular disc, variously lobed or lacinate, which is inserted at the mouth of the calyx-tube and is generally surrounded at its base by an outer fleshy annular disc; style 2-bifid, more or less connate at the base. *Fruit* a berry, enclosing 2 indehiscent cocci.

863

Briedelia montana, *Willd.; Brandis, For. Fl., 450; Gamble's Man. Timb., 357.*

Vern.—*Kargnalia, khaja, geia, kusi*, HIND.; *Geio*, NEPAL; *Kaisho*, ASS.; *Kurgnulia*, KUMAON; *Asáná*, MAR., CUTCH; *Asano*, BOMB., GUJ.; *Faturfoda*, GOA; *Vengé-maram, vengé*, TAM.; *Gundebingula, pantangi, ánem*, TEL.

Habitat.—A moderate-sized tree of the Sub-Himálaya, from Jhelum eastward, ascending to 4,000 feet; Oudh and Bengal.

Botanic Diagnosis.—"Branchlets and leaves wholly glabrous; bracteoles numerous, thinly membranous." (*Brandis.*)

DYE. 864

MEDICINE. 865

Dye.—Dr. **Dymock** thinks that the leaves might be used in tanning.

Medicine.—Reported to possess anthelmintic properties. Much used in Bombay and Goa as an astringent medicine. The bark "if soaked in water gives out much mucilage. The fibrous portion is very tough and strong." "Briedelia bark is well known as a valuable astringent in Western India." (*Surgeon-Major Dymock, Mat. Med., W. Ind., 589.*)

FODDER. 866

TIMBER. 867

Fodder.—The leaves are lopped for cattle fodder.

Structure of the Wood.—Grey, beautifully mottled; annual rings distinctly marked by darker and firmer wood on the outside of each ring. Weight 46 to 59 (?) lbs. per cubic foot.

It is very similar to that of **B. retusa,** and might be used for the same purposes.

868

Briedelia retusa, *Spreng.; Brandis, For. Fl., 449; Gamble's Man. Timb., 356.*

Syn.—B. CRENULATA, *Roxb.*, and B. SPINOSA, *Willd.; Roxb., Fl. Ind., Ed. C.B C., 705.*

Vern.—*Pathor, mark,* PB.; *Khája, kassi, gauli,* HIND.; *Kharaka, kaka,* KOL.; *Karika,* BHUMIJ; *Kanj, kaji,* KHARWAR; *Kúj,* MONGHYR; *Kadrúpala,* SANTAL; *Lamkana,* AJMERE; *Gauli,* GARHWAL; *Angnera,* BANSWARA; *Lamkana, angnera,* RAJPUTANA; *Geio,* NEPAL; *Pengji,* LEPCHA; *Nanda,* RAJBANSHI; *Katakuchi,* MECHI; *Kashi,* GARO; *Kamkúi,* CHITTAGONG; *Kasi, kosi,* URIYA; *Mullu-vengay, kamanji,* TAM.; *Koramánu, pedda-ánem, danki-bura, dudi máddi, koramadi, duriyamaddi,* TEL.; *Kassei,* GOND; *Karka,* KURKU; *Gúnjan, kati ain, asána,* MAR., BHIL; *Phatarphod, assana, asauna,* BOMB.; *Sun,* DUK.; *Asuna, gojé,* KAN.; *Adamarathu,* TINNEVELLY; *Tseichyee, seikgyi, tseikchyi, seikche,* BURM.; *Katta kaala, kat-takaala,* SINGH.

Habitat.—A large deciduous tree, with thorns on the bark of the young stems; found in the Sub-Himálayan tract, from the Chenab eastwards ascending to 3,600 feet; in Oudh, Bengal, Central and South India, and in Burma.

Botanic Diagnosis.—"Branchlets and under-side of the leaves tomentose; bracteoles few, coriaceous. Lateral nerves 15-20 pairs; calyx slightly enlarged in fruit." (*Brandis.*)

DYE. 869

MEDICINE. 870

Dye.—The bark is used in tanning.

Medicine.—The bark possesses medicinal properties similar to those of the preceding.

§ "Used as a liniment with gingelly oil in rheumatism." (*Surgeon-Major J. J. L. Ratton, M.D., Salem.*)

FOOD. 871

TIMBER. 872

Food.—The sweetish fruit is eaten, especially by wild pigeons. The leaves are cut to feed cattle and are said to free them from worms.

Structure of the Wood.—Sapwood small; heartwood grey to olive-brown, close-grained, durable, seasons well, and is moderately hard; the annual rings marked by concentric lines. It has a mottled grain and takes a beautiful polish.

It is used for cattle-yokes, agricultural implements, carts, and building. It stands well under water and is accordingly used for well-curbs.

873

B. stipularis, *Bl.; Gamble's Man. Timb., 356.*

Syn.—B. SCANDENS, *Roxb.; Fl. Ind., Ed. C.B.C., 706.*

Vern.—*Gourkassi,* URIYA; *Madlatáh, undergúpa,* OUDH; *Lilima,* NEPAL; *Dankibúra, siri-ánem* or *chiri-ánem,* TEL.; *Kihur, kohi,* ASS.; *Harinhara,* BENG.; *Sin-ma-no-pyin,* BURM.

Habitat.—A large, straggling or climbing shrub, met with in the Sub-Himálayan tract from the Jumna to Surba, ascending to altitude 2,000 feet; abundant in the Oudh forests, also in Bengal, Burma, South India the Malayan Peninsula, and Ceylon.

Botanic Diagnosis.—"Branchlets and under-side of the leaves tomentose; bracteoles few, coriaceous. Lateral nerves 8-12 pair. Calyx much enlarged in fruit." (*Brandis.*)

Structure of the Wood.—Greyish brown, moderately hard.

It is used for fuel in the Sunderbans.

TIMBER. 874

Briedelia tomentosa, *Bl.; Gamble's Man. Timb.,* 357.

875

Syn.—B. LANCEÆFOLIA, *Roxb., Fl. Ind., Ed. C.B.C.,* 706.

Vern.—*Sibri,* NEPAL; *Mantet,* LEPCHA; *Sirai, mindri,* BENG.

Habitat.—A small evergreen tree, met with in North-East Himálaya, ascending to 2,000 feet; in Eastern Bengal and in Burma.

Botanic Diagnosis.—Young branchlets pubescent or tomentose. Leaves small, glaucous, sparingly and minutely pubescent beneath.

Structure of the Wood.—Light olive-brown, hard, close-grained. Weight 64 lbs. per cubic foot.

TIMBER. 876

Brinjal, see **Solanum Melongena** *Linn.*

Brocoli, see **Brassica (oleracea) botrytis.**

BROMELIACEÆ.

877

A Natural Order of monocotyledons in which the ovary is generally inferior, occasionally only half-inferior and sometimes even altogether superior. They belong to the Cohort AMOMALES. Flowers distinct and hermaphrodite, regular, with a 2-seriate perianth. Calyx 3, green, the two posterior coherent; corolla 3, coloured, coherent and usually furnished with a nectariferous crest, spirally twisted in æstivation or rarely valvate, marcescent and again twisted with age. Stamens all perfect epigynous, perigynous, or hypogynous; filaments subulate and usually dilated at the base, free or connate and more or less adnate to the inner perianth-segments (corolla); anthers introrse 2-celled, basi- or dorsi-fixed, erect or incumbent. *Ovary* from position of the stamens must of course be completely inferior (in the Tribe BROMELIEÆ, *e.g.*, Ananas, &c.), half-inferior or superior (in Tribes PITCAIRNIEÆ and TILLANDSIEÆ). Ovules anatropous, numerous, 2-seriate at the inner angle of the cells, horizontal or ascending, rarely definite and pendulous from the top of the inner angle (Ananas). *Fruit* a 3-celled berry or a cepticidally 3-valved capsule, rarely loculicidal. Sometimes, as in the pine-apple, the individual fruits are coalesced into a succulent infruitescence crowned with a tuft of leaves. *Seeds* usually numerous, albumen farinaceous, with the embryo outside.

The BROMELIACEÆ are all tropical, American, and often epiphytic plants. In point of structure they are intermediate between the monocotyledon with a free, and those with an adherent, ovary.

As far as India is concerned, they are of little or no value, except the pine-apple, which is perhaps one of our most valuable fruits, and the fibre of which seems to have a good future before it (see **Ananas**).

BROMUS, *Linn.; Gen. Pl., III.,* 1200.

878

A genus of grasses containing about 40 species, chiefly natives of the temperate regions.

Sheath of leaf cut half way down. *Glumes* unequal, herbaceous, many-flowered, lower 1-veined, upper 3-5-veined. *Flowers* lanceolate-compressed, lower pale with a long awn (usually) founded on three veins, from below the tip. *Styles* below the top of the ovary. Nut furrowed, adhering to the pales.

Bromus arvensis, *Linn.; Duthie's Grasses,* 42; GRAMINEÆ.

Syn.—B. VERSICOLOR, *Poll.;* B. MULTIFLORUS, *Host.;* SERRAFALCS ARVENSIS, *Parl.*

FODDER. 879

Habitat.—The North-West Himálaya.

880

Bromus asper, *Linn.; Duthie's Grasses, 42.*

HAIRY-STALKED BROME GRASS.

Syn.—B. RAMOSUS, *Huds.;* B. MONTANUS, *Poll.;* B. HIRSUTUS, *Cart.;* FESTUCA ASPERA, *Mert. and Koch.*

Habitat.—A perennial grass found in North-West Himálaya.

Botanic Diagnosis.—Leaves broad, hairy. Panicle drooping, with long slightly divided branches; spikelets lanceolate; flowers remote, linear-lanceolate, lower pale hairy, 5-7-ribbed, lower flower twice the length of the upper glume and longer than its awn.

FODDER. 881

Fodder.—A good fodder grass for tracts sheltered by woods.

FODDER. 882

B. Schraderi, *Kunth.; Müeller, Extra-Trop. Pl., 53.*

PRAIRIE GRASS OF AUSTRALIA.

Syn.—CEROTOCHLOA PENDULA, *Schrad.*

Recently introduced for trial cultivation in the Botanical Gardens at Saharunpur and elsewhere. **Mr. Duthie** remarks: "**Mueller** describes this as one of the richest of all grasses, growing continuously and spreading readily from seeds, particularly on fertile and somewhat humid soil."

Broom, see **Cytisus scoparius,** *Linn.;* LEGUMINOSÆ.

883

BROUSSONETIA, *Vent.; Gen. Pl., III., 361.*

A genus of trees, with milky or opaline juice, containing 2 or 3 species, belonging to the Natural Order URTICACEÆ and the Tribe MOREÆ, natives of the Malaya, China, and Japan.

Leaves alternate (in **B. papyrifera** sometimes almost opposite), simple, petiolate, ovate-dentate, or when young 3-5-lobed, upper surface rough, under-surface soft tomentose, penniveined or at the base 3-costate; stipules lateral, membranaceous and deciduous. Inflorescence axillary, male in cylindrical catkins, female in compact, tomentose, round heads, with a greatly swollen receptacle; bracts interposed, truncate or clavate at the apex. *Flowers* diœcious (very much resembling those of **Morus**). *Male perianth* of 4 segments, free or connate at the base, membranous, valvate. Stamens 4, coiled up in bud, expanding with elasticity when mature. *Ovary* small, rudimentary. *Female perianth* ovoid, or tubular at the mouth, 3-4-dentate. Ovary included within the tubular perianth, stipitate, 1-locular; style subulate, entire (not bifid as in **Morus**). *Fruit* stipitate, girt at the base with the persistent perianth, drupaceous (in **Morus** the succulent sepals constitute the nutritious substance); mesocarp thin, except at the base and margins, where it is thick, succulent and edible, forming a forceps-like band which assists to eject the ripe seed. *Seed* solitary, finally separating from the endocarp; embryo incurved, subequal, oblong; radicle accumbent, ascending; albumen fleshy and thick, a layer protruded between the folds of the embryo.

The genus is named in honour of **P. N. V. Broussonet**, a naturalist who published in 1782 an account of the fishes of Barberry.

FIBRE. 884 | Paper. 885 | Burmese slates. 886 | Tapa-cloth. 887 | Karen Paper-cloth. 888

Broussonetia papyrifera, *Vent.; Kurz, For. Fl., Burm., II., 467;* URTICACEÆ.

THE PAPER-MULBERRY OR TAPA-CLOTH.

Vern.—*Malaing,* BURM.; *Aka kowzo, kename kowzo,* JAPAN.

Habitat.—A small tree, native of Japan, China, Polynesia, Siam, and said also to be wild in the Martaban hills.

Fibre.—The Japanese make paper from the bark of this tree, and the Burmese their curious papier-maché school slates (*Parabaik*). The Tapa-cloth of the South Sea Islands is made from it; also the Karens' mulberry paper-cloth.

FIBRE.

Perhaps no fibrous plant deserves to be more carefully investigated than this. Much time has apparently been wasted with experiments upon bamboo and rhea. It is probable that in both cases the experiments would have been much more profitable had they been directed to this fibre, with the view of discovering how far the paper-mulberry could economically be cultivated both as a paper-supply and as a new textile fibre. The Agricultural Department of India, during 1883, sent consignments of the seeds of this plant to British Burma. Plants were raised successfully and are reported (1884) to be growing vigorously. An attempt made during the rainy season to test the coppicing power of the plants at the Forest Garden of Tharawaddy was so successful that the portion coppiced could only be detected from the rest of the plantation by a close examination of the stools. **Dr. King**, in his annual report on the Botanical Gardens, Calcutta, for 1883-84, writes: "Some months ago I cut some branches of a paper mulberry tree (**Broussonetia papyrifera**) two years old, and had the bark removed. The latter was reported on by a paper-maker and pronounced, as I expected, an admirable paper material. Experience in this garden has already proved that this tree grows easily and rapidly in Lower Bengal, and I am assured by Mr. Maries, Superintendent of the Gardens of His Highness the Maharajah of Durbhunga, that it also grows well at Durbhunga—a much drier part of the province. If, therefore, villagers would take to growing this tree by the borders of their gardens and in the odd corners and scraps of ground in which Bengal abounds, there is a reasonable prospect that the province might produce in quantity one of the very best paper fibres known—a fibre at once strong and fine, and that has the great merit of requiring very little bleach. With the view of extending the cultivation of this tree, I am having thousands of young plants prepared for issue and for planting out in blank spots along the garden boundary."

"The Japanese are reported to propagate the plant very much as willows are grown in England. They use only the young shoots for the manufacture of paper. The stems are lopped into convenient pieces, and boiled until the bark separates from the wood. The dried bark is next moistened by soaking for a few hours in water. It is then scraped to remove superfluous matter, and thereafter boiled in wood-ashes until the fibres are thoroughly separated. After the boiling has been completed the fibre is beaten with wooden mallets until it is reduced to a paper pulp." (*Royle, Fibrous Pl.*, 342.) **Royle** points out that the process of paper-making described by **Kœmpfer**, as practised in Japan, so closely resembles the Nepal paper-making as to suggest that the practice was introduced to India through China.

The tapa or kapa paper-mulberry cloth already alluded to is in the South Sea Islands prepared in a somewhat similar way from the bark of this plant. The bark is soaked for a considerable time until it separates from the wood. It is then beaten out to the required degree of thinness. Mucilage from arrowroot is sometimes used both to join the pieces together and to give adhesiveness to the fabric. This is cut up into garments which are either worn plain or variously coloured and printed.

Prospects as a Paper Material.

889

Both as a future textile fibre and for the manufacture of paper, this is perhaps one of the most valuable fibres not at present being used by European commerce. In the Kew Report, 1879, p. 33, interesting information is given regarding this fibre. "A sample of the bark which came into the hands of **Mr. Routledge** is stated by him to be 'nearly, if not

Paper Material. 889

quite, the best fibre I have seen.' 'I must admit it is even superior to bamboo.' 'It requires very little chemicals, and gives an excellent yield—62·5 per cent. in the grey, *i.e.*, merely boiled, and 58 per cent. bleached." All this has been urged over and over again, but still the fibre does not take its true place. In a correspondence regarding bamboo as a supply for paper, **Dr. Brandis** expressed his opinion strongly in favour of paper-mulberry rather than bamboo. From a conviction that it would yet come to be appreciated, and in consideration of its slow growth, he recommended that a certain number of seedlings of the paper-mulberry should be planted out every year, in the forests where it could be grown, especially in Burma. The silk-worm can be fed upon the leaves of this plant, and the annual prunings of twigs to obtain a fresh flush of leaves for the silk-worm might be made to give a profitable return as a paper fibre.

890

Cultivation.—The following useful account of the propagation of the paper-mulberry as practised in Japan may be republished here: "It is propagated by layering, division of roots, cuttings, and by sowings, but the last method is slow and not usually practised.

"*Layering.*—In the latter part of March the ground is dug around the plant, light manure is applied, and the young twigs are then layered down in the ground, which has been previously dug. They are then covered with earth three inches thick, leaving only the tops of the twigs out of the ground. In the following spring, when small roots grow from the twigs, the layers are cut and planted in prepared ground, on small mounds about 18 inches apart, from which new shoots sprout in about ten days after planting. By September they often have reached a height of three feet.

"*Propagation by division of the roots.*—After the twigs have been cut for pulp, some of the mounds, on which grow suitable plants for propagation during the next year, are deeply covered with earth, and in the next spring season new shoots sprout from the hill. They are then taken up and the roots separated and planted in the nursery ground, and after three years they are transplanted to the fixed ground or field, and are fit for cutting after another year.

"*Method of planting and cultivation.*—The young plants may be planted on high ground, on mounds, or in fields, in March or April. In June they must be carefully weeded, the dry weeds being piled around each mound on which the plants are growing, for they make a good manure.

"The harvesting of the plants may take place any time during the season when the plants are deciduous (from September to February). The manner in which they are prepared for pulping is the same as for mulberry plants, as usually practised in Japan, and the average amount of the annual harvest of pulp has not yet been ascertained, as it differs considerably in different provinces.

"There are eleven or twelve varieties of the plant, besides which there are five varieties of the wild species, which are used for making paper of the inferior quality. The **Broussonetia Kazinoki** and **B. Kæmpferi** also belong to the same genus." (*Indian Forester, Vol. VIII., p. 48.*)

§ "Paper-Mulberry (**Broussonetia papyrifera**).—A packet of the seed of this useful plant was received from **Dr. King**, Director of the Calcutta Botanical Gardens, in October last. The seed was at once sown in pots and it germinated freely. The seedlings were kept in the seed-pots until this spring, when they were transplanted into nurseries. There are at present 1,240 plants in a healthy and thriving condition. This climate seems to be very suitable for them. Not a single seedling that germinated has died, either from the effects of cold, damp, or in the process of transplanting. The same can very seldom be said about many of our common timber trees. Being such an easily raised tree and so useful a one, I think I should not be wrong in recommending it to the attention of any person who may be in a position to extend its cultivation in Upper India. (*Annual Report, Botanic Gardens, Saharanpur 1881-82.*)

"The seedlings alluded to in my last report have done remarkably well. Should it ever be under consideration to cultivate this plant on a large scale, I can safely say that there need be no anxiety as to its not thriving to perfection in this climate. It is easily cultivated both by seed and from cuttings. Several of the plants are already in fruit. (*Annual Report on Saharanpur Garden for 1882-83.*" *Mr. J. F. Duthie.*)

Structure of the Wood.—Light-coloured, even-grained, not hard nor heavy.

TIMBER. 891

BROWNLOWIA, *Roxb.; Gen. Pl., I., 231.*

892

A genus of lofty trees belonging to the Natural Order TILIACEÆ; it comprises three species confined to tropical Asia.

Whole plant stellately pubescent or scaly. *Leaves* entire, 3-5-nerved. *Flowers* numerous, small, in large terminal or axillary panicles. Calyx bell-shaped, irregularly 3-5-fid. *Petals* 5, without gland. *Stamens* many, free, springing from a raised torus. *Staminodes* 5, within the stamens, opposite the petals and petaloid. *Anthers* sub-globose. Ovary 5-celled, each cell 2-ovulate; styles awl-shaped, slightly coherent; ovules ascending. *Carpels* ultimately nearly free, 2-valved, 1-seeded. *Albumen* none; cotyledons thick, fleshy.

Brownlowia elata, *Roxb.; Fl. Br. Ind., I., 381; Bot. Reg., t. 1472.*

893

Syn.—HUMEA ELATA, *Roxb.; Fl. Ind., Ed. C.B.C., 448.*

Vern.—*Masjot,* CHITTAGONG.

Habitat.—A lofty tree of the tidal forests of Chittagong and Tenasserim.

TIMBER. 894

B. lanceolata, *Benth.; Fl. Br Ind., I., 381.*

Habitat.—A tree of the tidal forests of the Sunderbuns, Arracan, and Tenasserim.

TIMBER. 895

B. peltata, *Benth.; Kurz, For. Fl. Burm., I., 153.*

Habitat.—A small tree of Tenasserim.

TIMBER. 896

Brucea Nima (?) quassioides, *Ham.*, see **Picrasma quassioides,** *Benth.*; SIMARUBEÆ.

BRUGUIERA, *Lam.; Gen. Pl., I., 679.*

897

A genus of trees belonging to the RHIZOPHOREÆ, comprising some eight species, natives of the tropics of the Old World.

Leaves opposite, coriaceous, oblong, entire, stipulate; peduncles axillary, recurved, one to many-flowered. *Calyx* 8-14-merous, adnate to the base of the ovary. *Petals* 8-14, oblong 2-fid, appendiculate, embracing the stamens which spring elastically from them when mature. *Stamens* 16-28, filaments filiform, anthers linear, mucronate, about as long as the filaments. *Ovary* 2-4-celled, included in the calyx-tube; cells 2-ovuled; style filiform; stigma 2-4-lobed, minute. *Fruit* turbinate, coriaceous, crowned with the calyx-limb, 1-celled and 1-seeded.

Bruguiera gymnorhiza, *Lam.; Fl. Br. Ind., II., 437;* RHIZOPHOREÆ.

898

One of the forms of the MANGROVE.

Syn.—B. RHEEDII, *Bl.* (*Beddome, c.*); RHIZOPHORA GYMNORHIZA; *Roxb., Fl. Ind., Ed. C.B.C., 390.*

Vern.—*Kakra, kankra,* BENG.; *Byubo,* BURM.

Habitat.—A small evergreen tree of the shores and tidal creeks of India, Burma, and the Andaman Islands.

TAN & DYE. Mangrove Bark. 899

Tan.—The bark is valuable, and with **Rhizophora mucronata,** *Lam.*, constitutes the tan known commercially as Mangrove bark (which see). It is a useful astringent, used also in dyeing black.

TIMBER. 900

Structure of the Wood.—Heartwood small, red, extremely hard. Weight 54 lbs. per cubic foot.

Used for firewood, house-posts, planks, and articles of native furniture.

901

BRUNELLA, *Linn.; Gen. Pl., II., 1203.*

A genus of perennial procumbent herbs belonging to the LABIATÆ; there are 2 to 3 species, natives of the temperate regions.

Leaves entire or inciso-dentate or even pinnatifid. Flowers 6 in a whorl, forming a dense terminal spike with two broad kidney-shaped bracts under each whorl. Calyx reddish purple, ultimately closed and compressed, upper lip flat, truncate, 3-toothed, lower bifid. Corolla ringent, upper lip concave, entire. Two inferior stamens, the longest anthers all 2-celled; filaments bifid, one branch barren.

902

Brunella (Prunella) vulgaris, *Linn.*

SELF-HEAL.

Vern.—*Aústakhadús*, PB.; *Ustúkhúdús*, SIND.

Habitat.—A small-branched, erect or creeping herb of the Himálaya, from 3,000 to 10,000 feet.

MEDICINE. 903

Medicine.—Regarded an expectorant and antispasmodic.

§ "The green leaves, smeared with castor oil and warmed over the fire, are applied externally to the anus in cases of painful piles." (*Surgeon-Major Thompson, M.D., Madras.*)

Brussels sprout, see **Brassica (oleracea) gemmifera.**

904

BRYONIA, *Linn.; Gen. Pl., I., 829.*

A genus of climbing herbs belonging to the Natural Order CUCURBITACEÆ. There are in all 12 species, inhabitants of the warm and temperate zones of the Old World.

Climbing herbs, scabrid and glabrous; tendrils 2-fid, in the Indian species. *Leaves* petioled, palmately 5-lobed or 3-5-angular. *Flowers* small, yellowish, males and females clustered in the same axils (in the Indian species shortly pedicelled). MALE: calyx-tube widely campanulate, 5-toothed; corolla 5-partite; stamens 3, inserted low down the calyx-tube; anthers free, two 2-celled, one 1-celled, cells curved or somewhat sigmoid, never quite conduplicate, connective, not produced; rudiment of ovary 0. FEMALE: calyx and corolla as in the male; ovary ovoid; style slender, 3-fid at the top, no disc at the base in the Indian species; ovules many, horizontal, placentas 3. *Berry* sphericai, indehiscent. *Seeds* not very many, oblong or ovoid, compressed.

The generic name *Bryonia* from βρυωνία and βρυώνη=to be full of, or to swell.

Bryonia callosa, *Rottl.*, syn. for **Cucumis trigonus,** *Roxb.*, which see.

B. epigæa, *Rottl.*, see **Corallocarpus epigæa,** *Hook. f.*

B. laciniosa, *Linn.; Fl. Br. Ind., II., 623; Wight, Ic., t. 500.*

THE BRYONY.

Vern.—*Gargú-narú*, HIND.; *Mala*, BENG.; *Kawale-che-dole*, BOMB.; *Nehoemaka*, MAL.; *Linga-donda*, TEL.

Habitat.—Throughout India, from the Himálaya to Ceylon.

MEDICINE. 905

Medicine.—"The whole plant is collected when in fruit for medicinal use. It is bitter and aperient, and is considered to have tonic properties." (*Dymock.*) Used as a medicine by the Santals.

FOOD. 906

Food:—The leaves are boiled and eaten as greens.

Bryonia pilosa, *Roxb.; Fl. Ind., III., 726, and*

B. rostrata, *Rottl.*, syn. for **Rhynchocarpa fœtida,** *Schrad.* 907

Vern.—*Kunkuma-donda, nága-donda,* TEL.

B. scabrella, *Linn. f.*, see **Mukia scabrella,** *Arn.*

B. umbellata, *Wall.*, see **Trichosanthes cucumerina,** *Linn.*

BRYOPHYLLUM, *Salisb.; Gen. Pl., I., 658.* 908

A genus of herbaceous perennials belonging to the CRASSULACEÆ: there are 4 species in tropical Africa, one of which extends throughout the tropical regions of the whole world.

Tall erect herbs. *Leaves* opposite, crenate. *Flowers* large, pendent, in spreading panicles with opposite branches. *Calyx* with a long inflated tube; lobes 4, short, valvate. *Corolla* with a campanulate tube and shortly 4-fid limb. *Stamens* 8, in two series, inserted on the middle of the corolla-tube. Hypogynous *scales* 4, obtuse. *Carpels* 4, free or connate at the base, attenuated into long styles; ovules numerous. *Follicles* 4, many-seeded.

The generic word is derived from *βρύω*, to be full of, or to burst forth; and *φύλλον*, a leaf, in allusion to the succulent nature of the leaves, and probably also because the leaves have the power of rooting by buds produced in the serrations on the margin of the leaves.

Bryophyllum calycinum, *Salisb.; Fl. Br. Ind., II., 413.* 909

Syn.—KALANCHOE PINNATA, *Pers.*; COTYLEDON RHIZOPHYLLA, *Roxb.*

Vern.—*Kóp-pátá,* BENG.; *Ahirávana-mahirávana, ghayamári,* BOMB., CUTCH. The Mohammedans call it *Zakhm-haiyát,* PERS., HIND.; *Zakhm-haiyát-ka-pattá,* DUK.; *Malai-kalli, rúna-kalli,* TAM.; *Síma-jamudu,* TEL.; *Ela-marunna, elamarunga, murikúti,* MALA.; *Lonná-hadakana-gidá,* KAN.; *Yoe-kiya-pin-ba,* BURM.

Habitat.—A succulent plant, with thick, fleshy leaves, from the crenulations of which, in contact with the ground, bulbules are produced which develope into new plants. Common throughout Bengal and the hotter moist parts of India to Ceylon and Malacca.

History.—According to **Roxburgh,** this plant was introduced into the Calcutta Botanic Gardens from the Moluccas. **Voigt** adds that it was brought by Lady Clive in 1799. As already stated, it is now found nearly over the greater part of the hot moist parts of India. In Lower Bengal it is one of the most abundant of gregarious herbs, and it has even spread through Assam and Sylhet to the valley of Mánipur. It is met with in fact throughout India, although less abundantly than in Bengal. **Dr. Dymock** appears to regard it as a native of the Deccan and of the Konkan, but the *Bombay Flora* (*Dalz. and Gibs., 1861*) gives it as merely common in the Warree country and near Belgaum. 910

Medicine.—"The leaves, slightly toasted, are used by the natives as an application to wounds, bruises, boils, and bites of venomous insects." "I have seen decidedly beneficial effects follow their application to contused wounds; swelling and discoloration were prevented, and union of the cut parts took place much more rapidly than it does with ordinary treatment by water-dressing." (*Dymock, p. 297.*)

MEDICINE. Leaves. 911

§ "Used in the form of poultice and powder for sloughing ulcers, it is a disinfectant." (*Surgeon W. Barren, Bhuj, Cutch.*)

BUCHANANIA, *Roxb.; Gen. Pl., I., 421.* 912

A genus of trees belonging to the Natural Order ANACARDIACEÆ; there are some 20 species, natives of tropical Asia, Australia, and the Pacific Islands.

Leaves alternate, petioled, simple, quite entire. *Panicles* terminal and axillary, crowded. *Flowers* small, white, hermaphrodite. *Calyx* short, 3-5-toothed or -lobed, persistent, imbricate. *Petals* 4-5, oblong, recurved, imbricate. *Disk* orbicular, 5-lobed. *Stamens* 8-10, free, inserted at the base of the disk. *Carpels* 5-6, free, seated in the cavity of the disk, one fertile, the rest imperfect; style short, stigma truncate; ovule 1, pendulous from a basal funicle. *Drupe* small, flesh scanty; stone crustaceous or bony, 2-valved. *Seed* gibbous, acute at one end; cotyledons thick; radicle superior.

The genus is named in honour of the late distinguished Indian botanist, **Dr. Buchanan Hamilton.**

913

Buchanania latifolia, *Roxb.; Fl. Br. Ind., II., 23.*

Vern.—*Piyár, piyál, piyála, chirónji* (the kernel), HIND.; *Chirónji* (the fruit), *piyál*, BENG.; *Chirauli* (the fruit), *chironji*, PB.; *Piál, payála, muriá, katbhilawa*, GARHWAL; *Piár, peira, pérrah*, OUDH; *Tarum*, KOL.; *Pial*, BHUMIJ; *Peea*, KHARWAR; *Taróp*, SANTAL; *Charu*, URIYA; *Achár, chár, chirónji* (the fruit), C. P.; *Sáráka, herka*, GOND; *Taro*, KURKU; *Sír*, BHIL; *Chár-ki-chároli* (the kernel), DUK.; *Piyál, chároli*, BOMB.; *Charwari*, HYDERABAD; *Mowda* or *katimango, marum, kat maá, aima, kátma-maram* (the plant), *kátma-payam* or *katma param* (the fruit), *kátma-parpu* (the kernel), TAM.; *Chara, charu mamudi, chinna mora, morli, chára-chettu* or *sára-chettu, chára-mámidi, járu-mámidi* (the plant), *chára-pandu* (the fruit), *chára-puppu* (the kernel), TEL.; *Nuskul, murkalu*, KAN.; *Kála maram*, MALA.; *Chároli*, GUJ., CUTCH; *Pyál-chár*, MAR.; *Piyála, chára, chirika*, SANS.; *Lonepho, lunbo, lamboben, lambo* or *lon-po, loneopomáa*, BURM.

References.—*Roxb., Fl. Ind., Ed. C.B.C., 365; Voigt, 272; Brandis, For. Fl., 127; Gamble, Man. Timb., 109; Kurz, For. Fl. Burm., I., 307; Beddome, t. 165; Dalz. and Gibs., Bomb. Fl., 52; Stewart's Pb. Pl., 45; Lisboa's Useful Pl., Bomb., 53; Drury's Us. Pl. Ind., 88.*

Habitat.—A tree, leafless only for a very short time. Found in the Sub-Himálayan tract from the Sutlej eastward, ascending to 2,000 feet; throughout India and Burma, common in the hotter and drier parts of the empire, and frequently associated with the *sál*, the *mahúa*, and the *dák*.

Properties and Uses—

GUM. 914

Gum.—A pellucid gum exudes from wounds on the stem (*Brandis*), more than half soluble in water, and is reported to resemble *Bassora Gum*. (See **Bassora** and also **Cochlospermum.**) It occurs in irregular broken fragments, brittle, pale, horn-coloured, tinged with brown, tasteless, soluble in water, except a small insoluble portion of basorine. It has been pronounced as having adhesive properties, similar to the inferior kinds of gum arabic, and as suitable for dressing textiles. The bark and the fruits furnish a natural varnish.

VARNISH. 915

TAN. Bark. 916

Tan.—The bark is used in tanning.

OIL. 917

Oil.—The kernels of the fruit yield an oil called *Chironji*, but owing to their being so much prized as a sweetmeat when cooked, this oil is rarely prepared. It is pale straw-coloured, limpid, sweet and wholesome. The kernels when broken readily yield this oil, 50 per cent. being obtained. (*Agri.-Hort. Soc. Jour. Ind., XII., 346.*)

MEDICINE. Gum. 918 Oil. 919

Medicine.—The gum is said to be administered in diarrhœa. The oil is used as a substitute for almond oil in native medicinal preparations and confectionery. It is also applied to glandular swellings of the neck. According to **Dr. Irvine** (*Medical Topography, Ajmir, 131*), the seed is very palatable and nutritious, especially when roasted; is used also in medicine and is considered heating. The fresh fruit is very agreeable.

Special Opinions.—§ "The fruits are said to be sweet and laxative. They are used to relieve thirst, burning of the body, and fever." (*Dr. U. C. Dutt, Serampore.*) "Used to improve the flavour of drugs in general." (*Surgeon W. Barren, Bhuj, Cutch.*)

Food.—"The kernel is a common substitute for almonds amongst the natives. It is largely used in sweetmeats. Its flavour is described as between that of the pistachio and the almond. It is eaten roasted with milk." (*Lisboa, Useful Plants of the Bombay Presidency, p. 150.*)

FOOD. Kernel. 920

The FRUIT is eaten by the hill tribes of Central India. Having first pounded them, along with the contained kernels, they dry them in the sun. As required, this is baked into a sort of bread and eaten.

Fruit. 921 Prepared bread. 922

"The forest tribes gather the seed and take out the kernel, which they exchange for grain, salt, and cloth. The kernel is an important article of trade, being largely used in native sweetmeats. Oil is also extracted from it." (*Bomb. Gaz., VII., 37.*)

§ "The fruit is sold in Bombay under the name of *chára-bhúr.*" (*Surgeon-Major W. Dymock, Bombay.*)

Fodder.—The LEAVES are said to be given as fodder (*Bomb. Gaz., X., 403.*)

FODDER. Leaves. 923

Structure of the Wood.—Greyish brown, moderately hard, with a small dark-coloured heartwood. It seasons well and is fairly durable if kept dry. Weight 30 to 36 lbs. per cubic foot.

TIMBER. 924

The wood "seasons well, is easily worked, and if kept dry is fairly durable." (*Bomb. Gaz., VII., 37.*) "The heartwood is hard, but the rest of the wood is poor. A seasoned cubic foot weighs 36 pounds." (*Bomb. Gaz., XV., 64.*)

It is used for boxes, bedsteads, bullock-yokes, doors, window-frames, tables, &c.

BUCKLANDIA, *Br.; Gen. Pl., I., 668.*

925

A tree attaining a height of 80 feet. *Leaves* alternate, cordate-ovate, accuminate, entire, long-petioled; stipules solitary or in pairs, large, oblong, coriaceous, deciduous. *Inflorescence* of 2-5-peduncled heads, at first enclosed between a pair of stipules; flowers adnate by their calyces, about 8 in a head, polygamous. *Calyx-tube* adnate to the ovary; limb 5-lobed. *Petals* in the ⚥ flower linear spathulate, fleshy, variable in number; in the ♀ rudimentary. *Stamens* 10-14 (in the ♀ none); filaments long. *Ovary* half inferior, 2-celled; styles 2, separate, soon divaricate; ovules 6 in each cell in two rows. *Capsule* nearly superior, woody, sub-globose, endocarp horny, showing a tendency to separate from the exocarp. *Seeds* 6 in each cell, oblong, trigonous; the upper wingless, solid, without any embryo, the lower one in each cell winged and fertile.

Bucklandia populnea, *R. Br.; Fl. Br. Ind., II., 429;* HAMAMELIDEÆ

926

Vern.—*Pipli*, NEPAL; *Singliang*, LEPCHA; *Dingdah*, KHASIA.

Habitat.—A large evergreen tree met with in the Eastern Himálaya, Khásia Hills, and hills of Martaban, from 3,000 to 8,000 feet.

Structure of the Wood.—Greyish brown, rough, moderately hard, close-grained, durable. Is very much used in Darjiling for planking and for door and window-frames.

TIMBER. 927

Buckthorn, see **Rhamnus catharticus,** *Linn.*, RHAMNEÆ.

Buck-wheat or **Brauk,** see **Fagopyrum esculentum,** *Mœnch.*

BUDDLEIA, *Linn.; Gen. Pl., II., 793.*

928

A genus of LOGANIACEÆ, comprising some 70 species, natives of the tropical and sub-tropical regions of Asia, America, and Africa.

Trees, shrubs, or herbs. *Leaves* opposite, entire or crenate, united by a stipulary line. *Cymes* dense, globose or corymbiform, axillary or in a thyrsoid terminal panicle. *Calyx* campanulate, 4-merous. *Corolla* urn-shaped; lobes 4, imbricate in bud. *Stamens* 4, on the corolla tube; anthers subsessile, ovate or oblong. Ovary 2-celled; style linear, clavate; ovules very many in each cell. *Capsule* septicidally 2-valved. *Seed* numerous, oblong or fusiform, testa usually loose or expanded into a wing or tail; albumen fleshy; embryo straight.

A genus named after **Adam Buddle**, a British botanist of some note.

929

Buddleia asiatica, *Lour.; Fl. Br. Ind., IV., 82.*

Syn.—B. NEEMDA, *Buch.; Roxb., Fl. Ind., Ed. C.B.C., 133; Wight, Ic., t. 894, 133.*

Vern.—*Bhati, dhaula, shiúntra,* KUMAON; *Bana,* SIMLA; *Newarpati,* NEPAL; *Pondám,* LEPCHA; *Nimda, budbhola,* CHITTAGONG; *Kyoung-miku,* BURM.

Habitat.—A large evergreen shrub of the Sub-Himálayan tract from the Indus eastward, ascending to 4,000 feet; Bengal, Burma, South India; chiefly found in second-growth forests, deserted village sites, and savannahs.

TIMBER. 930

Structure of the Wood.—Grey, moderately hard. Weight 44 lbs. per cubic foot.

931

B. Colvillei, *Hook. f.; Fl. Br. Ind., IV., 81.*

Vern.—*Puri singbatti,* NEPAL; *Pya-shing,* BHUTIA.

Habitat.—A small tree of the Eastern Himálaya, from 9,000 to 12,000 feet.

TIMBER. 932

Structure of the Wood.—Reddish brown, soft.

933

B. paniculata, *Wall.; Fl. Br. Ind., IV., 81.*

Syn.—B. CRISPA, *Benth.*

Vern.—*Spera wuna,* AFG.; *Dholtu, ghúttia, sodhera, sudhari,* N.-W. HIMALAYA; *Sinna,* NEPAL.

Habitat.—A large evergreen shrub of the Himálaya, from the Indus to Bhután, ascending to 7,000 feet.

TIMBER. 934

Structure of the Wood.—White, moderately hard, close-grained. Weight 41 lbs. per cubic foot.

Buffalo grass or **Gama grass,** see **Tripsacum dactyloides** (?).

Bullock's heart, see **Anona reticulata,** *Linn.*

935

BUPLEURUM, *Linn.; Gen. Pl., I., 886.*

A genus of UMBELLIFERÆ, comprising some 60 species, natives of Europe and temperate Asia. It is at once recognised from all its associates by its glabrous and glaucous, *entire,* thick leaves.

Glabrous *herbs* or shrubs. *Leaves* entire. *Umbels* compound; bracts and bracteoles foliaceous, setaceous, or 0. *Flowers* yellow or lurid, pedicelled or subsessile. *Calyx-teeth* 0. *Petals* obovate, emarginate. *Styles* short. *Fruit* laterally compressed, slightly constricted at the commissure; carpels terete or subpentagonal; primary ridges distinct, sometimes subulate, rarely obscure; secondary 0 or obscure; vittæ 1-3 between the primary ridges, rarely 0 or many; carpophore entire 2-fid or 2-partite. *Disc* depressed, rarely prominent in fruit. *Seed* terete, sometimes slightly grooved on the inner face.

The generic name is the Greek βούπλευρον—βοῦς (*bous*), the ox, and πλευρόν (*pleuron*), a rib, a name which probably indicates a resemblance to the curved lanceolate and entire leaves. Latin *Bupleuron,* Russ. *Buplewr,* It. *Bupleuro,* and Fr. *Bupleire.* They are generally known in English as the Hare's-ear.

Bupleurum falcatum, *Linn., var.* **marginata,** *Wall.; Fl. Br. Ind., II., 676.* 936

Vern.—*Kali sewar, sipil,* PB.

Habitat.—Met with in the mountainous tracts of Northern India, from 3,000 to 12,000 feet, extending from Kashmír to the Khásia Hills.

Medicine.—This, as also other members of the genus, is reputed to have stimulant properties. MEDICINE. 937

Food.—The root is said to be eaten. FOOD. 938

BURSERACEÆ. 939

"Balsamiferous trees or shrubs. *Leaves* alternate (very rarely opposite), imparipinnate or trifoliolate (very rarely unifoliolate), stipulate or ex-stipulate. *Inflorescence* racemose or paniculate. *Flowers* regular, small, hermaphrodite or often polygamous. *Calyx* free, 3-6-lobed, imbricate or valvate, often minute. *Petals* 3-6, distinct, rarely connate, imbricate or valvate. *Disk* annular or cupular, usually conspicuous, free or adnate to the base of the calyx. *Stamens* as many or twice as many as petals, inserted at the base or margin of the disk, equal or unequal; filaments free, rarely connate at the base, smooth; anthers dorsifixed, rarely adnate, 2-locular, dehiscing longitudinally. *Ovary* free, rarely 1-, more often 2-5-celled; style simple, stigma undivided or 2-5-lobed; ovules 2 or rarely 1 in each cell, anatropous, axile, usually pendulous, rarely ascending; micropyle superior, raphe ventral. *Fruit* drupaceous, indehiscent, containing 2-5 pyrenes, rarely pseudo-capsular and dehiscent. *Seeds* solitary, pendulous; testa membranous, albumen 0; cotyledons usually membranous, contortuplicate, rarely fleshy and planoconvex; radicle superior."

"*Drupe valvately dehiscent, pyrenes separating.*

Drupe trigonous . . . . . . 1. **Boswellia.**
Drupe broadly 3-winged . . . . . 2. **Triomma.**

Drupe indehiscent, pyrenes not separating.

Stamens 6-10—

Calyx 5-fid, urceolate. Disk clothing tube of calyx . . . . 3. **Garuga.**
Calyx 4-toothed, urceolate. Disk cupular. Flowers few, fasciculate . . . . . 4. **Balsamodendron.**

Stamens 8-10—

Calyx 4-6-toothed, small. Disk clothing base of calyx. Inflorescence paniculate . . . 5. **Protium.**
Calyx 4-6-lobed, imbricate. Disk annular . 6. **Bursera.**
Calyx usually 3-fid, valvate. Drupe ellipsoid, usually trigonous; style terminal . . . 7. **Canarium.**
Calyx 3-fid, valvate. Drupe usually gibbous; style lateral . . . . . . 8. **Santiria.**
Calyx 3-partite, large, valvate. Drupe globose 9. **Trigonochlamys.**

Stamens 5 . . . . 10. **Filicium.**"

(*Fl. Br. Ind., I.,* 527.)

There are about 160 species of Balsamiferous trees and shrubs belonging to this Natural Order in the world, all inhabitants of the tropical regions. In India there are 39 species referred to 10 genera. Of these, 34 or 87·2 per cent. are confined to the plains, 5 or 12·8 per cent. ascend to 5,000 feet in altitude, and none to higher altitudes. Their distribution over India is also striking: 28 or 71·8 per cent. occur in the eastern division of India, the majority being natives of Malacca. In South India 2 species occur, 2 in North India which also extend to Bombay, and another species confined to Sind; 5 species are general over the greater part of India.

940

BURSERA, *Linn.; Gen. Pl., I., 324.*

A genus of BURSERACEÆ, containing some forty species, mostly natives of tropical America.

Balsamiferous trees. *Leaves* alternate, imparipinnate, or rarely 1-foliolate. *Panicles* short, branched. *Flowers* hermaphrodite or polygamous. *Calyx* small, 4-6 partite or toothed, imbricate. *Petals* 4-6, short, patent at length reflexed, usually valvate. *Disk* annular, crenate. *Stamens* 8-12, nearly equal, inserted at the base of the disk. *Ovary* free, ovoid or sub-globose, 3-5-celled; style very short, stigma 3-5-lobed; ovules 2 in each cell. *Drupe* globose or ovoid, with 3-5 pyrenes.

This genus is named after **Joachim Burser**, a friend of **Caspar Bauhin**, Professor of Botany at Sara, Naples.

941

Bursera serrata, *Wall.; Fl. Br. Ind., I., 530.*

Syn.—LIMONIA PENTAGYNA, *Roxb.; Fl. Ind., Ed. C.B.C., 364.*

Vern.—*Murtenga*, ASS.; *Chitrika*, TEL. (on the Sircars); *Thadi-ben*, BURM.

Habitat.—A large, evergreen tree of the eastern moist zone of Bengal, Assam, Chittagong, and Burma.

TIMBER. 942

Structure of the Wood.—Hard sapwood light brown, heartwood red, close-grained, Weight 46 lbs. per cubic foot.

Good for furniture.

943

BUTEA, *Roxb.; Gen. Pl., I., 533.*

An elegant genus of LEGUMINOSÆ, with large orange-red flowers, containing three species, all Indian.

Erect trees or climbing shrubs, with 3-foliolate, stipellate leaves. *Flowers* densely fascicled, large, showy in axillary racemesi or terminal racemes or panicles. *Calyx* broadly campanulate; teeth short, deltoid. *Corolla* much exserted; petals nearly equal in length, the keel much curved, sub-obtuse or acute. *Stamens* diadelphous; anthers uniform. *Ovary* sessile or stalked, 2-ovuled; style filiform, curved, beardless, stigma capitate. *Pod* firm, ligulate, splitting round the single apical seed, the lowest part indehiscent.

A genus named after **John, Earl of Bute.**

944

Butea frondosa, *Roxb.; Fl. Br. Ind., II., 194.*

BUTEA GUM; BENGAL KINO; sometimes called the BASTARD TEAK.

Vern.—*Dhák, palás, tésú-ká-pér, kakria, kankrei, chichra*, HIND.; *Palas*, BENG.; *Chalcha*, BANDELKHAND; *Murut*, KOL.; *Murup*, SANTAL; *Pharsa*, BAIGAS; *Paras, faras*, BEHAR; *Palási, bulyettra*, NEPAL; *Lahokúng*, LEPCHA; *Palashu*, MECHI; *Porásu*, URIYA; *Chiúla, purohapalás, ckintá*, C. P.; *Murr*, GOND, KURKU; *Palása, khákára, khakha-o*, BOMB.; *Palás-ká-jhár, tésú-ká-jhar*, DUK.; *Khákará, khakhado, khákharnu-jháda*, GUJ.; *Khakar, pálás*, CUTCH; *Paras, palas, phalasá-cha-jhádá kakrácha-jháda*, MAR.; *Porasan, parasa, murukkan. puraishu, purashu, palásham*, TAM.; *Móduga, mohtu, tella módugu, móduga chettú, palá-shamu, pálásamu, palásamu, kimsukamu, motuku, pálás, modaga mardulu*, TEL.; *Muttuga, thorás, muttaga-mará, muttuga-gidá*, KAN.; *Plách-cha, murukka-maram*, MAL.; *Kinsuka, palása*, SANS.; *Darakhtepalah, palah*, PERS.; *Gasskeala* or *gaskoela, calukeale, káliya*, SINGH.; *Pouk, páv, pin*, BURM.

Habitat.—A moderate-sized, deciduous tree, found throughout India and Burma, extending in the North-West Himálaya as far as the Jhelum.

This is one of the most beautiful trees of the plains and lower hills of India. Although many of its properties are much appreciated by the natives, it must be admitted that the tree seems comparatively neglected.

"A waving, well-wooded country, set thick with bright scarlet flowering apple trees, gives some idea of many a Panch Mahál landscape when the *khákhra* is in bloom. In habit of growth it is not unlike the apple tree, and the leaves dropping when the flowers come, the top and outer branches stand out like sprays of unbroken scarlet. In the bud, the dark olive-green velvet of the calyx is scarcely less beautiful than the full flower." (*Bomb. Gaz., III., 199.*) Nearly every part of this interesting plant may be put to some useful purpose, and a little careful investigation seems all that is necessary to raise the gum at least to the position of an important commercial product.

THE GUM.

GUM. 945

Gum.—It yields naturally, or from artificial scars on the bark, a gum which is sold as "Bengal Kino" or *chúniá-gónd.* This occurs in the form of round tears, as large as a pea, often fragmentary, of an intense ruby colour and astringent taste. This gum may be purified by solution in water. It is translucent, but with age it darkens and becomes opaque. It is brittle, heat rendering it more so instead of melting it. It is generally known as *Kamarkas* in the bazars of the North-West Provinces, *Khákar-gónd* in Bombay, and *Chínyá-gónd, kinyá-gónd,* and *palás-kí-gónd* in Madras and some other places. In native medicine, Bengal kino is largely used as an astringent—a substitute for true kino. It is also employed in tanning.

946

Chemical Composition.—An aqueous solution of this gum by the action of persulphate of iron is changed into a dirty-green colour; a larger quantity occasioning a bright green precipitate. Acids throw down an orange or dirty-yellow pigment from the solution. A few drops of caustic potash change the colour to crimson, becoming grey with excess, until the whole of the colour is destroyed. Similar changes are effected by the action of caustic soda and ammonia. Carbonates of potash and of soda deepen the colour of the solution, but not so much as caustic potash. Metallic solutions, like acetate of lead, precipitate the whole of the colouring matter. Attempts have been made to fix these colours in the fibre of cotton, silk, wool, &c., with different mordants, but with very unsatisfactory results. (*Prof. Solly, in Journ. Royal Asiatic Society, 1838, and reproduced in several subsequent publications.*) **Roxburgh's** experiments with this gum are of sufficient importance to justify their being reproduced here: "This gum held in the flame of a candle swells and burns away slowly without smell or the least flame into a coal, and then into fine, light, white ashes. Held in the mouth it soon dissolves; its taste is strongly but simply astringent. Heat does not soften it, but rather renders it more brittle. Pure water dissolves it perfectly; the solution is of a deep clear red colour. It is in a great measure soluble in spirits, but this solution is paler, and a little turbid; the solution also becomes watery when spirit is added, and the spirituous more clear by the addition of water; diluted vitriolic acids render both solutions turbid and caustic; vegetable alkali changes the colour of the watery solution to a clean, deep, fiery blood-red. The spirituous it also deepens, but in a less degree. Sal marties changes the watery solution into a good durable ink."

Roxburgh pointed out that the Butea kino differs from the true kino in that it is more soluble and the solution more astringent in water than in spirit, while it is just the reverse with true kino. This property would thus admit of Butea kino being used where the presence of spirit was objectionable. According to the authors of the *Pharmacographia,* "This substance has a pure astringent taste, but no odour. It yielded us 1·8 per cent. of ash and contained 13·5 per cent. of water. Ether

GUM.

removes from it a small quantity of *pyrocatechin.* Boiling alcohol dissolves this kino to the extent of 46 per cent.; the solution, which is but little coloured, produces an abundant greyish-green precipitate with perchloride of iron, and a white one with acetate of lead. It may be hence inferred that a tannic acid, probably kino-tannic acid, constitutes about half the weight of the drug, the remainder of which is formed of a soluble mucilaginous substance which we have not isolated in a state of purity" (*p. 198*).

The Uses of the Gum may be said to be almost confined to medical science. As an astringent drug it is extensively used in India, and to a limited extent in Europe also. For industrial purposes the gum has made no progress, but there seems a good future for it both as a dye and tan. The natives of India are said to use it to precipitate and purify blue indigo. It is described as one of the best gums of the Central Provinces. In the Bombay Presidency it is said to be specially collected by the Náikdás. (*Bomb. Gaz., III., 109.*) From wounds in the bark a ruby-coloured astringent gum exudes, which loses colour by exposure, but it may be preserved by the gum being closely confined in a bottle. (*Lisboa, Useful Plants in the Bombay Presidency, 243.*) It is almost needless to republish the numerous references to this gum; suffice it to say that as the tree occurs throughout India, and grows at first rapidly, attaining its full size in little more than ten years, and requires no special care whatever, the supply of the gum and of the other products might be indefinitely extended should necessity arise. The so-called gum obtainable in our bazars is an exudation of a gummy sap from incisions on the bark. This hardens on exposure to the air into beautiful red-coloured transparent tears, which darken and become opaque with age unless kept in air-tight bottles. It would be interesting to have the timber of this tree treated in the same manner as in the preparation of Catechu, and it seems just possible, were this done, that a pure tanning extract might be obtained which would prove more suitable for industrial purposes than the gum at present met with in commerce. This is worthy of a trial, for, if found serviceable, the preparation of the extract might be combined with the separation of the bark fibre as a paper material. If at the same time the flowers as a dye-stuff could be made to give an additional return, Butea cultivation would become a profitable industry.

LAC. 947

Lac.—The lac insect is reared upon this tree in many parts of India—Chutia Nagpur, Central Provinces, the Deccan, Baroda, and Gujarát, &c. Commercially this is regarded as the second best quality of lac. (See **Lac.**)

DYE PROPERTIES.

DYE. Gum. 948 TAN. 949

Dye and Tan.—The GUM, as already stated, may be used both as a dye and tan, but, except in India, it is not in much demand for these purposes, and can hardly be viewed as a commercial product. By chemical actions special pigments and dyes may be prepared from it, which seems to deserve further and more careful attention. (See the account given under **Gum.**) As a tan the chief drawback to it seems to be the presence of so much gummy matter and a colouring agent mixed up with the tanning principle, the former of which retards its action. The following report by **Mr. Tiel** of an experiment with this gum as a tan is extracted from the *Journal of the Agri.-Horticultural Society of India, Vol. VIII., 25, for 1851:* "A piece of small calfskin, after being prepared in the usual way for the reception of tanning, was, on the 15th of July, immersed in a decoction of the '*dhák-palás,*' which was found readily soluble in cold water; the decoction was changed at intervals, as is usual, four times, each succession of liquor

The Tesu Dye.

TESU DYE.

being increased in strength to that which preceded it; each liquor was found to darken in colour in proportion to the time it was exposed to the action of the atmosphere. The skin during the process was constantly worked and attended to, and would, with like care, have been thoroughly tanned in five days with babúl bark. On the 21st July a piece of the skin was cut to dry out, to see if it really was tanned, and although it was highly coloured *through*, and had all the appearance of being thoroughly tanned, yet after being well washed, as is usual in currying, and dried out, it became as hard and as unpliable as a raw skin. Although it was highly coloured through, little or no tannin had combined with the skin. The tanning process was continued, adding gradually, as was required, more of the substance, until the 1st August, when another piece of the skin was tried by drying out, and with the same result as the first. From thence to the 15th and 25th August, respectively, when the whole two seers were consumed, the remainder of the skin was finally dried out, and found to be scarcely one-third tanned. It is therefore said to be of no use as a tanning substance, but it might, perhaps, be worthy of attention as a dyeing substance (for its colour seems very fast), or for tanning, could its astringent qualities, which are considerable, be easily deprived of so much colouring and gummy matters."

Tesu Flowers. 950

The FLOWERS, called *tésú, késú, kesuda*, or *palás-ké-phúl*, yield a brilliant but fleeting yellow dye, much used by the natives of India, especially during the *Hóli* festival. This is extracted either by expressing the coloured sap of the fresh flowers, or as a decoction or infusion from the dried flowers. The old leaves fade in February; the flush of new foliage appears in April and May, being preceded by a blaze of bright orange flowers, which at this season enliven the forests. No exact estimate of the number of trees, and therefore of the quantity of flowers which are annually produced, can be arrived at; but as the tree is one of the commonest plants in India, all over the drier undulations of the central plateau, the supply is practically unlimited. The flowers may be had for the gathering, as the annual production far exceeds the demand. They are collected in March and April, and as a rule are sun-dried. The petals are separated from the rest of the flower and preserved, and are in this condition sold, or when dry they are sometimes reduced to a powder. Simple immersion in water will extract the colour, but in some parts of the country the dye-stuff is boiled. The cloth to be dyed is sometimes boiled in the solution without the aid of any auxiliary or mordant. At other times the cloth, having been previously prepared by alum, lime, or ash, is then boiled with the colour, or again these substances are mixed with the dye-stuff and the cloth either boiled in the mixed dye and mordant solution, or left to steep in it for some time. The process of extraction of the dye is generally as follows. A given weight of dye is mixed thoroughly with twice as much water. After having been allowed to soak for some time, the mixture is boiled down to half its volume. It is then strained and allowed to cool. The cloth is either immersed in it before it cools, or is boiled with a required amount of the dye-solution, or is steeped in a cold dye-solution. The natives as a rule prefer the fleeting but brilliant yellow colour produced without the aid of any auxiliary, especially to dye the cloths worn at the *Hóli* festival; the fact of the colour being fleeting is viewed rather as an advantage than otherwise, since it can be got rid of after the festival is over. The addition of an alkali, as originally pointed out by the late **Dr. Roxburgh**, deepens the colour into orange and makes it at the same time a little less fleeting. Alum, lime, wood-ash or *sajjí-mátí* serves this purpose.

Powder. 951

Dye preparation. 952

Mordants. 953

Other vegetable substances are sometimes combined with the *tesu* in the production of yellow dyes, of which the following may be mentioned:

BUTEA frondosa.

Palas Flowers.

harsinghar **(Nyctanthes Arbor-tristis)**, *latkan* **(Bixa Orellana)**, *ál* or *aich* **(Morinda tinctoria)**, *haldí* or *turmeric* **(Curcuma longa)**, *baqam* **(Cæsalpinia Sappan)**, *gumbengfong* **(Plecospermum spinosum)**: the last mentioned is regarded specially useful as a silk dye. The presence of the *tésú* seems to improve the brilliancy of these dyes, but it is doubtful if its use can be recommended, since there seems no idea of its being in these combinations more durable than when used alone. Other vegetable substances are recommended to be combined with it, from some idea of their helping to make the colour less fleeting. The following are those most frequently used for this purpose: *hari* or *har* **(Terminalia chebula)**, *lódh* **(Symplocos racemosa)**. It is remarkable how extensively the bark of the last-mentioned plant is used in native dyes as an auxiliary intended to brighten and fix the colour. It is probable that the Mánipuri dye-auxiliary (**Garcinia pedunculata**) might with advantage be added to this list of dye-auxiliaries suitable for the *tésú* or *palás* dye. By combination with indigo, shades of green and light blue are sometimes produced. A grape-green *tésú, harsinghár* and indigo, with acidulated water, are used. To give light blue (the *lájwari* colour of Cawnpore), *tésú* alum, talc and indigo are used. (*Mr. Buck, Dyes and Tans of N.-W. P.*)

Auxiliaries 954
Combinations 955

Pigment. 956

A PIGMENT.—**Dr. Roxburgh** appears to be the only person who has experimented with a pigment extracted from these flowers. He says: "Amongst numberless experiments, I expressed a quantity of the juice of the fresh flowers, which was diluted with alum water, and rendered perfectly clear by depuration. It was then evaporated by the heat of the sun into a soft extract; this proves a brighter water-colour than any gamboge I have met with. It is now one year since I first used it, and it remains bright."

"Infusions of the dried flowers yielded me an extract very little, if anything, inferior to this last mentioned. They yield also a very fine durable yellow lake, and all these in a very large proportion." (*Roxb., Fl. Ind., Ed. C.B.C., 540.*)

THE TESU DYE. Abir. 957

An ABÍR.—**Voigt** says that the yellow dye obtained from the *tésú* flowers is used at the *Hólí* festival. This same fact is alluded to by several writers, but without describing the particular way in which it is used. Under the heading "ABÍR" will be found some further information, from which it will be seen that it serves as *gulál* for the flour of the singara-nut, but I am unable to discover the exact preparaiion which is used,—*e.g.*, a dry pigment or simply a powder of the petals mixed with the flour, or whether the decoction of the petals or a dye-solution is mixed with the flour just as it is being used. Probably in different parts of the country all three practices are followed.

Dr. Buchanan (in his *Statistical Account of Dinajpur*) says of this tree: "The flowers are not only offered to the gods, but in the festivals of spring serve to give a temporary yellow dye to the clothes of their votaries, on which account it is called *Vosonti*."

FIBRE.

IBRE. 958

Fibre.—Yields a strong fibre, said to be useful for paper-making and for cordage; also the young roots yield a strong fibre known as *chhoel*. This is made into ropes in Chutia Nagpur, Central Provinces, Oudh, Rájputana, and Bombay hill tracts, &c.; it is also used in some parts of India for making native sandals.

OIL. 959

Oil.—The seeds of this tree (*palás-pápṛá* of the bazar) yield a small quantity of bright, clear oil (by some authors called *Múdúga* oil); this is sometimes used medicinally.

MEDICINE.

MEDICINE. Gum. 960

Medicine.—THE GUM.—This is known as Bengal or Butea kino. Nearly the whole of the so-called kino of our bazars is this substance. **Dr. Waring** (in his *Bazar Medicines, p. 31*) remarks that this is of little moment, since it appears to be equally effectual. He says: "It is an excellent astringent, similar to Catechu, but being mild in operation it is better adapted for children and delicate females. The dose of the powdered gum is 10 to 30 grains, with a few grains of cinnamon." The addition of a little opium increases the efficacy.

Juice. 961

THE FRESH JUICE is used in phthisis and hæmorrhagic affections. It is also employed as an application to ulcers and relaxed sore-throat. As an astringent it is given in diarrhœa and dyspepsia. In the Konkan it is prescribed for fevers. "The use of the gum as an external astringent application is mentioned by **Chakradatta**; it is directed to be combined with other astringents and rock-salt. He recommends this mixture as a remedy for pterygium and opacities of the cornea." (*Dr. Dymock, Mat. Med., W. Ind., 187.*) **U. C. Dutt** informs us that the ancient Hindús used the gum as an external astringent only.

Seeds. 962

THE SEEDS.—Internally they are administered as an anthelmintic, but regarding the reliance which can be put upon their action considerable difference of opinion prevails. Some medical men think that they can be advantageously substituted for Santonine, while others view them as much less powerful. They have at the same time a warm purgative action, which often proves injurious to their anthelmintic property. They are, however, largely used in the treatment of round-worm. The following extract from **Dr. Waring's** *Bazar Medicines* will be found to give the leading facts regarding these seeds: "Butea seeds are thin, flat, oval or kidney-shaped, of a mahogany-brown colour, 1¼ to 1¾ inches in length, almost devoid of taste and smell. European experience has confirmed the high opinion held by the Mohammedan doctors as to their power in expelling *Lumbrici*, or *Round Worm*, so common amongst the natives of India. The seeds should be first soaked in water, and the testa, or shell, carefully removed; the kernel should then be dried and reduced to powder. Of this the dose is 20 grains thrice daily for three successive days, followed on the fourth day by a dose of castor oil. Under the use of this remedy, thus administered in the practice of **Dr. Oswald**, 125 lumbrici in one instance, and between 70 and 80 in another, were expelled. It has the disadvantage of occasionally purging, when its vermifuge properties are not apparent: in some instances also it has been found to excite vomiting and to irritate the kidneys, and though these ill effects do not ordinarily follow, yet they indicate caution in its employment." (*Bazar Medicines, Waring, pp. 31-32*). "In the *Bhavaprakása* the use of the seeds of the *Palása* as an aperient and anthelmintic is noticed; and they are directed to be beaten into a paste with honey for administration. **Sárangadhara** also recommends them as anthelmintic." (*Dr. Dymock.*) "Externally the seeds, when pounded with lemon-juice and applied to the skin, act as a rubefacient. I have used them successfully for the cure of the form of herpes known as Dhobie's itch." (*Surgeon-Major Dymock, p. 188.*) When made into a paste, they are used as a remedy for ringworm.

MEDICINE. Flowers. 963

THE FLOWERS are astringent, depurative, diuretic, and aphrodisiac; as a poultice they are used to disperse swellings and promote diuresis and the menstrual flow. They are given to *enciente* women in cases of diarrhœa, and are applied externally in orchitis.

Leaves. 964

THE LEAVES are described by the *Makhzan-ul-Adwiya* as astringent, tonic, and aphrodisiac, are used to disperse boils and pimples, and are given internally in flatulent colic, worms, and piles.

Bengal Kino.

Bark. 965

THE BARK, according to the *Hortus malabaricus*, is given in conjunction with ginger in cases of snake-bite.

The gum and other parts of **Butea superba**, *Roxb.*, are also used medicinally by the natives, being viewed as possessing the same properties as the corresponding parts of **B. frondosa.**

Charcoal. 966

Special Opinions.—§ "The charcoal from this plant was introduced by **Dr. T. W. Sheppard** in 1874, for bleaching the morphia manufactured at the Opium Factory, Ghazipur. It was selected after a series of experiments with the different forms of charcoal, its great advantage being its comparative freedom from saline matter; it can on this account be employed without any previous purification. Wood-charcoal possesses feebler decolorising powers than animal, but it had to be resorted to on account of the native prejudices against the use of bone-charcoal." (*Surgeon Warden, Professor of Chemistry, Calcutta Medical College.*)

MEDICINE 967

"I have tried the seeds of **B. frondosa** internally in numerous cases, and they are neither purgative nor febrifuge, at least not in one-drachm doses,—the largest quantity I have yet used. There is, however, no doubt that they are anthelmintic, at least to some extent. Administered in powder, morning and evening, for 2 or 3 days, and followed by a dose of some purgative, they generally expel from 1 to 3 or 4 round-worms; but failure is more frequent than success. That these seeds are not powerful enough to act always against the worms is proved by the expulsion of the latter in large numbers in many cases by the use of santonine, immediately after having failed with butea seeds. Both the kernel and the testa of the seeds possess the anthelmintic property. Dose of the powder for an adult, from 30 grains to 1 drachm. Four grains is an average dose for a child of 4 years.

"The inspissated juice of this plant (the butea kino of Indian commerce) is a good astringent, and as such is useful in all the complaints in which the true kino is indicated. It has been used in the same forms as those of the latter, but in somewhat larger doses,—*viz.*, from 15 to 40 grains." (*Honorary Surgeon Moodeen Sheriff, Khan Bahadur, Madras.*)

"This is a fairly useful anthelmintic and a good substitute for santonine, in some cases acting very well indeed. *Preparations.*—Powdered seeds, dose fifteen to thirty grains twice or thrice a day, followed by castor oil on the succeeding morning. The gum has been only lately used in this hospital, but as an astringent it is found to be a useful substitute for kino in the ordinary cases of diarrhœa and dysentery, of children especially. Preparations and doses, &c., are similar to kino." (*Apothecary J. G. Ashworth, Kumbakonam.*) "The leaves are astringent and used by the natives as a poultice to dispel tumorous hœmorrhoids, buboes, &c. The seeds are anthelmintic in doses of 20 grains. The gum is very astringent, and in doses of five grains most useful in checking serious diarrhœa. In large doses it is efficacious in hœmorrhage from the stomach and bladder. A strong solution of the gum is said to be a useful application for bruises and erysipalous inflammations." (*Surgeon R. A. Barker, Doomka.*) "In common use as an anthelmintic; dose for an adult, 20 grains of powdered seeds." (*Surgeon Mark Robinson, Coorg.*) "Root of this tree is used as an aphrodisiac by native physicians." (*Brigade Svrgeon S. M. Shircore, Moorshedabad.*) "This remedy exhibits its anthelmintic properties to perfection when used in the following formula: Powdered **Butea frondosa** gr. 3, powdered ginger gr. 1, santonine gr. $1\frac{1}{2}$; give for three successive nights, and follow by a dose of castor oil on the fourth morning." (*Honorary Surgeon Peter Anderson, Guntur, Madras.*) "The seeds are used in urinary diseases. In large doses the powder acts as a purgative, also as an anthelmintic. The flowers are used in the form of poultices in bladder diseases." (*Surgeon W. Barren, Bhuj,*

MEDICINE.

Cutch.) "Anthelmintic, doses 5 to 20 grains, used in round-worm." (*Choonna Lall, Civil Hospital Assistant, Jubbulpore.*) "Seeds are vermifuge; dose, powder 20 grains, three times a day, sometimes causes much vomiting and purging." (*Apothecary Thomas Ward, Madanapalli, Cuddapah.*) "The powdered gum is a useful astringent in chronic diarrhœa, pyrosis, and dyspepsia. The seeds are a powerful anthelmintic, especially in the case of round-worms." (*Brigade Surgeon J. H. Thornton, Monghyr.*) "Not nearly as efficacious as santonine." (*Surgeon-Major H. J. Hazlitt, Ootacamund, Nilgiri Hills.*) "Anthelmintic. The seeds are soaked in water, the testa removed, kernel dried and powdered; dose grains 20, three times a day for three successive days, followed by a dose of castor oil. They act effectually in expelling large numbers of Ascaris lumbricoides." (*Surgeon-Major A. F. Dobson, Bangalore.*) "The seeds are anthelmintic, used to expel round-worm, in doses of 10 grs. to 1 drachm of powder, according to age." (*Assistant Surgeon Shib Chunder Bhuttacharji, Chanda, Central Provinces.*) "The juice I have seen used by natives in dysentery and ringworm." (*Honorary Surgeon E. A. Morris, Negapatam.*)

FODDER. 968

Fodder.—The leaves are used as fodder for buffaloes and elephants. The leaves are regarded as a valuable manure.

TIMBER. 969

Structure of the Wood.—Dirty white, soft, not durable; no annual rings. It is said to be better under water, and so is used in North-West India for well-curbs and piles. Weight 30 to 40 lbs. per cubic foot.

It "grows to a height of about fifteen feet and seldom lives more than ten years." (*Bombay Gaz., VII., 40.*) "The wood is coarse and poor. A seasoned cubic foot weighs 33 lbs." (*Bombay Gaz., XV., Pt. I., p. 64.*)

SACRED USES.

SACRED. 970

Sacred and Domestic Uses.—From the name *palás* is said to be derived the name Plassey, the scene of Clive's famous victory. This beautiful tree is sacred to *Soma* (Moon); the wood is sacrificial, and is frequently mentioned in the Vedas.

DOMESTIC. 971

Dr. Buchanan, as already quoted, says the flowers are offered to the gods, and afford the yellow dye called *Vosonti*. The *palás* is sometimes represented as a sacred tree of the Buddhists. The word *palása* in Sanskrit means "leaf," but it has become a modern equivalent for the *dhák* (**Butea frondosa**), a tree which is supposed to be imbued with the immortalising *Soma*, the beverage of the gods. **Richard Folkard** (in his *Plant Lore*) says: "This tree is supposed to have sprung from the feather of a falcon imbued with the *Soma*." He adds: "The *palása* was much employed by the Hindús in religious ceremonies, particularly in one connected with the blessing of calves to ensure them proving good milkers." The triple leaves were deemed to typify, like the trident, the forked lightning, resembling the rod of Mercury, the Sanic and the Rowan rod. Speaking of these triple leaves **Mr. Lisboa** says: "The leaves of this plant are trifoliate; the middle leaflet is supposed to represent Vishnu, the left Brahma, and the right Shiv: hence its worship is enjoined in *Chaturmás Máhátma*. Hence also its use in the following three great ceremonies:—

SACRED USES.

Leaves as Plates. 972

"(1) The leaves are used as platters on the occasion of the investiture of the sacred thread, when a particular part of the ceremony, called *chewul* (that is, when the barber removes the last tuft of hair from the head of the child to be invested), is being performed.

Dry Twigs. 973

"(2) The dry twigs, under the designation of *samidhas*, are used for the feeding of *hom*, or sacred fire, in the ceremony which goes under the name of *nava grahas*, celebrated to secure the pacification of the nine planets (*nava*=nine, *grahas*=planets) on the occasion of *vástu shanti, i.e.,* entrance into a newly-built house, or one acquired from a non-Hindú.

Stem. 974

"(3) The stem is used as a staff on the day of *sodmúnj*, a part of the thread ceremony." (*Useful Plants of the Bombay Presidency, pp. 279-80.*)

Flowers. 975
Dishes. 976
Cups. 977

Dr. U. C. Dutt says the beautiful flowers were used as ear ornaments by the ancient Hindú women and were much admired by the poets. "The leaf-dishes used at caste-feasts are made by village Brahmans. Of two kinds,—plates, *patrávalis*, and cups, *dadiyás*,—the dishes are brought into Ahmedabad in bundles of 200 plates and 100 cups, and are sold according to size; the plates at from 3*d.* to 6*d.* (2-4 annas) the hundred, and the cups at from 1½*d.* to 2¼*d.* (1-1½ annas). Made of the dry leaves of the *khákhar* tree, **Butea frondosa,** fastened together with small slips of bamboo, they keep fit for use for two years. This industry is confined to the Daskroi villages near the city (Ahmedabad), where only they find a sale." (*Bombay Gaz., IV., 135.*)

978

Butea superba, *Roxb.; Fl. Br. Ind., II., 195.*

Vern.—*Latá-palásh*, BENG.; *Badúri*, SINGRAMPUR; *Chihúnt*, MONGHYR; *Nari murup*, SANTAL; *Palásí, pálásávela*, BOMB.; *Vél-khákar*, GUJ.; *Bél palás*, DUK.; *Yél-parás, palásavela*, MAR.; *Kodi-murukkam, kodi-palásham*, TAM.; *Tíge motku, tíge-móduga, tige-paláshama, báranki-chettu, tivva-máduga*, TEL.; *Balli-mut-taga*, KAN.; *Valliplách-cha, valli-murukka*, MAL.; *Latá-palása*, SANS.; *Samur*, GOND.; *Tunang*, KURKU; *Pouknway, poukgnwe*, BURM.

Habitat.—An extensive climber, scarcely differing from the preceding except in habit. Found in the forests of the Konkan, the Central Provinces, Central India, Rájputana, Bengal, Orissa, and Burma.

GUM. 979

Gum.—It yields a gum like that of **B. frondosa.**

DYE. 980

Dye.—The root is said to yield a red dye in Burma. "A sample of red dye-wood, appears to be the root of **Butea superba**, a well-known creeper whose scarlet blossoms are a conspicuous feature of the forest landscape in February and March." (*Indian Forester, Vol. X., p. 75.*) The information regarding this dye is exceedingly meagre; it is remarkable that no mention is made of the roots of the preceding species as affording a dye.

FIBRE. 981

Fibre.—An extensive climber, scarcely differing from **B. frondosa** except in habit. The roots, as also the young branches, afford a strong and useful fibre prepared in Chutia Nagpur, the Central Provinces, Central India, and Rájputana.

FODDER. 982

Fodder.—The leaves are regarded as a valuable fodder.

983

Butter.

Vern.—*Navanía*, SANS.; *Mákhan*, HIND, BENG., GUJ.; *Maska, loní*, MAR.

"The fatty portion of the milk of all mammalian animals is called 'Butter,' but the term in a commercial sense is restricted to that from the cow." (*Spons' Encyclop.*) In India butter is made from cows' or buffaloes' milk; it is chiefly used in the form of *ghí* or clarified butter. (See **Ghi.**)

984

BUXUS, *Linn; Gen. Pl., III., 266.*

Evergreen shrubs or undershrubs, with 4-sided branchlets and opposite, exstipulate leaves. *Flowers* monœcious, in axillary clusters. *Calyx* of male flower deeply 4-cleft, the segments opposite in pairs; of female flower, deeply 6-cleft, the segments in two circles of three each. *Stamens* 4, opposite the calyx-segments, inserted around a 4-sided rudimentary ovary. *Ovary* 3-celled, 3-cornered, with a flat top, the corners terminating in thick short styles, which alternate with the 3 inner calyx-segments. *Capsule* coriaceous, loculicidally 3-valved, each valve ending in 2 horns, being the valves of 2 of the styles, dissepiments attached to the valves. *Seeds* 3-6, trigonous.

Buxus sempervirens, *Linn.; Brandis, For. Fl., 447; Gamble's Man. Timb., 369;* EUPHORBIACEÆ

985

THE BOXWOOD TREE, *Eng;* BUCHSBAUM, *Germ.;* BUIS, *Fr.;* BOSSO, *It.*

Syn.—B. WALLICHIANA, *Baillon.*

Vern.—*Shanda laghúne,* AFG.; *Chikri,* KASHMIR; *Papri, papar, papur, paprang, shamshád, shumaj,* PB.

Habitat.—An evergreen shrub or small tree met with in the Suliman and Salt Ranges, North-West Himálaya, between 4,000 and 8,000 feet, in Bhután about 6,000 to 7,000 feet; but scattered in different parts of the Himálaya, chiefly on a calcareous soil and often in remote localities.

"The tree is, however, constant in one respect as regards its habitat, and thrives in moist and sheltered places, hugging the alluvial deposits along the banks of the perennial streams, and it does not thrive where it is exposed to winds, whether hot and dry, or cold and frosty. It also avoids the hot sides of the valley, and evidently prefers a north-west and northerly aspect." (*Report on the Boxwood supply in the Panjáb by Mr. Ribbentrop, "Indian Forester," XI., 26.*)

MEDICINE. 986

Medicine.—The wood is diaphoretic; leaves bitter, purgative, and diaphoretic, useful in rheumatism and syphilis. A tincture from the bark is used as a febrifuge.

FODDER. 987

Fodder.—Stewart says that goats will occasionally browse on the leaves, but other animals will not do so unless in times of dearth. They have been known to prove fatal to camels and cattle.

The leaves "are used in the south of France as manure for vineyards." (*Gamble.*)

TIMBER. 988

Structure of the Wood.—Yellowish white, hard, smooth, very close and even-grained. Annual rings distinctly marked by a narrow line without pores. Weight 55 to 65 lbs. per cubic foot. The rate of growth is variable, generally slow.

The following extract from **Mr. Gamble's** *Manual of Indian Timbers* will be found useful: "It is estimated that the cost per cubic foot of boxwood delivered at Saharanpur from the Kelso forest would be R1-8; its further cost by rail from Saharanpur to Bombay would be at least R1-8, or total R3 per cubic foot. Considering 1 cubic foot as weighing 60lb, we have the cost per ton as R112, which could only be just covered by receipts if the very best description of wood were sent down. There is consequently little likelihood of much trade in boxwood from the Himálayan forests."

Fancy Boxes. 989

Engraving slabs. 990

"The uses of boxwood are well known. In Europe it is used for engraving, turning, carving, and mathematical instruments. In the Himálaya, small boxes to contain butter, honey, tinder, snuff, &c., are made of it, and it is carved into combs. The boxwood to be used for engraving requires very careful and lengthened seasoning. On this subject and on the other requisite characters of boxwood for commercial purposes, the following extract from a letter of Messrs. J. Gardner and Sons, of Liverpool, to the Inspector-General of Forests, dated April 3rd, 1877, will give information:—

"The value of boxwood at Bombay of suitable texture for the English market, of which latter we can judge from a few sample pieces, will depend principally upon the quality.

"Wood from 2 to 4 inches diameter is required to be free from splits or cracks, otherwise, however free from knots and straight and round it may be, the value would not exceed £1 to £2 per ton, whilst if free from splits, round and straight, and with—

not exceeding one knot per foot in length	the value would probably be	£10 per ton,
exceeding 1 knot and not exceeding 2 knots		£7 10s. ,,
,, 2 knots ,, ,, 3 ,,		£5 ,,

—all knots or holes counted as such, however small.

BUXUS sempervirens.

Boxwood.

TIMBER.

"Wood 4 inches and upwards in diameter is preferred with one split rather than sound or with more than one split, any splits after the first reducing the value on account of the additional waste in working the same.

		Averaging per foot in length. 1 knot.	2 knots.	3 knots.
		£	£	£
The value of round and straight (1 split) averaging	4 to 5 inches diameter . .	6	4 10s.	3
	5 to 6 " " . .	9	6	3
	6 inches and upward diameter .	12	9	4 10s.

"If the splits are twisted more than 1 inch to the foot if small, 2 inches if medium size, and 3 inches to the foot length if large, the value is reduced one half.

"The above values will, of course, vary in accordance with the supply and demand for the various sizes and qualities.

"The most suitable texture of wood will be found growing upon the sides of mountains. If grown in the plains, the growth is usually too quick, and consequently the grain is too coarse; the wood of best texture being of slow growth and very fine in the grain.

"It should be cut down in the winter, and, if possible, stored at once in airy wooden sheds, well protected from sun and rain, and not to have too much air through the sides of the shed, more especially for the wood under 4 inches diameter.

"The boxwood also must not be piled upon the ground, but be well skidded under, so as to be kept quite free from the effects of any damp from the soil.

"After the trees are cut down, the longer they are left exposed, the more danger is there afterwards of the wood splitting more than is absolutely necessary during the necessary seasoning before shipment to this country.

"If shipped green there is great danger of the wood sweating and becoming mildewed during transit, which causes the wood afterwards to dry light and of a defective colour, and in fact rendering it of little value for commercial purposes.

"There is no occasion to strip the bark off, or to put cowdung or anything else upon the ends of the pieces to prevent their splitting.

"Boxwood is the nearest approach to ivory of any wood known, and will therefore probably gradually increase in value, as it, as well as ivory, becomes scarcer. It is now used very considerably in manufacturing concerns, but, on account of its gradual advance in price during the past few years, cheaper woods are in some instances being substituted.

"Small wood under 4 inches is used principally by flax-spinners for rollers and by turners for various purposes, rollers for rink-skates, &c., &c., and if free from splits is of equal value with the larger wood. It is imported here as small as 1½ inches in diameter, but the most useful sizes are from 2½ to 3½ inches, and would, therefore, we suppose, be from 15 to 30 or 40 years in growing, whilst larger wood would require 50 years and upwards at least,—perhaps we ought to say 100 years and upwards. It is used principally for shuttles for weaving silk, linen and cotton, and also for rule-making and wood-engraving. *Punch, The Illustrated London News, The Graphic,* and all the first-class pictorial papers use large quantities of boxwood."

Messrs. **Churchill and Sim**, reporting on some boxwood sent to them for sale in 1880, and which fetched 21 shillings per cwt., equivalent at 60lb per cubic foot to 11*s.* 1*d.*, or about R6 per cubic foot, say:—

"The pieces of boxwood were remarkably fine specimens, equal in quality to the best Abasia, and fetched a very high price, equivalent to £21 per ton. These logs were depreciated in value for ordinary purposes, owing to their having been squared, which was a mistake, as in that operation much valuable wood had been wasted, and when the bark is removed, a good protection to the log is destroyed. In the present state of the boxwood trade, and considering the fact that the supplies which have been coming forward for some time past are deteriorating in quality, from the action of the Turkish Government in closing the forests, and from other causes, the probability of a supply of this wood from India is a matter of considerable importance. The usual run of this wood would not, however, fetch the high price of this picked sample. The price realised cannot, however, be taken as any criterion, for whether supplies can be sent to this market, and sold at prices which will cover transit and freight, and then leave a profit, is very doubtful. Could this wood be regularly placed on the market at a moderate figure, there is no reason why a trade should not be developed in it."

Regarding the consignment of boxwood made to **Messrs. Gardner**

and Sons, the following extract from the Kew Report, 1881, will be found interesting :—

"**Messrs. Joseph Gardner and Sons,** the well-known timber merchants, wrote to **Dr. Brandis,** the Inspector-General of Forests in India, April 29th, 1881, on the subject: 'We bought the parcel (about 5 tons), landed ex *Strathmore* in London, at the high price of £30 per ton. At these high prices the consumption will be very limited indeed. Can you kindly inform us what the prospects are of securing any large quantities of this wood, say 5,000 to 10,000 tons, at about £10 per ton, in Liverpool or London? We are drawing our present supplies from Russia and Persia principally; but there are so many fiscal restrictions, and the wood is also inferior to your Indian shipments, that we should prefer drawing all our supplies from India. At anything like £30 per ton, only very small quantities can be used; at £10, however, it would probably be used very extensively for various purposes for which cheaper woods than boxwood are now used."

"To this communication **Dr. Brandis** replied, July 6:—

"'The boxwood resources of the country are very limited. There is no chance of such large supplies as from 5,000 to 10,000 tons being available from India'" (*p. 25*).

Mr. Ribbentrop, in *The Indian Forester,* reports on the Panjáb Boxwood as follows :—

"The cost in carriage between Saharunpur and Umballa is R0-1-9 per maund; deducting this from the above, I find that boxwood from Bashahr could be landed in London at R91-9-4 + R132 (carriage to Umballa)=R224 per ton. Taking the exchange at 1*s.* 7¾*d.*, this is an expense of £18 8*s.* 8*d.* per ton for the boxwood consigned to London.

"The price quoted at home is £24 per ton weight.

"It will thus be seen that by exporting the first-class wood we will realise, it is estimated, a net profit of R75 per ton, or R35 above the price which it would fetch in the local market.

"The net value of the Bushahr box forests, which are the only ones that need be taken into consideration as regards immediate exploitation, is therefore—

	R
700 tons, at R75 profit	=52,500
300 tons, „ 25 profit	= 7,500

"I roughly estimate that with strict protection and the gradual thinning out of dominant trees of other kinds, the present growing stock will replace the mature stock in 24 years, in which time the present mature timber may be extracted, reproduction being ensured at the same time." (*Indian Forester, XI., 28.*)

A good deal has been written regarding this subject, and a few interesting papers have appeared upon the possible future Indian trade in this most valuable timber. It is very much to be feared, however, that, both on account of the limited amount of boxwood and the heavy transport charges, India cannot to any very great extent become a source of supply to Europe. It will only pay to export selected pieces, and there does not appear to be any local demand for second-class boxwood. This seems to be the conclusion arrived at both in Europe and in India, and accordingly, in the Kew Report, already quoted, we find the subject of boxwood substitutes urged as more worthy of attention. "It is evident, therefore, that we cannot look to India to remedy the increasing dearth of boxwood. It would be obviously much to the advantage of any of our colonies that could send into the timber trade, in quantity, any wood which would be acceptable as a boxwood substitute." For a list of probable woods of this nature, see BOXWOOD SUBSTITUTES, page 518.

B. 990

Printed in the United States
By Bookmasters